METAL IONS
IN LIFE SCIENCES

VOLUME 3

The Ubiquitous Roles
of Cytochrome P450 Proteins

METAL IONS IN LIFE SCIENCES

edited by

Astrid Sigel,[1] Helmut Sigel,[1] and Roland K. O. Sigel[2]

[1] *Department of Chemistry*
Inorganic Chemistry
University of Basel
Spitalstrasse 51
CH-4056 Basel, Switzerland

[2] *Institute of Inorganic Chemistry*
University of Zürich
Winterthurerstrasse 190
CH-8057 Zürich, Switzerland

VOLUME 3

The Ubiquitous Roles of Cytochrome P450 Proteins

John Wiley & Sons, Ltd

Copyright © 2007   John Wiley & Sons Ltd, The Atrium, Southern Gate, Chichester,
West Sussex PO19 8SQ, England

Telephone   (+44) 1243 779777

Email (for orders and customer service enquiries): cs-books@wiley.co.uk
Visit our Home Page on www.wiley.com

All Rights Reserved. No part of this publication may be reproduced, stored in a retrieval system or transmitted in any form or by any means, electronic, mechanical, photocopying, recording, scanning or otherwise, except under the terms of the Copyright, Designs and Patents Act 1988 or under the terms of a licence issued by the Copyright Licensing Agency Ltd, 90 Tottenham Court Road, London W1T 4LP, UK, without the permission in writing of the Publisher. Requests to the Publisher should be addressed to the Permissions Department, John Wiley & Sons Ltd, The Atrium, Southern Gate, Chichester, West Sussex PO19 8SQ, England, or emailed to permreq@wiley.co.uk, or faxed to (+44) 1243 770620.

Designations used by companies to distinguish their products are often claimed as trademarks. All brand names and product names used in this book are trade names, service marks, trademarks or registered trademarks of their respective owners. The Publisher is not associated with any product or vendor mentioned in this book.

This publication is designed to provide accurate and authoritative information in regard to the subject matter covered. It is sold on the understanding that the Publisher is not engaged in rendering professional services. If professional advice or other expert assistance is required, the services of a competent professional should be sought.

The Publisher, the Editors and the Authors make no representations or warranties with respect to the accuracy or completeness of the contents of this work and specifically disclaim all warranties, including without limitation any implied warranties of fitness for a particular purpose. This work is sold with the understanding that the Publisher is not engaged in rendering professional services. The advice and strategies contained herein may not be suitable for every situation. In view of ongoing research, equipment modifications, changes in governmental regulations, and the constant flow of information relating to the use of experimental reagents, equipment, and devices, the reader is urged to review and evaluate the information provided in the package insert or instructions for each chemical, piece of equipment, reagent, or device for, among other things, any changes in the instructions or indication of usage and for added warnings and precautions. The fact that an organization or Website is referred to in this work as a citation and/or a potential source of further information does not mean that the Authors, the Editors, or the Publisher endorse the information the organization or Website may provide or recommendations it may make. Further, readers should be aware that Internet Websites listed in this work may have changed or disappeared between when this work was written and when it is read. No warranty may be created or extended by any promotional statements for this work. Neither the Publisher nor the Editors nor the Authors shall be liable for any damages arising herefrom.

Other Wiley Editorial Offices

John Wiley & Sons Inc., 111 River Street, Hoboken, NJ 07030, USA

Jossey-Bass, 989 Market Street, San Francisco, CA 94103-1741, USA

Wiley-VCH Verlag GmbH, Boschstr. 12, D-69469 Weinheim, Germany

John Wiley & Sons Australia Ltd, 42 McDougall Street, Milton, Queensland 4064, Australia

John Wiley & Sons (Asia) Pte Ltd, 2 Clementi Loop #02-01, Jin Xing Distripark, Singapore 129809

John Wiley & Sons Ltd, 6045 Freemont Blvd, Mississauga, Ontario L5R 4J3, Canada

Wiley also publishes its books in a variety of electronic formats. Some content that appears in print may not be available in electronic books.

Anniversary Logo Design: Richard J. Pacifico

British Library Cataloguing in Publication Data

A catalogue record for this book is available from the British Library

ISBN 978-0-470-01672-5

Typeset in 10/12pt Times by Integra Software Services Pvt. Ltd, Pondicherry, India
Printed and bound in Spain by Grafos S.A., Barcelona
This book is printed on acid-free paper responsibly manufactured from sustainable forestry in which at least two trees are planted for each one used for paper production.

The figure on the dustcover is part of Figure 2 of Chapter 6 by Andrew K. Udit, Stephen M. Contakes, and Harry B. Gray.

Historical Development and Perspectives of the Series
Metal Ions in Life Sciences

It is an old wisdom that metals are indispensable for life. Indeed, several of them, like sodium, potassium, and calcium, are easily discovered in living matter. However, the role of metals and their impact on life remained largely hidden until inorganic chemistry and coordination chemistry experienced a pronounced revival in the 1950s. The experimental and theoretical tools created in this period and their application to biochemical problems led to the development of the field or discipline now known as *Bioinorganic Chemistry, Inorganic Biochemistry*, or more recently also often addressed as *Biological Inorganic Chemistry*.

By 1970 *Bioinorganic Chemistry* was established and further promoted by the book series *Metal Ions in Biological Systems* founded in 1973 (edited by H.S., who was soon joined by A.S.) and published by Marcel Dekker, Inc., New York, for more than 30 years. After this company ceased to be a family endeavor and its acquisition by another company, we decided, after having edited 44 volumes of the *MIBS* series (the last two together with R.K.O.S.) to launch a new and broader minded series to cover today's needs in the *Life Sciences*. Therefore, the Sigels' new series is entitled

Metal Ions in Life Sciences

and we are happy to join forces in this new endeavor with a most experienced Publisher in the *Sciences*, John Wiley & Sons, Ltd, Chichester, UK.

The development of *Biological Inorganic Chemistry* during the past 40 years was and still is driven by several factors; among these are (i) the attempts to reveal the interplay between metal ions and peptides, nucleotides, hormones or vitamins, etc.; (ii) the efforts regarding the understanding of accumulation, transport, metabolism and toxicity of metal ions; (iii) the development and application of metal-based drugs; (iv) biomimetic syntheses with the aim to understand biological processes as well as to create efficient catalysts; (v) the determination of high-resolution structures of proteins, nucleic acids, and other biomolecules; (vi) the utilization of powerful spectroscopic tools allowing studies of structures and dynamics; and (vii), more recently, the widespread use of macromolecular engineering to create new biologically relevant structures at will. All this and more is and will be reflected in the volumes of the series *Metal Ions in Life Sciences*.

The importance of metal ions to the vital functions of living organisms, hence, to their health and well-being, is nowadays well accepted. However, in spite of all the progress made, we are still only at the brink of understanding these processes. Therefore, the series *Metal Ions in Life Sciences* will endeavor to link coordination chemistry and biochemistry in their widest sense. Despite the evident expectation that a great deal of future outstanding discoveries will be made in the interdisciplinary areas of science, there are still 'language' barriers between the historically separate spheres of chemistry, biology, medicine, and physics. Thus, it is one of the aims of this series to catalyze mutual 'understanding'.

It is our hope that *Metal Ions in Life Sciences* proves a stimulus for new activities in the fascinating 'field' of *Biological Inorganic Chemistry*. If so, it will well serve its purpose and be a rewarding result for the efforts spent by the authors.

Astrid Sigel, Helmut Sigel
Department of Chemistry
Inorganic Chemistry
University of Basel
CH-4056 Basel
Switzerland

Roland K. O. Sigel
Institute of Inorganic Chemistry
University of Zürich
CH-8057 Zürich
Switzerland

October 2005

Preface to Volume 3
The Ubiquitous Roles of Cytochrome P450 Proteins

Cytochrome P450 monooxygenases (P450s) own a dual functionality and are often referred to as 'mixed-function oxidases' because they possess both 'oxygenase' and 'oxidase' reactivities, meaning that they incorporate an oxygen atom from atmospheric dioxygen into a substrate molecule oxidizing it. The electron transfer functionalities of P450s also earned them the label of 'cytochrome'. The signature of all P450s is the heme Soret band at about 450 nm in the absorption spectrum of the Fe(II)-CO complex; this spectroscopic feature is diagnostic of a cysteinate residue bound *trans* to the carbon monoxide ligand.

This volume encompasses the breadth of research efforts focused on P450s from structure to function, including an appreciation of the diversity and complexity of the biotransformations they catalyze. The introductory Chapter 1, setting the scene, explores the many different levels at which these P450 enzymes (and their respective genes) have diverged in the process of evolution to yield the plethora of enzymes that are now termed 'P450 monooxygenases'. This broad and unprecedented reactivity of P450 enzymes which contain a heme iron center (most commonly iron(III)protoporphyrin IX) with a deprotonated cysteine side chain providing the fifth ligand to iron, has challenged the 'biomimetic community'. Indeed, much has been accomplished over the past five decades, e.g., regarding the understanding of the unusual activation of paramagnetic dioxygen by reductive oxygen cleavage, as is evident from Chapter 2 which provides an in-depth overview of structural and functional mimics of P450s.

The structures of P450 proteins and their molecular phylogeny are detailed in Chapter 3, together with the P450 nomenclature and classification. It is made clear in this account that the P450 protein fold is unique and highly conserved independent of the organism; astonishingly, the same fold is also used by some enzymes that catalyze non-P450 redox transformations. It is worthwhile to note that some P450s function at extremes of pH and heat, as was recently discovered with archaeans. The diversity of P450s is also reflected in aquatic species as outlined in Chapter 4 where expecially P450 activities in invertebrates are in the focus.

Chapter 5 examines the ability of electrochemical techniques to unravel fundamental aspects of the electron transfer process of P450 enzymes. This process is central to P450 catalysis and thus, in Chapter 6 extensive studies are summarized which have shown that interprotein electron transfer is facilitated by proper positioning, e.g., of the flavin mononucleotide- and heme-containing domains. Clearly, generation of a truly catalytic system that utilizes non-native redox cofactors in place of the reductase proteins is a 'holy grail' of P450 research.

The next four chapters center on mechanistic considerations. At first leakage reactions are considered which occur during the P450 catalytic cycle, followed by detailed evaluations of the structural basis for substrate recognition and catalysis, including the architecture of the active site. Roles of the secondary coordination sphere, i.e., at the proximal (thiolate) and the distal side (where activation of dioxygen and substrate binding occurs) are described next, as is the coordination to the heme iron of several small-molecule inhibitors, such as nitrogen monoxide (nitric oxide), carbon monoxide, cyanide, and imidazole.

P450-catalyzed hydroxylations and epoxidations, the biosynthesis of steroid hormones, and the catalyzed carbon–carbon bond cleavage are discussed in Chapters 11–13. The design and engineering of cytochrome P450 systems is detailed in Chapter 14 regarding the oxidation of non-natural substrates. Evidently, potential applications of altered P450s in the environmentally benign synthesis of chemical products and intermediates are expected in this research area.

In the terminating three chapters of *The Ubiquitous Roles of Cytochrome P450 Proteins* first the biotransformation of xenobiotics is comprehensively dealt with, i.e., the 'chemical defense' or response of an organism to foreign chemicals. Thereafter the metabolism of drugs by human P450 systems is described, an area of particular significance for drug development and finally also for daily life in the clinic, as is emphasized by a clinical pharmacist.

Astrid Sigel
Helmut Sigel
Roland K. O. Sigel

Contents

HISTORICAL DEVELOPMENT AND PERSPECTIVES OF THE SERIES	v
PREFACE TO VOLUME 3	vii
CONTRIBUTORS TO VOLUME 3	xvii
TITLES OF VOLUMES 1–44 IN THE *METAL IONS IN BIOLOGICAL SYSTEMS* SERIES	xxi
CONTENTS OF VOLUMES IN THE *METAL IONS IN LIFE SCIENCES* SERIES	xxiii

1 DIVERSITIES AND SIMILARITIES IN P450 SYSTEMS: AN INTRODUCTION 1
Mary A. Schuler and Stephen G. Sligar

1. Oxygenases: Mediators of Biochemical Diversity	2
2. P450 Superfamily: Diversity at the Sequence Level	3
3. Diversity of P450 Structures: Folds and Conformations for Functions	5
4. Diversity in P450 Mechanisms	6
5. Diversity in Regulation Across the Superfamily	13
6. Diversity in the Evolution of Common Metabolic Functions	16
7. Summary and Outlook	18
Acknowledgments	19
Abbreviations	19
References	19

2 STRUCTURAL AND FUNCTIONAL MIMICS OF CYTOCHROMES P450 27
Wolf-D. Woggon

1. Introduction	27
2. Iron Porphyrins Carrying a Thiolate or Modified Thiolate Ligand	31

	3. Structurally Remote P450 Mimics	40
	4. Concluding Remarks	52
	Acknowledgments	52
	Abbreviations	53
	References	53

3 STRUCTURES OF P450 PROTEINS AND THEIR MOLECULAR PHYLOGENY — 57
Thomas L. Poulos and Yergalem T. Meharenna

1. Introduction	58
2. P450 Evolution	59
3. P450 Families and Subfamilies	60
4. P450 Structures	62
5. Variation in P450 Function and Fold	81
6. Archaeon P450s	86
7. Summary and Conclusions	90
Acknowledgments	91
Abbreviations	91
References	91

4 AQUATIC P450 SPECIES — 97
Mark J. Snyder

1. Introduction. 'P450s Under the Surface'	98
2. Diversity of Aquatic Species	100
3. P450 Activities in Aquatic Invertebrates	101
4. Aquatic P450 Gene Families Identified	108
5. How Can We Use Information About P450s in Aquatic Species?	116
6. Conclusions and Outlook	120
Acknowledgments	121
Abbreviations	121
References	122

5 THE ELECTROCHEMISTRY OF CYTOCHROME P450 — 127
Alan M. Bond, Barry D. Fleming, and Lisandra L. Martin

1. Introduction	127
2. Redox Titration (Potentiometric Equilibrium) Measurements	131
3. Voltammetric (Dynamic) Measurements	139
4. Conclusions	150
Acknowledgments	151
Abbreviations	151
References	152

6 P450 ELECTRON TRANSFER REACTIONS   157
Andrew K. Udit, Stephen M. Contakes, and Harry B. Gray

    1. Introduction   158
    2. Catalytic Cycles   158
    3. Electron Tunneling Wires   171
    4. Concluding Remarks   180
       Acknowledgments   181
       Abbreviations   181
       References   181

7 LEAKAGE IN CYTOCHROME P450 REACTIONS IN RELATION TO PROTEIN STRUCTURAL PROPERTIES   187
Christiane Jung

    1. Introduction   188
    2. Protein Structural Parameters   191
    3. The Reaction Cycle of Cytochrome P450   195
    4. Protein Structural Parameters and Extent of Competitive Reactions 222
    5. Concluding Remarks   225
       Acknowledgments   226
       Abbreviations   226
       References   227

8 CYTOCHROMES P450 – STRUCTURAL BASIS FOR BINDING AND CATALYSIS   235
Konstanze von König and Ilme Schlichting

    1. Introduction   236
    2. Ligand Binding: Substrate Recognition and Access to the Distal Pocket 237
    3. Architecture of the Active Site of CYP101   239
    4. The Distal Acid-Alcohol Pair   242
    5. Experimental Characterization of Reaction Intermediates. Radiolysis as a Tool to Study Redox Reactions   250
    6. Crystal Structures of Oxy-Ferrous Complexes   253
    7. Mechanism: Summary, Conclusions, Speculations   258
       Acknowledgments   260
       Abbreviations   260
       References   261

9 BEYOND HEME-THIOLATE INTERACTIONS: ROLES OF THE SECONDARY COORDINATION SPHERE IN CYTOCHROME P450 SYSTEMS — 267
Yi Lu and Thomas D. Pfister

1. Overview of Cytochrome P450 Active Site Structure — 268
2. Secondary Coordination Sphere on the Proximal Side — 269
3. Secondary Coordination Sphere on the Distal Side — 275
4. Summary and Outlook — 280
   Acknowledgments — 281
   Abbreviations — 281
   References — 281

10 INTERACTIONS OF CYTOCHROME P450 WITH NITRIC OXIDE AND RELATED LIGANDS — 285
Andrew W. Munro, Kirsty J. McLean, and Hazel M. Girvan

1. Introduction. Interactions of Ligands and Substrates with P450 Enzymes: General Features — 286
2. Nitric Oxide and Its Interactions with P450s — 289
3. Interactions of Imidazoles and Substituted Imidazoles with P450s — 300
4. Other Ligands and Inhibitors of P450 Function — 305
5. Conclusions and Future Prospects — 310
   Acknowledgments — 310
   Abbreviations — 311
   References — 311

11 CYTOCHROME P450-CATALYZED HYDROXYLATIONS AND EPOXIDATIONS — 319
Roshan Perera, Shengxi Jin, Masanori Sono, and John H. Dawson

1. Introduction — 320
2. The Cytochrome P450 Enzymes — 322
3. Three-Dimensional Structures of the Active Sites of Cytochrome P450 Enzymes — 327
4. Role of the Cys Ligand: the Proximal Thiolate 'Push' and Distal Proton-Delivery — 333
5. Multiple Mechanisms of P450 Catalysis — 337
6. Multiple Oxidants in P450 Catalysis — 343
7. Two States Theory — 347
8. Influence of Substrate on the Spectral Properties and Reactivity of P450 Intermediates — 348
9. Formation and Reactivity of Transient P450 Oxygen Intermediates — 351

	10. Summary and Future Prospective	352
	Acknowledgments	353
	Abbreviations	353
	References	354

12 CYTOCHROME P450 AND STEROID HORMONE BIOSYNTHESIS 361

Rita Bernhardt and Michael R. Waterman

1.	Introduction	362
2.	Steroidogenic P450s	363
3.	Steroid Hormone Biosynthesis in the Adrenal Cortex	374
4.	Steroid Hormone Biosynthesis in the Gonads	379
5.	Extraadrenal and Extragonadal Steroidogenesis	382
6.	Outlook for the Future	385
	Acknowledgments	386
	Abbreviations	387
	References	387

13 CARBON-CARBON BOND CLEAVAGE BY P450 SYSTEMS 397

James J. De Voss and Max J. Cryle

1.	Introduction	398
2.	Cleavage Between Oxygenated Carbons	398
3.	Cleavage Alpha to Oxygenated Carbons	408
4.	Cleavage Alpha to Carbon Bearing Nitrogen	422
5.	Carbon-Carbon Bond Cleavage Involving Peroxides	423
6.	General Conclusions	429
	Acknowledgments	430
	Abbreviations	430
	References	430

14 DESIGN AND ENGINEERING OF CYTOCHROME P450 SYSTEMS 437

Stephen G. Bell, Nicola Hoskins,
Christopher J. C. Whitehouse, and Luet L. Wong

1.	Introduction	438
2.	Engineering Bacterial Cytochrome P450 Systems	444
3.	Engineering Mammalian Cytochrome P450 Enzymes	462
4.	Engineering Plant P450 Enzymes	467
5.	Conclusions and Outlook	468
	Abbreviations	469
	References	469

15 CHEMICAL DEFENSE AND EXPLOITATION. BIOTRANSFORMATION OF XENOBIOTICS BY CYTOCHROME P450 ENZYMES   477
Elizabeth M. J. Gillam and Dominic J. B. Hunter

   1. Introduction. Chemical Defense   478
   2. P450 Systems Involved in Chemical Defense   481
   3. Common Themes   529
   4. Industrial Applications of P450 Systems for Xenobiotic Decomposition   530
   5. Conclusions and Future Prospects   534
   Acknowledgments   535
   Abbreviations   535
   References   537

16 DRUG METABOLISM AS CATALYZED BY HUMAN CYTOCHROME P450 SYSTEMS   561
F. Peter Guengerich

   1. Introduction   562
   2. Importance of P450 Enzymes in Drug Metabolism   563
   3. Approaches to Predicting P450 Activity in Humans   565
   4. P450s Involved in Drug Metabolism   570
   5. Examples of Major Issues Involving Drug Metabolism by P450   576
   6. Summary   582
   Acknowledgments   583
   Abbreviations   583
   References   584

17 CYTOCHROME P450 ENZYMES: OBSERVATIONS FROM THE CLINIC   591
Peggy L. Carver

   1. Introduction   592
   2. Drug Interactions   593
   3. Drug Metabolism by Cytochrome P450 Enzymes   594
   4. Alterations in P450 Enzymes   596
   5. Active Transport of Drugs   603
   6. Enzyme-Transporter Cooperativity   603
   7. Use of Probes to Quantitate CYP Activity in Humans   604
   8. Prodrugs   604
   9. The Effect of Intravenous Versus Oral Administration on Drug Interactions   605
   10. Additional Factors Affecting Drug Interactions   606
   11. Herbal and Dietary Effects on CYP   608

12.	Interactions with Commonly Used Medications	610
13.	Beneficial Effects of Drug Interactions	611
14.	FDA Regulations Regarding CYP450-Mediated Drug Interactions	612
15.	Clinical Significance of Drug Interactions	613
16.	Summary	614
	Acknowledgment	615
	Abbreviations	615
	References	615

SUBJECT INDEX   619

Contributors

Numbers in parentheses indicate the pages on which the authors' contributions begin.

Stephen G. Bell Department of Chemistry, Inorganic Chemistry Laboratory, Oxford University, South Parks Road, Oxford, OX1 3QR, UK, <stephen.bell@chem.ox.ac.uk> (437)

Rita Bernhardt Institut für Biochemie, Universität des Saarlandes, P. O. Box 151150, D-66041 Saarbrücken, Germany, <ritabern@mx.uni-saarland.de> (361)

Alan M. Bond School of Chemistry, Monash University, Clayton, Victoria 3800, Australia, <alan.bond@sci.monash.edu.au> (127)

Peggy L. Carver University of Michigan College of Pharmacy, Clinical Sciences Department, 428 Church St., Ann Arbor, MI 48109-1065, USA, <peg@umich.edu> (591)

Stephen M. Contakes The Beckman Institute, California Institute of Technology, Pasadena, CA 91125, USA, <contakes@caltech.edu> (157)

Max J. Cryle School of Molecular and Microbial Sciences, University of Queensland, Brisbane, QLD 4072, Australia (397)

John H. Dawson Department of Chemistry and Biochemistry, and School of Medicine, University of South Carolina, Columbia, SC 29208, USA, <dawson@sc.edu> (319)

James J. De Voss School of Molecular and Microbial Sciences, University of Queensland, Brisbane, QLD 4072, Australia, <j.devoss@uq.edu.au> (397)

Barry D. Fleming School of Chemistry, Monash University, Clayton, Victoria 3800, Australia, <barry.fleming@sci.monash.edu.au> (127)

Elizabeth M. J. Gillam School of Biomedical Sciences, The University of Queensland, St. Lucia, Brisbane, Qld 4072, Australia, <e.gillam@uq.edu.au> (477)

Hazel M. Girvan Manchester Interdisciplinary Biocentre, School of Chemical Engineering and Analytical Science, University of Manchester, 131 Princess Street, Manchester M1 7DN, UK (285)

Harry B. Gray The Beckman Institute, California Institute of Technology, Pasadena, CA 91125, USA, <hbgray@caltech.edu> (157)

F. Peter Guengerich Department of Biochemistry and Center in Molecular Toxicology, Vanderbilt University School of Medicine, 638 Robinson Research Building, 23rd & Pierce Avenues, Nashville, TN 37232-0146, USA, <f.guengerich@vanderbilt.edu> (561)

Nicola Hoskins Department of Chemistry, Inorganic Chemistry Laboratory, Oxford University, South Parks Road, Oxford, OX1 3QR, UK (437)

Dominic J. B. Hunter School of Biomedical Sciences, The University of Queensland, St. Lucia, Brisbane, Qld 4072, Australia, <d.hunter1@uq.edu.au> (477)

Shengxi Jin Department of Chemistry and Biochemistry, University of South Carolina, Columbia, SC 29208, USA, <jinshengxi@gmail.com> (319)

Christiane Jung Max-Delbrück Center for Molecular Medicine, Research Group Protein Dynamics, D-13125 Berlin, Germany. Present address: KKS Ultraschall AG, Surface Treatment Division, Frauholzring 29, CH-6422 Steinen, Switzerland, <christiane_jung@hotmail.com> (187)

Yi Lu Department of Biochemistry and Department of Chemistry, University of Illinois at Urbana-Champaign, A322 Chemical and Life Science Building, Box 8-6, 600 S. Matthews Avenue, Urbana, IL 61801, USA, <yi-lu@uiuc.edu> (267)

Lisandra L. Martin School of Chemistry, Monash University, Clayton, Victoria 3800, Australia, <lisa.martin@sci.monash.edu.au> (127)

Kirsty J. McLean Manchester Interdisciplinary Biocentre, School of Chemical Engineering and Analytical Science, University of Manchester, 131 Princess Street, Manchester M1 7DN, UK (285)

Yergalem T. Meharenna Department of Biochemistry and Molecular Biology, Physiology and Biophysics, and Chemistry, University of California, Irvine, CA 92697-3900, USA, <ymeharen@uci.edu> (57)

CONTRIBUTORS

Andrew W. Munro Manchester Interdisciplinary Biocentre, School of Chemical Engineering and Analytical Science, University of Manchester, 131 Princess Street, Manchester M1 7DN, UK, <andrew.munro@manchester.ac.uk> (285)

Roshan Perera Department of Chemistry and Biochemistry, University of South Carolina, Columbia, SC 29208, USA, <perera@scripps.edu> (319)

Thomas D. Pfister Department of Biochemistry, University of Illinois at Urbana-Champaign, A322 Chemical and Life Science Building, Box 8-6, 600 S. Matthews Avenue, Urbana, IL 61801, USA (267)

Thomas L. Poulos Department of Biochemistry and Molecular Biology, Physiology and Biophysics, and Chemistry, University of California, Irvine, CA 92697-3900, USA, <poulos@uci.edu> (57)

Ilme Schlichting Department of Biomolecular Mechanisms, Max Planck Institute for Medical Research, Jahnstrasse 29, D-69120 Heidelberg, Germany, <ilme.schlichting@mpimf-heidelberg.mpg.de> (235)

Mary A. Schuler Departments of Cell and Developmental Biology, Biochemistry and Plant Biology, 190ERML, 1201 W. Gregory Drive, University of Illinois, Urbana, IL 61801, USA, <maryschu@uiuc.edu> (1)

Stephen G. Sligar Departments of Biochemistry, Chemistry and the College of Medicine, 116 Morrill Hall, MC-119, 505 S. Goodwin Avenue, Urbana, IL 61801, USA, <s-sligar@uiuc.edu> (1)

Mark J. Snyder Department of Clinical Pharmacology, University of California at San Francisco, Kenwood, CA 95452, USA. Contact address: P. O. Box 609, Kenwood, CA 95452, USA, <snyder181@comcast.net> (97)

Masanori Sono Department of Chemistry and Biochemistry, University of South Carolina, Columbia, SC 29208, USA, <sono@mail.chem.sc.edu> (319)

Andrew K. Udit Department of Chemistry, Occidental College, Los Angeles, CA 90041, USA, <udit@oxy.edu> (157)

Konstanze von König Department of Biomolecular Mechanisms, Max Planck Institute for Medical Research, Jahnstrasse 29, D-69120 Heidelberg, Germany, <konstanze.von.koenig@mpimf-heidelberg.mpg.de> (235)

Michael R. Waterman Department of Biochemistry, Vanderbilt University School of Medicine, Nashville, TN 37232-0146, USA, <michael.waterman@vanderbilt.edu> (361)

Christopher J. C. Whitehouse Department of Chemistry, Inorganic Chemistry Laboratory, Oxford University, South Parks Road, Oxford, OX1 3QR, UK (437)

Wolf-D. Woggon Department of Chemistry, Organic Chemistry, University of Basel, St. Johannsring 19, CH-4056 Basel, Switzerland, <wolf-d.woggon@unibas.ch> (27)

Luet L. Wong Department of Chemistry, Inorganic Chemistry Laboratory, Oxford University, South Parks Road, Oxford, OX1 3QR, UK, <luet.wong@chem.ox.ac.uk> (437)

Titles of Volumes 1–44 in the
Metal Ions in Biological Systems Series
edited by the SIGELs
and published by Dekker/Taylor & Francis

Volume 1: Simple Complexes
Volume 2: Mixed-Ligand Complexes
Volume 3: High Molecular Complexes
Volume 4: Metal Ions as Probes
Volume 5: Reactivity of Coordination Compounds
Volume 6: Biological Action of Metal Ions
Volume 7: Iron in Model and Natural Compounds
Volume 8: Nucleotides and Derivatives: Their Ligating Ambivalency
Volume 9: Amino Acids and Derivatives as Ambivalent Ligands
Volume 10: Carcinogenicity and Metal Ions
Volume 11: Metal Complexes as Anticancer Agents
Volume 12: Properties of Copper
Volume 13: Copper Proteins
Volume 14: Inorganic Drugs in Deficiency and Disease
Volume 15: Zinc and Its Role in Biology and Nutrition
Volume 16: Methods Involving Metal Ions and Complexes in Clinical Chemistry
Volume 17: Calcium and Its Role in Biology
Volume 18: Circulation of Metals in the Environment
Volume 19: Antibiotics and Their Complexes
Volume 20: Concepts on Metal Ion Toxicity
Volume 21: Applications of Nuclear Magnetic Resonance to Paramagnetic Species
Volume 22: ENDOR, EPR, and Electron Spin Echo for Probing Coordination Spheres
Volume 23: Nickel and Its Role in Biology
Volume 24: Aluminum and Its Role in Biology
Volume 25: Interrelations Among Metal Ions, Enzymes, and Gene Expression
Volume 26: Compendium on Magnesium and Its Role in Biology, Nutrition, and Physiology

Volume 27:	**Electron Transfer Reactions in Metalloproteins**
Volume 28:	**Degradation of Environmental Pollutants by Microorganisms and Their Metalloenzymes**
Volume 29:	**Biological Properties of Metal Alkyl Derivatives**
Volume 30:	**Metalloenzymes Involving Amino Acid-Residue and Related Radicals**
Volume 31:	**Vanadium and Its Role for Life**
Volume 32:	**Interactions of Metal Ions with Nucleotides, Nucleic Acids, and Their Constituents**
Volume 33:	**Probing Nucleic Acids by Metal Ion Complexes of Small Molecules**
Volume 34:	**Mercury and Its Effects on Environment and Biology**
Volume 35:	**Iron Transport and Storage in Microorganisms, Plants, and Animals**
Volume 36:	**Interrelations Between Free Radicals and Metal Ions in Life Processes**
Volume 37:	**Manganese and Its Role in Biological Processes**
Volume 38:	**Probing of Proteins by Metal Ions and Their Low-Molecular-Weight Complexes**
Volume 39:	**Molybdenum and Tungsten. Their Roles in Biological Processes**
Volume 40:	**The Lanthanides and Their Interrelations with Biosystems**
Volume 41:	**Metal Ions and Their Complexes in Medication**
Volume 42:	**Metal Complexes in Tumor Diagnosis and as Anticancer Agents**
Volume 43:	**Biogeochemical Cycles of Elements**
Volume 44:	**Biogeochemistry, Availability, and Transport of Metals in the Environment**

Contents of Volumes in the *Metal Ions in Life Sciences* Series

edited by the SIGELs
and published by John Wiley & Sons, Ltd, Chichester, UK
<http://www.wiley.com/go/mils>

Volume 1: Neurodegenerative Diseases and Metal Ions

1. The Role of Metal Ions in Neurology. An Introduction
   Dorothea Strozyk and Ashley I. Bush
2. Protein Folding, Misfolding, and Disease
   *Jennifer C. Lee, Judy E. Kim, Ekaterina V. Pletneva,
   Jasmin Faraone-Mennella, Harry B. Gray, and Jay R. Winkler*
3. Metal Ion Binding Properties of Proteins Related to Neurodegeneration
   *Henryk Kozlowski, Marek Luczkowski, Daniela Valensin, and
   Gianni Valensin*
4. Metallic Prions: Mining the Core of Transmissible Spongiform Encephalopathies
   David R. Brown
5. The Role of Metal Ions in the Amyloid Precursor Protein and in Alzheimer's Disease
   Thomas A. Bayer and Gerd Multhaup
6. The Role of Iron in the Pathogenesis of Parkinson's Disease
   *Manfred Gerlach, Kay L. Double, Mario E. Götz, Moussa B. H. Youdim,
   and Peter Riederer*
7. *In Vivo* Assessment of Iron in Huntington's Disease and Other Age-Related Neurodegenerative Brain Diseases
   George Bartzokis, Po H. Lu, Todd A. Tishler, and Susan Perlman
8. Copper-Zinc Superoxide Dismutase and Familial Amyotrophic Lateral Sclerosis
   Lisa J. Whitson and P. John Hart
9. The Malfunctioning of Copper Transport in Wilson and Menkes Diseases
   Bibudhendra Sarkar
10. Iron and Its Role in Neurodegenerative Diseases
    Roberta J. Ward and Robert R. Crichton

11. The Chemical Interplay between Catecholamines and Metal Ions in Neurological Diseases
    Wolfgang Linert, Guy N. L. Jameson, Reginald F. Jameson, and Kurt A. Jellinger
12. Zinc Metalloneurochemistry: Physiology, Pathology, and Probes
    Christopher J. Chang and Stephen J. Lippard
13. The Role of Aluminum in Neurotoxic and Neurodegenerative Processes
    Tamás Kiss, Krisztina Gajda-Schrantz, and Paolo F. Zatta
14. Neurotoxicity of Cadmium, Lead, and Mercury
    Hana R. Pohl, Henry G. Abadin, and John F. Risher
15. Neurodegerative Diseases and Metal Ions. A Concluding Overview
    Dorothea Strozyk and Ashley I. Bush
    Subject Index

Volume 2: Nickel and Its Surprising Impact in Nature

1. Biogeochemistry of Nickel and Its Release into the Environment
   Tiina M. Nieminen, Liisa Ukonmaanaho, Nicole Rausch, and William Shotyk
2. Nickel in the Environment and Its Role in the Metabolism of Plants and Cyanobacteria
   Hendrik Küpper and Peter M. H. Kroneck
3. Nickel Ion Complexes of Amino Acids and Peptides
   Teresa Kowalik-Jankowska, Henryk Kozlowski, Etelka Farkas, and Imre Sóvágó
4. Complex Formation of Nickel(II) with Sugar Residues, Nucleobases, Phosphates, Nucleotides, and Nucleic Acids
   Roland K. O. Sigel and Helmut Sigel
5. Synthetic Models for the Active Sites of Nickel-Containing Enzymes
   Jarl Ivar van der Vlugt and Franc Meyer
6. Urease. Recent Insights on the Role of Nickel
   Stefano Ciurli
7. Nickel Iron Hydrogenases
   Wolfgang Lubitz, Maurice van Gastel, and Wolfgang Gärtner
8. Methyl-Coenzyme M Reductase and Its Nickel Corphin Coenzyme F_{430} in Methanogenic Archaea
   Bernhard Jaun and Rudolf K. Thauer
9. Acetyl-Coenzyme A Synthases and Nickel-Containing Carbon Monoxide Dehydrogenases
   Paul A. Lindahl and David E. Graham
10. Nickel Superoxide Dismutase
    Peter A. Bryngelson and Michael J. Maroney
11. Biochemistry of the Nickel-Dependent Glyoxylase I Enzymes
    Nicole Sukdeo, Elisabeth Daub, and John F. Honek

12. Nickel in Acireductone Dioxygenase
    *Thomas C. Pochapsky, Tingting Ju, Marina Dang, Rachel Beaulieu,
    Gina Pagani*, and *Bo Ouyang*
13. The Nickel-Regulated Peptidyl-Prolyl *cis/trans* Isomerase SlyD
    Frank Erdmann and Gunter Fischer
14. Chaperones of Nickel Metabolism
    Soledad Quiroz, Jong K. Kim, Scott B. Mulrooney, and Robert P. Hausinger
15. The Role of Nickel in Environmental Adaptation of the Gastric Pathogen
    Helicobacter pylori
    *Florian D. Ernst, Arnoud H. M. van Vliet, Manfred Kist,
    Johannes G. Kusters, and Stefan Bereswill*
16. Nickel-Dependent Gene Expression
    Konstantin Salnikow and Kazimierz S. Kasprzak
17. Nickel Toxicity and Carcinogenesis
    Kazimierz S. Kasprzak and Konstantin Salnikow
    Subject Index

Volume 3: The Ubiquitous Roles of Cytochrome P450 Proteins
(this book)

Volume 4: Biomineralization. From Nature to Application
(tentative contents)

1. Crystals and Life. An Introduction
   Arthur Veis
2. Gene-Directed Crystal Growth Exemplified by the Biomineralization of Calcium Carbonate
   Fred H. Wilt and Christopher E. Killian
3. The Role of Enzymes in Biomineralization Processes
   Ingrid M. Weiss and Frédéric Marin
4. Metal-Bacteria Interactions at Both the Planktonic Cell and Biofilm Levels
   Ryan C. Hunter and Terry J. Beveridge
5. Biomineralization of Calcium Carbonate. The Interplay with Biosubstrates
   Amir Berman and Yael Levi-Kalisman
6. Sulfate-Containing Biominerals
   Fabienne Bosselmann and Matthias Epple
7. Oxalate Biominerals
   Enrique J. Baran and Paula V. Monje
8. Structural Control, Molecular Components, and Multi-Level Regulation of Biosilification in Diatoms
   Aubrey K. Davis, Kim Thamatrakoln, and Mark Hildebrand
9. Dynamics of Biomineralization and Biodemineralization
   Lijun Wang and George H. Nancollas
10. Mechanism of Mineralization of Collagen Based Connective Tissues
    Adele J. Boskey

11. Mammalian Enamel Formation
    Janet Moradian-Oldak and Michael Paine
12. Heavy Metals in the Jaws of Invertebrates
    Helga C. Lichtenegger
13. Ferritin. Biomineralization of Iron
    Elizabeth C. Theil
14. Molecular Biology and Magnetism of Magnetic Iron Minerals in Bacteria
    Richard B. Frankel, Sabrina Schuebbe, and Dennis Bazylinski
15. Mechanical Design of Biomineralized Tissues
    Peter Fratzl
16. Biominerals. Recorders of the Past
    Danielle Fortin, Susan Glasauer, and Sean Langley
17. Bio-Inspired Growth of Mineralized Tissue
    Darilis Suarez and William L. Murphy
18. Biomineralization of Novel Inorganic Materials for Application
    Helmut Cölfen and Markus Antonietti
19. Crystal Tectonics. Chemical Construction and Self-Organization
    Annie K. Powell
    Subject Index

Comments and suggestions with regard to contents, topics, and the like for future volumes of the series are welcome.

1

Diversities and Similarities in P450 Systems: An Introduction

Mary A. Schuler[1] and Stephen G. Sligar[2]

[1]Department of Cell and Developmental Biology, University of Illinois,
Urbana, IL 61801, USA
<maryschu@uiuc.edu>

[2]Department of Biochemistry, University of Illinois,
Urbana, IL 61801, USA
<s-sligar@uiuc.edu>

1. OXYGENASES: MEDIATORS OF BIOCHEMICAL DIVERSITY	2
2. P450 SUPERFAMILY: DIVERSITY AT THE SEQUENCE LEVEL	3
3. DIVERSITY OF P450 STRUCTURES: FOLDS AND CONFORMATIONS FOR FUNCTIONS	5
4. DIVERSITY IN P450 MECHANISMS	6
4.1. Diversity of Redox Partners	6
4.2. The Heme-Oxygen Catalytic Landscape	9
4.3. The Oxy and Peroxo Iron Intermediates	10
4.4. High-Valent Metal-Oxo Complexes	11
4.5. Uncoupling: Nature's Leakage Pathways	12
4.6. Other Heme-Thiolate Systems: Needs from a Mechanistic Viewpoint	12
5. DIVERSITY IN REGULATION ACROSS THE SUPERFAMILY	13
5.1. Transcriptional Regulation	13
5.2. Post-translational Regulation	15
6. DIVERSITY IN THE EVOLUTION OF COMMON METABOLIC FUNCTIONS	16
6.1. Hormone Biosynthesis	16
6.2. Xenobiotic Catabolism	17
6.3. Fatty Acid Hydroxylases: Bacteria to Mammals to Plants	17

7. SUMMARY AND OUTLOOK 18
   ACKNOWLEDGMENTS 19
   ABBREVIATIONS 19
   REFERENCES 19

1. OXYGENASES: MEDIATORS OF BIOCHEMICAL DIVERSITY

The introduction of oxygen into biochemical processes has had a profound effect on the evolution of life. An appreciation for this traumatic event was presented in a beautiful recent review on the linkages of gene development that occurred at this juncture [1]. Although the first important utilization of atmospheric dioxygen was perhaps through its use as a terminal electron acceptor in metabolic energy conversion, an equally important leap in complexity and diversity was appreciated when, in the mid-1950s, Osamu Hayaishi and Howard Mason discovered the oxygenases [2,3]. This advance changed the simplistic view of how Nature uses atmospheric dioxygen from that of a simple electron acceptor and pointed to the rich metabolic diversity allowed by the incorporation of atmospheric dioxygen into substrate molecules. The discovery, naming and mechanistic understanding of the first 'oxygenase' enzymes have provided wonderful opportunities and scientific impetus to understand the great diversity of these systems in the synthesis and catabolism of organic molecules.

Before describing their various levels of diversity, one must consider the prime biochemical similarity that categorizes nearly all of them within the class of 'oxidases' that use atmospheric dioxygen as a terminal electron acceptor and, hence, yield an oxidized substrate molecule. In technical terms, the 'oxygenases' that exist within this broader oxidase category are classed as 'mono-' or 'di-', depending on whether one or both atoms of atmospheric dioxygen are incorporated into their respective substrates. In their reaction cycles, the classic stoichiometry of monooxygenases represents a sort of 'half-way' point on the pathway for the full four-electron reduction of dioxygen to generate two molecules of water as is typical of the redox counting of cytochrome 'oxidases'. Positioned midstream in this pathway, monooxygenases require only two electrons and two protons to reductively cleave atmospheric dioxygen, producing only a single water molecule in the process while saving the second atom for the incorporation and formal oxidation of the organic substrate molecule. As a result of their dual functionality, cytochrome P450 monooxygenases (P450s) are often referred to as 'mixed-function oxidases' since they possess both 'oxygenase' and 'oxidase' reactivities. The electron transfer functionalities of P450s also earned them the label of 'cytochrome'.

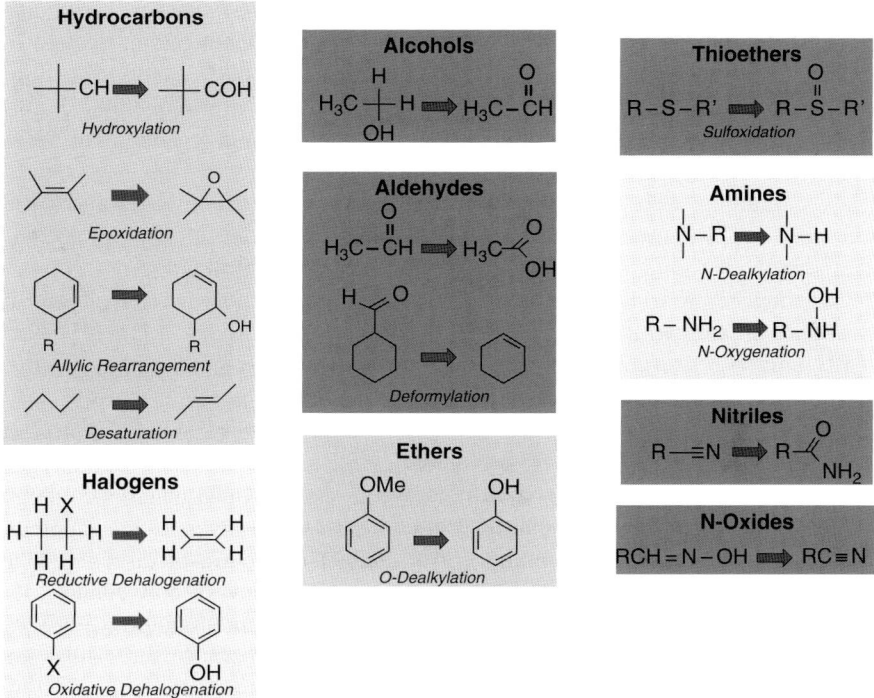

Figure 1. P450 chemistries.

This volume presents the breadth of research efforts focused on cytochrome P450 monooxygenases in their fullest, from structure through function to a deep appreciation of the diversity and complexity of the biotransformations that they catalyze (Figure 1). Historical perspectives on oxygenase discoveries over the past 50 years and mechanistic descriptions of their reaction cycles and metabolic transformations have been the subject of many other recent reviews [4–9]. Here, we focus on exploring the many different levels at which these enzymes (and their respective genes) have diverged in the process of evolution to yield the plethora of enzymes that are now termed 'P450 monooxygenases', even though they mediate a multitude of diverse reactions.

2. P450 SUPERFAMILY: DIVERSITY AT THE SEQUENCE LEVEL

Several years ago, realizing how many might be in store for future characterizations, researchers devised a nomenclature system for cytochrome monooxygenases (P450s) that designated sequences based on their degree of primary amino acid sequence identity [10,11]. In this system, the most highly related monooxygenase proteins were grouped into families whose members shared greater than 40% amino acid identity and designated with numbers (CYP1, CYP2, etc.) following the CYP (Cytochrome P450) designator used for all of these sequences. Families were divided into subfamilies whose members shared greater than 55% amino acid identity and designated with alphabetical characters (A, B, C, etc.). These subfamilies were further subdivided into individual loci designated with an additional set of numbers (CYP1A1, CYP1A2, CYP1A3, etc.). Without definitive genomic information demonstrating the existence of individual P450 loci, P450 sequences sharing more than 97% amino acid identity were designated as allelic variants with additional sets of numbers (v1, v2, etc.) [10,11]. The nearly universal acceptance of this highly structured nomenclature system provided an understandable index of the sequence relationships between the proteins found within a species as well as between the proteins found in the different kingdoms.

Stepping forward to the present, it is now widely appreciated that P450s exist in many bacteria and all archaea, fungi, and higher eukaryotes whose genomic DNA sequences have been completed. The numbers of full-length P450 genes existing in these species vary substantially in bacterial species from one in many species to 18–33 in some streptomycetes and 20–40 in some mycobacteria [12]. The numbers of full-length genes expand further in eukaryotic species, varying from three full-length P450 open reading frames (ORFs) in the *Saccharamyces cerevisiae* (budding yeast) genome to 34 in the *Chlamydomonas reinhardtii* genome, 46 in the *Apis mellifera* (honeybee) genome, 55 in the human genome, 71 in the *Physcomitrella patens* (moss) genome, 83 in the *Drosophila melanogaster* (fruitfly) genome, 100 in the *Anopheles gambiae* (mosquito) genome, 80 in the *Caenorhabditis elegans* (nematode) genome [11,13–16] (http://drnelson.utmem.edu, http://p450.antibes.inra.fr), 246 in the *Arabidopsis thaliana* genome [17–19], (http://Arabidopsis-P450.biotec.uiuc.edu, http://www.biobase.dk/P450/), and 356 in the *Oryza sativa* (rice) genome [13]. Multiply the number of organisms containing P450s by the number of full-length genes in any of them and there are indeed a large number of sequences within the phylogenetically diverse P450 superfamily. Present counts as of May 2006 include more than 5100 sequences (not counting alleles) in this massive superfamily (http://drnelson.utmem.edu). In some of the smaller plant genomes containing little repetitive DNA, they are estimated to represent 0.6% of the genome.

Genome-wide comparisons in some of these organisms with high P450 gene copy numbers have indicated that the degree of duplication and divergence in different P450 families and clans (their associated larger family groupings) are not constant. With the CYP51 family involved in sterol biosynthesis representing the only family common to fungi, animals, and plants [12,13,20,21], members of the plant-specific CYP71 family have proliferated to 52 members in *Arabidopsis* and 90 members in rice and members of the insect/animal-specific CYP4 family have proliferated from 4 members in the honeybee to 11 members in humans and 45 members in the mosquito (http://drnelson.utmem.edu; http://Arabidopsis-P450.biotec.uiuc.edu; http://p450.antibes.inra.fr). With P450 clans designated according to their family members with the lowest numeral, there are clearly some among the ten clans (CYP71) that have expanded substantially in plants compared with those that exist in animals.

With amino acid sequence identity representing one level of comparison among these many sequences, the organization of genes (as defined by the number and position of intron–exon junctions) within P450 families and subfamilies often supports the evolutionary relationships defined first by comparisons of these protein sequences. As an example of this organizational conservation, many of the 35 members of the *Arabidopsis* CYP71B subfamily have a single intron at the same place in their coding sequence, some have two introns at the same place and just two have no apparent introns (http://www.p450.kvl.dk; http://arabidopsis-p450.biotec.uiuc.edu). Scattered throughout the genome in clusters of duplicated P450 genes, all but one of the CYP71B loci with two introns are present within one tandem cluster while other clusters contain CYP71B loci with one intron. The same is true of the CYP71A subfamily where each in a cluster of six genes contains a single intron, each in three other sets of genes contain three introns and one more divergent gene contains four introns.

3. DIVERSITY OF P450 STRUCTURES: FOLDS AND CONFORMATIONS FOR FUNCTIONS

Despite these sequence diversities, P450 in many different families and many different organisms share a high degree of structural conservation in their secondary and tertiary folds [22–25] (see Chapter 3 in this volume). If one were to look at ribbon diagrams for the known P450 structures from across the room, all could immediately recognize the protein as belonging to the P450 superfamily. Moving closer, however, subtle variation in the positioning of secondary structure elements and the lengths of interconnecting loop regions contribute to the rich diversity of their catalytic sites and resulting specificities. Their commonality is manifested in a core structure of eleven α-helices (labeled A–K) and β-pleated sheets (labeled 1–4) surrounding the generally hydrophobic catalytic site buried within each protein. Variations in the lengths of the regions making up these core structures and in their intervening loops allow for these elements of secondary structure to create a diversity of three-dimensional active site structures. But, most important in our considerations of P450 catalytic site diversity is the fact that, within this core structure, comparatively small segments of the protein are involved in contacting the substrate and in the catalytic reaction cycle. These more limited regions include the loop between the B- and C-helices positioned over the heme (substrate recognition site 1 or SRS1 as originally described by Gotoh [26], the I-helix extending over the heme pyrrole ring B (SRS4), the amino-terminus of β-sheet 1–4 (SRS5) and the β-turn at the end of β-sheet 4 (SRS6). While quite variable in their individual sequences, alignments of 45 sequences representing each of the P450 subfamilies existing in *Arabidopsis* have indicated that no significant length variations exist in these internalized SRS regions [27]. Instead, some of the most prominent length variations occur in the previously mentioned α-helices and β-pleated sheets as well as external loop sequences and the loop between the F- and G-helices (between SRS2 and SRS3) that is involved in defining the substrate access and/or interactions with the endoplasmic reticulum membrane.

Alignments of the 20 crystal structures currently available for bacterial, fungal and mammalian P450s (listed in Chapter 3 of this volume [25]) have indicated that the most significant variations in backbone structure occur in three regions that have been designated as the B region (the loop between strand 5 of β-sheet 1, B′-helix, B-helix and B-C loop), the FG region (the C-terminus of the F-helix, the F-G loop and the N-terminus of the G-helix) and the β4 region (β-sheet 4) [28]. Structural backbone variations as well as side chain variations in these three regions as well as side chain variations in the previously mentioned SRS4 and SRS5 regions are the most likely contributors to diversity in substrate specificity among the P450s.

Many examples of the specificity differences conferred by very small variations in these SRS regions now exist in the naturally occurring differences between two closely related P450 proteins and in synthetically generated variations within a single P450. Examples of naturally occurring variations that lead to variations in substrate specificity are the mouse CYP2A4 and CYP2A5 sequences that mediate testosterone and coumarin hydroxylations, respectively, as the result of a limited number of amino acid variations in the SRS1, SRS2, SRS5, and SRS6 regions [29,30]. Other examples are the spearmint CYP71D15 and peppermint CYP71D18 sequences that differentiate between C6 and C3 limonene hydroxylations, respectively, based on a single amino acid variation in SRS5 (between the K helix and β1–4 strand) [31]. Examples of synthetically generated mutations that lead to variations in the substrate specificity of vertebrate and plant P450s are covered in several recent reviews [27,32,33].

4. DIVERSITY IN P450 MECHANISMS

4.1. Diversity of Redox Partners

The diversity of the P450 reactions can also be classified by the nature of the redox partners that introduce the two electrons required for the oxygenation of substrate. The systems carrying out this electron transfer reaction have been beautifully reviewed by Peterson and others [34–39]. Two major systems exist for those P450s that require an external feed from pyridine nucleotide. One of these utilizes a two-component electron transfer complex consisting of simplified FAD dehydrogenase and a small two-iron, two-sulfur redoxin to carry out the coupling of two-electron transfer to the sequential input of two redox equivalents needed by the P450 heme component [40]. The other utilizes a more complex flavoprotein, again with a FAD functioning as a hydride transfer catalyst interfacing with reduced pyridine nucleotide, but in this case, using a FMN prosthetic group cycling through a semiquinone to function as the two-to-one electron transformer.

In some bacterial systems, such as CYP102 (P450BM3) from *Bacillus megaterium*, the diflavin and heme catalytic domains are found linked into a single polypeptide [41]. Regardless of the nature of coupling to pyridine nucleotide, the P450 cycle needs a single electron to reduce the protein so that atmospheric dioxygen can bind and form the ferrous dioxygen complex. This intermediate was characterized first in the microbial P450CAM (CYP101) protein by Peterson, Gunsalus, and colleagues in the early 1970s [42,43] and has more recently been stabilized in the human CYP3A4 protein through incorporation into nanoscale, soluble phospholipid bilayers [44]. Interestingly, recent work from the Ortiz de Montellano laboratory has revealed further diversity in the provision of electron input in the case of CYP119 from the thermophilic *Sulfolobus solfataricus* [45]. Here, in addition to a temperature-stable iron-sulfur protein, a novel 2-oxoacid-ferredoxin oxidoreductase that utilizes pyruvic acid rather than NAD(P)H as the source of reducing equivalents couples to this system. In the future, one may expect to see additional variations in the mechanisms of providing the electrons needed for the classic P450 reaction cycle.

As reviewed in McLean et al. [46], additional diversity exists in some systems that do not use atmospheric dioxygen and two electrons as co-substrates and instead use the reduced dioxygen product, peroxide, as input to provide both the oxygen nucleus and the redox equivalents. One of the more unusual bacterial P450s in this category is *Bacillus subtilis* CYP152A1 that catalyzes hydroxylation of its long-chain fatty acid substrates directly using hydrogen peroxide [47]. Another is *Fusarium oxysporum* CYP55A1 (P450NOR) that catalyzes the hydroxylation of nitric oxide into nitric oxide by reduction of the P450 with NADH [48]. Some of the unusual eukaryotic P450s in this category are those in the plant CYP74A subfamily (allene oxide synthases) in which a hydroperoxide in the substrate is subsequently rearranged to form a reactive allene oxide that is subsequently converted to jasmonic acid [49,50]. Others in this unusual category are the plant CYP74B subfamily proteins (hydroperoxide lyases) and CYP74D subfamily proteins (divinyl ether synthases) that break down 13- and 9-carbon fatty acid hydroperoxides, respectively, into shorter signaling molecules and fungal defense compounds [51–53] and the mammalian CYP5A1 protein (thromboxane synthases) that catalyzes the isomerization of prostaglandin H2, yielding thromboxane A2 [54]. Interestingly, nonsteroidal anti-inflammatory drugs (NSAID) that interfere with the cyclooxygenase activity of prostaglandin endoperoxide H synthase (PGHS) and subsequent production of prostaglandins in mammals also competitively block allene oxide synthases and the subsequent production of jasmonic acid in plants [55].

In organellar compartments, such as animal mitochondria and plant chloroplasts, the electron transfer components are most similar to the bacterial FAD dehydrogenases and two-iron, two-sulfur redoxins. In the inner mitochondrial membranes of mammalian cells, adrenodoxin (Adx) and NADPH-dependent adrenodoxin reductase (AdR) provide electrons to CYP11A and CYP11B proteins, which are involved in cholesterol side chain cleavage and modification, and other P450s localized within the mitochondria [37]. In *Schizosaccharomyces pombe* (fission yeast), an iron sulfur protein (etp1) shares enough structural similarity with mammalian AdR that it can substitute for its activity in heterologous expression systems [56]. The pathogenic bacterium, *Mycobacterium tuberculosis*, also contains an electron transfer component (FprA) that is chemically and structurally related to mammalian AdR [57,58]. In plant chloroplasts, ferredoxin (Fd) and NADPH-dependent ferredoxin reductase (FNR), which are electron transfer components of the photosynthetic electron transfer chain [59], provide electrons to P450s normally localized within this organelle as well as heterologous plant, bacterial and mammalian P450s targeted to this organelle by genetic engineering [60,61]. In several unusual cases, some bacteria utilizing ferredoxin-like proteins have them fused in-frame with P450 coding sequences allowing them to attain optimal electron transfer rates within a single protein. Examples of this include a *Methylococcus capsulatus* CYP51 protein that is fused to a 3Fe-4S ferredoxin and a *Rhodococcus* sp. CYP116B2 that is fused to a dioxygenase reductase and a 2Fe-2S ferredoxin center [62–64].

In the cytosolic compartments of vertebrate, insect and plant cells, membrane-bound P450s found in the endoplasmic reticulum utilize NADPH-dependent P450 reductases and, sometimes, NADH-dependent cytochrome b_5 reductase/cytochrome b_5 complexes. Compared with the diversity of P450s in these organisms, their electron transfer partners are few in number and reasonably well conserved. In the organisms where complete genomic sequences are available, P450 reductase is encoded by one gene in the human, *C. elegans* (nematode), *A. gambiae* (mosquito), and *D. melanogaster* genomes and three genes in the *A. mellifera* (honeybee) genome (Genbank accessions). For reasons that are as yet unclear, multiple NADPH-dependent P450 reductase genes exist in most higher plant genomes with two identified in the sequenced *A. thaliana* and *Oryza sativa* (rice) genomes and two and three identified in *Helianthus tuberosus* (artichoke) and *Populus* sp. (poplar) cDNA collections, respectively [65,66]. Sequence comparisons among these indicate that the *Musca domestica* (housefly) P450 reductase most commonly used for heterologous expression in insect cell systems is 84% identical to *D. melanogaster* P450 reductase, 76% identical to *A. gambiae* P450 reductase and 66% identical to *Bombyx mori* (silkworm) P450 reductase and 55–56% identical to vertebrate P450 reductases. In comparison, the *Arabidopsis* P450 reductases are 61% identical to one another, 65–70% identical to the artichoke P450 reductases, 55–66% identical to the rice P450 reductases and 40–41% to vertebrate and fruitfly P450 reductases; the rice P450 reductases share approximately the same degree of relatedness to each other (63% identical) as the *Arabidopsis* P450 reductases share to one another. Cytochrome b_5 and NADH-dependent cytochrome b_5 reductase genes in these organisms have similarly low copy numbers in the vertebrate and insect genomes, higher copy numbers in plant genomes and high degrees of conservation among all.

Relevant to some of the post-translational regulatory mechanisms moderating P450 activities discussed in post-translational regulation (Section 5.2.), several of these electron transfer partners are subject to phosphorylation events that can modulate their activities.

4.2. The Heme-Oxygen Catalytic Landscape

Many very recent review articles have appeared which document the development of the current understanding of the P450 monooxygenase catalytic mechanism [7,67,68]. The most typical discussion of the reaction cycles of the cytochrome P450s utilizes a cyclic reaction path that begins with the protein substrate-free and the heme iron in the ferric state with the five d-electrons in a low spin ($S = 1/2$) configuration. This low-spin state is formed due to a large ligand field contributed by the axial thiolate 'fifth' ligand contributed by a cysteine residue and an axially coordinated 'sixth' water molecule.

The overall goal of any mechanistic understanding is the description of the intermediate states of heme, oxygen and substrate as well as the electrons/protons needed to link these states into a reaction cycle. As described earlier, the diversity of catalytic specificities in P450s exists because of the variations in particular sets of active site residues that have allowed these enzymes to evolve with varying degrees of plasticity. In some cases, such as the enzymes involved in the regiospecific hormone oxygenations occurring in humans, insects and plants, the catalytic site provides for a great deal of specificity and selectivity. In other cases, such as the human hepatic and insect midgut enzymes involved in xenobiotic detoxifications, the catalytic sites appear to have selectivities for broad classes of compounds, but are rather promiscuous with respect to the organic structures within particular classes. In both cases, however, it now appears that a complementarity between substrate and the active site geometry, as well as any linked induced conformational changes permitted, can result in displacement of the axial water with a resultant weakening of the ligand field and conversion of the protein to the high spin ($S = 5/2$) electronic configuration. This complementarity of substrate and active site displacing the water heme ligand has a substantial functional significance. Since the next step in the reaction cycle involves a ferric–ferrous reduction of the heme iron and the ferrous iron must be in the five coordinated high-spin state for subsequent dioxygen binding, a change in coordination of the ferric iron to the high-spin electronic configuration will facilitate electron transfer through a change in the redox potential of the metal center [69,70].

4.3. The Oxy and Peroxo Iron Intermediates

In the P450s studied to date, the oxy-ferrous state is only quasi-stable, and decays to the ferric resting state with the release of superoxide very rapidly. In the membrane-bound systems where substrate turnover is slower, this autoxidation reaction is thought to be responsible for the ultimate production of cytotoxic reactive oxygen species (ROS). A second electron input from the redox donor would then yield a two-electron reduced dioxygen heme center, a formal ferric-peroxo or ferric-hydroperoxo state. Although postulated for many decades based on simple electron counting, this state was only recently observed directly via low-temperature cryoradiolytic reduction in a series of seminal papers by Hoffman and collegues [71–73]. The ferric-peroxo state represents a critical branch point in the diversity of P450 reactions as indicated in Figure 2. The now classic P450 reaction cycle involves specific protonation of the peroxo anion to form the hydroperoxo and a subsequent proton delivery to result in a heterolytic scission of the O–O bond of heme-bound dioxygen. As envisioned by Groves

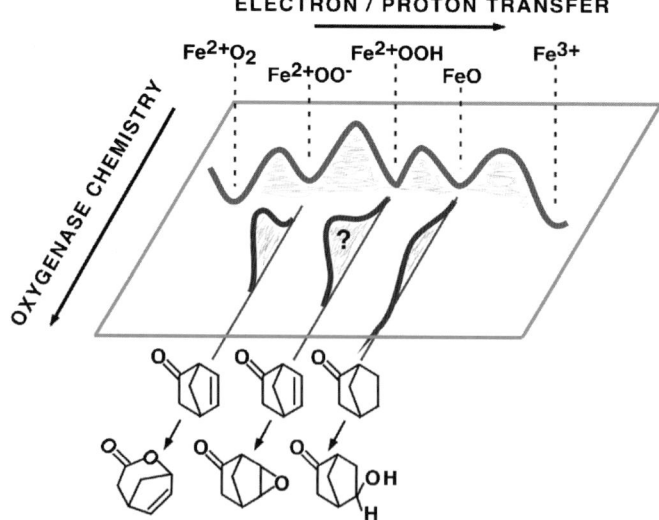

Figure 2. Schematic of the cytochrome P450 reactivity landscape.

(see Section 4.4. below) nearly three decades ago (and further reported in this volume), substrate oxygenation occurs through a step-wise hydrogen abstraction and radical recombination event in the enzyme active site.

However, in principle, each of the intermediates shown in Figure 2 could be active in substrate metabolism. The ferrous oxygenated state could potentially operate as a nucleophile since early Mössbauer spectroscopy from the Debrunner laboratory demonstrated that the iron was more ferric-like and the well-known 'alpha-effect' of the proximal oxygen lone pairs should push electron density out toward the distal oxygen atom [74]. Despite many attempts to show that a P450 ferrous-oxy state could catalyze ester hydrolysis or hemiacetal formation, no such reactivity has been demonstrated. However, a recent publication suggests that the simple one-electron reduced dioxygen complex can indeed serve as a catalyst in CYP2E1-mediated deboronation of bortezomib, a potent inhibitor of the 26S proteosome [75]. The input of a second reducing equivalent would form the peroxo anion, an even more potent nucleophile. Indeed, a peroxo heme adduct has been suggested to serve as the catalytic intermediate in the final step in the aromatization of the A-ring in estrogen biosynthesis [76] as well as in the production of nitric oxide by nitric oxide synthase [77].

4.4. High-Valent Metal-Oxo Complexes

Following formation of a hydroperoxo intermediate in the P450 reaction cycle, a second proton delivery will result in cleavage of the oxygen–oxygen bond and the formation of a higher-valent metal-oxo complex that is at the redox level of the peroxidase 'Compound I'. While a 'Compound I' state has been characterized structurally and spectroscopically in several of the peroxidases, it has yet to be observed in the dioxygen-dependent reaction cycle of P450. However, by focusing on the substrate and the resultant stereochemistry of carbon center functionalization, Groves and co-workers made the seminal discovery of a step-wise hydrogen abstraction mechanism [78,79]. Coon and colleagues first demonstrated that, under unfavorable substrate oxygenation conditions, a reaction cycle intermediate could be reduced by two additional redox equivalents to form a second water molecule [80], and the Sligar laboratory used isotope effects to show that commitment to oxygenase catalysis and water production shared a common intermediate [81]. These indirect tools have provided strong evidence for the existence of a higher-valent metal-oxo state in P450 reaction mechanisms.

The extremely hot 'Fe=O' oxidant could easily oxidize an unactivated alkane as well as perform easier heteroatom, alkene, or allylic oxygenations. The natural question that emerges when one contemplates the great diversity of substrates and metabolic profiles is whether product distribution and substrate specificity are controlled by protein constraints, inherent substrate site reactivities and/or specific isozyme requirements. White and colleagues have examined these parameters in the context of the bacterial CYP101 protein [82], but in more complex P450 active site architectures, there remains the possibility that complex homotropic and heterotropic cooperativity can alter the observed metabolic profiles [33]. This diversity of substrate recognition is of paramount importance to human health in that, in many cases, observed drug–drug interactions and determination of effective therapeutic dose are dictated by P450 turnover rates. The application of *in vitro* studies for the prediction of drug–drug interactions *in vivo* has recently been reviewed [83] and more detailed analyses, including the possible time-dependent interaction between multiple P450s has been presented for CYP2D6 [84] and CYP3A4 [85].

4.5. Uncoupling: Nature's Leakage Pathways

The complexity of the enzymatic cycle and the presence of several reactive intermediates along the reaction coordinate gives rise to another intrinsic feature of the P450 mechanism, which is the 'uncoupling' or leaking of reducing equivalents into nonproductive pathways capable of producing cytotoxic reactive oxygen species such as superoxide and peroxide. The overall efficiency of converting the consumption of electrons from pyridine nucleotide oxidation to the critical active intermediate(s) necessary for substrate metabolism depends on the kinetic partitioning between the commitment to catalysis of a particular P450 intermediate versus the dissociation of the corresponding iron-superoxide or iron-peroxide complex. Through a number of studies, this efficiency or uncoupling ratio has been shown to be sensitive to the hydration of the distal site of the heme and accessibility to solvent. The influence of the protein structure and dynamics on these leakage pathways is reviewed by Jung [86] (see Chapter 7 of this volume). Other factors which strongly modulate uncoupling include the substrate structure and mobility within the active center of the protein and can also be systematically probed using point mutations of the distal amino acids [87–90].

4.6. Other Heme-Thiolate Systems: Needs from a Mechanistic Viewpoint

There are several other classes of heme proteins which contain thiolate as a proximal heme axial ligand such as the nitric oxide synthases (NOS), chloroperoxidase, cystathionine β-synthase, and the sensor proteins CooA and eIF2a kinase. A brief overview of the properties of these has recently been presented by Omura [91]. An important difference between the enzymes and sensor proteins listed above is in the fact that the former retain their thiolate proximal ligand in the reduced ferrous state and show the definitive Soret band at 442–450 nm when saturated with carbon monoxide (CO) as opposed to the 'sensor' proteins wherein the thiolate is displaced by CO or nitric oxide (NO) in the reduced state. It is beyond the scope of this introductory chapter to provide a discussion of these other heme-thiolate systems although a complete understanding of the bioinorganic mechanisms of the P450 oxygenases benefits from discoveries in these related enzymes.

5. DIVERSITY IN REGULATION ACROSS THE SUPERFAMILY

5.1. Transcriptional Regulation

Layered on the previously mentioned structural diversity in catalytic site residues is transcriptional diversity in the range of tissues and stimuli capable of expressing individual P450 genes (except in organisms containing a single constitutively expressed P450). Because of this transcriptional diversity, even P450 loci coding for highly similar proteins have potential for mediating different physiological functions with individual P450s being expressed in one tissue and not the next. In mammalian systems, examples of this transcriptional diversity exist in the vertebrate CYP1A1 and CYP1A2 genes where CYP1A1 is not expressed at any detectable constitutive level, but highly induced by arylhydrocarbons in many tissues and CYP1A2 is constitutively expressed in liver and further induced by arylhydrocarbons only in liver [92–94]. Other examples exist in the numerous members of the human CYP2 family that are expressed in adults at distinctly higher levels in liver (CYP2E1), lung (CYP2S1), thymus (CYP2U1), etc. Displaying distinct developmental variations, these same loci are expressed in fetal tissues at varying levels that are sometimes higher and sometimes lower than observed in adult tissues [95].

In the insect world, examples of differentially regulated sets of related P450 transcripts are fewer in number, primarily because many of these have been cloned only in recent times. Those that display distinct developmental and tissue-specific expression patterns are the *D. melanogaster* mitochondrial CYP302A1, CYP314A1, and CYP315A1 proteins (also designated as the Halloween genes *dib, shd* and *sad*) which mediate the 22-, 20- and 2-hydroxylations, respectively, on the ecdysteroid nucleus [96]. Another microsomal CYP306A1 protein (*D. melanogaster phm*) mediates the 25-hydroxylation on the ecdysteroid nucleus [97,98]. Consistent with their role in early ecdysone synthesis, *dib, sad* and *phm* are expressed early in larval development in the prothoracic gland cells of the larval ring gland. And, *shd* that has a function in later ecdysone modifications, is expressed later in larval development and in multiple tissues. Another *D. melanogaster* CYP307A1 (designated as *spo*) in this pathway has an expression pattern suggesting that it has a role in early embryogenesis outside of prothoracic glands, but its exact function has not yet been defined [99].

The larger complements of P450 genes in plant genomes as well as the large-scale cDNA/EST and microarray projects being carried out in

Arabidopsis and *Oryza* (http://www.arabidopsis.org/; http://rgp.dna.affrc.go.jp/; http://www.tigr.org/tdb/e2k1/osa1/) have provided many more examples of differentially regulated transcripts within individual P450 subfamilies. Examples of this exist in the 5-member CYP86A subfamily that contains the functionally characterized CYP86A1, CYP86A2, CYP86A4, CYP86A7, and CYP86A8 involved in fatty acid hydroxylations [100–102], the 37-member CYP71B subfamily that contains the genetically characterized CYP71B15 in camalexin synthesis [103] and 36 uncharacterized members and the 17-member CYP71A subfamily whose functions are completely uncharacterized. Transcript profiling by microarray analyses as well as RT-PCR analyses have indicated that each of the members within these subfamilies is independently regulated, with some being expressed exclusively in one or another tissue and others being constitutively expressed in all tissues, albeit to different levels [104]. Enumerations of the full-length cDNAs existing for each of the 246 full-length cDNAs in *Arabidopsis* [104] make this point eminently clear with several loci in these particular subfamilies represented by 5–7 full-length cDNAs, others represented by 3–4 full-length cDNAs and yet others not represented by any cDNAs or ESTs.

Apart from tissue and developmental cues triggering transcription, individual P450 loci in animals and plants are capable of responding to varying sets of chemical inducers encountered in their dietary sources or through exposure to environmental toxins. Among the best characterized of these transcriptional activating molecules are polycyclic aromatic hydrocarbons, such as 3-methylcholanthrene (3-MC), benzo[α]pyrene, β-naphthoflavone and 2,3,7,8-tetrachlorodibenzo-*p*-dioxin (TCDD, dioxin) and the drug phenobarbital (PB). In mammalian systems, the signal transduction cascades and the range of genes activated by these two groups of compounds differ: arylhydrocarbons induce expression of genes in the CYP1A subfamily whereas phenobarbital and its related compounds induce expression of genes in the CYP2A, CYP2B, CYP2C, and CYP2D subfamilies [105,106]. In addition to these extensively studied xenobiotic inducers, many of the compounds encountered in plant food sources or used to preserve food materials are capable of inducing mRNAs for vertebrate and insect phase I (P450) and phase II (glutathione S-transferase) detoxicative activities. Among these, furanocoumarins represent a much-studied group of plant defense compounds that are capable of inducing CYP1A1, CYP1A2, and CYP2B1 transcripts in rats [107,108] and CYP6B and CYP9A subfamily transcripts in insects [109–115]. Other larger groups of compounds that are capable of activating vertebrate antioxidant response cascades include antioxidants (e.g., the common food preservative *tert*-butylhydroquinone) [116], natural α,β-unsaturated aldehydes (e.g., *trans*-2-hexanal) [117] and simple phenolic compounds (e.g., caffeic acid) [118].

In addition to these examples of transcriptional induction of individual genes, a small number of P450 loci are subject to transcriptional suppression. One of the most prominent examples of this phenomenon is the vertebrate male-specific

CYP2C11 that is down-regulated in response to arylhydrocarbons, such as 3-MC and TCDD without changing the half-life of its transcript [119,120]. Other examples of P450 suppression are reviewed in Lee and Riddick [119].

5.2. Post-translational Regulation

Studies on a variety of vertebrate P450s have reported that some are post-translationally modified by phosphorylation (CYP2B1, CYP2B4, CYP2E1, CYP11A1, CYP17A1), glycosylation (CYP11A1, CYP19A1), nitration (CYP4A subfamily), and ubiquitination (CYP3A4, CYP2B1) [122–124]. Evidence indicates that, in the cases of CYP2B1, CYP2B4 and CYP2E1, phosphorylation of a serine located on the P450 surface in the C-helix is effected by cAMP-dependent protein kinase A (PKA). Modification at these sites, which occur at analogous positions in these three proteins, provides for the rapid post-translational repression of these particular enzymes and significant reduction in the synthesis of their downstream components. In the case of CYP2E1, phosphorylation has been shown to act as a switch inactivating the microsomal enzyme immediately with kinetics much more rapid than if its activity was regulated by transcriptional repression or protein degradation [123]. Phosphorylation at this site has also been shown to decrease the proportion of CYP2E1 targeted to the endoplasmic reticulum membrane, allowing some to be targeted into the mitochondria using this protein's cryptic organellar targeting sequence [125].

In the case of mitochondrial CYP11A1, phosphorylation of serines and threonines is effected by cAMP-dependent protein kinase C (PKC) resulting in the activation of this particular enzyme's activity [126]. And, in the case of CYP17A1, phosphorylation of serines and threonines by PKA results in the activation of this enzyme's activity [127]. In other cases, such as CYP3A4, CYP2B1, and CYP2E1, ubiquitination targets these proteins to degradation pathways [128,129] through mechanisms that are distinct from the classical ubiquitin-dependent degradation pathways [130].

According to our present understanding of this post-translational regulatory mechanism, low-level ubiquitination of the CYP3A proteins appears to be linked to extended activation of their expression and followed by formation of unusual high-molecular-weight CYP3A4-ubiquitin aggregates in microsomal membranes free of cytosolic components. The fact that these aggregates are subsequently degraded by the cytosolic components and blocked from forming in the presence of substrates suggests a link between ubiquitination and substrate stabilization of this particular group of vertebrate P450s [130,131].

The effects of some other post-translational modifications are less well characterized. In the case of CYP19A1, glycosylation has been shown to occur on a residue within the N-terminal signal sequence that is capable of directing insertion of this P450 across the ER membrane in the types of insect cells used for heterologous expression [132], but the significance of this modification is not yet clear. These examples serve to illustrate the sorts of modifications affecting individual P450 activities and are not meant to be comprehensive. Further information on the range of vertebrate P450s affected by these modifications are available in Aguiar et al. [124]. Little is yet known about the extent of post-translational modification occurring on insect and plant P450s.

With some of these modifications impacting the activities of the P450 monooxygenases on which they occur, it is worth noting that the extent and types of post-translational modification vary substantially among the heterologous expression systems currently being used for production and analysis of P450 activities.

6. DIVERSITY IN THE EVOLUTION OF COMMON METABOLIC FUNCTIONS

Despite their obvious differences in physical appearance, many organisms maintain a set of common metabolic processes that include the synthesis of hormones, modification of fatty acids, catabolism of xenobiotics, signaling molecules, etc., mediated in different ways by highly divergent P450s. Several recent reviews detail the range of P450-mediated reactions currently known in bacteria and fungi [12], insects [96,133,134], plants [18,19,104,135], and humans [33]. The broad range of P450s mediating common functions are exemplified in the following three categories of reactions.

6.1. Hormone Biosynthesis

Leading to production of sterols that serve as essential structural components of membranes and as precursors for steroid hormones, CYP51 sequences exist in all P450-containing organisms [12,13,20,136]. Testosterone and estradiol, which represent well-characterized examples of the sex-specific steroid hormones occurring in vertebrates, are derived from their cholesterol precursor via CYP11A, CYP17A, and CYP19A subfamily members that are partitioned between the mitochondria and endoplasmic reticulum [33]. Subsequent modifications are mediated by CYP2A, CYP2B, CYP2C, and CYP3A subfamily members located in the endoplasmic reticulum. Brassinosteroids, which represent the plant equivalents of these vertebrate steroid hormones, are synthesized by multiple P450-mediated modifications on the phytosterol skeleton derived from cholesterol [137–139]. Several members in the plant CYP85A, CYP90A, CYP90B, CYP90C, and CYP90D subfamilies mediate these modifications in a biochemical grid that interconnects various intermediates in the synthesis of castasterone and brassinolide, the biologically active forms [140–146]. Ecdysone, which is the insect steroid hormone, is synthesized by an array of P450s in the CYP302A, CYP306A, CYP307A, CYP314A and CYP315A subfamilies that were previously mentioned in Section 5.1.

Interestingly, some of these P450-mediated modifications as well as other independent P450-mediated modifications on these steroidal hormones lead to their inactivation. Here, examples include the CYP3A-mediated 6β-, 16β-, and 11β-hydroxylations of testosterone in vertebrates [33,147], the CYP734A1-mediated 26-hydroxylation of brassinolide in plants as well as the CYP72C-mediated hydroxylation of brassinolide (at an undefined position) [148–151]. P450s also mediate the inactivation of ecdysone in insects via a 26-hydroxylation [152].

6.2. Xenobiotic Catabolism

With many P450s known to mediate xenobiotic catabolism in vertebrates, an appreciation of diversity and inter-organismal comparisons are best focused by addressing P450s involved in catabolism of plant toxins encountered to varying degrees by the vertebrates and insects that ingest them. Inactivations of furanocoumarins, which were also previously mentioned in transcriptional regulation (Section 5.1), are mediated by the CYP3A4 proteins in humans and closely related enzymes in other vertebrates [33] and CYP6B subfamily proteins as well as the CYP321A1 protein in insects [153–157]. Vertebrate and insect P450s involved in detoxification of this class of compounds attack the double bond of the furan ring as well as methoxy and other substituents on the furanocoumarin core structure [158–160]. Owing to the dimensions of their catalytic sites, P450s in other families and subfamilies are often targets for inactivation by furanocoumarins with, for example, human CYP2A6 being inhibited by binding of this compound in its narrowly constrained catalytic site [161,162].

6.3. Fatty Acid Hydroxylases: Bacteria to Mammals to Plants

Fatty acid hydroxylations are mediated by a wide range of P450s in vastly different subfamilies in different organisms. In the broad set of activities catalyzed by fatty acid hydroxylases, those that catalyze the hydroxylation of the terminal methyl group on aliphatic fatty acid chains are designated as 'ω-hydroxylases' while those that catalyze hydroxylations on internal carbons are designates as 'in-chain hydroxylases'. Heterologous expressions of these enzymes have indicated they differ in their preferences for fatty acid chain length and degree of saturation and substitution on these chains.

In bacteria, one of the first enzymes characterized was the previously mentioned *B. megaterium* CYP102A1 (P450BM3) that hydroxylates C12 to C18 fatty acids. The more recently *B. subtilis* CYP102A2 and CYP102A3 proteins have been shown to have similar preferences for long-chain fatty acids [163].

In the blue-green alga *Anabaena variabilis*, the CYP110 protein mediates hydroxylation of long-chain saturated and unsaturated fatty acids [164]. In the alkane-assimilating yeast *Candida maltosa*, CYP52A subfamily proteins mediate hydroxylations of C12 to C16 fatty acids with varying preferences and efficiencies [165]. In animals, several of the many CYP4A subfamily members have been shown to hydroxylate fatty acids [166,167]. In comparison with the fatty acid hydroxylases existing in bacteria and plants, several of these mammalian fatty acid hydroxylases are unusual in that they covalently ligate the heme to the backbone of the I-helix at a glutamic acid positioned four residues prior to the (D/E)T in the course of their hydroxylation reactions [168]. In insects, only one protein, *D. melanogaster* CYP6A8, has been shown to mediate any sort of fatty acid hydroxylation [169]. In plants, members of many P450 subfamilies have been shown to catalyze ω-hydroxylations, in-chain hydroxylations and epoxidations of medium- and long-chain fatty acids [170]. The range of plant-specific P450 families involved include the CYP81B, CYP86A, and CYP94A families as well as the most recently identified CYP76A and CYP709C families [100–102,171–178]. The breadth of P450 families involved in fatty acid transformations in these organisms exists in striking contrast with the widely conserved CYP51 family involved in core sterol synthesis.

7. SUMMARY AND OUTLOOK

Looking across the tree of life, it is amazing that one finds cytochrome P450s represented in all life forms. Interestingly, common to these various life forms is a bipartite functionality. One major function of these monooxygenases is in xenobiotic metabolism and detoxification. Here, one finds the use of these monooxygenases in reactions ranging from the metabolism of pharmaceuticals in humans [33,179] to the removal of toxic substances in insects and plants. A second major role in all life forms finds the P450s functioning in the biosynthesis of hormones and signaling molecules, again providing critical functions in mammals, insects and plants.

It is clear that, despite their many differences, their similarities and existing technologies available for transferring genes from organism to the next are now allowing researchers with clear insight into the machinations of these enzymes to engineer P450s for many biotechnological applications. As summarized by Bernhardt [180], recombinant P450s have already been used for improving production of pharmaceuticals in bacterial systems and enhancing synthetic processes in plants with many remaining to be tested. Looking ahead in this ever-expanding field of research is exciting with work on P450 diversity providing the tools for many future biotechnological applications.

ACKNOWLEDGMENTS

The authors thank Dr Ilia Denisov and Mr Sanjeewa Rupasinghe for scientific contributions and Ms Anu Murphy and Ms Kara Sandfort for compiling references. Research on insect P450s is supported by National Institutes of Health R01 GM071826 (MAS), on plant P450s is supported by National Science Foundation grant NSF2010 MCB 0115068 (MAS), and on bacterial and mammalian P450s is supported by National Institutes of Health R37 GM31756 and R37 GM33775 (SGS).

ABBREVIATIONS

Adr	adrenodoxin reductase
Adx	adrenodoxin
FAD	flavin adenine dinucleotide
Fd	ferredoxin
FMN	flavin mononucleotide
FNR	NADPH-dependent ferredoxin reductase
3-MC	3-methylcholanthrene
NAD(P)H	nicotinamide adenine dinucleotide (phosphate), reduced
NOS	nitric oxide synthase
NSAID	nonsteroidal anti-inflammatory drug
ORF	open reading frame
PB	phenobarbital
PGHS	prostaglandin endoperoxide H synthase
PKA	cAMP-dependent protein kinase A
PKC	cAMP-dependent protein kinase C
ROS	reactive oxygen species
TCDD	2,3,7,8-tetrachlorodibenzo-p-dioxin

REFERENCES

1. J. Raymond and D. Segre, *Science, 311*, 1764–1767 (2006).
2. O. Hayaishi, M. Katagiri, and S. Rothberg, *J. Am. Chem. Soc., 77*, 5450–5451 (1955).
3. H. S. Mason, L. Fowlks, and E. Peterson, *J. Am. Chem. Soc., 77*, 2851–2914 (1955).
4. M. Sono, M. P. Roach, E. D. Coulter, and J. H. Dawson, *Chem. Rev., 96*, 2841 (1996).
5. F. P. Guengerich, *Chem. Res. Toxicol., 14*, 611–650 (2001).
6. F. P. Guengerich, *Curr. Drug Metab., 2*, 93–115 (2001).
7. I. G. Denisov, T. M. Makris, S. G. Sligar, and I. Schlichting, *Chem. Rev., 105*, 2253–2277 (2005).

8. T. M. Makris, I. Denisov, I. Schlichting, and S. G. Sligar, in *Cytochrome P450: Structure, Mechanism, and Biochemistry* (P. R. Ortiz de Montellano, ed.), 3rd edn, Kluwer Academic/Plenum Publishers, New York, 2005, pp. 149–182.
9. T. M. Makris, K. von Koenig, I. Schlichting, and S. G. Sligar, *J. Inorg. Biochem.*, 100, 507–518 (2006).
10. D. R. Nelson, T. Kamataki, D. J. Waxman, F. P. Guengerich, R. W. Estabrook, R. Feyereisen, F. J. Gonzalez, M. J. Coon, I. C. Gunsalus, O. Gotoh, K. Okuda, and D. W. Nebert, *DNA Cell Biol.*, 12, 1–51 (1993).
11. D. R. Nelson, L. Koymans, T. Kamataki, J. J. Stegeman, R. Feyereisen, D. J. Waxman, M. R. Waterman, O. Gotoh, M. J. Coon, R. W. Estabrook, I. C. Gunsalus, and D. W. Nebert, *Pharmacogenetics*, 6, 1–41 (1996).
12. S. L. Kelly, D. E. Kelly, C. J. Jackson, A. G. S. Warrilow, and D. C. Lamb, in *Cytochrome P450: Structure, Mechanism, and Biochemistry* (P. R. Ortiz de Montellano, ed.), 3rd edn, Kluwer Academic/Plenum Publishers, New York, 2005, pp. 585–617.
13. D. R. Nelson, M. A. Schuler, S. M. Paquette, D. Werck-Reichhart, and S. Bak, *Plant Physiol.*, 135, 756–772 (2004).
14. D. R. Nelson, *Phytochem. Rev.*, in press (2006).
15. N. Tijet, C. Helvig, and R. Feyereisen, *Gene*, 262, 189–198 (2001).
16. C. Claudianos, H. Ranson, R. M. Johnson, S. Biswas, M. A. Schuler, M. R. Berenbaum, R. Feyereisen, and J. G. Oakeshott, *Insect. Mol. Biol.*, 15, 615–636 (2006).
17. S. M. Paquette, S. Bak, and R. Feyereisen, *DNA Cell Biol.*, 19, 307–317 (2000).
18. D. Werck-Reichhart, S. Bak, and S. Paquette, in *The Arabidopsis Book* (C. R. Somerville and E. M. Meyerowitz, eds), American Society of Plant Biologists, Rockville, MD, 2002, doi/10.1199/tab.0028, http://www.aspb.org/publications/arabidopsis
19. M. A. Schuler and D. Werck-Reichhart, *Annu. Rev. Plant Biol.*, 54, 629–667 (2003).
20. N. Debeljak, M. Fink, and D. Rozman, *Arch. Biochem. Biophys.*, 409, 159–171 (2003).
21. M. R. Waterman and G. I. Lepesheva, *Biochem. Biophys. Res. Commun.*, 338, 418–422 (2005).
22. S. E. Graham and J. A. Peterson, *Arch. Biochem. Biophys.*, 369, 24–29 (1999).
23. E. F. Johnson and C. D. Stout, *Biochem. Biophys. Res. Commun.*, 338, 331–336 (2005).
24. T. L. Poulos and E. F. Johnson, in *Cytochrome P450: Structure, Mechanism, and Biochemistry* (P. R. Ortiz de Montellano, ed.), 3rd edn, Kluwer Academic/Plenum Publishers, New York, 2005, pp. 87–111.
25. T. L. Poulos and Y. T. Meharenna, in *The Ubiquitous Roles of P450 Proteins*, Vol. 3 of *Metal Ions in Life Sciences* (A. Sigel, H. Sigel, and R. K. O. Sigel, eds), John Wiley & Sons, Ltd, Chichester, UK, 2007, pp. 57–96.
26. O. Gotoh, *J. Biol. Chem.*, 267, 83–90 (1992).
27. S. Rupasinghe and M. A. Schuler, *Phytochem. Rev.*, in press (2006).
28. J. Baudry, S. Rupasinghe, and M. A. Schuler, *Protein Eng. Design Selec.*, 19, 345–353 (2006).

29. R. L. P. Lindberg and M. Negishi, *Nature, 339*, 632–634 (1989).
30. M. Negishi, M. Iwasaki, R. O. Juvonen, T. Sueyoshi, T. A. Darden, and L. G. Pedersen, *Mutat. Res., 350*, 43–50 (1996).
31. M. Schalk and R. Croteau, *Proc. Natl. Acad. Sci. USA, 97*, 11948–11953 (2000).
32. T. L. Domanski and J. R. Halpert, *Curr. Drug Metab., 2*, 117–137 (2001).
33. F. P. Guengerich, in *Cytochrome P450: Structure, Mechanism, and Biochemistry* (P. R. Ortiz de Montellano, ed.), 3rd edn, Kluwer Academic/Plenum Publishers, New York, 2005, pp. 377–530.
34. M. J. Hintz and J. A. Peterson, *J. Biol. Chem., 256*, 6721–6728 (1981).
35. P. W. Roome and J. A. Peterson, *Arch. Biochem. Biophys., 266*, 32–40 (1988).
36. P. W. Roome and J. A. Peterson, *Arch. Biochem. Biophys., 266*, 41–50 (1988).
37. A. V. Grinberg, F. Hannemann, B. Schiffler, J. Muller, U. Heinemann, and R. Bernhardt, *Proteins, 40*, 590–612 (2000).
38. I. F. Sevrioukova, H. Li, and T. L. Poulos, *J. Mol. Biol., 336*, 889–902 (2004).
39. M. J. I. Paine, N. S. Scrutton, A. W. Munro, A. Gutierrez, G. C. K. Roberts, and C. R. Wolf, in *Cytochrome P450: Structure, Mechanism, and Biochemistry* (P. R. Ortiz de Montellano, ed.), 3rd edn, Kluwer Academic/Plenum Publishers, New York, 2005, pp. 115–148.
40. M. B. Murataliev, R. Feyereisen, and F. A. Walker, *Biochim. Biophys. Acta, 1698*, 1–26 (2004).
41. L. O. Narhi and A. J. Fulco, *J. Biol. Chem., 261*, 7160–7169 (1986).
42. J. A. Peterson, Y. Ishimura, and B. W. Griffin, *Arch. Biochem. Biophys., 149*, 197–208 (1972).
43. J. D. Lipscomb, S. G. Sligar, M. J. Namtvedt, and I. C. Gunsalus, *J. Biol. Chem., 251*, 1116–1124 (1976).
44. I. G. Denisov, Y. V. Grinkova, B. J. Baas, and S. G. Sligar, *J. Biol. Chem., 281*, 23313–23318 (2006).
45. A. V. Puchkaev and P. R. Ortiz de Montellano, *Arch. Biochem. Biophys., 434*, 169–177 (2005).
46. K. J. McLean, M. Sabri, K. R. Marshall, R. J. Lawson, D. G. Lewis, D. Clift, P. R. Balding, A. J. Dunford, A. J. Warman, J. P. McVey, A. M. Quinn, M. J. Sutcliffe, N. S. Scrutton, and A. W. Munro, *Biochem. Soc. Trans., 33*, 796–801 (2005).
47. I. Matsunaga, A. Ueda, N. Fujiwara, T. Sumimoto, and K. Ichihara, *Lipids, 34*, 841–846 (1999).
48. Y. Shiro, M. Fujii, T. Iizuka, S. Adachi, K. Tsukamoto, K. Nakahara, and H. Shoun, *J. Biol. Chem., 270*, 1617–1623 (1995).
49. W. C. Song, C. D. Funk, and A. R. Brash, *Proc. Natl. Acad. Sci. USA, 90*, 8519–8523 (1993).
50. N. Tijet and A. R. Brash, *Prostaglandins Other Lipid Meditat., 68–69*, 423–431 (2002).
51. K. Matsui, M. Shibutani, T. Hase, and T. Kajiwara, *FEBS Lett., 394*, 21–24 (1996).
52. A. Itoh and G. A. Howe, *J. Biol. Chem., 276*, 3620–3627 (2001).
53. A. N. Grechkin, *Prostaglandins Other Lipid Meditat., 68–69*, 457–470 (2002).
54. L. H. Wang, A. L. Tsai, and P. Y. Hsu, *J. Biol. Chem., 276*, 14737–14743 (2001).
55. Z. Pan, B. Camara, H. W. Gardner, and R. A. Backhaus, *J. Biol. Chem., 273*, 18139–18145 (1998).

56. M. Bureik, B. Schiffler, Y. Hiraoka, F. Vogel, and R. Bernhardt, *Biochemistry, 41*, 2311–2321 (2002).
57. R. T. Bossi, A. Aliverti, D. Raimondi, F. Fischer, G. Zanetti, D. Ferrari, N. Tahallah, C. S. Maier, A. J. Heck, M. Rizzi, and A. Mattevi, *Biochemistry, 41*, 8807–8818 (2002).
58. K. J. McLean, N. S. Scrutton, and A. W. Munro, *Biochem. J., 372*, 317–327 (2003).
59. D. Ohta and M. Mizutani, *Front. Biosci., 9*, 1587–1597 (2004).
60. D. P. O'Keefe, J. M. Tepperman, C. Dean, K. J. Leto, D. L. Erbes, and J. T. Odell, *Plant Physiol., 105*, 473–482 (1994).
61. T. Lacour and H. Ohkawa, *Biochim. Biophys. Acta., 1433*, 87–102 (1999).
62. C. J. Jackson, D. C. Lamb, T. H. Marczylo, A. G. Warrilow, N. J. Manning, D. J. Lowe, D. E. Kelly, and S. L. Kelly, *J. Biol. Chem., 277*, 46959–46965 (2002).
63. G. A. Roberts, G. Grogan, A. Greter, S. L. Flitsch, and N. J. Turner, *J. Bacteriol., 184*, 3898–3908 (2002).
64. D. J. Hunter, G. A. Roberts, T. W. Ost, J. H. White, S. Muller, N. J. Turner, S. L. Flitsch, and S. K. Chapman, *FEBS Lett., 579*, 2215–2220 (2005).
65. M. Mizutani and D. Ohta, *Plant Physiol., 116*, 357–367 (1998).
66. D. K. Ro, J. Ehlting, and C. J. Douglas, *Plant Physiol., 130*, 1837–1851 (2002).
67. P. R. Ortiz de Montellano (ed.), *Cytochrome P450, Structure, Mechanism, and Biochemistry*, 3rd edn, Kluwer Academic/Plenum Publishers, New York, 2005.
68. S. G. Sligar, T. M. Makris, and I. G. Denisov, *Biochem. Biophys. Res. Commun., 338*, 346–354 (2005).
69. S. G. Sligar, *Biochemistry, 15*, 5399–5406 (1976).
70. M. Fisher and S. G. Sligar, *J. Am. Chem. Soc., 107*, 5018–5019 (1985).
71. R. Davydov, T. M. Makris, V. Kofman, D. E. Werst, S. G. Sligar, and B. M. Hoffman, *J. Am. Chem. Soc., 123*, 1403–1415 (2001).
72. R. Davydov, V. Kofman, H. Fujii, T. Yoshida, M. Ikeda-Saito, and B. M. Hoffman, *J. Am. Chem. Soc., 124*, 1798–1808 (2002).
73. T. M. Makris, R. Davydov, I. G. Denisov, B. M. Hoffman, and S. G. Sligar, *Drug Metab. Rev., 34*, 691–708 (2002).
74. M. Sharrock, P. G. Debrunner, C. Schulz, J. D. Lipscomb, V. Marshall, and I. C. Gunsalus, *Biochim. Biophys. Acta, 420*, 8–26 (1976).
75. J. Labutti, I. Parsons, R. Huang, G. Miwa, L.-S. Gan, and J. S. Daniels, *Chem. Res. Toxicol., 19*, 539–546 (2006).
76. M. Akhtar, M. Calder, D. Corina, and J. Wright, *Biochem. J., 201*, 569–580 (1982).
77. H.-G. Korth, R. Sustmann, C. Thater, A. R. Butler, and K. U. Ingold, *J. Biol. Chem., 269*, 17776–17779 (1994).
78. J. T. Groves, G. A. McClusky, R. E. White, and M. J. Coon, *Biochem. Biophys. Res. Commun., 81*, 154–160 (1978).
79. J. T. Groves and C. C.-Y. Wang, *Curr. Opin. Chem. Biol., 4*, 687–695 (2000).
80. L. D. Gorsky, D. R. Koop, and M. J. Coon, *J. Biol. Chem., 259*, 6812–6817 (1984).
81. W. M. Atkins and S. G. Sligar, *Biochemistry, 27*, 1610–1616 (1988).
82. R. E. White, M. B. McCarthy, K. D. Egeberg, and S. G. Sligar, *Arch. Biochem. Biophys., 228*, 493–502 (1984).
83. R. S. Obach, R. L. Walsky, K. Venkatakrishnan, J. B. Houston, and L. M. Tremaine, *Clin. Pharmacol. Ther., 78*, 582–592 (2005).

84. K. Ito, D. Hallifax, R. S. Obach, and J. B. Houston, *Drug Metab. Dispos., 33*, 837–844 (2005).
85. A. Galetin, K. Ito, D. Hallifax, and J. B. Houston, *J. Pharmacol. Exp. Ther., 314*, 180–190 (2005).
86. C. Jung, in *The Ubiquitous Roles of P450 Proteins*, Vol. 3 of *Metal Ions in Life Sciences* (A. Sigel, H. Sigel, and R. K. O. Sigel, eds), John Wiley & Sons, Ltd, Chichester, UK, 2007, pp. 187–234.
87. P. J. Loida and S. G. Sligar, *Protein Eng., 6*, 207–212 (1993).
88. P. J. Loida and S. G. Sligar, *Biochemistry, 32*, 11530–11538 (1993).
89. M. Budde, M. Morr, R. D. Schmid, and V. B. Urlacher, *Chembiochem., 7*, 789–794 (2006).
90. J. P. Clark, C. S. Miles, C. G. Mowat, M. D. Walkinshaw, G. A. Reid, S. N. Daff, and S. K. Chapman, *J. Inorg. Biochem., 100*, 1075–1090 (2006).
91. T. Omura, *Biochem. Biophys. Res. Commun., 338*, 404–409 (2005).
92. F. J. Gonzalez and Y. H. Lee, *FASEB J., 10*, 1112–1117 (1996).
93. J. P. Whitlock, Jr., *Annu. Rev. Pharmacol. Toxicol., 39*, 103–125 (1999).
94. Y. Fujii-Kuriyama and J. Mimura, *Biochem. Biophys. Res. Commun., 338*, 311–317 (2005).
95. D. Choudhary, I. Jansson, I. Stoilov, M. Sarfarazi, and J. B. Schenkman, *Arch. Biochem. Biophys., 436*, 50–61 (2005).
96. L. I. Gilbert, *Mol. Cell. Endocrinol., 215*, 1–10 (2004).
97. R. Niwa, T. Matsuda, T. Yoshiyama, T. Namiki, K. Mita, Y. Fujimoto, and H. Kataoka, *J. Biol. Chem., 279*, 35942–35949 (2004).
98. J. T. Warren, A. Petryk, G. Marques, J. P. Parvy, T. Shinoda, K. Itoyama, J. Kobayashi, M. Jarcho, Y. Li, M. B. O'Connor, C. Dauphin-Villemant, and L. I. Gilbert, *Insect Biochem. Mol. Biol., 34*, 991–1010 (2004).
99. T. Namiki, R. Niwa, T. Sakudoh, K. Shirai, H. Takeuchi, and H. Kataoka, *Biochem. Biophys. Res. Commun., 337*, 367–374 (2005).
100. I. Benveniste, N. Tijet, F. Adas, G. Philipps, J.-P. Salaün, and F. Durst, *Biochem. Biophys. Res. Commun., 243*, 688–693 (1998).
101. K. Wellesen, F. Durst, F. Pinot, I. Benveniste, K. Nettesheim, E. Wisman, S. Steiner-Lange, H. Saedler, and A. Yephremov, *Proc. Natl. Acad. Sci. USA, 98*, 9694–9699 (2001).
102. H. Duan and M. A. Schuler, *Plant Physiol., 137*, 1067–1081 (2005).
103. N. Zhou, T. L. Tootle, and J. Glazebrook, *Plant Cell, 11*, 2419–2428 (1999).
104. M. A. Schuler, H. Duan, M. Bilgin and S. Ali, *Phytochem. Reviews*, in press (2006).
105. D. J. Waxman, *Arch. Biochem. Biophys., 369*, 11–23 (1999).
106. T. Sueyoshi and M. Negishi, *Annu. Rev. Pharmacol. Toxicol., 41*, 123–143 (2001).
107. J. H. Gwang, *Cancer Lett., 109*, 115–120 (1996).
108. A. Baumgart, M. Schmidt, H. J. Schmitz, and D. Schrenk, *Biochem. Pharmacol., 69*, 657–667 (2005).
109. H. Prapaipong, M. R. Berenbaum, and M. A. Schuler, *Nucleic Acids Res., 22*, 3210–3217 (1994).
110. C.-F. Hung, H. Prapaipong, M. R. Berenbaum, and M. A. Schuler, *Insect Biochem. Mol. Biol., 25*, 89–99 (1995).
111. C.-F. Hung, T. L. Harrison, M. R. Berenbaum, and M. A. Schuler, *Insect Mol. Biol., 4*, 149–160 (1995).

112. J. L. Stevens, M. J. Snyder, J. F. Koener, and R. Feyereisen, *Insect Biochem. Mol. Biol.*, *30*, 559–568 (2000).
113. W. Li, M. R. Berenbaum, and M. A. Schuler, *Insect Biochem. Mol. Biol.*, *31*, 999–1011 (2001).
114. R. A. Petersen, A. R. Zangerl, M. R. Berenbaum, and M. A. Schuler, *Insect Biochem. Mol. Biol.*, *31*, 679–690 (2001).
115. X. Li, M. R. Berenbaum, and M. A. Schuler, *Insect Biochem. Mol. Biol.*, *11*, 343–351 (2002).
116. F. Shahidi, *Nahrung.*, *44*, 158–163 (2000).
117. R. B. Tjalkens, S. W. Luckey, D. J. Kroll, and D. R. Petersen, *Arch. Biochem. Biophys.*, *359*, 42–50 (1998).
118. A. K. Jaiswal, R. Venugopal, J. Mucha, A. M. Carothers, and D. Grunberger, *Cancer Res.*, *57*, 440–446 (1997).
119. C. Lee and D. S. Riddick, *Biochem. Pharmacol.*, *59*, 1417–1423 (2000).
120. A. Bhathena, C. Lee, and D. S. Riddick, *Drug Metab. Dispos.*, *30*, 1385–1392 (2002).
121. D. S. Riddick, C. Lee, A. Bhathena, Y. E. Timsit, P. Y. Cheng, E. T. Morgan, R. A. Prough, S. L. Ripp, K. K. Miller, A. Jahan, and J. Y. Chiang, *Drug Metab. Dispos.*, *32*, 367–375 (2004).
122. B. Oesch-Bartlomowicz and F. Oesch, *Biol. Chem.*, *383*, 1587–1592 (2002).
123. B. Oesch-Bartlomowicz and F. Oesch, *Arch. Biochem. Biophys.*, *409*, 228–234 (2003).
124. M. Aguiar, R. Masse, and B. F. Gibbs, *Drug Metab. Rev.*, *37*, 379–404 (2005).
125. M. A. Robin, H. K. Anandatheerthavarada, G. Biswas, N. B. Sepuri, D. M. Gordon, D. Pain, and N. G. Avadhani, *J. Biol. Chem.*, *277*, 40583–40593 (2002).
126. I. Vilgrain, G. Defaye, and E. M. Chambaz, *Biochem. Biophys. Res. Commun.*, *125*, 554–561 (1984).
127. L. H. Zhang, H. Rodriguez, S. Ohno, and W. L. Miller, *Proc. Natl. Acad. Sci. USA*, *92*, 10619–10623 (1995).
128. K. K. Korsmeyer, S. Davoll, M. E. Figueiredo-Pereira, and M. A. Correia, *Arch. Biochem. Biophys.*, *365*, 31–44 (1999).
129. M. A. Correia, S. Sadeghi, and E. Mundo-Paredes, *Annu. Rev. Pharmacol. Toxicol.*, *45*, 439–464 (2005).
130. R. C. Zangar, A. L. Kimzey, J. R. Okita, D. S. Wunschel, R. J. Edwards, H. Kim, and R. T. Okita, *Mol. Pharmacol.*, *61*, 892–904 (2002).
131. A. L. Kimzey, K. K. Weitz, F. P. Guengerich, and R. C. Zangar, *Biochemistry*, *42*, 12691–12699 (2003).
132. O. Shimozawa, M. Sakaguchi, H. Ogawa, N. Harada, K. Mihara, and T. Omura, *J. Biol. Chem.*, *268*, 21399–21402 (1993).
133. J. G. Scott and Z. Wen, *Pest Manag. Sci.*, *57*, 958–967 (2001).
134. R. Feyereisen, in *Comprehensive Molecular Insect Science*, Vol. 4, (L. I. Gilbert, K. Latrou, and S. S. Gill, eds), Elsevier, Oxford, 2005, pp. 1–77.
135. K. A. Nielsen and B. L. Moller, in *Cytochrome P450: Structure, Mechanism, and Biochemistry* (P. R. Ortiz de Montellano, ed.), 3rd edn, Kluwer Academic/Plenum Publishers, New York, 2005, pp. 553–583.
136. R. Bernhardt and M. R. Waterman, in *The Ubiquitous Roles of P450 Proteins*, Vol. 3 of *Metal Ions in Life Sciences* (A. Sigel, H. Sigel, and R. K. O. Sigel, eds), John Wiley & Sons, Ltd, Chichester, UK, 2007, pp. 361–396.

137. G. J. Bishop and C. Koncz, *Plant Cell, 14*, S97–S110 (2002).
138. S. Fujioka and T. Yokota, *Annu. Rev. Plant Biol., 54*, 137–164 (2003).
139. S. Choe, *Plant Physiol., 126*, 539–548 (2006).
140. M. Szekeres, K. Németh, Z. Koncz-Kálmán, J. Mathur, A. Kauschmann, T. Altmann, G. P. Rédei, F. Nagy, J. Schell and C. Koncz, *Cell, 85*, 171–182 (1996).
141. S. Choe, B. P. Dilkes, S. Fujioka, S. Takatsuto, A. Sakurai, and K. A. Feldmann, *Plant Cell, 10*, 231–243 (1998).
142. G. J. Bishop, T. Nomura, T. Yokota, K. Harrison, T. Noguchi, S. Fujioka, S. Takatsuto, J. D. Jones, and Y. Kamiya, *Proc. Natl. Acad. Sci. USA, 96*, 1761–1766 (1999).
143. Y. Shimada, S. Fujioka, N. Miyauchi, M. Kushiro, S. Takatsuto, T. Nomura, T. Yokota, Y. Kamiya, G. J. Bishop, and S. Yoshida, *Plant Physiol., 126*, 770–779 (2001).
144. Y. Shimada, H. Goda, A. Nakamura, S. Takatsuto, S. Fujioka, and S. Yoshida, *Plant Physiol., 131*, 287–297 (2003).
145. T.-W. Kim, J.-Y. Hwang, Y.-S. Kim, S.-H. Joo, S. C. Chang, J. S. Lee, S. Takatsuto, and S. K. Kim, *Plant Cell, 17*, 2397–2412 (2005).
146. T. Nomura, T. Kushiro, T. Yokota, Y. Kamiya, G. J. Bishop, and S. Yamaguchi, *J. Biol. Chem., 280*, 17873–17879 (2005).
147. M. H. Choi, P. L. Skipper, J. S. Wishnok, and S. R. Tannenbaum, *Drug Metab. Dispos., 33*, 714–718 (2005).
148. M. M. Neff, S. M. Nguyen, E. J. Malancharuvil, S. Fujioka, T. Noguchi, H. Seto, M. Tsubuki, T. Honda, S. Takatsuto, S. Yoshida, and J. Chory, *Proc. Natl. Acad. Sci. USA, 96*, 15316–15323 (1999).
149. E. M. Turk, S. Fujioka, H. Seto, Y. Shimada, S. Takatsuto, S. Yoshida, M. A. Denzel, Q. I. Torres, and M. M. Neff, *Plant Physiol., 133*, 1643–1653 (2003).
150. M. Nakamura, T. Satoh, S. Tanaka, N. Mochizuki, T. Yokota, and A. Nagatani, *J. Exp. Bot., 56*, 833–840 (2005).
151. N. Takahashi, M. Nakazawa, K. Shibata, T. Yokota, A. Ishikawa, K. Suzuki, M. Kawashima, T. Ichikawa, H. Shimada, and M. Matsui, *Plant J., 42*, 13–22 (2005).
152. D. R. Williams, M. J. Fisher, and H. H. Rees, *Arch. Biochem. Biophys., 376*, 389–398 (2000).
153. C.-F. Hung, M. R. Berenbaum, and M. A. Schuler, *Insect Biochem. Mol. Biol., 27*, 377–385 (1997).
154. J.-S. Chen, M. R. Berenbaum, and M. A. Schuler, *Insect Mol. Biol., 11*, 175–186 (2002).
155. Z. Wen, L. Pan, M. R. Berenbaum, and M. A. Schuler, *Insect Biochem. Mol. Biol., 33*, 937–947 (2003).
156. X. Li, J. Baudry, M. R. Berenbaum, and M. A. Schuler, *Proc. Natl. Acad. Sci. USA., 101*, 2939–2944 (2004).
157. M. Sasabe, Z. Wen, M. R. Berenbaum, and M. A. Schuler, *Gene, 338*, 163–175 (2004).
158. G. W. Ivie, in *Light-Activated Pesticides* (J. R. Heitz and K. R. Downum, eds), Vol. 339 of *American Chemical Society Symposium Series*, Washington, 1987, pp. 216–230.

159. J. K. Nitao, M. Berhow, S. M. Duval, D. Weisleder, S. F. Vaughn, A. Zangerl, and M. R. Berenbaum, *J. Chem. Ecol.*, *29*, 671–682 (2003).
160. W. Mao, S. Rupasinghe, A. Zangerl, M. A. Schuler, and M. R. Berenbaum, *Insect Mol. Biol.*, *15*, 169–179 (2006).
161. J. Maenpaa, R. Juvonen, H. Raunio, A. Rautio, and O. Pelkonen, *Biochem. Pharmacol.*, *48*, 1363–1369 (1994).
162. J. K. Yano, M. H. Hsu, K. J. Griffin, C. D. Stout, and E. F. Johnson, *Nature Struct. & Mol. Biol.*, *12*, 822–823 (2005).
163. M. C. Gustafsson, O. Roitel, K. R. Marshall, M. A. Noble, S. K. Chapman, A. Pessegueiro, A. J. Fulco, M. R. Cheesman, C. von Wachenfeldt and A. W. Munro, *Biochemistry*, *3*, 5474–5487 (2004).
164. S. Torres, C. R. Fjetland and P. J. Lammers, *BMC Microbiol.*, *5*, 16–27 (2005).
165. U. Scheller, T. Zimmer, E. Kargel, and W. H. Schnuck, *Arch. Biochem. Biophys.*, *328*, 245–254 (1996).
166. A. E. Simpson, *Gen. Pharmacol.*, *28*, 351–359 (1997).
167. R. T. Okita and J. R. Okita, *Curr. Drug. Metab.*, *2*, 265–281 (2001).
168. K. R. Henne, K. L. Kunze, Y.-M. Zheng, P. Christmas, R. J. Soberman, and A. E. Rettie, *Biochemistry*, *40*, 12925–12931 (2001).
169. C. Helvig, N. Tijet, R. Feyereisen, F. A. Walker, and L. L. Restifo, *Biochem. Biophys. Res. Commun.*, *325*, 1495–1502 (2004).
170. J.-P. Salaün and C. Helvig, *Drug Interact.*, *12*, 261–283 (1995).
171. F. Cabello-Hurtado, Y. Batard, J.-P. Salaün, F. Durst, F. Pinot, and D. Werck-Reichart, *J. Biol. Chem.*, *273*, 7260–7267 (1998).
172. N. Tijet, C. Helvig, F. Pinot, R. Le Bouquin, A. Lesot, F. Durst, J. P. Salaun, and I. Benveniste, *Biochem. J.*, *332*, 583–589 (1998).
173. R. LeBouquin, F. Pinot, I. Benveniste, J. P. Salaün, and F. Durst, *Biochem. Biophys. Res. Commun.*, *261*, 156–162 (1999).
174. R. LeBouquin, M. Skrabs, R. Kahn, I. Benveniste, J. P. Salaün, L. Schreiber, F. Durst, and F. Pinot, *Eur. J. Biochem.*, *268*, 3083–3090 (2001).
175. R. A. Kahn, R. LeBouquin, F. Pinot, I. Benveniste and F. Durst, *Arch. Biochem. Biophys.*, *391*, 180–187 (2001).
176. F. Xiao, S. M. Goodwin, Y. Xiao, Z. Sun, D. Baker, X. Tang, M. A. Jenks, and J. M. Zhou, *EMBO J.*, *23*, 2903–2913 (2004).
177. S. Kandel, M. Morant, I. Benveniste, E. Blee, D. Werck-Reichhart, and F. Pinot, *J. Biol. Chem.*, *280*, 35881–35889 (2005).
178. K. Tamaki, H. Imaishi, H. Ohkawa, K. Oono, and M. Sugimoto, *Biosci. Biotechnol. Biochem.*, *69*, 406–409 (2005).
179. P. B. Danielson, *Curr. Drug Metab.*, *3*, 561–597 (2002).
180. R. Bernhardt, *J. Biotechnol.*, *124*, 128–145 (2006).

2

Structural and Functional Mimics of Cytochromes P450

Wolf-D. Woggon
Department of Chemistry, University of Basel,
St. Johannsring 19, CH-4056 Basel, Switzerland
<wolf-d.woggon@unibas.ch>

1. INTRODUCTION	27
2. IRON PORPHYRINS CARRYING A THIOLATE OR MODIFIED THIOLATE LIGAND	31
3. STRUCTURALLY REMOTE P450 MIMICS	40
3.1. Iron Porphyrins	40
3.2. Manganese Porphyrins	41
3.3. Ruthenium Porphyrins	46
3.4. Various Electron-Deficient Metal Porphyrins	48
4. CONCLUDING REMARKS	52
ACKNOWLEDGMENTS	52
ABBREVIATIONS	53
REFERENCES	53

1. INTRODUCTION

The discovery of cytochromes P450 dates back more than 50 years [1]. Today hundreds of monooxygenases are known that are involved in the metabolism of drugs, e.g., in human liver, and in the production and transformation of hormones and fatty acids in glands [2].

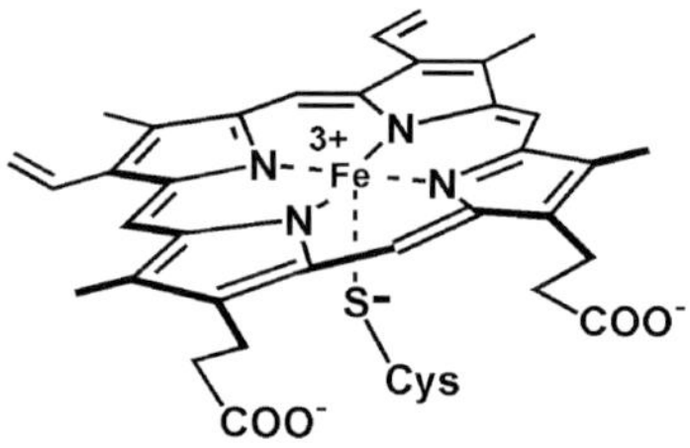

Figure 1. The cofactor of cytochromes P450.

Cytochromes P450 all contain the same cofactor **1** (Figure 1), a non-covalently bound iron(III)protoporphyrin IX to which a deprotonated cysteine provides the fifth ligand to iron. The presence of this iron complex and its sophisticated and broad reactivity in substrate oxidations by means of reductive oxygen cleavage made the system a challenging target to prepare model compounds as structural analogues and functional mimics in order to gain insight into the mechanisms of P450 proteins. This endeavor is one of the first examples of enzyme model research that united various disciplines of chemistry. It is the aim of this account to review some highlights of a joint, and often competitive effort by inorganic, organic and physical chemists who synthesized/used enzyme models as a source to understand heme-thiolate monooxygenase catalysis. In general, the natural structure **1** has been simplified in various ways; the peripheric methyl and vinyl substituents were omitted as well as in most cases the difficult-to-handle thiolate ligand. Instead, for most of the complexes the *meso*-positions are substituted by phenyl rings in order to prevent oxidation at the *meso*-C atoms and further to provide a handle for functionalization. In some cases the β-pyrrole positions have been substituted with methyl groups and butyl chains to render the complexes more soluble in organic solvents and to restrict conformational mobility of the phenyl rings at the *meso*-positions.

Last but not least, many groups which more recently prepared quite elaborate active site analogues of P450 took advantage of experimental procedures and concepts in porphyrin chemistry developed in the 1970s and 1980s to obtain oxygen binding heme models, namely by Baldwin et al. (*capped* porphyrins, **2** [3]), Battersby et al. (*strapped* porphyrins, **3** [4]), Collmann et al. (*picket-fence*, **4** [5]) and Momenteau et al. (*basket-handle* porphyrins, **5** [6], Figure 2).

Since many results reviewed in this account were concerned with the synthesis of intermediates of the catalytic cycle of cytochromes P450, for clarity today's established mechanism (referring to cytochrome P450$_{cam}$) is shown in Figure 3, ignoring the dispute about alternative oxidants.

The cofactor chemistry of cytochromes P450 is characterized by the binding and unusual activation of paramagnetic oxygen through reductive oxygen cleav-

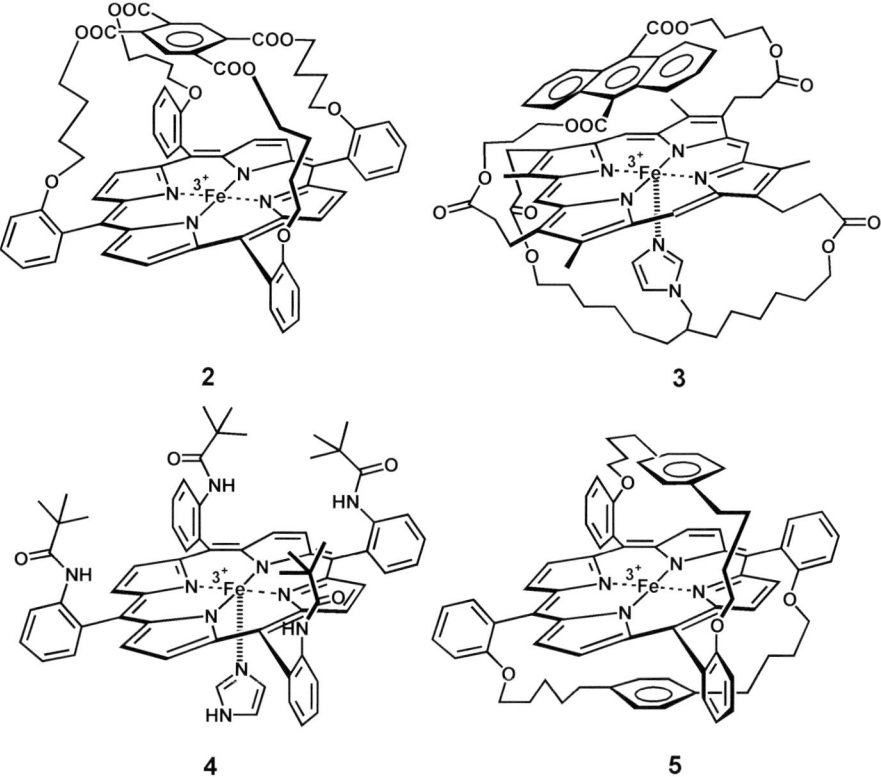

Figure 2. Examples of *face-protected* porphyrins.

age. In the absence of substrate the resting state of P450$_{cam}$ **6** is a predominatly low-spin Fe(III)porphyrin, displaying $E_0 = -290$ mV that renders reduction by the redox protein putidaredoxin (Pdx) to iron (II) thermodynamically unfavorable (Pdx, $E_0 = -235$ mV). Only when the water cluster occupying the sixth coordination position is replaced by camphor the complex becomes high-spin and E_0 shifts to -170 mV (see **7** in Figure 3), such that reduction to iron(II), **8**, and subsequent binding of oxygen can occur. The resulting diamagnetic oxygen complex **9** is then reduced to **10** by a slow, rate-contributing electron transfer from the redox protein.

The peroxo complex **10** is protonated twice, such that water becomes a leaving group and the short-lived, very reactive oxo iron(IV) porphyrin radical cation **11** (Cpd I) is produced. As shown for camphor this intermediate is capable to hydroxylate non-activated positions and it is believed that this oxidant is also the reactive species in all other P450 reactions, e.g., epoxidations, hetero atom oxidations, diol cleavages and aromatic and benzylic hydroxylations [7,8].

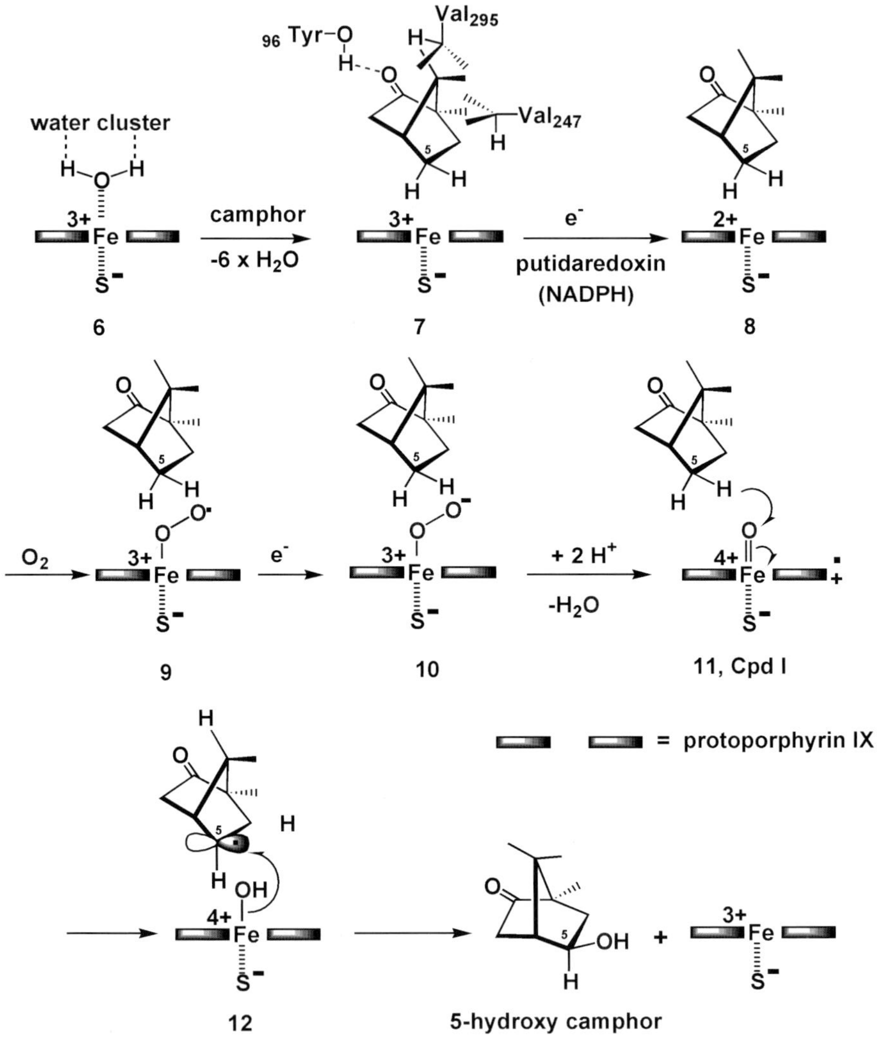

Figure 3. Catalytic cycle of cytochrome P450$_{cam}$.

Whereas today most of these reactions can be catalyzed by synthetic and specifically designed chiral metal complexes [9], the regioselective and enantioselective hydroxylation of non-activated positions remains an unsolved problem and has attracted organic and inorganic chemists ever since cytochrome P450 was discovered as Nature's paradigm.

2. IRON PORPHYRINS CARRYING A THIOLATE OR MODIFIED THIOLATE LIGAND

The earliest examples of P450 models **13** [10], **14** [11], and **15** [12] which were designed to display a thiolate coordination to iron confirmed the notion that natural P450 enzymes contain this unusual ligand system (Figure 4). For **13** an EPR spectrum was obtained similar to the high-spin state of the E·S complex of P450 (see **7**), and **15**, after reduction to iron(II), was shown to bind CO to yield **16** which displayed a split Soret band (at ~ 380 and 450 nm) characteristic of the corresponding natural P450 state.

Using **14** protonated at sulfur did not yield a 'hyper' absorption spectrum (split Soret band) after reduction and CO addition, indicating that the natural P450 cofactor contains indeed a mercaptide bound to iron. Due to the non-covalent attachment of the thiophenolate in **13** and the flexibility of and the accessibility to the aryl thiolate of **14** and the alkyl mercaptide of **15**, these complexes were valuable spectroscopical models of cytochrome P450, however, they were not useful to carry out P450-type reactions.

Subsequently the target of the 'second generation' of heme-thiolate enzyme models was designed to obtain a sterically congested thiolate ligand 'forced' into coordination with iron (Figure 5). This aim was accomplished using the basket-handle approach [6] for the synthesis of the doubly-bridged iron porphyrin

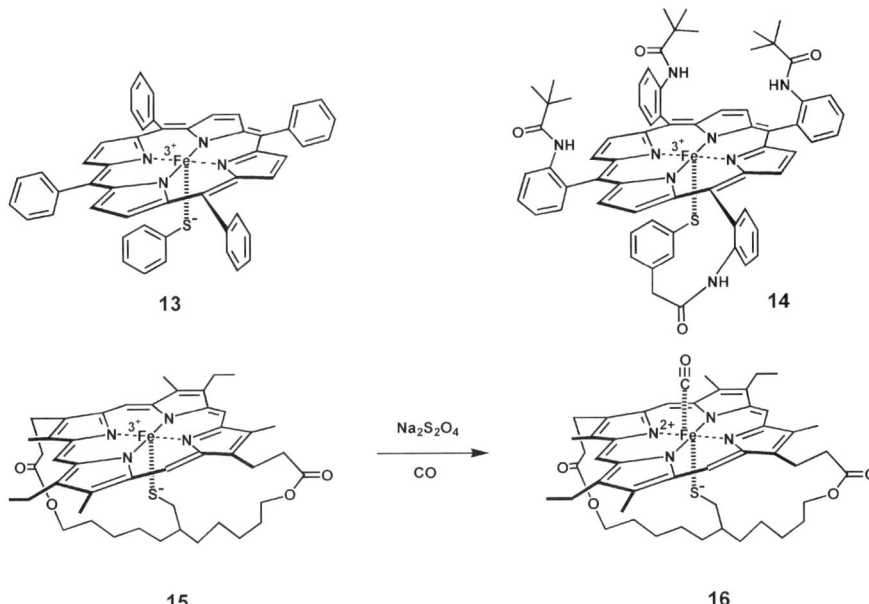

Figure 4. P450 models carrying a thiolate ligand.

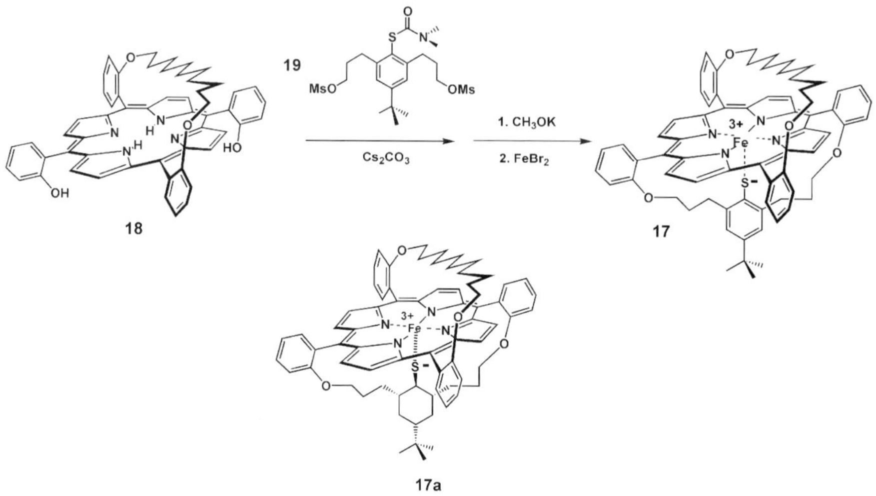

Figure 5. Doubly-bridged active-site analogues **17** and **17a**.

17 [13]. For this purpose the mono-bridged porphyrin **18** was condensed with the protected sulfur bridge **19**, prepared in eight steps from 4-*tert* butyl phenol. Then iron was inserted, the S-protecting group was removed and **17** obtained in acceptable overall yield of 5% from *o*-methoxybenzaldehyde.

Complex **17** is a high-spin iron(III) system which has the alkane bridge as a substrate covalently attached. The P450-like reactivity of the compound was shown by employing the 'shunt pathway', i.e., the direct access of Cpd I (see **11** from **7** by means of PhIO). Indeed, reaction of **17** with F_5PhIO gave hydroxylation of a non activated C-H bond of the alkane bridge [14]. Further, **17** could be reduced with $Na_2S_2O_4$ in the presence of ^{13}CO to furnish the six-coordinate complex with a characteristic 457 nm absorption and a ^{13}CO resonance at 206 ppm (200 ppm for ^{13}CO-P450$_{cam}$ [15], 197 ppm for ^{13}CO-**16** [12]). Accordingly, **17** is an analogue of the E·S complex of cytochrome P450 because it displays similar spectroscopic features and the inherent reactivity of the native system.

One significant difference between **17** and **7** became apparent: It was very difficult to reduce Fe(III) to Fe(II) quantitatively, indicating a greater stability of the higher oxidation state. This was confirmed by cyclic voltammetry, **17** displayed a completely reversible redox behavior at $E_0 = -607$ mV (DMF, sat. LiBr) [16] which is more than 400 mV more negative than that of E·S P450$_{cam}$. When the thiophenolate of **17** was replaced by an alkyl thiolate (see **17a** [17]), the redox potential shifted to even more negative values $E_0 = -714$ mV. These observations suggested a reduced charge density at the thiolate of native cytochrome P450, confirming an earlier proposal [15]. It was subsequently discovered in X-ray structures of several P450 proteins such as P450$_{cam}$, P450$_{terp}$, P450$_{eryF}$, and P450$_{BM-3}$ that the cysteine derived thiolate ligand is hydrogen bonded to

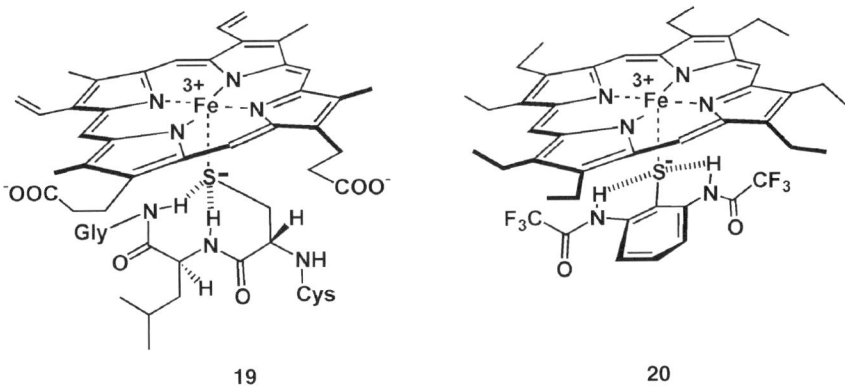

Figure 6. Hydrogen bonding of the thiolate ligand in P450$_{cam}$ **19** and the model complex **20**.

two amino acids of the protein backbone (see **19** in Figure 6) and hence, its charge donating capacity to iron is smaller than in model complex **17** and **18** for example.

This particular proximal coordination sphere of P450 was elegantly mimicked by the high-spin octaethyl porphyrin iron(III) **20** [18] which exhibits $E_0 = -350$ mV, a redox potential about 300 mV more positive than for a complex in which the ligand of **20** is replaced by Ph–S$^-$. Thus, through complementary experiments with P450 proteins and enzyme mimics it has been possible indeed to associate the value of E_0 of the resting state of P450$_{cam}$ with the electron-donating character of the thiolate ligand.

Another feature of the resting state of P450$_{cam}$ is the unusual iron(III) low-spin state which is the second parameter which renders **6** 'resting' and unreactive towards reductive oxygen cleavage. As already noted in [10] and confirmed for active site analogues **17** and **18**, all heme-thiolate models with a free sixth coordination site are high-spin and only change to low-spin if strong ligands such as imidazole or CH$_3$O$^-$ are coordinating to the distale site. It should be noted that in cases where a low-spin character of five-coordinate heme-thiolate models was reported [19] most likely an undisclosed CH$_3$O$^-$ binds to iron deriving from the sulfur deprotection under basic conditions.

Since for **6** one water molecule of a cluster coordinates to iron(III) (see **21** in Figure 7), the question was raised whether water coordination would be sufficient to stabilize the low-spin state. To investigate this problem the iron(III) porphyrin **22** was synthesized, which has a thiolate and one water molecule coordinating to iron. Evidence for water coordination rests on the three-pulse ESEEM and Davis ENDOR spectra of **22** and the corresponding D$_2$O and H$_2$^{17}O complexes which revealed the coupling between the high-spin iron and H, D, and ^{17}O [20]. A spin change of **22** to low-spin was only observed when the water ligand was

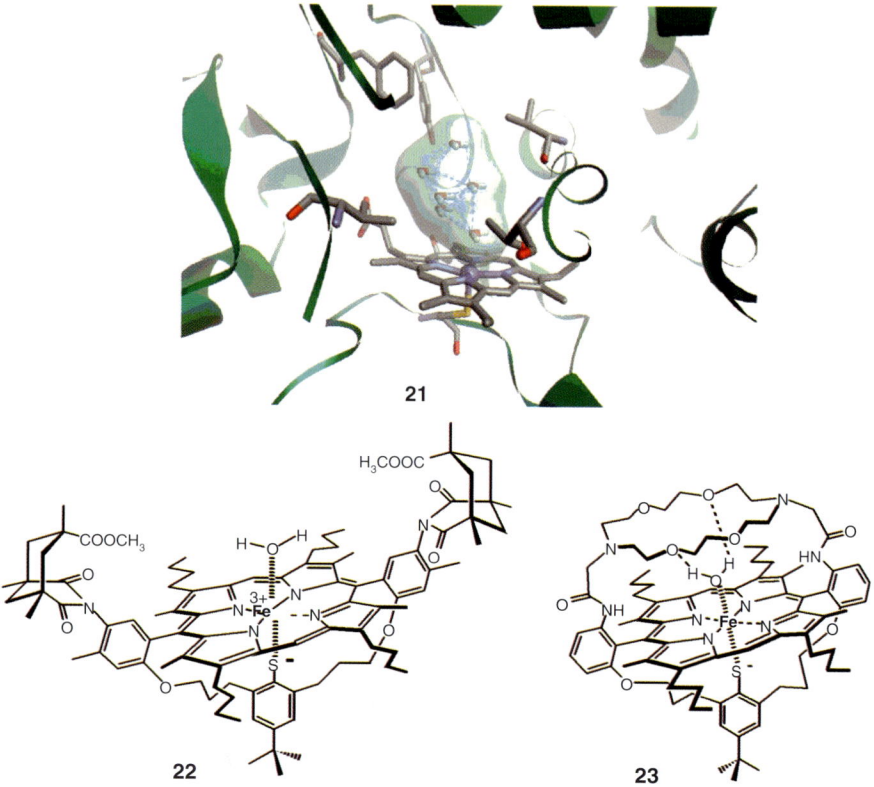

Figure 7. The resting state of cytochrome P450$_{cam}$ **21** (adapted from an X-ray structure [21]) and corresponding active site analogues **22** and **23**.

exchanged for 1,2-dimethyl imidazole. Accordingly the coordination of a thiolate and one water ligand doesn't stabilize the low-spin state.

From these experiments it was concluded that the crown-capped porphyrin **23** would be a model more close to the resting state of cytochrome P450$_{cam}$. The complex **23** was synthesized and EPR studies pursued with **23** and with the corresponding complex without water. The latter iron porphyrin displayed a high-spin character, as expected, whereas **23** indeed showed a low-spin EPR. In contrast, the Ba^{2+} crown-ether complex remained high-spin. Accordingly, hydrogen bonding of the water in **23** to the crown-ether oxygens creates a 'HO$^-$'-type ligand, coordinating to iron at the distal site. It is intriguing to speculate that the water cluster in natural P450$_{cam}$ (see **21**), is polarized such that 'HO$^-$' rather than water actually binds to iron, hence stabilizing the low-spin state [22,23].

In view of the fact that several cytochromes P450 contain a 'masked' thiolate ligand, $-SO_3^-$ was believed to be a suitable new fifth ligand because (i) only

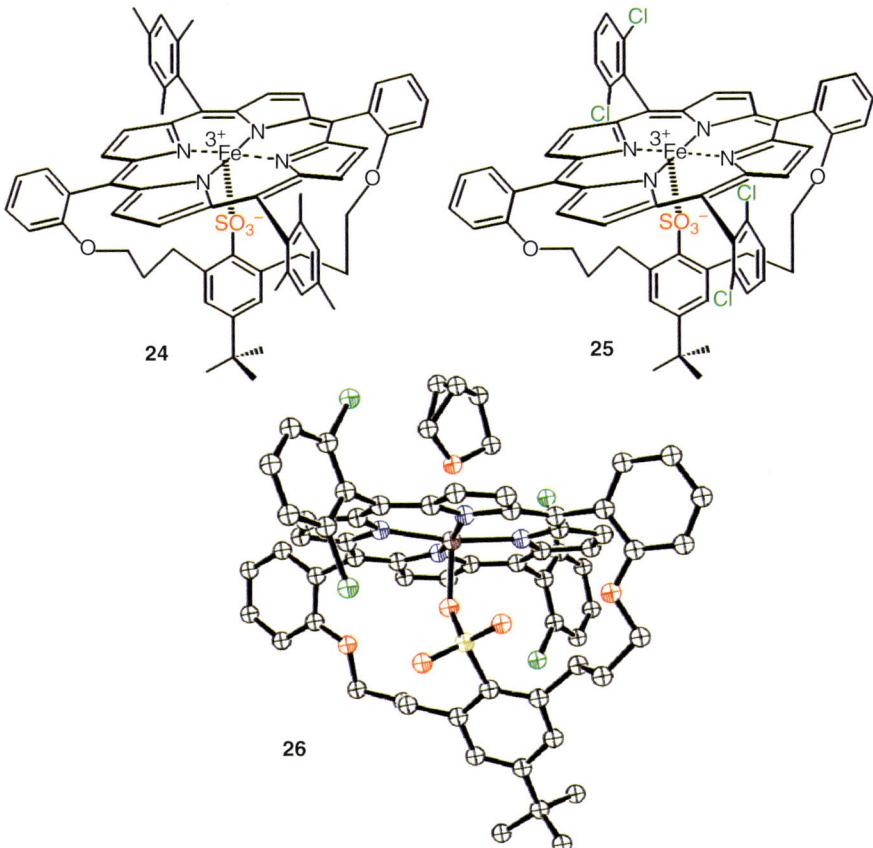

Figure 8. A new family of P450 enzyme mimics.

one-third of a negative charge interacts with the iron orbitals, (ii) E_0 of such a system was expected to be more positive, and (iii) in contrast to the thiolate the SO_3^- ligand should be stable in the presence of co-oxidants. Two iron complexes **24** [24] and **25** [25] were synthesized using the strategy employed for **17** (Figure 8).

Both compounds display the expected anodic shifts of redox potentials, i.e., $E_0(\mathbf{24}) = -290\,\text{mV}$ and $E_0(\mathbf{25}) = -210\,\text{mV}$, and hence, are electrochemically very close to the resting state of P450$_{cam}$, confirming the significance of the electron donating character of the fifth ligand in cytochromes P450 and its model compounds. The spin state of the latter, however is different, due to the absence of the water cluster **24** and **25** are both high-spin, even in coordinating solvents such as methanol.

DFT calculations predicted that one oxygen of the SO_3^- group coordinates to iron and further revealed that Cpd I derived from **24** displays a reaction profile for allylic oxidation and epoxidation of propene which is more or less indistinguishable from the iron porphyrin with a thiolate ligand [25]. The coordination mode of the SO_3^- ligand was confirmed by X-ray analysis [26] (see **26**). The high-spin complex **24** was shown to be a valuable model for NO binding to the resting state and to the E·S complex of $P450_{cam}$ in methanol and toluene, respectively. In contrast to the non-coordinating toluene in methanol one solvent molecule occupies the distal coordination site, mimicking water binding of the resting state. Activation parameters for NO binding to **24** and **24**-MeOH determined by various stopped-flow techniques are in good agreement with those obtained for the corresponding states of $P450_{cam}$ and certain differences such as for k_{off} could be explained by the absence of the protein environment in the model systems [27]. From both compounds the corresponding Cpd I equivalents could be generated and with **25** some reaction characteristic of cytochrome P450 could be catalyzed: epoxidations, N-dealkylations, diol cleavage, aromatic hydroxylation, and the oxidation of non-activated C–H groups [26]. The last two reactions were observed in a radical clock experiment with **27** in organic solvents, unbiased by any influence from P450's substrate binding site, aimed to determine the lifetime of possible radical intermediates [28].

Oxidation of the substrate **27** gave three products **28**, **29**, and **30** (Figure 9) [28]. Using the known rate of rearrangement of radical **31** → **32** (Figure 10) in the same solvent, the lifetime $\tau = 625\,\text{fs}$ of an intermediate radical cluster **33** could be calculated from the product ratio **29:30** = 9:1. Since this lifetime is more than three times higher than τ determined for the same substrate oxidized by cytochrome P450 and higher than the lifetime of a transition state, it was inferred that **33** is an intermediate on the reaction coordinate (Figure 10). Further it was concluded that Cpd I catalyzes a concerted but non-synchronous insertion of oxygen into the C–H bond and the difference between model reactions in

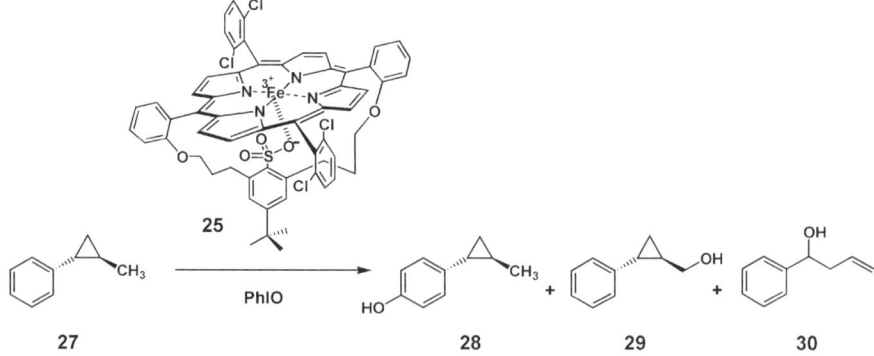

Figure 9. Catalytic oxidation using the enzyme mimic **25**.

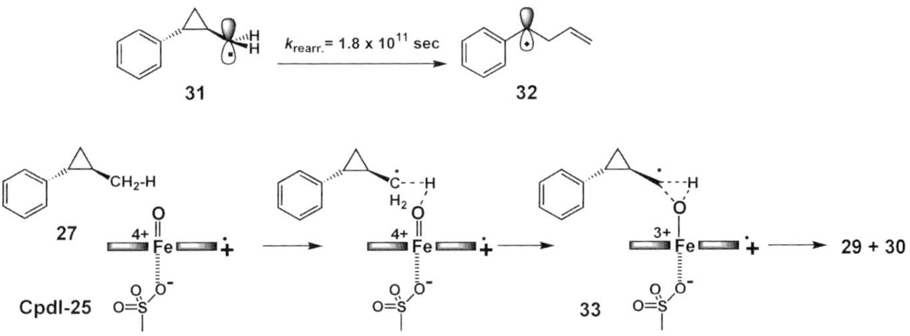

STRUCTURAL AND FUNCTIONAL MIMICS OF CYTOCHROMES P450

Figure 10. Proposed mechanism of oxygen insertion by Cpd I into a C–H bond.

solution and oxidations with P450 proteins is due to steric interactions between the substrate radical and the protein which slows down the rate of rearrangement in the substrate binding site compared with k_{rearr} in solution.

The stability of the thiolate ligand with respect to decomplexation from iron and/or oxidation in the presence of co-oxidants, however, was also accomplished through steric congestion different from the concepts mentioned above.

The 'thiolate-tail' iron(III) complex **34** (Figure 11) [29] has been reported to be relatively stable, presumably because k_{off} for decomplexation is slowed down due to steric bulk of the 'picket fences' and for the same reason the thiolate is well protected against oxidation by co-oxidants generally used in the 'shunt pathway'. Comparing the reactions of **34** with those of Fe(TPP)Cl, both in the presence of PhCOOOH, suggested that **34** acts via heterolytic O–O bond cleavage whereas Fe(TPP)Cl supports homolytic O–O bond scission. In agreement with these results are O-demethylation experiments with the substrate 1,4-dimethoxy

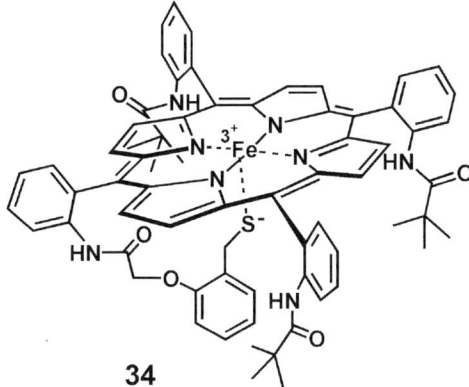

Figure 11. A 'picket-fenced' thiolate porphyrin.

benzene, **34** and PhCOOOH. In contrast to Fe(TPP)Cl, for **34** ^{18}O incorporation from the ^{18}O-labeled oxidant was not observed and using CD$_3$O-Ph-OCH$_3$ a $k_\mathrm{H}/k_\mathrm{D} = 11.7$ was determined which is identical to the isotopic effect calculated from incubations with rat liver microsomes [29]. Further, various substrate oxidations, e.g., adamantane→1-adamantanol, revealed a significantly increased catalysis for **34** versus Fe(TPP)Cl [23], confirming the importance of the thiolate ligand [24].

The 'twin-coronet' iron porphyrin **35** [30] is another sophisticated P450 mimic, displaying very efficient **'shielding' of the** thiolate ligand (Figure 12). Interestingly, **35** showed a low-spin **EPR** and an **extremely** negative $E_0 = -950\,\mathrm{mV}$ [30]. These data suggest that a **six-coordinate iron(III)** complex was obtained; presumably due to the final, basic **hydrolysis of** the S-protecting group, one of the phenolic OH groups (red) was **deprotonated** and hence, coordinates to iron stabilizing the low-spin state.

On reaction of **35** with one equivalent of KO$_2$ in the presence of oxygen iron(III) was reduced to iron(II) (see **36**) which was shown to bind O$_2$ and CO reversibly [31] (Figure 13). The oxygen adduct **37** was remarkably stable, suggesting H-bonding of the end-on oxygen to one of the OH groups of the BINOL cavity. This feature renders **37** an excellent model of the corresponding state of P450$_\mathrm{cam}$ **38** for which the end-on bound oxygen is stabilized by an ordered water molecule and by the side chain OH group of Thr$_{252}$ [32].

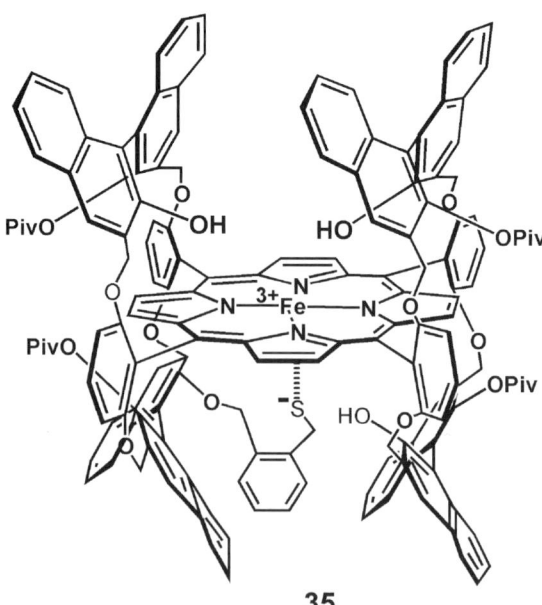

Figure 12. A binaphtyl-shielded P450 mimic.

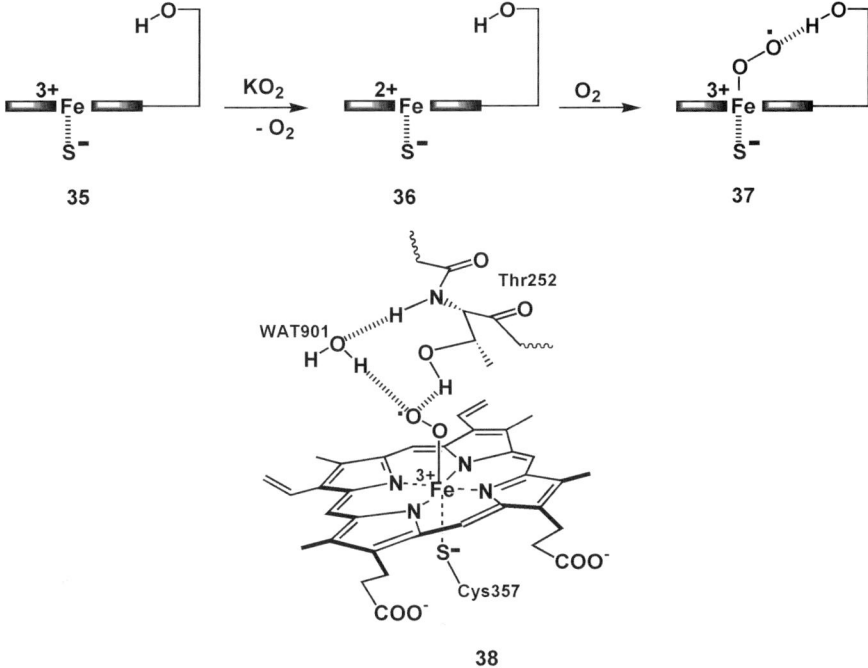

Figure 13. Stabilization of the oxygen adduct **37**.

The 'twin coronet' porphyrin [33] was then applied as a mimic of prostaglandin H synthase (e.g., PGSH-1) which catalyzes the addition of oxygen to arachidonic acid in the following sequence: by H• abstraction from a doubly allylic activated C–H bond, i.e., removal of the 13-*pro*-S hydrogen of arachidonic acid, a stabilized radical is generated which is subsequently and stereospecifically trapped by O_2 to yield a hydroperoxide that is reduced to the corresponding alcohol as the final product.

In contrast to other P450s H• abstraction from the substrate is not directly mediated by the oxo iron(IV) porphyrin radical cation (Cpd I), but by the tyrosinyl radical of Tyr385, 10 Å apart from the heme, generated by electron transfer to Cpd I.

The complex **35** was shown to catalyze this event using **39**, a truncated analogue of arachidonic acid, as a substrate (Figure 14). In the model reaction the Cpd I equivalent **40** was produced from **35** by means of mCPBA, intramolecular electron transfer from the naphtyl OH group to the porphyrin gave the napht-O• radical **41** which abstracted H• from **39**. The resulting allyl radical was trapped by $^{18}O_2$ to give the desired (hydroperoxide) alcohol **42** with quantitative ^{18}O-incorporation; the TON was calculated as 16.5.

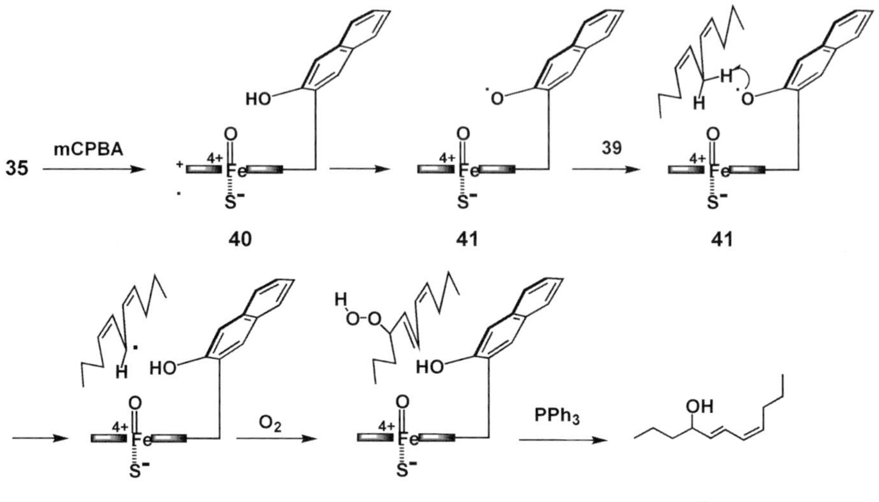

Figure 14. Reaction of **35** as an active-site model of PGSH.

3. STRUCTURALLY REMOTE P450 MIMICS

The development of P450 mimics devoid of the thiolate ligand has been an attractive target since all the difficulties regarding the stability of the thiolate coordination could be circumvented. Ever since the 'shunt-pathway', i.e., the generation of Cpd I with 'O' donors from simple iron(III) porphyrins such as TPPs was discovered in 1979 [34], this research area has grown steadily, now encompassing thousands of publications. In the course of this development various substituted porphyrin chromophores were used, complexing metal ions such as iron(III), manganese(III), and ruthenium(II). All of these ions form high-valent oxidation states that can be produced under various conditions in the presence of PhIO, H_2O_2, peroxynitrite, mCPBA, tert-butyl hydroperoxide, and N-oxides.

3.1. Iron Porphyrins

The characterization of the oxo iron(IV) porphyrin radical cation as a model of Cpd I (see **11**) was first accomplished by treating **43** (TMPFeCl) with mCPBA to yield **44** (Figure 15), displaying UV, EPR, Mössbauer, and NMR spectra in agreement with several data available from Cpd I species of horseradish peroxidase (HRP) and chloroperoxidase (CPO). It was further demonstrated that **44** exchanges its oxygen with $H_2^{18}O$ and is an excellent oxidizing agent converting norbornene **45** to the epoxide **46** [35].

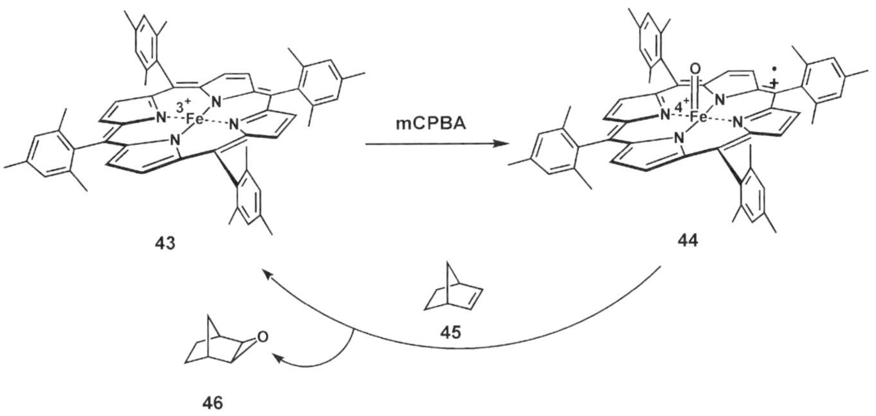

STRUCTURAL AND FUNCTIONAL MIMICS OF CYTOCHROMES P450

Figure 15. The first characterization of a model complex of Cpd I.

The conclusion that this species is the only reactive intermediate of P450-catalyzed reactions was a breakthrough in P450 research, has inspired the P450 community ever since and even became a dogma that was only questioned recently [36,37]. Subsequently, using *meso*-tetraphenyl porphyrin iron (TPPFeCl) it was shown that PhIO is an equally suitable reagent to generate a Cpd I equivalent that is capable of epoxidizing various Z-configured olefins. In contrast, the *E*-diastereoisomers were bad substrates which in some cases did not convert at all. This observation has been explained due to steric hindrance of the substrate with the porphyrin plane and it was suggested that the olefin approaches the Fe=O bond in a side-on fashion [38]. Asymmetric epoxidation of simple olefins employing chiral, vaulted binaphtyl iron porphyrins such as **47** and PhIO (Figure 16) have been accomplished in modest to bad yields and enantiomeric excess (ee) up to 72% [39]. With the same system ethylbenzene afforded 1-phenylethanol with 40% ee [40] and sulfoxidation of alkyl aryl sulfides was achieved with 48% ee.

3.2. Manganese Porphyrins

Exchanging Fe(III) for Mn(III) in porphyrins such as TPP (see **48** in Figure 17), proved to be a very good choice since the corresponding high-valent **49** gave significantly better yields in hydrocarbon oxidations than a corresponding iron porphyrin catalyst [41]. In this context it was found that both Mn(IV)=O and Mn(V)=O (see **49**) are competent oxidants of olefins, but display different reactivities (Figure 17). In contrast to the Mn(IV)=O complex Mn(V)=O exchanges the oxo oxygen atom very rapidly with water, and its pyridine adduct (see **50**), reacts much faster with olefins under retention of configuration [42].

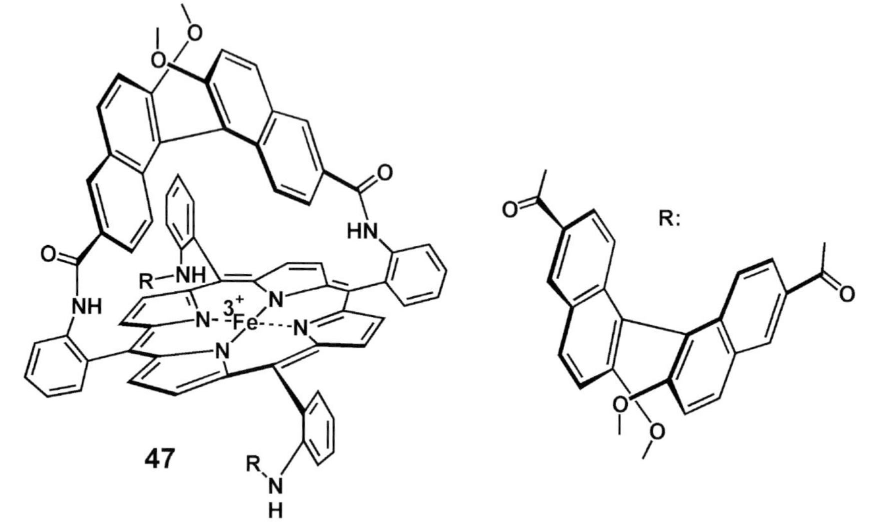

Figure 16. Vaulted chiral porphyrins for enantioselective reactions.

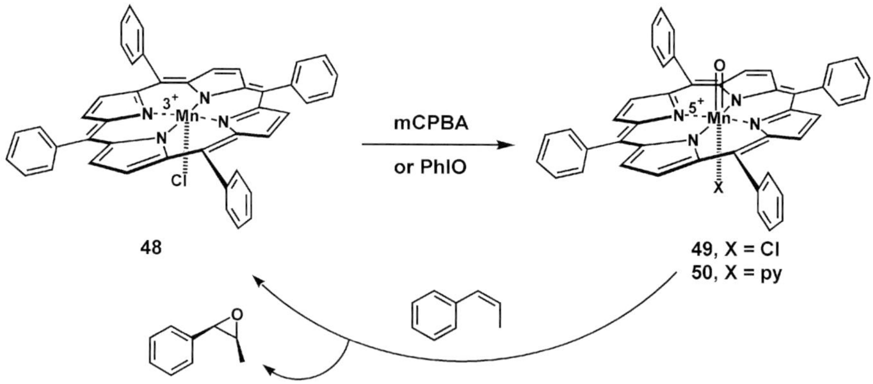

Figure 17. Sterospecific epoxidation by means of oxo manganese porphyrins.

In addition, several aerobic processes for olefin epoxidation have been described employing manganese porphyrins. The general concept consists of using a mixture of, e.g., **48**, O_2, and a reducing agent such as $NaBH_4$ [43], colloidal Pt/H_2 [44], or ascorbate in a biphasic system [45]. In all cases the structure of the oxygen transfer reagent is not known, the procedures are, however, rather efficient due to the stability of the Mn porphyrin systems under these conditions. For example, the water-soluble complex **51**/colloidal $Pt/H_2/O_2$ (Figure 18) gave

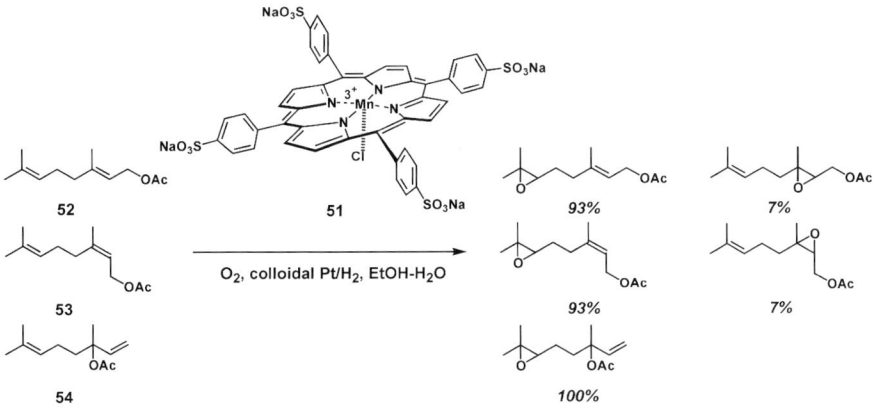

Figure 18. Regio- and stereoselective epoxidation of terpenes.

regio- and stereoselective epoxidation of terpenes such as **52**, **53**, and **54** with turnovers between 10^3 and 10^4 [44].

Besides containing reactive centers such as metal cofactors, enzymes carry binding sites to orient the substrate in an adequate fashion. This is particularly important for enantiospecific C–H bond cleavage catalyzed by cytochromes P450 (see Figure 3). In this context the attachment of substrate recognition sites to the periphery of Mn(III) porphyrins has been a major breakthrough. Due to their lipophilic cavity and their hydrophilic surface, β-cyclodextrins are particularly suited for binding hydrophobic substrates which are the most frequent targets of P450 oxidations. In general, mono-β-cyclodextrins bind substrates that fit into the cavity of 7.8 Å depth with association constants $< 10^4$ M^{-1}. Hence, it can be expected that substrates that bind to two or more β-cyclodextrins display binding constants between 10^6 and 10^8 M^{-1}. This concept was demonstrated *inter alia* by determining $K_a = 3.3 \times 10^6$ M^{-1} for the binding of cholesterol **55** to the dimeric β-cyclodextrin **56** (Figure 19) [46].

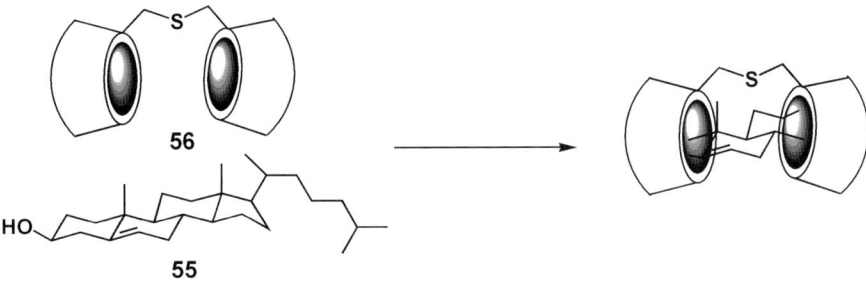

Figure 19. Substrate recognition by dimeric β-cyclodextrins.

Attaching two or four β-cyclodextrins to the *meso*-phenyl groups of a Mn(III) porphyrin (see **57** in Figure 20), epoxidation of stilben-type substrates such as **58** → **59** [47] and benzylic hydroxylation of diphenyl ethane substrates **60** → **61** [48] could be selectively catalyzed in the presence of PhIO with turnovers up to 10^2 (Figure 20). It is important to note that all substrates contain end groups that are known to form good inclusion complexes with β-cyclodextrin such as $PhNO_2$ (see **58**) and Ph-*tert*-butyl (see **60**). This fact and the structural similarity between substrates and products, in particular coordination of the epoxide or the benzylic alcohol to Mn(III) poses the principal problem of product inhibition. Further, the thioether linkages to β-cyclodextrins are most likely oxidized to sulfoxides under reaction conditions. The reactivity of the supramolecular system towards hydroxylation of non-activated C-H bonds was shown by stereospecific conversion of the androstanediol derivative **62** → **63** (Figure 21), this oxidation at C-6 of the steroid documents a remarkably selective procedure, nearly impossible to accomplish with other methods [48].

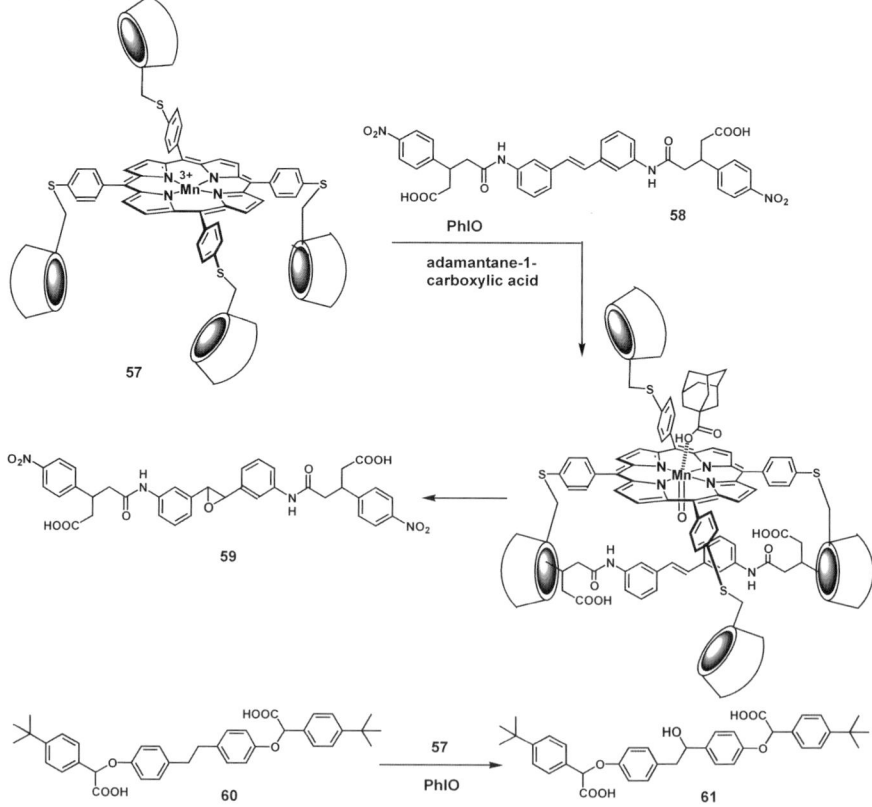

Figure 20. P450-like reactions with β-cyclodextrin-attached Mn porphyrins.

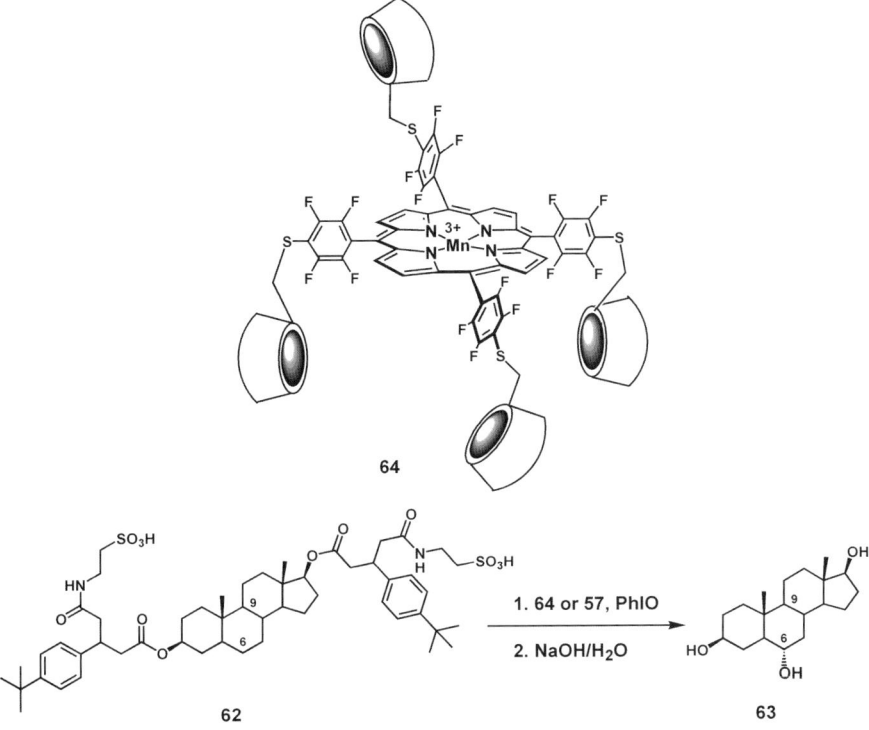

Figure 21. Regio- and stereospecific steroid hydroxylation.

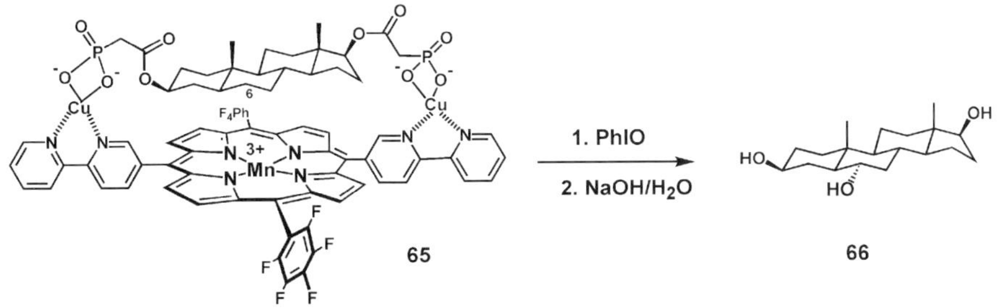

Figure 22. Substrate recognition through complex formation.

More recently the system has been optimized in various ways to allow for example the hydroxylation at C-9 of a steroid. Extending the concept mentioned above, the stability against autoxidation and the reactivity of the oxo manganese intermediate was improved by introducing electron-withdrawing tetrafluorophenyl groups at the *meso*-positions of the porphyrin chromophore (see **64** in Figure 21) [49].

Alternatively, substrate recognition by the catalyst was accomplished by Cu^{2+} coordination rather than by cyclodextrins (Figure 22). For this purpose the Mn(III) porphyrin contains two *meso*-bipyridyl units and the substrate two terminal phosphate groups that all together form an excellent *bis*-Cu^{2+} complex of substrate and catalyst such that hydroxylation occurs exclusively at C-6α see **65 → 66** [50].

3.3. Ruthenium Porphyrins

High-valent oxo ruthenium compounds have been known for decades as valuable reagents for catalytic oxidations. Thus, it was only logical to investigate whether ruthenium porphyrins can be used as P450 enzyme mimics. Ru(II) porphyrins such as **67** containing four *meso*-mesityl substituents that prevent μ-oxo dimer formation are readily oxidized with PhIO to yield the diamagnetic dioxo Ru(VI) porphyrin **68** [51]. The same compound can also be obtained as the final product when **67** is exposed to air; during this transformation the mono-oxo Ru(IV) complex **69** was observed and spectroscopically characterized (Figure 23) [52].

Ru complexes such as **67**, however, display a mayor advantage over corresponding complexes with Co, Fe, Mn, or Os [53] because within this group they are the only complexes which react with pyridine N-oxides such as **70** to yield highly active dioxo Ru(VI) porphyrins **68**. Pyridine N-oxides are superior to N-oxides of, e.g., 4-cyano-*N,N*-dimethylaniline [54], that can be used for Fe(III) porphyrins to give Cpd I analogues. The latter yields 4-cyano-*N,N*-dimethylaniline which by itself is a substrate and hence, subsequently demethylated.

With the system **67** + **70** turnovers of up to 17 000 have been accomplished in the stereospecific epoxidation of simple olefins [53]. Very efficient catalytic

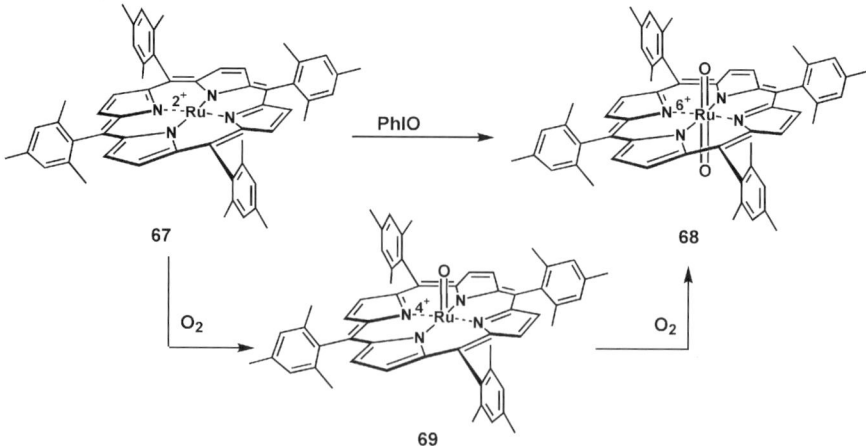

Figure 23. Formation of dioxo ruthenium porphyrins.

oxidation of hydrocarbons, for example adamantane **71** → **72** + **73** (turnover 11 000), was also achieved in the presence of **70** and HBr (Figure 24).

The reactivity of O=Ru=O porphyrins towards double bonds was also employed in supramolecular systems (Figure 25). The water-soluble *bis*-β-cyclodextrin Ru porphyrin **74** [55] binds the carotenoid **75** ($K_a = 8.3 \times 10^6 M^{-1}$ of the corresponding Zn complex), extracted from a hexane phase, to produce an inclusion complex **76** which in the presence of *tert*-butyl hydroperoxide (TBHP) gave the corresponding dioxo complex **77** that catalyzed the regiospecific epoxidation and subsequent cleavage of the central C-(15) − C-(15′) double bond

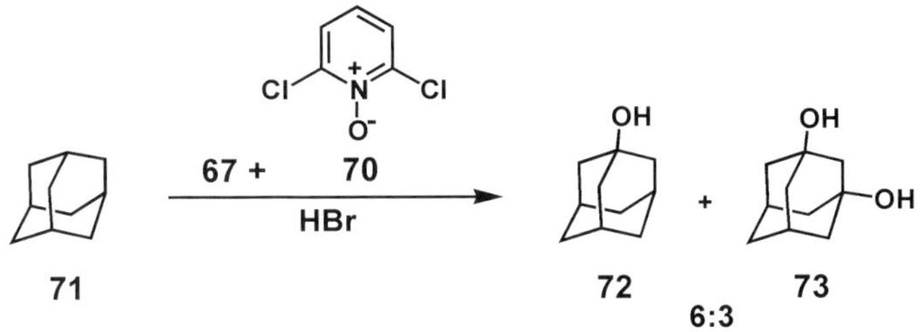

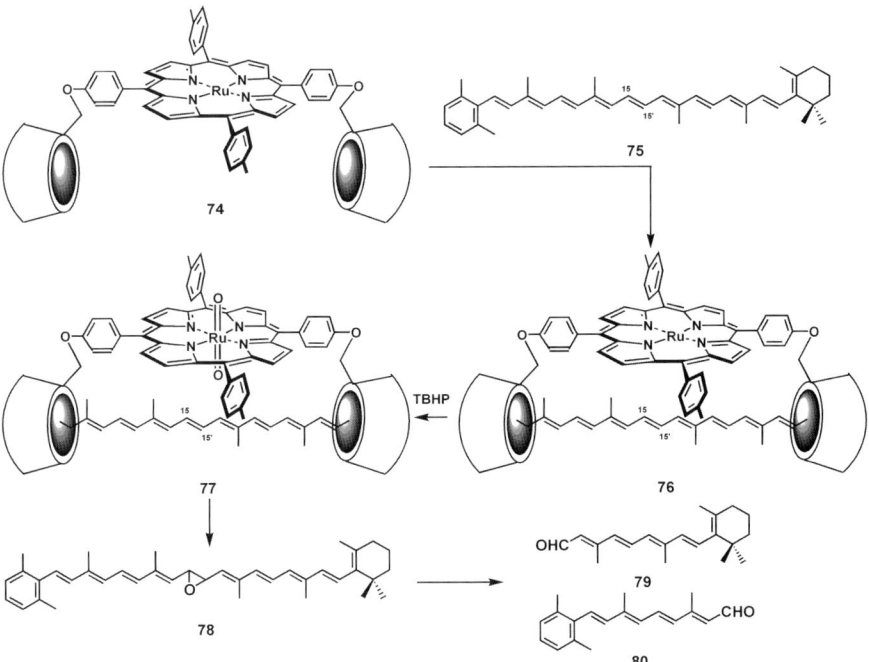

Figure 24. Catalytic oxidation of non-activated C–H bonds.

Figure 25. Regiospecific epoxidation of carotenoids catalyzed by a Ru porphyrin.

of the substrate (see **78 → 79 + 80** in Figure 25). This system serves well as a mimic of the enzyme 5,15′-β-carotene monooxygenase which produces retinal **79** in tissues of higher organisms [56,57].

3.4. Various Electron-Deficient Metal Porphyrins

Most co-oxidants used in excess to generate catalytic P450 enzyme mimics bleach the porphyrin chromophore in minutes at room temperature. Destruction of the prophyrin is most likely an electrophilic reaction at either the *meso-* or the β-pyrrole positions [58,59].

The attachment of phenyl groups carrying electron withdrawing substitutents at *meso*-carbons is expected to: (i) improve the stability of complexes; (ii) shift the redox potential anodically, i.e., it stabilizes Fe(II) over Fe(III); (iii) render the corresponding Cpd I analogue more reactive, and (iv) should provide enough steric hindrance to prevent μ-oxo dimer formation. In case the β-pyrrole positions are not substituted such that the π-system of the phenyl rings is not completely decoupled from the porphyrin all these effects are expected to be substantial and beneficial for catalysis.

Introducing electron-withdrawing groups such as Cl, Br or NO_2 into the β-pyrrole positions will certainly contribute to the stability of the porphyrin against electrophilic attack and will also enhance the reactivity of Cpd I [60]. However, the bulky halogens also lead to severe distortion of the macrocycle, i.e., to various *saddle-shaped* porphyrin conformations, causing dramatic batochromic shifts in the UV-Vis (to ∼ 450 nm) spectrum. This substitution also gives rise to anodic shifts of E_0, however, less significant than expected because the energy of the HOMO is raised.

One of the first complexes adopting the former approach was **81** carrying four *meso*-2,6-dichlorophenyl groups (Figure 26). This complex proved to be an excellent catalyst in the presence of F_5PhIO for the epoxidation of olefins such as norbornene **45** with 300 turnovers/s and a total turnover of 10 000 [61]; for alkane hydroxylation, however, **81** was less efficient.

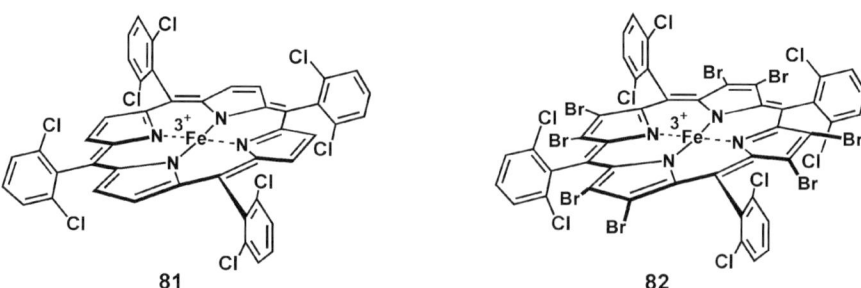

Figure 26. Polyhalogenated porphyrins.

The development of a bromination procedure (NBS) of the β-pyrrole positions of corresponding Zn porphyrins [62] led to the synthesis of the very stable complex **82** which was a reactive catalyst for the hydroxylation of non-activated C–H bonds.

Attaching SO_3H groups to the *meso*-phenyl units of electron deficient porphyrins provides further advantages. Complexes such as **83** (Figure 27) become water-soluble [60] and effectively mimic the enzyme lignin peroxidase (see **84** → **85** + **86** in Figure 27) and the SO_3H functionality allows, for example, binding to a poly(vinylpyridinium) polymer such that the catalyst becomes immobilized by Coulomb interactions and the pyridine of the polymer acts as fifth ligand to iron (see **87** in Figure 28), or manganese [63].

Electronic activation of porphyrins was further accomplished by replacing the β-pyrrole hydrogens of the Mn complex corresponding to **81** with NO_2 groups [64]. These catalysts even work well in the presence of H_2O_2 and preferentially hydroxylate aromatic compounds. The best results were obtained when 1–4 NO_2 groups were introduced, this is most likely related to the fact that

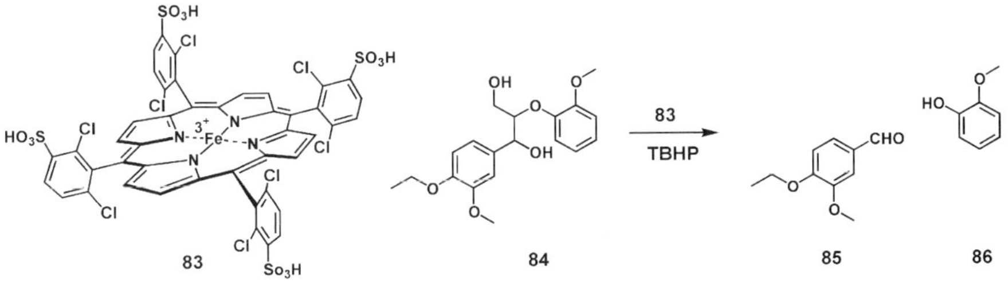

Figure 27. Cleavage of a lignin-type substrate with a water soluble electron deficient porphyrin.

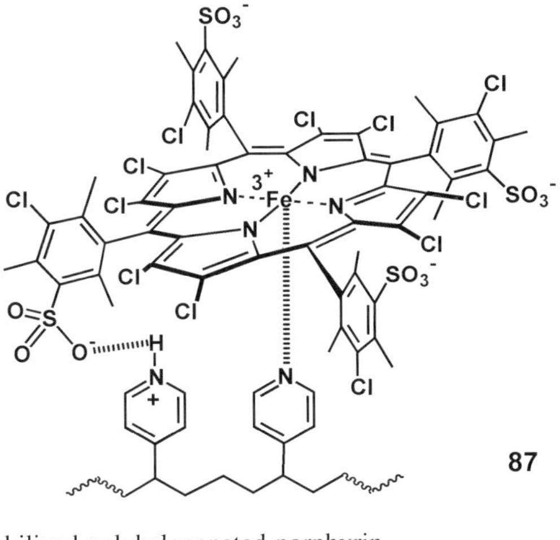

Figure 28. Immobilized polyhalogenated porphyrin.

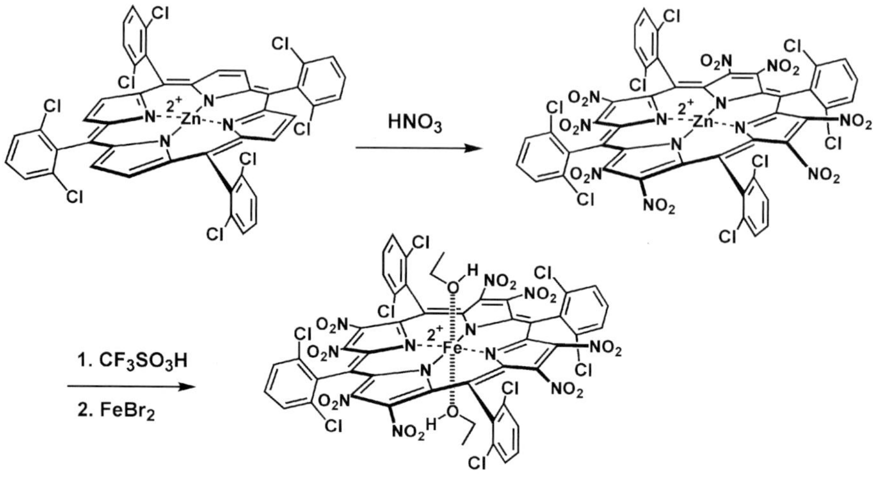

Figure 29. Preparation of polynitrated iron porphyrins.

the redox potential of the couple Mn(II)/Mn(III) (and also Fe) shifts dramatically towards positive values with increasing number of NO_2 groups. Hence, it seemed that with more than five NO_2 groups Mn(II) cannot be oxidized. This was somewhat confirmed in a more recent investigation of the iron porphyrin **88** (Figure 29) containing 8 NO_2 substituents [65]. The reversible one-electron oxidation of Fe(II) to Fe(III) occurs at $+1.03$ V which readily explains the stability of Fe(II), even in the presence of air. Interestingly, the hexacoordinate *bis*-EtOH adduct **88** is a high-spin Fe(II) system and according to X-ray structure analysis the porphyrin is completely planar. Only if the same compound

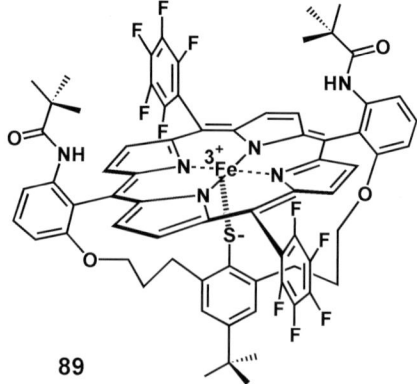

Figure 30. A P450 and CPO mimic containing electron-withdrawing substituents.

is crystallized from wet acetone, the hexacoordinate *bis*-H_2O iron(II)porphyrin becomes completely distorted and of intermediate spin.

The system **89** that combines electron-withdrawing pentafluorophenyl units and a thiolate coordinating to iron also displayed interesting features (Figure 30) [66]. The reduced form of **89** was found to bind CO reversibly showing a split Soret with a UV maximum at 448 nm. The redox potential was determined $E_0 = -134$ mV identical to the value of the heme-thiolate protein chloroperoxidase. It was demonstrated that the compound behaved as a valuable P450/CPO equivalent for it catalyzed the epoxidation of Z-olefins as well as the chlorination of activated C–H bonds with turnovers up to 1500.

Using an electron-deficient manganese porphyrin and a rhodium pyridyl complex both embedded in a membrane setup (see **90** in Figure 31), it has been

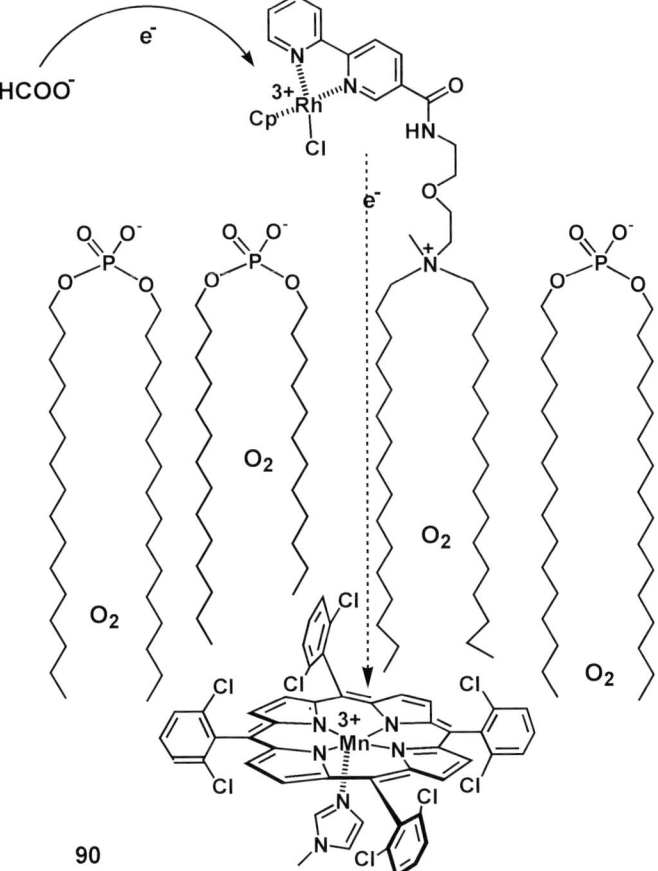

Figure 31. A vesicle construct containing various components of the native P450 system.

possible to assemble all the principal components of the P450 system [67]. The rhodium formate unit resembles the P450 reductase which shuffles electrons to the manganese porphyrin to initiate reductive oxygen cleavage generating O=Mn(V) as the reactive species. Though the manganese porphyrin is quite stable under these conditions the total turnover for the epoxidation of α-pinene is a modest 360 and rather dependent on the choice of the vesicle's head group.

4. CONCLUDING REMARKS

The 'biomimetic P450 community' has focused on creating systems that would help to understand the broad and unprecedented reactivity of P450 enzymes. In conjunction with enzymatic studies this goal has been accomplished to a great extent. However, though the principal P450 reactivity could be established with systems particularly designed [7,8,68] and promising routes to prepare Cpd I have been recently discovered [69], no real practical and competitive lab procedures above milligram scale have emerged from these investigations, neither regarding simple epoxidations nor regio- and stereospecific 'O' insertion into C–H bonds. The hydroxylation of non-activated C–H bonds requires a strong oxidant such as Cpd I and everything electronically similar to Cpd I would probably react in a radical type reaction by H• abstraction, and hence any selectivity is difficult to achieve in a synthetic setup. Further, most of the efforts have failed to make the porphyrin chromophore more stable in the presence of an excess of co-oxidant required to generate Cpd I-like species. Hence there is a need to invent new methods to create oxo metal species under conditions that are tolerated by the ligand [69].

Nature has developed rather rigid substrate binding sites, e.g., for $P450_{cam}$, that through proper orientation of the substrate allow for exposing the one and only C–H bond to the metal oxo bond. On the other hand, flexible, distal pockets exist in various P450s which is seen in the phenomenon of metabolic switching and much more pronounced in drug-metabolizing P450s such as CYP3A4 [2,70], which turn over structurally very diverse substrates and/or hydroxylate one compound at different positions. In order to mimic the protein's acceptance of substrate diversity *and* selectivity of oxidation a self-organizing supramolecular system is the only solution. At present, however, no simply accessible and versatile metal porphyrins containing such supramolecular substrate recognition site are available.

ACKNOWLEDGMENTS

Contributions from the author's lab were financially supported by the Swiss National Science Foundation and F. Hoffmann-La Roche AG.

ABBREVIATIONS

Cpd I	compound I
CPO	chloroperoxidase
DFT	density functional theory
DMF	dimethylformamide
ee	enantiomeric excess
ENDOR	electron nuclear double resonance
ESEEM	electron spin-echo envelope modulation
HOMO	highest occupied molecular orbital
HRP	horseradish peroxidase
mCPBA	*m*-chloroperbenzoic acid
NBS	*N*-bromosuccinimide
Pdx	putidaredoxin
PGSH	prostaglandin H synthase
Ph	phenyl
TBHP	*tert*-butyl hydroperoxide
TMP	tetra mesityl porphyrin
TON	turnover number
TPP	tetraphenyl porphyrin

REFERENCES

1. T. Omura and R. Sato, *J. Biol.Chem.*, 237, 1375–1376 (1962).
2. P. Guengerich, in *Cytochrome P450, Structure Mechanism and Biochemistry* (P. Ortiz de Montellano, ed.), 3rd edn, Kluwer Academic, New York, 2005, 377–463.
3. J. W. Sparapany, M. J. Crossley, J. E. Baldwin, and J. A. Ibers, *J. Am. Chem. Soc.*, 110, 4559–4564 (1988).
4. A. R. Battersby, S. A. J. Bartholomew, and T. Mitta, *J. Chem. Soc., Chem. Commun.*, 1291–1293 (1983).
5. J. P. Collman, R. R. Gagne, C. A. Reed, T. R. Halbert, G. Lang, and W. T. Robinson, *J. Am. Chem. Soc.*, 97, 1427–1439 (1975).
6. M. Momenteau, J. Mispelter, B. Loock, and E. Bisagni, *J. Chem. Soc., Perkin Trans. 1*, 189–196 (1983).
7. W.-D. Woggon, *Topics in Current Chemistry*, 184 (Bioorganic Chemistry), 39–96 (1997).
8. B. Meunier, S. P. de Visser, and S. Shaik, *Chem. Rev.* 104, 3947–3980 (2004).
9. *Comprehensive Asymmetric Catalysis I-III* (E. N. Jacobsen, A. Pfaltz, and H. Yamamoto, eds), Springer, Berlin 1999.
10. J. P. Collman, T. N. Sorell, and B. M. Hoffman, *J. Am. Chem. Soc.*, 97, 913–914 (1975).
11. J. P. Collmann and S. Groh, *J. Am. Chem. Soc.*, 104, 1391–1403 (1982).
12. A. R. Battersby, W. Howson, and A. D. Hamilton, *J. Chem. Soc., Chem. Commun.*, 1266–1268 (1982).

13. B. Staeubli, H. Fretz, U. Piantini, and W.-D. Woggon, *Helv. Chim. Acta, 70*, 1173–1193 (1987).
14. H. Patzelt and W.-D. Woggon, *Helv. Chim. Acta, 75*, 523–530 (1992).
15. T. G. Traylor, T. C. Mincey, and A. P. Berzinis, *J. Am. Chem. Soc., 103*, 7084–7089 (1981).
16. W.-D.Woggon, *Nachr. Chem. Tech. Lab., 36*, 890–895 (1988).
17. H. Aissaoui, S. Ghirlanda, C. Gmuer, and W.-D. Woggon, *J. Mol. Catalysis A: Chemical, 113*, 393–402 (1996).
18. N. Ueyama, N. Nishikawa, Y. Yamada, T.-A. Okamura, and A. Nakamura, *J. Am. Chem. Soc., 118*, 12826–12827 (1996).
19. T. Higuchi, U. Shinobu, and M Hirobe, *J. Am. Chem. Soc., 112*, 7051–7053 (1990).
20. H. Aissaoui, R. Bachmann, A. Schweiger and W.-D. Woggon, *Angew. Chem., Int. Ed., 37*, 2998–3002 (1998).
21. T. L. Poulos, B. C. Finzel, and A. J. Howard, *Biochemistry, 25*, 5314–5322 (1986).
22. M. Lochner, M. Meuwly, and W.-D. Woggon, *Chem. Commun., 12*, 1330–1332 (2003).
23. M. Lochner, L. Mu, and W.-D. Woggon, *Adv. Synthesis and Catalysis, 345*, 743–765 (2003).
24. D. N. Meyer and W.-D. Woggon, *Chimia, 59*, 85–87 (2005).
25. S. Kozuch, T. Leifels, D. Meyer, L. Sbaragli, S. Shaik, and W.-D. Woggon, *Synlett, 4*, 675–684 (2005).
26. D. Meyer, T. Leifels, L. Sbaragli, and W.-D. Woggon, *Biochem. Biophys. Res. Commun., 338*, 372–377 (2005).
27. A. Franke, N. Hessenauer-Ilicheva, D. Meyer, G. Stochel, W.-D. Woggon, and R. van Eldik, submitted.
28. L. Sbaragli and W.-D.Woggon, *Synthesis, 9*, 1538–1542 (2005).
29. Y. Urano, T. Higuchi, M. Hirobe, and T. Nagano, *J. Am. Chem. Soc., 119*, 1208–1209 (1997).
30. F. Tani, M. Matsu-ura, S. Nakayama, M. Ichimura, N. Nakamura, and Y. Naruta, *J. Am. Chem. Soc., 123*, 1133–1142 (2001).
31. F. Tani, M. Matsu-ura, S. Nakayama, and Y. Naruta, *Coord. Chem. Rev., 226*, 219–226 (2002).
32. I. Schlichting, J. Berendzen, K. Chu, A. M. Stock, S. A. Maves, D. E. Benson, R. M. Sweet, D. Ringe, G. A. Petsko, and S. G. Sligar, *Science, 287*, 1615–1622 (2000).
33. E. Matsui, Y. Naruta, F. Tani, and Y. Shimazaki, *Angew. Chemie, Int. Ed., 42*, 2744–2747 (2003).
34. J. T. Groves, T. E. Nemo, and R. S. Myers, *J. Am. Chem. Soc., 101*, 1032–1033 (1979).
35. J. T. Groves, R. C. Haushalter, M. Nakamura, T. E. Nemo, and B. J. Evans, *J. Am. Chem. Soc., 103*, 2884–2886 (1981).
36. R. E. P. Chandrasena, K. P Vatsis, M. J. Coon, P. F Hollenberg, and M. Newcomb, *J. Am. Chem. Soc., 126*, 115–126 (2004).
37. F. Ogliaro, S. P. de Visser, S. Cohen, P. K. Sharma, and S. Shaik, *J. Am. Chem. Soc., 124*, 2806–2817 (2002).
38. J. T. Groves and T. E. Nemo, *J. Am. Chem. Soc., 105*, 5786–5791 (1983).
39. J. T. Groves and P. Viski, *J. Org. Chem., 55*, 3628–3634 (1990).
40. J. T. Groves and P. Viski, *J. Am. Chem. Soc., 111*, 8537–8538 (1989).

41. J. T. Groves, W. J. Kruper, and R. C. Haushalter, *J. Am. Chem. Soc.*, *102*, 6377–6380 (1980).
42. J. T. Groves and M. K. Stern, *J. Am. Chem. Soc.*, *110*, 8628–8638 (1988).
43. I. Tabushi and N. Koga, *J. Am. Chem. Soc.*, *101*, 6456–6458 (1979).
44. I. Tabushi and K. Morimitsu, *J. Am. Chem. Soc.*, *106*, 6871–6872 (1984).
45. D. Mansuy, M. Fontecave, and J. F. Bartoli, *J. Chem. Soc., Chem. Commun.*, *6*, 253–254 (1983).
46. R. Breslow and B. Zhang, *J. Am. Chem. Soc.*, *118*, 8495–8496 (1996).
47. R. Breslow, X. Zhang, R. Xu, M. Maletic, and R. Merger, *J. Am. Chem. Soc.*, *118*, 11678–11679 (1996).
48. R. Breslow, Y. Huang, X. Zhang, and J. Yang, *Proc. Natl. Acad. Sci. USA*, *94*, 11156–11158 (1997).
49. J. Young and R. Breslow, *Angew. Chem. Int. Ed.*, *39*, 2692–2694 (2000).
50. Z. Fang and R. Breslow, *Organic Letters*, *8*, 251–254 (2006).
51. J. T. Groves and R. Quinn, *Inorg. Chem.*, *23*, 3844–3846 (1984).
52. J. T. Groves and K. H. Ahn, *Inorg. Chem.*, *26*, 3831–3833 (1987).
53. T. Higuchi and M. Hirobe, *J. Mol. Catalysis A: Chemical*, *113*, 403–422 (1996).
54. C. M. Dicken, F.-L. Lu, M. W. Nee, and T. C. Bruice, *J. Am. Chem. Soc. 107*, 5776–5789 (1985).
55. R. R. French, P. Holzer, M. G. Leuenberger, and W.-D. Woggon, *Angew. Chem., Int. Ed.*, *39*, 1267–1269 (2000).
56. A. Wyss, G. Wirtz, W.-D. Woggon, R. Brugger, M. Wyss, A. Friedlein, H. Bachmann, and W. Hunziker, *Biochem. Biophys. Res. Commun.*, *271*, 334–336 (2000).
57. M. G. Leuenberger, C. Engeloch-Jarret, and W.-D. Woggon, *Angew. Chem., Int. Ed.*, *40*, 2614–2617 (2001).
58. P. R. Ortiz de Montellano, *Acc. Chem. Res.*, *31*, 543–549 (1998).
59. D. Kumar, S. P. De Visser, and S. Shaik, *J. Am. Chem. Soc.*, *127*, 8204–8213 (2005).
60. D. Dolphin, T. G. Traylor, and L. X. Xie, *Acc. Chem. Res.*, *30*, 251–259 (1997).
61. P. S. Traylor, D. Dolphin, and T. G. Traylor, *J. Chem. Soc., Chem. Commun.*, 279–280 (1984).
62. T. G. Traylor and S. Tsuchiya, *Inorg. Chem.*, *26*, 1338–1339 (1987).
63. S. Campestrini and B. Meunier, *Inorg. Chem.*, *31*, 1999–2006 (1992).
64. J.-F. Bartoli, V. Mouries-Mansuy, K. Le Barch-Ozette, M. Palacio, P. Battioni, and D. Mansuy, *J. Chem. Soc. Chem. Commun.*, 827–828 (2000).
65. K. M. Barkigia, M. Palacio, Y. Sun, M. Nogues, M. W. Renner, F. Varret, P. Battioni, D. Mansuy, and J. Fajer, *Inorg. Chem.*, *41*, 5647–5649 (2002).
66. H.-A. Wagenknecht, C. Claude, and W.-D. Woggon, *Helv. Chim. Acta, 81*, 1506–1520 (1998).
67. A. P. H. J. Schenning, J. H. Spelberg, D. H. W. Hubert, M. C. Feiters, and R. J. M. Nolte, *Chem. Eur. J.*, *4*, 871–880 (1998).
68. W.-D. Woggon, *Acc. Chem. Res.*, *38*, 127–136 (2005).
69. R. Zhang, R. E. P. Chandrasena, E. Martinez, J. H. Horner, and M. Newcomb, *Org. Lett.*, *7*, 1193–1195 (2005).
70. A. Chougnet, C. Stoessel, and W.-D. Woggon, *Bioorg. Med. Chem. Lett.*, *13*, 3643–3645 (2003).

3

Structures of P450 Proteins and Their Molecular Phylogeny

Thomas L. Poulos and Yergalem T. Meharenna
Department of Biochemistry and Molecular Biology, Physiology and Biophysics, and Chemistry
University of California, Irvine, CA 92697-3900, USA
<poulos@uci.edu>
<ymeharen@uci.edu>

1. INTRODUCTION	58
2. P450 EVOLUTION	59
3. P450 FAMILIES AND SUBFAMILIES	60
3.1. Naming of P450s	60
4. P450 STRUCTURES	62
4.1. Conservation of the P450 Fold	62
4.2. Domain Structure	64
4.3. Cysteine Ligand Loop	65
4.4. Heme–Propionate Interactions	67
4.5. Oxygen Binding Region	68
4.6. Conformational Changes Required for Substrate Binding	70
4.7. Electron Transfer Site	74
4.8. Membrane Binding	76
4.9. Oligomeric Structure	77
5. VARIATION IN P450 FUNCTION AND FOLD	81
5.1. Enzymes with a P450 Function but Different Fold	81
5.2. Enzymes with the P450 Fold but Different Function	83
5.3. Mechanistic Variations	85
6. ARCHAEON P450s	86

7. SUMMARY AND CONCLUSIONS	90
ACKNOWLEDGMENTS	91
ABBREVIATIONS	91
REFERENCES	91

1. INTRODUCTION

The various genome projects have revolutionized the way we approach biological research, and near the top of the list of research areas most effected is P450. A relatively short time ago (early 1980s) the cloning and sequencing of a new P450 was a major event. Things have changed. In a breathtakingly short period of time we know how many P450 genes a diverse array of organisms have. We humans have only 63 while *Drosophila* has 83 and *Arabidopsis* has 286. Why? Solving this problem will require the much more difficult and time-consuming task of what is sometimes called functional proteomics, or in this case, working out the functional P450ome for each organism. The information in hand should also contribute to our understanding on the evolutionary relatedness of P450s. Already it is clear that the P450 fold is unique and highly conserved, independent of organism. However, the same fold is used by enzymes that do not catalyze P450 chemistry.

Another important consideration in P450 phylogeny is the broadly defined functional properties used to classify P450s. One of the most clearly defined functional divisions of P450s is between those involved in xenobiotic detoxification and those involved in the production of essential metabolic intermediates. The first group includes the well-known drug-metabolizing P450s that are not substrate selective. The second group includes mitochondrial steroid hydroxylases and various bacterial P450s that are highly specific since these must produce critical intermediates essential for survival. Which came first, the detoxifying P450s or the specific P450s?

While the functional P450ome will take some time to work out, the progress in P450 structural biology has progressed rapidly owing to at least three factors. First, the various genome projects have helped to identify new P450s that would no doubt have taken much longer to discover using pre-genomic era approaches. Second, heterologous recombinant expression has become more routine with an ever-increasing array of methodologies. Third, the ground breaking work on how to express and crystallize membrane bound P450s [1] has encouraged others to pursue the solution of a diverse range of membrane-associated P450s. The goal of solving the structure of all human P450s now appears to be a realistic possibility. In addition, the rapid accumulation of P450 structures should help in assigning a specific function to a new P450. Comparison of a structure with unknown function to a more well characterized P450 should greatly aid in identifying or at least narrowing the possible substrate(s) of the unknown. Given the increasing ease of P450 structure determinations, solving a new structure may become competitive in terms of time and resources with more traditional genetic and biochemical methods for determining function. This is similar to the rationale and justification for the current large-scale structural genomics projects where it is hoped that the fold of an uncharacterized gene product will help to assign function.

What follows is a current overview on P450 structures. We begin in Sections 2 and 3 with a brief primer on P450 evolution and nomenclature, the later being a seemingly Byzantine system that is nonetheless absolutely critical if one is to have any hope of understanding P450 evolution, including the functional connection between various P450s from different species. In Section 4 we focus on the details of P450 structure and how the same P450 fold has been used to catalyze non-P450 redox chemistry. Section 5 centers on enzymes with the P450 fold, but different function and enzymes with P450-like functions, but different fold, and finally Section 6 provides an overview on recently discovered archaea P450s.

2. P450 EVOLUTION

Geologic [2] and molecular biological [3] evidence suggest that eukaryotes and eocyte prokaryotes are immediate relatives, indicating that any gene occurring in both eubacteria and eukaryotes probably arose from a common ancestral gene more than 3.5 billion years ago [4]. As noted in Section 1, all P450s are either highly specific and are required for the production of specific metabolic intermediates or exhibit much less selectivity and are critical for detoxification. Since the majority of bacterial P450s are selective, it follows that early P450s were selective and that the decrease in specificity evolved later as more complex organisms had to contend with a more diverse and hostile environment. Thus, adaptive diversification of P450 genes to environmental variables such as the need to detoxify plant phytoalexins and other environmental chemicals resulted in an enormous increase in the number of mammalian P450 genes over 300–400 million years [5]. Therefore, P450 enzymes probably originated during early evolution to provide the organism with oxidative and peroxidative metabolism of important endogenous molecules, as well as the breakdown of environmental chemicals utilized for energy. Subsequently, P450 enzymes in eukaryotes diversified, in response to the adverse effects of dietary pressures and other environmental chemicals [5].

An important question is the identity of the oldest P450 that gave rise to mammalian P450s. One view holds that the specific lanosterol 14 α-demethylase, CYP51, is the precursor to all mammalian P450s [6–10]. This is based primarily on CYP51-like bacterial P450s being the closest homologs to a mammalian P450 in terms of sequence identity. Moreover, its sterol 14 α-demethylation reaction is evolutionarily conserved. This, of course, also assumes that mammalian P450s arose from a bacterial ancestor. Similar P450s in bacteria and mammals also could be due to convergent as opposed to divergent evolution. An alternate view was proposed by Rezen et al. [11] where CYP51 was acquired by lateral gene transfer. Thus the question on the bacterial precursor to mammalian P450s remains open. The most recent discovery of archaea P450s raises some additional possibilities on the origins of mammalian P450s.

3. P450 FAMILIES AND SUBFAMILIES

3.1. Naming of P450s

Although the main topic of this chapter is structure, it is important to first explain how P450s are named and classified. This was not so critical when there were only a few P450 crystal structures. However, more structures are being determined and to understand how these structures are related functionally and evolutionarily requires a basic understanding of P450 nomenclature. When it became apparent that the number of P450 genes and proteins is very large, it also became necessary to develop a nomenclature system [12–16]. There are two naming systems: one for genes and one for proteins. Naming a P450 gene include the root symbol CYP for cytochrome P450 ('cyp' for mouse and Drosophila). The cDNA and genes are italicized, followed by an Arabic numeral designating the P450 family and a letter indicating the subfamily (when two or more exist), and an Arabic numeral representing the individual gene within the subfamily, e.g., '*CYP2C8*' or '*CYP19*' ('*cyp2c8*' and '*cyp19*', respectively, in mouse).

The mRNA and enzyme in all species including mouse should include all capital letters without italics, e.g., 'CYP2C8' or 'CYP19'. 'P' ('ps' in mouse and *Drosophila*) after the gene number is used to denote a pseudogene. Thus, the *CYP2C8* describes CYP family 2 member in the C subfamily, and isoform 8. Approximately 20 CYP isoforms among the CYP 1, 2, 3, and 4 subfamilies metabolize drugs. Each isoform is a product of a separate gene and these genes within the same subfamily can share considerable sequence homology. Where a single gene demonstrates variations, it is termed polymorphic. Polymorphic enzymes are produced from corresponding polymorphic gene variants and may possess quite different enzyme activities.

The different polymorphic variants carry a further number following an asterisk. For example, CYP2D6*1 is the commonly occurring variant of CYP2D6. Genes within a given family appear to have the same number of exons and similar intron–exon boundaries [12]. Nebert and Nelson [12] thus recommend comparing intron–exon organization in addition to standard sequence alignments in drawing a conclusion about P450 evolution. There are intermediate levels such as B-class and E-class [17] as well as clans (http://drnelson.utemem.edu/CytochromeP450.html) defined for the superfamily and family of P450s.

The naming of the protein as opposed to the gene follows a similar set of rules. CYPs exhibiting greater than 40% identity at the amino acid level belong to the same family. This is done by determining the percentage identity of the representative protein sequences aligned from each family and subfamily. This percentage only reflects a comparison of overlapping portions of the sequences while gaps and unmatched ends are not counted in the overall length. If the new sequence is at least 40% identical to any other sequence, then the new sequence belongs to that family. However, there are more than a half-dozen exceptions to this rule [15]. The designation for P450 families is CYPXXX, where CYP = cytochrome P450 and XXX = an integer range assigned to a family. For example, P450cam is designated CYP101 and belongs to the bacterial family. This family includes all CYPs between CYP101 and CYP199.

Once the family or subfamily is identified, the new sequence is compared with that of all other members in the group. If the new sequence is only a few amino acids different from a known sequence, it is given the same name. If the sequence is a new member of the family or subfamily, it is given the next available number. New gene families have been given names associated with the enzymatic activity when known. Sometimes new families are given names based on their evolutionary association with previously named families. Fungal, yeast, and other lower eukaryote families start at CYP51 and may continue to CYP69. Plant gene families start at CYP71 and may continue to CYP99. Bacterial families start at CYP101. When any of these sets becomes full, it is recommended continuing at CYP201 for animals, CYP501 for lower eukaryotes, CYP701 for plants, and CYP1001 for bacteria [12].

Sequences that are greater than 55% identical are in the same subfamily designated by a letter after the family number. Thus CYP11A belongs to family 11 and subfamily A. Inclusion of more distant species within the same subfamily lowers this value to >46%, and this value changes as the database continues to expand [16]. For instance, the cholesterol side chain cleavage enzyme, steroid 11-β-monoxygenase and aldosterone synthase are all assigned to the CYP11 family because they share greater or equal to 40% sequence identity. Steroid 11-β-monoxygenase and aldosterone synthase also share more than 55% sequence identity and thus both are assigned to CYP11B1 and CYP11B2, respectively. However, CYP11B1 and CYP11B2 share less than 55% sequence identity with the cholesterol side chain cleavage enzyme which is designated CYP11A [18].

4. P450 STRUCTURES

4.1. Conservation of the P450 Fold

As of February 2006 there were 26 unique P450 structures available (Table 1) in the Protein Data Bank. This structural data-base now is sufficiently large to safely conclude that the overall P450 fold shown in Figure 1 is conserved, independent of phylogenetic origin. This point can be illustrated by comparing the first prokaryotic P450 structure solved, P450cam, and the first mammalian

Table 1. List of unique P450 structures deposited in the Protein Data Bank as of February 2006.

P450 crystal structures		PDB id	CYP	Resolution	Date	Reference
1.	Heme/FMN-binding domains of the Cytochrome P450 BM3	1BVY	102	2.03 Å	1998	[51]
2.	Cytochrome P450 cam	1CPP	101	2.60 Å	1985	[96]
3.	Cytochrome P450 terp	1CPT	108	2.30 Å	1994	[97]
4.	Cytochrome P450 2C5	1DT6	2C5	3.00 Å	2000	[1]
5.	Cytochrome P450 14α-sterol demethylase with 4-phenylimidazole bound	1E9X	51	2.10 Å	2000	[43]
6.	Cytochrome P450: 119 with 4-phenylimidazole bound	1F4T	119	1.93 Å	2000	[46]
7.	Cytochrome P450 2D6	2F9Q	2D6	3.00 Å	2005	[98] (to be published)
8.	Cytochrome P450 154c1	1GWI	154c1	1.92 Å	2002	[99]
9.	Cytochrome P450 BM3	2HPD	102	2.00 Å	1993	[100]
10.	Cytochrome P450 bs beta with fatty acid bound	1IZO	152a1	2.10 Å	2002	[81]
11.	Cytochrome P450 oxyb	1LFK	165B3	2.20 Å	2002	[101]
12.	Cytochrome P450 Mt2	1N40	121	1.06 Å	2002	[102]
13.	Cytochrome P450 175a1	1N97	175a1	1.80 Å	2002	[93]
14.	Cytochrome P450 154a1	1ODO	154a1	1.85 Å	2003	[103]
15.	Cytochrome P450 2C9	1OG2	2C9	1.85 Å	2003	[104]
16.	Cytochrome P450 eryf	1OXA	107A1	2.10 Å	1995	[105]
17.	Cytochrome P450 epok with epothilone d-bound	1PKF	167A1	2.10 Å	2003	[106]
18.	Cytochrome P450 2B4	1PO5	2B4	1.60 Å	2003	[61]
19.	Cytochrome P450 2C8	1PQ2	2C8	2.70 Å	2004	[107]
20.	Cytochrome P450 nor	1ROM	55A1	2.00 Å	1997	[75]
21.	Cytochrome P450 158a2	1S1F	158a2	1.50 Å	2004	[108]

22. Cytochrome P450 cin with 1,8-cineole bound	1T2B	176A	1.70 Å	2004	[49]
23. Cytochrome P450 3A4	1TQN	3A4	2.05 Å	2004	[55]
24. Cytochrome P450 st	1UE8	hypothetical P450	3.00 Å	2003	[109]
25. Cytochrome P450 OxyC	1UED	165C4	1.90 Å	2003	[110]
26. Cytochrome P450 2A6 with coumarin bound	1Z10	2A6	1.90 Å	2005	[111]

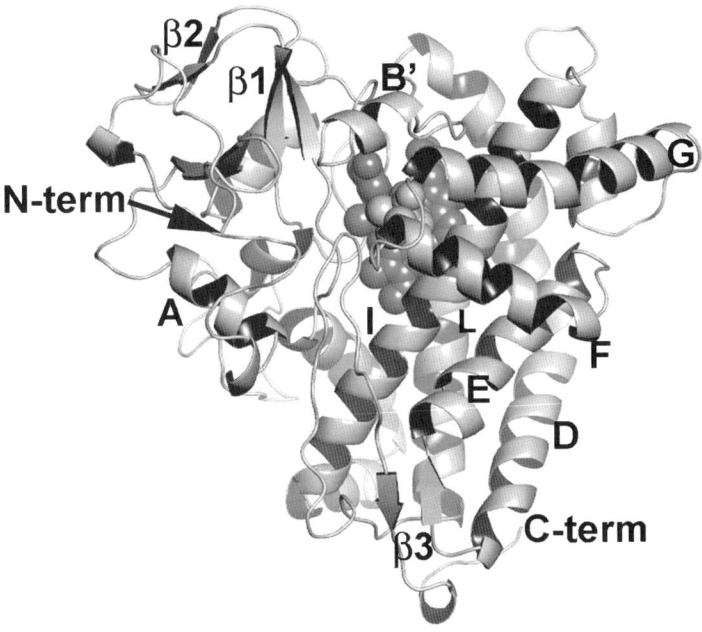

Figure 1. Crystal structure of P450cam showing the key elements of secondary structure.

membrane-bound P450 structure solved, P4502C5, which also represent two of the most distantly related P450s. In general microsomal P450s are larger than prokaryotic P450s. Ignoring for the moment the N-terminal region of microsomal P450s that help anchor the protein to the microsomal membrane, the main difference in size is due to longer surface loops in the microsomal P450s. Both P450s have about the same number of helical segments, 20 in P450cam and 21 in P4502C5, although the helices tend to be longer in the microsomal protein with a total helical content of 279 residues in P4502C5 compared to 205 in P450cam.

A typical way of comparing proteins is to carry out a full least-squares superimposition of Cα atoms. Such a comparison between P450cam and P4502C5 gives an rms deviation of Cα atoms of 2.1 Å for 195 common residues. If distance criteria are relaxed to include 304 common residues the rms deviation increases to 2.8 Å. The proximal and distal surfaces of the heme, defined by the L- and I-helices, respectively, are the two regions most highly conserved. However, such similarity is often masked by a full least-squares fit. In our P450cam/P4502C5 comparison a 10 residue section of the I-helix that defines the oxygen binding site around the key Thr252 (P450cam numbering), the average distance in Cα atoms is 2.2 Å while the average for an 11-residue stretch defining the Cys ligand loop region is 2.4 Å. Based on this type of analysis, one would conclude that the I- and L-helices are, on average, no more similar than any other elements of structure. Such comparisons based on a full least-squares fit between proteins can be deceptive and not give a clear picture on the similarity in active site regions. For example, if the hemes of P450cam and P4502C5 are superimposed rather than the entire protein then the I-helix oxygen binding region average distance decreases from 2.2 to 1.8 Å while the Cys ligand loop decreases from 2.4 to 1.2 Å. A 1.2 Å difference is getting close to the limits of error in moderate-resolution X-ray structures and indicates that the Cys ligand loop region is not just similar, but identical. The reason for the large difference between a global fit and a fit between just the hemes is that the protein further from the active site exhibits much greater variability than those segments closest to the heme. As might be expected, the I- and L-helices, especially those regions near the heme, are very similar in all P450s relative to the heme. Thus it is important to select an appropriate frame of reference in comparing P450 structures. An overall fit certainly is important for global comparisons, but for active site comparisons, the frame of reference should be narrowed to the confines of the active site and the obvious reference point for P450s is the heme.

4.2. Domain Structure

P450s are normally described as consisting of a β domain and helical domain. Domains often are associated with independent folding units. For example, peroxidases clearly fold into N- and C-terminal helical-rich domains with the heme sandwiched between helices from each domain (Figure 2).

However, P450s do not fold into clearly divided N- and C-terminal domains (Figure 2). Both the I- and L-helices which bracket the heme are in the C-terminal half of the molecule while a majority of the β-structure is also in the C-terminal half, but is situated far from the heme. The majority of the N-terminal half of the molecule surrounds the inner core of C-terminal residues closest to the heme. The N-terminus begins on the β-sheet side (left-hand side, Figure 1) then crosses over at the B'-helix to the helical domain (right-hand side, Figure 1). After the I-helix, the polypeptide crosses back to the left-side β domain before crossing back one more time to the right-side helical domain. This type of zig-zag folding

STRUCTURES AND MOLECULAR PHYLOGENY OF P450 PROTEINS

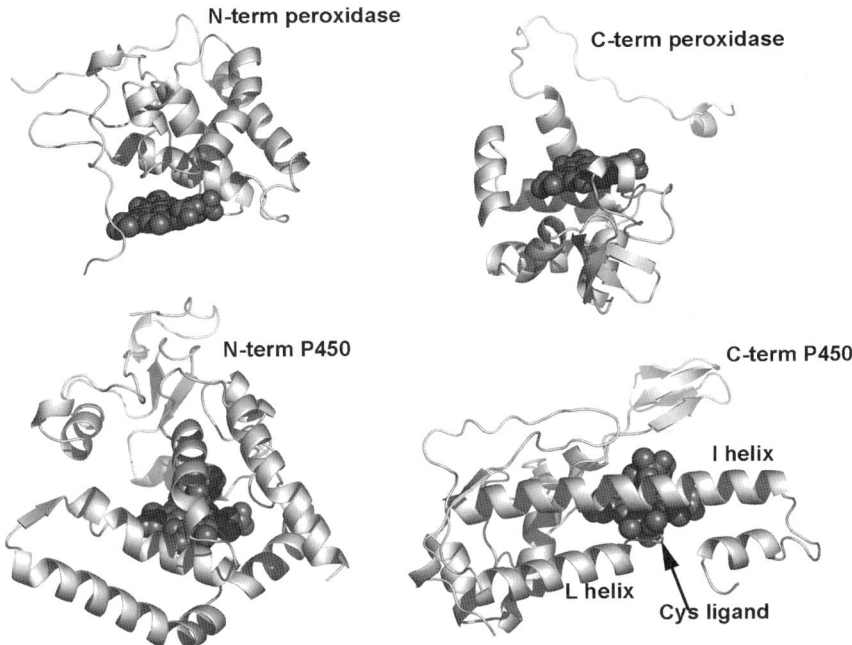

Figure 2. The top diagram shows the N- and C-terminal halves of typical heme peroxidases. Peroxidases fold into two distinct N- and C-terminal domains with the heme sandwiched between these domains. P450s exhibit a very different folding pattern. Note that the C-terminal half of the molecule is situated close to and surrounds the heme while the N-terminal half packs around the inner C-terminal half.

pattern places restrictions on how P450s fold. Clearly P450s cannot fold as the protein is synthesized on the ribosome, especially since the C-terminal half forms the inner core. The entire polypeptide must first be released from the ribosome before proper folding can be achieved.

4.3. Cysteine Ligand Loop

It generally is agreed that the Cys ligand and its surrounding protein environment plays a critical and unique role in the control of P450 chemistry. However, all proteins with Cys–Fe bonds share with P450s common structural features surrounding the Fe ligation environment, even though many of these proteins have no functional or structural connection to P450. First consider the details of the Cys ligand in P450s. In P450s the protein architecture surrounding the proximal Cys ligand is strictly conserved. In all P450s whose structures are known, the peptide NH of the Cys ligand accepts an antiparallel β-strand-type H-bond from a peptide carbonyl O atom as shown in Figure 3. This region

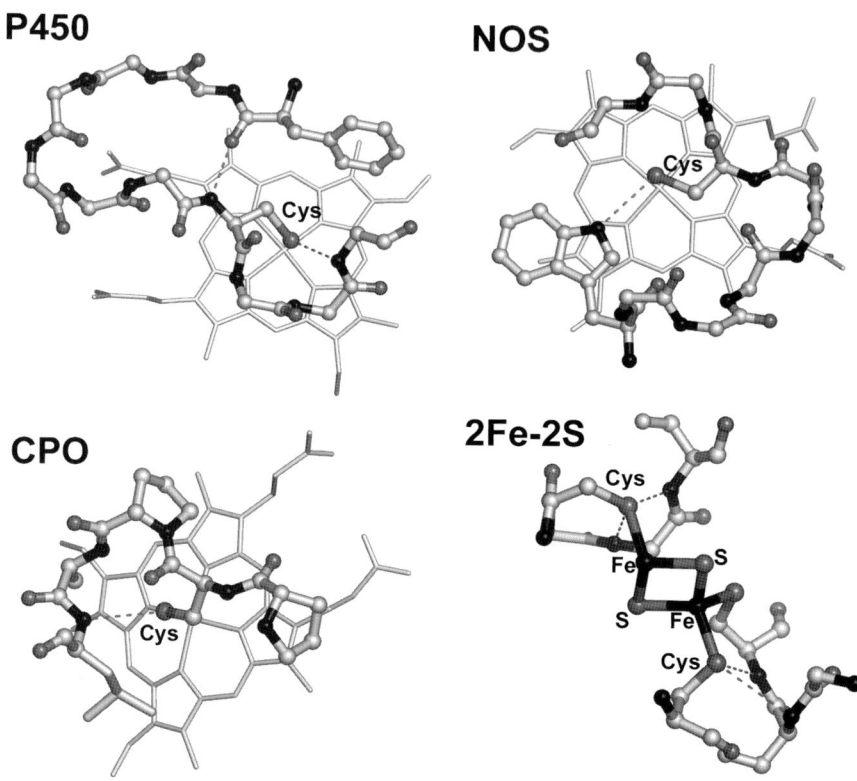

Figure 3. Models showing the H-bonding pattern to the sulfur ligand in four different iron-sulfur proteins. All 4 have H-bonds between peptide NH groups and the sulfur ligand.

defines an antiparallel β-bulge which forms a tight structure that places the Cys sulfur atom in a position where it is surrounded by peptide NH groups that have no H-bond acceptors other than the thiolate ligand.

One of these NH groups forms a good H-bond with the Cys sulfur. Placement of a sulfur iron ligand near peptide NH groups is a characteristic feature found in other heme thiolate enzymes. Chloroperoxidase (CPO) has no structural homology to P450, yet here, too, the Cys sulfur ligand H-bonds with a peptide NH very much the same as in P450 [19]. Nitric oxide synthase (NOS) is similar as well, except here the Cys ligand H-bonds with the indole ring NH group of a conserved Trp residue [20–22]. This similarity also extends to non-heme iron-sulfur ferredoxin-like proteins that contain 2, 3, and 4 iron-sulfur clusters (Figure 3). The importance of the S-HN hydrogen bonds in iron-sulfur proteins was first brought to light from the rubredoxin and ferredoxin crystal structures back in 1975 [23]. As a result, much of the insights regarding the local protein environment surrounding the Cys ligand derive from non-heme Fe-S proteins.

Both experimental and theoretical approaches have been utilized to shed further light on the functional relevance of Cys-HN H-bonds. The general idea is that the conserved Cys-HN H-bonded motif modulates the iron redox potential. Without such H-bonds the iron will exhibit a redox potential well below that of physiologically important reductants, thus trapping the iron center in the oxidized, Fe(III) state. Providing H-bond donors to the sulfur attenuates the negative charge on the sulfur ligand which increases the redox potential, thus enabling the system to be reduced by NAD(P)H and other biological reductants. Work with model metallo-thiolate systems supports this picture [24–26] as do theoretical studies with iron-sulfur proteins [27].

The relevance of the Cys-HN H-bonds extends beyond simple redox potential control to the O_2 activation process and subsequent hydroxylation reactions. While it has not yet been possible to directly study the active hydroxylating species in P450s owing to its apparent very short lifetime, modern density functional theory (DFT) coupled with molecular mechanics approaches have provided some important insights. By analogy with peroxidases, the intermediate generally thought to carry out hydroxylation in P450s is compound I, $Por^{+\bullet}Fe(IV)=O$, where $Por^{+\bullet}$ is the porphyrin radical. The analogy with peroxidases, however, is limited because compound I in peroxidases is relatively long-lived while in P450s compound I has never been directly observed in the normal catalytic cycle. The obvious difference between these two enzyme systems is that peroxidases use His as a heme ligand while P450s use Cys. Initially it was thought that the nature of the ligand would help promote cleavage of the O–O bond. In peroxidases, however, the His ligand can be replaced with either Glu or Gln and the rate of compound I formation is not dramatically altered [28,29]. Even so, the nature of the ligand does play a critical role in determining the reactivity of $Por^{+\bullet}Fe(IV)=O$. DFT calculations show that the thiolate group contains significant radical character in P450 compound I [30,31]. However, as soon as an H-bond donor to the thiolate sulfur is added to the model significant radical character shifts to the porphyrin [31] thus changing the chemical reactivity of compound I [32,33] to produce a better hydroxylating agent. As the new C–O and O–H bonds are formed with the Fe(IV)=O oxygen, electron density must be transferred from the ferryl, Fe(IV)=O, oxygen. The porphyrin radical 'hole' provides the required electron sink with the H-bonded thiolate ligand, as opposed to a His ligand found in peroxidases, stabilizing the ferryl-to-porphyrin radical charge transfer [34].

4.4. Heme–Propionate Interactions

Theoretical studies also have shed light on a heretofore ignored structural element of P450 chemistry which is the heme propionates. In all P450s and most heme enzymes whose structures are known, the heme propionates are tied down by ion pairs or H-bonds, which are generally viewed as one way of anchoring the heme in place. Thus, the role of the heme propionates would appear to be purely structural. This view was broadened when the manganese peroxidase crystal structure was solved since one heme propionate was found to serve as a Mn(II) ligand. Electron transfer from Mn(II) to the porphyrin or Fe(IV)=O center in compound I is very likely mediated along the propionate group. Recent DFT calculations further broaden the possible role of the heme propionates in shaping the electronic structure of compound I [35,36]. Significant spin density is delocalized on the heme propionates in the absence of any ion pairs or H-bonds which decreases the amount of porphyrin radical character. However, ion pairs/H-bonds to the heme propionates reverse this effect and increase the amount of porphyrin radical character which, in turn, generates a more potent hydroxylating agent.

In all known P450 structures at least one heme propionate is strongly paired with at least one basic amino acid side chain. This in itself is not too informative, but an examination of the known peroxidase structures reveals an interesting pattern consistent with the heme propionates playing a role in the reactivity of compound I. All heme peroxidases whose structures are known, except cytochrome *c* peroxidase (CCP), form a porphyrin cation radical in compound I. Nearly all these peroxidases also have ion pairs between the heme propionates and Lys or Arg. One outlier is CCP which neither forms a porphyrin radical nor has the propionates tied up in ion pairs. The obvious way to test these ideas is to remove or add ion pairs to heme propionates and study the effects on porphyrin radical stability. One study along these lines has been published. Ascorbate peroxidase forms a porphyrin radical in compound I [37]. Removing a key Arg group ion paired with the heme propionate does decrease the stability of the porphyrin radical [38].

4.5. Oxygen Binding Region

One conserved feature of the I-helix relevant to function is the local rupture in the normal α-helical H-bonding pattern near the highly conserved Thr252 (P450cam numbering, Figure 4). This local widening of the I helix enables solvent to be positioned into the helical groove. It now is known that this helical distortion and associated solvent structure is critical for the O_2 activation process [39] and is covered in more detail in Chapter 8 of this book. Thr252, however, is not conserved. For example, P450eryF has Ala while P450cin has Asn, yet both structures have the same I-helix distortion as found in P450cam. Thus the 'conserved' Thr252 is not important in the I-helix distortion. Thr252 is thought to stabilize the Fe(III)-OOH hydroperoxy intermediate in P450cam [40] which may also be applicable to most P450s but certainly not all. Since P450eryF has an Ala in place of Thr252, some other group must compensate for the missing Thr252 OH group. It now appears that a substrate OH serves this function or, possibly, serves as a direct proton donor to dioxygen [41,42].

As important as the I-helix is for substrate binding and O_2 activation, there now is an interesting example of a unique I-helix conformation. In CYP51 the I-helix breaks just on the N-terminal side of the I-helix Thr residue (Figure 5).

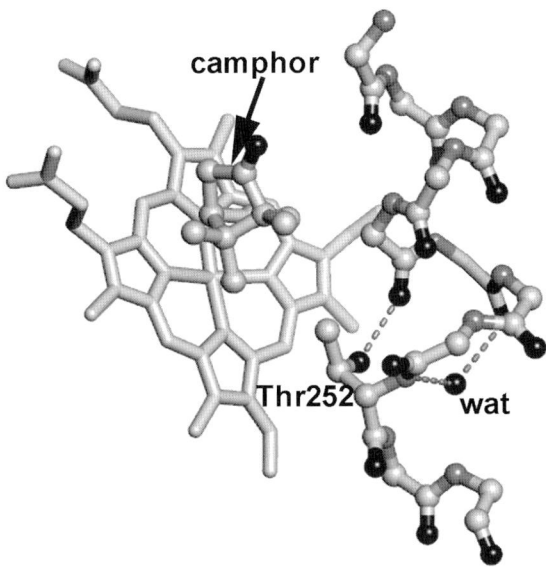

Figure 4. The I-helix in P450cam. The local disruption in the normal helical H-bonding pattern enables the side chain OH of Thr252 to donate an H-bond to the peptide carbonyl O atom of Gly248. This widening in the I-helix leaves a space that is taken up by water which makes up for some of the missing helical H-bonds.

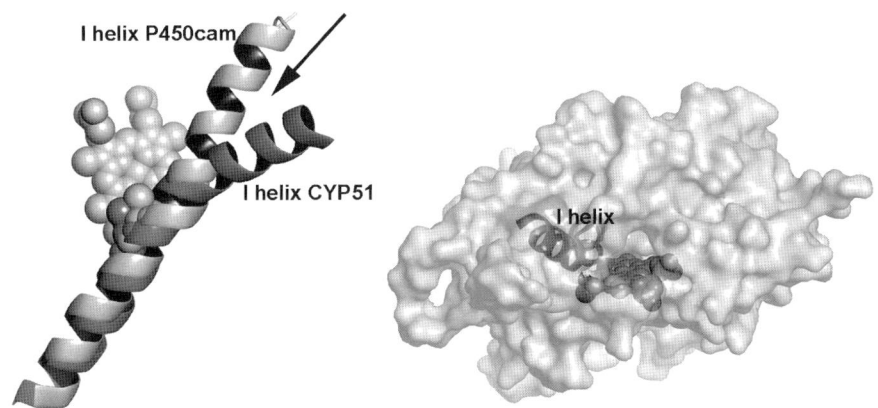

Figure 5. Crystal structure of CYP51 (1E9X [43]). Compared to P450cam, the I-helix is kinked. As shown on the right, this bending of the I-helix provides a new opening to the active site approximately parallel to the heme.

The main consequence of this large bend in the I-helix is to create a new opening to the active site which raises possibilities of novel entry and exit pathways for substrate and products [43]. Despite the break in the I-helix, CYP51 retains the local widening of the helix similar to P450cam.

4.6. Conformational Changes Required for Substrate Binding

Perhaps one of the key questions underlying P450 structure and function is how the same basic fold adapts to accommodate the requirements for substrate binding and catalysis. The seemingly endless array of P450 substrates include large, small, charged, neutral, and polar molecules. Some P450s are highly specific such as most prokaryotic P450s and mitochondrial P450s participating in steroid metabolism while others, such as microsomal drug-metabolizing P450s, are promiscuous with regard to the types of substrates oxidized. In looking at other enzyme systems that exhibit the same fold and chemistry, but work on different substrates, proteases come to mind and especially the serine proteases. Trypsin and chymotrypsin have the same fold and catalytic residues, yet trypsin cleaves only at Lys and Arg while chymotrypsin cleaves at hydrophobic side chains. This is achieved by modification of a recognition pocket, which requires fairly minor changes in structure to alter specificity. Because the substrate is a polymer, selectivity is achieved by binding sites removed from the catalytic residues. It also is clear that large-scale conformational changes such as large domain motions are not required in proteases. The active site forms a relatively rigid scaffold to which the substrate binds. Thus the catalytic site remains unchanged and only the specificity pockets spatially removed from the catalytic site need change.

The analogy with proteases or, for that matter, any enzyme system that operates on polymers breaks down once we go beyond the same fold and chemistry. P450s work on comparatively small and often rigid substrates. As a result, the catalytic site where bonds are made and broken cannot be neatly separated from those regions of the active site that define specificity. Thus, P450s neither have nor can they have separate specificity pockets, as do proteases. Moreover, the requirements of the oxygen activation chemistry necessitates burying the heme and oxygen binding site within the confines of the protein and thus the active site of P450s do not define an open cleft at the molecular surface as with many other enzymes. This further requires that P450s be flexible structures in order to allow substrate entry followed by closure of the active site.

The first clear-cut picture of the required open/close motion to become available was based on a comparison of the P450BM3 substrate-free and -bound complexes [44]. The loop connecting the F- and G-helices defines the entry to the active site and in the substrate-free structures the F- and G-helices are positioned such that the active site remains open. When substrate binds the F- and G-helices slide as a unit over the relatively stationary I-helix, thus closing the active site (Figure 6). Other P450s also experience similar open/close motions.

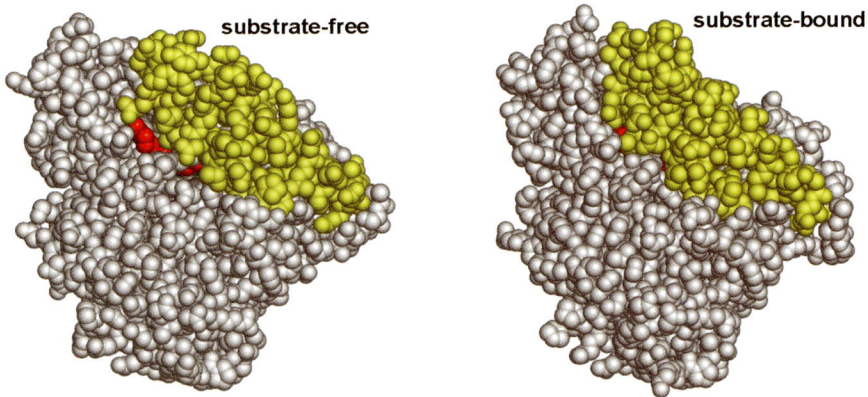

Figure 6. Structure of P450BM3 in the substrate-free and -bound forms. The F- and G-helices are yellow while the substrate is red.

To date the largest range of motions observed is with P4502B4 where some residues change position by as much as 18 Å [45].

In addition to a simple two-state open/close motion there also is the possibility for induced fit where the same active site can adopt a different shape depending on what substrate or ligand is bound. The first example of an induced-fit-type process was CYP119 where the structure was solved with two different ligands bound, phenylimidazole and imidazole (Figure 7) [46]. When the smaller imidazole binds to the heme iron, the loop connecting the F- and G-helices dips into the active site where direct contacts are made with the ligand. This requires a disruption of the C-terminal end of the F-helix. The energetic cost of breaking helical hydrogen bonds is partially compensated for by new intramolecular contacts formed between the protein and imidazole. This is a clear example of the protein shaping itself around the ligand and further underscores the importance of the F- and G-helices in substrate binding. Such adaptation, however, cannot occur unless there are compensating forces that drive conformational changes. In the CYP119 example, helical H-bonds and other favorable nonbonded interactions which are lost when imidazole binds are gained back by new favorable ligand–protein contacts as well as new intramolecular protein contacts. Obviously, CYP119 was not designed to adapt to these types of inhibitors and what was found in the crystal structures was a serendipitous discovery. However, it is intriguing to consider the possibility that those P450s that lack specificity such as the drug-metabolizing P450s also shape themselves around substrates of different sizes and shapes. If so, then in addition to requiring open/close motions there must also be compensating forces that recover the energy lost when secondary structure and other favorable nonbonded interactions are disrupted when a substrate binds and the active site adapts. Thus, the information content included in the P450 fold may be far more complex than simple open/close motions.

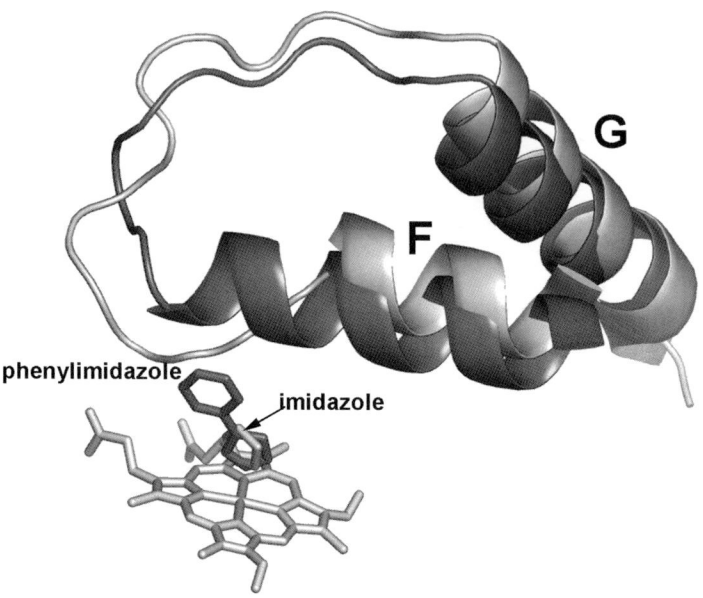

Figure 7. Structure of CYP119 in a complex with both the imidazole (light model) and phenylimidazole (dark model). In the imidazole complex the C-terminal end of the F helix unfolds which increases the length of the F/G loop, thus allowing the loop to extend closer to the imidazole for additional stabilizing contacts. This is an example of 'induced' fit.

There now are 13 unique structures of P450s with substrate bound, although with some P450s there are multiple structures with inhibitors and substrates bound. In most cases the substrates are very different in size, shape, and polarity. Thus, it is not surprising that there are some dramatic changes in elements of secondary structure contacting the substrate, especially the B'-helix region. However, site-directed mutagenesis studies have shown that very few changes in sequence can lead to rather large changes in specificity. One of the earliest examples centered on $P450_{COH}$ (CYP2A5) and $P450_{15\alpha}$ CYP2A4 [47]. $P450_{COH}$ oxidizes coumarin while $P450_{15\alpha}$ oxidizes testosterone yet these two P450s differ at only 11 amino acids. Mutagenesis work [47] showed that changing only one amino acid in $P450_{COH}$ to the corresponding residue in $P450_{15\alpha}$ enables $P450_{COH}$ to oxidize testosterone. Considering that coumarin and testosterone are very different substrates, this remains a dramatic and most unexpected result.

It might then be anticipated that P450s with very similar substrates will have very similar active sites with only a few amino acid substitutions required to alter specificity. The one pair of prokaryotic P450s that best fits this type of comparison is P450cam and P450cin. P450cin is from the soil bacterium *Citrobacter braakii* which can use cineole as a carbon source [48]. As shown in Figure 8, the substrates for these two P450s are very similar. Indeed, both

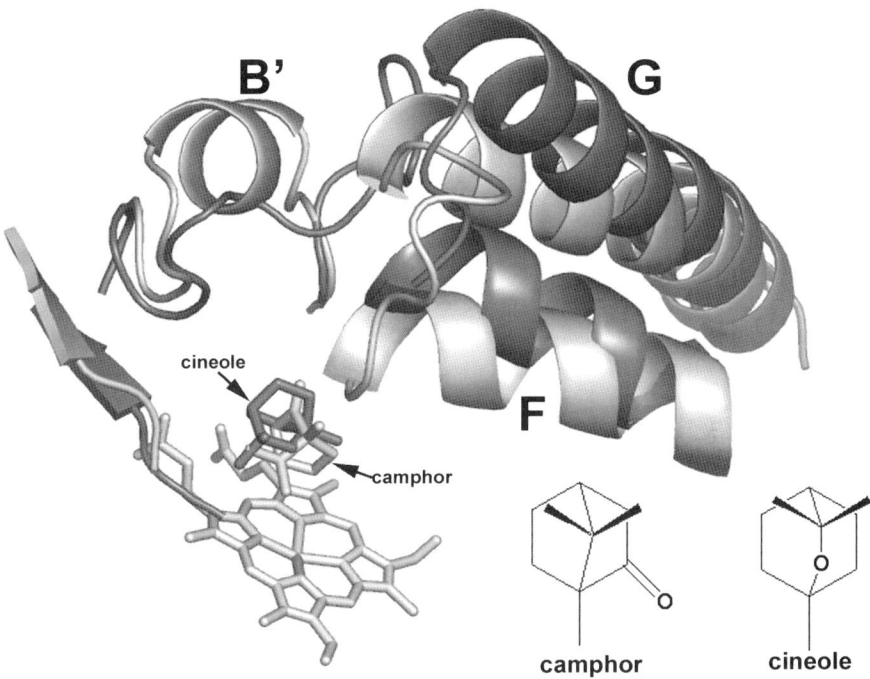

Figure 8. A comparison of the P450cam (light model) and P450cin (dark model) active site topographies. Differences in active site architecture is unexpectedly large considering the close similarity in the structure of the substrates. Note that P450cin has no B'-helix.

camphor and cineole have the same number of carbon and oxygen atoms except camphor has a ketone oxygen while cineole has an ether oxygen. Given the similarity in substrates, it might have been anticipated that P450cin and P450cam would share very similar active site structures despite the relatively low sequence homology, 27% identity and 46% similarity [48]. The crystal structure, however, of P450cin shows that the active site architecture is surprisingly different from P450cam. For example, P450cin does not have a B'-helix, a key substrate contact point in many P450s. As shown in Figure 8, the topography around the B'-helix region and the F/G loop is substantially different between these P450s. Even so, as might be expected from the similar size and shape of substrates, the active site volume is essentially the same, 256 Å^3 for P450cin and 264 Å^3 for P450cam [49]. This suggests that there may be a number of different active site structures that can accommodate the same or nearly the same substrates. This contrasts sharply with our earlier example of $P450_{COH}$ and $P450_{15\alpha}$.

One critical difference in comparing the $P450_{COH}/P450_{15\alpha}$ pair with the P450cam/P450cin pair is that P450cin and P450cam are highly specific as they must be since these P450s are essential components for the organism in using their respective substrates as carbon sources. This is not the case with microsomal P450s which are much less selective. In addition, the relaxed specificity of microsomal P450s is often paid for by higher levels of uncoupling where reducing equivalents are funneled off as H_2O_2 rather than hydroxylated product. P450cam and P450cin cannot afford such wasteful consumption of reducing equivalents. Thus, it would appear that a small handful of amino substitutions would not be enough to flip specificity between P450cam and P450cin without leading to higher levels of uncoupling and/or much lower product production. The most obvious reason for why P450cam and P450cin are so different is that each derives from a genetically distinct organism and according to the current procedures for constructing P450 phylogenetic trees, they are not closely related. This could be an example of convergent evolution where different active site topographies evolved to bind very similar substrates. It thus would be very interesting to find a camphor-metabolizing P450 from a soil organism other than *P. putida*. One might expect to find a substantially different active site topography, yet the same high level of specificity.

4.7. Electron Transfer Site

A potentially important difference between microsomal P450s and prokaryotic P450s is the presumed electron transfer site. The weight of the evidence supports the view that redox partners dock on the proximal side of the protein where the P450 heme lies closest to the molecular surface (Figure 9). Since P450cam uses a $(FeS)_2$ ferredoxin, putidaredoxin, as the donor while microsomal

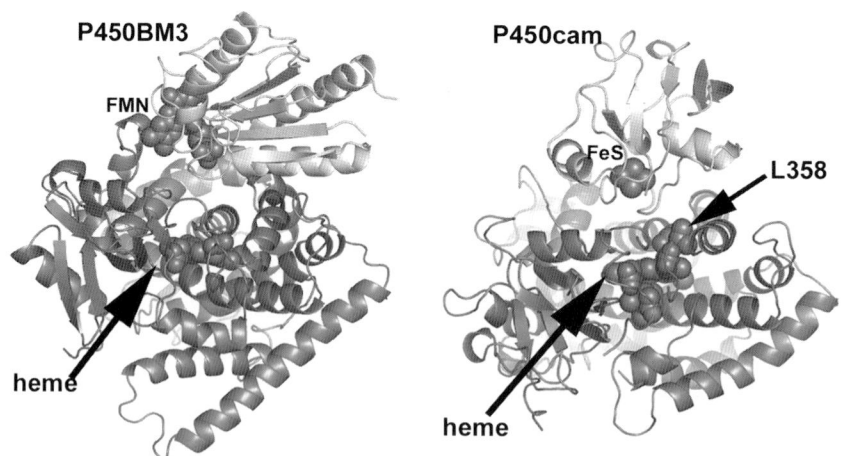

Figure 9. On the left is the structure of the complex formed between P450BM3 and its FMN domain taken from Sevrioukova et al. [51]. On the right is the hypothetical model of the complex formed between P450cam and its redox partner, putidaredoxin [52]. The docking site on the proximal surface of the P450 provides the closest approach of the heme to the donating cofactor, FMN or $(Fe-S)_2$.

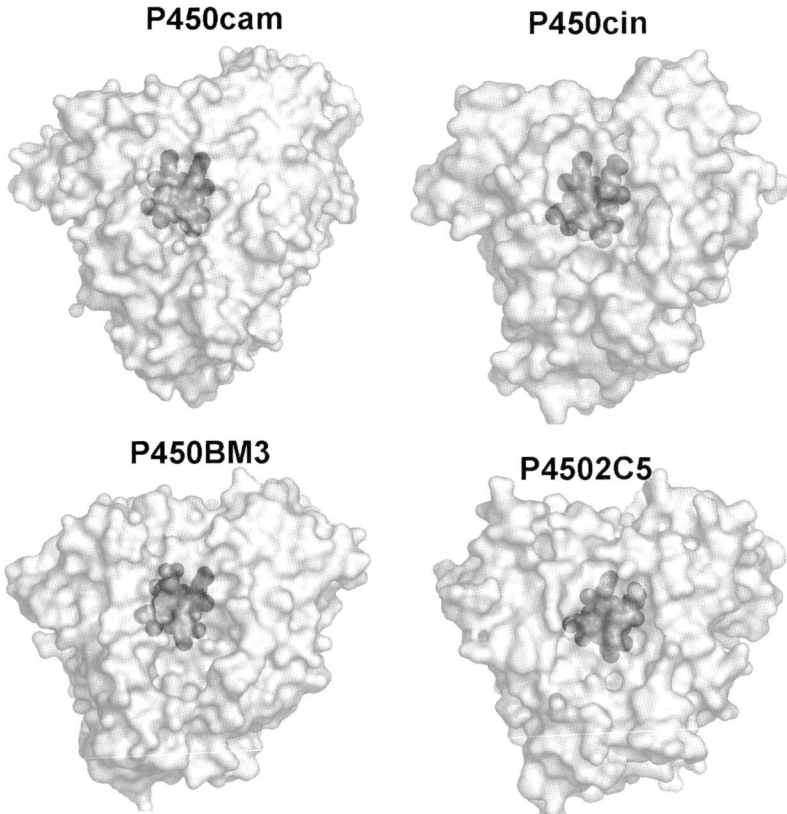

Figure 10. Molecular surface models viewed along the proximal surface where redox partners are thought to bind. P450BM3 and P4502C5 have a deeper more concave docking site while both P450cam and P450cin have a more shallow pocket. It incorrectly was anticipated that P450cin would more closely resemble P450BM3 and P4502C5 since P450cin also uses a flavodoxin-like FMN protein as its redox partner.

P450s use the FMN module of P450 reductase, it is expected that the docking sites might be substantially different. For this type of comparison it is best to use P450BM3 as the microsomal model P450. Although P450BM3 is a bacterial enzyme, it more closely resembles mircosomal P450s in both structure and function, including the use of the reductase FMN module as the electron donor. The main difference is that in P450BM3 the heme and reductase domains are fused as a single polypeptide chain [50]. Moreover, the structure of P450BM3 complexed with the FMN domain is known [51] and hypothetical models of the P450cam-putidaredoxin consistent with several experimental approaches are available [52] (Figure 9). A comparison of P450 proximal docking sites is shown in Figure 10. In both P450BM3 and P4502C5 the docking site just above the heme is broader and deeper defining a more concave site than in P450cam.

Also shown in Figure 10 is P450cin. P450cin is a bacterial P450 that utilizes FMN- and FAD-containing redox proteins with the FMN protein, called cindoxin, serving as the electron donor to the P450 [48]. Thus, P450cin belongs to the P450BM3 family with respect to electron transfer partners with the exception that the FMN and FAD proteins are separate polypeptides. Given the similarity in redox partners, one might expect P450cin to more closely resemble P450BM3/P4502C5 than P450cam. However, based on a detailed structural comparison [49], just the opposite was found. As shown in Figure 10, the docking surface in P450cin is shallow as in P450cam and not as concave as in P450BM3. Thus, the electron transfer complex formed between P450cin and its redox partner, cindoxin, must differ from what is seen in the P450BM3-FMN X-ray structure. This also means that the structural features of the presumed electron transfer site will not be very useful in predicting the type of electron transfer partner utilized by any particular P450.

4.8. Membrane Binding

Although much of what has been discussed thus far centers on similarities between all P450s, a remaining important unknown is how interactions between the membrane and P450 effects structure, although there is a reasonably clear picture of how P450 is oriented relative to the membrane [53]. Atomic force microscopy studies indicate that the P450 extends about 35 Å above the membrane which means that in addition to the N-terminal membrane anchor, other regions of the protein must also be embedded in the membrane [54]. Piecing together these various observations lead to one possible model of P450-membrane interaction (Figure 11). In this model the F/G loop and substrate access channel are oriented toward the membrane. With the entry to the active site oriented toward the membrane one can envision lipophilic substrates partitioning between the membrane and active site pocket.

P4503A4, the P450 responsible for the metabolism of about half of all drugs, is expected to have an adaptable active site. It is capable of adjusting to the size and shape of a diverse range of substrates, indicating that the F/G loop area should be flexible. A unique feature of the P4503A4 structure in this region is a Phe cluster (Figure 12) in the vicinity of what would be the F/G loop in other P450s [55,56]. Instead of a meandering F/G loop, P4503A4 has two additional helices, which have been termed the F'- and G'-helices. Notice that the F-helix has a break before continuing as the F'-helix which helps to complete part of the Phe cluster. Molecular dynamics simulations show that this region of P4503A4 centered on Phe215 (Figure 12) is quite flexible [57] and that in the substrate-free form the Phe cluster is easily disrupted. Since the F/G loop region is expected

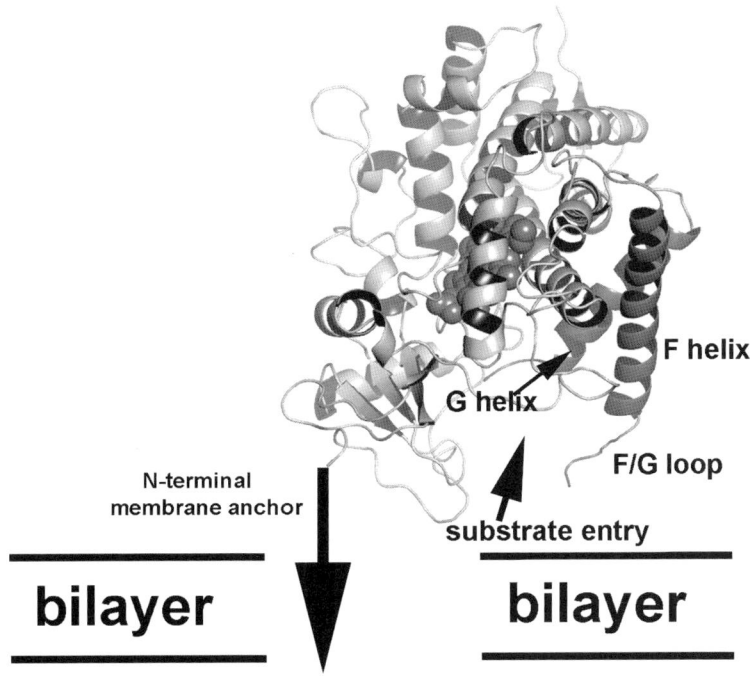

Figure 11. Schematic diagram showing a probable orientation of P4502C5 (1DT6, [1]) relative to the membrane bilayer.

to be close to the membrane, it is interesting to consider the possibility that the Phe cluster might help anchor P4503A4 to the membrane in a more open conformation than observed in the crystal structure.

4.9. Oligomeric Structure

We normally think of P450s as being monomeric. However, there now are examples of multimeric forms of P450s. P450BM3 was the first example of a prokaryotic P450 where it was clearly shown that higher-order structures form in solution [58]. More recently it has been found that the specific activity of P450BM3 markedly decreases at low enzyme concentrations, suggesting that a dimeric form favored at higher concentrations is active while the monomer is inactive [59]. If P450BM3 is active only as a dimer, there are some interesting parallels with nitric oxide synthase (NOS). Like P450BM3, NOS is catalytically self-sufficient with the FMN/FAD reductase fused to the C-terminal end of the heme domain. NOS forms a tight dimer between the heme domains and the current view on how electrons are transferred is that the monomer A reductase

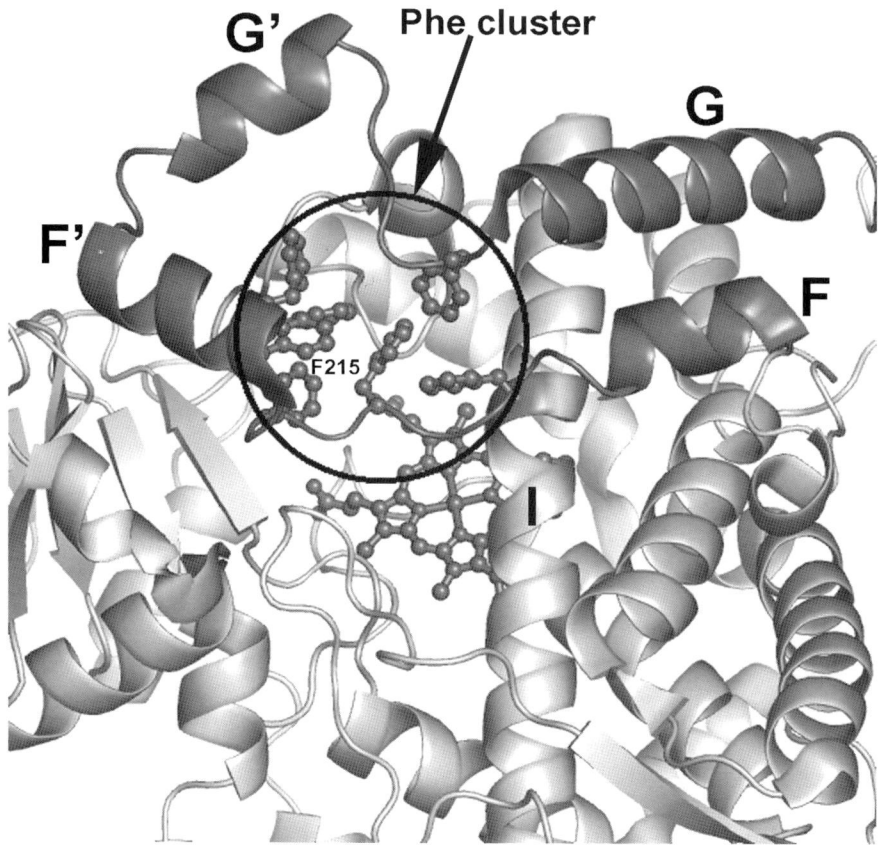

Figure 12. Crystal structure of P4503A4 (1TQN, [55,56]) showing the Phe cluster in P4503A4 situated just above the substrate binding pocket.

reduces the heme of monomer B [60]. Such 'cross-talk' could explain why P450BM3 is not active as a monomer. The available structural data on P450BM3 allow for some estimates on the structural feasibility of such a model.

Of special relevance is the fact that various crystal forms of P450BM3 heme domain form the same dimer in the crystalline asymmetric unit, even though the space groups and cell dimensions differ, suggesting that the dimer observed in the crystal is the physiologically important dimer. One such crystal structure was of the complex formed between the heme and FMN domains which very likely mimics the physiologically important electron transfer complex [51]. The crystal structure of this complex was derived from P450BM3 missing the FAD module and thus consisted of the heme and FMN domains. However, during the course of crystallization the linker connecting the heme and FMN domains was proteolyzed and although the crystal contained two molecules of heme domain in the asymmetric unit, there was only one FMN domain. What remains unknown is whether or not the FMN domain derived from the heme domain of monomer A or B. If we assume that P450BM3 behaves similar to NOS and that FMN from monomer A reduces the heme of monomer B, we can ask if the missing heme-FMN linker can span the distance between the heme and FMN domains when FMN monomer A is docked to the electron transfer site on the monomer B heme domain.

Figure 13 shows the crystal structure of the heme-FMN dimer with the relevant N- and C-termini labeled. The distance between the indicated C-terminus of heme and N-terminus of the FMN domains is about 40 Å. The structure is missing a 17 residue section of the linker. Since a completely extended polypeptide has a spacing of 3.6 Å between Cα atoms, it is quite reasonable for the linker to connect the ends of the heme and FMN domains when the FMN domain of monomer A is docked to the heme domain of monomer B. This, of course, does not prove that such electron transfer cross-talk occurs in P450BM3, but from the limited information at hand, this scenario is at least structurally feasible and can explain why P450BM3 is inactive as a monomer.

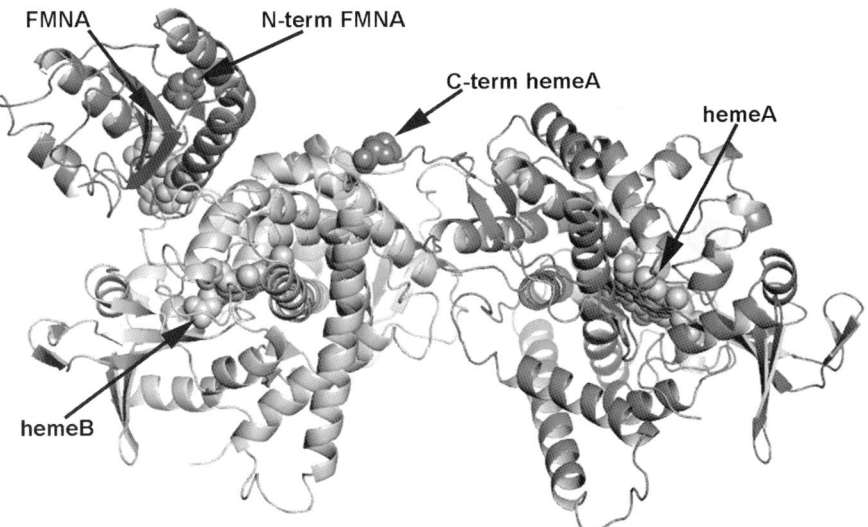

Figure 13. Model of the P450BM3-FMN domain complex. In the crystal structure there are two heme domains, but only one FMN domain per asymmetric unit. Labeling of the domains assumes that the FMN (FMNA dark model) is connected to the hemeA domain (also dark model). Although the linker connecting the hemeA and FMNA domains is not visible in the electron density map, the linker is long enough for FMNA to bind to the hemeB.

Perhaps the structurally most fascinating oligomeric architecture of a P450 so far known is that of P4502B4 [45,61]. As shown in Figure 14, the heme site is totally exposed owing to a very large displacement of the F- and G-helical regions. This enables a segment of a symmetry-related molecule in the crystal asymmetric unit to dip into the active site where His226 of molecule B coordinates the heme iron of molecule A and vice versa. Of course it can be argued that the dimeric structure is not physiologically relevant or is a serendipitous consequence of crystallization. However, spectral and analytical ultracentrifugation studies are consistent with the same dimer found in the crystal structure forming in solution [61]. Moreover, the crystal lattice will not force the protein to adopt a conformation that is not accessible in solution given appropriate conditions [62]. Shortly after this initial finding Scott et al. [45] solved the structure of P4502B4 complexed with an inhibitor bound

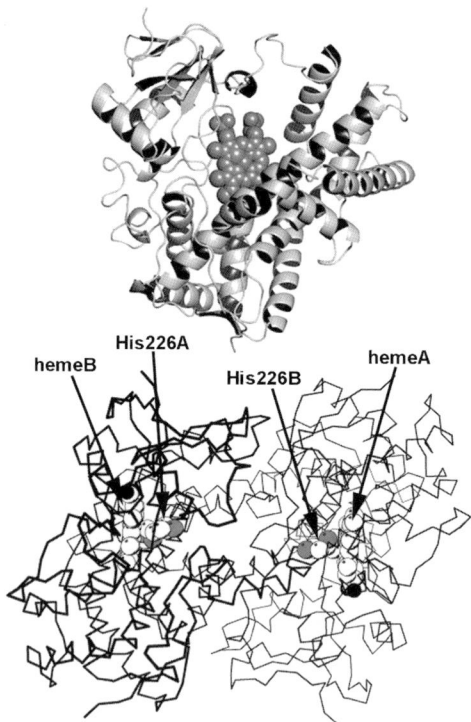

Figure 14. Molecular models of P4502B4 [61]. The top diagram is looking along the distal surface and shows that the active site pocket is wide open. The lower diagram shows how the dimers interact. His226 of molecule A penetrates into the distal pocket of molecule B where it coordinates to the heme iron while His226 of molecule B does the same in molecule A.

in the active site. Compared with the dimeric structure, the active site closes, resulting in the movement of some segments relative to the dimeric structure as large as 18 Å. What is most remarkable is that there is essentially no change in secondary structure. That is, helices involved in the open/close motion move as a unit without any disruption in helical structure. These and other studies strongly suggest that the P450 fold is unusually flexible and capable of undergoing extremely large open/close motions. This no doubt is relevant to function, especially for the nonselective drug metabolizing P450s that must bind a wide variety of substrates with very different sizes and shapes.

5. VARIATION IN P450 FUNCTION AND FOLD

5.1. Enzymes with a P450 Function but Different Fold

Enzymes that catalyze P450-like monooxygenation reactions are not confined to enzymes that exhibit the P450 fold. One of the most notable examples is chloroperoxidase (CPO) from the fungus *Caldariomyces fumago* [63]. CPO consists of 299 residues and forms a nearly all-helical structure (Figure 15). The structure is unique and bears no similarity to other known crystal structures. The biological role of CPO is to catalyze peroxide-dependent chlorination reactions in the formation of the natural product caldaryomycin (1,2-dichloro-2,5-dihydroxy cyclopentane) [63]. In this reaction CPO operates like a typical peroxidase to form Compound I [64–67] whose oxidizing equivalents are used to oxidize Cl$^-$ to form HOCl as the chlorinating agent. CPO, however, also exhibits P450 and

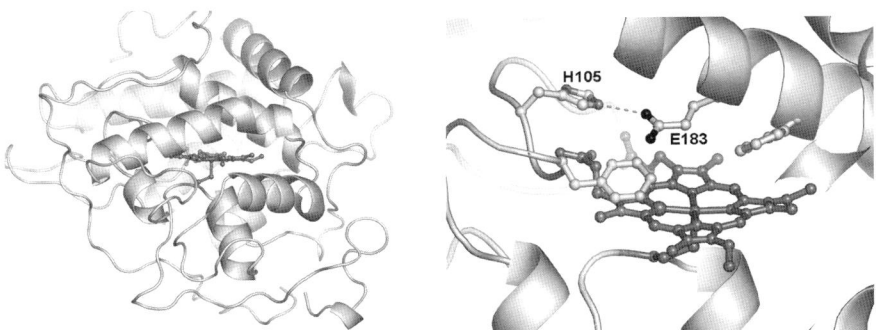

Figure 15. Crystal structure of chloroperoxidase (1CPO). The active site view on the right highlights interactions around the peroxide binding site. Glu183 is thought to be the acid–base catalytic group required for heterolytic cleavage of the peroxide O–O bond.

catalase activity and as such is one of the most diverse of the known heme enzymes. The structure of CPO [19] revealed that this unusual enzyme is a P450-peroxidase functional hybrid, even though the overall structure bears no similarity to either peroxidases or P450. Like P450, CPO has a Cys ligand which accepts an H-bond from a peptide NH exactly as in other thiolate-Fe proteins. As a result, the spectroscopic properties of CPO, especially the Fe(II)-CO complex, is the same as P450 [68]. However, unlike P450s the distal pocket where peroxide binds is polar (Figure 15) with Glu183 situated directly adjacent to the peroxide binding site. Glu183 very likely serves the same function as the acid–base catalytic His in other peroxidases [69].

Nitric oxide synthase also is a well characterized heme-thiolate enzyme that exhibits no structural homology with P450s, yet catalyzes a similar reaction. NOS catalyzes the hydroxylation of L-arginine to citrulline and NO. The first step in the reaction shown below, where $R-NH_2$ is an L-Arg guanidinium nitrogen atom, is a classic P450 monooxygenation reaction:

$$R-NH_2 + O_2 + NADPH + H^+ \rightarrow R-NH-OH + NAD^+ + H_2O$$

$$RNH-OH + 0.5NADPH + 0.5H^+ + O_2 \rightarrow NO + 0.5NAD^+ + H_2O + \text{L-citrulline}$$

Again, the only structural property shared with P450 is the heme-thiolate ligand which accepts an H-bond although in NOS the H-bond donor is the indole ring nitrogen of a Trp residue rather than a peptide NH group. NOS is a homodimer that has a much more open and accessible active site than P450s. While P450s can undergo rather large open-close motions [44,45] especially when substrates bind, the NOS active site is rigid and remains open, possibly because a more rigid β-sheet structure dominates the active site architecture rather than helices as in P450s. As a result, NOS faces a problem in having the oxygen activation machinery accessible to bulk solvent which can lead to the nonproductive formation of peroxide rather than hydroxylated product. P450s solve this problem by sequestering the active site from the bulk solvent and by providing a network of H-bonded solvent molecules that shuttle protons to the iron-linked dioxygen [39,40]. NOS does not have this option since the most important product of the reaction, NO, must rapidly diffuse out of the active site where it can serve its important signaling functions.

One possibility for how NOS solves the problem of an open active site is to tightly couple proton and electron transfer to the dioxygen complex, Fe(II)-OO^{2-}, to give first the hydroperoxy Fe(III)-OOH^- and then the dihydroperoxy, Fe(III)-OOH_2 intermediate [70]. Fe(III)-OOH_2 rapidly undergoes O–O bond cleavage to give the active hydroxylating intermediate, Fe(IV)=O. To achieve rapid electron/proton transfer NOS uses the cofactor, tetrahydrobiopterin (BH4), as the electron donor to the dioxygen complex [71]. As shown in Figure 16, BH4 directly interacts with the heme so the distance between donor

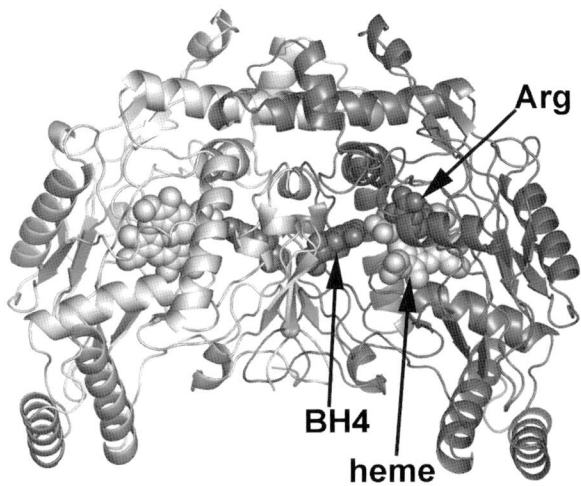

Figure 16. The dimeric heme domain of endothelial nitric oxide synthase (eNOS). The location of the substrate (L-Arg), heme, and cofactor BH4 are indicated. Unlike P450s, the NOS active site pocket is open when substrate is bound.

and acceptor is short, thus allowing a possibly concerted reduction and protonation of Fe(II)-OO. BH4 itself could serve as a proton donor or, possibly, an ordered water molecule which directly H-bonds to the N atom of NO complexed to the Fe(II) NOS heme [72] which, of course, assumes that NO is a reasonable mimic of dioxygen. As a result neither peroxy intermediate, Fe(III)-OO$^-$ nor Fe(III)-OOH, builds up, thus minimizing the chances for uncoupling to release H_2O_2. In sharp contrast, the one known structure of a P450 electron transfer complex [51] shows the donor flavin and acceptor heme to be about 18 Å apart.

5.2. Enzymes with the P450 Fold but Different Function

There now are a growing number of examples of enzymes with the P450 fold, but exhibit a different function. The first such P450 discovered was P450nor from the denitrifying fungus *Fusarium oxysporium* [73,74] that catalyzes the following reaction:

$$2NO + NAD(P)H + H^+ \rightarrow N_2O + NAD(P)^+ + H_2O$$

Although the crystal structure of P450nor [75] shows that this enzyme belongs to the P450 family of enzymes, the reaction does not require the participation of another redox protein and the iron atom remains in the Fe(III) state. Therefore,

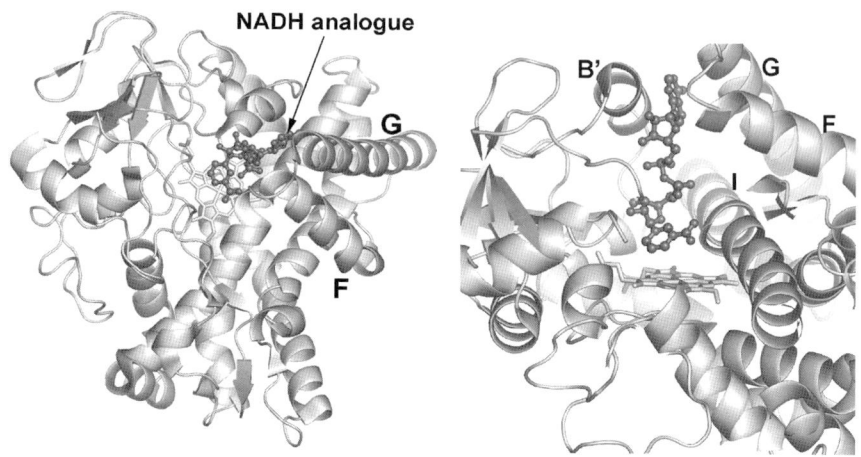

Figure 17. The crystal structure of P450nor [76] showing the location of the NADH analogue.

NAD(P)H serves as a direct electron donor to NO. If NAD(P)H is to serve as a direct electron donor then the NAD(P)H should bind close to the iron-linked NO. While the structure of P450nor shows sufficient room for NADPH, solving the structure of the P450nor-NAD(P)H complex has been plagued by the relatively low affinity of the enzyme for the cofactor [76].

Mutation of residues 73 and 75 in the B'-helix to Gly, however, considerably improved cofactor binding [77] enabling the crystal structure of P450nor in a complex with the NADH analogue nicotinic acid adenine dinucleotide to be solved [76] (Figure 17). As predicted the cofactor binds near the iron-linked NO, thus supporting a direct hydride transfer mechanism. The cofactor enters the active site near the F and G helices and especially the loop connecting these helices as in other P450s. This region of P450s is quite flexible and can undergo a considerable open/close motion in order to allow substrate entry. P450nor works in a similar fashion except the active site is adapted to bind the NADH cofactor rather than P450 substrates.

Another group of P450s with a similar fold, but different function, utilize H_2O_2 to oxidize long-chain fatty acids at the α- or β-positions [78–80]. Therefore, these enzymes do not redox cycle between Fe(III)/Fe(II) heme iron, but instead the Fe(III) iron reacts directly with H_2O_2. As a result there is no redox partner and in this regard, these P450s are similar to P450nor. As expected, the crystal structure of P450BSβ exhibits the classic P450 fold [81] (Figure 18). The fatty acid substrate binds in the access channel, similar to the way in which fatty acids bind to P450BM3. A major difference, however, is that P450BSβ has Arg242 in the I-helix that provides H-bonding interactions to the carboxyl group of the fatty acid (Figure 18).

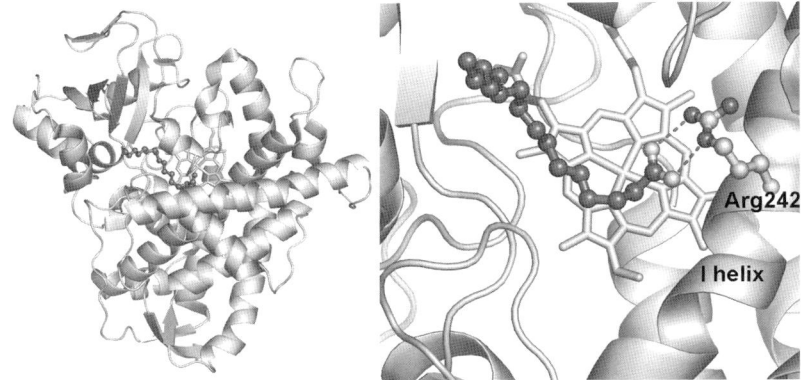

Figure 18. Crystal structure of P450BSβ (1IZO, [80]). Note that the carboxyl end of the fatty acid is deep in the active site where it H-bonds with Arg242 in the I-helix.

5.3. Mechanistic Variations

A major mechanistic difference between traditional P450 monooxygenase chemistry and peroxidases is that peroxidases utilize a strategically positioned acid/base catalytic group that transfers a proton from the iron-linked peroxide O atom to the leaving peroxide O atom. Most heme peroxidases use His [82] although others appear to use Glu [19,83]. In most P450s and nitric oxide synthases, however, amino acids do not serve as acid–base catalysts, but instead water serves as the direct proton donor to the iron-linked dioxygen [39]. One possible reason for this difference is that in peroxidases the substrate, peroxide, enters with protons while oxygen does not. Thus, in peroxidases it is necessary to move the proton from the iron-linked O atom to the leaving O atom. To accomplish this task specifically and rapidly requires a suitably positioned amino acid side chain. In contrast, the P450-oxy complex must be protonated from solvent and thus, the P450 active site is arranged such that a network of H-bonded solvent molecules operates as the proton delivery machine. P450BSβ offers an interesting variation on these general trends. Since the substrate is H_2O_2 there should be an acid–base catalytic side chain as in other peroxidases. Nevertheless, P450BSβ does not provide such a group. Instead, the proposed mechanism uses the substrate carboxyl to serve as the acid–base catalyst [81]. Here the carboxyl group serves the same proton-shuttle function as the His in other peroxidases. Such substrate-assisted catalysis is not limited to P450BSβ since P450eryF is thought to use a substrate OH group as a direct proton donor to dioxygen [42,84].

Two common threads that tie together these 'non-traditional' P450s are that the redox state of the iron remains unchanged and that redox partners are not required. It is generally thought that redox partners bind on the proximal side of the molecule where the Cys-ligand face of the heme lies closest to the surface. Indeed, the one known structure of a P450 redox complex shows docking between the exposed end of the FMN module and the proximal surface of the P450. These complexes tend to be dominated by electrostatic interactions and hence, the molecular surfaces between the donor and accepter are electrostatically complementary with the P450 serving as the electropositive partner, leading to a distinct asymmetric distribution of charge. The electrostatic surface potential of P450nor and P450BSβ, however, are quite different from other P450s with a much more even distribution of charge. It is interesting to consider the possibility that such differences in charge distribution may be a characteristic structural feature that separates those P450s requiring a redox partner from those that do not. That is, those that are designed to bind redox partners have a more electropositive proximal surface. Such information could be useful in the era of structural genomics where structures are solved prior to much or any knowledge of function. A recent example is Cyp119, the first P450 identified from a hyperthermophile [85]. Although the function of CYP119 was unknown at the time the structure was solved, the structure [46,86] exhibited the typical asymmetric charge distribution, so one would predict that CYP119 has a redox partner. The close structural similarity to P450BM3 also led to the prediction that CYP119 might be a fatty acid hydroxylase. After the crystal structure was solved the redox partners were discovered and CYP119 was found to be a fatty acid hydroxylase that uses a ferredoxin as the direct electron donor [87].

6. ARCHAEON P450s

On rare occasions the scientific community is humbled by the realization that our cherished notions of life on Earth are incorrect. Until the late 1970s the world was neatly divided into prokaryotes and eukaryotes. In the 1970s it became clear that this classification was totally inadequate [88] and that there is a third group, called archaea, which look like bacteria, but are biochemically and genetically a distinct form of life. Archaeans thrive in some of the most extreme conditions of pH and heat on Earth. The best studied are those that live near deep sea vents where temperatures reach over 100 °C. Since the early Earth was inhospitable to a majority of current life forms owing to extremes of temperature, it is natural to ask if current day archaea are direct descendents of some of the earliest single-cell organisms. Fortunately the peculiar chemistry required of archaea in order to survive in extreme environments enables a piecing together of a chemical fossil record. Archaea membranes are composed of isoprenoid units rather than the more traditional fatty acids. These are linked to the glycerol backbone via ether linkages rather than ester linkages found in most other membranes, thus leading to greater stability to the extremes of pH and heat where some archaea thrive.

The unique isoprenoid content of archaea provides the basis for the chemical fossil record, which leads to estimates of life on Earth as long ago as 3.8 billion years. If these chemical deposits correlate with archaea, then archaea do indeed represent a very old life form. Of course it remains unclear if the biochemistry of current day archaea are a true reflection of life billions of years ago, but the prospects are intriguing.

Given these peculiar properties of archaea it was quite surprising when the first archaea P450, CYP119, was discovered, quite by accident, in 1996 [85]. Wright et al. [85] were in the process of cloning the thymidylate synthase gene from the acidothermophilic archaeon *Sulfolobus sulfataricus* and found sequences consistent with a P450. Based on the P450 nomenclature this P450 was termed CYP119. *Sulfolobus sulfataricus* grows optimally at pH 3.5 and 85 °C [89] so CYP119 was the first thermostable P450 to be identified. Subsequent cloning, expression, and purification illustrated that CYP119 has a melting point of 90 °C compared to $\simeq$ 55 °C for P450cam [90,91].

In this era of structural genomics it is not uncommon to purify an enzyme without knowing its function and CYP119 is one example. However, a BLAST search shows that CYP119 is most similar at 35% sequence identity to the fatty acid hydroxylase P450BM3, CYP102, from *B. megaterium*. The crystal structure of CYP119 was solved not too long after its discovery, but showed no more pronounced similarity to P450BM3 than other prokaryotic P450s [46,86] (Figure 19). A major problem in deciphering the biological role of CYP119 is that, to reconstitute CYP119 hdyroxylase assay requires redox partners that can survive the extreme of temperature optimal for CYP119. While the hunt for such redox partner was underway, it was found that fatty acids do bind to CYP119 to give the traditional low-to-high-spin spectral shift, but only at elevated temperatures (unpublished observations). Fortunately, the redox partners for CYP119 now have been identified [92]. The direct electron donor to CYP119 is 7Fe cluster ferredoxin which is not too unusual. However, the electron donor to the ferredoxin is a 2-oxoacid-ferredoxin oxidoreductase that utilizes pyruvic acid rather than NAD(P)H as the source of reducing equivalents which is unusual and the first non-NAD(P)H utilizing redox partner found for any P450. The discovery of the CYP119 redox partners has enabled the development of high-temperature assays and CYP119 was found to be a fatty acid hydroxylase [92] as expected. Even though turnover is not fast with lauric acid as the substrate, of the order of 15–20 min^{-1} and it remains to be seen if fatty acids are the true endogenous substrates, the way is now open to further probe the role of CYP119 in *Sulfolobus sulfataricus*.

Soon after the structure of CYP119 was solved, a second thermophilic P450 structure was solved, that of CYP175A1 from *Thermus thermophilus* [93]. Here the high degree of sequence and structural similarity to P450BM3 clearly suggested that CYP175A1 might be a fatty acid hydroxylase. Blasco et al. [94] then found that P450175A1 can hydroxylate β-carotene. Whether or not this is the

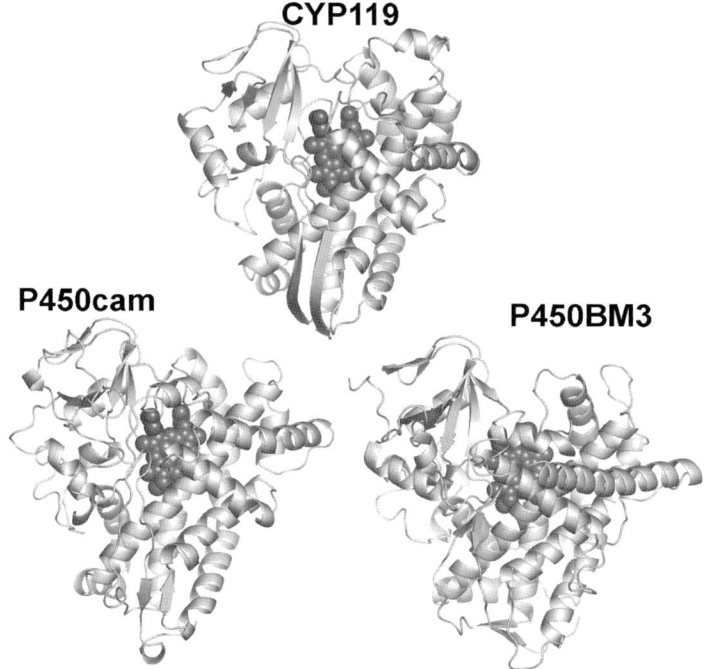

Figure 19. Crystal structures of CYP119, P450cam, and P450BM3. The hemes are shown as spacefilled atoms.

physiologically important substrate remains to be seen, but this does indicate that CYP175A1 is able to oxidize long-chain hydrocarbons.

The discovery of thermal-stable P450s is exciting from the perspective of utilizing P450s for practical industrial application such as the synthesis of important organic intermediates. Thermal stability often translates to increased stability in other hostile environments such as extremes of pH and organic solvents. One approach to the practical utilization of thermal-stable P450s is to either engineer the known thermophilic P450s to catalyze the chemistry of interest or introduce those structural features responsible for heat stability into one's favorite, but relatively unstable protein. Thus, it is important to uncover those structural features responsible for thermal stability in CYP119 and CYP1751A1.

Both CYP119 and CYP1751A1 are smaller, with fewer long loops than other P450s. Indeed, CYP1751A1 has been described as P450BM3 without the loops. The more interesting structural feature possibly related to thermal stability in CYP119 is an extensive 39-Å-long aromatic cluster running along one side of the molecule (Figure 20). This type of extensive aromatic network is not present in other known P450 structures. Mutagenesis of selected residues in the cluster does support the view that the aromatic cluster is an important structural feature

STRUCTURES AND MOLECULAR PHYLOGENY OF P450 PROTEINS

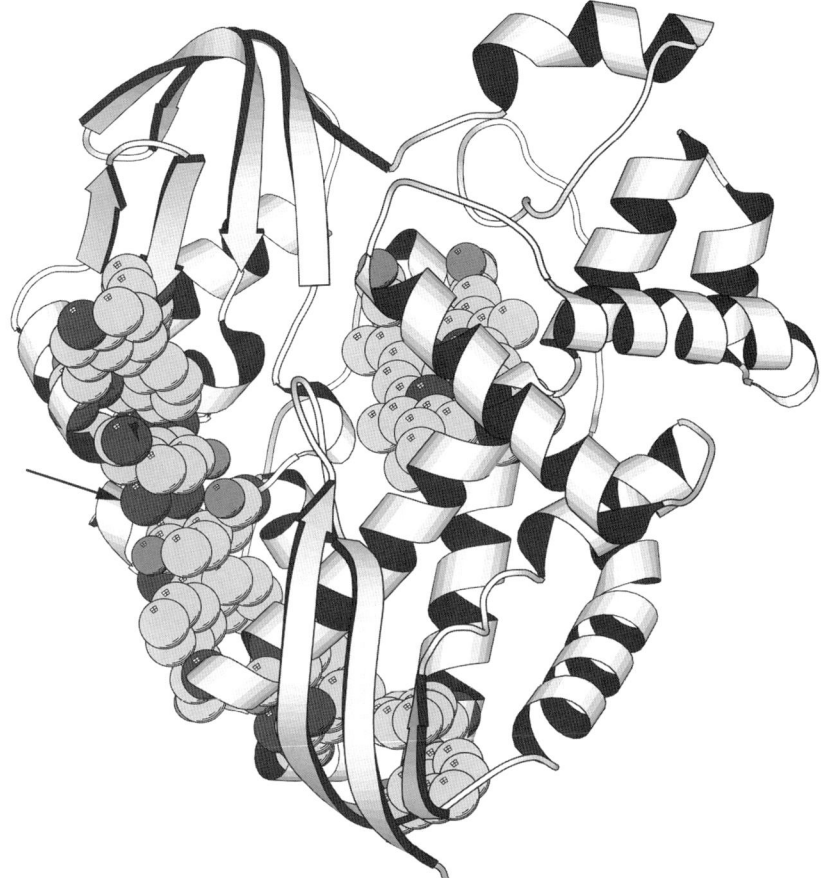

Figure 20. CYP119 crystal structure. The arrow indicates the center of an extended aromatics cluster unique to CYP119 that helps contribute to CYP119 thermal stability.

that helps to confer additional thermal stability [95]. However, CYP1751A1 does not have a similar aromatic cluster. Instead it appears that CYP1751A1 may derive enhanced thermal stability from networks of salt bridges that contain at least three residues [93]. In both CYP119 and P450175A1 about 60% of the salt bridges consist of three or more residues while the closest non-thermophilic P450 is P450cam at about 37% [93]. It thus appears that thermal stable P450s have recruited very much the same structural features that confer stability to other thermal stable enzymes unrelated to P450s. The additional intriguing aspect of thermal-stable P450s is whether or not these P450s also can undergo large open/close motions as in other P450s. If so, an interesting question to be answered is how it is that such conformational gymnastics are compatible with extremes of temperature and pH?

7. SUMMARY AND CONCLUSIONS

The rate at which new P450 structures are being solved is increasing owing to improved heterologous expression systems, a better understanding of how to handle membrane-bound P450s, and a growing sequence data base that has enabled the identification of heretofore unknown P450s. All P450s, independent of their origins, share a common three-dimensional fold. The P450 fold appears to be conformationally quite flexible since there now are growing examples of very large open/close motions that control access to the active site and some hints of induced fit where the protein 'shapes' itself around ligands of different sizes and shapes. The P450 fold also can accommodate non-P450 chemistry such as peroxidase activity and substrate transformations where no external electron source is required.

The list of novel P450-like proteins is very likely to grow as more previously unkown P450s are discovered from the many genome projects now complete or close to completion. In addition to novel P450s, the list of redox partners utilized by various P450s has grown, including those with the redox partner covalently attached. It no longer is possible to neatly divide P450s into those that use a single FMN/FAD reductase and those that use a two-component system consisting of $(Fe-S)_2$ and FAD proteins. There remains much to be learned about the interactions between P450s and redox partners.

Finally, we can expect the current wave of activity in P450 crystallography to have a major impact in three general areas:

1. Given the current interest in structural genomics coupled with technological advances, it is no longer an unrealistic goal to expect the structures of all human P450s to be solved as well as P450s from other important organisms such as *Mycobacterium tuberculosis*. There is substantial hope that this will dramatically improve the success rate of the structure-based drug design efforts targeted to specific P450s. The one cautionary note is that the observed flexibility of certain P450s makes structure-based inhibitor/drug design a particularly challenging problem. If the P450 of interest is quite flexible, the drug designer must take into account such flexibility which is often quite difficult to do in most computer-based efforts. There is room for optimism that clever new algorithms will take into account P450 flexibility. Recognizing the limitations of current drug design approaches as applied to P450s, however, is the first step.
2. A long-sought practical goal in P450 research is to harness the power of P450 chemistry in the production of industrially and medically useful intermediates. The recent discovery of thermal-stable P450s should help in this regard if

thermal-stable P450s with commercially interesting reactions can be found or if those structural features responsible for thermal stability can be introduced into an otherwise unstable, but industrially useful P450s. The near future should see a growth in this area of research.

3. Finally, one of the most interesting basic research outcomes of P450 structural genomics is to work out the function of the newly discovered P450s. What exactly are P450s doing in archaea, *C. elegans*, etc.? The structures should prove to be a great aid in at least narrowing the search for substrates and hence, identifying function.

ACKNOWLEDGMENTS

Work in our lab was supported by NIH grants GM33688 and GM057353.

ABBREVIATIONS

BH4	tetrahydrobiopterin
BLAST	basic local alignment search tool
caldariomycin	1,2-dichloro-2,5-dihydroxycyclopentane
CPO	chloroperoxidase
CYP51	lanosterol 14-α-demethylase
DFT	density functional theory
FAD	flavin adenine dinucleotide
FMN	flavin mononucleotide
NADPH	nicotinamide adenine dinucleotide phosphate, reduced
NOS	nitric oxide synthase
rms	root-mean-square

REFERENCES

1. P. A. Williams, J. Cosme, V. Sridhar, E. F. Johnson, and D. E. McRee, *Mol. Cell,* 5, 121–132 (2000).
2. A. H. Knoll, *Science,* 256, 622–627 (1992).
3. M. C. Rivera and J. A. Lake, *Science,* 257, 74–76 (1992).
4. W. F. Loomis, *Four Billion Years: An Essay on the Evolution of Genes and Organisms,* Sunderland, Mass., Sinauer Assoc., Inc., 1988.
5. F. J. Gonzalez and D. W. Nebert, *Trends in Genetics,* 6, 182–186 (1990).
6. Y. Yoshida, M. Noshiro, Y. Aoyama, T. Kawamoto, T. Horiuchi, and O. Gotoh, *J. Biochem. (Tokyo),* 122, 1122–1128 (1997).
7. D. R. Nelson, *Arch. Biochem. Biophys.,* 369, 1–10 (1999).
8. N. Debeljak, S. Horvat, K. Vouk, M. Lee, and D. Rozman, *Arch. Biochem. Biophys.,* 379, 37–45 (2000).

9. D. Debeljak, M. Fink, and D. Rozman, *Arch. Biochem. Biophys.*, *409*, 159–171 (2003).
10. Y. Yoshida, Y. Aoyama, M. Noshiro, and O. Gotoh, *Biochem. Biophys. Res. Commun.*, *273*, 799–804 (2000).
11. T. Rezen, N. Debeljak, D. Kordis, and D. Rozman, *J. Mol. Evol.*, *59*, 51–58 (2004).
12. D. W. Nebert and D. R. Nelson, *Methods Enzymol.*, *206*, 3–11 (1991).
13. D. W. Nebert and F. J. Gonzalez, *Ann. Rev. Biochem.*, *56*, 945–993 (1987).
14. D. W. Nebert, D. R. Nelson, and R. Feyereisen, *Xenobiotica*, *19*, 1149–1160 (1989).
15. D. R. Nelson, T. Kamataki, D. J. Waxman, F. P. Guengerich, R. W. Estabrook, R. Feyereisen, F. J. Gonzalez, M. J. Coon, I. C. Gunsalus, O. Gotoh, K. Okuda, and D. W. Nebert, *DNA Cell Biol.*, *12*, 1–51 (1993).
16. D. R. Nelson, L. Koymans, T. Kamataki, J. J. Stegeman, R. Feyereisen, D. J. Waxman, M. R. Waterman, O. Gotoh, M. J. Coon, R. W. Estabrook, I. C. Gunsalus, and D. W. Nebert, *Pharmacogenetics*, *6*, 1–42 (1996).
17. O. Gotoh, *Cytochrome P450* (T. Omura, Y. Ishimura, and Y. Fujii-Kuriyama, eds), 2nd edn, Kodansha, Tokyo, 1993.
18. F. Berisha, *An essay submitted in partial fulfillment of the requirements for the degree of MRes Modelling Biological Complexity*, University College London, 2004.
19. M. Sundaramoorthy, J. Terner, and T. L. Poulos, *Structure*, *3*, 1367–1377 (1995).
20. C. S. Raman, H. Li, P. Martaske, V. Kral, B. S. S. Masters, and T. L. Poulos, *Cell*, *95*, 939–950 (1998).
21. B. R. Crane, A. S. Arvai, D. K. Ghosh, C. Wu, E. D. Getzoff, D. J. Stuehr, and J. A. Tainer, *Science*, *279*, 2121–2126 (1998).
22. T. O. Fischmann, A. Hruza, X. D. Niu, J. D. Fossetta, C. A. Lunn, E. Dolphin, A. J. Prongay, P. Reichert, D. J. Lundell, S. K. Narula, and P. C. Webe, *Nature Struct. Biol.*, *6*, 233–242 (1999).
23. E. Adman, K. D. Watenpaugh, and L. H. Jensen, *Proc. Natl. Acad. Sci. USA*, *72*, (1975).
24. N. Ueyama, T. Terakawa, M. Nakata, and A. Nakamura, *J. Am. Chem. Soc.*, *105*, 7098–7102 (1983).
25. N. Ueyama, T. Okamura, and A. Nakamura, *J. Am. Chem. Soc.*, *114*, 8129–8137 (1992).
26. N. Ueyama, N. Nishikawa, Y. Yamada, T. Okamura, and A. Nakamura, *J. Amer. Chem. Soc.*, *118*, 1286–1287 (1996).
27. R. Langen, G. M. Jensen, U. Jacob, P. J. Stephens, and A. Warshel, *J. Biol. Chem.*, *267*, 25625–25627 (1992).
28. K. Choudhury, M. Sundaramoorthy, J. M. Mauro, and T. L. Poulos, *J. Biol. Chem.*, *267*, 25656–25659 (1992).
29. K. Choudhury, M. Sundaramoorthy, A. Hickman, T. Yonetani, E. Woehl, M. F. Dunn, and T. L. Poulos, *J. Biol. Chem.*, *269*, 20239–20249 (1994).
30. M. Green, *J. Amer. Chem. Soc.*, *121*, 7939–7940 (1999).
31. F. Ogliaro, S. Cohen, S. P. de Visser, and S. Shaik, *J. Amer. Chem. Soc.*, *122*, 12892–12893 (2000).
32. R. Weiss, D. Mandon, T. Wolter, A. X. Trautwein, M. Müther, E. Bill, A. Gold, K. Jayaraj, and J. Terner, *J. Biol. Inorg. Chem.*, *1*, 377–383 (1996).

33. N. Suzuki, T. Higuchi, Y. Urano, and K. Kikuchi, *J. Amer. Chem. Soc., 121*, 11571–11572 (1999).
34. M. Filatov, N. Harris, and S. Shaik, *J. Chem. Soc., Perkin Trans. 2*, 399–410 (1999).
35. V. Guallar, M.-H. Baik, S. J. Lippard, and R. A. Friesner, *Proc. Natl. Acad. Sci. USA, 100*, 6998–7002 (2003).
36. J. C. Schoneboom, S. Cohen, H. Lin, S. Shaik, and W. Thiel, *J. Am. Chem. Soc., 126*, 4017–4034 (2004).
37. W. R. Patterson, T. L. Poulos, and D. B. Goodin, *Biochemistry, 34*, 4342–4345 (1995).
38. T. P. Barrows and T. L. Poulos, *Biochemistry, 44*, 14062–14068 (2005).
39. I. Schlichting, J. Berendzen, K. Chu, A. M. Stock, S. A. Maves, D. E. Benson, R. M. Sweet, D. Ringe, G. A. Petsko, and S. G. Sligar, *Science, 287*, 1615–1622 (2000).
40. S. Nagano and T. L. Poulos, *J. Biol. Chem., 280*, 31659–31663 (2005).
41. J. R. Cupp-Vickery, H. Li, and T. L. Poulos, *Proteins, 20*, 197–201 (1994).
42. S. Nagano, J. R. Cupp-Vickery, and T. L. Poulos, *J. Biol. Chem., 280*, 22102–22107 (2005).
43. L. M. Podust, T. L. Poulos, and M. R. Waterman, *Proc. Natl. Acad. Sci. USA, 98*, 3068–3073 (2001).
44. H. Li and T. L. Poulos, *Nat. Struct. Biol., 4*, 140–146. (1997).
45. E. E. Scott, M. A. White, Y. A. He, E. F. Johnson, C. D. Stout, and J. R. Halpert, *J. Biol. Chem., 279*, 27294 – 27301 (2004).
46. J. K. Yano, L. S. Koo, D. J. Schuller, H. Li, P. R. Ortiz de Montellano, and T. L. Poulos, *J. Biol. Chem., 275*, 31086–31092 (2000).
47. R. L. Lindberg and M. Negishi, *Nature, 339*, 632–634 (1989).
48. D. B. Hawkes, G. W. Adams, A. L. Burlingame, P. R. Ortiz de Montellano, and J. J. De Voss, *J. Biol. Chem., 277*, 27725–27732 (2002).
49. Y. T. Meharenna, H. Li, B. H. David, D. B. Hawkes, A. G. Pearson, J. De Voss, and T. L. Poulos, *Biochemistry, 43*, 9487–9494 (2004).
50. L. O. Narhi and A. J. Fulco, *J. Biol. Chem., 262*, 6683–6690 (1987).
51. I. F. Sevrioukova, H. Li, H. Zhang, J. A. Peterson, and T. L. Poulos, *Proc. Natl. Acad. Sci. USA, 96*, 1863–1868 (1999).
52. T. Pochapsky, T. A. Lyons, S. Kazanis, T. Arakaki, and G. Ratnaswamy, *Biochimie, 78*, 723–733 (1996).
53. Y. Ohta, S. Kawato, H. Tagashira, S. Takemori, and S. Kominami, *Biochemistry, 31*, 12680–12687 (1992).
54. T. H. Bayburt and S. G. Sligar, *Proc. Natl. Acad. Sci. USA, 99*, 6725–6730 (2002).
55. J. K. Yano, M. R. Wester, G. A. Schoch, K. J. Griffin, C. D. Stout, and E. F. Johnson, *J. Biol. Chem., 279*, 38091–38094 (2004).
56. P. A. Williams, J. Cosme, D. M. Vinkovic, A. Ward, H. C. Angove, P. J. Day, C. Vonrhein, I. J. Tickle, and H. Jhoti, *Science, 305*, 683–686 (2004).
57. H. Park, S. Lee, and J. Suh, *J. Amer. Chem. Soc., 127*, 13634–13642 (2005).
58. S. D. Black and S. T. Martin, *Biochemistry, 33*, 12056–12062 (1994).
59. R. Neeli, H. M. Girvan, A. Lawrence, M. J. Warren, D. Leys, N. S. Scrutton, and A. W. Munro, *FEBS Lett., 579*, 5582–5588 (2005).

60. U. Siddhanta, A. Presta, B. Fanparallel, D. Wolan, D. L. Rousseau, and D. J. Stuehr, *J. Biol. Chem.*, 273, 18950–18958, (1998).
61. E. E. Scott, Y. A. He, M. R. Wester, M. A. White, C. C. Chin, J. R. Halpert, E. F. Johnson, and C. D. Stout, *Proc. Natl. Acad. Sci. USA*, 100, 13196–13201 (2003).
62. T. L. Poulos, *Proc. Natl. Acad. Sci. USA*, 100, 13121–13122 (2003).
63. D. R. Morris and L. P. Hager, *J. Biol. Chem.*, 241, 1763–1768 (1966).
64. M. M. Palcic, R. Rutter, T. Araiso, L. P. Hager, and H. B. Dunford, *Biochem. Biophys. Res. Commun.*, 94, 1123–1127 (1980).
65. L. P. Hager, D. L. Doubek, R. M. Silverstein, J. H. Hargis, and J. C. Martin, *J. Amer. Chem. Soc.*, 94, 4364–4366 (1972).
66. R. Rutter, M. Valentine, M. P. Hendrich, L. P. Hager, and P. G. Debrunner, *Biochemistry*, 22, 4769–4774 (1983).
67. R. Rutter, L. P. Hager, H. Dhonau, M. Hendrich, M. Valentine, and P. G. Debrunner, *Biochemistry*, 23, 6809–6816 (1984).
68. P. F. Hollenberg and L. P. Hager, *J. Biol. Chem.*, 248, 2630–2633 (1973).
69. M. Sundaramoorthy, J. Terner, and T. L. Poulos, *Chem. Biol.*, 5, 461–473 (1998).
70. T. L. Poulos, *Drug Metab. Dispos.*, 33, 10–18 (2005).
71. A. R. Hurshman, C. Krebs, D. E. Edmondson, B. H. Huynh, and M. A. Marletta, *Biochemistry*, 38, 15689–15696 (1999).
72. H. Li, C. S. Raman, P. Martasek, B. S. S. Masters, and T. L. Poulos, *Biochemistry*, 40, 5399–5406 ((2001).
73. H. Shoun, Y. Sudo, Y. Seto, and T. Beppu, *J. Biochem. (Tokyo)*, 94, 1219–1229 (1983).
74. H. Shoun and T. Tanimoto, *J. Biol. Chem.*, 266, 11078–11082 (1991).
75. S. Y. Park, H. Shimizu, S. Adachi, A. Nakagawa, I. Tanaka, K. Nakahara, H. Shoun, E. Obayashi, H. Nakamura, T. Iizuka, and Y. Shiro, *Nat. Struc. Biol.*, 4, 827–832 (1997).
76. R. Oshima, S. Fushinobu, F. Su, L. Zhang, N. Takaya, and H. Shoun, *J. Mol. Biol.*, 342, 207–217 (2004).
77. L. Zhang, T. Kudo, N. Takaya, and H. Shoun, *J. Biol. Chem.*, 277, 33842–33847 (2002).
78. C. Cambillau, I. Matsunagaa, A. Uedaa, N. Fujiwarab, T. Sumimotoa, and K. Ichihara, *Lipids*, 34, 841–846 (1999).
79. I. Matsunaga, N. Yokotani, O. Gotoh, E. Kusunose, M. Yamada, and K. Ichihara, *J. Biol. Chem.*, 272, (1997).
80. I. Matsunaga, T. Sumimotoa, A. Ueda, E. Kusunosea, and K. Ichihara, *Lipids*, 35, 365–371 (2000).
81. D.-S. Lee, A. Yamada, H. Sugimoto, I. Matsunaga, H. Ogura, K. Ichihara, S.-i. Adachi, S.-Y. Park, and Y. Shiro, *J. Biol. Chem.*, 278, 9761–9767 (2003).
82. T. L. Poulos and J. Kraut, *J. Biol. Chem.*, 255, 8199–8205 (1980).
83. H. Shimizu, D. J. Schuller, W. N. Lanzilotta, S. Sundaramoorthy, D. M. Arciero, A. B. Hooper, and T. L. Poulos, *Biochemistry*, 40, 13483–13490 (2001).
84. J. R. Cupp-Vickery, O. Han, C. R. Hutchinson, and T. L. Poulos, *Nature Struc. Biol.*, 3, 632–637 (1996).
85. R. L. Wright, K. Harris, B. Solow, R. H. White, and P. J. Kennelly, *FEBS Letters*, 384, 235–239 (1996).

86. S. Y. Park, K. Yamane, S. Adachi, Y. Shiro, S. A. Maves, K. E. Weiss, and S. G. Sligar, *J. Inorg. Biochem., 91*, 491–501 (2002).
87. A. V. Puchkaev and P. R. Ortiz de Montellano, *Arch. Biochem. Biophys., 434*, 169–177 (2005).
88. C. R. Woese and G. E. Fox, *Proc. Natl. Acad. Sci. USA, 74*, 5088–5090 (1977).
89. G. Schafer, *Biochim. Biophys. Acta, 1277*, 163–200 (1996).
90. L. S. Koo, C. E. Immoos, M. S. Cohen, P. J. Farmer, and P. R. Ortiz de Montellano, *J. Amer. Chem. Soc., 124*, 5684–5691 (2002).
91. M. A. McLean, S. A. Maves, K. E. Weiss, S. Krepich, and S. G. Sligar, *Biochem. Biophys. Res. Comm., 252*, 166–172 (1998).
92. A. V. Puchkaev and P. R. Ortiz de Montellano, *Arch. Biochem. Biophys., 434*, 169–177 (2005).
93. J. K. Yano, F. Blasco, H. Li, R. D. Schmid, A. Henne, and T. L. Poulos, *J. Biol. Chem., 278*, 608–616 (2003).
94. F. Blasco, I. Kauffmann, and R. D. Schmid, *Appl. Microbiol. Biotechnol., 64*, 671–674 (2004).
95. A. V. Puchkaev, L. S. Koo, and P. R. Ortiz de Montellano, *Arch. Biochem. Biophys., 409*, 52–58 (2003).
96. T. L. Poulos, B. C. Finzel, I. C. Gunsalus, G. C. Wagner, and J. Kraut, *J. Biol. Chem., 260*, 16122–16130 (1985).
97. C. A. Hasemann, K. G. Ravichandran, J. A. Peterson, and J. Deisenhofer, *J. Mol. Biol., 236*, 1169–1185 (1994).
98. P. Rowland, F. E. Blaney, M. G. Smyth, J. J. Jones, V. R. Leydon, A. K. Oxbrow, C. J. Lewis, M. G. Tennant, S. Modi, D. S. Eggleston, R. J. Chenery, and A. M. Bridges, to be published.
99. L. Podust, Y. Kim, M. Arase, B. Neely, B. Beck, H. Bach, D. Sherman, D. Lamb, S. Kelly, and M. Waterman, *J. Biol. Chem., 278*, 12214–12221 (2003).
100. K. G. Ravichandran, S. S. Boddupalli, C. A. Hasermann, J. A. Peterson, and J. Deisenhofer, *Science, 261*, 731–736 (1993).
101. K. Zerbe, O. Pylypenko, F. Vitali, W. Zhang, S. Rouset, M. Heck, J. W. Vrijbloed, D. Bischoff, B. Bister, R. D. Sussmuth, S. Pelzer, W. Wohlleben, J. A. Robinson, and I. Schlichting, *J. Biol. Chem., 277*, 47476–47485 (2002).
102. C. G. Mowat, D. Leys, K. J. McLean, S. L. Rivers, A. Richmond, A. W. Munro, M. Ortiz Lombardia, P. M. Alzari, G. A. Reid, S. K. Chapman, and M. D. Walkinshaw, *Acta Crystallogr., D 58*, 704–705 (2002).
103. L. M. Podust, H. Bach, Y. Kim, D. C. Lamb, M. Arase, D. H. Kelly, S. L. Sherma, and M. R. Waterman, *Protein Sci., 13*, 255 (2004).
104. P. A. Williams, J. Cosme, A. Ward, H. C. Angove, D. Matak-Vinkovic, and H. Jhoti, *Nature, 424*, 464–468 (2003).
105. J. R. Cupp-Vickery and T. L. Poulos, *Nature Struc. Biol., 2*, 144–153 (1995).
106. S. Nagano, H. Li, H. Shimizu, C. Nishida, H. Ogura, P. R. Ortiz de Montellano, and T. L. Poulos, (2003) submitted.
107. G. A. Schoch, J. K. Yano, M. R. Wester, K. J. Griffin, C. D. Stout, and, E. F. Johnson, *J. Biol.Chem., 279*, 9497–9503 (2004).
108. B. Zhao, F. P. Guengerich, A. Bellamine, D. C. Lamb, M. Izumikawa, L. Lei, L. M. Podust, M. Sundaramoorthy, J. A. Kalaitzis, L. M. Reddy, S. L. Kelly,

B. S. Moore, D. Stec, M. Voehler, J. R. Falck, T. Shimada, and M. R. Waterman, *J. Biol. Chem., 280*, 11599–11607 (2005).
109. Y. Oku, A. Ohtaki, S. Kamitori, N. Nakamura, M. Yohda, H. Ohno, and Y. Kawarabayasi, *J. Inorg. Biochem., 98*, 1194–1199 (2004).
110. O. Pylypenko, F. Vitali, K. Zerbe, J. A. Robinson, and I. Schlichting, *J. Biol. Chem., 278*, 46727–46733 (2003).
111. J. K. Yano, M. H. Hsu, K. J. Griffin, C. D. Stout, and E. F. Johnson, *Nat. Struct. Mol. Biol., 12*, 822–823 (2005).

4

Aquatic P450 Species*

Mark J. Snyder

Department of Clinical Pharmacology, University of California at San Francisco,
Kenwood, CA 95452, USA
<snyder181@comcast.net>

1.	INTRODUCTION. 'P450s UNDER THE SURFACE'	98
2.	DIVERSITY OF AQUATIC SPECIES	100
	2.1. Invertebrates	100
	2.2. Fish	100
	2.3. Mammals	100
3.	P450 ACTIVITIES IN AQUATIC INVERTEBRATES	101
	3.1. Background to Biochemical Activities	101
	3.2. Xenobiotic Effects on CYP Enzyme Activities and Gene Induction	102
	3.2.1. Cnidarians	102
	3.2.2. Annelids (Polychaetes)	103
	3.2.3. Molluscs	104
	3.2.4. Arthropods (Crustaceans)	105
	3.2.5. Echinoderms	108
4.	AQUATIC P450 GENE FAMILIES IDENTIFIED	108
	4.1. Invertebrate P450 Families	108
	4.2. Fish P450 Families	111
	4.3. Mammal P450 Families	115

*This review is dedicated to the memory of Eva P. Mulder.

5. HOW CAN WE USE INFORMATION ABOUT P450s
   IN AQUATIC SPECIES? 116
   5.1. Invertebrate Species Physiology and Ecology 116
   5.2. Environmental Monitoring/Ecosystem Health 118
   5.3. Potential Monitoring of Toxicants Present in Food 119
6. CONCLUSIONS AND OUTLOOK 120
   ACKNOWLEDGMENTS 121
   ABBREVIATIONS 121
   REFERENCES 122

1. INTRODUCTION. 'P450s UNDER THE SURFACE'

Organisms produce many different detoxification enzymes and other proteins in response to xenobiotic exposures. Cytochrome P450s (CYP) are one of the major phase I-type classes of enzymes found in terrestrial and aquatic organisms, ranging from bacteria to vertebrates. CYPs metabolize a wide variety of substrates including endogenous molecules (e.g., fatty acids, eicosanoids, steroids) and xenobiotics (e.g., hydrocarbons, pesticides, drugs). Aquatic species occupy every aspect of the environment, from above the surface (intertidal) to below the sediments. They have extremely diverse physiologies and are exposed to a vast array of potential toxicants as a result existing in variable environments. Aspects of aquatic species' cytochrome P450 enzymes have been studied for the last 30 years. In a few phyla, P450 activities have been measured and are responsive to xenobiotic exposures. Until the last 20 years, there was little progress in identifying P450 gene diversity in aquatic species. Molecular biology tools have greatly advanced this field and will lead to the identification of as much or more diversity for this protein superfamily as described in terrestrial organisms. Recent work has expanded our knowledge of the CYP superfamily, and new developments will rapidly advance the usefulness of these genes into such disciplines as biomarker research.

Cytochrome P450 monooxygenase enzymes comprise a widely distributed protein superfamily. One recent published accounting gives more than 750 sequences belonging to more than 107 different families [1a]. Many more CYP genes have been described in the last several years and a running total can be found in regular updates (see D. R. Nelson's web page [1b]). Phylogenetic analysis of CYP diversity suggests that a common ancestor to all present day P450 forms existed prior to the evolution of eukaryotes [2].

When CYPs or a subcellular preparation containing them (microsomes and/or mitochondria) are reduced, they give a characteristic absorbance peak at 450 nm in the presence of carbon monoxide [3]. Functions of P450s in the metabolism of endogenous compounds (e.g., steroids, fatty acids, eicosanoids) and xenobiotics (e.g., dietary plant chemicals, various aromatic hydrocarbons (PAH, AH), polychlorinated biphenyls (PCB), insecticides, drugs) have been extensively studied in the last 30 years [4] and include hydroxylation, epoxidation, oxidative deamination, S-, N-, and O-dealkylations, and dehalogenation. Though end results of P450 reactions are most often more hydrophilic and presumably more excretable, some products of P450 metabolism are more reactive, leading to cellular damage and in certain cases to the initiation of carcinogenesis [5].

To avoid confusion that typically occurs with investigator-specific gene naming terminology, the Cytochrome P450 Nomenclature Committee has described the systematic organization of one of the largest protein families [6]. A single CYP protein is generally coded by one gene, and assignment to a particular family (e.g., CYP1) is based primarily on $\geq 40\%$ overall amino acid sequence identity. Further division to a particular subfamily (e.g., CYP1A, the first member being CYP1A1...CYP1A$_N$) is based on at least 55% overall sequence identity. A higher-order classification scheme for this protein superfamily has been developed by which CYP families are grouped into CLANs, an extra step necessitated by the ever growing number of published CYP sequences [1].

For all but a few organisms, the true number of P450 genes is unknown. The mouse and human have 102 and 57 described CYP genes, respectively [7]. Bacteria may have from 0 to 11 different CYP genes, while the current champion in this regard is a plant (rice), *Oryza sativa*, with 323 different CYP proteins produced in its lifetime. The high number of CYP genes in plants may reflect the outcome of plant–animal 'warfare' as an evolutionary driving mechanism. In response to herbivory, plants may be forced to evolve new noxious secondary products to stave off insect herbivores [8].

From 1988 to the present, more than 38 000 published articles are retrievable using the keyword 'P450' in a search using the Pubmed database. The vast majority of this literature is concentrated on biochemical and toxicological aspects in higher vertebrates. In the marine environment, the best-studied member of the cytochrome P450 superfamily is CYP1A1, the major form induced by dioxins, PAHs and PCBs. Antisera have been produced against the CYP1A1 of a number of fish species and successfully used to detect changes in enzyme protein levels by immunoblotting and enzyme-linked immunosorbant assay (ELISA) [9–11]. This promising strategy has been utilized only to a limited degree in the study of other xenobiotic-affected CYP forms in marine phyla, especially invertebrates as will be discussed below.

The majority of this review will be devoted to aquatic invertebrate CYPs. I will discuss marine vertebrates (fish and mammals) in the context of providing a summary of what is currently known about CYP gene families. Fish P450 biochemistry and molecular biology deserves additional review and is too considerable to do justice to the field in this chapter.

2. DIVERSITY OF AQUATIC SPECIES

2.1. Invertebrates

Aquatic invertebrates are found in all the world's freshwater, estuarine, and marine environments. Their diversity rivals or exceeds that of terrestrial insects. They can be found in the most polluted aquatic habitats worldwide and comprise the lowest to highest levels of the food chain. As such, particular species can be sentinels of ecosystem health. An example of this is the 'mussel watch' program that consists of repeated sampling of established mollusc beds and analyses of tissues for pollutant (and algal toxin load prior to market sale). Some species are capable of existing in resting stages (cysts) for tens or more years without signs of life (for example, some crustaceans). The numbers and family relationships of aquatic invertebrate P450 families to mammals are currently unknown. The vast differences in aquatic invertebrate physiology will be undoubtedly highlighted in equally diverse families of P450s as more information is uncovered. Much of this review will focus on aquatic invertebrate P450, though the diversity of CYP genes in fish and aquatic mammals will also be discussed.

2.2. Fish

Fish live in a myriad of aquatic environments. They can be found at the deepest depths of the ocean, underground cave systems, and high-altitude lakes and streams. Fish live in subfreezing temperatures, at pressures of >14 tons per square centimeter, and in environments of extremely low oxygen. Some species live within relatively small areas of tens of meters or less their entire lives (e.g., killifish). Others can migrate thousands of miles each year (for example, tunas, sharks). As potential biomonitoring organisms, some can be considered as representative of the health of large ecosystems and others, the health of a few hundred meters of a particular river. The vertebrate relationship of fish to mammals is also reflected in conserved steroid-related and toxicant-catabolizing P450 families as will be discussed below.

2.3. Mammals

The smallest total biomass of the animals discussed here are the marine mammals. This group includes whales, dolphins, porpoises, seals, sea lions, otters, and manatees/dugongs. I include only those groups that live predominately in water. This group is interesting for several reasons with regard to P450s. Some species live primarily in the open ocean (e.g., whales, dolphins) and P450 markers in these animals could be considered reflective of the health (pollution) of the ecosystem as a whole. Others live mainly close to 'shore' (otters, manatees, beluga whales) and P450s in these species could be construed as indicative of the health of heavily human-impacted local environments. Levels of pesticides in whale blubber, and PCB's in beluga whale tissues are examples of the former and latter, respectively. The physiology of marine mammals is similar to humans and it would be expected that P450 families involved in normal life functions such as steroid production and catabolism are similar as well.

3. P450 ACTIVITIES IN AQUATIC INVERTEBRATES

3.1. Background to Biochemical Activities

Khan et al. [12] described the initial finding of P450-type activity in marine invertebrates. P450-type enzymatic activities have since been reported in arthropods (crustaceans), annelids (polychaetes), cnidarians, molluscs, poriferans, platyhelminths, and echinoderms (see [13–17] for extensive reviews of early literature on marine invertebrate P450s). CYP distribution within various marine invertebrate phyla falls into the following patterns. In molluscs, P450 is highest in the digestive gland and also found in blood cells, gills, foot, and gonads [14]. For echinoderms, CYP activities are also highest in digestive tissue (pyloric ceca) and also found in gonads and haemal plexus [18]. Similarly, crustaceans have the highest total P450 protein in the hepatopancreas, but also show high activity in green gland, gonads, and stomach [13].

Though P450 enzymes are found in aquatic invertebrate phyla, there has been much confusion about the relevance of these proteins in the metabolism of toxicants. Total P450 protein and associated enzymatic activities are usually measured at levels ten-fold lower than that found in mammals [15]. Some studies have failed to show P450 activity inductions following xenobiotic exposures [19–21]. The presence of endogenous inhibitors in microsomal preparations has been suggested as a major reason for many of the reported problems with aquatic invertebrate P450 activity measurements (discussed in [13]). Strong evidence from many species (best studied in the lobster *Panulirus argus* [22]) also suggests that preparation of marine invertebrate microsomes for the detection of P450 enzymatic activities results in the destruction of NADPH-cytochrome P450 reductase, the enzyme is responsible for the transfer of electrons necessary for P450 activity. Closely related to this issue (though in terrestrial invertebrates), insect gut contents have been found to directly inhibit P450 activities. An inhibitory factor released from gut tissue as a result of feeding and/or the effects of digestive enzyme(s) themselves on P450 proteins is the hypothesized cause [23].

Another documented confounding issue in prior marine invertebrate P450 studies is the 'one-electron oxidation or transfer' problem. Organic hydroperoxides are capable of catalyzing P450 reactions without NADPH present in both mammals and aquatic invertebrates. The possibility of longer-lived peroxidation in crustacean systems than in mammals suggests a different heme environment protecting against hydroperoxide-P450 damage. Similar peroxidative effects are seen in mussel and sea star P450 systems *in vitro* [24,25]. Increases in oxidation in the C6 position of benzo[a]pyrene (BaP), leading to quinone formation with peroxidation *in vivo* is postulated as one of the common 'one-electron transfer' mechanisms noted in many aquatic invertebrates [25,26]. The low oxygen environments in intertidal and sedimentary habitats of aquatic versus terrestrial environments have been postulated as the driving mechanism favoring the substantial increases in xenobiotic metabolism with organic peroxides [25,27]. Also, NADPH-cytochrome P450 reductase may not be fully required for oxidation of P450 substrates in aquatic environments [13] since NADH can be utilized by P450s in various marine invertebrates with cytochrome b_5 as donor [28,29]. P450 enzyme activity is better supported by NADH in both crayfish and sea stars [25,30], while activity with NADPH as electron donor is higher in sea anemone [27]. Similar confounding data exist for a wide range of aquatic invertebrate phyla for which some representative species have higher BaP hydroxylase activity better supported without added NADPH [29]. The reasons for the variations in P450 activities with different electron donors and subcellular environments will hopefully be resolved by some dedicated biochemistry.

3.2. Xenobiotic Effects on CYP Enzyme Activities and Gene Induction

3.2.1. Cnidarians

The phylum Cnidaria has received attention from only a few investigators with regard to detoxification mechanisms. Cnidarian total microsomal P450 protein is low at about 50 pmol per mg protein versus other aquatic invertebrates such as crayfish that vary between 300–700 pmol per mg protein [31]. Benzo[a]pyrene metabolism by anemone, *Bunodosoma cavernata* has been studied by Winston et al. [27] who found both NADPH- and NADH-dependent BaP metabolism by anemone microsomes. This is in agreement with other marine invertebrate microsomal activity observations for which NADH [25,32], NADPH [27], or neither [33] demonstrated increased requirements for efficient *in vitro* BaP metabolism. Organic hydroperoxide substitution of NADPH or NADH in the microsomal incubations also supported BaP metabolism. Gassman and Kennedy [34] demonstrated P450 enzyme activity in the coral *Favia fragrum*, and higher BaP hydroxylase was found in animals from sites of increasing pollutant concentrations.

Immunological cross-reactivity among several anemone species proteins has been shown against both mammalian and fish anti-P450 antisera [34]. The need to research the biochemical adaptive mechanisms (including the potential involvement of detoxification enzymes such as P450s) of these organisms in much greater detail is indicated by the frequent occurrence of cnidarians around coral reefs and in nearshore environments considered critical to monitor by many. As discussed further below, responses to stresses such as anthropogenic chemicals in cnidarians and other aquatic invertebrate phyla may be useful biomarkers that provide clues of potential harm to aquatic habitats.

3.2.2. Annelids (Polychaetes)

Most of the early work on detoxification enzymes of aquatic members of this phylum (polychaetes) was done in the laboratory of R. Lee in the late 1970s to the early 1980s (see review in [15]). Only a few studies have been performed on CYPs in this phylum in the past decade or so. Whole-polychaete PAH metabolism following placement in contaminated sediment aquaria [36,37] was clearly induced in the worm, *Capitella* sp. Both fluoranthrene conversion to polar metabolites and depuration rates were elevated after 1–2 days of exposure. These results support the idea of inducible P450-based detoxification mechanisms in polychaetes that mediate the worm responses to hydrocarbon exposures.

In the last few years, new information has been developed on the identification and expression of new polychaete CYP genes. The phylogenetic relationships of these new genes are discussed in Section 4. CYP4AT1 and CYP331A1 in *Capitella capitata* are induced by hydrocarbon exposures in a differential manner [38]. Both BaP and fluoranthrene induce CYP4AT1 by two-fold, while 3-methylcholanthrene (3-MC) slightly inhibits expression. CYP331A1 is unaffected by BaP and fluoranthrene and is slightly induced by 3-MC.

Two new CYP4 family-related genes were found in *Nereis virens* [39]. Exposure to BaP had no effect on either CYP, while one was induced two-fold by clofibrate and three-fold by benz[a]anthracene. In a further study, the new genes were identified as CYP4BB1 and CYP342A1 [40]. Both CYPs were expressed in a bacterial system and found to metabolize pyrene to a single metabolite, 1-hydroxypyrene at rates of 58–71 pmol/min/mg CYP protein. Site directed mutagenesis of the axial heme binding cysteine for serine in either CYP abolished pyrene metabolism. This work strongly indicates that members of the CYP4 and related CYP families have detoxifying metabolic activities in aquatic invertebrates.

The distribution of polychaetes is particularly high in areas heavily impacted by anthropogenic chemicals in nearshore environments. Clearly, as for cnidarians, this phylum deserves more attention with regard to detoxification enzymes such as P450s.

3.2.3. Molluscs

Aspects of mollusc hydrocarbon metabolism in molluscs have received considerable attention in the last 15 years. Increased BaP activity was measured in visceral microsomes from atrazine-exposed snails, *Lymnaea palustris* [41]. Mytilid bivalves have been the focus of a greater volume of work. *Mytilus galloprovincialis* was exposed to phenobarbital, clofibrate, or 3-MC followed by measurements of digestive gland microsome total P450 protein, laurate hydroxylase, and BaP activities [42]. 3-MC elevated all three measurements, while phenobarbital elevated only laurate hydroxylase (LAH) activity, and clofibrate was without effect. CYP1A1-like immunoreactivity of mussel microsomal proteins increased following 10 day exposure to PCBs without any change in CYP1A1-like mRNA expression [43]. P450-mediated hydrocarbon metabolism was markedly shifted with PCB exposure. Mussels sampled from field sites contaminated by multiple pollutants led to the observations that a CYP1A-like immunoreactive protein was elevated, while a CYP3A-like protein was reduced in digestive glands [44].

Reduced DNA adduct formation in *M. edulis* was correlated with inhibition of P450 activity in the presence of BaP in isolated digestive gland cells [45]. β-naphthoflavone (βNF) treatment of octopus, *Octopus pallidus*, resulted in decreased P450, EROD (ethoxyresorufin *O*-deethylase) and ECOD (ethoxycoumarin *O*-deethylase) activities of digestive gland microsomes [46]. Injections of octopus with PCB elevated ECOD activity only. P450 activity inhibition (ECOD) has been documented in *M. galloprovincialis* [47] with phenobarbital exposure.

Vertebrate steroids have been found in aquatic invertebrates in studies over the past few decades [17]. Recent work has shown that aromatase activities are seasonally modulated with reproductive periods in scallop, *Patinopectin yessoensis* [48]. Aromatase activities are related to tissue estradiol-17-β and both enzyme and steroid are immunolocalized in the testis and ovary of scallops. Mollusc digestive gland microsomes show altered testosterone and estradiol metabolism when incubated in the presence of the environmental endocrine disruptor tributyltin (TBT) [49–51]. This *in vitro* inhibition was also found in whole animal TBT incubation experiments, supporting the theory of sex steroid alterations as contributing to the phenomenon of induced imposex (masculinization of females) in gastropod molluscs. TBT inhibits mollusc NADPH cytochrome P450 reductase *in vitro* in a dose-dependent manner [50], with obvious negative impacts on the ability of P450 enzymes to metabolize substrates. Imposex is greatly stimulated when molluscs are treated directly with testosterone, and simultaneous treatment with inhibitors of androgen receptors blocks TBT-induced imposex [52]. The inhibition of P450-mediated steroid metabolism by TBT, although not the only mechanism involved in mollusc imposex induction [53], has resulted in population-level reproductive abnormalities. Future work on the interaction of xenobiotics with metabolizing systems in marine organisms such as P450 enzymes is obligated by these findings in marine invertebrates.

With the use of heterologous immunological and molecular probes, CYP1A1-, CYP3A-, CYP4A-, and CYP11A-like proteins were identified in *M. edulis* digestive glands [54,55]. Mussel CYP1A1 homolog expression appears to have some relationship with seasonal BaP activity patterns. Interestingly, the mussel digestive gland BaP activity peaks during the Summer while ECOD activity is highest during the late Fall/early Winter and total P450 peaks in both Summer and Winter months [56]. In contrast, there is a decrease in total digestive gland P450 in *M. galloprovincialis* from late Winter/late Spring [57] coincident with dramatic declines in body burdens of a number of different xenobiotics. Partial purification of a *M. edulis* digestive gland P450 protein was achieved using a several step protocol involving detergent solubilization and polyethylene glycol fractionation of microsomes followed by several chromatographic steps (affinity, ion-exchange, and hydroxylapatite). The protein gave a strong signal on Western blots probed with fish CYP1A1 antibody [58]. Further chromatography steps led to the identification of three different fractions, each with significant BaP activity. In another study with the freshwater mussel *Unio tumidus*, no evidence for the presence of CYP1A was found using reverse transcription PCR experiments with several degenerate primer pairs [59].

The first identified aquatic invertebrate CYP family was CYP10 from the snail with no known function to date [72]. The second mollusc CYP gene, CYP4Y1, was identified from *M. galloprovincialis* digestive gland [60]. CYP4Y1 expression was inhibited by increasing concentrations of βNF added in static exposures. Similarly, mussels were exposed to microbially degraded weathered oil obtained by incubation of natural Santa Barbara oil seep microbials mats with weathered crude oil. Degraded oil exposures at levels normally found in nearshore mussel habitats in Southern California (1–17 ppm) significantly reduced CYP4Y1 mRNA levels in a dose-dependent manner [61]. Hydrocarbon exposure reduced CYP enzyme activities or specific isoform levels in several other molluscs including *O. pallidus, Chryptochiton stelleri*, and *M. edulis* [26,46,47,62]. A partial cDNA for an additional CYP4 family member was recently identified (not yet officially named) in the mussel *U. tumidus* [59] and there is no information about potential inducers.

A new CYP family, CYP320, was recently identified in the tropical freshwater snail *Biomphalaria glabrata*, the intermediate host for the parasite *Schistosoma mansoni* that causes schistosomiasis [63]. CYP320A1 is found in multiple tissues, including ovitestis, mantle, and nephridium, but is currently without a clear function.

3.2.4. *Arthropods (Crustaceans)*

BaP metabolism has been reported by crayfish, *Procambrus clarkii*, tissues [32] and was apparently limited by low P450 reductase activity in hepatopancreas and green gland. This resulted from substantially greater metabolite formation in the presence of organic hydroperoxides that directly transfer electrons and oxygen to the P450 heme. Hepatopancreas BaP activity demonstrated a preference for NADH while green gland preferred NADPH as electron donor. The utility of crustacean P450 enzyme activities as useful biomarkers of exposure in the field has been confirmed [64]. Elevated BaP activity was found in both laboratory-exposed crabs, *Carcinus aestuarii*, and those from more contaminated field sites. Similar field data was obtained with crayfish, *P. clarkii*, using 7-ethoxyresorufin *O*-deethylase activity to monitor the effects of pesticide spraying [65]. Grass shrimp, *Palaemonetes* sp., had elevated P450 activity from sites more heavily contaminated by hydrocarbons [66]. Dioxin injection into male crayfish, *Pacifastacus leniusculus*, elevated total P450 by two-fold 7 days following a single treatment that resulted in 25% mortality [67]. Lower BaP activity was found in barnacles (*Balanus eburneus*) exposed to increasing concentrations of α-naphthoflavone [28]. Induction of a CYP3A-like P450 was identified following tributyltin ingestion in crabs, *Callinectes sapidus* [51,68].

Several studies from a single group have shown that xenobiotic exposures of a fresh water crustacean (*Daphnia magna*) may interrupt P450-mediated metabolism of steroids [69–71]. Though assays of microsomal P450 activities proved difficult because of many of the aforementioned biochemical problems, whole Daphniid exposures showed that alterations in the metabolism of [^{14}C]-testosterone were found following exposure to PB, malathion, βNF, and the endocrine disruptor diethylstilbesterol [69,70]. These data suggest that aquatic invertebrate P450 steroid metabolizing activities may be potentially valuable markers to test the effects of endocrine disrupting chemicals. P450 alterations were identified in the blue crab, *C. sapidus*, following exposure to TBT that is known to affect mollusc endocrine systems, as previously mentioned [68]. These changes included alterations in hepatopancreas microsomal testosterone metabolism, and increased levels of a CYP3A-like protein were found following TBT exposure.

CYP2L1 was the first identified marine invertebrate P450 from the hepatopancreas of the spiny lobster, *P. argus* [73]. This same laboratory has previously demonstrated that partially purified microsomal fractions containing this protein are capable of metabolizing a host of substrates including aminopyrine, benzo[a]pyrene, 7-ethoxycoumarin, and the steroids testosterone and progesterone, but not the molting hormone ecdysone [22,74]. The natural endogenous substrates for the CYP2L1 gene are currently unknown, although microsomes from a yeast heterologous expression system efficiently metabolized testosterone and progesterone [75]. An antibody directed against the spiny lobster CYP2L1 has cross-reactivity against microsomal proteins of related lobsters and several vertebrates, but not against a variety of other decapod crustaceans [76]. This antibody has unknown utility for CYP protein measurements in other marine invertebrate species. In a study of a wide variety of aquatic species, only crustaceans among the invertebrates (not molluscs, annelids or echinoderms) were found to have metabolic profiles of PCB congeners suggestive of the presence of CYP1- and CYP2 family-like P450s [77]. The presence of specific P450s capable of these particular biochemical activities currently remains unclear.

CYP45 was identified in the hepatopancreas of the American lobster, *Homarus americanus* [78] and expression coincided with changes in hemolymph molting hormone levels (ecdysteroids) [79]. Lobster hepatopancreas is the major organ for the metabolism of endogenous and ingested ecdysteroids [80,81] and injection of the active molting hormone (20-hydroxyecdysone) elevated hepatopancreas CYP45 expression [78]. These results suggested that CYP45 may be involved in either further ecdysteroid metabolism or in other aspects of the physiological events that culminate in molting. Two additional CYPs were identified in hepatopancreas of the crab, *Carcinus maenas* [82]. CYP4C39 was not inducible by ecdysteroids or a variety of xenobiotic exposures. The crab CYP330A1 was inducible by molting hormone injections similar to CYP45 in *H. americanus*. Partial genes for other hepatopancreas P450s from *H. americanus*, CYP4C18, and from a penaeid shrimp (*Penaeus setiferus*), CYP4C17, were also described [60]. Expression of lobster CYP4C18 gene appears unaffected by ecdysteroid treatment [83]. Future work will undoubtedly uncover the specific CYPs involved in steroid molting hormone production and degradation in crustaceans and other aquatic invertebrates.

An anti-CYP45 polyclonal antibody was developed and exposure of juvenile *H. americanus* to phenobarbital or heptachlor resulted in 2- to 15-fold elevations in hepatopancreas CYP45 protein levels [84]. This confirmed that the increased CYP45 gene expression following treatment by these xenobiotics was definitively associated with increases in the CYP45 protein [78]. Similarly, *C. maenas* CYP330A1 expression was elevated by low levels of phenobarbital or BaP, while higher exposure levels were without effect [82].

The CYP4 family degenerate primer RT-PCR strategy [83] was also employed to isolate partial CYPs from the water flea *Daphnia pulex* [85]. There were 4 new CYP4C, two CYP4AN, and one CYP4AP family members. Two *D. pulex* populations exhibited differences in CYP4C32 and CYP4AP1 expression that correlated with the polyphenol richness of the vegetation around their aquatic habitat. Dietary polyphenol exposure resulted in different CYP induction patterns, suggesting the possible involvement of CYP4s in subalpine *D. pulex* population differentiation related to polyphenol levels of the environmental vegetation.

In addition to the xenobiotic-related involvement of CYP4 family members in aquatic/marine invertebrates, one study has examined a potential role of CYP4 in molting physiology in crayfish, *Orconectes limosus*. CYP4C15 was identified from the Y-organs, the site of crustacean molting hormone biosynthesis [86]. It is expressed only in the Y-organs and levels are highest during premolt when the organ is most active in ecdysteroid synthesis, suggesting a potential role in molting physiology, similar to hepatopancreas-localized lobster CYP45 and crab CYP330A1 described previously.

3.2.5. Echinoderms

The metabolism of BaP by echinoderm microsomal preparations has been studied in detail with many similarities to the previous work on other aquatic invertebrate phyla [87–89]. *In vitro* studies of the requirements for NADPH or NADH and the effects of added organic hydroperoxide were performed. Organic peroxide and NADH promoted greater BaP hydroxylase metabolism by sea star pyloric ceca microsomes than NADPH. Both NADPH and peroxide supported the increase in BaP hydroxylase activity after BaP exposure, but peroxide was more efficient as found for other marine invertebrates. Treatment of sea stars, *Asterias rubens*, with PCB or BaP elevated BaP hydroxylase turnover rate (product formed per time, per amount of P450 protein), while BaP exposure elevated both BaP hydroxylase activity and turnover rate [87–89]. Coplanar PCB induced the levels of an immunoreactive CYP1A-like protein by 73 fold in *A. rubens* while a noncoplanar congener was without effect [90]. Exposure of sea stars to either PCB or BaP also inhibited P450 pregnenolone hydroxylase activity [89]. Seasonal variations in sea star total P450 have also been identified [91] with the highest P450 enzyme protein levels found just prior to the beginning of reproductive period associated with increases in steroid metabolism.

There is additional immunological evidence for the presence of P450 enzymes belonging to the CYP1, CYP2, and CYP3 families in sea stars [87]. The first echinoderm CYP genes were identified from digestive tissues of the sea urchin *Lytechinus anamesus*, belonging to the CYP4 family, CYP4C19, and CYP4C20 [60]. No information is currently available about the expression patterns of these two members of the CYP4 family in sea urchins.

4. AQUATIC P450 GENE FAMILIES IDENTIFIED

4.1. Invertebrate P450 Families

A listing of aquatic invertebrate and vertebrate CYP genes was gleaned from a combination of database searches from D. R. Nelson's website [1b], the NCBI protein database, and current literature. Aquatic invertebrate CYP genes are given in Table 1. We can expect the total number of CYP genes in any aquatic invertebrate species to be related to the total found in nematodes (74 genes), mosquitoes (105 genes), fruit fly (84 genes), and sea squirt (urochordate, 97 genes).

Table 1. Cytochrome P450 genes identified in aquatic invertebrates.

P450 Gene	Species (common name)	P450 Gene	Species (common name)
CYP1C1	*Hydra magnipapillata* (Hydra)[a] *Strongylocentrotus purpuratus* (Purple sea urchin)[a]	CYP4FA1	*Hydra magnipapillata* (Hydra)[a] *Coscinasterias muricata* (Australian sea star)
CYP2L1,2	*Panulirus argus* (Spiny lobster)	CYP5A1	*Strongylocentrotus purpuratus* (Purple sea urchin)[a]
CYP2 (various)	*Strongylocentrotus purpuratus* (Purple sea urchin)[a]	CYP9	*Reniera* sp. (Sponge)
CYP3A (various)	*Strongylocentrotus purpuratus* (Purple sea urchin)[a]	CYP10A1	*Lymnaea stagnalis* (Pond snail)
CYP4C15,24	*Orconectes limosus* (Spinycheek crayfish)	CYP20	*Reniera* sp. (Sponge)
CYP4C16	*Penaeus setiferus* (White shrimp)		*Hydra magnipapillata* (Hydra)[a]
CYP4C17	*Haliotis rufescens* (Red abalone)		
CYP4C18	*Homarus americanus* (American lobster)	CYP27A,B	*Reniera* sp. (Sponge)
CYP4C19	*Lytechinus anamesus* (White sea urchin)		
CYP4C20	*Lytechinus anamesus* (White sea urchin)	CYP30A1	*Mercenaria mercenaria* (Northern quahog)
CYP4C29	*Cherax quadricarinatus* (Redclaw crayfish)		
CYP4C30	*Coscinasterias muricata* (Australian sea star)	CYP38A1	*Suberite domuncula* (Sponge)
CYP4C31,32,33,34	*Daphnia pulex* (Water flea)		
CYP4C39	*Carcinus maenas* (Green shore crab)	CYP45	*Homarus americanus* (American lobster)
CYP4C,4F	*Reniera* sp. (Sponge)	CYP49A1	*Reniera* sp. (Sponge)
CYP4V5,7	*Strongylocentrotus purpuratus* (Purple sea urchin)[a]		
CYP4Y1	*Mytilus galloprovincialis* (Blue mussel)	CYP51	*Reniera* sp. (Sponge) *Strongylocentrotus purpuratus* (Purple sea urchin)[a]

Table 1. (Continued).

P450 Gene	Species (common name)	P450 Gene	Species (common name)
CYP4AL1	*Mytilus edulis planulatus* (Australian mussel)	CYP320A1	*Biomphalaria glabrata* (Freshwater tropical snail)
CYP4AM1	*Plebidonax deltoides* (Goolwa cockle)	CYP322A1	*Perna canaliculus* (Green lipped mussel)
CYP4AN1,2,3	*Daphnia pulex* (Water flea)	CYP323A1	*Saccostrea glomerata* (Sydney rock oyster)
CYP4AP1,2	*Daphnia pulex* (Water flea)	CYP330A1	*Carcinus maenas* (Green crab)
CYP4AT1	*Capitella capitata* (Gallery worm)	CYP331A1	*Capitella capitata* (Gallery worm)
CYP4BB1	*Nereis virens* (Clamworm)	CYP342A1	*Nereis virens* (Clamworm)

[a] Partial sequences with significant identity to the indicated P450 family and subfamily.

Since the last review on this topic in 2000 [17], twelve new CYP gene families have been described in aquatic invertebrates. Surprisingly, two partial CYP1C1-like sequences found in *Hydra magnipapillata* and *Strongylocentrotus purpuratus* (urchin) are now known. Much effort has gone towards the discovery of CYP1 genes in this group of organisms without success (see previous discussion on antibodies and molecular probes). CYP2 is now represented in *S. purpuratus*, extending the finding of the CYP2L subfamily in the spiny lobster, *P. argus* [75,76,92]. The CYP2L amino acid sequence is approximately 36% identical to mammalian CYP2 family members. Three more related genes, CYP2L2 and two closely related partial sequences were recently identified in this species by the same laboratory [92]. A new family, CYP330A1, was identified in the crab *C. maenas* with the highest degree of identity belonging to the CYP2 family at 35–37% [82]. An interesting new family, CYP320A1, was found in the snail *B. glabrata* with some homology to the CYP2 and CYP17 families, but with no clear relationship or functional clues [63]. Mining of the sea urchin genome has thus far provided many partial CYP sequences in addition to the above belonging to the following gene subfamilies: CYP3A, CYP4V5, CYP4V7, CYP5A1, and possibly a CYP51-like gene.

The CYP4 family has the most representation in aquatic invertebrates to date, a similar finding to the extensive CYP4 membership present in the insects [17,60,93,94]. In fact, the CYP4 and CYP6 families each comprise about 26% of the 83 total complete CYP genes in *D. melanogaster* [95]. Some of the new findings of new CYP4 family members in aquatic invertebrates stem undoubtedly from the use of similar degenerate CYP4 primers and the ease of RT-PCR cloning first championed in the mid-1990s [83,93]. The CYP4C subfamily is found in crustaceans, echinoderms, sponges, and molluscs [1b,

60,82,85,86]. The CYP4 family has a number of other subfamilies represented in a variety of aquatic invertebrates [1b,38,40,60.] An additional family with significant relationship to CYP4 (~ 35%) is CYP342A1 from the polychaete *N. virens* [40].

The CYP6-related P450 families are another story. Brown et al. [96] identified a new CYP family, CYP30, in female gonadal tissue from the clam *Mercenaria mercenaria*. Although its function is unknown, this sequence shares significant homology with vertebrate CYP3 and invertebrate CYP6, CYP9, and CYP25 families [1]. Several other aquatic invertebrate CYPs share significant homology with the vertebrate CYP3 family: CYP45 from the lobster *H. americanus* [60], CYP331A1 from the polychaete *C. capitata* [38], and a CYP9-like sequence from the sponge *Reniera* sp. [1b]. Vertebrate CYP3 family members are usually expressed in digestive and respiratory systems [97]. CYP3A genes and related CYP family members are thought to be strongly adaptive to changing environmental pressures, have the potential for broad substrate specificity, and appear to be involved in resistance to xenobiotics [98]. Aquatic invertebrate CYP genes related to these families will undoubtedly be interesting and important for environmental monitoring, as discussed further below.

Surprisingly, Hydra has no representatives in any known mitochondrial CYP family. The CYP genes responsible for several of the biosynthetic steps of insect ecdysteroid hormones are located in the mitochondria [99]. Since crustaceans, and potentially members of other aquatic invertebrate phyla, also use these steroid hormones for similar physiological functions, mitochondrial CYP families will likely be found in aquatic invertebrates. Vertebrate steroids have been identified in hemolymph of aquatic invertebrates and associated with reproductive functions in molluscs as indicated by the previously described TBT story. Other evidence suggests that vertebrate neuroactive steroids such as allopregnanolone and pregnenolone sulfate may function to regulate crustacean locomotor activity [100]. Such physiological functions for steroids previously thought active only for vertebrates argues for the potential involvement of mitochondrial CYPs in aquatic invertebrate steroid metabolism.

The extensive physiological similarities between insects and crustaceans suggest that at least some of the eight *Drosophila* mitochondrial CYPs [95] may be found in aquatic invertebrates. Some of these mitochondrial CYPs may also be involved in detoxification responses such as CYP12A4 in *Drosophila* that is responsible for resistance to the pesticide lufenuron [101].

4.2. Fish P450 Families

Considerable efforts have greatly advanced the database of fish CYP genes (Table 2). The P450 database [1b] details the likelihood of 55 CYP genes for the Japanese pufferfish, *Takifugu rubripes*, and 81 CYP genes for the Zebrafish, *Danio rerio*. The fish CYP1 family currently has three subfamilies: A, B, and C.

Table 2. Cytochrome P450 genes identified in fish.

P450 Gene	Species (common name)	P450 Gene	Species (common name)
CYP1A1[a]	*Oncorhynchus mykiss* (Trout)	CYP2K9,10, 11,12,13,14,15	*Takifugu rubripes* (Japanese pufferfish)
	Dicentrarchus labrax (European sea bass)	CYP2K16,17, 18,19,20,21,22	*Danio rerio* (Zebrafish)
	Salmo salar (Salmon)		
CYP1A3	*Oncorhynchus mykiss* (Trout)	CYP2M1	*Oncorhynchus mykiss* (Trout)
CYP1A9	*Anguilla anguilla* (European eel)		
	Anguilla japonica (Japanese eel)	CYP2N1,2	*Fundulus heteroclitus* (Killifish)
		CYP2N3	*Stenotomus chrysops* (Scup)
CYP1B1	*Platichthys flesus* (European flounder)	CYP2N4,5,6,7,8	*Chaetodon* sp. (Butterfly fish)
	Danio rerio (Zebrafish)	CYP2N9,10,11	*Takifugu rubripes* (Japanese pufferfish)
	Cyprinus carpio (Common carp)	CYP2N13	*Danio rerio* (Zebrafish)
	Takifugu rubripes (Japanese pufferfish)		
	Tetraodon nigroviridis (Freshwater pufferfish)	CYP2P1,2,3	*Fundulus heteroclitus* (Killifish)
	Anguilla japonica (Japanese eel)	CYP2P4,5	*Takifugu rubripes* (Japanese pufferfish)
	Oreochromis niloticus (Nile tilapia)	CYP2P6, 7,8,9,10	*Danio rerio* (Zebrafish)
	Pleuronectes platessa (Plaice)	CYP2P11	*Micropterus salmoides* (Large mouth bass)
CYP1B2	*Cyprinus carpio* (Common carp)		
CYP1C1	*Stenotomus chrysops* (Scup)	CYP2R1	*Danio rerio* (Zebrafish)
	Takifugu rubripes (Japanese pufferfish)	CYP2R1,2,3	*Takifugu rubripes* (Japanese pufferfish)
	Danio rerio (Zebrafish)		
	Cyprinus carpio (Common carp)	CYP2U1	*Takifugu rubripes* (Japanese pufferfish)
	Tetraodon nigroviridis (Freshwater pufferfish)		*Danio rerio* (Zebrafish)
	Anguilla japonica (Japanese eel)		

AQUATIC P450 SPECIES

CYP1C2	*Stenotomus chrysops* (Scup)	CYP2V1	*Danio rerio* (Zebrafish)
	Takifugu rubripes (Japanese pufferfish)		
	Danio rerio (Zebrafish)	CYP2X1	*Ictalurus punctatus* (Catfish)
	Cyprinus carpio (Common carp)	CYP2X2,3,4,5	*Takifugu rubripes* (Japanese pufferfish)
	Tetraodon nigroviridis (Freshwater pufferfish)	CYP2X6,7,8,9, 10,11,12	*Danio rerio* (Zebrafish)
	Fundulus heteroclitus (Killifish)		
		CYP2Y1,2	*Takifugu rubripes* (Japanese pufferfish)
CYP2K1v1,2,3	*Oncorhynchus mykiss* (Trout)	CYP2Y3,4	*Danio rerio* (Zebrafish)
CYP2K2	*Fundulus heteroclitus* (Killifish)		
CYP2K3,4,5	*Oncorhynchus mykiss* (Trout)	CYP2Z1,2	*Takifugu rubripes* (Japanese pufferfish)
CYP2K6,7,8	*Danio rerio* (Zebrafish)	CYP2Z3,4	*Danio rerio* (Zebrafish)
CYP2AA1,2,3,4, 5,6,7,8,9	*Danio rerio* (Zebrafish)	CYP7C1	*Takifugu rubripes* (Japanese pufferfish)
			Danio rerio (Zebrafish)
CYP2AD1	*Takifugu rubripes* (Japanese pufferfish)		
CYP2AD2,3,6	*Danio rerio* (Zebrafish)	CYP7D1	*Danio rerio* (Zebrafish)
CYP2AD4	*Oryzias latipes* (Medaka)		
CYP2AD5	*Gasterosteus aculeatus* (Stickleback)	CYP8A1	*Danio rerio* (Zebrafish)
		CYP8A1,2	*Takifugu rubripes* (Japanese pufferfish)
CYP3A27	*Oncorhynchus mykiss* (Trout)		
CYP3A30	*Fundulus heteroclitus* (Killifish)	CYP8B1,2	*Takifugu rubripes* (Japanese pufferfish)
CYP3A38,40	*Oryzias latipes* (Medaka)	CYP8B1,2,3	*Danio rerio* (Zebrafish)
CYP3A45	*Oncorhynchus mykiss* (Trout)		
CYP3A48,49,50	*Takifugu rubripes* (Japanese pufferfish)	CYP11A1	*Oncorhynchus mykiss* (Trout)
CYP3A48,49,50	*Takifugu rubripes* (Japanese pufferfish)	CYP11A1	*Oncorhynchus mykiss* (Trout)
CYP3A56	*Fundulus heteroclitus* (Killifish)		*Takifugu rubripes* (Japanese pufferfish)

Table 2. (Continued).

P450 Gene	Species (common name)	P450 Gene	Species (common name)
CYP3A65	*Danio rerio* (Zebrafish)		*Dasyatis americana* (Southern stingray)
CYP3A68,69	*Micropterus salmoides* (Largemouth bass)	CYP11A1,2	*Danio rerio* (Zebrafish)
CYP3B1,2	*Fundulus heteroclitus* (Killifish)	CYP11B1	*Danio rerio* (Zebrafish)
CYP3C1	*Danio rerio* (Zebrafish)	CYP17	*Oncorhynchus mykiss* (Trout)
CYP3D1	*Takifugu rubripes* (Japanese pufferfish)		*Oryzias latipes* (Medaka)
			Squalus acanthias (Dogfish)
CYP4F7	*Dicentrarchus labrax* (European sea bass)		*Ictalurus punctatus* (Catfish)
CYP4F28	*Takifugu rubripes* (Japanese pufferfish)	CYP17A1,2	*Takifugu rubripes* (Japanese pufferfish)
CYP4F43	*Danio rerio* (Zebrafish)		*Danio rerio* (Zebrafish)
CYP4T1	*Oncorhynchus mykiss* (Trout)	CYP19	*Ictalurus punctatus* (Catfish)
CYP4T2	*Dicentrarchus labrax* (European sea bass)		*Oncorhynchus mykiss* (Trout)
CYP4T5	*Takifugu rubripes* (Japanese pufferfish)		*Carassius auratus* (Goldfish)
CYP4T8	*Danio rerio* (Zebrafish)		*Oryzias latipes* (Medaka)
			Tilapia nilotica (Tilapia)
CYP4V6,7	*Danio rerio* (Zebrafish)		*Haplochromis burtoni* (Astatotilapia)
			Paralichthys olivaceus (Japanese flounder)
CYP5A1	*Danio rerio* (Zebrafish)	CYP19A1,2	*Takifugu rubripes* (Japanese pufferfish)
CYP7A1	*Danio rerio* (Zebrafish)		*Danio rerio* (Zebrafish)
	Takifugu rubripes (Japanese pufferfish)		
CYP20	*Danio rerio* (Zebrafish)	CYP27A1,2,3	*Takifugu rubripes* (Japanese pufferfish)
	Takifugu rubripes (Japanese pufferfish)	CYP27A.a.b.c.d	*Danio rerio* (Zebrafish)
	Tetraodon nigroviridis (Japanese pufferfish)		
	Salmo salar (Atlantic salmon)	CYP27B1	*Danio rerio* (Zebrafish)
			Takifugu rubripes (Japanese pufferfish)

CYP21	*Takifugu rubripes* (Japanese pufferfish)		
	Danio rerio (Zebrafish)	CYP27C1	*Takifugu rubripes* (Japanese pufferfish)
	Tetraodon nigroviridis (Japanese pufferfish)		*Danio rerio* (Zebrafish)
CYP24	*Takifugu rubripes* (Japanese pufferfish)	CYP39	*Danio rerio* (Zebrafish)
	Danio rerio (Zebrafish)		
		CYP46A1,2	*Danio rerio* (Zebrafish)
CYP26A1	*Takifugu rubripes* (Japanese pufferfish)	CYP46A1,2,3	*Takifugu rubripes* (Japanese pufferfish)
	Danio rerio (Zebrafish)		
		CYP51A1	*Takifugu rubripes* (Japanese pufferfish)
CYP26B1	*Takifugu rubripes* (Japanese pufferfish)		*Danio rerio* (Zebrafish)
	Danio rerio (Zebrafish)		

[a] Denotes presence of CYP1A in additional species without subfamily designation.

Similar to the situation of the extensive CYP4 family in insects and other invertebrates, the CYP2 family appears to be the largest CYP family in fishes. There are twelve CYP2 subfamilies currently described in fish. Fish have four CYP3 subfamilies, and three CYP4 subfamilies. The following additional CYP families have been identified in fish to date: CYP5, CYP7, CYP8, CYP11, CYP17, CYP19, VYP20, CYP21, CYP24, CYP26, CYP27, CYP39, CYP46, and CYP51. Seventeen of eighteen mammalian CYP families are found in fish, as described for the pufferfish *T. rubripes* [102]. The only mammalian CYP family absent in fish appears to be CYP39.

4.3. Mammal P450 Families

Marine mammal CYP genes thus far identified belong to only four families, CYP1, CYP2, CYP3, and CYP4 (Table 3). Many of these sequences have been added by one research group analyzing a cDNA library from Minke whale liver, *Balaenoptera acutorostrata* [103]. It should be expected that the marine mammal CYP families will be very similar to that found in terrestrial mammals. The total number could be in the range of that found in mouse (102 genes), fish (81 genes), and human (57 genes).

Table 3. Cytochrome P450 genes identified in marine mammals.

P450 Gene	Species (common name)	P450 Gene	Species (common name)
CYP1A1	*Balaenoptera acutorostrata* (Minke whale)	CYP2C	*Balaenoptera acutorostrata* (Minke whale)
	Phocoenoides dalli (Dall's porpoise)		
	Eumetopias jubatus (Stellar sealion)	CYP2E	*Balaenoptera acutorostrata* (Minke whale)
	Phoca largha (Spotted seal)		
	Phoca fasciata (Ribbon seal)		
	Phoca groenlandica (Harp seal)	CYP3A32	*Balaenoptera acutorostrata* (Minke whale)
	Halichoerus grypus (Grey seal)	CYP3A33	*Phocoenoides dalli* (Dall's porpoise)
		CYP3A34	*Eumetopias jubatus* (Stellar sea lion)
CYP1A2	*Eumetopias jubatus* (Stellar sealion)	CYP3A35	*Phoca largha* (Spotted seal)
	Phoca fasciata (Ribbon seal)	CYP3A36	*Phoca fasciata* (Ribbon seal)
	Phoca groenlandica (Harp seal)		
	Halichoerus grypus (Grey seal)	CYP4A	*Balaenoptera acutorostrata* (Minke whale)
CYP1B1	*Stenella coeruleoalba* (Stripped dolphin)		

5. HOW CAN WE USE INFORMATION ABOUT P450s IN AQUATIC SPECIES?

5.1. Invertebrate Species Physiology and Ecology

Evidence that changes in P450 enzyme activities and/or gene expression are important in the survival of marine invertebrates to xenobiotics is not definitively supported by the current literature. Many of the original problems in the interpretation of detoxification enzyme changes following xenobiotic exposures for insect P450 enzymes [104] are similar to those involving marine invertebrate enzymes. We now have data on P450 enzyme, and/or molecular data on gene and protein level, and activity changes in different species with exposure to various xenobiotics. Most studies use model P450 enzyme substrates with *in vitro* microsomal assays following exposure of the organism to a different xenobiotic. The question still remains about what can be interpreted about other types of enzymatic and/or gene activity alterations that are presumably not directed against the inducer itself?

Still required to answer these points in aquatic organisms are the following:

1. Clear evidence is still needed to show that P450 enzyme induction in aquatic invertebrates results in the increased metabolism of the inducer, a point solvable by heterologous expression of specifically induced P450 cDNAs and the study of their particular *in vitro* metabolic capabilities. For example, this approach has been successful with various members of the insect CYP6 family (for example, [105–107]). With the new progress in the identification of marine invertebrate CYP genes, expression and metabolic characterization of these enzymes should be the next steps.
2. There remains a paucity of information on natural variations in particular CYP gene activities with life stage, nutritional state, reproduction, etc. It has been fairly simple to demonstrate changes in CYP protein activity and expression following exposure to xenobiotics, but what roles do various P450 enzymes play in normal metabolism, growth, and reproduction for example? At present, there are few such studies among the vast number of different aquatic invertebrate phyla.

Recent discussions have appeared in the literature relating current molecular advances in our knowledge of insect detoxification enzymes, such as P450s, to real-world problems. There is a significant amount of evidence for specialist-herbivore insect P450s acting as key innovations that allow increased resource utilization, i.e., increased exploitation efficiency on toxic host plants [108,109]. Conservation of particular amino acids within the hypothesized substrate recognition sites for furanocoumarin-metabolizing P450s of the CYP6B family are thought to be key features in the evolution of members of papilioinid caterpillars [109,110].

Aquatic organism responses to natural dietary chemicals from different flora and fauna have also been reviewed [111,112]. A wide variety of compounds are known to be feeding deterrents and toxicants to various marine vertebrate and invertebrate consumers. What role(s) do the various detoxification mechanisms play, such as cytochrome P450s, in supporting the ability of particular consumers to metabolize and excrete toxic marine compounds from their diets? Specific P450 activities are critical for many specialist-feeding herbivorous insects in allowing them to feed on plants containing compounds toxic to other species [108,113]. It is likely that similar detoxification mechanisms will be found in specialist feeders among the huge number of marine invertebrate species. This area is ripe for new investigations.

5.2. Environmental Monitoring/Ecosystem Health

Most of the work in the field of aquatic organism cytochrome P450 has related to hierarchy of organism to ecosystem health. The focus of the clear majority of this work in the past 30 years was either single-species exposure to one or multiple chemicals in the laboratory or multiple individuals sampled from one or more 'contaminated' versus control field sites. Tissues were then routinely sampled (usually fish livers or invertebrate digestive tissues such as crustacean hepatopancreas or mollusc digestive gland) and extracted for CYP enzyme assay, or immunoblotting or ELISA using specific CYP antisera. Since the advent of molecular biology techniques, CYP cDNAs and/or genomic clones have been isolated and sequenced. Probes were then developed for gene expression analyses by Northern blotting or more recently, real-time PCR for more accurate mRNA level determinations. These are the current tools for the measurement of alterations in CYP enzyme activity and gene expression under different scenarios. Soon gene microarrays and proteomics will become easier and cheaper to perform, with the hope that complete protein response profiles will provide quick and definitive information about the state of organism/species/ecosystem health.

The potential development of an array of biomarkers, including CYPs, will provide the maximum amount of organism health information when coupled to cellular and whole-organism responses. An example of cellular responses would be genotoxicity tests such as micronucleus and Comet assays [114,115]. For example, whole-organism responses have been examined coincident to a set of biomarker assays in transplanted mussels [116]. Tissue lesions and parasite load were correlated with PAH and PCB levels in the contaminated sites and with tissue levels of heat-shock proteins, acetylcholin esterase, CYP enzyme activity, P-glycoprotein, metallothioneins, and DNA adducts. The 'medical' approach is similar to that used by physicians in treated humans and we should adopt such extended informational profiles to aid in the diagnosis of organism and ultimately, ecosystem health.

Although xenobiotic resistance is discussed in detail in Chapter 15, this topic is germane to the aquatic environment and several points need to be brought out here. It has become widely accepted that high levels of P450 activities are associated with insect herbivores feeding on plants that contain high concentrations of toxic secondary metabolites [109]. Elevated expression of P450s belonging to the CYP6 family in both *Drosophila* and *Musca domestica* (housefly) is also associated with metabolic resistance to certain insecticides. The expression of insect CYP6A1 and CYP6A2 is regulated by *trans*-acting factors (from other chromosomes). Mutation(s) in this factor(s) are thought to result in resistance to both organophosphate and carbamate insecticides [117].

Rapid insect resistance to pesticides has involved alterations in the regulation of detoxification gene expression (e.g., P450, glutathione S-transferase), amplification of gene number (e.g., glutathione S-transferase, esterase, P-glycoprotein), and gene mutations (e.g., acetylcholine esterase, *Bacillus thuringiensis* toxin receptor) [118]. Resistant strains of *Drosophila* can exhibit 50- to 100-fold higher basal cytochrome P450 enzyme activities and gene expression patterns versus controls [98,119]. *Drosophila* CYP4E2 is overexpressed several fold in DDT-resistant strains [120], and housefly CYP6A1 and CYP6D1 are ten-fold overexpressed by insecticide-resistant strains [121,122]. CYP6A2 constitutive expression is nearly absent versus highly expressed in underproducer and overproducer *Drosophila* strains [123]. Lowered expression appears regulated by mutated negative regulatory elements that lead to overexpression of particular CYP genes [117,123]. Daborn et al. [98] have demonstrated at a molecular level the first example of a mutation in a P450 gene that results in insecticide resistance. The high level of DDT resistance found in many strains of *Drosophila melanogaster* resulted from 100-fold upregulation of just a single specific P450 enzyme (CYP6G1). The root cause is the insertion of a transposable element into the CYP6G1 promoter. We know so much about the spread of xenobiotic resistance in terrestrial insects, yet nothing at all is known about resistance in aquatic species. Indeed, judging by the incredible increases in the cases of insecticide resistance worldwide [124], and the extent of watersheds and oceans as final dumping grounds for these chemicals, such resistance mechanisms must be in play in aquatic species.

5.3. Potential Monitoring of Toxicants Present in Food

Bivalve molluscs are the only aquatic food source constantly monitored for the presence of toxicants, and this occurs most frequently during the Summer months when algal blooms are more common. Mussels, oysters, and scallops retain high levels of algal toxins (for example, red-tide saxitoxin). Toxins are monitored by a sampling program involving animal exposures, and results are reported as numbers of mouse units [125].

Organohalogenated chemicals such as PCBs, insecticides, and flame retardants are found in seafood consumed worldwide [126]. Corresponding levels of these contaminants can be measured in human consumers at elevated levels in some ethnic groups [127]. The persistence of such chemicals in fish has been documented decades following cessation of use [128].

As alternatives to chemical analyses, biomarker measurements could be used to assess aquatic food contaminant levels. Genotoxicity measures using Comet and micronucleus assays have been suggested to monitor aquatic food levels of PAHs when fed to rats [115]. The potential use of particular CYP measurements of selected aquatic food species could provide additional useful monitoring tools to certify the safety of foods prior to human consumption. The availability of sufficient information on fish P450s is approaching a level where CYP monitoring could be considered for this purpose.

6. CONCLUSIONS AND OUTLOOK

This review has discussed advances in the P450 involving aquatic animals, with the focus on marine invertebrates. Many of the advances occurred due to advances in molecular biology, especially RT-PCR procedures. This approach circumvents many of the problems with P450 enzyme measurements and biochemical purifications in aquatic invertebrates. A series of genome sequencing projects for aquatic organisms is ongoing and will lead to definitive identification of all the CYP genes and pseudogenes in each type of aquatic organism. There is much promise in extending molecular approaches to other types of detoxification enzymes. Among the most promising are the glutathione S-transferase and esterase enzyme families, both of which are multigenic in many organisms (for example, [129]). Similar to the P450 field, relatively little is known about the numbers and types of glutathione S-transferases, esterases, hydrolases, and other detoxification genes in marine invertebrates.

Human gene polymorphisms, especially in CYP2 family members, are known to result in tremendous differences in an individual's xenobiotic metabolism abilities [130,131]. Analyses of human CYP2D6 variation in the European population resulted in the finding of almost 50 mutations that could be grouped according to the metabolic abilities of each single allele [132]. We should expect such differences in aquatic species (especially invertebrates), and primary evidence in support of this notion is derived from the many well-known cases of insecticide resistance in laboratory and field insect populations. One possible consequence of such variation is that xenobiotic inputs may select for individuals with the specific detoxification pathways to survive and reproduce under the selective pressure(s) of one or more anthropogenic chemicals. This could result in the loss of genetic variation within a population, or perhaps, in the loss of an entire population or species from a region with significant xenobiotic contamination(s). For example, the advent of molecular genetic approaches to examine such variations may have useful impacts for the analyses of mussel CYP genes as an informative part of a 'mussel watch' program, on a local, regional or even worldwide scale.

The diversity of life under the surface is also reflected in the diversity of relationships between species. What do we know of the interactions between microbial and algal products and to P450 family evolution in related consumers? The plethora of pigments, offensive and defensive compounds, pheromones, and yet-to-be-discovered metabolic pathways in diverse aquatic and marine species will probably be reflected in equally diverse P450 and other detoxification enzyme functions. Is the 'aquatic warfare' versus 'plant warfare' hypotheses comparison valid in the aquatic world? Sponge, tunicate and hemichordate DNA arrays and/or sequenced genomes may help provide the suite of identified detoxification genes needed to examine these questions. Novel P450 families and P450-specific chemical reactions will emerge as new species and species interrelationships are identified. Only investigator disinterest and the ever-shrinking availability of grant funding limit these endeavors.

This review has included a history of CYP research in aquatic animals, with the emphasis on invertebrates. There is much to learn, and the following is a brief list of interesting and perhaps critical points/questions to answer for those working in this field:

- Deciphering what the magnitudes of changes in biomarkers actually means to individual cells to whole organisms. What algorithms/models will be developed to integrate this information?
- How can we tease out the most informative biomarkers in the most informative 'indicator species' in the most critical environments? What P450's will be useful?
- Does knowledge of model organism P450 translate into useful information for aquatic species? What are the relationships between cell signaling pathways and alterations in P450 expression? What intracellular factors affect individual P450 activities within cells?

ACKNOWLEDGMENTS

This work was supported over the years with grants from the National Science Foundation, the National Sea Grant College Program, the US Environmental Protection Agency, and the US Department of Agriculture. I wish to thank the following for their scientific expertise that helped mold my interests in P450 research: Dr R. Feyereisen, Dr M. Cohen, Dr J. Andersen, and Dr G. Cherr. I also thank Dr J. Clegg who encouraged me to work at the Bodega Marine Laboratory and Dr E. S. Chang who provided laboratory space to pursue aquatic P450 research.

ABBREVIATIONS

AH	aromatic hydrocarbon
BaP	benzo[a]pyrene
βNF	β-naphthoflavone
CYP	cytochrome P450
DDT	dichlorodiphenyltrichloroethane
ECOD	ethoxycoumarin O-deethylase
ELISA	enzyme-linked immunosorbant essay
EROD	ethoxyresorufin O-deethylase
LAH	laurate hydroxylase
3-MC	3-methylcholanthrene

NADH	nicotinamide adenine dinucleotide, reduced
NADPH	nicotinamide adenine dinucleotide phosphate, reduced
NCBI	National Center for Biotechnology Information
PAH	polyaromatic hydrocarbon
PB	phenobarbital
PCB	polychlorinated biphenyl
RT-PCR	real-time polymerase chain reaction
TBT	tributyltin

REFERENCES

1. (a) D. R. Nelson, *Comp. Biochem. Physiol. 121C*, 15–22 (1998); (b) web page: http://drnelson.utmem.edu/CytochromeP450.html
2. D. R. Nelson and H. W. Strobel, *Mol. Biol. Evol.*, *4*, 572–593 (1987).
3. T. Omura and R. Sato, *J. Biol. Chem.*, *239*, 2370–2379 (1964).
4. F. J. Gonzalez, *Pharmacol. Rev.*, *40*, 243–288 (1989).
5. J.-Y. Hong and C. S. Yang, *Environ. Health Perspect.*, *105*, 759–762 (1997).
6. D. W. Nebert, M. Adesnik, M. J. Coon, R. W. Eastabrook, F. J. Gonzalez, F. P. Guengerich, I. C. Gunsalus, E. F. Johnson, B. Kemper, W. Levin, I. R. Phillips, I. Sato, and M. R. Waterman, *DNA*, *6*, 1–11 (1987).
7. D. R. Nelson, D. C. Zeldin, S. M. G. Hoffman, L. J. Matais, H. M. Wain, and D. W. Nebert, *Pharmacogenetics*, *14*, 1–18 (2004).
8. F. J. Gonzalez and D. W. Nebert, *Trends Gen.*, *6*, 182–186 (1990).
9. J. J. Stegeman and P. J. Kloepper-Sams, *Environ. Health Perspect.*, *71*, 87–95 (1987).
10. A. Goksoyr, *Sci. Total Environ.*, *101*, 255–262 (1991).
11. A. Goksoyr and L. Forlin, *Aquat. Toxicol.*, *22*, 287–312 (1992).
12. M. A. Q. Khan, A. Kamal, R. J. Wolin, and J. Runnels, *Bull. Environ. Contam. Toxicol.*, *8*, 219–228 (1972).
13. M. O. James, *Xenobiotica*, *19*, 1063–1076 (1989).
14. D. R. Livingstone, M. A. Kirchin, and A. Wiseman, *Xenobiotica*, *19*, 1041–1042 (1989).
15. D. R. Livingstone, in *Advances in Comparative Environmental Physiology* (R. Gilles, ed.), Vol. 7, Springer-Verlag, Berlin, 1991, pp. 46–185.
16. M. O. James, and S. M. Boyle, *Comp. Biochem. Physiol.*, *121C*, 157–172 (1998).
17. M. J. Snyder, *Aquat. Toxicol.*, *48*, 529–547 (2000).
18. P. J. den Besten, H. J. Herwig, E. G. van Donsalaar, and D. R. Livingstone, *Mar. Biol.*, *107*, 171–177 (1990).
19. J. F. Payne, *Mar. Pollut. Bull.*, *5*, 112–116 (1977).
20. S. C. Singer, P. E. Marsh, F. Gonsoulin, and R. Lee, *Biol. Bull.*, *153*, 377–386 (1980).
21. M. O. James and P. J. Little, *Comp. Biochem. Physiol.*, *78C*, 241–245 (1984).
22. M. O. James, *Arch. Biochem. Biophys.*, *282*, 8–17 (1990).
23. S. M. Valles and S. J. Yu, *J. Econ. Entomol.*, *89*, 1508–1512 (1996).

24. P. J. den Besten, S. C. M. O'Hara, and D. R. Livingstone, *Mar. Environ. Res., 34*, 309–313 (1992).
25. P. J. den Besten, P. Lemaire, and D. R. Livingstone, *Xenobiotica, 24*, 989–1001 (1994).
26. J. J. Stegeman, *Mar. Biol., 89*, 21–30 (1985).
27. G. W. Winston, M. H. Mayeaux, and L. M. Heffernan, *Mar. Environ. Res., 45*, 89–100 (1998).
28. J. J. Stegeman and H. B. Kaplan, *Comp. Biochem. Physiol., 68C*, 55–61 (1981).
29. M. Solé and D. R. Livingstone, *Comp. Biochem. Physiol. C, 141*, 20–31 (2005).
30. C. S. E. Jewell and G. W. Winston, *Comp. Biochem. Physiol., 92B*, 329–339 (1989).
31. L. M. Heffernan and G. W. Winston, *Comp. Biochem. Physiol., 121C*, 371–383 (1998).
32. C. S. E. Jewell, M. H. Mayeaux, and G. W. Winston, *Comp. Biochem. Physiol., 118C*, 369–374 (1997).
33. D. R. Livingstone, P. Garcia-Martinez, J. J. Stegeman, and G. W. Winston, *Biochem. Soc. Trans., 16*, 779 (1988).
34. N. J. Gassman and C. J. Kennedy, *Bull. Mar. Sci., 50*, 320–330 (1992).
35. L. M. Heffernan, R. P. Ertl, J. J. Stegeman, D. R. Buhler, and G. W. Winston, *Mar. Environ. Res., 45*, 353–357 (1996).
36. V. E. Forbes, T. L. Forbes, and M. Holmer, *Mar. Ecol. Prog. Ser., 132*, 63–70 (1996).
37. V. E. Forbes, M. S. Andreassen, and L. Christensen, *Environ. Toxicol. Chem., 20*, 1012–1021 (2001).
38. B. Li, H. C. Bisgaard, and V. E. Forbes, *Biochem. Biophys. Res. Commun., 325*, 510–517 (2004).
39. K. F. Rewitz, C. Kjellerup, A. Jorgennsen, C. Petersen, and O. Andersen, *Comp. Biochem. Physiol. C, 138*, 89–96 (2004).
40. A. Jorgensen, L. J. Rasmussen, and O. Andersen, *Biochem. Biophys. Res. Commun., 336*, 890–897 (2005).
41. W. Baturo and L. Lagadic, *Environ. Toxicol. Chem., 15*, 771–781 (1996).
42. X. Michel, J.-P. Salaun, F. Galgani, and J.-F. Narbonne, *Mar. Environ. Res., 38*, 257–273 (1994).
43. D. R. Livingstone, C. Nasci, M. Solé, L. Da Ros, S. C. M. O'Hara, L. D. Peters, V. Fossato, A. N. Wootton, and P. S. Goldfarb, *Aquat. Toxicol., 38*, 205–224 (1997).
44. J. P. Shaw, L. D. Peters, and J. K. Chipman, *Mar. Environ. Res., 58*, 649–653 (2004).
45. C. L. Mitchelmore, C. Birmelin, J. K. Chipman, and D. R. Livingstone, *Aquat. Toxicol., 41*, 193–212 (1998).
46. D. M. Y. Cheah, P. F. A. Wright, D. A. Holdway, and J. T. Ahokas, *Aquat. Toxicol., 33*, 201–214 (1995).
47. A. Galli, D. Del Chiero, R. Nieri, and G. Bronzetti, *Mar. Biol., 100*, 69–73 (1988).
48. M. Osada, H. Tawarayama, and K. Mori, *Comp. Biochem. Physiol. B, 139*, 123–128 (2004).
49. M. J. J. Ronis and A. Z. Mason, *Mar. Environ. Res., 42*, 161–166 (1996).
50. Y. Morcillo and C. Porte, *Trends Anal. Chem., 17*, 109–116 (1998).

51. E. Oberdörster, D. Rittscoff, and P. McClellan-Green, *Mar. Pollut. Bull.*, *36*, 144–151 (1998).
52. J. Oehlmann, E. Stroben, U. Schulte-Oehlmann, B. Bauer, P. Fioroni, and B. Markert, *Fresenius J. Anal. Chem.*, *354*, 540–545 (1996).
53. M. M. Santos, L. F. Castro, M. N. Viera, J. Micael, R. Morabito, P. Massanisso, and M. A. Reis-Henriques, *Comp. Biochem. Physiol. C*, *141*, 101–109 (2005).
54. A. N. Wootton, C. Herring, A. J. Spry, A. Wiseman, D. R. Livingstone, and P. S. Goldfarb, *Mar. Environ. Res.*, *39*, 21–26 (1995).
55. A. N. Wootton, P. S. Goldfarb, P. Lemaire, S. C. M. O'Hara, and D. R. Livingstone, *Mar. Environ. Res.*, *42*, 1–4 (1996).
56. M. A. Kirchin, A. Wiseman, and D. R. Livingstone, *Comp. Biochem. Physiol.*, *101C*, 81–91 (1992).
57. M. Solé, C. Porte, and J. Albaigés, Environ. *Toxicol. Chem.* *14*, 157–164 (1995).
58. C. Porte, P. Lemaire, L. D. Peters, and D. R. Livingstone, *Mar. Environ. Res.*, *39*, 27–31 (1995).
59. S. Chaty, F. Rodius, and P. Vasseur, *Aquat. Toxicol.*, *69*, 81–94 (2004).
60. M. J. Snyder, *Biochem. Biophys. Res. Commun.*, *249*, 187–190 (1998).
61. M. J. Snyder, E. Girvetz, and E. P. Mulder, *Arch. Environ. Contam. Toxicol.*, *41*, 22–29 (2001).
62. D. Schlenk and D. R. Buhler, *Biochem. Biophys. Res. Commun.*, *163*, 476–480 (1989).
63. A. E. Lockyer, L. R. Noble, D. Rollinson, and C. S. Jones, *Mem. Inst. Oswaldo Cruz*, *100*, 259–262 (2005).
64. M. C. Fossi, C. Savelli, and S. Casini, *Comp. Biochem. Physiol.*, *121C*, 321–331 (1998).
65. C. Porte and E. Escartin, *Comp. Biochem. Physiol. 121C*, 333–338 (1998).
66. E. Oberdörster, M. Martin, C. F. Ide, and J. A. McLachlan, *Arch. Environ. Contam. Toxicol.*, *37*, 512–518 (1999).
67. C. M. Ashley, M. G. Simpson, D. M. Holdich, and D. R. Bell, *Aquat. Toxicol.*, *35*, 157–169 (1996).
68. E. Oberdörster, D. Rittscoff, and P. McClellan-Green, *Aquat. Toxicol.*, *41*, 83–100 (1998).
69. W. S. Baldwin and G. A. LeBlanc, *Environ. Toxicol. Chem.*, *13*, 1013–1021 (1994).
70. W. S. Baldwin, D. L. Milam, and G. A. LeBlanc, *Environ. Toxicol. Chem.*, *14*, 945–952 (1995).
71. W. S. Baldwin, S. E. Graham, D. Shea, and G. A. LeBlanc, *Ecotoxicol. Environ. Safety*, *39*, 104–111 (1998).
72. Y. Teunissen, W. P. M. Geraerts, H. Van Heerikhuizen, R. J. Planta, and J. Joosse, *J. Biochem.*, *112*, 249–252 (1992).
73. M. O. James, S. M. Boyle, H. G. Trapido-Rosenthal, W. C. Smith, R. M. Greenberg, and K. T. Shiverick, *Arch. Biochem. Biophys.*, *329*, 31–38 (1996).
74. M. O. James and K. T. Shiverick, *Arch. Biochem. Biophys.*, *233*, 1–9 (1984).
75. S. M. Boyle, M. P. Popp, C. W. Smith, R. M. Greenberg, and M. O. James, *Mar. Environ. Res.*, *46*, 25–28 (1998).
76. S. M. Boyle and M. O. James, *Mar. Environ. Res.*, *42*, 1–6 (1996).
77. J. F. Brown, *Mar. Environ. Res.*, *34*, 261–266 (1992).

78. M. J. Snyder, *Arch. Biochem. Biophys., 358*, 271–276 (1998).
79. M. J. Snyder and E. S. Chang, *Gen. Comp. Endocrinol., 81*, 133–145 (1991).
80. M. J. Snyder and E. S. Chang, *Gen. Comp. Endocrinol., 83*, 118–131 (1991).
81. M. J. Snyder and E. S. Chang, *Gen. Comp. Endocrinol., 85*, 286–296 (1992).
82. K. F. Rewitz, B. Styrishave, and O. Andersen, *Biochem. Biophys. Res. Commun., 310*, 252–260 (2003).
83. M. J. Snyder, J. A. Scott, J. F. Andersen, and R. Feyereisen, *Meth. Enzymol., 272*, 304–312 (1996).
84. M. J. Snyder and E. P. Mulder, *Aquat. Toxicol., 55*, 177–190 (2001).
85. P. David, C. Dauphin-Villemant, A. Mesneau, and J. C. Meyran, *Mol. Ecol., 12*, 2473–2480 (2003).
86. S. Aragon, S. Claudinot, C. Blais, M. Maibeche, and C. Dauphin-Villemant, *Insect Biochem. Mol. Biol., 32*, 153–159 (2002).
87. P. J. den Besten, J. M. L. Elenbaas, J. R. Maas, S. J. Dielman, H. J. Herwig, and P. A. Voogt, *Aquat. Toxicol., 20*, 95–110 (1991).
88. P. J. den Besten, J. R. Maas, D. I. Zandee, P. A. Voogt, and D. R. Livingstone, *Comp. Biochem. Physiol., 110C*, 165–168 (1991).
89. P. J. den Besten, P. Lemaire, D. R. Livingstone, B. Woodin, J. J. Stegeman, H. J. Herwig, and W. Seinen, *Aquat. Toxicol., 26*, 23–40 (1993).
90. B. Danis, S. Goriely, P. Dubois, S. W. Fowler, V. Flamand, and M. Warnau, *Aquat. Toxicol., 69*, 371–383 (2004).
91. P. J. den Besten, *Comp. Biochem. Physiol., 121C*, 139–146 (1998).
92. S. M. Boyle, R. M. Greenberg, and M. O. James, *Mar. Environ. Res., 46*, 57–60 (1998).
93. J. A. Scott, F. H. Collins, and R. Feyereisen, *Biochem. Biophys. Res. Commun., 205*, 1452–1459 (1994).
94. R. Feyereisen, *Annu. Rev. Entomol., 44*, 507–533 (1999).
95. N. Tijet, C. Helvig, and R. Feyereisen, *Gene, 262*, 189–198 (2001).
96. D. J. Brown, G. C. Clark, and R. J. Van Beneden, *Comp. Biochem. Physiol., 121C*, 351–360 (1998).
97. A. G. McArthur, T. Hegelund, R. L. Cox, J. J. Stegeman, M. Liljenberg, U. Olsson, P. Sundberg, and M. C. Celander, *J. Mol. Evol., 57*, 200–211 (2003).
98. P. J. Daborn, J. L. Yen, M. R. Bogwitz, G. Le Goff, E. Feil, S. Jeffers, N. Tijet, T. Perry, D. Heckel, P. Batterham, R. Feyereisen, T. G. Wilson, and R. H. ffrench-Constant, *Science, 297*, 2253–2256 (2002).
99. A. Petryk, J. T. Warren, G. Marqués, M. P. Jarcho, L. I. Gilbert, J. Kahler, J.-P. Parvy, Y. Li, C. Dauphin-Villemant, and M. B. O'Connor, *Proc. Natl. Acad. Sci. USA, 100*, 13773–13778 (2003).
100. M. J. Snyder and H. V. S. Peeke, *Neurosci. Lett., 313*, 65–68 (2001).
101. M. R. Bogwitz, H. Chung, L. Magoc, S. Rigby, W. Wong, M. O'Keefe, J. A. McKenzie, P. Batterham, and P. J. Daborn, *Proc. Natl. Acad. Sci. USA, 102*, 12807–12812 (2005).
102. D. R. Nelson, *Arch. Biochem. Biophys., 409*, 18–24 (2003).
103. E. Y. Kim, H. Iwata, Y. Fujise, and S. Tanabe, *Mar. Environ. Res., 58*, 495–498 (2004).
104. F. Gould, *Ecol. Entomol., 9*, 29–34 (1984).

105. J. F. Andersen, J. G. Utermohlen, and R. Feyereisen, *Biochemistry, 33*, 2171–2177 (1994).
106. R. Ma, M. R. Berenbaum, and M. A. Schuler, *Arch. Biochem. Biophys., 310*, 332–340 (1994).
107. B. C. Dunkov, V. M. Guzov, G. Mocelin, F. Shotkoski, A. Brun, M. Amichot, R. H. ffrench-Constant, and R. Feyereisen, *DNA Cell Biol., 16*, 1345–1356 (1997).
108. M. R. Berenbaum, C. Favret, and M. A. Schuler, *Am. Nat. Suppl., 148*, S139–S155 (1996).
109. M. R. Berenbaum, *J. Chem. Ecol., 28*, 873–896 (2002).
110. X. Li, J. Baudry, M. R. Berenbaum, and M. A. Schuler, *Proc. Natl. Acad. Sci. USA, 101*, 2939–2944 (2004).
111. M. E. Hay, in *Ecological Roles of Marine Natural Products* (V. J. Paul, ed.), Comstock Publishing Associates, Ithaca, NY, 1992, pp. 93–118.
112. J. R. Pawlik, *Chem. Rev., 93*, 1911–1922 (1993).
113. M. J. Snyder, and J. I. Glendinning, *J. Comp. Physiol. Part A, 179*, 255–261 (1996).
114. K. Saotome, and M. Hayashi, *Mutagenesis, 18*, 73–76 (2003).
115. S. Lemiere, C. Cossu-Leguille, A. Bispo, M.-J. Jourdain, M.-C. Lanhers, D. Burnel, and P. Vasseur, *Environ. Toxicol. 19*, 387–395 (2004).
116. N. Bodin, T. Burgeot, J. Y. Stanisiere, G. Bocquene, D. Menard, C. Minier, I. Boutet, A. Amat, Y. Cherel, and H. Budzinski, *Comp. Biochem. Physiol. C, 138*, 411–427 (2004).
117. R. Feyereisen, *Toxicol. Lett., 82:83*, 83–90 (1995).
118. R. H. French-Constant, P. J. Daborn, and G. Le Goff, Trends Gen., 20, 163–170 (2004).
119. M. Amichot, A. Brun, A. Cuany, G. De Souza, T. Le Mouel, J. M. Bride, M. Babaoult, J. P. Salaun, R. Rahmani, and J. B. Berge, *Comp. Biochem. Physiol., 121C*, 311–319 (1998).
120. M. Amichot, A. Brun, A. Cuany, C. Helvig, J. P. Salaun, F. Durst, and J. B. Bergé, in *Cytochrome P450 8th International Conference* (M. C. Lechner, ed.). John Libbey Eurotext, Paris, 1994, pp. 689–692.
121. F. A. Cariño, J. F. Koerner, F. W. Plapp, and R. Feyereisen, *Insect Biochem. Molec. Biol., 24*, 411–418 (1994).
122. J. G. Scott, *J. Pesticide Sci., 21*, 241–245 (1996).
123. S. M. Dombrowski, R. Krishnan, M. Witte, S. Maitra, C. Diesing, L. C. Waters, and R. Ganguly, *Gene, 221*, 69–77 (1998).
124. I. Denholm, G. J. Levine, and M. S. Williamson, *Science, 297*, 2222–2223 (2002).
125. J. J. Sullivan, M. G. Simon, and W. T. Iwaoka, *J. Food Sci., 48*, 1312–1314 (1983).
126. A. G. Smith and S. D. Gangolli, *Food Chem. Toxicol., 40*, 767–779 (2002).
127. D. C. Cole, J. Sheeshka, E. J. Murkin, J. Kearney, F. Scott, L. A. Ferron, and J. P. Weber, *Arch. Environ. Health, 57*, 494–509 (2002).
128. D. R. Luellen, G. G. Vadas, and M. A. Unger, *Sci. Total Environ.*, in press (2006)
129. M. J. Snyder and D. R. Maddison, *DNA Cell Biol., 16*, 1373–1384 (1997).
130. H. K. Kroemer and M. Eichelbaum, *Life Sci., 56*, 2285–2298 (1995).
131. J.-Y. Hong and C. S. Yang, *Environ. Health Perspect., 105*, 759–762 (1997).
132. D. Marez, M. Legrand, N. Sabbagh, J.-M. Lo Guidice, C. Spire, J.-J. Lafitte, U. A. Meyer, and F. Broly, *Pharmacogenetics, 7*, 193–202 (1997).

5

The Electrochemistry of Cytochrome P450

Alan M. Bond, Barry D. Fleming, and Lisandra L. Martin

School of Chemistry, Monash University, Clayton, Victoria 3800, Australia
<alan.bond@sci.monash.edu.au>
<barry.fleming@sci.monash.edu.au>
<lisa.martin@sci.monash.edu.au>

1. INTRODUCTION		127
2. REDOX TITRATION (POTENTIOMETRIC EQUILIBRIUM) MEASUREMENTS		131
2.1. Methods		131
2.2. Discussion of Potentiometrically Determined Reversible Potentials		133
3. VOLTAMMETRIC (DYNAMIC) MEASUREMENTS		139
3.1. Theoretical Considerations		140
3.2. Practical Considerations		143
3.3. Electrode Modifications		144
3.4. Discussion of Voltammetrically Determined Reversible Potentials		145
4. CONCLUSIONS		150
ACKNOWLEDGMENTS		151
ABBREVIATIONS		151
REFERENCES		152

1. INTRODUCTION

Cytochromes P450 are redox enzymes found widely across nature in many living organisms. To function catalytically they require electrons, which they receive from an electron transport chain starting with NAD(P)H. Ultimately, the electrons are used to form a high-energy oxidizing species that, most notably, can allow the enzyme to efficiently hydroxylate an unactivated C–H bond.

The electroactive component of cytochrome P450 is an iron-protoporphyrin IX (heme). The heme is uniquely positioned within a three-dimensional network of one long folded polypeptide chain, as shown in Figure 1. Generally, in a substrate-free, low-spin conformation there are six coordination sites around the iron atom: the four lateral sites are occupied by nitrogen atoms of the tetrapyrrole macrocycle, the proximal site by a sulfur atom of a thiolate (cysteine) group, and the distal site by a weakly bound water molecule. As will be discussed later, it is this primary coordination environment and the factors influencing it, in addition to the dielectric setting of the heme, that contribute significantly to the level of electronic energy required to activate the enzyme.

As described in the generalized P450 catalytic cycle (Figure 2), two electrons are required in the overall catalytic process. After binding of substrate, the first electron is used to reduce the iron in the heme group, initiating the binding of the co-substrate, O_2. The second electron then reduces the oxygenated heme, inevitably leading to what is commonly regarded as 'Compound I', the active oxidant.

The scope of this review is confined to describing the first one-electron reduction process and any chemical steps which may be directly coupled to this process, e.g., substrate and proton binding (see highlighted sections in Figure 2). The overall process of interest can be conveniently described in electrochemical

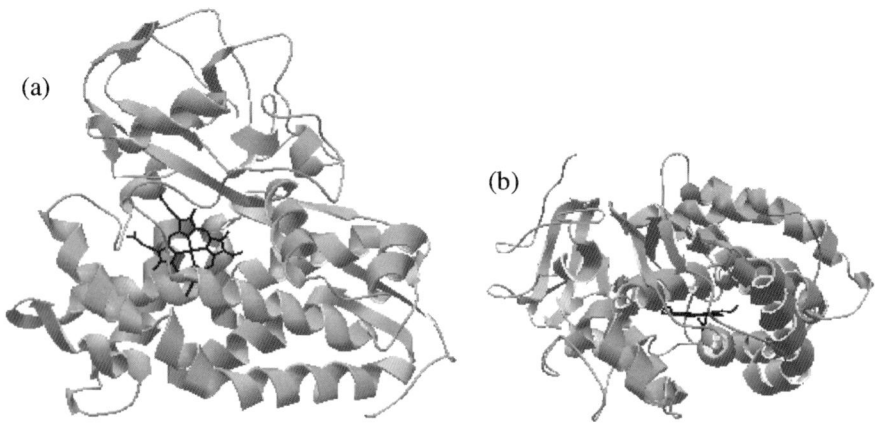

Figure 1. A ribbon representation of the P450cam structure looking (a) into the distal face, and (b) along the plane of the heme (with distal face at the bottom). Images generated with Swiss-PDBViewer software using crystal structure data contained in PDB file 1AKD.

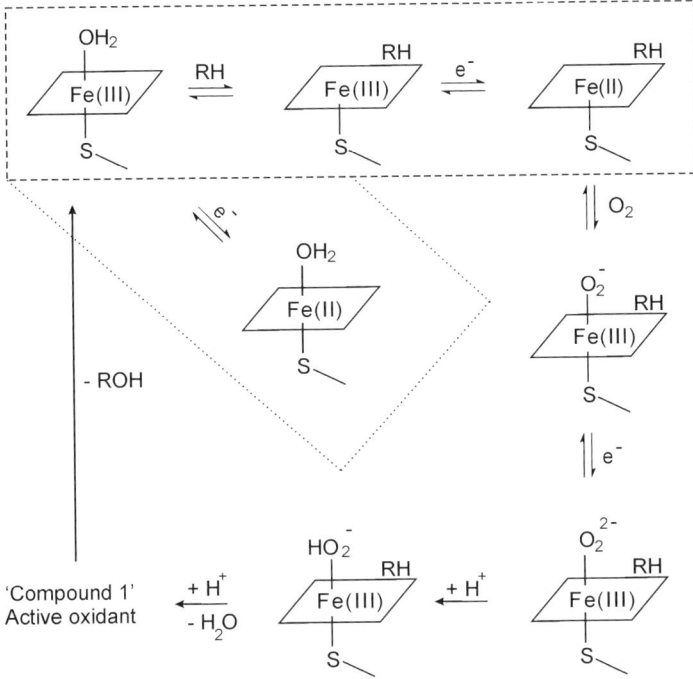

Figure 2. Generalized P450 catalytic cycle. The region of interest in this review is contained in the dotted area, with another important step in the electrochemistry of P450 shown in the dashed area.

terms under equilibrium (potentiometric or zero-current conditions) via the half-cell reaction (Equation 1),

$$[\text{cyt P450, Fe(III)}] + e^- \xrightleftharpoons{E^0} [\text{cyt P450, Fe(II)}] \quad (1)$$

where E^0 is the reversible potential versus a reference electrode, and Fe(III) and Fe(II) refer to the oxidized and reduced forms of the cytochrome P450 (cyt P450) heme respectively (with or without substrate bound in the active site pocket). Thus, Equation (1) is a generic representation of the P450 redox process that does not consider the phase of the enzyme (solution or solid), differences in the protonation levels of the oxidized or reduced forms, or the presence or absence of substrate. Traditionally in electrochemical investigations of redox processes, the reference half-cell reaction that completes the electrochemical cell is assumed to be the standard hydrogen electrode (SHE, Equation 2).

$$H^+_{(aq)} + e^- \rightleftharpoons 1/2\, H_{2(g)} \quad (2)$$

The reference half-cell reaction is assumed to be reversible under all operating conditions and assigned a value of 0 Volt under standard conditions (25 °C, unit activity of $H^+_{(aq)}$ and $H_{2(g)}$). In practice, for reasons of experimental convenience, the reference electrode is either a Ag/AgCl or saturated calomel electrode (SCE) whose potential is assumed to be accurately known versus SHE. In this review all reversible potential values are corrected where necessary and reported versus SHE (potential differences between the SHE and normal hydrogen electrode, NHE, are neglected). On the basis of assuming the reference half-cell has a potential of 0 V, the P450 process (Equation 1) can be characterized by a standard potential, E^0 measured as voltage versus SHE. Provided E^0 values are reported versus SHE, then E^0 is directly dependent on the difference between the free energies of the oxidized and reduced states of the enzyme (Equation 3),

$$\Delta G^0 = -nFE^0 = G_{Fe(II)} - G_{Fe(III)} \tag{3}$$

since ΔG for the reference H^+/H_2 process is defined as zero under standard conditions. ΔG^0 in turn may be related to the equilibrium constant, K, by the relationship (Equation 4)

$$\Delta G^0 = -RT \ln K \tag{4}$$

where, as shown in Equation (5),

$$K = \frac{\left(a_{H_{2(g)}}\right)^{1/2} a_{Fe(II)}}{a_{H^+_{(aq)}} a_{Fe(III)}} \tag{5}$$

Experimentally, standard conditions are essentially impossible to obtain for enzymes so a reversible formal potential E^0_f rather than E^0 is reported in this review. Differences between E^0 and E^0_f arise from terms such as junction potentials, non-unity activities, and non-standard temperatures [1,2]. Hence, the value of E^0_f may be used as a relative guide to the thermodynamic driving force required to 'activate' the enzyme, and as such its magnitude is important in understanding the catalytic process.

Determining a value for E^0_f has been accomplished traditionally via Nernst titrations under equilibrium conditions (thermodynamic reversibility). Initially, E^0_f values were obtained predominately in this manner. More recently, voltammetric approaches (that are inherently simpler) have been used to obtain E^0_f values for P450 enzymes. In potentiometric methods, no current flows and hence the cell is in a true equilibrium state. In voltammetric methods, current flows and hence, the enzyme half-cell reaction (Equation 1) is not truly at equilibrium. Rather, this process is under kinetic, not thermodynamic, control, and is now defined according to Butler–Volmer theory [2] by Equation (6),

$$[\text{cyt P450, Fe(III)}] + e^- \xrightleftharpoons{(E^0_f, k^0_f, \alpha)} [\text{cyt P450, Fe(II)}] \tag{6}$$

where k_f^0 is the standard rate constant at E_f^0, and α the charge transfer coefficient. Often k_f^0 is sufficiently fast that the reaction appears to be in equilibrium on the timescale of the voltammetric measurement. In this case E_f^0 can be measured directly from voltammetric data. If not, simulations must be employed to model the voltammetric behavior and hence, used to derive the thermodynamic and kinetic parameters on the basis of experimentally determined voltammograms.

In this review we are interested in providing a comparison of the E_f^0 values obtained between potentiometric and voltammetric methods, as well as those obtained within each method. Interestingly, much variation is observed in both cases. This implies that questions arise, such as: Why is this so? What do the values obtained by each method actually mean? Can values be directly compared with each other? Is the value measured (typically *in vitro*) truly representative of the *in vivo* enzyme? If k_f^0 values can be obtained, what is their significance?

2. REDOX TITRATION (POTENTIOMETRIC EQUILIBRIUM) MEASUREMENTS

Redox titration measurements in various forms have for over 30 years been utilized to characterize the redox properties of cytochrome P450 systems.

2.1. Methods

There are three key factors common to all P450 redox titration measurements:

(a) a means to reduce (and oxidize) the enzyme;
(b) a measure of the P450 redox state;
(c) a measure of the reversible potential, $\left[E_f^0\right]_{soln}$ for the solution-phase half-cell reaction as shown in Equation (7)

$$[\text{cyt P450, Fe(III)}]_{\text{solution}} + e^- \underset{}{\overset{\left[E_f^0\right]_{soln}}{\rightleftharpoons}} [\text{cyt P450, Fe(II)}]_{\text{solution}} \quad (7)$$

The reduction (or oxidation) of the P450 can be accomplished chemically, enyzmatically, photochemically or electrochemically, the latter typically in the presence of suitable redox mediators. $\left[E_f^0\right]_{soln}$ is commonly measured potentiometrically, or in some cases, calculated on the basis of the equilibrium state of a reaction with a dye whose $\left[E_f^0\right]_{soln}$ value is known. Electronic and EPR spectral methods have mainly been used to determine the state of the reduction–oxidation of the P450 enzyme.

The Nernst equation describing a potentiometric titration is given by Equation (8)

$$E = [E^0]_{soln} + \frac{RT}{F} \ln \left\{ \frac{a_{Fe(II)}}{a_{Fe(III)}} \right\} \tag{8}$$

where E is the measured potential referred to the SHE scale, and $a_{Fe(II)}$ and $a_{Fe(III)}$ are the activities of the reduced and oxidized P450 respectively. The simplified form of the Nernst equation is given in Equation (9),

$$E = [E_f^0]_{soln} + \frac{RT}{F} \ln \left\{ \frac{[Fe(II)]}{[Fe(III)]} \right\} \tag{9}$$

where activities are replaced by concentration and activity coefficients are incorporated into the $[E_f^0]_{soln}$ value.

In principle, the protonation levels of the Fe(III) and Fe(II) forms of P450 need not be identical, in which case the reversible potential will have a pH dependence. In the case of P450cam, Sligar and Gunsalus [3] have introduced a square scheme, as shown in Equation (10):

$$[\text{cyt P450, H}^+, \text{Fe(III)}] \xrightleftharpoons[\pm e^-]{[E_f^0]_1} [\text{cyt P450, H}^+, \text{Fe(II)}]$$

$$K_1 \updownarrow \pm H^+ \qquad\qquad \pm H^+ \updownarrow K_2 \tag{10}$$

$$[\text{cyt P450, Fe(III)}] \xrightleftharpoons[\pm e^-]{[E_f^0]_2} [\text{cyt P450, Fe(II)}]$$

At equilibrium, the observed reversible potential, $[E_f^0]_{soln}$, can be expressed as follows (Equation 11)

$$[E_f^0]_{soln} = [E_f^0]_1 + \frac{RT}{F} \ln \left\{ \frac{[H^+] + K_1}{[H^+] + K_2} \right\} \tag{11}$$

Clearly, if $K_1 = K_2$, then $[E_f^0]_{soln} = [E_f^0]_1$. An analogous relationship applies for the equilibrium constants relevant to substrate binding [4]. Thus, if substrate binding to both the Fe(III) and Fe(II) forms is equal (same equilibrium constant) then $[E_f^0]_{soln}$ would be unaffected by the presence of substrate.

Equation (9) therefore relates the measured solution potential E to the proportion of the oxidized and reduced heme present, and also contains terms associated

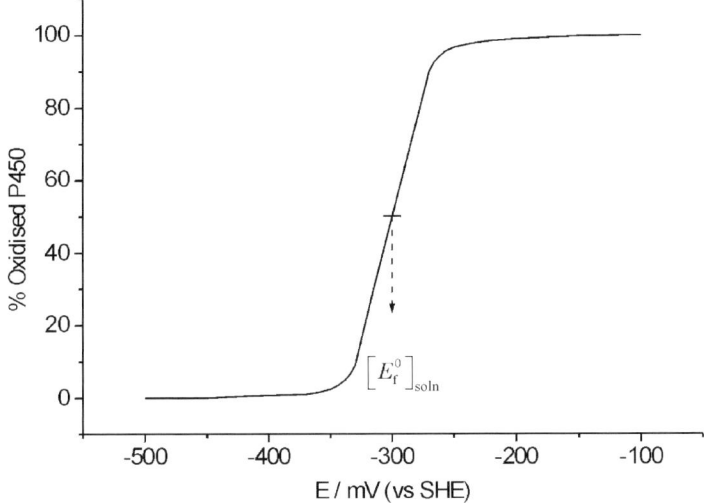

Figure 3. Example of a Nernst titration curve for a one-electron redox couple with a $[E_f^0]_{soln}$ value of -300 mV versus SHE.

with proton and substrate reactions coupled with electron transfer, and junction potentials. Such contributions may be incorporated knowingly or unknowingly in the reported values of $[E_f^0]_{soln}$. Clearly, the pH value must be stated when $[E_f^0]_{soln}$ values are reported. As shown in Figure 3, a plot of the oxidation–reduction state against E, when [Fe(III)]=[Fe(II)], will yield $[E_f^0]_{soln}$ [5]. A well-written practical and theoretical description of this technique can be found elsewhere [1]. The following section will describe how the application of such techniques has been used to determine factors important in the modulation of the P450 reversible potential.

2.2. Discussion of Potentiometrically Determined Reversible Potentials

The $[E_f^0]_{soln}$ values for a variety of isolated, substrate-free P450 enzymes in aqueous buffered solutions (pH 7–7.5) obtained by redox titration methods are summarized in Table 1 from most negative to least negative. The solution conditions (buffer, added electrolyte, or other additives) and methodology used are also summarized. Clearly, the reported values vary by in excess of 200 mV, ranging from -427 to -182 mV. These significant differences in the $[E_f^0]_{soln}$ values could highlight the diversity present within the secondary structure of the P450 superfamily, despite P450 enzymes maintaining a highly conserved primary structural motif. For example, the least negative $[E_f^0]_{soln}$ value reported for P450cin is believed to be the result of a more hydrophobic distal coordination

Table 1. $[E_f^0]_{soln}$ values for isolated substrate-free cytochrome P450 enzymes in aqueous buffered solutions (pH 7–7.5) determined by redox titration methods ($T = 20$–$25\,°C$).

P450	Family	$[E_f^0]_{soln}$/mV vs SHE	Solution conditions	Method*	Reference
BM3	bacterial	-427 ± 4	0.1 M KPhos	Chem. Abs. Pot.	[11]
		(-368 ± 6)			[12]
scc	bovine	-412 ± 2	0.1 M KPhos, 20% glycerol, 1m M EDTA, 0.1 m M DTT	Enzym. EPR Dye	[13]
RhF	bacterial	-380 ± 4	0.1 M Tris-HCl, 0.5 M KCl	E'chem. Abs. Pot.	[14]
BioI	bacterial	-330 ± 5	n/s	Chem. Abs. Pot.	[15]
LM	rat	-329	0.1 M KPhos	Chem. Abs. Pot.	[16]
LM	rat	-326	0.1 M KPhos	Chem. Abs. Dye	[16]
PB-B	rabbit	-319 ± 4	0.1 M Kphos, 0.01 M EDTA, 5–12% glycerol	Photochem. Fluor. Dye	[17]
nor	fungal	-307	0.1 M KPhos, 0.05 M EDTA	Photochem. Abs. Pot.	[18]
cam	bacterial	-306 ± 10	0.05 M Tris/HCl, 0.2 M KCl	E'chem. Abs. Pot.	[19]
βNF/ISF-G	rat	-304	0.1 M KPhos, 0.01 M EDTA, 5–12% glycerol	Photochem. Fluor. Dye	[17]
cam	bacterial	-303 ± 10	n/s	Enzym. Abs. n/s	[20]
LM	rat	-300	0.05 M NaPhos, 0.01 M EDTA, 25% (v/v) glycerol	Photochem. Abs. Dye	[21]
cin	bacterial	-182	n/s	Chem. Abs. Pot.	[6]

* Methods are designated, in order from top to bottom: method of P450 reduction, method of measuring the state of the P450, and method for measuring the solution potential; n/s = not stated, Abs. = absorbance, Chem. = chemical, Dye = dye equilibrium, E'chem. = electrochemical, Enzym. = enzymatic, EPR = electron spin resonance, Fluor. = fluorescence, Photochem. = photochemical, Pot. = potentiometric

environment [6]. The significant contribution made by the dielectric constant of the region surrounding the heme on the value of the redox potential is well documented [7–10].

It is also important to note that the experimental method, pH, buffer, electrolyte, ionic strength and other solution additives could also contribute to the $[E_f^0]_{soln}$ value reported. In particular, the discussions above show that $[E_f^0]_{soln}$ values are a function of pK_a values, so that more data is needed on the pH dependence of P450 electron transfer to consequently distinguish different factors that contribute to $[E_f^0]_{soln}$ values. Furthermore, improvements in the isolation, expression and purification systems for P450 enzymes have occurred in recent times. Thus, the more recent potentiometric data would be in all likelihood at P450 purity levels that were not possible 20–30 years ago. A new systematic collection of potentiometrically determined $[E_f^0]_{soln}$ values obtained as a function of pH with highly purified P450 samples is needed for the full significance of the present literature to be completely understood.

In most cases, the value of $[E_f^0]_{soln}$ has been observed to become less negative when a substrate is bound in the active site pocket. Table 2 shows the change in reduction potential values, $\Delta[E_f^0]_{soln}$ where Equation (12)

$$\Delta[E_f^0]_{soln} = [E_f^0]_{soln(substrate)} - [E_f^0]_{soln(substrate-free)} \qquad (12)$$

holds for a selection of P450 enzymes when a substrate is present. The general explanation given for this potential shift is that the distal coordinated H$_2$O molecule is displaced, and/or the level of hydration in the active site is reduced upon substrate binding [22]. The resultant lower dielectric environment destabilizes the Fe(III) state relative to the Fe(II) state, thereby shifting the value of $[E_f^0]_{soln}$ to less negative potentials (reduction of Fe(III) to Fe(II) becomes easier). The regulatory effect of the ligand field stabilization energy has also been cited as an important contributing factor [20].

However, the overall effect on $[E_f^0]_{soln}$ is governed by the nature and extent of the relative substrate binding (see Equations 10 and 11). The level of steric crowding generated by the bound substrate is a key determinant of whether the

Table 2. $\Delta[E_f^0]_{soln}$ values for some cytochrome P450 enzymes in the presence of substrate as determined from redox potentiometry titrations.

P450	Substrate	$\Delta[E_f^0]_{soln}$/mV	References
cam	d-camphor	133 ± 10	[19,20,23]
BioI	palmitoleic acid	131 ± 2	[15]
BM3	arachidonate	129 ± 10	[11,12]
scc	cholesterol	107 ± 2	[13]
cin	1,8-cineole	12 ± 2	[6]

aqua ligand remains coordinated or not [22]. The less negative value of $\left[E_f^0\right]_{soln}$ has been correlated with a change in spin state, from a low-spin, hexacoordinate form to a high-spin, pentacoordinate form [4,23]. It has also been considered that even when substrate binding does occur, resulting in a less negative value of $\left[E_f^0\right]_{soln}$ and spin-state change, the heme iron can remain hexacoordinate [22]. Raag and Poulos [22] hypothesized that this would be consistent with a mechanism whereby a coordinated OH^- molecule in the substrate-free enzyme is protonated to become H_2O in order to stabilize the heme within the lowered dielectric environment. Such a mechanism could also contribute to the pH dependence of $\left[E_f^0\right]_{soln}$ values.

Instances have been reported whereby little change in the value of $\left[E_f^0\right]_{soln}$ was observed when typical substrates were bound [6,16]. The authors point out that, despite a spin-state change, the lack of any significant potential change can be explained by the substrate having similar affinities for both the reduced and oxidized form of the enzyme (analogous to the situation where $K_1 = K_2$ in Equation 11). P420cam, the inactive form of P450cam, was also shown to have a similar reversible potential when camphor was or was not present – a reflection of the enzyme's inability to bind the substrate [24]. Interestingly, the $\left[E_f^0\right]_{soln}$ value in this case was $-211 \pm 10\,mV$, nearly 100 mV more positive than the P450cam value. Significant structural changes were proposed, in particular, substitution of the proximal thiolate ligand in the reduced form.

Microsomal P450s naturally exist bound to phospholipid membranes. Replicating such conditions experimentally can be quite difficult. However, when several solubilized P450 enzymes (membrane binding domain removed) were interacted with model dilauroyl-GPC membranes, little change in the value of $\left[E_f^0\right]_{soln}$ was observed when compared with the solubilized form [16,17]. This indicates that membrane–protein interactions, important for enzyme orientation, stability and catalytic activity, may have little impact on the heme redox cofactor environment [25,26].

The $\left[E_f^0\right]_{soln}$ values shown in Table 1 are for isolated P450 heme domains that have been reduced by electrons supplied by, in the majority of cases, a non-natural reductant. So, for P450s with relatively exposed heme cofactors, the $\left[E_f^0\right]_{soln}$ values measured may not necessarily be indicative of the physiologically active enzyme given that electron transfer usually occurs when the enzyme interacts with a partner protein(s) as shown in Figure 4. The $\left[E_f^0\right]_{soln}$ value for the isolated heme domain from P450RhF, was found to be 21 and 43 mV less negative than that observed for the heme-FMN complex and intact holoenzyme, respectively [14,27]. Similarly, the $\left[E_f^0\right]_{soln}$ value for P450scc was 32 mV less negative then that reported for P450scc when complexed with adrenodoxin (both in the presence of substrate) [28]. Protein–protein inter-domain interactions are likely to result in small structural changes that have some affect on the heme microenvironment. However, the authors suggest that increased solvent exposure upon removal of the reductase domain contributes to the increased $\left[E_f^0\right]_{soln}$

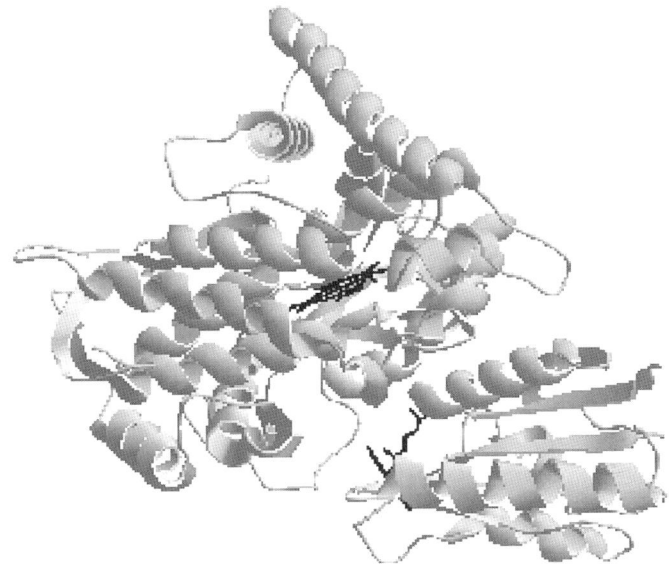

Figure 4. A ribbon representation of the P450BM3 heme domain complexed to the FMN reductase domain. Generated with Swiss-PDBViewer software using crystal structure data contained in PDB file 1BVY.

value [14]. Typically, the shift to less negative $\left[E_\mathrm{f}^0\right]_\mathrm{soln}$ values is commensurate with a decrease in the dielectric environment of the heme, the opposite of that implied by the authors. It is possible that uncomplexing the heme domain may effectively result in constriction of the substrate channel, and hence reduced solvent exposure to the heme cofactor. Further structural analysis of complexed and uncomplexed heme domain would be useful in this case. Interestingly, the $\left[E_\mathrm{f}^0\right]_\mathrm{soln}$ values observed for free and putidaredoxin-complexed camphor-bound P450cam were similar [23]. In this case, it is likely that the impact of substrate binding on $\left[E_\mathrm{f}^0\right]_\mathrm{soln}$ significantly outweighed that of the complexation with its electron transfer partner.

Protein engineering (site-directed mutagenesis) of P450 enzymes has led to valuable insights into factors that lead to modulation of the P450 reversible potential [11,19,29–34]. Morishima and co-workers have highlighted the critical role of the proximal hydrogen-bonding network; in particular how it affects the electron donating ability of the thiolate ligand [29,30,33]. The $\left[E_\mathrm{f}^0\right]_\mathrm{soln}$ values for a series of P450cam mutant enzymes, L358P, Q360L, Q360P, Q360E, and Q360K were all less negative than that of the wild-type enzyme (camphor-bound). In each case, removal of an amide proton from the pocket surrounding the cysteine residue resulted in an increased electron density on the thiolate, aiding the so-called 'push' effect, and destabilization of the reduced heme state.

A similar function for the proximal H-bonding network on the reversible potential of heme enzymes and P450 model complexes has been reported [35,36]. A theoretical consideration of the effect of changes in hydrogen bonding, as well as ligand replacement and dielectric screening on the 'push' effect has also been reported [37].

Significant thermodynamic control of the heme domain of P450BM3 was revealed by a series of F393 mutations [11,31,38,39]. A phenylalanine residue located near the heme-binding thiolate ligand is highly conserved within many P450s. Substitution of this phenyalanine residue with either histidine (F393H) or alanine (F393A) resulted in significantly less negative values of $[E_f^0]_{soln}$ for the substrate-free heme domain (95 and 115 mV respectively), whereas substitution with the more sterically bulky tryptophan residue (F393W) yielded a mutant with an $[E_f^0]_{soln}$ value 53 mV more negative than that of the wild-type enzyme. Despite the changes in $[E_f^0]_{soln}$ values being correlated to the steric nature of the residues, structural evidence did not fully support this as being the main contributing factor. Rather, an assortment of contributions was proposed due to the proximity of residue 393 to the heme, in particular the vinyl groups, and the thiolate ligand. However, despite the less negative $[E_f^0]_{soln}$ values and concomitant faster electron transfer with the F393H and F393A mutants, steady-state turnover rates were significantly reduced.

Altering the hydrophobicity of the active site pocket of P450cam by replacing a tyrosine residue with phenylalanine (Y96F) resulted in a 58 mV less negative value of $[E_f^0]_{soln}$ compared with the wild-type enzyme [19]. It was suggested that the increased hydrophobic nature of the active site destabilized the hydrogen-bond network associated with the active site, and facilitated the protonation of the coordinated water molecule (as discussed above). Also shown with P450cam, was that changing the proximal (coordinated) cysteine residue to a histidine (C357H), resulted in a less negative reversible potential (albeit lower than expected given the mixed spin-state) [34]. Neutralising the positively charged arginine residue (R112) at the proximal surface of P450cam by site-directed mutagenesis led to more negative values of $[E_f^0]_{soln}$ [32]. Despite being a surface amino acid, a significant impact (25–60 mV) on the heme reversible potential was still observed. This was the result of a combination of the less positive electrostatic field of the protein, a reduction in the charge withdrawal from one of the heme propionate groups, and a 'knock-on' reduction in the 'push' effect of the thiolate — all resulting in an increased stability of the Fe(III) state.

Potentiometric data relating to proton coupling to the P450 reversible potential is very limited. In fact, only two reports (both for bacterial systems) could be found in the present literature [3,6]. The first, a thorough and pioneering effort, was back in 1979 by Sligar and Gunsalus [3]. They clearly showed that a proton was preferentially binding upon reduction of P450cam (in the presence of camphor). For ease of comparison with the other more recent study and voltammetric data, we have taken their original potentiometric data (Figure 3 [3])

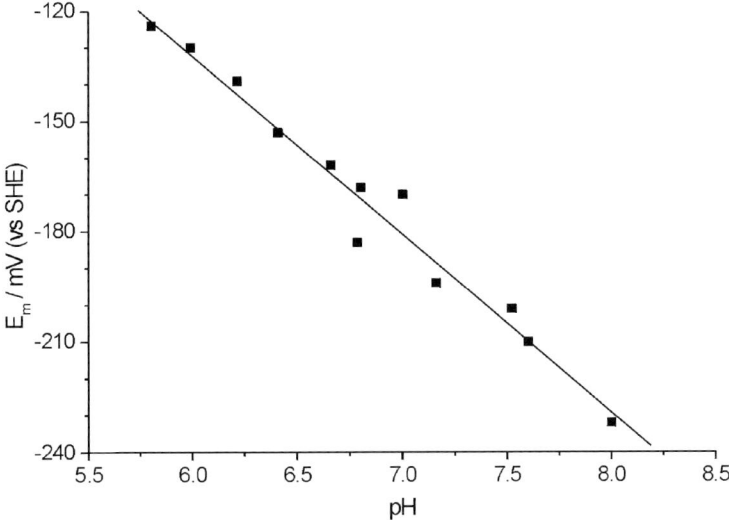

Figure 5. pH dependence of the value of $\left[E_f^0\right]_{soln}$ for P450cam in the presence of camphor. Data compiled from Figure 3 of [3] and plotted here to illustrate the dependence of the redox potential on the approximate pH values. The solid line is included as guide only.

and converted it to approximate pH values, as shown in Figure 5. The approximately $-59\,mV$ per pH unit change observed in this case was similar to that observed for P450cin [6]. This trend for P450cin was observed both in the presence and absence of substrate. It would be interesting to see whether this is a trait of the whole P450 superfamily, and hence more studies like this are encouraged.

3. VOLTAMMETRIC (DYNAMIC) MEASUREMENTS

Voltammetric techniques have been used extensively to evaluate the electron transfer properties of redox proteins (thermodynamic, kinetic, and mechanistic information) [40–43]. The protein may either be in solution, and hence diffuse to/from the electrode surface, or be confined to the electrode surface, whereby electron transfer takes place, as shown in Figure 6. In practice, solution experiments with large redox enzymes such as P450s have not been highly successful. This has been attributed to poor communication between the electrode and protein due to the combination of electrode fouling by proteinaceous material and/or the deeply buried redox-active site within the protein shell preventing electron transfer [40,42,44]. As will be discussed in the following section, it has become quite apparent that voltammetrically determined reversible potential values are

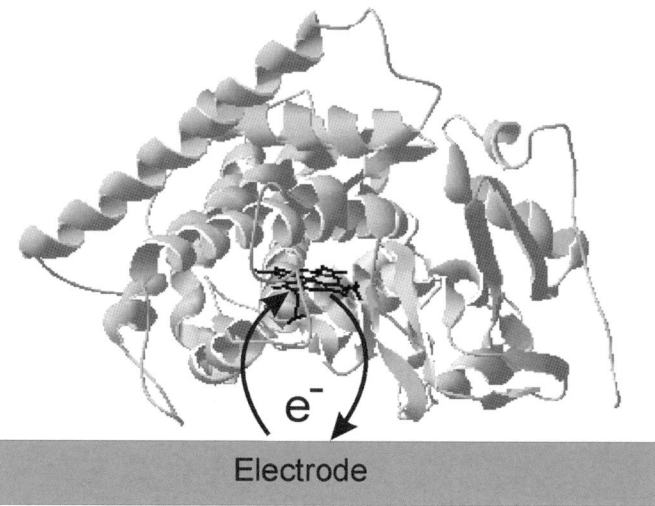

Figure 6. Schematic diagram of P450BM3 heme domain undergoing electron transfer at an electrode surface. This could be representative of a molecule diffusing to the electrode surface, or one that is confined by some means at the electrode surface.

not necessarily the same as the $\left[E_f^0\right]_{soln}$ values determined by classical potentiometric methods.

3.1. Theoretical Considerations

In a potentiometric experiment, two electrodes are present; the indicating electrode, which is in direct contact with the test solution, and the reference electrode that is separated from the test solution by a salt bridge.

Typically, the cell used in a voltammetric experiment has a working electrode where the P450 reaction of interest takes place, a reference electrode as used in potentiometry, and an auxiliary electrode. When used with a potentiostated form of instrumentation, the three-electrode system minimizes uncompensated resistance (R_u) effects (Ohmic IR_u drop).

In a voltammetric experiment, a time (t)-dependent potential difference (working versus reference electrode) is applied and the current I measured as a function of potential (time). The resultant plot of I versus E (or t) is referred to as a voltammogram. Commonly, the technique of cyclic voltammetry is employed in P450 studies. In this I–E–t format, the potential commences at an initial value E_i, is swept at a scan rate v in one direction until the potential direction is reversed at the switching potential E_s, as illustrated in Figure 7. Often the sequence is repeated for several cycles in multiple potential cycling experiments.

ELECTROCHEMISTRY OF CYTOCHROME P450

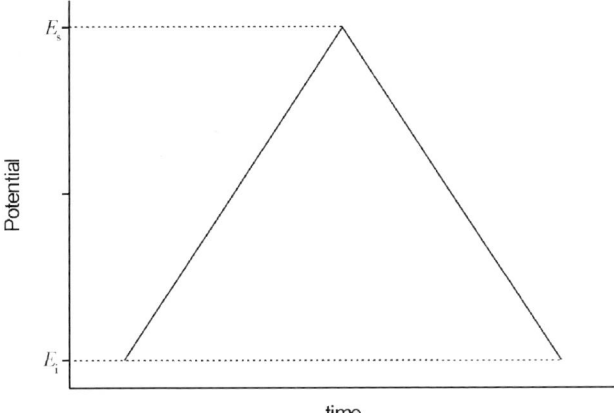

Figure 7. Example of a typical potential waveform used in a dc cyclic voltammetric experiment.

Figure 8 shows the asymmetric peak shape response expected under transient conditions at a macrodisk electrode (bulk transport to the electrode is controlled by linear diffusion) when both the reactant and product are soluble so that the reaction in Equation (13) applies:

$$[\text{cyt P450, Fe(III)}] + e^- \xrightleftharpoons{[E_f^0]_{\text{soln}},[k_f^0]_{\text{soln}},\alpha_{(\text{soln})}} [\text{cyt P450, Fe(II)}] \qquad (13)$$

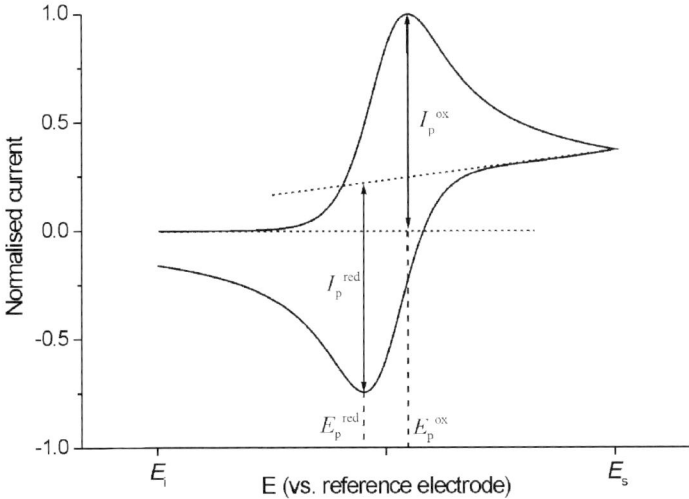

Figure 8. Typical cyclic voltammogram obtained for a reversible one-electron process.

This class of reaction is characterized by reduction and oxidation peak potentials (E_p^{red} and E_p^{ox}) and peak currents (I_p^{red} and I_p^{ox}) and should lead to an $[E_f^0]_{soln}$ value that is almost the same as the value obtained potentiometrically. For a reversible process (large $[k_f^0]_{soln}$ value) the midpoint potential E_m gives a good approximation of $[E_f^0]_{soln}$ since

$$E_m = \frac{(E_p^{ox} + E_p^{red})}{2} \approx [E_f^0]_{soln} \qquad (14)$$

Unfortunately, little P450 solution-phase voltammetric data has been reported. Predominately, voltammetric studies with P450 have been undertaken with surface-confined enzymes. In an ideal thin-layer film of P450 (sub- to monolayer coverage), the electron transfer process can be described by the electrode reaction

$$[\text{cyt P450, Fe(III)}]_{\text{thin film}} + e^- \underset{}{\overset{[E_f^0]_{\text{thin film}}, [k_f^0]_{\text{thin film}}, \alpha_{(\text{thin film})}}{\rightleftharpoons}} [\text{cyt P450, Fe(II)}]_{\text{thin film}} \qquad (15)$$

In this diffusionless process, a reversible response would produce the symmetrical peak shaped voltammogram shown in Figure 9 (large $[k_f^0]_{\text{thin film}}$ value) for values $E_p \simeq [E_f^0]_{\text{thin film}}$. In contrast, a process with slow electrode kinetics on the time scale given by the scan rate would exhibit the nonsymmetrical shape illustrated in Figure 9. In this case,

$$E_m = \frac{(E_p^{ox} + E_p^{red})}{2} \approx [E_f^0]_{\text{thin film}} \qquad (16)$$

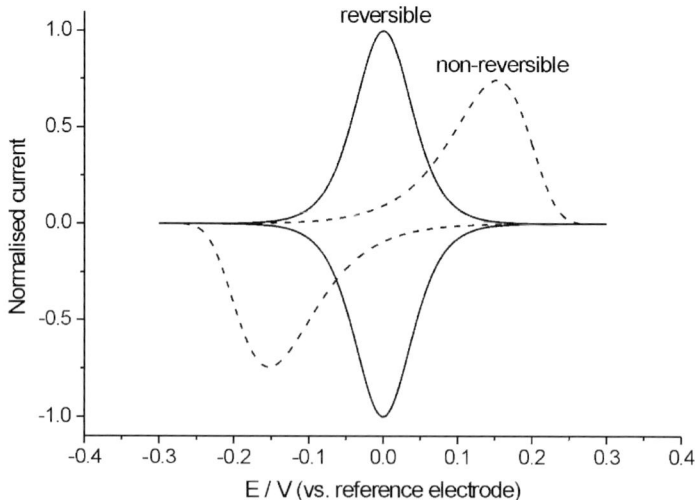

Figure 9. An example of a reversible (solid line) and nonreversible (dashed line) cyclic voltammogram for an ideal one-electron thin-film reaction.

but comparison of simulated and experimental voltammograms would produce a better estimate of $[E_f^0]_{thin\,film}$ and $[k_f^0]_{thin\,film}$ (and α as well).

Again, truly ideal, thin-film voltammetry associated with monolayer coverage is rarely encountered with P450 voltammetry. The work of Rusling et al. [44–46] has best shown the nuances of voltammetry with surface confined P450s, which is represented by the generic Equation (17),

$$[\text{cyt P450, Fe(III)}]_{surf} + e^- \xrightleftharpoons[]{[E_f^0]_{surf},[k_f^0]_{surf},\alpha_{(surf)}} [\text{cyt P450, Fe(II)}]_{surf} \quad (17)$$

where $[E_f^0]_{surf}$, $[k_f^0]_{surf}$, and $\alpha_{(surf)}$ are the reversible formal potential, standard rate constant, and charge transfer coefficient for the surface-confined P450 electron transfer process, respectively. Typically, $[E_f^0]_{surf}$ is again approximated by E_m, as shown in Equation (18):

$$E_m = \frac{(E_p^{ox} + E_p^{red})}{2} \approx [E_f^0]_{surf} \quad (18)$$

However, the extent to which the value of E_m has direct thermodynamic significance in the case of surface confined voltammetry is often unclear.

References [2,47,48] should be consulted for more detailed descriptions of the voltammetric theory and experimental details.

3.2. Practical Considerations

Typical cyclic voltammograms for a P450 enzyme confined at an electrode surface are shown in Figure 10. The potential at which the peak of the reduction current E_p^{red} and oxidation current E_p^{ox} occur are used to calculate the midpoint potential, and hence $[E_f^0]_{surf}$, as shown in Equation (18). In the case illustrated, a $[E_f^0]_{surf}$ value of 17 ± 5 mV was obtained for P450BM3. The inset shows a voltammogram that highlights some of the often-encountered difficulties with P450 voltammetry, especially the low Faradaic current when compared with the background non-Faradaic current. A large background effectively means an increased error in the $[E_f^0]_{surf}$ value, given the difficulty of accurate location of the position of the reduction and oxidation peaks (even with background subtraction procedures). With suitable voltammograms, mechanistic and kinetic information can also be gained from such experiments by varying the rate at which the potential is varied with time.

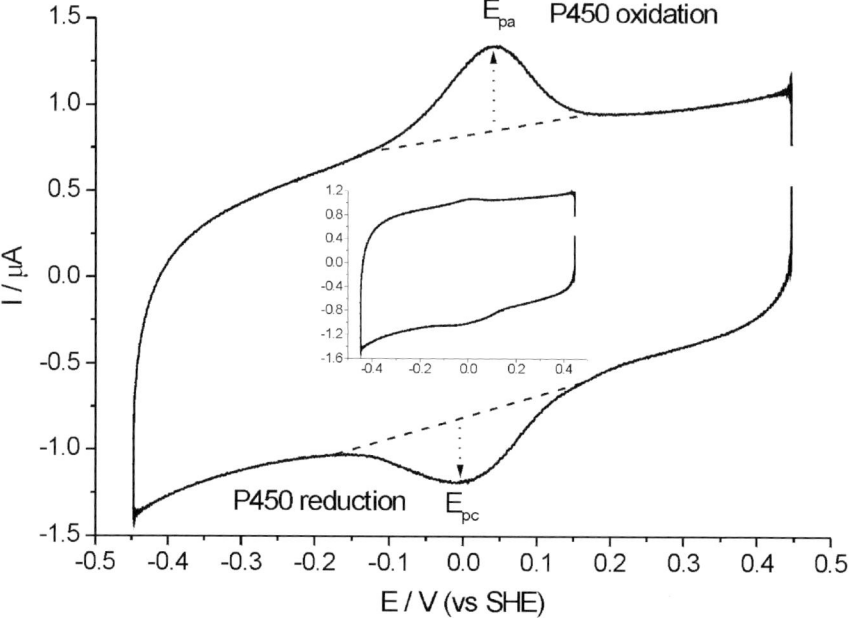

Figure 10. Cyclic voltammograms for a surface confined P450 enzyme. The main voltammogram highlights the parameters used to calculate $[E_f^0]_{surf}$, whereas the inset highlights the typical low Faradaic-background current situation prevalent with P450 voltammetric responses. Actual experiments were performed with cytochrome P450BM3 heme domain confined in a biomembrane-like layer of didodecyldimethylammonium bromide (DDAB) at an edge-plane graphite (EPG) electrode.

3.3. Electrode Modifications

A variety of methods for confining P450 enzymes at an electrode surface have been attempted. The design and features of such functional P450 films, with particular applications to biosensing and bioreacting has been described recently [48–51]. Our interest here is to highlight those approaches that have been used to suitably probe the thermodynamics of the underlying electron transfer process. There have been several reports of P450 enzymes being incorporated into electrochemically driven, whether directly or indirectly, catalytic processes that result in substrate turnover or sensing [52–62]. These are obvious end points or applications of P450 electrochemistry that are highly desirable, yet in these cases the fundamental electrochemistry was not reported or had insufficient detail and hence will not be mentioned here.

Also, determination of a reliable $[E_f^0]_{surf}$ value from voltammetric data requires suitably distinguishable reduction and oxidation peaks from which peak potentials can be determined. There are some instances where the reported $[E_f^0]_{surf}$

values must be treated with caution, given the significant difficulties in recognizing these peak parameters from the observed voltammograms (even in light of background subtraction procedures). As such, data of this nature are not considered in the following discussions.

3.4. Discussion of Voltammetrically Determined Reversible Potentials

The first report of the direct electrochemistry of a P450 enzyme was that of P450cam with an unmodified edge-plane graphite electrode [63]. The favorable response was based on the electrostatic interaction between negatively charged groups associated with electrode and positive charges located at the surface of the protein. The $\left[E_f^0\right]_{surf}$ values given in the absence and presence of camphor (-285 ± 11 mV and -149 ± 10 mV), were remarkably consistent with that reported previously by potentiometric methods. The diffusion-controlled response was reportedly reliable and reproducible only when freshly purified protein was present. It was also implied that the strictly anaerobic atmosphere ($[O_2] < 1.5$ ppm) was also important (as has been shown since by others). Follow-up work in the same laboratory with unmodified edge-plane graphite electrodes reported comparable $\left[E_f^0\right]_{surf}$ values for camphor-bound wild-type and mutant P450cam enzymes (under similar conditions, albeit based on only one scan with a scan rate 200 times faster than the earlier work) [64]. It is interesting to note though that attempts by others to reproduce this type of unmodified electrode procedure with other P450 enzymes have generally not proven successful. A weak response was observed for P450 2E1 when adsorbed at a bare glassy carbon (GC) electrode, and showed a substrate-free $\left[E_f^0\right]_{surf}$ value of -90 ± 5 mV (significantly more positive than might be expected given potentiometric values for similar mammalian P450s) [65]. Whilst formation of a Langmuir–Blodgett film of P450scc which was deposited directly onto an indium–tin oxide (ITO) electrode indicated some level of Faradaic response, presumably from the heme, the values quoted for the anodic and cathodic peak positions seemed inconsistent with the data provided [61,66].

By far the most convincing reports on the 'electrochemistry of P450s' have been when some form of electrode modification(s) has occurred. In essence, this has usually taken the form of the P450 enzyme being entrapped, confined or attached to the electrode surface by some means.

Of these methods, the most popular given its ease of use, is trapping the P450 in a multilayered biomembrane-like film. The molecules used to form such films consist of insoluble (in water) surfactants or lipids, such as DDAB, DDAPSS and DMPC. Despite their relative insolubility in water, their amphiphilic nature causes them to form into temperature- and concentration-dependent lamellar phases. Under the experimental conditions used in most electrochemical studies, the molecules form a liquid crystalline phase, within which the enzyme can remain stable, diffuse (albeit less so than in water) and as has been shown, can undergo reversible electron transfer. Several P450 enzymes have been confined to an electrode surface in such a manner and voltammetric measurements performed.

Table 3 summarizes the $[E_f^0]_{surf}$ values measured for P450 enzymes confined to an electrode surface in a film of DDAB. Some differences in $[E_f^0]_{surf}$ values between P450s are observed, which may be related to the varied structure and amino acid environments in the vicinity of the heme in each case. However, despite the similar nature of the DDAB layers formed, even slight experimental differences with parameters such as pH, electrolyte, temperature, buffer, reference electrodes, method of midpoint determination, etc., may contribute to the value of the observed $[E_f^0]_{surf}$. It is therefore difficult to confidently assign the main contributing factors and make any concrete conclusions about these differences in the $[E_f^0]_{surf}$ values. What is immediately obvious though is that $[E_f^0]_{surf}$ values acquired by this method are significantly more positive compared with those from potentiometric titrations — this important point will be discussed in more detail later. The method itself still has some questions about how it affects the integrity of the enzyme [51]. For instance, no significant changes in $[E_f^0]_{surf}$ have been observed when in the presence of substrate, unlike that observed with potentiometric titrations. Reasons for this may be that the enzyme active site has been altered by the interaction with the DDAB and hence does not allow substrate binding to occur, the DDAB lipid/membrane may 'sequester' the substrate and hinder it from binding, or the substrate does bind but interactions with the membrane alter the redox cofactor (low dielectric environment,

Table 3. Voltammetrically measured $[E_f^0]_{surf}$ values for P450 enzymes when confined in a film of DDAB at an electrode surface (T = 22 ± 3 °C).

P450	Origin	Electrode	$[E_f^0]_{surf}$/mV vs SHE	Reference
2C9	human	EPG	−110 ± 5	[67]
2E1	human	GC	−85 ± 5	[65]
2C18	human	EPG	−73 ± 5	[67]
C17	bovine	EPG	−54 ± 4	[68]
st	bacterial	PFC	−53	[69]
cin	bacterial	EPG	−52 ± 5	[6]
C17	human	EPG	−48 ± 4	[68]
2C9	human	EPG	−41 ± 4	[70]
2C19	human	EPG	−23 ± 5	[67]
119	bacterial	BPG	−17 ± 5	[71]
BM3	bacterial	EPG	−9 ± 4	[72]
C17	porcine	EPG	−3 ± 4	[68]
cam	bacterial	PG	22 ± 6	[45]

more hydrophobic, water exclusion) causing the large positive increase in $\left[E_f^0\right]_{surf}$ which masks any increase relating to the substrate. However, changes in $\left[E_f^0\right]_{surf}$ as a result of CO binding to the reduced heme have been observed, commensurate with that observed with potentiometric data (albeit of a smaller magnitude) [16]. Also, absorption spectra for P450cam when present in a vesicle dispersion, either in solution or confined at a surface, still revealed the characteristic 450 nm absorption upon reduction and CO binding [45].

Variations to this approach have been to include glutaraldeyde and BSA to help cross-link the enzyme when in the DDAB film [73]. The same group extended this approach further by incorporation of the P450/DDAB film into a methyltrimethoxysilane sol-gel, enabling enhanced responses in organic solvents [74,75]. More recently, it was revealed that changing the counter-ion from Br$^-$ (DDAB) to PSS$^-$ (DDAPSS) did not significantly alter the $\left[E_f^0\right]_{surf}$ observed for CYP102 [72,76].

Another entrapment or confining method that has resulted in a P450 voltammetric response being observed was accomplished with an alternate layer-by-layer approach. This approach is characterized by the alternate adsorption of a polyion and a P450 enzyme, such that a multi-layered film containing stable, ordered, active enzyme is created. The polyion used to obtain the most favourable Faradaic response from P450cam, when an MPS modified gold electrode was used, was PSS [77]. On PG electrodes, alternate P450cam layers with PEI also showed significant Faradaic responses [46]. Comparison of the $\left[E_f^0\right]_{surf}$ values obtained from these two procedures will be made in the proceeding section. Other P450s to have been incorporated into such layers include P450 1A2 (PSS) [78] and P450 3A4 (PDDA) [79].

The use of clay-modified electrodes has also revealed encouraging results. A significant adsorbed response was observed for P450cam at a sodium montmorillonite-modified GC electrode [80]. A positive shift in $\left[E_f^0\right]_{surf}$ of 80 and 86 mV was observed in the presence of either CO or metapyrone respectively, highlighting the heme origin of the response. Inclusion of a non-ionic detergent, Tween 80, in the clay-modified GC electrode procedure was shown to dramatically improve the electron transfer to P4502B4 [81]. The promoting effect of the Tween 80 was related to its ability to monomerize the microsomal membranous P450 sample. P4502B4 was also observed to adsorb to an MPA-modified Au electrode, with an electrochemical response characterized by an $\left[E_f^0\right]_{surf}$ value of −23 mV [82].

A novel approach was recently described whereby the heme domain of P450BM3 was wired to a basal plane graphite (BPG) electrode by interaction between a pyrene moiety engineered onto the surface of the protein with the graphite electrode surface [83]. One reported positive attribute of this method was the fast electron transfer kinetics relative to other voltammetric methods.

In order to make sufficiently relevant comparisons of $\left[E_f^0\right]_{surf}$ values between methods, ideally the same P450 sample would have to be used (in addition to

Table 4. Voltammetrically determined $\left[E_f^0\right]_{\text{surf}}$ values for cytochrome P450cam on different modified electrodes under similar solution conditions (pH 7.2 ± 0.2).

Modifier	Electrode	$\left[E_f^0\right]_{\text{surf}}$/ mV vs SHE	Reference
DDAB	EPG	22	[45]
MPS-(PEI/PSS)	Au	−6	[77]
DDAB/BSA/glutaraldehyde	GC	−19	[73]
(PEI)$_6$	BPG	−74	[46]
DMPC	EPG	−121	[45]
Clay	GC	−139	[80]
Solution (potentiometry)	n/a	-303 ± 10	[20]

other experimental conditions). As yet, no systematic comparison of methods has been undertaken. Table 4 shows a collection of reported $\left[E_f^0\right]_{\text{surf}}$ values for P450cam derived under different electrode conditions. The significant variation in the value of $\left[E_f^0\right]_{\text{surf}}$ is immediately obvious; both when compared between modification techniques (up to 160 mV) and even more so with the potentiometrically determined value (up to 325 mV). It would appear then that the electrode and any electrode modifiers (e.g., large cations, polyelectrolytes, surfactants, lipids) can affect the thermodynamics (and kinetics) of the electron transfer process.

For cast films of synthetic surfactants or lipids that contain P450 enzymes or other heme-containing proteins, the value of $\left[E_f^0\right]_{\text{surf}}$ for the protein was found to be dependent on the nature of the lipid [44,45,84,85]. As shown in Table 4, both DDAB and DMPC resulted in less negative reversible potential values for P450cam when compared with the potentiometrically determined value, but the magnitude of these increases was significantly different (325 and 182 mV respectively) [45]. This positive shift in the measured reversible potential of P450s, and those similarly observed for other heme proteins, has been ascribed to interactions between the lipid and the protein; which may result in a partial unfolding (or enforced refolding) of the protein structure, and hence a change in the level of solvent exposure to the heme, and perhaps even a change in spin-state. The possibility of electrical double-layer effects on the potential experienced by the protein at the film/electrode interface have also been proposed [45,85]. The largely hydrophobic environment of a multilayered surfactant film is dramatically different from that experienced by the enzyme when in a buffered aqueous solution in which potentiometric titrations are performed [70]. This is very likely to impact on the dielectric environment of the heme, and hence bring about changes in the reduction potential, given the increased stability of the Fe(II) relative to the Fe(III) state.

In solution, the interaction between cytochrome P450cam and the clay mineral, sodium montmorillonite, resulted in no measurable structural changes [80].

However, when adsorbed at a clay-modified electrode, a $\Delta\left[E_f^0\right]_{surf}$ of 164 mV was observed, compared with the value determined in solution. In this case, it was surmised that the adsorption process leads to partial dehydration of the environment near the heme.

Surface-enhanced resonance Raman spectroscopy (SERRS) was used to identify some conformational and redox potential changes for P450cam when it was immobilized on Ag particles coated with different length self-assembled monolayers (SAMs) [86]. The changes observed were related to the strength of the coordinative bonds of the axial ligands, dehydration of the enzyme, and the electric-field-induced destabilization of the ferrous state.

Electrode- and electrode-modifier-induced reversible potential changes have been reported for other heme- and non-heme-containing proteins [84,85,87–90]. Heme exposure to solvent, or more significantly in these cases, lack of exposure, has been shown to be a key factor that affects the observed reversible potential. For cytochrome b_5, the complex formed between the protein and a polylysine-modified electrode surface was found to not only neutralize the charge on the exposed heme propionate, but effectively to lessen the dielectric environment at the heme by the exclusion of water [88,90]. These changes resulted in an up to 100 mV increase in the measured value of $\left[E_f^0\right]_{surf}$, making it energetically more favorable for heme reduction. Complexation between the anionic proteins, rubredoxin and ferredoxin, and positively charged polyions or multivalent ions, whether in solution or at an electrode surface, also resulted in shifts to less negative potentials for these proteins [89].

As with $\left[E_f^0\right]_{soln}$, the value of $\left[E_f^0\right]_{surf}$ for P450s has also been shown to depend on pH. Figure 11 shows the effect of pH on the $\left[E_f^0\right]_{surf}$ value determined for human and bovine P450arom when confined in DDAB at an EPG electrode. The trend of approximately -59 mV change per pH unit has been identified for a number of P450s, including P450cam [44,46], P450cin [6], P450BM3 [83], P450 2C9 [67,70], P450 2C18, and P450 2C19 [67]. This confirms that even when confined at an electrode surface, protonation of a species near the heme occurs upon reduction. The identity of the ferrous protonation site is not yet confirmed, but given the high pK_a values ($pK_a > 10$) generally inferred by the available voltammetric data, OH^- is the likely base that gets protonated upon reduction [67].

The solution ionic strength has also been shown to affect the value of $\left[E_f^0\right]_{surf}$ [68,70]. In the presence and absence of substrate, the $\left[E_f^0\right]_{surf}$ values for both P450 2C9 (human) and P450 C17 (porcine) became less negative by about 20 mV when the ionic strength was increased 10-fold from 0.1 to 1.0 M potassium phosphate. This implies that at the higher ionic strength, the P450 Fe(II) form is stabilized over the Fe(III) form.

The kinetics of some surface-confined P450 electron transfer reactions have been determined; with $\left[k_f^0\right]_{surf}$ values varying from approximately 2 to 650 s^{-1} [45,65,70,72,76,81,83]. Most of these values have been based on the

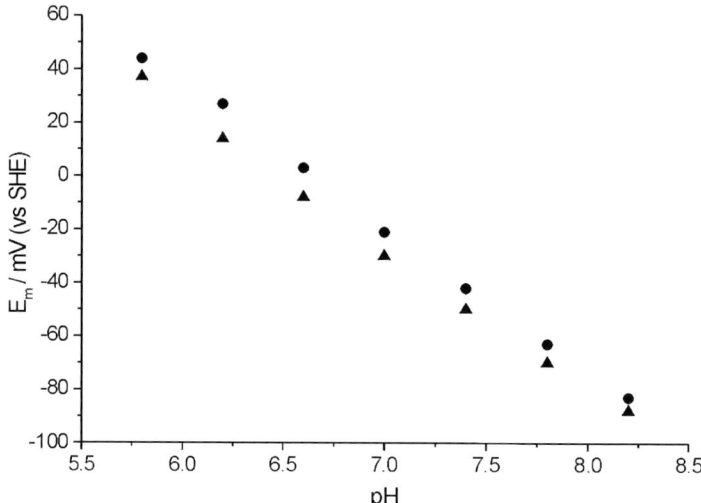

Figure 11. $[E_f^0]_{surf}$ values for human (circles) and bovine (triangles) P450arom as a function of pH when confined in a biomembrane-like layer of DDAB at an EPG electrode.

thin-film electrochemical model proposed by Laviron [91]. There has been some contention over the applicability of this model to surface-confined P450s, on the basis that the layers formed by these confining procedures may not necessarily result in diffusionless thin-layer voltammetric behavior [92]. In addition, the varied approaches used to immobilize the different P450s (and experimental conditions) make it difficult to compare $[k_f^0]_{surf}$ values.

4. CONCLUSIONS

The electrochemistry of cytochrome P450 consists of data derived from potentiometric titrations and voltammetric experiments. To date the focus has essentially been on the collection of a 'redox potential' value, without much thought given to the meaning of this value. This has led to difficulties when comparisons between P450s and different experimental conditions are attempted.

More potentiometric studies are encouraged; $[E_f^0]_{surf}$ values obtained by this approach can be an accurate, reliable and comparable, particularly if the enzyme sample is known to be pure, a systematic, consistent methodology is taken, and protocols are carefully thought through.

Ideally, voltammetric data $[E_f^0]_{surf}$ should corroborate that obtained by potentiometry $[E_f^0]_{soln}$. However, it is quite frustrating and disappointing that the straightforward voltammetric experiment does not meet this expectation. The different symbols used in this review for the potential values obtained by each method are for good reasons: they are not the same! As shown, the $\left[E_f^0\right]_{surf}$ values are significantly dependent on the method of surface immobilization. We would therefore recommend that the best form of comparison would consist of using, if possible, a similar immobilization approach for all the different P450s. Future work should also be aimed at understanding the factors contributing to the E_f^0 values, as well as more effort to obtaining thin films (monolayers), as the data obtained from trapping methods currently used are much harder to understand. Kinetic studies with respect to P450s are in their infancy, particularly from voltammetric experiments. Given that k_f^0 applies specifically to the value of $\left[E_f^0\right]_{surf}$, very little can be concluded at this stage.

ACKNOWLEDGMENTS

Financial support from the Australian Research Council in the form of a Discovery Grant and Federation Fellowship to A.M.B., and a Monash University Research Fund Postdoctoral Fellowship to B.D.F. are gratefully acknowledged.

ABBREVIATIONS

BPG	basal plane graphite
BSA	bovine serum albumin
DDAB	didodecyldimethlammonium bromide
DDAPSS	didodecylammonium poly(styrenesulphonate)
DMPC	dimyristoylphosphatidylcholine
DTT	dithiothreitol
E^0	reversible standard potential (under standard conditions)
E_f^0	reversible formal potential (under non-standard conditions)
$\left[E_f^0\right]_{soln}$	reversible formal potential for a solution-phase redox process
$\left[E_f^0\right]_{surf}$	reversible formal potential for a surface-confined redox process
EDTA	ethylenediamine-N,N,N',N'-tetraacetate
EPG	edge-plane graphite
FMN	flavin mononucleotide
GC	glassy carbon
GPC	glyceryl-3-phosphorylcholine
ITO	indium–tin oxide
k_f^0	rate constant
$\left[k_f^0\right]_{soln}$	rate constant for a solution-phase redox process
$\left[k_f^0\right]_{surf}$	rate constant for a surface-confined redox process
$\left[k_f^0\right]_{thin\,film}$	rate constant for a thin-film redox process
MPA	3-mercaptopropionic acid
MPS	3-mercapto-1-propanesulfonic acid
NHE	normal hydrogen electrode

PDDA	poly(dimethyldiallylammonium chloride)
PG	pyrolytic graphite
Phos	phosphate
PSS	poly(styrenesulphonate)
R_u	uncompensated resistance
SAM	self-assembled monolayer
SCE	saturated calomel electrode
SERRS	surface-enhanced resonance Raman spectroscopy
Tris	tris-(hydroxymethyl)-aminomethane

REFERENCES

1. P. L. Dutton, *Methods Enzymol., 54*, 411–435 (1978).
2. A. J. Bard and L. R. Faulkner, *Electrochemical Methods: Fundamentals and Applications*, John Wiley & Sons, Inc., New York, 2001.
3. S. G. Sligar and I. C. Gunsalus, *Biochemistry, 18*, 2290–2295 (1979).
4. S. G. Sligar, *Biochemistry, 15*, 5399–5406 (1976).
5. In potentiometric titrations it has been commonly denoted, E_0' or E_m, where 'm' stands for the midpoint, given that non-standard conditions are typically used.
6. K.-F. Aguey-Zinsou, P. V. Bernhardt, J. J. De Voss, and K. E. Slessor, *Chem. Commun., 3*, 418–419 (2003).
7. R. J. Kassner, *J. Am. Chem. Soc., 95*, 2674–2677 (1973).
8. A. K. Churg and A. Warshel, *Biochemistry, 25*, 1675–1681 (1986).
9. G. R. Moore, G. W. Pettigrew, and N. K. Rogers, *Proc. Natl. Acad. Sci. USA, 83*, 4998–4999 (1986).
10. D. F. V. Lewis and P. Hlavica, *Biochim. Biophys. Acta, 1460*, 353–374 (2000).
11. T. W. B. Ost, C. S. Miles, A. W. Munro, J. Murdoch, G. A. Reid, and S. K. Chapman, *Biochemistry, 40*, 13421–13429 (2001).
12. S. N. Daff, S. K. Chapman, K. L. Turner, R. A. Holt, S. Govindaraj, T. L. Poulos, and A. W. Munro, *Biochemistry, 36*, 13816–13823 (1997).
13. D. R. Light and N. R. Orme-Johnson, *J. Biol. Chem., 256*, 343–350 (1981).
14. D. J. B. Hunter, G. A. Roberts, T. W. B. Ost, J. H. White, S. Muller, N. J. Turner, S. L. Flitsch, and S. K. Chapman, *FEBS Lett., 579*, 2215–2220 (2005).
15. R. J. Lawson, D. Leys, M. J. Sutcliffe, C. A. Kemp, M. R. Cheesman, S. J. Smith, J. Clarkson, W. E. Smith, I. Haq, J. B. Perkins, and A. W. Munro, *Biochemistry, 43*, 12410–12426 (2004).
16. F. P. Guengerich, D. P. Ballou, and M. J. Coon, *J. Biol. Chem., 250*, 7405–7414 (1975).
17. F. P. Guengerich, *Biochemistry, 22*, 2811–2820 (1983).
18. Y. Shiro, M. Fujii, Y. Isogai, S.-i. Adachi, T. Iizuka, E. Obayashi, R. Makino, K. Nakahara, and H. Shoun, *Biochemistry, 34*, 9052–9058 (1995).
19. V. Reipa, M. P. Mayhew, M. J. Holden, and V. L. Vilker, *Chem. Commun., 4*, 318–319 (2002).
20. M. T. Fisher and S. G. Sligar, *J. Am. Chem. Soc., 107*, 5018–5019 (1985).

21. S. G. Sligar, D. L. Cinti, G. G. Gibson, and J. B. Schenkman, *Biochem. Biophys. Res. Commun., 90*, 925–932 (1979).
22. R. Raag and T. L. Poulos, *Biochemistry, 28*, 917–922 (1989).
23. S. G. Sligar and I. C. Gunsalus, *Proc. Natl. Acad. Sci. USA, 73*, 1078–1082 (1976).
24. S. A. Martinis, S. R. Blanke, L. P. Hager, S. G. Sligar, G. Hui Bon Hoa, J. J. Rux, and J. H. Dawson, *Biochemistry, 35*, 14530–14536 (1996).
25. E.-S. Wu and C. S. Yang, *Biochemistry, 23*, 28–33 (1984).
26. K.-H. Kim, T. Ahn, and C.-H. Yun, *Biochemistry, 42*, 15377–15387 (2003).
27. G. A. Roberts, A. Celik, D. J. B. Hunter, T. W. B. Ost, J. H. White, S. K. Chapman, N. J. Turner and S. L. Flitsch, *J. Biol. Chem., 278*, 48914–48920 (2003).
28. J. D. Lambeth and S. O. Pember, *J. Biol. Chem., 258*, 5596–5602 (1983).
29. S. Yoshioka, S. Takahashi, K. Ishimori, and I. Morishima, *J. Inorg. Biochem., 81*, 141–151 (2000).
30. S. Yoshioka, T. Tosha, S. Takahashi, K. Ishimori, H. Hori, and I. Morishima, *J. Am. Chem. Soc., 124*, 14571–14579 (2002).
31. T. W. B. Ost, J. P. Clark, C. G. Mowat, C. S. Miles, M. D. Walkinshaw, G. A. Reid, S. K. Chapman, and S. Daff, *J. Am. Chem. Soc., 125*, 15010–15020 (2003).
32. M. Unno, H. Shimada, Y. Toba, R. Makino, and Y. Ishimura, *J. Biol. Chem., 271*, 17869–17874 (1996).
33. T. Tosha, S. Yoshioka, H. Hori, S. Takahashi, K. Ishimori, and I. Morishima, *Biochemistry, 41*, 13883–13893 (2002).
34. K. Auclair, P. Moenne-Loccoz, and P. R. Ortiz de Montellano, *J. Am. Chem. Soc., 123*, 4877–4885 (2001).
35. T. L. Poulos, *J. Biol. Inorg Chem., 1*, 356–359 (1996).
36. A. Dey, T.-a. Okamura, N. Ueyama, B. Hedman, K. O. Hodgson, and E. I. Solomon, *J. Am. Chem. Soc., 127*, 12046–15053 (2005).
37. F. Ogliaro, S. P. de Visser, and S. Shaik, *2002, 91*, 554–567 (2002).
38. T. W. B. Ost, A. W. Munro, C. G. Mowat, P. R. Taylor, A. Pesseguiero, A. J. Fulco, A. K. Cho, M. A. Cheesman, M. D. Walkinshaw, and S. K. Chapman, *Biochemistry, 40*, 13430–13438 (2001).
39. Z. Chen, T. W. B. Ost, and J. P. M. Schelvis, *Biochemistry, 43*, 1798–1808 (2004).
40. L.-H. Guo and H. A. O. Hill, *Adv. Inorg. Chem., 36*, 341–375 (1991).
41. A. M. Bond, *Inorg. Chim. Acta., 226*, 293–340 (1994).
42. F. Armstrong, H. A. O. Hill, and N. J. Walton, *Acc. Chem. Res., 21*, 407–413 (1988).
43. F. A. Armstrong and G. S. Wilson, *Electrochim. Acta, 45*, 2623–2645 (2000).
44. J. F. Rusling, *Acc. Chem. Res., 31*, 363–369 (1998).
45. Z. Zhang, A.-E. F. Nassar, Z. Lu, J. B. Schenkman, and J. F. Rusling, *J. Chem. Soc., Faraday Trans., 93*, 1769–1774 (1997).
46. B. Munge, C. Estavillo, J. B. Schenkman, and J. F. Rusling, *ChemBioChem, 4*, 82–89 (2003).
47. A. M. Bond, *Broadening Electrochemical Horizons*, Oxford University Press, New York, 2002.
48. J. F. Rusling and Z. Zhang, in *Biomolecular Films: Design, Function and Applications* (J. F. Rusling, ed.), Marcel Dekker Inc., New York, 2003, pp. 1–64.
49. V. V. Shumyantseva, T. V. Bulko, and A. I. Archakov, *J. Inorg. Biochem., 99*, 1051–1063 (2005).

50. N. Bistolas, U. Wollenberger, C. Jung, and F. W. Scheller, *Biosens. Bioelectron.*, 20, 2408–2423 (2005).
51. A. K. Udit and H. B. Gray, *Biochem. Biophys. Res. Commun.*, 338, 470–476 (2005).
52. F. Scheller, R. Renneberg, W. Schwarze, G. Strnad, K. Pommerening, H.-J. Prumke, and P. Mohr, *Acta Biol. Med. Germ.*, 38, 503–509 (1979).
53. K. M. Faulkner, M. S. Shet, C. W. Fisher, and R. W. Estabrook, *Proc. Natl. Acad. Sci. USA*, 92, 7705–7709 (1995).
54. R. W. Estabrook, K. M. Faulkner, M. S. Shet, and C. W. Fisher, *Methods Enzymol.*, 272, 44–51 (1996).
55. V. Reipa, M. P. Mayhew, and V. L. Vilker, *Proc. Natl. Acad. Sci. USA*, 94, 13554–13558 (1997).
56. N. Sugihara, Y. Ogoma, K. Abe, Y. Kondo, and T. Akaike, *Polym. Adv. Technol.*, 9, 307–313 (1998).
57. V. V. Shumyantseva, T. V. Bulko, S. A. Alexandrova, N. N. Sokolov, R. D. Schmid, T. Bachmann, and A. I. Archakov, *Biochem. Biophys. Res. Commun.*, 263, 678–680 (1999).
58. M. P. Mayhew, V. Reipa, M. J. Holden and V. L. Vilker, *Biotechnol. Prog.*, 16, 610–616 (2000).
59. V. V. Shumyantseva, T. V. Bulko, T. T. Bachmann, U. Bilitewski, R. D. Schmid, and A. I. Archakov, *Arch. Biochem. Biophys.*, 377, 43–48 (2000).
60. A. K. Udit, F. H. Arnold, and H. B. Gray, *J. Inorg. Biochem.*, 98, 1547–1550 (2004).
61. C. Paternolli, M. Antonini, P. Ghisellini, and C. Nicolini, *Langmuir*, 20, 11706–11712 (2004).
62. N. Sultana, J. B. Schenkman, and J. F. Rusling, *J. Am. Chem. Soc.*, 127, 13460–13461 (2005).
63. J. Kazlauskaite, A. C. G. Westlake, L.-L. Wong, and H. A. O. Hill, *Chem. Commun.*, 18, 2189–2190 (1996).
64. K. K.-W. Lo, L.-L. Wong, and H. A. O. Hill, *FEBS Lett.*, 451, 342–346 (1999).
65. A. Fantuzzi, M. Fairhead, and G. Gilardi, *J. Am. Chem. Soc.*, 126, 5040–5041 (2004).
66. C. Nicolini, V. Erokhin, P. Ghisellini, C. Paternolli, M. K. Ram, and V. Sivozhelezov, *Langmuir*, 17, 3719–3726 (2001).
67. A. Shukla, E. M. Gillam, D. J. Mitchell, and P. V. Bernhardt, *Electrochem. Commun.*, 7, 437–442 (2005).
68. D. L. Johnson, A. J. Conley, and L. L. Martin, *J. Mol. Endocrinol.*, 36, 349–359 (2006).
69. Y. Oku, A. Ohtaki, S. Kamitori, N. Nakamura, M. Yohda, H. Ohno, and Y. Kawarabayasi, *J. Inorg. Biochem.*, 98, 1194–1199 (2004).
70. D. L. Johnson, B. C. Lewis, D. J. Elliot, J. O. Miners, and L. L. Martin, *Biochem. Pharmacol.*, 69, 1533–1541 (2005).
71. L. S. Koo, C. E. Immos, M. S. Cohen, P. J. Farmer, and P. R. Ortiz de Montellano, *J. Am. Chem. Soc.*, 124, 5684–5691 (2002).
72. B. D. Fleming, Y. Tian, S. G. Bell, L.-L. Wong, V. Urlacher, and H. A. O. H. Hill, *Eur. J. Biochem.*, 270, 4082–4088 (2003).
73. E. I. Iwuoha, S. Joseph, Z. Zhang, M. R. Smyth, U. Fuhr, and P. R. Ortiz de Montellano, *J. Pharm. Biomed. Anal.*, 17, 1101–1110 (1998).
74. E. Iwuoha and M. R. Smyth, *Biosens. Bioelectron.*, 12, 53–75 (1997).

75. E. Iwuoha, S. Kane, C. O. Ania, M. R. Smyth, P. R. Ortiz de Montellano, and U. Fuhr, *Electroanalysis, 12*, 980–986 (2000).
76. A. K. Udit, N. Hindoyan, M. G. Hill, F. H. Arnold, and H. B. Gray, *Inorg. Chem., 44*, 4109–4111 (2005).
77. Y. M. Lvov, Z. Lu, J. B. Schenkman, X. Zu, and J. F. Rusling, *J. Am. Chem. Soc., 120*, 4073–4080 (1998).
78. C. Estavillo, Z. Lu, I. Jansson, J. B. Schenkman, and J. F. Rusling, *Biophys. Chem., 104*, 291–296 (2003).
79. S. Joseph, J. F. Rusling, Y. M. Lvov, T. Friedberg, and U. Fuhr, *Biochem. Pharmacol., 65*, 1817–1826 (2003).
80. C. Lei, U. Wollenberger, C. Jung, and F. W. Scheller, *Biochem. Biophys. Res. Commun., 268*, 740–744 (2000).
81. V. V. Shumyantseva, Y. D. Ivanov, N. Bistolas, F. W. Scheller, A. I. Archakov, and U. Wollenberger, *Anal. Chem., 76*, 6046–6052 (2004).
82. G. S. Xuan, S. Jung, and S. Kim, *Bull. Korean Chem. Soc., 25*, 165–166 (2004).
83. A. K. Udit, M. G. Hill, V. G. Bittner, F. H. Arnold, and H. B. Gray, *J. Am. Chem. Soc., 126*, 10218–10219 (2004).
84. A.-E. F. Nassar, Y. Narikiyo, T. Sagara, and N. Nakashima, *J. Chem. Soc., Faraday Trans., 91*, 1775–1782 (1995).
85. Z. Zhang and J. F. Rusling, *Biophys. Chem., 63*, 133–146 (1997).
86. S. Todorovic, C. Jung, P. Hildebrandt, and D. H. Murgida, *J Biol Inorg Chem, 11*, 119–127 (2006).
87. F. A. Tezcan, J. R. Winkler, and H. B. Gray, *J. Am. Chem. Soc., 120*, 13383–13388 (1998).
88. M. Rivera, R. Seetharaman, D. Girdhar, M. Wirtz, X. Zhang, X. Wang, and S. White, *Biochemistry, 37*, 1485–1494 (1998).
89. Z. Xiao, M. J. Lavery, A. M. Bond, and A. G. Wedd, *Electrochem. Commun., 1*, 309–314 (1999).
90. M. Wirtz, V. Oganesyan, X. Zhang, J. Studer, and M. Rivera, *Faraday Discuss., 116*, 221–234 (2000).
91. E. Laviron, *J. Electroanal. Chem., 101*, 19–28 (1979).
92. M. J. Honeychurch, *Langmuir, 14*, 235–237 (1998).

6

P450 Electron Transfer Reactions

Andrew K. Udit,[1] *Stephen M. Contakes,*[2] *and Harry B. Gray*[2]

[1]Department of Chemistry, Occidental College, Los Angeles, CA 90041, USA
[2]Beckman Institute, California Institute of Technology, Pasadena, CA 91125, USA
<udit@oxy.edu>
<contakes@caltech.edu>
<hbgray@caltech.edu>

1. INTRODUCTION	158
2. CATALYTIC CYCLES	158
2.1. Native Mechanism	158
2.2. Electron Transfer Pathways	160
2.2.1. P450BM3: A Bacterial Model for Mammalian Systems	160
2.3. Dioxygen Reduction	163
2.4. Biocatalysis: The P450 Holy Grail	165
2.4.1. Mediators	167
2.5. Electrochemical Heme Oxidation	168
3. ELECTRON TUNNELING WIRES	171
3.1. Metal-Diimine Wires	171
3.2. P450-Wire Conjugates	174
3.3. Wire-Mediated P450 Reduction	178
3.4. Wire-Mediated P450 Oxidation	179
3.5. Electronic Coupling	180
4. CONCLUDING REMARKS	180
ACKNOWLEDGMENTS	181
ABBREVIATIONS	181
REFERENCES	181

Metal Ions in Life Sciences, Volume 3   Edited by Astrid Sigel, Helmut Sigel and Roland K. O. Sigel
© 2007 John Wiley & Sons, Ltd

1. INTRODUCTION

The heme-thiolate cytochromes P450 (P450) perform regio- and stereospecific reactions under physiological conditions [1–3]. The hallmark reaction catalyzed by these enzymes is catalytic hydroxylation of hydrocarbons:

$$R\text{–}H + NAD(P)H + H^+ + O_2 \rightarrow R\text{–}OH + NAD(P)^+ + H_2O \qquad (1)$$

The vast P450 synthetic repertoire also includes alkene epoxidation, heteroatom (N,S) oxidation, dealkylation, and (anaerobic) dehalogenation [4]. *In vivo* activities include hormone biosynthesis, xenobiotic elimination, drug (de-)activation, and carcinogenesis [5], while potential *in vitro* applications range from chemical synthesis [6] to biosensing [7].

The signature of all P450s is the heme Soret band at ~ 450 nm in the absorption spectrum of the Fe^{II}–CO complex; this spectroscopic feature is diagnostic of a cysteinate bound *trans* to the carbon monoxide ligand [8]. Thiolate ligation to the iron center is proposed to be critical for P450 catalysis, facilitating heterolytic cleavage of dioxygen for subsequent substrate oxidation [9]. Indeed, mutating the P450 axial cysteine to another amino acid abolishes the unique chemistry catalyzed by P450 [10].

Central to P450 catalysis is electron transfer (ET). Catalysis is initiated with the first ET event, while the overall rate is limited by the second ET. Much effort has been expended to investigate the factors that govern ET to and from P450 heme centers in order to understand the catalytic mechanism and ultimately harness P450 oxidation chemistry for *in vitro* applications.

2. CATALYTIC CYCLES

2.1. Native Mechanism

The generally accepted catalytic cycle for P450 is shown in Figure 1 [2]. The resting state is a six-coordinate low-spin Fe^{III} (E^o between -330 and -370 mV versus NHE, depending on the P450), with porphyrin nitrogen donor atoms occupying the four equatorial positions and cysteine (proximal) and water (distal) in the axial positions. Catalysis is initiated by substrate binding in the protein active site with displacement of the axial water, thereby converting the six-coordinate low-spin Fe^{III} heme to a five-coordinate high-spin complex. Substrate binding is accompanied by a change in heme potential (substrate-dependent, but generally greater than $+100$ mV) that has been shown to initiate ET to the heme from native reductase proteins [11]. One-electron reduction to Fe^{II} allows dioxygen to bind to the iron, forming an Fe^{II}–O_2 complex. The rate-limiting step in the cycle is a second one-electron reduction followed by protonation that

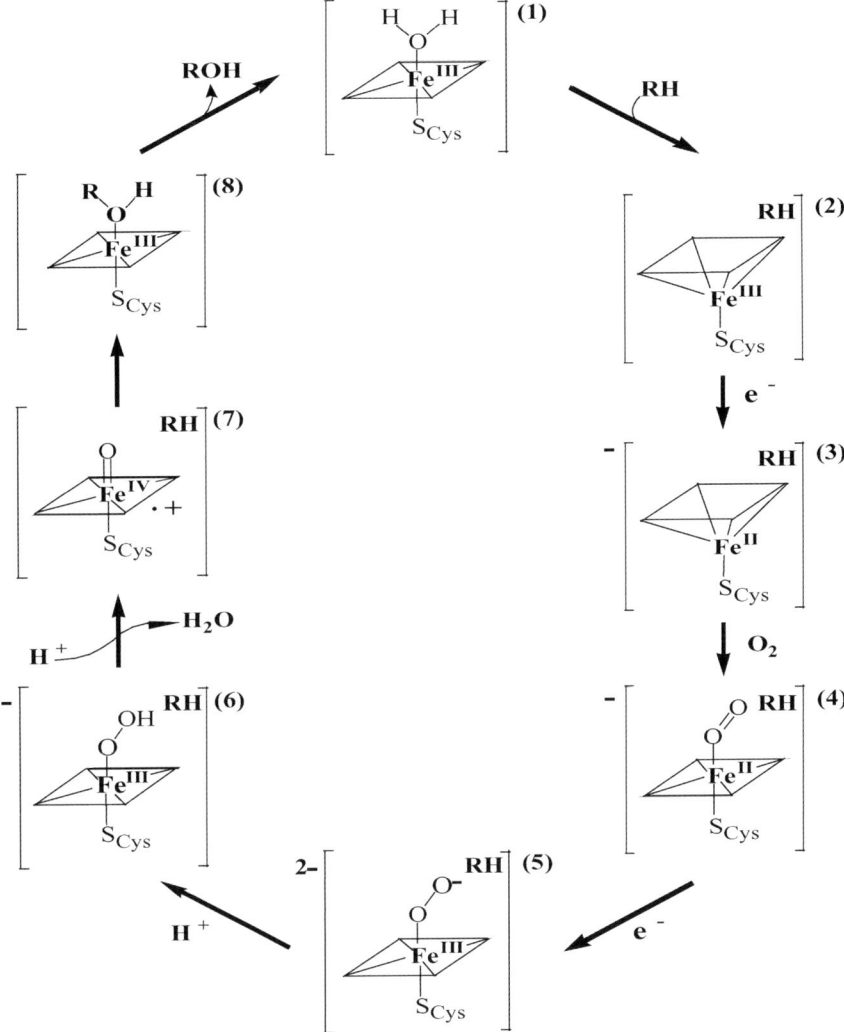

Figure 1. The P450 catalytic cycle.

generates the peroxy complex, Fe^{III}–OOH^-. This species is further protonated to yield a molecule of water and a high-valent iron-oxo active intermediate referred to as compound I. The precise nature of this iron-oxo complex is not definitively known, although most likely it is an Fe^{IV}=O unit coupled to a porphyrin cation radical [12]. In the standard catalytic cycle, compound I oxygenates substrate and regenerates the P450 resting state.

An alternative catalytic pathway termed the 'peroxide shunt' also exists. In this pathway, reduced dioxygen is supplied in the form of peroxide, which can enter the catalytic cycle and directly generate the Fe^{III}–OOH^- complex (**6** in Figure 1). This pathway has been discussed in detail elsewhere [13,14].

2.2. Electron Transfer Pathways

P450s can be divided into two general categories, mammalian and bacterial, with key differences in their ET machinery. While bacterial forms are soluble proteins that receive electrons from cytosolic ferredoxins (Fe-S proteins) and ferredoxin reductases, mammalian proteins by contrast are membrane-bound and are reduced by membrane-associated flavin-containing reductases. Additionally, the two classes also display low sequence identity (e.g., 16% amino acid identity between human cytochrome P4501A2 and bacterial P450BM3 [15]); in fact, even closely related P450s are known to be somewhat divergent in sequence (e.g., 63% amino acid identity between *Bacillus* P450 homologs 102A1 and 102A2 [16]). Despite these differences, P450s share a common catalytic mechanism governed by ET (Figure 1): two electrons are delivered to the P450 heme by native reductase proteins in a highly controlled manner, resulting in the activation of dioxygen for subsequent substrate oxidation.

2.2.1. P450BM3: A Bacterial Model for Mammalian Systems

P450 from *Bacillus megaterium* (P450BM3) is typical of bacterial isoforms — it is water soluble, relatively stable, and exhibits rapid turnover rates — yet it utilizes a mammalian-like diflavin reductase and has higher sequence identity with eukaryotic forms. Indeed, the primary amino acid sequence of the P450BM3 reductase and heme domains are approximately 35 and 30% identical with their eukaryotic microsomal counterparts [17]. The similarity of P450BM3 to mammalian P450s has made this protein an attractive model for eukaryotic enzymes [18], particularly since crystal structures of the P450BM3 heme domain complexed with the FMN portion of the reductase are available [19]. Despite advances in crystallization that have produced some mammalian P450 structures (e.g., P4503A4 [20], P4502C5 [21], P4502C9 [22]), analogous interdomain structures are not available for the eukaryotic isoforms. Such structures would certainly shed light on heme domain-reductase interactions that facilitate ET.

P450BM3 (119 kDa) consists of two domains: (1) a reductase domain at the carboxy end of the polypeptide containing FAD and FMN cofactors is responsible for mediating ET from NADPH in solution to the heme domain; and (2) an amino-terminal heme domain that contains Fe^{III}-protoporphyrin IX [23]. ET in P450BM3 begins with NADPH in solution ($E° = -340$ mV versus NHE) transferring hydride to the FAD in the reductase domain [11].

Subsequent transfer of an electron to the second reductase flavin (FMN) results in formation of the semiquinone forms of each flavin. Reduction of the heme by the FMN semiquinone can only occur with substrate bound, which effects a low- to high-spin transition in the heme and shifts the heme potential ($\Delta E° > 100\,\text{mV}$). A second FAD electron is then transferred to FMN, which in turn reduces the heme a second time according to the scheme in Figure 1.

The crystal structure of the P450BM3 heme domain complexed with the FMN domain (Figure 2) provides key insights into the native ET pathway between FMN (blue) and the heme (red). As shown in the structure, the heme is deeply buried within the protein matrix, making it difficult for an external source to reduce the iron. Thus, proper positioning and docking of the FMN domain directly underneath the heme is likely to be crucial for efficient ET. The FMN isoalloxazine ring has its methyl groups pointing directly to the heme binding loop (residues 382–400, highlighted in yellow in Figure 2); without the heme domain, this part of the FMN would be solvent-exposed. Thus, the methyl groups are expected to act as a bridge, providing an interdomain pathway for ET. Indeed, distance calculations put the C7-methyl group only 4 Å from the amide nitrogen of Gln^{387} [19].

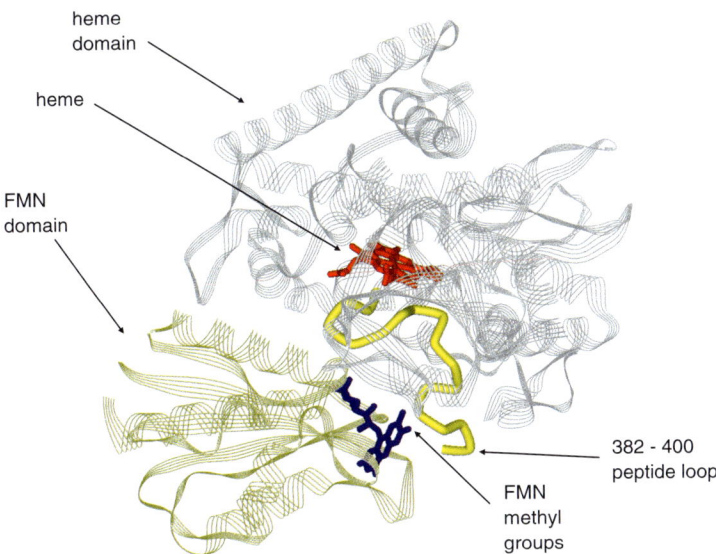

Figure 2. Structure of the complex between the cytochrome P450BM3 heme and native FMN domains. Highlighted are the heme (red) and FMN (blue) cofactors and the intervening 382–400 peptide loop (yellow).

Unlike the complex between cytochrome P450 from *Pseudomonas putida* (P450CAM) and putidaredoxin (Pdx) (the prototypal bacterial P450 system), where electrostatic forces are the main contributors to the docking interactions between the two ET domains [24], P450BM3 likely uses a combination of forces to bring together its ET partners. Electrostatic interactions between the two P450BM3 domains are relatively weak: analysis of the crystal structure reveals two direct hydrogen bonds, one salt bridge (between His^{100} and Glu^{494}), and several water-mediated contacts [19]. Given that the contact interface spans some 967 Å^2, these electrostatic interactions contribute minimally to docking. Most likely, the electrostatic forces act in concert with the short peptide linker between the heme and FMN domains to orient the two partners for ET. In fact, substrate oxidations catalyzed by separately expressed reductase and heme domains exhibit poor turnover rates in the presence of NADPH [25]. The striking difference in catalytic rate between P450BM3 and mammalian enzymes is attributable in part to the absence of a covalent linkage between heme and reductase domains in the latter: while P450BM3 typically displays rates on the order of 1000 turnovers per minute [26], the mammalian systems are two to three orders of magnitude slower (e.g., $\sim$ 18 turnovers per minute for P4502B1-mediated oxidation of benzphetamine) [27]. Note also that the FMN to heme ET rate for P450BM3 is of the order of $\sim 200\,s^{-1}$ [28].

The positioning of P450BM3 domains and cofactors suggests that Gln^{387} plays an important role in mediating interdomain ET. Support for this proposal comes from experiments in which the heme domain had been covalently tethered to a ruthenium(II) diimine in the single-surface cysteine mutant N387C [29]. Very rapid ET rates from electronically excited Ru^{II} to the heme suggest that the complex is well coupled to the iron center (2.5×10^6 and $4.6 \times 10^5 \, s^{-1}$ with and without substrate, respectively).

With the photochemical work as a guide, we designed electrochemical experiments aimed at exploiting the same surface cysteine mutant [30]. In these experiments, *N*-(1-pyrene)iodoacetamide covalently tethered to the P450BM3 heme domain surface thiolate was used to anchor the enzyme to a basal plane graphite electrode. Electrochemically wiring the enzyme in this way converts the electrode to an artificial reductase. Analysis of cyclic voltammetry experiments ($Fe^{III/II}$ redox couple at $-340\,mV$ versus Ag/AgCl) yielded a standard rate constant (k°, $\Delta G^\circ = 0$) of $\sim 650\,s^{-1}$ [30]. This ET rate is among the highest reported for a P450-electrode system, underscoring the relatively strong electronic coupling to the heme achieved through amino acid 387. ET rates for photochemical heme reduction extrapolated to zero driving force (assuming $\lambda = 0.8\,eV$ [31]) are 280 and $3300\,s^{-1}$ for substrate-free and substrate-bound P450BM3, values that are similar to k° obtained from the electrochemical experiments. Our finding of similar rates by two different methods utilizing the same surface position for probe attachment suggests that interactions at the surface in the immediate vicinity of Gln^{387} facilitate interdomain ET in the native system.

2.3. Dioxygen Reduction

The P450 catalytic mechanism (Figure 1) involves the reduction of heme-bound dioxygen to water. This can occur by the native pathway in the presence of substrate (forming compound I, represented as $Fe^V=O$ in Figure 3) to give oxidized substrate and regenerate the $Fe^{III}-OH_2$ resting state. Other routes, termed 'uncoupled pathways' (or simply 'uncoupling'), result in the consumption of reducing equivalents to form one-, two-, or four-electron reduced species of dioxygen without substrate oxidation (hence, product formation is uncoupled from the consumption of electrons).

The mechanism of dioxygen reduction begins with formation of the ferrous-oxy (or ferric-superoxide) complex, $Fe^{II}-O_2$. Uncoupling in the catalytic cycle can occur at this point, resulting in release of superoxide. However, compared with the other species in the catalytic cycle, the ferric-superoxide complex is relatively long-lived (autooxidation at $\sim 0.05\ s^{-1}$ for P450CAM $Fe^{II}-O_2$ at 20 °C,

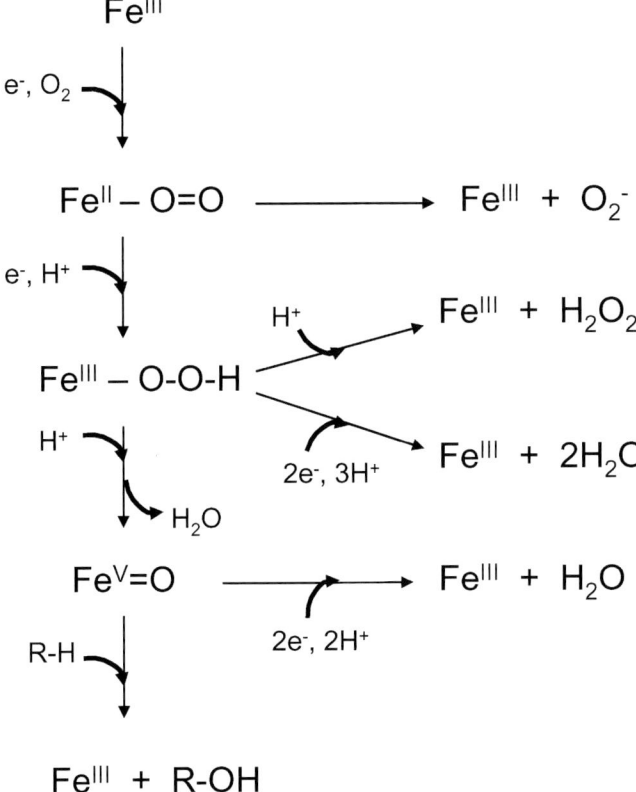

Figure 3. Dioxygen reduction and substrate oxygenation catalyzed by P450.

pH 7) [32], so this pathway is a minor contributor to uncoupling. In the presence of additional reducing equivalents, the ferric-superoxide complex accepts a second electron and a proton, forming a ferric hydroperoxide, Fe^{III}–O–OH.

Once formed, the ferric hydroperoxide can decay in several ways. First, protonation of the oxygen distal to the heme iron results in release of water and formation of a high-valent ferryl species, compound I. In the presence of an appropriately positioned substrate, oxygen rebound occurs, producing oxygenated substrate and Fe^{III} heme. An alternative pathway is oxidase activity: once compound I is formed, if there is no substrate at the active site (e.g., an exogenous molecule or a protein side chain) and additional electrons are available, ferryl heme reduction by two electrons yields water and the Fe^{III} heme. Thus, oxidase activity results in consumption of four electrons (in the native system, two equivalents of NAD(P)H), four protons, and one molecule of dioxygen, yielding two molecules of water [6,33,34].

Protonation of the ferric-hydroperoxide at the oxygen proximal to the heme iron results in formation of hydrogen peroxide, which can dissociate from the heme and regenerate the Fe^{III} resting state. This is the mode of uncoupling most often observed in unproductive P450 reactions; it is seen frequently in P450s that display poor reaction efficiency [34]. In these instances, access of the ferric hydroperoxide to water (a proton donor) is often implicated as the primary reason for uncoupling via this route [2]. This is best illustrated in experiments using wild-type P450 with a series of substrates [35]: molecules considered to be the best substrates effect complete heme dehydration (every low-spin six-coordinate substrate-free (hydrated) heme is converted to a high-spin five-coordinate substrate-bound iron center). Reactions with these good substrates are generally fully coupled and give 100% yields (i.e., all electrons are used to generate oxygenated substrates). By contrast, poor substrates (or mutations that affect substrate binding) only give partial heme dehydration (mixed-spin hemes).

Clearly, active-site water molecules play different roles in P450 mechanisms. While water molecules are required for proton delivery to the distal oxygen to yield compound I, excessive active-site hydration can lead to uncoupling through protonation of the ferric hydroperoxide proximal oxygen followed by peroxide release. Crystallographic work by Nagano and Poulos has shed considerable light on this matter [36]. Their study with P450CAM has shown that active site Thr^{252} – the conserved Thr in P450s – is critical: once thought to donate a proton to the distal oxygen of the ferric hydroperoxide, Thr^{252} in fact appears to stabilize the Fe^{III}-OOH unit by acting as a hydrogen-bond acceptor. It was proposed that this interaction increases the proton affinity of the distal oxygen, promoting protonation by active-site water molecules to produce water and compound I from the ferric hydroperoxide.

The dioxygen reactions detailed above — native substrate oxidation, oxidase activity, superoxide release, and peroxide formation — are the canonical pathways for reduction of dioxygen by P450. In addition, recent electrochemical investigations [30,37] have revealed yet another pathway. Analysis of rotating-disk electrode experiments and chemical fluorescence assays have shown that P450-electrodes can catalyze four-electron reduction of dioxygen to water. A possible intermediate in these reactions is compound I, in which case the P450-electrodes should oxidize substrates. However, the absence of substrate oxidation could mean that four-electron reduction of dioxygen does not involve the formation of ferryl species: instead of following one of the canonical pathways, two-electron reduction of the ferric-peroxy complex could produce water directly (Figure 3). Precedent for this can be found in prior work with cobalt porphyrins, where it was shown that either two- or four-electron reduction of dioxygen could be effected, with the pathway of choice controlled by the electronic nature of the active complex [38]. Apparently, the key to forming water from dioxygen is to increase the ET rate such that reduction of the metal-peroxy complex occurs faster than dissociation of peroxide from the active center. With a P450-electrode, by operating at a potential more negative than the $Fe^{III/II}$ redox couple, the P450 ferric-peroxy complex is rapidly reduced, yielding water as the only product.

2.4. Biocatalysis: The P450 Holy Grail

The various pathways for dioxygen reduction underscore the complexity of the P450 catalytic cycle. ET, proton transfer, dioxygen activation to compound I, and protein conformational changes are all required in the proper combination in order to achieve substrate oxidation. A major goal in the field is to demonstrate P450 oxygenase activity *in vitro*; indeed, the development of functional P450 biocatalytic systems to solve medical [5], pharmacological [39], chemical synthetic [40], and bioremediation [41] problems is one of the holy grails of modern biological inorganic chemistry.

Expression and purification protocols for various mammalian and bacterial P450s are widely available [3], as are assays for P450 activity [42,43], leaving ET (and productive activation of dioxygen) as the primary hurdle standing in the way of a functional P450 system for biosensing and biocatalysis. *In vivo*, the ultimate source of electrons for P450 catalysis is NAD(P)H; this requirement for NAD(P)H is perhaps the greatest roadblock to commercial use of P450s, owing to its high cost, relative instability, and inefficient recovery protocols. As a result, several investigations have been launched with the goal of supplying P450 with the electrons it requires for substrate oxidation [44]. A convenient method for regenerating NAD(P)H would involve an electrode. Sadly, electrochemical regeneration of NAD(P)H proceeds through two one-electron transfer steps, a process that competes unfavorably with a pathway that forms biologically inactive dimers from one-electron reduced forms of the cofactor [45]. Chemical reductants for NAD(P)H have the additional drawback of either not being reusable or requiring regeneration themselves (which would typically involve a negatively polarized electrode that could lead to NAD(P)H dimers) [46]. Enzymatic cofactor recycling systems utilizing purified proteins provide an efficient way of regenerating NAD(P)H *in vitro* [47], coupling NAD(P)H oxidation by P450 with NAD(P)$^+$ reduction by another enzymatic system (e.g., formate dehydrogenase). However, these additional proteins complicate matters (e.g., issues regarding optimal reaction conditions and the certainty of unanticipated side reactions). Finally, *in vivo* recycling systems that exploit whole-cell machinery for NAD(P)H regeneration suffer because of poor reaction rates, in addition to the usual problems associated with whole-cell reactions (e.g., poor cell permeability of substrates/products, poor product recovery) [48].

Direct electrochemical reduction is an attractive means of providing P450 with reducing equivalents, thereby replacing the native ET machinery (e.g., NAD(P)H, reductases). However, achieving direct electrochemistry of large proteins without mediators is a formidable challenge. The intervening peptide medium results in extremely weak electronic coupling between redox cofactors and the electrode surface, rendering heterogeneous ET too slow to measure [31,49,50]; this is especially true for P450, where the heme is deeply buried in the folded polypeptide. Thus, we and others have pursued ways to increase heme-electrode electronic couplings, as can be accomplished using electrodes with carefully prepared surfaces: methodologies include use of polished edge-plane graphite [51], surfactant assemblies on basal-plane graphite [41,52,53], mercaptan films on gold [54], polyion layers on carbon [55,56], covalent surface modification [30], clay on carbon [57], and rhodium-graphite electrodes [58]. In each instance voltammetry was possible, allowing in-depth characterization of P450 heme (Fe$^{III/II}$ and Fe$^{II/I}$) redox couples. Voltammetric data and the resulting interpretations of experiments using these techniques are discussed by Bond et al. in Chapter 5 of this volume.

While successful at achieving ET, many of the surface-modified electrodes are not useful for substrate oxidation, while those that are catalytically active function at impractical rates and display poor coupling efficiencies. In addition to the issues with dioxygen activation (discussed above), surface-confined P450-electrodes suffer from two additional problems. First, confining P450s (and other proteins) to electrode surfaces raises questions regarding protein fold and stability [59]; indeed, voltammetry of electrode-confined P450 typically reveals altered heme potentials ($\Delta E°$ up to +300 mV relative to solution), suggesting a perturbed heme environment [52,53]. Second, decreased polypeptide flexibility resulting from adsorption and/or protein aggregation on the electrode surface may hinder substrate access and restrict conformational changes that are necessary for catalysis. The latter is particularly important, since it has been shown that P450s undergo large conformational changes in order to bind substrates [60]. Also, subtle structural changes that alter the hydrogen-bonding network in the vicinity of the heme are implicated as necessary perturbations to the heme electronic structure for subsequent substrate oxidation [61].

2.4.1. Mediators

We and others have investigated P450 bioelectrochemical systems in which small molecule mediators in solution shuttle electrons between an electrode and the enzyme (Figure 4). These systems have been shown to catalyze substrate oxidation, which is not surprising, as the enzyme retains its native solution properties during the mediated electrochemical reaction. Critical to this method is finding an appropriate small molecule that will efficiently mediate the ET reaction.

Estabrook and co-workers discovered a promising system utilizing a platinum electrode and a water-soluble cobalt(III) sepulchrate cage complex ($E^{\circ}(Co^{III/II}) = -550$ mV versus Ag/AgCl) as the electron shuttle [62]. Mediated catalysis was demonstrated with a variety of mammalian and bacterial P450s, with rates approaching that of NAD(P)H-driven systems: as an example, reactions with P450BM3 and lauric acid yielded turnover rates of 110 min^{-1} (compared with 900 min^{-1} for reactions with NADPH) [62]. Following this work, cobaltocene was used as a scaffold to construct a suitable mediator for bioelectrochemical catalysis with P450BM3. [63] A dicarboxy derivative (1,1'-dicarboxycobaltocenium hexafluorophosphate, $E^{\circ}(Co^{III/II}) = -830$ mV versus Ag/AgCl at pH 7) was synthesized with the intention of improving water solubility and disfavoring aggregation, two problems encountered with the sepulchrate system. The synthesized mediator was capable of reducing both FAD and FMN in the P450BM3 reductase domain as well as ferric iron in the heme domain, while electrolysis reactions with the holo protein resulted in lauric acid hydroxylation (16.5 turnovers per minute). Interestingly, catalytic reactions occurred with just the P450BM3 heme domain and the metallocene (1.8 turnovers per minute). Unfortunately, neither system could overcome the problem of direct dioxygen reduction by the reduced form of the mediator. For cobaltocene, this problem was evident from the coupling efficiency, which revealed that only 2% of the total current passed resulted in product formation.

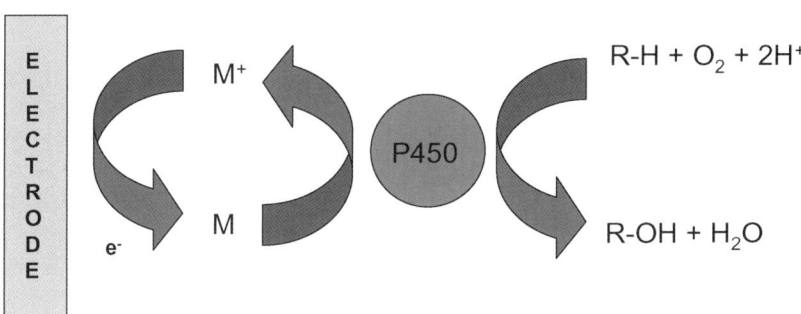

Figure 4. Mediated electrocatalytic P450 system.

Vilker et al. reported an electrochemical system with P450CAM using Pdx – the native P450CAM ET partner – as the electrochemical mediator [64]. Direct reduction of Pdx with an antimony-doped tin oxide electrode initiated the ET cascade that ultimately resulted in dioxygen activation and camphor hydroxylation. An average turnover rate of $36\,min^{-1}$ and 2600 total turnovers were reported. Notably, reactor performance was optimal when the concentration of Pdx was approximately two orders of magnitude larger than that of P450CAM, suggesting that either electrode to Pdx or Pdx to P450CAM ET was inefficient (likely the former, given that Pdx and P450CAM are native redox partners). Once again, the co-substrate dioxygen proved problematic. The working electrode had to be screened with a platinum mesh to disproportionate peroxide formed from direct dioxygen reduction, while the necessary dioxygen for catalysis was generated *in situ* through water oxidation at another electrode.

Clearly, the key obstacle in the mediated systems is dioxygen reduction, resulting in futile redox cycling and production of reactive oxygen species. Still, using mediated electrochemical reduction is a simple and effective way of constructing a functional P450 bioelectrochemical catalytic system. The challenge is to find an electrochemical mediator that readily transfers electrons to P450, while displaying minimal reactivity with dioxygen in solution. One candidate is $[Cp^*Rh(bpy)(H_2O)]Cl_2$ (Rh), which has been shown to function as a recyclable hydride transfer mediator, capable of regenerating nicotinamide- (NADH, NADPH) and isoalloxazine-based (FAD, FMN) cofactors in solution [65–67]. To the best of our knowledge, Rh activity with P450 has not been demonstrated, although it represents an attractive possibility: in addition to more closely mimicking the native P450 system (hydride transfer for Rh versus ET for prior P450-mediator systems [62,63]), it would have the advantage of minimizing uncoupling, as Rh regeneration was found to proceed efficiently in the presence of dioxygen [65]. This point was demonstrated in reactions using Rh to reduce flavins in the catalytic cycle of styrene monooxygenase, which displayed a coupling efficiency of 60% while catalyzing substrate turnover at rates comparable to the native system [67].

2.5. Electrochemical Heme Oxidation

The ability to access and characterize compound I is central to understanding P450 function. While several investigations have attempted to observe such high-valent species in the catalytic cycle [61,68–70], none has led to definitive characterization of compound I. A novel approach now being explored involves direct oxidation of the P450 heme. Section 3.4 of this chapter discusses methodologies using molecular wires to generate oxidized P450 species; complementing those studies are electrochemical investigations aimed at generating high-valent heme complexes.

Regarding electrocatalysis, specific problems that arise when running a P450-electrode system cathodically – despite the type of P450-electrode system being considered (e.g., mediated, surface-confined) – include uncoupling leading to futile redox cycling (through P450-catalyzed one-, two-, and four-electron reduction of dioxygen). One way to overcome the disadvantages of the cathodic reaction would be to generate ferryl species directly through electrochemical heme oxidation, essentially running the catalytic cycle (Figure 1) in reverse, thereby allowing both observation and characterization of catalytically active species. In addition, running the cycle in reverse opens the way for dioxygen- and NAD(P)H-independent substrate turnover.

Oxidation of a P450 water-ligated Fe^{III} heme should produce ferryl species directly (Figure 5). In fact, in work with chloroperoxidase [71] and horseradish peroxidase [72], one- and two-electron oxidation of the Fe^{III}–OH_2 heme resulted in rapid conversion to the ferryl species Fe^{IV}=O–H (compound II) and Fe^{V}=O (compound I). If similar ferryl species could be generated in P450, they should be able to oxygenate substrates. Prior work by Groves and Gilbert with model Fe^{III}-porphyrin complexes is a case in point: in these experiments electrochemical one-electron oxidation produced ferryl species capable of epoxidation [73].

Direct electrochemical oxidation of a P450 heme would likely occur through two one-electron steps. Thus, compound II would be generated first, followed by oxidation to give compound I. The oxidizing strength of potential ferryl species can be estimated from thermodynamic calculations. Mayer [74] has proposed that the key determinant for C–H bond activation by metal-oxo complexes is the strength of the resulting $(M^{(n-1)+})$O–H bond following hydrogen-atom abstraction: if the O–H bond (D(O–H)) in the (formally one-electron-reduced)

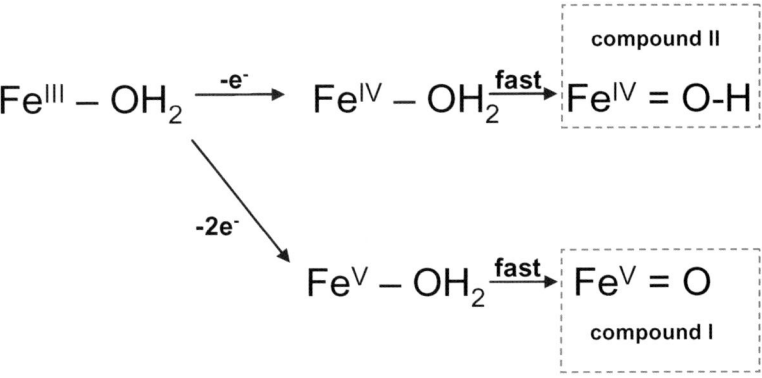

Figure 5. Generation of high-valent P450s.

$$Fe^{IV}=O-H + \bar{e} \rightleftharpoons Fe^{III}-O-H$$

$$1/2\, H_2 \rightleftharpoons H^+ + \bar{e}$$

$$Fe^{III}-O-H + H^+ \rightleftharpoons Fe^{III}-OH_2$$

$$H\bullet \rightleftharpoons 1/2\, H_2$$

$$\overline{Fe^{IV}=O-H + H\bullet \rightleftharpoons Fe^{III}-OH_2}$$

Figure 6. Thermodynamic cycle for estimating the oxidizing strength of P450 compound II.

'rebound' complex is stronger than the C–H bond (D(C–H)) being broken, the reaction can proceed. D(O–H) for the reaction

$$M^{n+}=O + H^\bullet \rightleftharpoons M^{(n-1)+}-OH \quad (2)$$

can be estimated using a simple thermodynamic cycle (Figure 6). An empirical expression for D(O–H) [75] is based on this cycle:

$$D(O–H)\,(\text{kcal mol}^{-1}) = 23.06 \times [E°(Fe^{IV/III})] + 1.37 \times pK_a(Fe^{III}-OH_2) + 57 \quad (3)$$

Green et al. [9] used this approach to estimate that D(O–H) in compound II of chloroperoxidase is 98 kcal mol^{-1}, a value that would enable the corresponding compound I to oxidize most alkanes (D(C–H) $\sim$ 95–99 kcal mol^{-1}).

Voltammetry experiments using P450CAM confined within didodecyldimethylammonium bromide (DDAB) surfactant films on carbon electrodes produced a species that appeared to be an oxidized heme [76]. Rapid scanning (> 30 V/s at room temperature, or > 1 V/s at 4 °C) yielded two redox couples, Fe$^{III/II}$ ($E_{1/2} = -227$ mV versus Ag/AgCl, pH 8) and another couple **E** at positive potentials ($E_{1/2} = 830$ mV). Voltammetry at constant ionic strength and variable pH revealed that **E** was pH dependent, shifting -47 ± 7 mV/pH unit. Since redox reactions involving porphyrin macrocycles are relatively insensitive to solution conditions [77], **E** is not likely to be a porphyrin-derived ET reaction. In addition, voltammetry in the presence of imidazole appeared to quench **E**, which rules against the likelihood of **E** being an amino acid redox couple. The remaining possibility is an Fe-centered redox reaction, suggesting that **E** represents the heme Fe$^{IV/III}$ redox couple.

With water as the axial ligand, it is possible that one-electron oxidation of the P450 heme would give compound II (Fe^{IV}=O–H). Turnover experiments in the P450CAM-DDAB system at high potentials (0.9 V versus Ag/AgCl, pH 8) produced P450-dependent sulfoxidation of thioanisole. Interestingly, experiments involving other typical P450 substrates (alkanes, alkenes) yielded no products. Assuming production of a ferryl species (**E**), it would appear that the Fe^{IV}=O–H unit is not powerful enough to perform reactions more challenging than thioanisole oxidation (within the electrochemical film system). This result is not surprising, in view of the fact that the catalytically active species in the canonical P450 mechanism, compound I, is one full oxidation state higher than **E**. Using equation (3), D(O–H) for Fe^{III}–OH_2 in the electrochemical P450-DDAB film system (taking $E^\circ = 1.14$ V versus NHE in DDAB and pK_a for Fe^{III}–$OH_2 = 9$) is estimated to be ~ 94 kcal mol^{-1}. Clearly, an alkane cannot be oxidized by the corresponding electrochemically generated ferryl species.

3. ELECTRON TUNNELING WIRES

3.1. Metal-Diimine Wires

Metal-diimine wires (often referred to as sensitizer-tethered substrates, or electron tunneling wires) have been developed to trigger redox reactions at buried enzyme active sites [78]. Earlier work revealed that photogenerated $[Ru(bpy)_3]^{3+}$ reacts with horseradish peroxidase and microperoxidase-8 to generate high-valent heme intermediates (compounds I and II) on sub-microsecond timescales [79]. Similar attempts to use $[Ru(bpy)_3]^{3+}$ to generate high-valent heme intermediates in P450s were unsuccessful, presumably due to the inefficiency of ET from the buried heme through the protein backbone [78]. Wires circumvent this problem by providing a direct ET pathway between $[Ru(bpy)_3]^{3+}$ and a buried enzyme active site.

Electron tunneling wires contain a ligand with affinity for a buried active site, which is tethered to a photoactive metal-diimine 'head group' that can bind at the enzyme surface (Figure 7). The tether, residing in the substrate access channel, provides an electron tunneling pathway for injection or abstraction of electrons to and from the active site. The ligand and tether can be optimized to bind selectively to the enzyme of interest. Most wires reported to date have employed Ru^{II} or Rh^I diimine complexes as head groups. Although originally designed as photoredox triggers, wires have been used as structural probes [60,80,81], fluorescent sensors [82,83], enzyme inhibitors [84], and electrochemical sensors [85]. Wires also have been developed for enzymes other than P450, including inducible nitric oxide synthase [86,87], *Arthrobacter globiformis* amine oxidase [84,85], myeloperoxidase [88], and lipoxygenase [89].

The flash-quench method can trigger wire-mediated ET to and from buried enzyme active sites, as shown schematically in Figure 8. Photoexcitation into the $[Ru(bpy)_3]^{2+}$ metal-to-ligand charge-transfer band generates an excited state

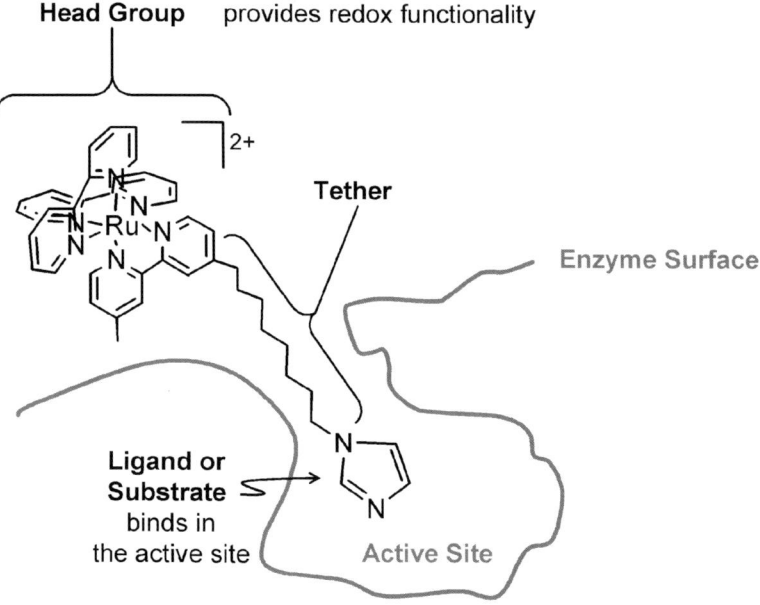

Figure 7. Electron tunneling wires contain a substrate or ligand tethered to a redox-active head group.

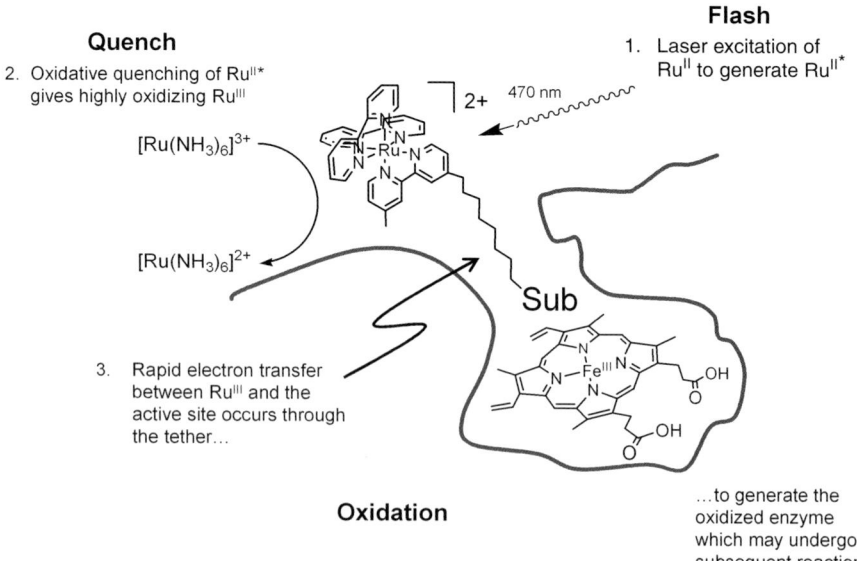

Figure 8. Wires can be used to deliver electrons or holes to buried active sites. Based on ideas developed in [78].

Table 1. Reduction Potentials.

Couple	$(V, NHE)^a$	Ref.
P450CAM $Fe^{3+/2+}$ (water bound)	~ -0.3	[11]
P450CAM $Fe^{3+/2+}$ (substrate bound)	~ -0.15	[11]
$[Ru((CF_3)_2bpy)_2(bpy)]^{3+/2+}$	+1.87 (AN)	[93]
$[Ru(bpy)_3]^{3+/2+}$	+1.26	[91]
	+1.53 (AN)	
$[Ru(bpy)_2(phen)]^{3+/2+}$	+1.52 (AN)	[94]
$[Ru(4,4'-Me_2bpy)_3]^{3+/2+}$	+1.10	[92]
$[Ru(4,4'-5,5'Me_4bpy)_3]^{3+/2+}$	+0.80	[92]
$[Ru(bpy)_3]^{3+/2+*}$	−0.62	[91]
$[Ru(bpy)_2(phen)]^{3+/2+*}$	−0.67	[91]
$[Ru(4,4'-Me_2bpy)_3]^{3+/2+*}$	−0.61	[91]
$[Ru(4,4'-5,5'Me_4bpy)_3]^{3+/2+*}$	−0.75	[91]
$[Ru(bpy)_3]^{2+*/+}$	+1.08	[91]
$[Ru(bpy)_2(phen)]^{2+*/+}$	+1.02	[91]
$[Ru(4,4'-Me_2bpy)_3]^{2+*/+}$	+0.91 (AN)	[91]
$[Ru(4,4'-5,5'Me_4bpy)_3]^{2+*/+}$	+0.80 (AN)	[91]
$[Ru((CF_3)_2bpy)_2(bpy)]^{2+/+}$	−0.59 (AN)	[93]
$[Ru(bpy)_3]^{2+/+}$	−1.28	[91]
	−1.07 (AN)	
$[Ru(4,4'-Me_2bpy)_3]^{2+/+}$	−1.37	[92]
$[Ru(4,4'-5,5'Me_4bpy)_3]^{2+/+}$	−1.36	[92]
$[Re(CO)_3(bpy)Im]^{2+/+}$	+1.33	[92]
$[Re(CO)_3(bpy)Im]^{2+/+*}$	−0.96	[90]
$[Re(CO)_3(bpy)Im]^{+*/0}$	+0.97	[90]
$[Re(CO)_3(bpy)Im]^{+/0}$	−1.34	[92]

^{a} All values refer to aqueous solution unless otherwise noted.

that is both a better oxidant and reductant than the ground-state complex (Table 1). In certain cases, this excited state may oxidize or reduce the active site directly, but more commonly ET is triggered by flash-quench with an added redox agent, which generates a powerful oxidant or reductant, $[Ru(bpy)_3]^{3+}$ or $[Ru(bpy)_3]^+$, respectively (Figure 9). The reversible oxidative quenchers $[Ru(NH_3)_6]^{3+}$ and methyl viologen, and the irreversible oxidative quencher $[Co(NH_3)_5Cl]^{2+}$ are used in many experiments. Reductive quenchers include p-methoxy-N, N-dimethylaniline (pMDMA) and ascorbate. As can be seen from Table 1, redox potentials of Ru^{II} and Re^{I} wires vary widely.

Molecular wires can in principle be used to photogenerate P450 intermediates both by reductive and oxidative processes (Figure 10). Reduction of ferric P450 to the ferrous state will be discussed in Section 3.3 and oxidation to high-valent states will be discussed in Section 3.4. More ambitious schemes could in principle be used to produce other species: for example, the ferric-superoxide (or $Fe^{III}-O_2$)

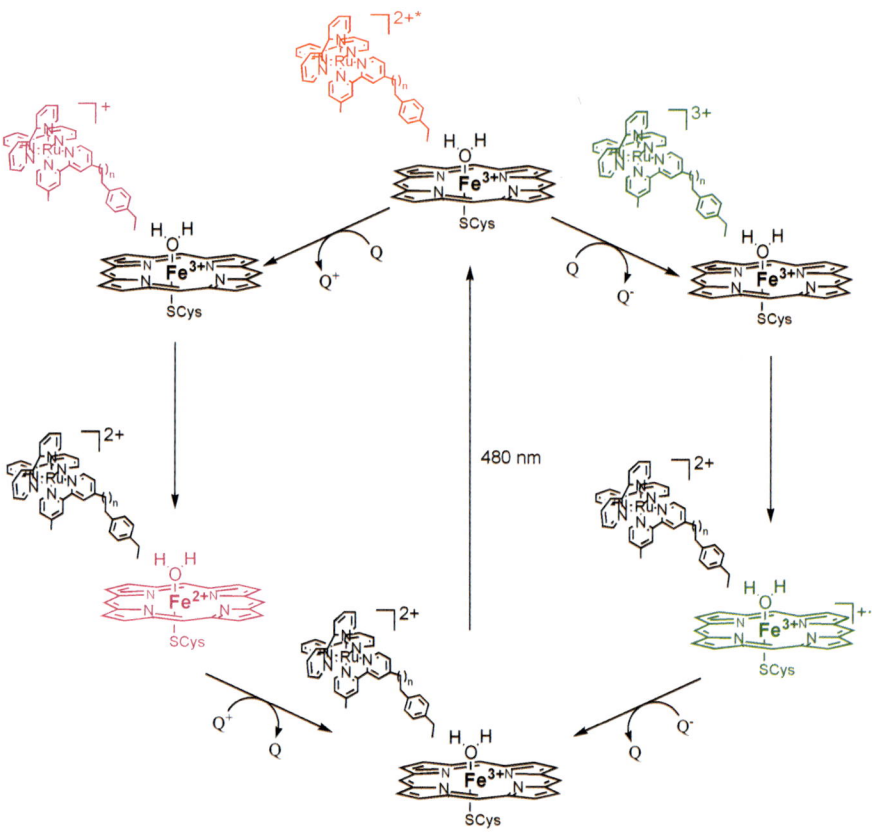

Figure 9. The flash-quench method for photogeneration of P450CAM redox states using Ru^{II} wires. Q = quenching agent (reproduced from [78] by permission of Wiley-VCH, Weinheim).

intermediate (**5** in Figure 1) could theoretically be produced in a flow system and photoreduced to give a ferric hydroperoxide species.

3.2. P450-Wire Conjugates

Sixteen Ru^{II}-diimine wires developed for P450CAM are shown in Figure 11. The structures and ET properties of P450-wire conjugates vary widely, depending on the nature of the tethers and terminal groups. Interestingly, while these wires bind to P450CAM with micromolar to nanomolar dissociation constants (Table 2), they do not bind P450BM3 [98]. Most likely, this remarkable selectivity is a consequence of the differing substrate profiles for the two enzymes: while P450CAM shows maximal activity with adamantyl-like substrates, P450BM3

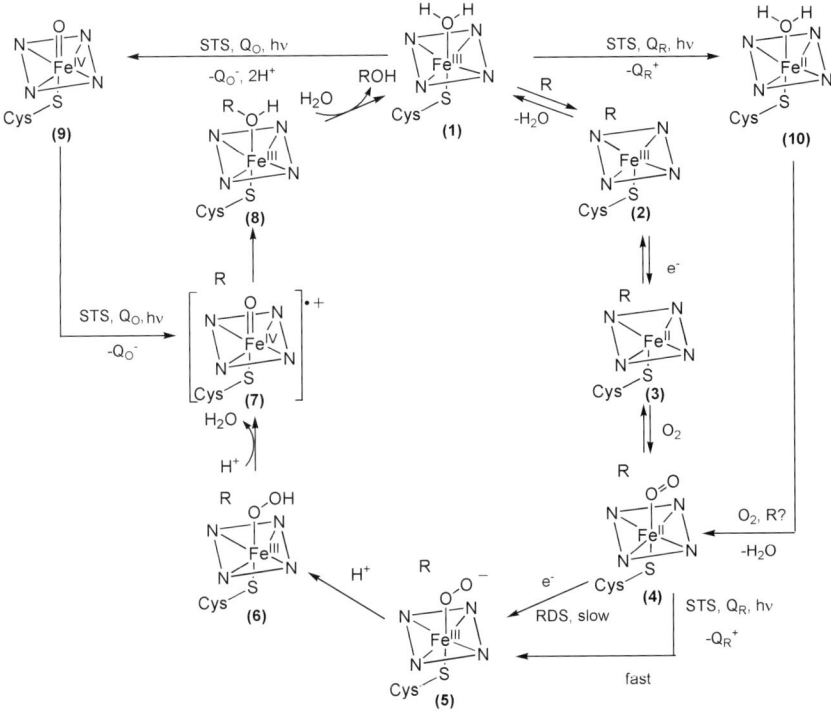

Figure 10. Photogeneration of redox states in the cytochrome P450 catalytic cycle. Q_O: oxidative quencher, Q_R: reductive quencher, RDS: rate-determining step, STS: a wire with photoactive head group (typically $[Ru(bpy)_3]^{2+}$ or $[Re(CO)_3(phen)Im]^+$).

prefers long chain fatty acids. Further, P450BM3 has an arginine (Arg^{47}) at the mouth of the substrate access channel, which has been proposed to interact electrostatically with the carboxyl group of the native fatty acid substrates, thereby stabilizing the enzyme-substrate complex [99]. By contrast, this same Arg residue would interact unfavorably with positively charged wire head groups (Figure 11).

Imidazole-terminated metal-diimine wires directly ligate the heme Fe, as evidenced by a shift in the Soret peak from 416 to 420 nm [78,100]. Notably, the spectral shift indicates that water is not fully displaced from the active site [95–97]. Wires terminated by hydrophobic substrates are incapable of coordinating the heme directly. When these wires bind P450CAM, water may or may not be displaced from the active site. Alkyl-tethered wires with ethylbenzene substrates do not produce significant changes in heme ligation or spin state on binding [95], while alkyl and perfluorobiphenyl-tethered adamantyl wires cause partial displacement of coordinated water from the active site [95–97].

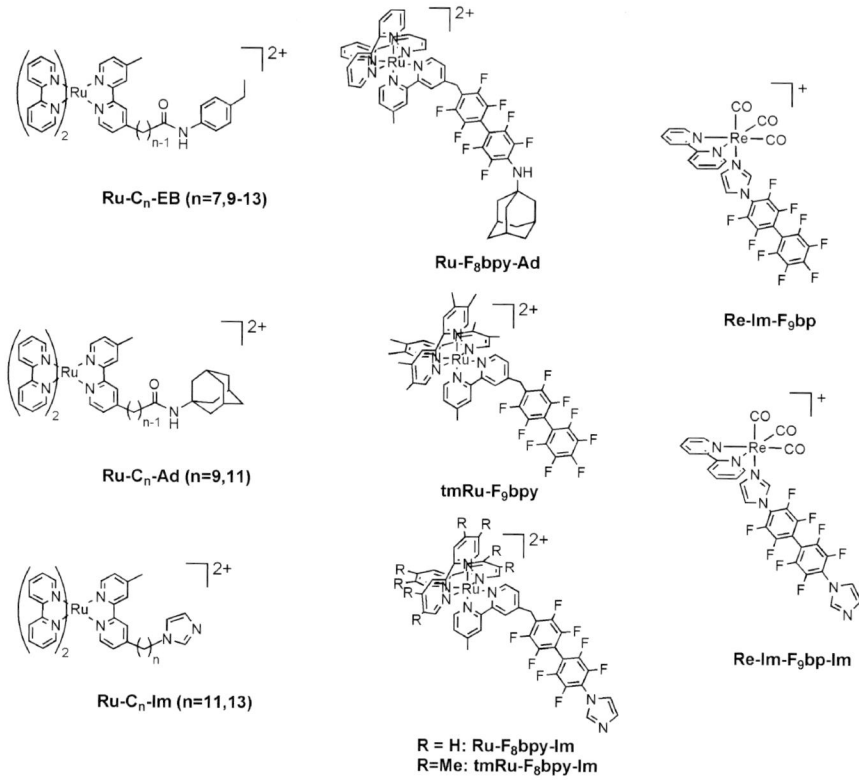

Figure 11. Ru^{II} and Re^{I} wires that bind cytochrome P450CAM. tm: tetramethyl, EB: ethylbenzene.

Crystal structures have been reported for conjugates between P450CAM and wires terminated with hydrophobic groups [60,81–83]. Several of these structures are shown in Figure 12 along with that of P450CAM with camphor in the active site. As can be seen from Figure 12a, native P450CAM possesses a 'closed' structure in which there is no clear path to the buried heme. In its wire complexes, P450CAM is forced into an 'open' conformation, involving displacement of the P450CAM F, G, and I helices [60,81]. The I helix bulge and the associated catalytically relevant Asp^{251} and Thr^{252} residues occupy positions analogous to those observed on dioxygen binding [60]. It is likely that these 'open' conformations correspond to the structures of transient intermediates populated when substrates enter and products leave the active site [60].

Interestingly, P450-wire conjugates in the 'open' conformation show catalytic activity. When P450:$[Ru-C_9-Ad]^{2+}$ was treated with NADH in the presence of Pdx and Pdx reductase, the reductant was consumed at a rate of $8 \pm 2\,\mu mol/min$ per μmol P450, while product assays showed 10% formation

Table 2. P450CAM-wire conjugates.

Wire[a]	K_d (μM)	d_{M-Fe} (Å)	Comment	Ref.
[Ru-F$_8$bpy-Ad]$^{2+}$	0.077 ± 0.011	22.1[b] 21.8[c]	[d]	[96,100]
[tmRu-F$_9$bpy]$^{2+}$	2.1 ± 0.5	17.0[b]	[d]	[96,100]
[Ru-F$_8$bpy-Im]$^{2+}$	3.7 ± 0.5	18.1[b]	Direct photoreduction; 30% Fe^{II} produced; $k_f = 4.4 \times 10^6$ s^{-1}; $k_b \sim$ undetermined	[96,100]
[tmRu-F$_8$bpy-Im]$^{2+}$	0.48 ± 0.18	18.1[b]	Direct photoreduction; 74% Fe^{II} produced; $k_f = 2.8 \times 10^7$ s^{-1}; $k_b \sim 1.7 \times 10^8$ s^{-1}	[96,100]
[Ru-C$_{11}$-Im]$^{2+}$	> 50		[e]	[95]
[Ru-C$_{13}$-Im]$^{2+}$	4.1	21.2	[d]; ET after reductive quench; $k_f = 2 \times 10^4$ s^{-1}	[78,95]
[Ru-C$_9$-Ad]$^{2+}$	0.4 0.19(Λ) 0.09(Δ)	21.4	[e]	[95]
[Ru-C$_{11}$-Ad]$^{2+}$	0.6	21.0	[d]; ET after reductive quench with pMDMA yields [P$_{Cys}$-Fe^{II}(H$_2$O)]$^-$ $k_f = 2 \times 10^4$ s^{-1}	[78,95]
[Ru-C$_7$-EB]$^{2+}$	6.5	19.5	[d]; ET after reductive quench with pMDMA yields [P$_{Cys}$-Fe^{II}(H$_2$O)]$^-$ $k_f \sim 10^3$ s^{-1}	[95]
[Ru-C$_9$-EB]$^{2+}$	0.7	19.4	[d]; ET after reductive quench with pMDMA yields [P$_{Cys}$-Fe^{II}(H$_2$O)]$^-$	[95]
[Ru-C$_{10}$-EB]$^{2+}$	0.9	19.9	[d]; ET after reductive quench with pMDMA yields [P$_{Cys}$-Fe^{II}(H$_2$O)]$^-$	[95]
[Ru-C$_{11}$-EB]$^{2+}$	0.9	20.1	[d]; ET after reductive quench with pMDMA yields [P$_{Cys}$-Fe^{II}(H$_2$O)]$^-$; $k_f = 2 \times 10^4$ s^{-1} ET after oxidative quench with [Co(NH$_3$)$_5$Cl]$^{2+}$ gives [P$_{Cys}$-Fe^{IV}(HO)]? or a porphyrin radical; $k_f = 6 \times 10^3$ s^{-1}	[78,95]
[Ru-C$_{12}$-EB]$^{2+}$	1.5	20.5	[d]; ET after reductive quench with pMDMA yields [P$_{Cys}$-Fe^{II}(H$_2$O)]$^-$	[95]
[Ru-C$_{13}$-EB]$^{2+}$	1.7	20.6	[d]; ET after reductive quench with pMDMA yields [P$_{Cys}$-Fe^{II}(H$_2$O)]$^-$	[95]
[Re-Im-F$_8$bp-Im]$^+$	0.6 ± 0.5		[d]	[86,97]
[Re-Im-F$_9$bp]$^+$	22 ± 8		[d]	[86,97]

[a] See Figure 11.
[b] Distances were determined from transient emission studies using the Förster relationship.
[c] Distance was determined by X-ray crystallography.
[d] Direct photoinduced ET was not observed.
[e] Photochemical data are not available.

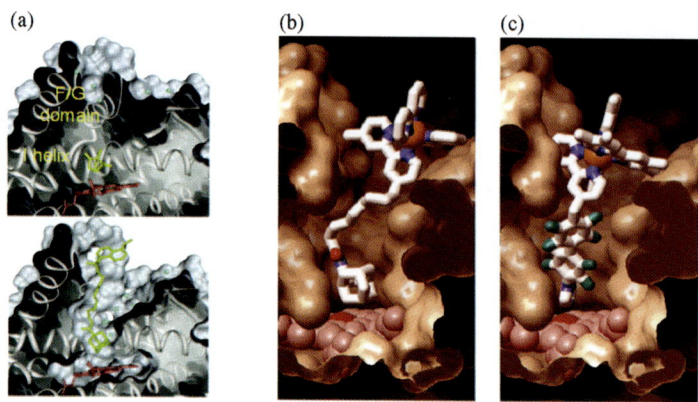

Figure 12. Cutaway views of the crystallographically determined structures of P450CAM wire complexes: (a) Comparison of the closed P450CAM conformation in its complex with camphor (top) and the 'open' conformation adopted in its complex with a dansylated adamantyl wire (bottom) (reproduced by permission from [60]). (b) Structure of the P450CAM:Ru-C$_9$-Ad:P450CAM^{2+} complex. (c) Structure of the P450CAM:[tmRu-F$_8$bpy-Im]$^{2+}$ complex (reproduced by permission of [100]).

of the hydroxylated wire, [Ru-C$_9$-Ad-OH]$^{2+}$ [95]. Control experiments with 2-adamantylacetamide revealed a turnover rate of $\sim 90 \pm 20\,\mu$mol/min per μmol P450. The lower rate observed for P450:[Ru-C$_9$-Ad]$^{2+}$ conjugate could be due to conformational changes near Asp^{251} and Thr^{252} associated with [Ru-C$_9$-Ad]$^{2+}$ binding.

3.3. Wire-Mediated P450 Reduction

Studies to date have focused on photoreduction of ferric P450CAM. In conjugates with alkyl-tethered imidazole wires, ET to the heme is much slower than intrinsic decay or Förster energy transfer quenching [78,95]. The [Ru(bpy)$_3$]$^{2+}$ excited-state lifetimes in these conjugates allow generation of [Ru(bpy)$_3$]$^{3+}$ or [Ru(bpy)]$^+$ in flash-quench experiments; this was successfully exploited to produce ferrous P450CAM on sub-millisecond timescales [78]. Laser photoexcitation of a P450CAM:[Ru-C$_{13}$-Im]$^{2+}$ conjugate in the presence of pMDMA resulted in nanosecond formation of [Ru-C$_{13}$-Im]$^+$ followed by Ru$^I \to$ Fe^{III} ET with $k_{ET} = 2 \times 10^4\,\text{s}^{-1}$. Similar experiments using P450CAM conjugates with [Ru-C$_{11}$-EB]$^{2+}$ and [Ru-C$_{11}$-Ad]$^{2+}$ also showed heme photoreduction, as evidenced by a Soret shift from 417 to 390 nm, which was attributed to formation of the previously unobserved [Fe^{II}-(OH$_2$)]. The Ru$^I \to$ Fe^{III} rate constant for [Ru-C$_{11}$-EB]$^+$ and [Ru-C$_{11}$-Ad]$^+$ complexes was identical with that for [Ru-C$_{13}$-Im]$^+$, $2 \times 10^4\,\text{s}^{-1}$, indicating that heme ligation is not required for efficient wire-mediated ET. In flash-quench experiments with [Ru-C$_7$-EB]$^{2+}$, where there is a heme-wire gap, the Ru^I to Fe^{III} ET rate is an order of magnitude lower [95].

Our study of wire-mediated heme photoreduction was extended to include light-induced turnover in the presence of dioxygen [95]. A hydroxylated adamantyl wire, [Ru-C$_{11}$-AdOH]$^{2+}$, was detected in aerated solutions of P450CAM:[Ru-C$_{11}$-Ad]$^{2+}$, pMDMA, and catalase (to scavenge any H$_2$O$_2$ produced) that had been subjected to steady-state irradiation ($\lambda > 450$ nm, 4 °C). Potentially, this method represents yet another way of achieving cofactor-independent *in vitro* biocatalysis.

Perfluorobiphenyl-tethered wires with [Ru(bpy)$_3$]$^{2+}$ and [Ru(Me$_4$bpy)$_3$]$^{2+}$ head groups exhibit different ET behavior from their aliphatic counterparts. A key difference is that imidazole-terminated perfluorobiphenyl-tethered wires directly photoreduce P450CAM [100]. Pulsed-laser (355 nm) irradiation of the P450CAM:[tmRu-F$_8$bpy-Im]$^{2+}$ conjugate forms Fe^{II}-P450 in high yield (~74%) on the nanosecond timescale. The rate constants for forward and back ET are 2.8×10^7 and ~1.7×10^8 s^{-1}. Notably, the forward ET rate is ~1000-fold greater than that observed for the aliphatic wires, while the extremely fast back ET is near the coupling-limited rate [100].

Experiments with [Ru-F$_8$bpy-Im]$^{2+}$ also show nanosecond formation of Fe^{II}-P450CAM in ~30% yield with $k_f = 4.4 \times 10^6$ s^{-1} (k_b undetermined); this is six times slower than that for [tmRu-F$_8$bpy-Im]$^{2+}$, as expected since the driving force for the latter wire is greater by ~0.13 eV. Semiclassical theory predicts that there should be a four-fold rate enhancement in the reaction with [Ru(4,4'-5,5'Me$_4$bpy)$_3$]$^{2+*}$ [100]. Interestingly, the heme in P450CAM complexes with perfluorobiphenyl-tethered wires [Ru-F$_8$bpy-Ad]$^{2+}$ and [tmRu-F$_9$bpy]$^{2+}$ is not reduced upon photoexcitation, likely because there is no through-bond coupling pathway in these systems.

3.4. Wire-Mediated P450 Oxidation

Wires terminated with groups that do not coordinate the heme can in principle be used for photogeneration of catalytically relevant P450 intermediates. In experiments aimed at photooxidation of ferric P450 to compound II, we irradiated P450CAM bound to [Ru-C$_{11}$-EB]$^{2+}$ in the presence of an irreversible oxidative quencher, [Co(NH$_3$)$_5$Cl]$^{2+}$ [78]. Photogeneration of [Ru-C$_{11}$-EB]$^{3+}$ ($E° = 1.26$ V versus NHE) was followed by heme oxidation, as evidenced by the formation of a product (heme Soret at 390 nm) with $k_{ET} = 6 \times 10^3$ s^{-1}. In our initial report, we speculated that the oxidized species was either a porphyrin cation radical or Fe^{IV}-OH$_2$, noting that a blue-shifted Soret was consistent with porphyrin π-cation radical spectra [78]. Subsequent work has established that the oxoiron(IV) intermediate of chloroperoxidase compound II has a p$K_a > 8.2$ [9]. Based on this finding, it is likely that the oxidized intermediate is actually a protonated ferryl species, $Fe^{IV}=O-H$. We found that tyrosine radical formation competes with heme oxidation and that the yield of oxidized product can be increased in reactions with P450CAM mutant Y96F [95]. Interestingly, although the heme decays back to its ferric resting state without appreciable decomposition, oxidized heme is only observed in the presence of $[Co(NH_3)_5Cl]^{2+}$ [78].

3.5. Electronic Coupling

A recent theoretical study has elucidated the factors promoting electronic couplings in P450CAM-wire complexes [101]. Molecular trajectories for wire docking in the P450CAM active-site channel were computed using molecular dynamics methods. Donor-acceptor electronic couplings calculated using the Hartree-Fock method and Koopmans' theorem are consistent with the relative ET rates for perfluorobiphenyl-bridged wires. More importantly, the theoretical work shows that the couplings mediated by the perfluorobiphenyl tethers are dominated by superexchange through extra electron states, while those through hydrocarbon tethers are dominated by hole-mediated superexchange interactions. Furthermore, a small number of wire conformations with strong donor-acceptor couplings accounts for the observed ET rates, especially for wires that interact with the heme through weak van der Waals interactions.

4. CONCLUDING REMARKS

Much of P450 research is driven by (1) the search for catalytically active species and (2) the desire to harness P450 activity for a wide array of applications. We have learned a great deal about the catalytic mechanism in work thus far, and it is likely that *in vitro* oxidative catalysis with P450 – the so-called Holy Grail – will be realized in the near future. In our work, we have found that a suitably modified electrode is the simplest and most cost-effective way of replacing the P450 ET machinery. A promising system for large-scale biocatalysis would incorporate a small molecule electrochemical mediator. The key is to find a mediator that: (1) can participate in rapid and reversible ET with the electrode and enzyme; (2) can be electrochemically reduced at moderate potentials (to avoid solvent reduction, and possibly other side reactions); and (3) does not readily reduce dioxygen in solution. Studies of P450 ET fundamentals will eventually lead to catalytic systems that meet these requirements.

Our work with electron tunneling wires has demonstrated the feasibility of designing enzyme-selective molecules that can be used to photochemically induce redox reactions in P450. However, definitive characterization of catalytically relevant P450 intermediates has yet to be realized using this methodology. Nevertheless, the observation of reductive turnover in the presence of dioxygen indicates that flash-quench photoreduction of P450-wire conjugates with transiently generated ferrous-oxy P450CAM is a promising avenue for future work. We also plan to develop wires to probe the ET properties of other P450s, as well as design and synthesize P450 isoform-specific inhibitors for *in vivo* applications.

ACKNOWLEDGMENTS

Our work is supported by NIH, NSF, NSERC, HHMI, the Ellison Medical Foundation, and the Arnold and Mabel Beckman Foundation.

ABBREVIATIONS

Ad	adamantyl
AN	acetonitrile
bp	biphenyl
bpy	2,2′-bipyridine
Cp	cyclopentadienyl
DDAB	didodecyldimethylammonium bromide
EB	ethylbenzene
ET	electron transfer
FAD	flavin adenine dinucleotide
FMN	flavin mononucleotide
Im	imidazole
Me	methyl
NADPH	nicotinamide adenine dinucleotide phosphate, reduced
NHE	normal hydrogen electrode
P450	cytochrome P450
P450BM3	cytochrome P450 from *Bacillus megaterium*
P450CAM	cytochrome P450 from *Pseudomonas putida*
Pdx	putidaredoxin
phen	1,10-phenanthroline
pMDMA	p-methoxy-N,N-dimethylaniline
Rh	$[Cp^*Rh(bpy)(H_2O)]Cl_2$
tm	tetramethyl

REFERENCES

1. D. F. V. Lewis, *Guide to Cytochromes P450: Structure and Function*, Taylor and Francis, New York, 2001, pp. 256.
2. P. R. Ortiz de Montellano, ed., *Cytochrome P450: Structure, Mechanism, and Biochemistry*, 2nd edn., Plenum Press, New York, 1995, pp. 652.

3. I. R. Phillips and E. A. Shephard, in *Methods in Molecular Biology*, Vol. 107, Humana Press, Totowa, 1998, pp. 482.
4. M. Sono, M. P. Roach, E. D. Coulter, and J. H. Dawson, *Chem. Rev.*, 96, 2842–2883 (1996).
5. F. P. Guengerich, *Mol. Interv.*, 3, 194–204 (2003).
6. A. Glieder, E. T. Farinas, and F. H. Arnold, *Nat. Biotechnol.*, 20, 1135–1139 (2002).
7. S. Joseph, J. F. Rusling, Y. M. Lvov, T. Friedberg, and U. Fuhr, *Biochem. Pharmacol.*, 65, 1817–1826 (2003).
8. C.-A. Yu and I. Gunsalus, *J. Biol. Chem.*, 249, 106–106 (1974).
9. M. T. Green, J. H. Dawson, and H. B. Gray, *Science*, 304, 1653–1656 (2004).
10. S. Yoshioka, S. Takahashi, H. Hori, K. Ishimori, and I. Morishima, *Eur. J. Biochem.*, 268, 252–259 (2001).
11. S. Daff, S. Chapman, K. Turner, R. Holt, S. Govindaraj, T. Poulos, and A. Munro, *Biochemistry*, 36, 13816–13823 (1997).
12. S. Shaik, D. Kumar, S. P. de Visser, A. Altun, and W. Thiel, *Chem. Rev.*, 105, 2279–2328 (2005).
13. G. D. Nordblom, R. E. White, and M. J. Coon, *Arch. Biochem. Biophys.*, 175, 524–533 (1976).
14. E. G. Hrycay, J. A. Gustafsson, M. Ingleman-Sundberg, and L. Ernster, *Biochem. Biophys. Res. Comm.*, 66, 209–216 (1975).
15. V. Sieber, C. A. Martinez, and F. H. Arnold, *Nat. Biotechnol.*, 19, 456–460 (2001).
16. C. R. Otey, J. J. Silberg, C. A. Voigt, J. B. Endelman, G. Bandera, and F. H. Arnold, *Chem. Biol.*, 11, 309–318 (2004).
17. K. G. Ravichandran, S. S. Boddupalli, C. A. Hasemann, J. A. Peterson, and J. Deisenhofer, *Science*, 261, 731–736 (1993).
18. A. W. Munro, D. G. Leys, K. J. McLean, K. R. Marshall, T. W. B. Ost, S. Daff, C. S. Miles, S. K. Chapman, D. A. Lysek, C. C. Moser, C. C. Page, and P. L. Dutton, *Trends Biochem. Sci.*, 27, 250–257 (2002).
19. I. F. Sevrioukova, H. Li, H. Zhang, J. A. Peterson, and T. L. Poulos, *Proc. Natl. Acad. Sci. USA*, 96, 1863–1868 (1999).
20. P. A. Williams, J. Cosme, D. M. Vinkovic, A. Ward, H. C. Angove, P. J. Day, C. Vonrhein, I. J. Tickle, and H. Jhoti, *Science*, 305, 683–686 (2004).
21. P. A. Williams, J. Cosme, V. Sridhar, E. F. Johnson, and D. E. McRee, *Mol. Cell*, 5, 121–131 (2000).
22. P. A. Williams, J. Cosme, A. Ward, H. C. Angove, D. M. Vinkovic, and H. Jhoti, *Nature*, 424, 464–468 (2003).
23. L. O. Narhi, and A. J. Fulco, *J. Biol. Chem.*, 262, 6683–6690 (1987).
24. A. E. Roitberg, M. J. Holden, M. P. Mayhew, I. V. Kurnikov, D. N. Beratan, and V. L. Vilker, *J. Am. Chem. Soc.*, 120, 8927–8932 (1998).
25. S. S. Boddupalli, T. Oster, R. W. Estabrook, and J. A. Peterson, *J. Biol. Chem.*, 267, 10375–10380 (1992).
26. L. O. Narhi and A. J. Fulco, *J. Biol. Chem.*, 261, 7160–7169 (1986).
27. J. R. Reed and P. F. Hollenberg, *J. Inorg. Biochem.*, 93, 152–160 (2003).
28. A. W. Munro, S. Daff, J. R. Coggins, J. G. Lindsay, and S. K. Chapman, *Eur. J. Biochem.*, 239, 403–409 (1996).

29. I. F. Sevrioukova, C. E. Immoos, T. L. Poulos, and P. J. Farmer, *Isr. J. Chem., 40*, 47–53 (2000).
30. A. K. Udit, M. G. Hill, V. G. Bittner, F. H. Arnold, and H. B. Gray, *J. Am. Chem. Soc., 126*, 10218–10219 (2004).
31. H. B. Gray and J. R. Winkler, *Annu. Rev. Biochem., 65*, 537–561 (1996).
32. J. D. Lipscomb, S. G. Sligar, M. J. Namtvedt, and I. C. Gunsalus, *J. Biol. Chem., 251*, 1116–1124 (1976).
33. S. Kadkhodayan, E. D. Coulter, D. M. Maryniak, T. A. Bryson, and J. H. Dawson, *J. Biol. Chem., 270*, 28042–28048 (1995).
34. L. L. Wong, C. G. Westlake, and D. P. Nickerson, *Structure and Bonding, 88*, 175–207 (1997).
35. P. C. Cirino and F. H. Arnold, *Adv. Synth. Catal., 344*, 1–6 (2002).
36. S. Nagano and T. L. Poulos, *J. Biol. Chem., 280*, 31659–31663 (2005).
37. A. K. Udit, N. Hindoyan, M. G. Hill, F. H. Arnold, and H. B. Gray, *Inorg. Chem., 44*, 4109–4111 (2005).
38. F. C. Anson, C. Shi, and B. Steiger, *Acc. Chem. Res., 30*, 437–444 (1997).
39. F. P. Guengerich, *Nat. Rev. Drug Discov., 1*, 359–366 (2002).
40. M. W. Peters, P. Meinhold, A. Glieder, and F. H. Arnold, *J. Am. Chem. Soc., 125*, 13442–13450 (2003).
41. E. Blair, J. Greaves, and P. J. Farmer, *J. Am. Chem. Soc., 126*, 8632–8633 (2004).
42. U. Schwaneberg, C. Otey, P. C. Cirino, E. Farinas, and F. H. Arnold, *J. Biomol. Screen., 6*, 111–117 (2001).
43. R. L. Walsky and R. S. Obach, *Drug Metab. Dispos., 32*, 647–660 (2004).
44. J. Stubbe and W. A. van der Donk, *Chem. Rev., 98*, 705–762 (1998).
45. C. O. Schmakel, S. V. Santhanam, and P. J. Elving, *J. Am. Chem. Soc., 97*, 5083–5092 (1975).
46. R. Wienkamp and E. Steckhan, *Angew. Chem. Int. Ed., 21*, 782–783 (1982).
47. K. Seelbach, B. Riebel, W. Hummel, M.-R. Kula, V. I. Tishkov, A. M. Egorov, C. Wandrey, and U. Kragl, *Tetrahedron Lett., 37*, 1377–1380 (1996).
48. S. Schneider, M. Wubbolts, D. Sanglard, and B. Witholt, *Appl. Environ. Microbiol., 64*, 3784–3790 (1998).
49. H. B. Gray and J. R. Winkler, *Quart. Rev. Biophys., 36*, 341–372 (2003).
50. F. M. Hawkridge and I. Taniguchi, in *The Porphyrin Handbook* (K. M. Kadish, K. M. Smith, and R. Guilard, eds.), Vol. 8, Academic Press, San Diego, 2000, pp. 191–262.
51. J. Kazlauskaite, A. C. G. Westlake, L.-L. Wong, and H. A. O. Hill, *Chem. Commun.*, 2189–2190 (1996).
52. A. K. Udit, W. Belliston-Bittner, E. C. Glazer, Y. H. L. Nguyen, J. M. Gillan, M. G. Hill, M. A. Marletta, D. B. Goodin, and H. B. Gray, *J. Am. Chem. Soc., 127*, 11212–11213 (2005).
53. C. E. Immoos, J. Chou, M. Bayachou, E. Blair, J. Greaves, and P. J. Farmer, *J. Am. Chem. Soc., 126*, 4934–4942 (2004).
54. A. Fantuzzi, M. Fairhead, and G. Gilardi, *J. Am. Chem. Soc., 126*, 5040–5041 (2004).
55. Y. M. Lvov, Z. Lu, J. B. Schenkman, X. Zu, and J. F. Rusling, *J. Am. Chem. Soc., 120*, 4073–4080 (1998).

56. B. Munge, C. Estavillo, J. B. Schenkman, and J. F. Rusling, *ChemBioChem,* 4, 82–89 (2003).
57. C. Lei, U. Wollenberger, C. Jung, and F. W. Scheller, *Biochem. Biophys. Res. Comm.,* 268, 740–744 (2000).
58. V. V. Shumyantseva, S. Carrara, V. Bavastrello, D. J. Riley, T. V. Bulko, K. G. Skryabin, A. I. Archakov, and C. Nicolini, *Biosens. Bioelectron.,* 21, 217–222 (2005).
59. M. T. de Groot, M. Merkx, and M. T. M. Koper, *J. Am. Chem. Soc.,* 127, 16224–16232 (2005).
60. A.-M. A. Hays, A. R. Dunn, R. Chiu, H. B. Gray, C. D. Stout, and D. B. Goodin, *J. Mol. Biol.,* 344, 455–469 (2004).
61. M. C. Glascock, D. P. Ballou, and J. H. Dawson, *J. Biol. Chem.,* 280, 42134–42141 (2005).
62. R. Estabrook, K. Faulkner, M. Shet, and C. Fisher, *Methods Enzymol.,* 272, 44–51 (1996).
63. A. K. Udit, F. H. Arnold, and H. B. Gray, *J. Inorg. Biochem.,* 98, 1547–1550 (2004).
64. V. Reipa, M. P. Mayhew, and V. L. Vilker, *Proc. Natl. Acad. Sci. USA,* 94, 13554–13558 (1997).
65. F. Hollman, B. Witholt, and A. Schmid, *J. Mol. Catal. B: Enzym.,* 19–20, 167–176 (2002).
66. F. Hollman, A. Schmid, and E. Steckhan, *Angew. Chem. Int. Ed.,* 40, 169–171 (2001).
67. F. Hollman, P.-C. Lin, B. Witholt, and A. Schmid, *J. Am. Chem. Soc.,* 125, 8209–8217 (2003).
68. R. Davydov, T. M. Markris, V. Kofman, D. E. Werst, S. G. Sligar, and B. M. Hoffman, *J. Am. Chem. Soc.,* 123, 1403–1415 (2001).
69. T. Egawa, H. Shimada, and Y. Ishimura, *Biochem. Biophys. Res. Comm.,* 201, 1464–1469 (1994).
70. I. Schlichting, J. Berendzen, K. Chu, A. M. Stock, S. A. Maves, D. E. Benson, R. M. Sweet, D. Ringe, G. A. Petsko, and S. G. Sligar, *Science,* 287, 1615–1622 (2000).
71. J. Berglund, T. Pascher, J. R. Winkler, and H. B. Gray, *J. Am. Chem. Soc.,* 119, 2464–2469 (1997).
72. T. Egawa, D. A. Proshlyakov, H. Miki, R. Makino, T. Ogura, T. Kitagawa, and Y. Ishimura, *J. Biol. Inorg. Chem.,* 6, 46–54 (2001).
73. J. T. Groves and J. A. Gilbert, *Inorg. Chem.,* 25, 123–125 (1986).
74. J. M. Mayer, *Acc. Chem. Res.,* 31, 441–450 (1998).
75. F. G. Bordwell, J. Cheng, G. Ji, A. V. Satish, and X. Zhang, *J. Am. Chem. Soc.,* 113, 9790–9795 (1991).
76. A. K. Udit, M. G. Hill, and H. B. Gray, *J. Inorg. Biochem.,* 100, 519–523 (2006).
77. M. A. Phillippi, E. T. Shimomura, and H. M. Goff, *Inorg. Chem.,* 20, 1322–1325 (1981).
78. J. J. Wilker, I. J. Dmochowski, J. H. Dawson, J. R. Winkler, and H. B. Gray, *Angew. Chem. Int. Ed.,* 38, 90–92 (1999).
79. D. W. Low, J. R. Winkler, and H. B. Gray, *J. Am. Chem. Soc.,* 118, 117–120 (1996).

80. I. J. Dmochowski, A. R. Dunn, J. J. Wilker, B. R. Crane, M. T. Green, J. H. Dawson, S. G. Sligar, J. R. Winkler, and H. B. Gray, *Method. Enzymol.*, *357*, 120–133 (2002).
81. A. R. Dunn, I. J. Dmochowski, A. M. Bilwes, H. B. Gray, and B. R. Crane, *Proc. Natl. Acad. Sci. USA*, *98*, 12420–12425 (2001).
82. A. R. Dunn, A.-M. A. Hays, D. B. Goodin, C. D. Stout, R. Chiu, J. R. Winkler, and H. B. Gray, *J. Am. Chem. Soc.*, *124*, 10254–10255. (2002).
83. I. J. Dmochowski, B. R. Crane, J. J. Wilker, J. R. Winkler, and H. B. Gray, *Proc. Natl. Acad. Sci. USA*, *96*, 12987–12990 (1999).
84. S. M. Contakes, G. A. Juda, D. B. Langley, N. W. Halpern-Manners, A. P. Duff, A. R. Dunn, H. B. Gray, D. M. Dooley, J. M. Guss, and H. C. Freeman, *Proc. Natl. Acad. Sci. USA*, *102*, 13451–13456 (2005).
85. C. R. Hess, G. A. Juda, D. M. Dooley, R. N. Amii, M. G. Hill, J. R. Winkler, and H. B. Gray, *J. Am. Chem. Soc.*, *125(24)*, 7156–7157 (2003).
86. W. Belliston-Bittner, A. R. Dunn, Y. H. L. Nguyen, D. J. Stuehr, J. R. Winkler, and H. B. Gray, *J. Am. Chem. Soc.*, *127*, 15907–15915 (2005).
87. A. R. Dunn, W. Belliston-Bittner, J. R. Winkler, E. D. Getzoff, D. J. Stuehr, and H. B. Gray, *J. Am. Chem. Soc.*, *127*, 5169–5173 (2005).
88. S. M. Contakes and H. B. Gray, unpublished results.
89. T. Holman, V. Kenyon, Y. Nyugen, S. M. Contakes, and H. B. Gray, unpublished results.
90. J. Miller, Ph.D. Thesis, California Institute of Technology (Pasadena, CA), 2002.
91. D. M. Roundhill, *Photochemistry and Photophysics of Metal Complexes*, Plenum Press, New York, 1994, pp. 356.
92. K. Kalyanasundaram, *Photochemistry of Polypyridine and Porphyrin Complexes*, Academic Press, London, 1992, pp. 626.
93. M. Furue, K. Maruyama, T. Oguni, M. Naiki, and M. Kamachi, *Inorg. Chem.*, *31*, 3792–3795 (1992).
94. R. J. Staniewica, R. F. Sympson, and D. G. Hendricker, *Inorg. Chem.*, *16*, 2166–2171 (1977).
95. I. J. Dmochowski, PhD Thesis, California Institute of Technology (Pasadena, CA), 2000.
96. A. R. Dunn, PhD Thesis, California Institute of Technology (Pasadena, CA), 2003.
97. W. Belliston Bittner, PhD Thesis, California Institute of Technology (Pasadena, CA), 2005.
98. A. Udit and H. B. Gray, unpublished results.
99. M. Noble, C. Miles, S. Chapman, D. Lysek, A. Mackay, G. Reid, R. Hanzlik, and A. Munro, *Biochem. J.*, *339*, 371–379 (1999).
100. A. R. Dunn, I. J. Dmochowski, J. R. Winkler, and H. B. Gray, *J. Am. Chem. Soc.*, *125*, 12450–12456 (2003).
101. I. V. Kurnikov, A. R. Dunn, H. B. Gray, and M. A. Ratner, submitted.

7

Leakage in Cytochrome P450 Reactions in Relation to Protein Structural Properties

Christiane Jung*

Max-Delbrück Center for Molecular Medicine, Research Group Protein Dynamics
D-13125 Berlin, Germany
<christiane_jung@hotmail.com>

1.	INTRODUCTION	188
2.	PROTEIN STRUCTURAL PARAMETERS	191
	2.1. Experimental Approaches	191
	2.2. Theoretical Models and Definitions	192
3.	THE REACTION CYCLE OF CYTOCHROME P450	195
	3.1. Step 1. Substrate Binding and High-Spin/Low-Spin State Equilibrium	196
	3.2. Step 2. Delivery of the First Electron	202
	3.3. Step 3. The P450-CO Complex as a Probe for Heme Pocket Properties	203
	3.3.1. Spectral Properties of P450-CO. Sensor for the Heme Pocket Polarity	203
	3.3.2. Kinetics of CO Binding. Sensor for Heme Pocket Dynamics	208
	3.4. Step 3. The First Branch Point: The P450-O_2 Complex	211
	3.5. Step 4. The Second Branch Point: The P450 Peroxo and Hydroperoxo Complexes	213

---

*Present address: KKS Ultraschall AG, Surface Treatment Division, Frauholzring 29, CH-6422 Steinen, SZ, Switzerland

3.6. Step 5. The Third Branch Point: The P450 Iron-Oxo
  Intermediate 215
   3.6.1. The Oxidase Reaction 216
   3.6.2. Formation of Protein Radicals 217
 3.7. Side Reactions Connected with Inactivation or Modification
  of P450 221
4. PROTEIN STRUCTURAL PARAMETERS AND EXTENT OF
 COMPETITIVE REACTIONS 222
 4.1. Leakage Due to Hydrogen Peroxide Formation 222
 4.2. Leakage Due to Protein Radical Formation 224
5. CONCLUDING REMARKS 225
 ACKNOWLEDGMENTS 226
 ABBREVIATIONS 226
 REFERENCES 227

1. INTRODUCTION

Recent observations of protein radicals produced by the iron-oxo intermediate in thiolate heme proteins [1] initiated the idea that protein radical formation might be one of the diverse side reactions (leakage) in cytochrome P450 enzymes (P450) during turnover in the natural reaction cycle. How relevant are side reactions for the function and what are the structural conditions under which the P450 cycle leaks?

Usually, the activity of P450 is characterized according to its main reaction or main substrate. However, in many cases side reactions (Table 1) occur because of the disturbed interplay of the heme and the protein. Therefore, P450 catalysis is a complex phenomenon although the net reaction (1) looks very simple.

$$O_2 + 2\ H^+ + 2\ e^- + RH \rightarrow ROH + H_2O \qquad (1)$$

A substrate RH with an inert carbon-hydrogen bond (C-H) reacts with O_2 if two protons and two electrons are present. The O-O bond is cleaved and one oxygen atom is inserted into the substrate to produce alcohol while the other oxygen atom produces water.

Because of spin conservation rules for a chemical reaction the triplet state of O_2 does not react with a singulet state of the substrate without catalysis. For taking place of reaction (1) in P450 systems two reaction cycles have to interplay: the electron delivery cycle and the P450 cycle (Figure 1). In the electron delivery cycle the electrons are provided from NAD(P)H to redox proteins (flavin proteins) which transfer the electrons directly or via other redox proteins

Table 1. Side products formed as result of leakage or inactivation in the P450 cycle.

Side product	Branch point in the P450 cycle	Examples	References
P420	at different steps in the cycle, mainly after step 2: first reduction	all P450s known so far	[8,57,85,161–166]
O_2^-	at step 3: autoxidation of the O_2 complex	almost all P450	[111,112,167–172]
H_2O_2	at step 3: autoxidation of the dioxygen complex and dismutation of O_2^- at step 4: peroxo and hydroperoxo complex	microsomal P450s mitochondrial P450s bacterial P450s P450cam and mutants P450eryF	[111–113,115] [114,178] [88,173–176] [177]
H_2O	at step 5: iron-oxo complex	microsomal P450s bacterial P450s	[111–113,115] [173–176]
Protein radicals (Tyr, Trp)	at step 5: iron-oxo complex	P450scc P450cam P450BMP iNOSox, nNOSox	[137] [15] [149] [150]
Heme covalent coupling to protein	via step 5: iron-oxo complex	CYP4 family P450cam	[153,154] [180]
Heme adduct with aldehydes	via step 4: hydroperoxo complex	CYP2B4 CYP102	[155] [179]
Heme loss	via step 4: hydroperoxo complex	diverse P450s	[156]

(iron-sulfur proteins, cytochrome b_5) to P450. In the P450 cycle O_2 and the substrate RH come together to form the product ROH. Neglecting the electronic reactivity of the substrate itself the efficiency of product formation can be diminished by five main protein-based phenomena. The electrons are not or inefficiently transported to P450 because of (i) bypassing the electrons from the delivery cycle to other systems, (ii) by blocking the electron transport within this cycle, and (iii) by improper interaction between the redox protein with P450. When the electrons have entered the P450 cycle the product formation is influenced (iv) by possible inhibition phenomena or (v) by formation of side products. The latter one is called uncoupling or leakage of the P450 cycle and is in the focus of this chapter.

One may ask what are the forces and the time regime for driving the P450 cycle and uncoupling (leakage) reactions? The insertion of an oxygen atom into a carbon-hydrogen bond according to the net reaction (1) should proceed

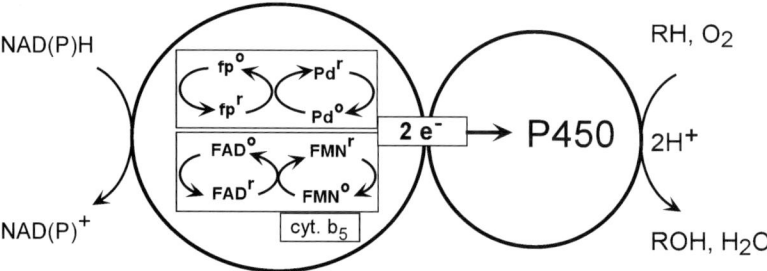

Figure 1. The electron delivery cycle (left) and the P450 reaction cycle (right) of the monooxygenase systems cytochrome P450. Depending on the source of the system either the NADH/flavin (fp)/iron-sulfur (Pd) redox protein (upper box) or the NADPH/FAD/FMN redox protein (lower box) chain is operating. Cytochrome b_5 is involved for the second electron transfer step in some P450s (lower box).

spontaneously because the free enthalpy ΔG is negative ($\sim -384\,\text{kJ/mol}$) [2]. The insertion of the oxygen atom into the substrate would require the splitting of the O-O and C-H bonds which both need an activation free enthalpy $\Delta G^{\#}$ in the order of $+(418$ to $460\,\text{kJ/mol})$ in the absence of a catalyst. A catalyst does not change the driving force and should not change the final product but diminishes the activation enthalpy to speed up the reaction. The activation enthalpy for P450 catalyzed substrate hydroxylations is in the order of 38–71 kJ/mol [2]. This fundamental statement refers to each elementary step. However, if several parallel and consecutive elementary reactions are possible, the outcome of the total net reaction is often very sensitive to small structural changes of the reactant and the protein-based catalyst. This can drive the process in different directions within the energy landscape of the chemical system.

The heme as prosthetic group in P450 enzymes is held in a pocket build up by the surrounding amino acids of the protein (see Figure 2 in Section 2.1.). The iron as catalytic center is coordinated by the four nitrogen atoms of the porphyrin ring, the negatively charged sulfur of a cysteine as the so-called axial proximal or 5th ligand and a ligand of different nature as distal or 6th ligand depending on the state along the reaction cycle. O_2 as one reactant is bound as distal ligand to the iron and the substrate reactant is located in proximity to the bound O_2 [3,4]. Water molecules from the solvent penetrate the protein, enter and leave the heme pocket and interfere with substrate and iron ligand binding and product release. The dynamic behavior of the protein structure which determines or modifies these processes plays a significant role also for the chemical reactions taking place at the heme. This chapter shows which structural properties of the P450 protein in its interaction with substrates and dioxygen are or may be relevant for the leakage of the P450 reaction cycle. Most of the statements focus on P450cam from the soil bacterium *Pseudomonas putida* (CYP101) which is the best studied P450 enzyme so far.

2. PROTEIN STRUCTURAL PARAMETERS

2.1. Experimental Approaches

In recent years the crystal structure of many P450 proteins have been solved demonstrating the great similarity of the global and secondary structures independently of the source of the enzyme and of the fact that the amino acid sequence identity is less than 20% for most of the proteins [5]. Despite this progress in structure analysis by crystallography the information is, however, limited because only snap shots of the three-dimensional position of the main atoms in the crystal are obtained.

The recently developed technique of cryo-crystallography [3] has significantly improved the insight into structures of intermediate complexes. Nevertheless, energetic states of the electronic or nuclear system, which might be relevant for reactivity, are not accessible. Spectroscopic studies are therefore necessary. The data obtained by spectroscopic methods, however, cannot be directly related to three-dimensional coordinates without assumption of theoretical models which describe properties of the chromophor under consideration. These models may be a qualitative description or a quantitative calculation of electronic and structural parameters as recently reviewed [6].

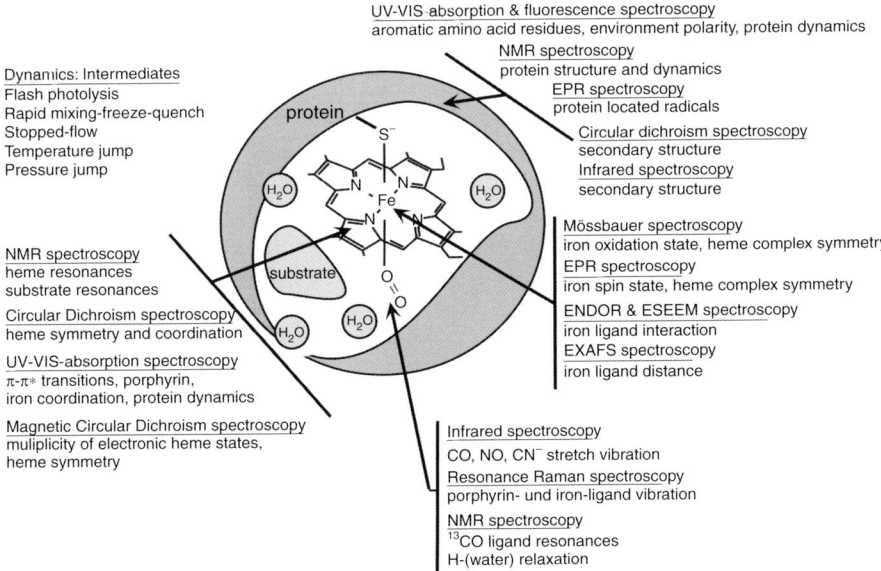

Figure 2. Overview about spectroscopic techniques used to study the structural behavior of cytochromes P450.

Figure 2 shows the most important spectroscopic techniques applied to P450 enzymes. The relevant parts of the heme protein which are captured by these techniques are highlighted by arrows. The secondary structure of the protein has been studied by CD [7] and FTIR [8–10]. NMR was applied to determine also changes in the three-dimensional structure of P450cam in solution when it interacts with the redox protein putidaredoxin [11,12]. The electronic transitions of aromatic amino acid residues, in particular tyrosine and tryptophan, are well studied by UV-visible absorption and fluorescence spectroscopy and have been used to monitor the polarity of the chromophor environment [13,14]. Intermediate amino acid radicals have recently been detected by EPR [1,15]. The extensively used UV-visible spectroscopy in the range of 350 nm to 700 nm captures the electronic $\pi-\pi^*$ transitions of the porphyrin [16] which are modulated by the axial heme iron ligands and by polar effects of surrounding amino acids influencing the porphyrin π-system and the peripheral substituents. The circular dichroism of these transitions in the absence (CD) and presence of a magnetic field (MCD) is used to characterize the symmetry and energetic degeneracy of the electronic states of the heme and represent a fingerprint for ligand complexes under study [16–19].

NMR has also been applied to study the heme properties, the electronic parameters of axial ligands as well as of substrates bound in the heme pocket [20,21]. NMR studies performed on the CO complex have been correlated with stretch vibration frequencies obtained by infrared spectroscopy (FTIR) to show specific effects of the substrate on the ligand [22]. Resonance Raman spectroscopy is a powerful technique to characterize the porphyrin normal vibrations whose frequency do strongly depend on the oxidation and spin state of the heme iron [23–25]. Iron-ligand bond vibration modes and ligand stretch modes are also accessible by this technique. The oxidation and spin states of the heme iron are also reflected in Mössbauer data [26,27]. EPR [16], ESEEM [28], ENDOR [29], and EXAFS [30,31] techniques give a detailed insight into the iron oxidation and spin states, the nearest iron environment and iron-ligand distances. The few cited references are not a complete reflection of the tremendous work in this field. Many of these techniques have been applied on P450 already more than 20 years ago and the data of these studies on the different P450s have been summarized over the years in several excellent reviews and books [16,32–34]. Some of these techniques are now used in combination with stopped-flow, rapid-mixing freeze-quench, flash photolysis or other time-resolved approaches such as temperature or pressure jump techniques.

2.2. Theoretical Models and Definitions

Most of the steps in the P450 reaction cycle involve binding and dissociation phenomena which can be rationalized by the sketch of an energy landscape shown in Figure 3 originally established for the rebinding phenomena of CO

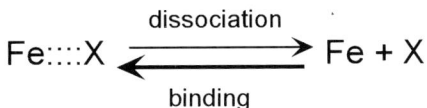

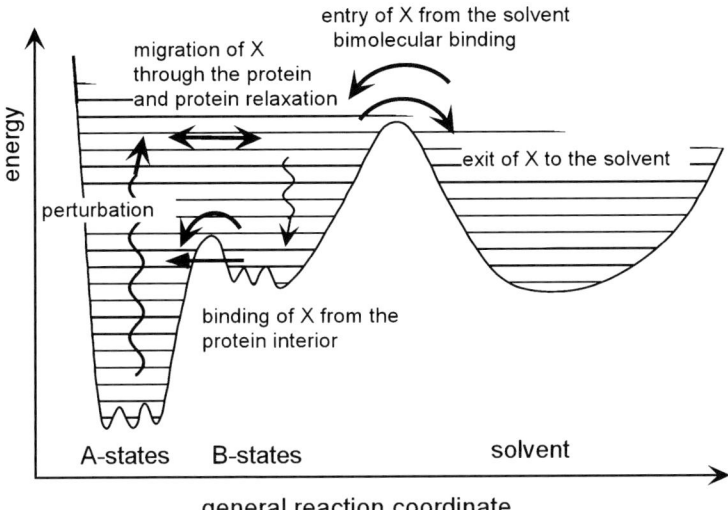

Figure 3. Generalized energy landscape for the binding and dissociation of small molecules X such as iron ligands, substrates and water molecules. The left global minimum (A states) corresponds to the nearest heme environment and the right broad minimum to the bulk solvent.

flash photolysis in heme proteins [35]. Because P450 with its molecular weight of around 50 kDa represents a large molecule, a protein interior and a solvent environment can be differentiated. When a small molecule X, which can either be the substrate, an iron ligand or just a water molecule, binds to or close to the heme iron (active center) then three steps can be differentiated: (i) entry of X into the protein from the bulk solvent; (ii) migration of X through the protein whereby it may be transiently bound to different sites in the protein (B states) and (iii) binding of X to the site with the highest energetic stability (A states). If any kind of perturbation is applied to the protein the small molecule X can dissociate from A, migrate within the protein or even exit back into the bulk solvent. These steps are always accompanied by analogous steps for disordered solvent molecules which penetrate the protein in a different extent and interfere or are accompanied with the binding/dissociation of X.

Studies under high hydrostatic pressure turned out to be a good tool to analyze these binding phenomena [36]. The pressure P affects only the volume V of the system to be analyzed and the water molecules in the protein are important entities reflected in the reaction volume ΔV and the activation volume $\Delta V^{\#}$ (Equation 2).

$$\left|\frac{\partial \ln K}{\partial P}\right|_T = -\frac{\Delta V}{R \cdot T}; \quad \left|\frac{\partial \ln k}{\partial P}\right|_T = -\frac{\Delta V^{\#}}{R \cdot T} \qquad (2)$$

K and k are the equilibrium constant and the rate constant, respectively. T and R represent the temperature and the gas constant, respectively. The change of the volume with pressure at constant temperature yields the isothermal compressibility (Equation 3) where V_0 is the averaged volume as reference value. The compressibility can be determined from spectroscopic studies assuming that the spectroscopic signal $v(R_s, P)$ (i.e., the absorption frequency) changes when a polar group approaches the chromophore at a distance R_s [37,38]. The value of *const* depends on the kind of spectroscopic signal used.

$$\beta = -\frac{1}{V_0}\left[\frac{\partial V}{\partial P}\right]_T; \quad \beta \equiv const \cdot \frac{1}{v_0}\left[\frac{\partial v(R_s, P)}{\partial P}\right]_T \qquad (3)$$

The absorption frequency can be separated in three contributions (Equation 4).

$$v(R_s, P) = v_{vac} + v_s(R_0, P=0) + \alpha \cdot P \qquad (4)$$

v_{vac} is the absorption frequency in the absence of the solvent ('vacuum'). $v_s(R_0, P=0)$ is the change of the absorption frequency caused by the interaction with the solvent molecules in a distance R_0 at the pressure $P=0$ kbar ('solvent shift'). α is the slope which can be experimentally determined from the change of the band frequency with pressure. As seen below the 'solvent shift' is related to the high-spin state content of the heme iron. The sum of the two first terms in Equation (4) corresponds to v_0 in Equation (3).

The compressibility β is related to the averaged volume fluctuation $<V^2>$ (Equation 5). k_B is the Boltzman constant. If ordered or disordered water

$$<V^2> = k_B \cdot T \cdot \beta \cdot V \qquad (5)$$

molecules are significantly involved in a process then an antagonistic behavior of the effect of osmotic pressure of the solvent (Π, Equation 6) and the hydrostatic pressure (P) should be observed [39]. X_{H_2O} and V_{H_2O} are the mole fraction and the mole volume of water in the solvent, respectively. The model behind is that the protein is regarded as water penetrable 'membrane' which can release water molecules to balance the water concentration gradient to the solvent (osmotic pressure effect) while hydrostatic pressure pushes water molecules into the protein.

$$\Pi = R \cdot T \cdot \ln\left[\frac{X_{H_2O}}{V_{H_2O}}\right] \qquad (6)$$

The rate of water influx into the protein can only indirectly be determined by the change of the population of a protein state which is assigned to a particular content of water. For P450 the low-spin state (*LS*) and the high-spin state (*HS*) are the relevant states to be considered (see below). The slope of the loss of the high-spin-state content ΔHS with the temperature change ΔT (from 297 K to 180 K within 10 min) is interpreted as a water influx rate k_{H_2O} (Equation 7) for the heme pocket.

$$k_{H_2O} = \left[\frac{\Delta HS}{\Delta T \cdot 10 \text{ min}} \right] \quad (7)$$

The inverse value of k_{H_2O} has been named as rigidity factor of the heme pocket [40]. If this rigidity factor is related to the mobility of the substrate bound in the heme pocket then a correlation with the *B* factor (Equation 8) for the substrate in the crystal structure should exist. Δx is the deviation of the substrate from a particular reference point.

$$B = 8\pi^2 <\Delta x^2> \quad (8)$$

Electron transfer is an essential process for P450 function. According to the Marcus theory the electron transfer rate depends on three parameters: (i) the difference between the redox potential of the electron donor and acceptor (ΔE) which is related to the free enthalpy (driving force) $\Delta G = -n \cdot F \cdot \Delta E$ where F is the Faraday constant and n the number of transferred electrons; (ii) the reorganization energy λ needed to reach the conformational state from which the electron can transfer and (iii) the distance R_d between the electron donor and the electron acceptor. Equation (9) shows the relation of these three parameters to the electron transfer rate applied for proteins [41].

$$\log_{10} k_{et} = 15 - 0.6 \cdot R_d - 3.1 \cdot (\Delta G + \lambda)^2 / \lambda \quad (9)$$

All other parameters or terms used in this chapter correspond to the common definitions found in textbooks of physical chemistry.

3. THE REACTION CYCLE OF CYTOCHROME P450

The P450 reaction cycle (Figure 4) starts with binding the substrate to the heme protein (step 1). In step 2 the heme iron is reduced by electron delivery from NAD(P)H via redox proteins (flavin protein, iron-sulfur protein). In a subsequent step 3 the reduced heme iron binds molecular oxygen. The O_2 complex accepts a second electron with synchronous or subsequent attachment of a proton to the oxygen ligand (step 4). This intermediate dioxygen species quickly decomposes by splitting the O-O bond and releasing a water molecule. Thus, a highly reactive

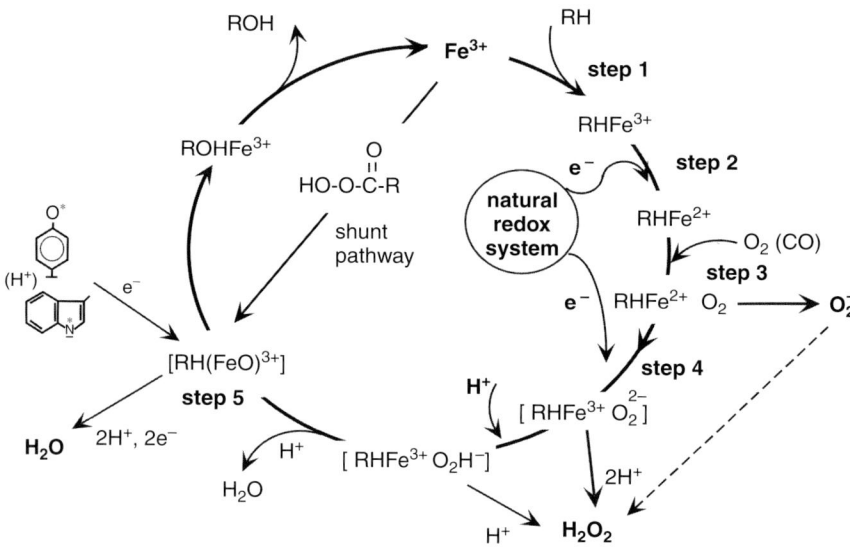

Figure 4. The cytochrome P450 reaction cycle with main steps 1 to 5 and the uncoupling side reactions (leakage).

iron-oxo species (compound I) is formed (step 5), which is assumed to hydroxylate the substrate. Finally, the product is released from the heme pocket and the resting state of the enzyme is reformed for the next cycle. These main steps are similar for all P450 enzymes. However, each step is far from being simple and several branch points for side reactions exist leading to the leakage of the cycle (Table 1).

3.1. Step 1. Substrate Binding and High-Spin/Low-Spin State Equilibrium

At the beginning of the cycle the heme iron is in the oxidized form (ferric, Fe^{3+}) and exists in two states which differ in the total spin of the five electrons in the d-orbitals of the iron. This so-called low-spin/high-spin state equilibrium is one of the most important structural properties in P450 enzymes because many other properties of the protein along the P450 reaction cycle show surprisingly a good relation to the high-spin content which is the fraction of the protein in this spin state.

The heme iron in the low-spin state has a total spin of $S = 1/2$ and corresponds to the $^2T_2$ term in the ligand field theory [16,42]. Early NMR solvent-proton relaxation studies [43,44] suggested that a water molecule should coordinate to the iron as 6th ligand or exist very close to it. This suggestion has been confirmed by ESEEM spectroscopy [28] and the crystal structure of P450cam [45] showing that the heme iron is six-coordinated. A water molecule of a cluster with six water molecules is the distal ligand. Therefore, the formation of the low-spin state requires that water molecules from the solvent or closest heme environment have to enter the heme pocket for binding to the iron (see Figure 3 in Section 2.2). The low-spin state is characterized by EPR g-values of 2.42, 2.25, and 1.91 [46,47]. In the high-spin state the total spin is $S = 5/2$ and the corresponding term is $^6A_1$. The distal coordination position is not occupied in the high-spin state if produced by camphor binding to P450cam [48]. The EPR g-values are 3.95, 7.80, and 1.78 [47].

Although the spin states are well characterized by EPR the equilibrium parameters for the population of these states are not adequately mirrored because the measurements have to be done at very low temperatures (20 K). It has been found that the fraction of the spin states at low temperatures strongly depends on the cooling rate [49]. UV-visible spectroscopy turned out to be much more convenient to characterize the spin state equilibrium at all temperatures. The π–π^* transitions of the porphyrin are strongly modulated by the coordination sphere and the oxidation state of the iron. The low-spin state reveals the Soret band at $\sim$417 nm and the visible Q-bands at $\sim$529 nm and $\sim$568 nm while the Soret band in the high-spin state appears at $\sim$392 nm with weak bands at $\sim$507 nm and 634 nm [16,50]. For most of the P450 enzymes the iron exists in the low-spin state for the substrate-free protein. When a substrate binds to P450 the low-spin state is completely or partially shifted to the high-spin state. For example, for P450cam an almost 100% conversion from the low-spin to the high-spin state is observed when the natural substrate 1R-camphor binds [50]. However, camphor analogues only partially convert the heme iron to the high-spin state [51]. Many other P450-substrate complexes from different organisms behave very similar [52,53]. There are, however, also examples where no high-spin shift at all upon substrate binding is observed although turnover is detected [54].

At the beginning the finding of fractional high-spin content was very puzzling because it has been shown that the substrate binds in a 1:1 ratio to the enzyme as seen in the crystal structure [55] and concluded from FTIR studies [56]. The remaining low-spin state population could therefore not result from a fraction of P450 which has not bound a substrate. In addition, it has been discovered already very early that this low-spin/high-spin state equilibrium also exists in the absence of a substrate and can significantly be modulated by temperature, pressure, and pH of the solvent. Temperature decrease and pressure increase lead to a higher population of the low-spin state [57]. A lower pH value shifts the equilibrium in substrate-free P450cam to the high-spin state [58]. Mutation of specific amino acid residues can also change the spin state to high-spin even if no substrate is bound, i.e., substrate-free Y96F shows a high-spin content of 25% [59].

According to the Tanabe-Sugano diagram for the energy levels of d_5 electron systems [60] the high-spin state $^6A_1$ would be the ground state for weak ligand field strengths while the low-spin state $^2T_2$ becomes the ground state for strong ligand field strengths [16]. A thermal spin state equilibrium can appear close to the crossing region of both energy levels either by thermal population of the excited state at constant ligand field or by thermal change of the ligand field induced by variation of the iron-ligand bond, the ligand itself or the electric field around the complex. The electric field effect has been discussed based on quantum chemical calculations [61] showing that only in the presence of an electrostatic field around the heme the low-spin state becomes the ground state. Recent QM/MM calculations show that the energetic gap between the low-spin and high-spin states is strongly dependent on the protein environment [62].

However, the change of the iron-ligand bond has been shown to be most relevant for P450 [63]. Based on crystal structure data the ligand field of the distal ligand can be changed either by removing the water cluster and forming a 5-coordinated heme complex (100% high-spin, i.e., camphor binding in P450cam) or by changing the protonation state of a singly bound water molecule. The latter possibility has been suggested from the finding that, i.e., the norcamphor complex of P450cam, which reveals a high-spin content of ca. 46%, has a water molecule bound at the distal position with an occupancy of 0.97 [55]. It has been argued that OH^- would produce a strong ligand field and therefore induce the low-spin state while H_2O is a weak ligand inducing the high-spin state. In this case the protonation equilibrium of the water molecule at the sixth ligand position would be the origin for the spin state equilibrium. The loss of five of the six water molecules in the cluster may change the pK_a of the remaining water molecule.

However, it is not clear whether the high occupancy seen in the crystal is also realized for the protein in solution and whether the slight differences seen in the EPR g-values for the low-spin state of the substrate-free (1.91, 2.26, 2.45), camphor (1.91, 2.25, 2.45) [47], norcamphor (1.91, 2.24, 2.42, unpublished), and norbornane (1.91, 2.24, 2.41, unpublished) complexes measured under identical conditions from the same stock protein and at the same EPR spectrometer can be interpreted in terms of a different protonation state of the coordinated water molecule. The presence of a water molecule at the sixth coordination site even in a substrate-bound protein has also been discussed for P4502B4 based on DFT calculations and used to explain the fractional high-spin state content [64]. Other theoretical studies reviewed in [6] have been employed to study the specific coordination properties of a water molecule as sixth ligand. Whatever the detailed situation in the different P450 proteins is, the formation of the high-spin state is connected with release of water molecules from the heme pocket even when one water molecule still remains close to the heme iron while the low-spin state is accompanied by the entry of water molecules. Studies on the antagonistic effect of hydrostatic and osmotic pressure on the high-spin/low-spin state equilibrium

(Equation 10) support this conclusion [65].

$$\text{low-spin state} \underset{k_{-1}}{\overset{k_1}{\rightleftharpoons}} \text{high-spin state} \qquad (10)$$

In conclusion, the spin state equilibrium reflects an averaged residence time of water molecules in the heme pocket and its vicinity. It is therefore a parameter describing the dynamic behavior of the protein structure and the substrate mobility in the heme pocket [55].

Two important findings for a set of substrate-P450cam complexes with different populations of the high-spin state at standard conditions, i.e., 20 °C, 0.1 MPa, (named as 'initial high-spin content', HS_i) strongly support this conclusion:

(i) In temperature jump experiments the rate constant k_{-1} (Equation 10) for the relaxation back to the low-spin state depends on HS_i [66] (Figure 5A).

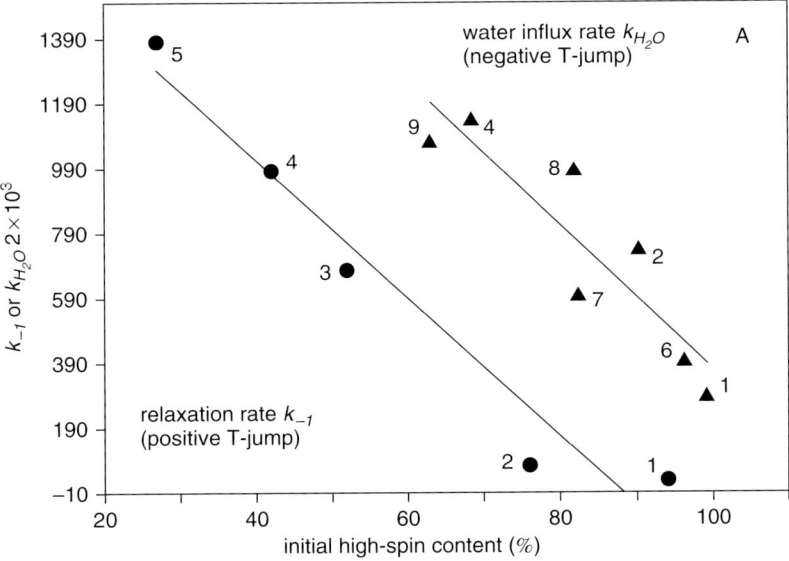

Figure 5. A: The relaxation constants for the high-spin to the low-spin state transition of P450cam for different substrate complexes as function of the initial high-spin content. (●) k_{-1} (Eq. 10) after a positive temperature jump (10 °C to 13 °C) [66] and (▲) k_{H_2O} (water influx rate, Eq. 7) after a negative temperature jump (297 K to 180 K) within 10 minutes [40] scaled by 2000 to fit in the same plot with k_{-1}. B: The reaction volume for the pressure-induced high-spin to low-spin state transition for different substrate complexes of P450cam in dependence on the initial high-spin content of the complexes [40]. Substrates: 1, 1R-camphor; 2, 1R-camphorquinone; 3, fenchone; 4, norcamphor; 5, TMCH; 6, 1S-camphor; 7, 1S-camphorquinone, 8, camphane; 9, norbornane.

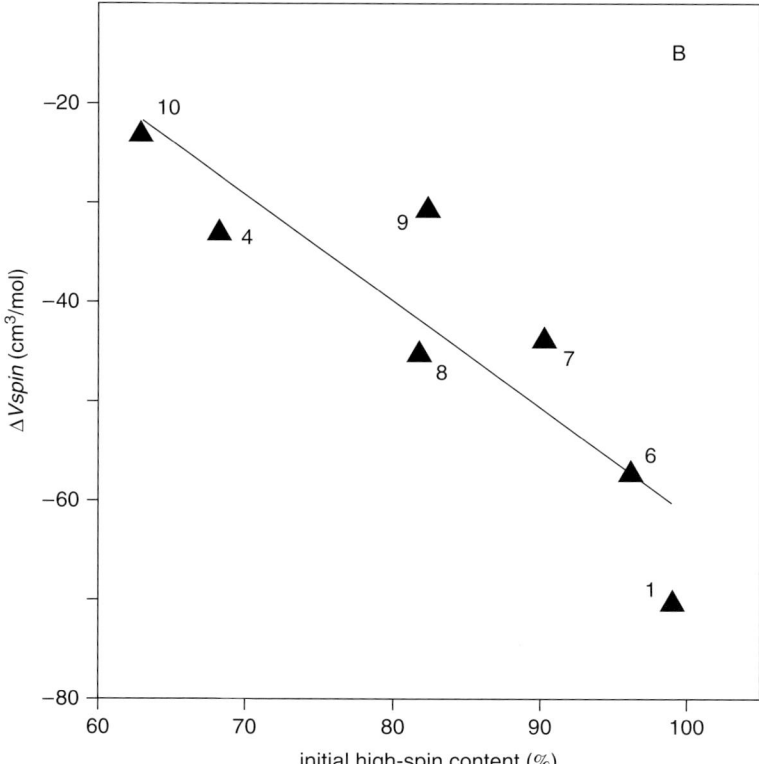

Figure 5. (Continued).

An analogous correlation is observed for a set of P450cam-substrate complexes which are quickly cooled to 180 K (negative temperature jump). The high-spin content trapped at 180 K depends on the structure of the substrate complex [40]. The water influx rate (Equation 7) determined from these experiments reveals the same trend. An increasing relaxation rate, respective water influx rate, correlates with a decreasing initial high-spin content (Figure 5A). The more water molecules are initially in the heme pocket (low initial high-spin content) the faster the water molecules return back into the heme pocket after stopping the temperature perturbation.

(ii) A similar behavior is found for the reaction volume (Equation 2) of the high-spin/low-spin state equilibrium determined from high-pressure experiments. The absolute value of the reaction volume for the pressure-induced low-spin state formation increases with increasing the initial high-spin state content HS_i in a set of P450-substrate complexes (Figure 5B) [40,57]. This observation is explained by a hydration model. If P450 exists in the high-spin state, most of the active site water molecules, present in the heme pocket in the absence of a substrate, are replaced by the substrate after its binding. When pressure is applied, water molecules are pressed into the protein again which increases the probability that a water molecule coordinates to the heme iron and produces the low-spin state. The more water molecules exist initially in the protein, the fewer water molecules have to be pressed into the protein to get the low-spin state. It has been argued from FTIR studies [56] that the substrate is not squeezed out from the heme pocket under pressure which would of course induce a population of the low-spin state just because of substrate removal.

For P450cam it was found that this hydration phenomenon depends on the structure of the substrate (Figure 6) and reflects the mobility of the substrate within the heme pocket [40,55].

The natural substrate 1R-camphor is held in the optimal orientation in the active site by the hydrogen bond of its keto group and the hydroxyl group of the amino-acid residue Tyr96, and by hydrophobic contacts of its methyl groups C-8, C-9 to Val295 and Asp297 in the β_3 sheet, and of the methyl group C-10 to Val247 in the I-helix and Thr185 in the F-helix [48]. In particular the hydrophobic contacts make the main contribution to the low substrate mobility

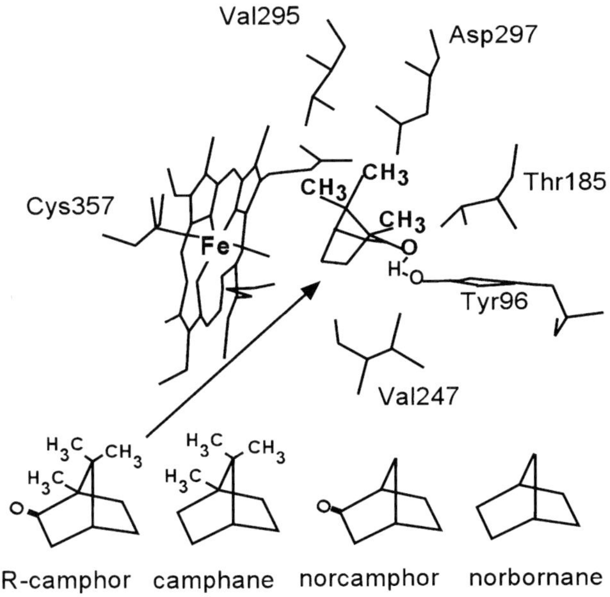

Figure 6. The heme active site of P450cam indicating the amino acid residues which are relevant for the binding of camphor (top) and the structure of camphor analogues (bottom).

and high rigidity of the heme pocket [40]. The loss of the methyl groups in the substrate analogues leads to a more mobile substrate binding mode which facilitates the access of the heme pocket for water molecules. According to the binding model (Figure 3) this means that substrate binding and water exit to the solvent or migration within the protein are tightly coupled. Three different routes for substrate entry and exit have been proposed based on molecular dynamics simulations [67]. The route 2a in P450cam which involves the B/C loop, the F/G loop and the beta-1 hairpin, is also the dominant route in other bacterial P450s [68].

Obviously, the substrate seems to use special pathways to approach the heme environment and these pathways might be similar in different P450s. However, whether the water molecules do also use the same or specific pathways is not known. A channel of three water molecules (numbers 687, 523, and 566) linking Thr252 in the I-helix groove of the active center and the buried Glu366 close to the protein surface has been found in the crystal structure of camphor-bound P450cam. This water channel is perturbed when camphor analogues are bound or the substrate-free protein is formed. In these cases the water molecules are removed or dislocated and may contribute to ordered and disordered water molecules in the heme pocket [55]. A contribution of this water channel to the proton transfer needed for dioxygen activation has been however ruled out by mutation of Glu366 to methionine. In this mutant the hydroxylation activity is not significantly changed compared to the wild type protein [69].

3.2. Step 2. Delivery of the First Electron

The next step in the reaction cycle is the reduction of the heme iron to the iron(II) state (reduced form). In the natural cycle the electron is provided from NAD(P)H via electron transfer steps between redox proteins (Figure 1). The redox potential of the participating proteins, the protein-protein recognition (predominantly by electrostatic interactions) and the kind of pathways for intra-protein electron transport are the relevant parameters determining the efficiency of the reduction of the heme iron. The latter one is the most uncertain parameter, difficult to study experimentally and actually only accessible in detail by theoretical methods such as used, i.e., for the HARLEM pathway program [70].

It has been experimentally demonstrated for several P450s that the standard redox potential of the P450 protein correlates with its high-spin state content when different substrate complexes of a particular P450 are compared [50,63]. This relation may be explained by the fact that the reduced heme iron is in the five-coordinated high-spin ($S = 2$) state [3,71,72]. This means that there is no water molecule bound to the iron. The cleavage of this water molecule from the heme iron before reduction would significantly facilitate the electron transfer. For example, the shift of the redox potential from $-300\,mV$ for substrate-free P450cam (low-spin) to $-170\,mV$ for camphor-bound P450cam (high-spin) [50] would correspond to a free energy of interaction between water and the heme iron of ~ 3 kcal/mol which is in the order of the relative energy E(1/2)–E(5/2) of ~ 4 kcal/mol between a water-ligated heme low-spin state and a water-unligated high-spin state estimated for P4502B4 using DFT calculations [64]. Further theoretical studies show that the reduction of the water unligated 5-coordinated heme complex in P450 is more exothermic than the ligated one by 10.7 kcal/mol [6]. The main reason is the negatively charged sulfur ligand which makes P450 a bad electron acceptor compared to heme complexes with, for instance, an imidazole as proximal ligand. So, release of the water molecule from the heme iron may gate the electron transfer.

It has been found for P450cam [73] as well for other P450s, such as P4502B4 [52], that the reduction rate also increases with increasing high-spin state content. Other P450s seem not to follow this relation [74]. However, the reduction of the P450 heme iron in the natural reaction cycle is a step where the two reaction cycles of the monooxygenase system meet (Figure 1). Several parameters are therefore interlinked which make it difficult to separate different effects. At least for the soluble P450cam the relation of the reduction parameters to the high-spin content is clear indicating that the hydration state of the heme pocket in the oxidized protein has consequences for the next step in the P450 reaction cycle.

3.3. Step 3. The P450-CO Complex as a Probe for Heme Pocket Properties

The reduced heme iron is able to bind molecular oxygen as the next step in the cycle to form an unstable O_2 complex (see below). Because of this instability other small iron ligands which form a more stable complex have been used to characterize the reduced P450 state and to model the O_2 complex. The most important ligand is carbon monoxide. It is clear from the orbital hybridization for CO and O_2 in binding to the iron that CO cannot model all properties of O_2, i.e., the binding geometry [9]. The crystal structure of the CO and the O_2 complexes of P450cam have confirmed this difference [3,4,75,76].

However, it turned out that CO is an excellent probe to study polar/electrostatic properties and the dynamics of the heme pocket [56,77] which both are relevant for the leakage phenomenon. Therefore, the properties of the CO complex will be discussed in some detail.

3.3.1. Spectral Properties of P450-CO. Sensor for the Heme Pocket Polarity

The best known spectral property of the CO complex is its strongly red-shifted Soret band from normally 420 nm (i.e., for myoglobin-CO) to 450 nm which was the basis for the discovery of this enzyme and giving its name cytochrome P450. The origin for this spectral effect is the strong overlap of the 3p-orbitals of the negatively charged sulfur of the cysteine axial ligand with the iron 3d-orbitals and the porphyrin π-molecular orbitals. Quantum chemical calculations (π-INDO) indicated that this overlap, together with the strong electronic push effect of the negatively charged sulfur, leads to an orbital mixing of the weak 'normal' UV electronic transitions $e_g(\pi_{axial}) \rightarrow b_{1u}^*$ (porphyrin) (around 340 nm) and the strong 'normal' Soret electronic transitions $a_{1u}, a_{2u} \rightarrow e_g^*$ of the porphyrin (around 420 nm) resulting in a synchronous intensity borrowing and red-shift of both bands [78].

Alternatively, a new transition from the sulfur $3p_z$ orbital to the porphyrin e_g^* orbitals, not seen in myoglobin models, (IEHT calculations [79]) or the split of mixed transitions of porphyrin $a_{2u}, a_{1u} \rightarrow e_g^*$, $d_{x2-y2} \rightarrow e_g^*$ and $d_\pi \rightarrow d_{xy}^*$ ('split' Soret band, INDO/CI calculations [80]) have also been proposed to be the origin. Semi-empirical quantum chemical calculations revealed a partial negative net charge of the sulfur ligand of approximately -0.5 to -0.6 [78]. Such partial net charge can be produced by a hydrogen bond between the sulfur ligand and a neighboring amino acid residue as quantum chemical calculations using the CNDO method suggest [78]. Indeed, the crystal structure of P450cam shows that there is a fine-tuned hydrogen bond network within a cage formed by the neighboring amino acid residues Leu358, Gln360, Phe360, and Leu356 for P450cam [48]. Several studies showed the importance of this hydrogen bond network for the function of P450 [81–84]. One can imagine that this arrangement is sensitive to structural changes induced by different perturbations such as temperature, pressure, pH, and specific reagents. Indeed, it is well known that the characteristic absorption band at 450 nm is easily moved back to 420 nm ('normal' Soret band) [85]. This so-called cytochrome P420 is probably not a homogeneous conformation and different structural changes have been discussed (protonation of the negatively charged sulfur, replacement of the cysteine by histidine [6,78]). The reaction volume for the conversion of P450-CO to P420-CO is in the range of $-50 \, cm^3/mol$ to $-160 \, cm^3/mol$ [37,57]. Although the term P420 is originally defined only for the CO complex, characteristic spectral changes associated with P420 formation have also been observed for the oxidized protein. For a set of substrate analogues bound to the oxidized form of P450cam it was found that the reaction volume for P420 formation depends on the high-spin content. Increasing absolute values of $-80 \, cm^3/mol$ for the substrate-free protein (100% low-spin) to $-220 \, cm^3/mol$ for the camphor or adamantanone complex (100% high-spin) were observed [57]. These large values indicate that P420 formation is accompanied with large conformational changes and reorganisation of solvent molecules. Considering the diverse observations of P420 formation in almost all P450 systems (Table 1) this conformational conversion must be regarded as one possible side reaction which diminishes activity and should be considered in the context of the leakage phenomenon of the P450 reaction cycle.

The position of the Soret band in the CO complex varies for the different P450s between 448 nm and 452 nm [16]. For a long time it was assumed that for a particular P450 protein the position of the Soret band in the CO complex is not altered by binding substrates or changing solvent conditions. However, detailed studies on P450cam revealed that this assumption is not justified. It turned out that in substrate complexes, which produce the high-spin state in the oxidized protein, the Soret band in the CO complex appears at a lower wavelength compared to those which are low-spin complexes in the oxidized form (Figure 7A) [37]. The shift is small but significant. When the position of the Soret band is expressed in energy-proportional units, such as the wavenumber (cm^{-1}), there is an increase of the wavenumber with the increasing initial high-spin content for a set of substrate complexes of P450cam [37]. The shift from pure low-spin to 100% high-spin state population is in the order of $100\,cm^{-1}$. The Soret band in the CO complex shifts to lower wavenumbers when high hydrostatic pressure is applied (Figure 7A). The isothermal compressibility calculated from the slope

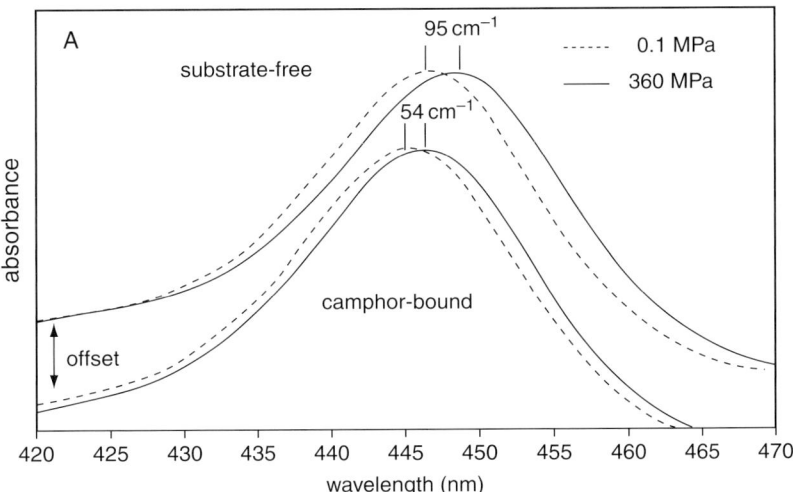

Figure 7. **A**: The Soret band peak of substrate-free and camphor-bound P450cam-CO at 0.1 MPa and 360 MPa at 4 °C, 100 mM potassium phosphate buffer, pH 7, 60% glycerol [37]. **B**: The CO stretch vibration frequency (expressed in wave number unit cm^{-1}) of different substrate complexes of P450cam-CO [56] as function of the initial high-spin content produced in the ferric form. **C**: Isothermal compressibility determined from the pressure-induced Soret band shift (Eq. 3 with $const = 1$) [36,37] of P450cam-CO substrate complexes. **D**: Isothermal compressibility determined from the pressure-induced shift of the CO stretch vibration band (Eq. 3 with $const = 1$) [88,89] of P450cam-CO substrate complexes. Substrates: 1, substrate-free; 2, adamantane; 3, norbornane; 4, norcamphor; 5, TMCH; 6, fenchone; 7, 1R-camphor; 8, dimethylallyl borneol ether; 9, bromocamphor; 10, camphane; 11, allylborneol ether; 12, 1S-camphor.

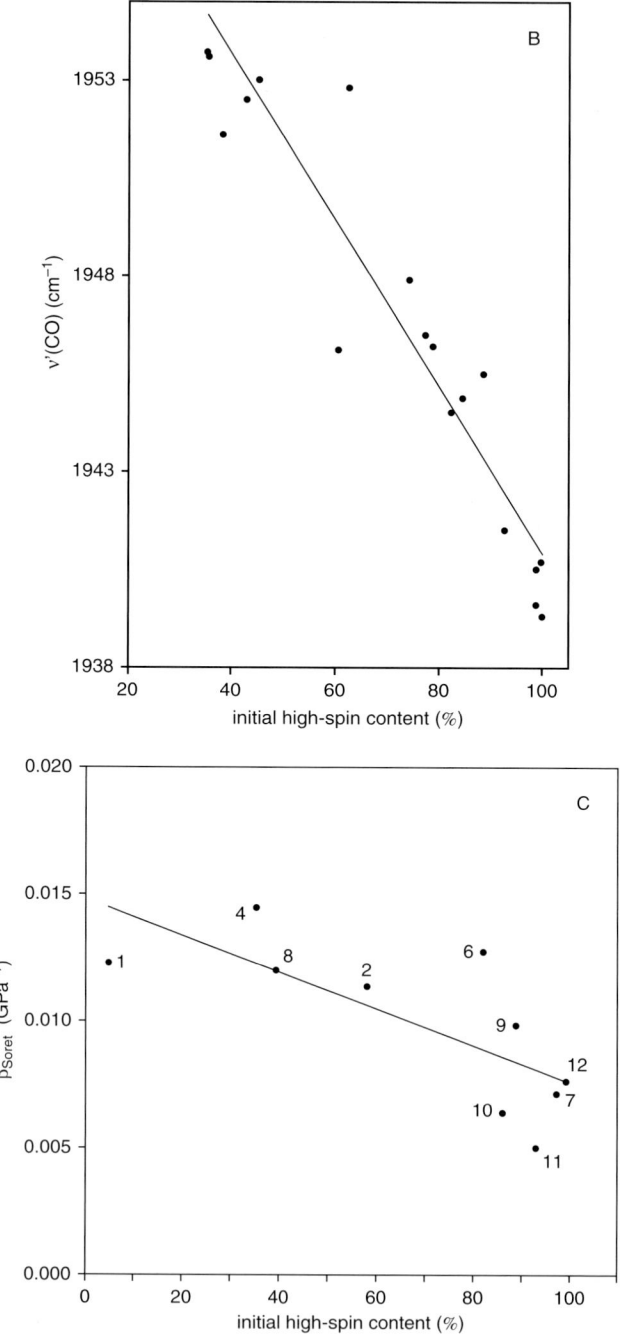

Figure 7. (Continued).

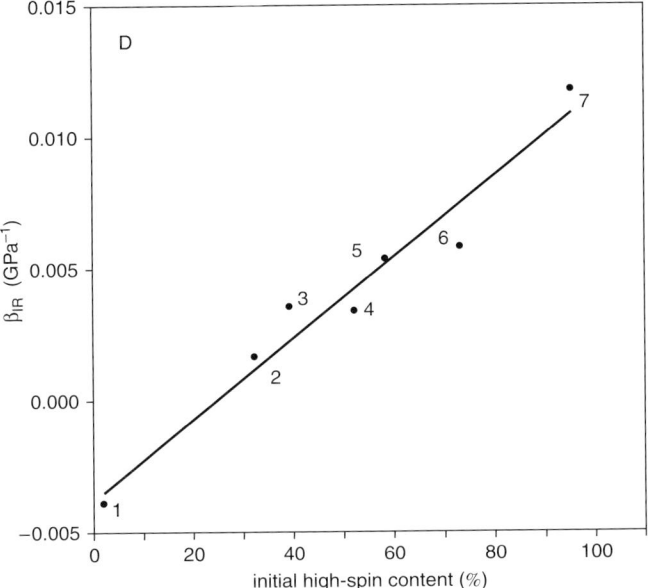

Figure 7. (Continued).

of the pressure-induced shift (Equation 3) depends on the initial high-spin content HS_i (Figure 7C). The model explaining this relation is that the distance between the chromophor (heme) and a polar group (i.e., water molecule) shortens when the protein is compressed [37,38]. This 'solvent shift' must therefore be related to the high-spin state content. Based on these studies it is concluded that P450 substrate complexes with a lower initial high-spin content (respective more water in the heme pocket) show a more compressible protein structure. This property is related to a higher mobility of the substrate within the heme pocket [40].

The quantum chemical studies mentioned above showed that the push effect of the proximal sulfur ligand leads to an increase of the electron density in the porphyrin-π-system (which is related to the red shift of the Soret band) as well as in the π-system of the CO ligand (increased iron π-backdonation) [86]. In particular this increased iron π-backdonation to the CO ligand led assume that the CO stretch vibration frequency $\nu(CO)$ of P450 proteins should generally appear at lower values compared to the frequencies in heme proteins with a histidine as fifth iron ligand. However, this assumption could not be verified experimentally by FTIR measurements [9] when focusing only on P450. The reason is that other effects of amino acids forming the heme pocket and the nature of the bound substrate influence the CO dipole from the distal side (push-pull effect). The push effect of the fifth ligand, however, exists and is reflected in a down-shift of the trend line in the plot of the CO stretch mode frequency versus the frequency of the iron-CO stretch vibration of different heme proteins [9,22,87].

Furthermore, for a large set of substrates for P450cam it was found that the frequency of the CO stretch vibration at room temperature and ambient pressure correlates with the initial high-spin content produced by the substrates in the oxidized form of P450cam [56] (Figure 7B). The lower the initial high-spin content the higher the CO stretch vibration frequency appears. In addition, high-pressure FTIR studies showed that the slope of the pressure-induced shift of the CO stretch mode frequency does also depend on the initial high-spin content [88]. Surprisingly, however, the compressibility extracted from this slope showed an inverse dependency on the initial high-spin content (Figure 7D) compared with the compressibility determined from the pressure effect on the Soret band (Figure 7C) discussed above for the same set of substrates [89].

The above puzzling finding could be explained by a model considering direct and distal side effects of water molecules in the heme pocket on ν(CO). The partial negative net charge at the terminal oxygen atom of the CO ligand (−0.197 [86]) leads to an electrostatic interaction with a positive electrostatic potential produced by the protein on the distal side resulting in a red-shift of the CO stretch mode band for complexes with 100% high-spin state content compared to a distal side without such positive electrostatic potential. In addition, the almost linear Fe-CO geometry allows a good polarization of the CO ligand and makes it very sensitive to even weak electrostatic fields. With an increasing number of water molecules in the heme pocket, this electrostatic interaction is disturbed due to compensation of the positive electrostatic potential. Therefore, the CO stretch mode frequency must increase again as observed [88]. It should be already noted here that this conclusion is important for understanding the leakage phenomenon discussed below.

3.3.2. Kinetics of CO Binding. Sensor for Heme Pocket Dynamics

Similar to the O_2 complex the bond of the CO ligand to the iron can be broken by light absorption of the heme [90]. When the light is switched off the ligand travels back to the heme iron (Figure 3). Usually, the binding affinity is larger for CO by a factor of ten compared to O_2 for the same P450 protein [91] but the binding is slower (k_{on} $\sim 10^4$–10^5 $M^{-1}s^{-1}$, in the presence of a substrate) compared to the O_2 ligand (k_{on} > $10^6 M^{-1}s^{-1}$) (Table 2). This difference has been employed in the photo-action spectrum to proof the participation of P450 in substrate conversion [92].

From the two major rebinding processes (geminate rebinding from the protein interior [35,93] at temperatures lower than 180 K and rebinding from the solvent above 180 K) (Figure 3) only the latter one is relevant for the leakage phenomenon discussed in this chapter. Flash photolysis and stopped-flow studies should show similar results what is also the case. For a set of substrate complexes

Table 2. Rate constants for CO and O_2 binding for selected P450 proteins as examples.

P450	Substrate	T (°C)	k (M^{-1}s^{-1}) × 10^4	References
O_2 binding				
CYP2B4	benzphetamine	15	250	[169]
P450LM4	-	10	50	[182]
P450scc	cholesterol, SF	25	530	[181]
P450scc	22R-hydroxycholesterol, SF	25	6.2	[181]
P450scc	20,22-dihydroxycholesterol, SF	25	3.2	[181]
CO binding				
P450terp	substrate-free, SF	5	360	[170]
P450terp	+ terpineol, SF	5	2	[170]
P450BM3	substrate-free, SF	5	160	[170]
P450BM3	substrate-free, SF	20	420	[170]
P450BM3	+ palmitate, SF	5	130	[170]
P450BM3	+ palmitate, SF	20	390	[170]
P450cam	substrate-free; SF	5	29.5 (46%); 297 (54%)	[94]
P450cam	substrate-free, F-FTIR	26.8	158 (54%); 133 (8%); 381 (31%)	[77]
P450cam	substrate-free, F	20	850	[95]
P450cam	substrate-free, F	20	732	[96]
P450cam	norbornane, SF	3.8	332	[94]
P450cam	norbornane, F-FTIR	26.8	344	[77]
P450cam	norcamphor, SF	4.5	381	[94]
P450cam	norcamphor, F-FTIR	26.8	341	[77]
P450cam	norcamphor, F	20	1000	[95]
P450cam	adamantane, F	20	1300	[95]
P450cam	R-camphor, SF	5.6	3	[94]
P450cam	R-camphor, F-FTIR	26.8	9.8	[77]
P450cam	R-camphor, F	20	10	[95]
P450cam	R-camphor, F	20	6.8	[96]
P450cam	camphane, SF	4.8	1.6 (52%); 7.8 (48%)	[94]
P450cam	camphane, F-FTIR	26.8	10.4 (61%); 49.7 (25%)	[77]
P450cam	bromocamphor, F	20	55	[95]
P450cam	TMCH, F	20	75	[95]
P450cam	fenchone, F	20	150	[95]
P450scc	substrate-free	25	20	[181]
P450scc	cholesterol, SF	25	22	[181]
P450scc	20α-hydroxycholesterol, SF	25	16	[181]
P450scc	22R- hydroxycholesterol, SF	25	0.69	[181]
P450scc	20,22-dihydroxycholesterol, SF	25	0.38	[181]

F: flash photolysis with UV-vis detection; F-FTIR: flash photolysis with FTIR detection; SF: stopped flow with UV-vis detection.

of P450cam two groups could be separated based on stopped-flow and flash photolysis studies under high pressure [94,95]. The one group of substrate complexes shows a fast CO binding and is characterized by a positive activation volume (Equation 2) for the CO binding [norcamphor, norbornane and adamantane (class II-substrates) and the substrate-free protein]. The other group shows a negative activation volume and a slow CO binding kinetics [1R-camphor, camphane, fenchone, 3-endo-bromocamphor and 3,3,5,5-tetramethylcyclohexanone (TMCH), class I substrates].

For the first group the initial high-spin content HS_i is low while for the second group almost 80–100% high-spin content is observed. As explained in detail in [94] a positive activation volume for the binding process indicates a high solvent accessibility of the heme pocket. All class I substrates possess methyl groups (Figure 6) while class II substrates do not. It was concluded that the methyl groups of the substrate are the relevant structural entities which modulate significantly the CO binding properties of P450cam. Disturbing these interactions leads to a higher substrate mobility and accessibility of the heme pocket for water. One might conclude that the slow CO binding kinetics for the class I substrate-complexes of P450cam is caused by a high energy (ΔH) needed to loosen the contact of the methyl groups to the protein. This is, however, not the case as is concluded from the values for the activation enthalpy $\Delta H_{on}^{\#}$ and activation entropy $\Delta S_{on}^{\#}$ of CO binding in substrate-free ($\Delta H_{on}^{\#} = 61.9\,\mathrm{kJ/mol}$, $\Delta S_{on}^{\#} = +98.7\,\mathrm{JK^{-1}mol^{-1}}$, $\Delta G_{on}^{\#} = 34.44\,\mathrm{kJ/mol}$; 5°C) and 1R-camphor-bound P450cam ($\Delta H_{on}^{\#} = 31.8\,\mathrm{kJ/mol}$, $\Delta S_{on}^{\#} = -43.6\,\mathrm{JK^{-1}mol^{-1}}$, $\Delta G_{on}^{\#} = 43.93\,\mathrm{kJ/mol}$; 5°C) determined from flash photolysis studies [96]. The energetic contribution of the entropic term ($-T \cdot \Delta S^{\#}$) to the free enthalpy of activation $\Delta G_{on}^{\#}$ is the reason which makes CO binding in substrate-free P450cam faster than in the presence of camphor.

The large positive activation entropy in substrate-free P450cam may indicate that the CO molecule travels along many pathways to the heme iron. Along each pathway, however, many contacts (e.g., contacts to many water molecules) have to be broken, reflected in the large positive activation enthalpy. The higher flexibility and stronger compressibility of the structure in substrate-free and the class-II substrate complexes of P450cam discussed above is in agreement with this view. This explanation may also hold for other P450 proteins for which a much slower CO binding in the presence of a substrate compared to the substrate-free protein has been observed (Table 2).

In summary, the studies on the oxidized P450 and the CO complex suggest: (i) The accessibility of the heme pocket for water molecules is the fundamental structural property behind the high-spin/low-spin state equilibrium in the oxidized P450 as well as behind the compressibility and binding dynamics in the reduced P450-CO complex. Thus, the hydration phenomenon is not only restricted to the oxidized protein but is also relevant for the reduced protein state in the next step in the P450 reaction cycle. (ii) The presence of water molecules in the heme pocket disturbs polar or electrostatic contacts between the terminal oxygen atom of the ligand and amino acid residues on the distal side. It will be shown below how relevant these conclusions for the leakage phenomenon are.

3.4. Step 3. The First Branch Point: The P450-O_2 Complex

At the beginning of the spectroscopic characterization of P450 it was expected that the O_2 complex should also show a red-shifted Soret band at 450 nm in the UV-visible spectrum as observed for the CO complex. However, this is not the case. The Soret band appears at ~420 nm while the UV band is seen at 354 nm and the Q band at 553 nm with a shoulder at ~585 nm [19,97]. This 'normal' position of the Soret band has been explained by quantum chemical calculations which simply showed that the stronger electronegativity of O_2 compared to CO leads to withdrawing electron density from the porphyrin π-system to the dioxygen ligand. This withdrawal partially compensates the push effect of the negatively charged sulfur ligand to the porphyrin π-system. Because π-electron density and Soret band position are related to each other, this means that the Soret band shifts back to the normal wavelength position [78,86]. There is a controversial interpretation of the spectrum based on INDO-CI calculations which indicate that even in the O_2 complex the Soret band is 'split' similar to the CO complex induced by the admixture of sulfur $3p_z$ orbitals to porphyrin molecular orbitals [98].

Although O_2 can only bind to the reduced iron to give a diamagnetic complex, the electron density distribution does not correspond to Fe^{2+}-O_2 but resembles more the Fe^{3+}-O_2^- resonance form. This phenomenon is well known for hemoglobin and myoglobin [99] and has also been observed for P450 and verified by Mössbauer data [72], resonance Raman studies [100], and quantum chemical calculations [6,86]. It should be noted that this resonance form does not mean that there is an electron spin separation as one might conclude from magnetic susceptibility measurements on oxyhemoglobin published almost 30 years ago [101]. This analysis showed a significant deviation from diamagnetism and let open what the electronic ground state is. Quantum chemical calculations [99] however showed that the electronic ground state is diamagnetic. However, there is an electronic triplet state where the spins are separated and which is energetically not far from the ground state and can partially be populated.

A similar situation is valid for P450, too [6]. This state is ~1 kcal/mol above the ground state [98] and can be populated and should lead to the splitting of the oxygen-iron bond with releasing O_2^- and leaving iron(III) behind. The negatively charged terminal oxygen atom (-0.442[16]; -0.291[98]) of the bound O_2 ligand could also accept a proton from the local environment and could leave the ligand as HO_2. However, it has been found that predominantly O_2^- is produced [102] which can in the presence of two protons dismutate to hydrogen peroxide. The release of O_2^- and the reforming of the Fe(III) state is termed as autoxidation.

Table 3. Autoxidation rate of the O_2 complex for selected P450 proteins as examples.

P450	Conditions	T (°C)	k (s^{-1})	References
2B4 (LM2)	buffer	2	20; 2.5; 0.25	[168]
2B4 (LM2)	liposomes	2	15	[168]
LM4	buffer	10	0.99; 0.24	[171]
2B4	buffer, substrate-free	15	0.96; 0.13; 0.016	[169]
2B4	buffer, benzphetamine	15	0.13; 0.048	[169]
P450BMP	arachidonic acid	5	0.032	[170]
P450BMP	arachidonic acid	20	0.22	[170]
P450terp	substrate-free	20	0.017	[170]
	substrate-free	5	0.00071	[170]
P450119	substrate-free	5	0.08	[172]
P450cam	substrate-free	2	0.0018	[167]
P450cam	substrate-free	10	0.0045	[167]
P450cam	substrate-free	20	0.0140	[167]
P450cam	camphor	2	0.00055	[167]
P450cam	camphor	10	0.00125	[167]
P450cam	camphor	20	0.0039	[167]
P450scc	cholesterol	25	0.13	[181]
P450scc	22R-hydroxycholesterol	25	0.028	[181]
P450scc	20,22-dihydroxycholesterol	25	0.013	[181]
CYP158A2	flaviolin	23	0.042	[183]

So, the O_2 complex with its autoxidation reaction is the first real branch point along the P450 reaction cycle. The autoxidation rate varies significantly between the different P450s and is strongly dependent on the temperature (Table 3). The O_2 complex can be stabilized for some minutes at low temperatures [19]. It has also been observed that the substrate-free protein or substrate-complexes with a lower high-spin state content autoxidize faster than the complexes with 100% high-spin state content.

The crystal structure of the O_2 complex of camphor-bound P450cam has been recently solved [3,4]. Two water molecules on the distal side mediate a hydrogen bond network between Thr252-OH, Gly248 and the terminal oxygen atom of the O_2 ligand. One water molecule directly hydrogen bonds to the terminal oxygen atom of the O_2 ligand. Thr252-OH is also in a sufficient distant to assist in hydrogen bonding. Mutation of Thr252 to alanine keeps these water molecules in place. In contrast, the mutation of Asp251, next to Thr252, to Asn leads to a structural change hindering that these water molecules can be stabilized at these positions. The crystal structure of the O_2 complex of the erythromycin-bound P450eryF, has also been solved very recently [103]. This protein has an alanine instead of threonine at the analogous position of Thr252 in the I-helix of P450cam. A threonine at this position has been found in many P450s and assumed to be important for stabilizing a hydrogen bond network in the I-helix groove around the O_2 ligand. This function has been taken over in P450eryF by the 5-OH group of the substrate erythromycin which donates a hydrogen bond to the terminal oxygen atom of the bound ligand.

3.5. Step 4. The Second Branch Point: The P450 Peroxo and Hydroperoxo Complexes

The delivery of the second electron to the O_2 complex is a very specific process which initiates the actual O-O bond cleavage process. It has been shown for P450cam already very early [46,104] that putidaredoxin (Pdx) as the redox partner does not only deliver the first and second electron but functions also as an effector molecule for the oxygen activation. Indeed, in the meantime many studies demonstrated specific structural changes, induced by P450cam-Pdx interaction. In particular the hydrogen bond network at the proximal heme side is affected which stabilizes the charge of the thiolate ligand [11,25,105,106]. Based on rapid-double-mixing stopped-flow experiments under single turnover conditions it was recently suggested that binding of reduced Pdx initiates a lowering of the negative charge of the thiolate ligand reflected in a blue-shift of the Soret band of the P450-O_2 complex from 418 nm to 414 nm. This modified P450-O_2 complex with the less electron-rich thiolate ligand can be reduced easier [106].

The second electron which has reached the heme distributes over the whole porphyrin complex. Already earlier semi-empirical quantum chemical calculations [16] showed that after reduction of the O_2 complex without previous attachment of a proton to the terminal oxygen atom, this additional electron would be predominantly localized in the porphyrin π-system. The π-bond order for the O-O bond and the Fe-O bond would slightly change in opposite directions (from 0.496 to 0.444 and from 0.457 to 0.473 for O-O and Fe-O, respectively). However, when a proton has bound to the terminal oxygen atom then the second electron would predominantly be localized at the oxygen. In this case the O-O bond order is drastically diminished and becomes even antibonding (-0.054) while the Fe-O bond order does only slightly decrease (0.435). In contrast, when performing the same calculations with imidazole as proximal iron ligand the decrease of the Fe-O bond order is much more pronounced and the O-O bond order is low but remains still bonding after reduction of the protonated O_2 complex. DFT calculations of the O_2 complex and its reduced forms also clearly indicate that single protonation of the distal oxygen atom significantly weakens the O-O bond while the second protonation would simultaneously transform to the iron-oxo species without any energetic barrier. The protonation of the proximal oxygen would lead to hydrogen peroxide formation [107]. So, the electronic push effect of the negatively charged sulfur ligand is reflected rather in this specific behavior at protonation than in enforcing a charge gradient along the axial bonding axis. Studies with more elaborated theoretical methods performed in the recent years come to a similar conclusion [6].

Although the reduced nonprotonated O_2 complex (ferric peroxo complex) has been included as one step in the P450 reaction cycle since at the beginning of the P450 research it was not clear whether this complex does really exist. Recent studies with artificial reduction of the O_2 complex by γ-irradiation at cryogenic temperatures and EPR and ENDOR spectroscopic analysis of P450cam wild type and mutants showed that under nonequilibrium conditions such a complex ($g_\perp \sim 2.25$) is indeed formed [108,109]. In the wild type enzyme a significant amount of the ferric peroxo complex has been detected at 6 K which converts to the ferric hydroperoxo complex ($g_\perp \sim 2.3$) already when heated to 77 K. The Thr252Ala mutant behaves similarly. However, in the Asp251Asn mutant the ferric peroxo species is already the primary intermediate at elevated temperature indicating that proton transfer to the terminal oxygen atom is delayed. Stepwise annealing the wild type samples bound with the camphor substrate to temperatures above the glass temperature of the solvent/protein mixture at about 200 K converted the hydroperoxo intermediate directly via three intermediates in the sequence of $g_\perp \sim 2.6$, $g_\perp \sim 2.5$, and $g_\perp \sim 2.48$ to the hydroxy-camphor product complex and then to the ferric low-spin resting state ($g_\perp \sim 2.41$). An analogous annealing study performed by Nyman and Debrunner [110] gave similar results. Although the appearance of the 5-exo-hydroxy-camphor product complex suggests that the O-O bond was cleaved the expected iron-oxo-porphyrin intermediate was not detected.

The hydroperoxo complex is the second branch point at which the P450 reaction cycle can leak or uncouple leading to the formation of hydrogen peroxide instead of the water molecule produced from the terminal oxygen atom after O-O bond cleavage in the main route. Uncoupling hydrogen peroxide formation has been observed for almost all P450 systems (Table 1) [111–115]. This leakage phenomenon is generally explained by a disturbed active site or improper binding of the substrate in the active center of P450.

The most detailed mechanistic insight is given again for P450cam and has been extensively reviewed [4,116–119]. In the ferric P450cam, Thr252-OH in the I-helix groove donates a hydrogen bond to the peptide C=O of Glu248. Upon reduction and O_2 binding Asp251 undergoes a flip which initiates the break of the hydrogen bond between Thr252 and Glu248 and allows water molecules to enter the heme pocket. Thr252-OH is now able to hydrogen bond to the terminal oxygen atom of the O_2 ligand. Originally it was thought that this hydrogen bond is important for the specific proton transport because mutation to alanine, glycine or cysteine at position 252 drastically decreased the camphor hydroxylation activity without significantly reducing the NADH oxidation but increasing significantly the H_2O_2 formation [69].

Interestingly, a similar observation was made for P450BM3 when Thr268 in the I-helix was replaced by alanine [120]. Replacing Thr252 in P450cam by polar amino acid residues such as serine or asparagine restored significantly the original hydroxylation activity. A missing or changed hydrogen bond network between

Thr252, water and terminal oxygen atom was proposed to disturb the specific proton transfer to the O_2 ligand and facilitate uncoupling. QM/MM calculations showed that in the Thr252Ala mutant the water network is broken and the water molecules are more mobile [62]. However, this conclusion came recently in question because the crystal structure of the O_2 complex of the Thr252Ala mutant of P450cam showed that the two water molecules and the hydrogen bond network were still in place [4] excluding a role of Thr252 as proton transfer mediator. However, both water molecules are absent when Asp251 is mutated to Asn. In this case the camphor hydroxylation activity as well as the NADH oxidation are almost abolished [121]. This indicates that the second proton is transferred via the catalytic water molecules. So, the function of Thr252-OH is believed to stabilize the hydroperoxo complex and mutation will destabilize this complex and facilitate the release of hydrogen peroxide. What would this conclusion mean for the uncoupling observed when different substrates are bound to P450? The structural origin is not explored for most of the P450 enzymes. For a large set of substrate complexes of P450cam, however, it was found that the percentage amount of consumed O_2 which is converted to H_2O_2 depends on the high-spin content produced by the binding of the particular substrate [88]. It turned out that with decreasing high-spin content the extent of uncoupling is increased. It was suggested that the increased mobility of the substrate which facilitates the accessibility of the heme pocket for water molecules is the relevant property inducing uncoupling. The parameters which reflect this scenario are summarized below in Section 4.

Initiated by studies on substrate oxidations with CYP2B4 and CYP2E1 [122,123] the peroxo and hydroperoxo complexes have been discussed to be potential oxidizing species. Recent studies on P450cam and its Thr252Ala mutant in their reaction with the natural substrate camphor and alkene derivatives also lead to the suggestion that the hydroperoxo complex may be the oxidizing species in epoxidation reactions [124]. For a detailed discussion the reader is referred to recent reviews [118,125]. However, it is generally assumed that the iron-oxo intermediate is the most relevant oxidizing species which is formed after cleavage of the O-O bond.

3.6. Step 5. The Third Branch Point: The P450 Iron-Oxo Intermediate

The O-O bond can principally split in two ways. The homolytic cleavage would yield a OH radical and leave the second reduction equivalent in the iron-oxo species. In the terminology of peroxidases this heme species corresponds to compound II. For the heterolytic cleavage mode OH^- or H_2O in the presence of a proton are formed and the heme complex remains formally as $[Fe^{3+}\text{-}O]$ unit with oxygen as electrophilic species. However, the real electron density distribution over the whole complex resembles more a high-valent iron-oxo porphyrin π-cation radical [Fe(IV)-O] – P$^+$ suggested from studies of metal porphyrin studies and peroxidases [126,127] where this intermediate is called compound I. The heterolytic splitting is assumed to be the relevant mode in the natural reaction cycle for most of the P450 enzymes. However, P450 reactions are so complex [128] that homolytic cleavage might also be considered in particular cases.

So far, the iron-oxo compound I intermediate could not be trapped in the natural reaction cycle of P450. However, in analogy to chloroperoxidase from *Caldariomyces fumago* (CPO), which is also a thiolate heme protein, it is suggested that compound I in P450 should be characterized by such an iron(IV)-porphyrin π-cation radical species as discussed above [129,130]. Because of the lack of experimental data a tremendous amount of theoretical work has been published over the years which describes the electronic structure of the putative compound I of P450 and its mode of interaction with the substrate to produce the hydroxylated product which has been reviewed recently [6,131] and will not be further discussed here.

The iron-oxo intermediate is the branch point for at least two side reactions: the formation of a water molecule (oxidase reaction) and the formation of protein radicals.

3.6.1. The Oxidase Reaction

The formation of the water molecule from the iron-oxo intermediate has not been directly observed so far. However the existence of this oxidase reaction has been concluded from analysis of the stoichiometry between NAD(P) oxidation and O_2 consumption [111,112,132]. According to equation (1) the molecular ratio between the extent of NAD(P)H oxidation, O_2 consumption, and formation of hydroxylated substrate (R-OH) should be 1:1:1. There are, however, many examples where the ratio of R-OH formation to NAD(P)H oxidation is significantly lower than one [111] indicating that the redox equivalents of NAD(P)H are channelled to other products. These other products are O_2^- and H_2O_2 as discussed above which can be assayed by different techniques [133]. To a certain extent free NAD(P)H without an enzyme can produce O_2^- and H_2O_2 due to dismutation. Correcting the enzymatic activity data for these reactions, the ratio [NAD(P)H]/[O_2] should be one and the ratio [H_2O_2]/[O_2] reflects the fraction of H_2O_2 formed from O_2 within the P450 cycle via the hydroperoxo complex at the second branch point. If the ratio [NAD(P)H]/[O_2] exceeds one then water formation is indicated. The contribution of water formation is estimated from the ratio ([NAD(P)H] – ([H_2O_2] + [R-OH]))/([O_2] – ([H_2O_2] + [R-OH])) [132]. This ratio approaches a value of two if all NAD(P)H and O_2 are consumed to produce water because water formation from dioxygen requires 4 electrons and 4 protons.

To form a water molecule from the iron-oxo complex the two additional electrons and protons must reach the oxygen atom bound to the iron and one must assume that an unspecific transfer via internal solvent molecules is the relevant mechanism. Indeed, it has been shown for the iron-oxo intermediate in microperoxidases and heme model complexes [134] as well as for horseradish peroxidase [135] that the iron bound oxygen can principally exchange with bulk water. The classical example for the oxidase side reaction is the uncoupling of NADPH and O_2 consumption of P450LM in the presence of perfluorocyclohexane [132,136]. Many other examples have been reviewed [111].

3.6.2. Formation of Protein Radicals

Although radical formation during P450 reactions has been observed earlier [137] detailed studies of a radical formation were only recently reported [1,27]. The original motivation of these studies was to characterize the oxidation and spin states of the supposed iron-oxo species by spectroscopic techniques. Since the early beginning of P450 research many attempts have been made to trap this intermediate in the natural reaction cycle. However, this was without any success. The reason is that this species is so short-lived and does not accumulate under any conditions applied so far that it is not possible to analyze it. It has been estimated [118] from the data of the cryo-radiolytically produced species [109] that the hydroperoxy complex reacts with a rate constant on the order of $1000\,s^{-1}$, meaning that the pseudo-first-order rate constant for the camphor hydroxylation by the transient iron-oxo species should be larger than $1000\,s^{-1}$ to hinder any accumulation of the iron-oxo intermediate. Therefore, another way of trapping this intermediate has been applied. Formally, the reaction of oxidized P450 with external oxidants such as peracids or peroxides which carry already the two redox equivalents should lead to the same intermediate (so-called shunt pathway, Figure 4). For more than 30 years [138] it has been observed that the reaction of P450 with peracids, peroxides, and iodosobenzene can lead to substrate conversion similar to the natural cycle. This observation may therefore justify to trap the iron-oxo intermediate by the shunt pathway for further spectroscopic characterization. These studies resulted, however, in controversial findings.

For these experiments CPO was taken as reference because this enzyme possesses a thiolate proximal ligand like P450 and shows a reaction cycle which formally corresponds to the shunt pathway in P450. When CPO is mixed in stopped-flow experiments with an oxidant such as H_2O_2, peroxy acetic acid or meta-chloro-perbenzoic acid (mCPBA) a green-colored compound I is formed which shows a UV-visible absorption spectrum after several milliseconds characterized by a broad Soret band around 370 nm and a weak but significant long-wavelength band at ~ 680 nm [139,140]. This spectrum is characteristic for a porphyrin π-cation radical, very well known for iron porphyrin model complexes [127]. EPR (9.6 GHz) and Mössbauer spectroscopic studies on CPO samples obtained by rapid-mixing with peroxy acetic acid and freeze-quenching [141] identified compound I as a Fe(IV)=O (spin: S = 1) and porphyrin-π-cation radical (spin: S' = 1/2) where both spins couple antiferromagnetically. The EPR-spectrum of compound I in CPO produced with peracids has g-values of $g_\parallel = 2$ and $g_\perp = 1.75$. The Mössbauer study gives an isomer shift of $\delta = 0.14$ mm/s and a quadrupole splitting of $\Delta E_Q = 1.02$ mm/s.

Stopped-flow experiments with oxidants similar to those described for CPO have also been performed on P450cam (CYP101) [142] and on the thermostable CYP119 [143] with the shortest time for detection after 6 ms and 2 ms, respectively, and using mCPBA as oxidant. In contrast to CPO, the spectrum of an intermediate was only detectable as the result of a single-value decomposition analysis. It turned out that 2–3% of the spectral changes represent a pattern resembling that observed for CPO compound I. This result has been taken as a proof that a porphyrin π-cation radical is formed also in P450. Earlier analogous stopped-flow studies, however, without single-value decomposition analysis revealed spectra which did not match the spectra discussed above [144–146]. Recently, the kinetics of the formation of an intermediate in the reaction of H_2O_2 as well as mCPBA with sub-stoichiometric oxidant concentrations has been studied with P450cam in the absence and presence of substrates [147] indicating that a protonation/deprotonation equilibrium of presumably Tyr96 should assist the formation of the intermediate. Very recently, it has also been argued based on stopped-flow studies with UV-visible detection that the pH value might significantly affect the kinetics and therefore the steady state concentration of a compound I species in P450cam [148]. Considering only the results of the stopped-flow experiments one might conclude that P450 behaves like CPO. However, doubt came up with rapid-mixing/freeze-quench experiments using peracids as external oxidant first with P450cam [15,27,47] and later on with other thiolate heme proteins [1,149,150].

These studies showed that the iron in P450cam and P450BMP is indeed in the iron(IV) state with the characteristic Mössbauer parameters of the doublet with an isomer shift of $\delta = 0.13$ mms^{-1} and a quadrupole splitting of $\Delta E_Q = 1.94$ mms^{-1} for both proteins (Figure 8A). The isomer shift is similar to the value observed for CPO ($\delta = 0.14$ mms^{-1}) [141] and the value obtained from the recent theoretical study [131]. The higher quadrupole splitting of P450 compared to CPO ($\Delta E_Q = 1.02$ mms^{-1}) may indicate differences in the symmetry of the heme complex. This specific difference in the symmetry could not be correctly simulated by theoretical methods ($\Delta E_Q = 0.64$–0.67 mms^{-1} [131], and $\Delta E_Q = 0.35$–0.64 mms^{-1} [151]). However in contrast to CPO, the EPR spectra of the freeze-quenched intermediate in P450cam and P450BMP did not show any signal which would indicate a porphyrin π-cation radical. Instead, an EPR signal for an organic radical which overlaps with the signals of the natural low-spin heme iron has been found [27]. Using high-frequency EPR at 94 GHz, 190 GHz, and 285 GHz EPR and site-directed mutagenesis [15] it was possible to assign the

EPR signals in P450cam to specific tyrosines. EPR g-values of $g_y = 2.0044$, $g_z = 2.0022$ and a distribution of $g_x = 2.0078$ to 2.0064 were obtained which are fingerprints for tyrosine radicals in a polar environment with heterogeneous hydrogen bonding. Simulations of the 94 GHz EPR signature (Figure 8B) in wild-type P450cam led to the assignment of Tyr96 as radical site. This was confirmed by mutating this tyrosine to phenylalanine what significantly changed the EPR radical signature (Figure 8B) but did not remove a radical signal.

Simulations of this EPR spectrum indicated that Tyr75 is the radical site in the Y96F mutant. When Tyr75 is also replaced by phenylalanine the typical tyrosine radical signal is removed. However, the double mutant shows still a

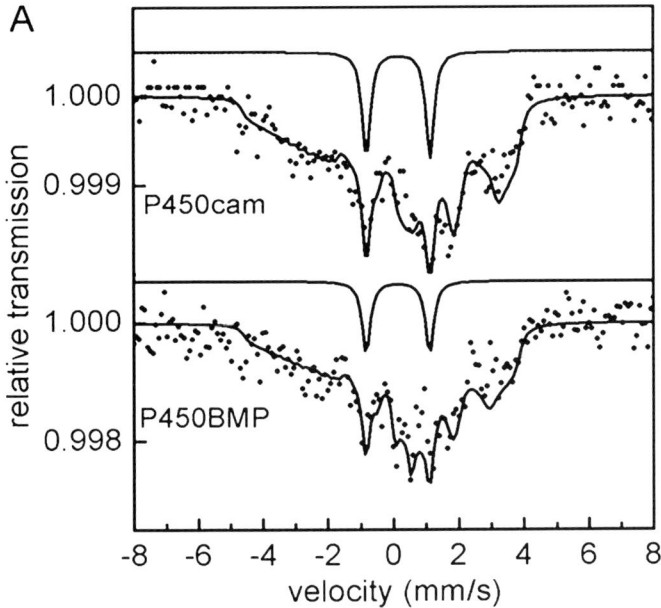

Figure 8. **A**: Mössbauer spectra of the freeze-quenched ($-110\,°C$) intermediate of wild-type ^{57}Fe-P450cam and ^{57}Fe-P450BMP after 8 ms reaction with peroxy acetic acid at room temperature. The solid doublet spectrum above the experimental spectrum corresponds to the simulated spectrum of the Fe(IV)-species. **B**: 94 GHz EPR spectra of the freeze-quenched ($-110\,°C$) intermediate of P450cam, wild-type and Y96F mutant and P450BMP, wild-type, after 8 ms reaction with peroxy acetic acid at room temperature. The residual spectrum for P450BMP was obtained from the subtraction of the scaled P450cam, wild-type spectrum from the spectrum for P450BMP. **C**: The amount of freeze-quenched ($-110\,°C$) intermediate formed by reaction of wild-type P450cam with peroxy acetic acid at room temperature and at different reaction times given for the tyrosine radical, the Fe(IV) species and the unknown Fe(III) high-spin species. All spectra in A, B, and C are constructed from original data published in [1,15,27,47,149,152].

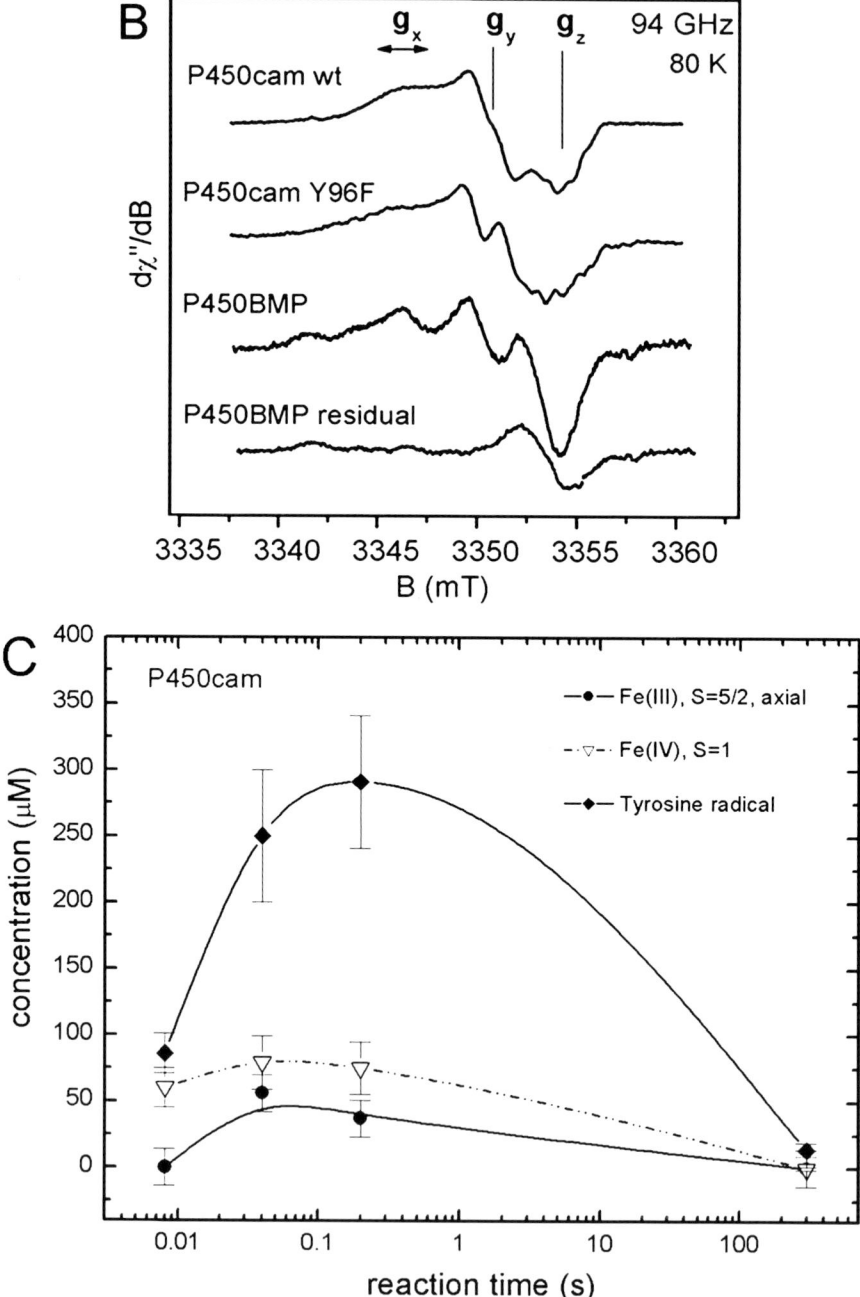

Figure 8. (Continued).

very weak signal with a g value of 2.003 which could not be assigned so far [15]. Also for P450BMP an EPR spectrum for the freeze-quenched intermediate was observed which revealed a β-proton hyperfine tensor consistent with a tyrosine radical (Figure 8B). The spectrum of P450BMP had a similar overall shape as that of wild-type P450cam. Tyr305 and Tyr334 in P450BMP both have side chain orientations, consistent with the observed β-proton hyperfine tensors. However, compared to the P450cam spectrum a strong negative signal component on the high-field side was additionally observed which is by far too intense for a g_z component of a tyrosine radical alone (Figure 8B) [149]. A fitting analysis suggested that this negative signal originates from a tryptophan radical (g = 2.0027) which amounted to 30% of the total signal intensity. All three tryptophans in the neighborhood of the heme (Trp90, Trp96, Trp367) have side chain orientations leading to hyperfine tensors for the β-protons, which would be consistent with the spectral width of this residual signal.

For wild-type P450cam the intermediate was freeze-quenched after different reaction times (Figure 8C). At 8 ms the amount of the tyrosine radical scales approximately with the amount of Fe(IV) observed. This correlation is lost already after 40 ms. The detailed analysis of the 9.6 GHz EPR spectra revealed that an additional axial ferric high-spin intermediate is increasingly formed during the time course before the original low-spin complex of the start sample was recovered [152]. The time course of the formation and decay of the radical signal reveals a maximum indicating that it is a real intermediate.

All these freeze-quench experiments and most of the stopped-flow studies were performed with a substrate-free protein because it was observed that the substrate (camphor in P450cam) blocks the access of the heme pocket for the external oxidant [1,47]. The amount of radical and iron(IV) formed in the P450cam intermediate is about 15% (8 ms reaction) of the total amount of substrate-free protein but only 0.4% in the presence of camphor. However, the fraction of tyrosine radical formed was again increased when substrates with a higher mobility (i.e., norcamphor (1.2%) or norbornane (1.6–2%)) were bound in the heme pocket, which allowed a better heme pocket accessibility (rigidity factors: 1R-camphor (6.81), norcamphor (1.75), and norbornane (1.86) [40]).

3.7. Side Reactions Connected with Inactivation or Modification of P450

Although inactivation and modification of P450 should not be classified in terms of leakage or uncoupling of the P450 cycle it nevertheless leads to a deviation or branching off from the normal cycle. Formation of P420 understood as loss of the proximal ligand (see above) can certainly occur at all steps of the P450 cycle and depends on various parameters which will be hard to generalize. One cannot exclude that most of the P420 formation observed is the result of the unnatural environment of the isolated and purified proteins under *in vitro* conditions.

Different heme modifications have been observed. Covalent coupling of the heme to the protein can occur when a carboxylate group is close to a methyl group substituent of the porphyrin. In an autocatalytic process this methyl group becomes hydroxylated probably activated via a ferryl species. The resulting alcohol group forms an ester bond to the glutamic acid residue in case of CYP4A [153]. This kind of heme modification has been observed in several P450s when a conserved carboxylate group close to a methyl group is found [154]. Heme adducts from aldehydes in a natural NADPH/O_2 or mCPBA driven P450 reaction have been found whereby it is still unclear which oxygen species is relevant [155]. Heme degradation initiated by heme loss from the protein by destruction of the heme-protein contacts has been described [156]. The mechanism is still unclear but the heme degradation is assumed to be caused by hydrogen peroxide released during the uncoupling from the hydroperoxo complex (second branch point). Several other reactions related to P450 modification due to uncoupling have been recently reviewed [157]. A more detailed description of these processes is not within the scope of this chapter.

4. PROTEIN STRUCTURAL PARAMETERS AND EXTENT OF COMPETITIVE REACTIONS

The structural and spectroscopic characterization of the complexes at the different steps along the reaction cycle (Figure 4) shows that the P450-based catalysis is a very complex phenomenon and that the accessibility to the heme pocket or heme environment for water molecules seems to play a significant role. The detailed mechanism is often discussed on the basis of crystal structure data which restrict explanations to local geometrical reorientations of amino acid side chains and/or geometry of bound ligands during reaction. The thermodynamic and kinetic behavior of the complexes in solution might, however, also be significantly affected by disordered water molecules which are not resolved in the crystal. Two leakage phenomena will be high-lighted as summary.

4.1. Leakage Due to Hydrogen Peroxide Formation

It turned out that the high-spin/low-spin state equilibrium in the ferric form of P450 reflects a structural property of the protein which is relevant for each step in the cycle. This property is the accessibility of the heme pocket for water. These water molecules may not be strictly ordered but rather disordered. Figure 9 shows the cross correlation of the extent of H_2O_2 formation observed for diverse wild-type and mutant P450cam substrate complexes with the logarithm of the spin state equilibrium constant. Although there is scatter in the data the overall trend is clear. The lower the high-spin state content (or the higher the low-spin state content) the more hydrogen peroxide in the uncoupling reaction is

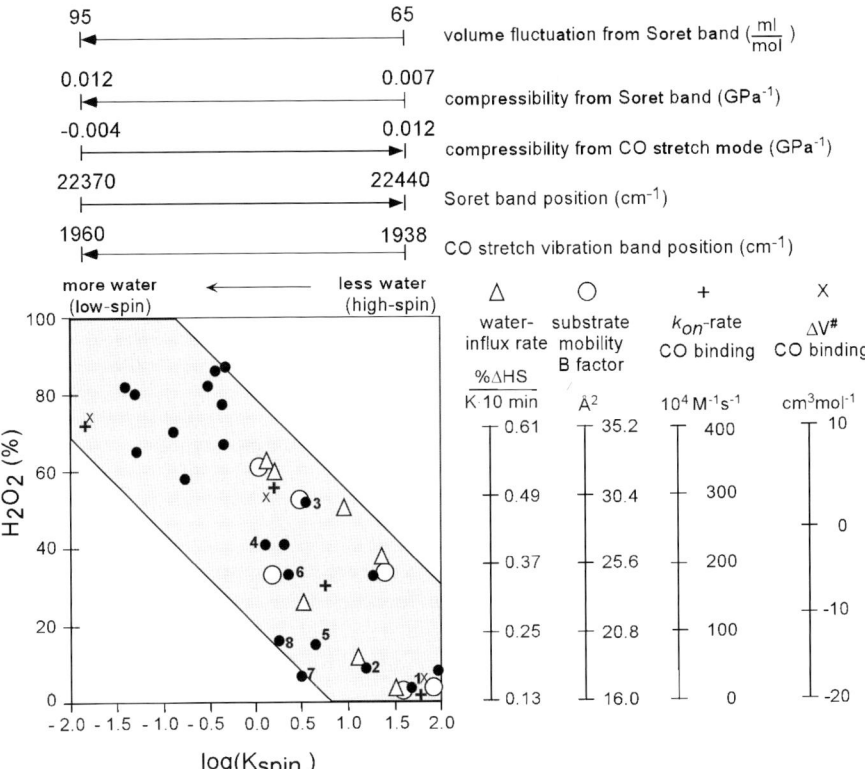

Figure 9. The summarizing cross-correlation between the percentage amount of H_2O_2 formed from O_2 consumed in the natural reaction cycle of P450cam and mutants bound with different substrates which induce a different high-spin content in the P450 ferric form with several protein structural parameters. The detailed experimental activity data are given in [88].

produced. Comparing this finding with the structural parameters extracted from the different experiments it means that the more water molecules exist in the heme pocket or heme pocket vicinity the easier H_2O_2 is formed. This is reflected in a higher water influx rate and a higher B factor of the substrate reflecting its mobility in the heme pocket. This dynamic behavior is also reflected in the increasing *on*-rate and a more positive activation volume for CO binding when the water accessibility is increased. The heme pocket respective the protein becomes more compressible with increasing water content, or in other words the volume fluctuation determined from the Soret band increases.

The behavior of the CO stretch mode reflects a local structural change on the distal side where the important hydrogen bond network around Thr252 in the oxidized and the O_2 complex of P450cam was observed in the crystal structure. Increasing water entry into the heme pocket disturbs specific structural arrangements in this hydrogen bond network on the distal side which changes the electrostatic field around the CO dipole respective O_2 ligand. This changed electrostatic potential seems to be the reason that the second proton needed for cleavage of the O-O bond in the hydroperoxo complex is not specifically transported to the terminal oxygen for water formation but may attach to the proximal oxygen atom. In this case the splitting of the Fe-O bond is faster than the O-O bond cleavage leading to the release of H_2O_2.

4.2. Leakage Due to Protein Radical Formation

Although the tyrosine and tryptophan radicals are observed in the shunt pathway where the iron-oxo intermediate is produced with external oxidants such radical formation cannot be ruled out to occur also in the natural reaction cycle. The observation of the formation of tyrosine and tryptophan radicals in the iron-oxo species leads to the conclusion that intramolecular electron transfer from the protein residues to the heme occurs if a porphyrin radical is initially produced in the reaction with the oxidant (Figure 4). According to equation (9) two factors are important for radical formation: (i) the difference in the redox potential between tyrosine or tryptophan and the iron-oxo species which includes also the energetic contribution for stabilization of the protein radical by the protein environment and (ii) the distance between the heme and the tyrosine or tryptophan as radical site.

Although the redox potential of the compound I/compound II couple for P450 is not known from direct measurements it has been indirectly estimated from a study of the rates of oxidation of substituted N,N-dimethylanilines by liver microsomal P450PB-B to be in the range of 1.85 V (vis. SCE) (~ 2 V (vis. NHE)) [158]. The redox potentials for the tyrosine phenoxy/neutral tyrosine phenoxy radical pair and the tryptophan side chain neutral indolyl/indolyl radical pair have been determined to be ~ 0.94 V (NHE) and 1.05 V (NHE), respectively [159]. Even considering that the electrostatic environment around the amino acid residues may slightly modify their redox potential [160] the large difference of approximately 1 V gives a driving force of $\Delta G = 1$ eV. With this value an electron transfer distance of ~ 23 Å can be estimated from equation (9) using the reorganization energy of ~ 1 eV obtained for P450PB-B [158] and a rate of $20\,s^{-1}$ and $26\,s^{-1}$ estimated from the 15% and 19% radical formation within 8 ms for P450cam and P450BMP, respectively [1]. Interestingly, the distance between Tyr96-OH to the heme edge and the iron is 7.6 Å and 9.4 Å, respectively (Figure 10). For Tyr75 the distances are 7.6 Å and 11.8 Å, respectively. For P450BMP the distances for Tyr334-OH und Tyr305-OH are 12.2 Å und 13.7 Å (heme edge), respectively. Trp96-N is in a distance of 7.8 Å to the heme edge and 2.8 Å to the propionic acid side chain. These distances point to a smaller driving force of approximately 0.25 eV or the difference between ΔG and λ should be ~ 0.75 eV according to the plot given in [41].

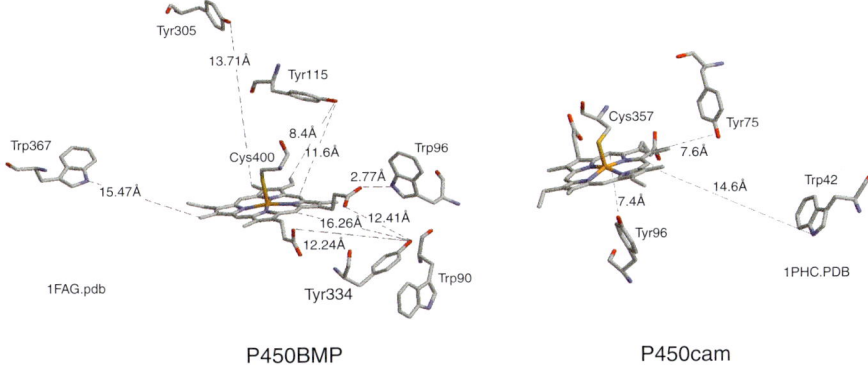

Figure 10. The location of tyrosines and tryptophans which are closest to the heme in P450cam (1PHC.PDB) and P450BMP (1FAG.PDB).

Although these estimations are very rough, they show that protein radical formation in P450 is a feasible process which could also occur in the natural cycle and compete with substrate hydroxylation if the substrate is not optimal concerning fit in the active site or chemistry of the oxygen insertion into the carbon-hydrogen bond.

P450 is not an exception but fits in the rule concerning protein radical formation at the compound I level. For many other heme proteins including nitric oxide synthase [150] tyrosine as well as tryptophan radical formation induced by electron transfer to an intermediate porphyrin compound I species have been observed in the last 10 years and extensively studied by different techniques as recently reviewed [1].

5. CONCLUDING REMARKS

Since the beginning of P450 research more than 35 years ago, the uncoupling of the P450 reaction cycle from its normal route has always been observed, in particular in microsomal systems. There seems to be general agreement on the explanation that the improper fit of the substrate in the heme pocket and/or the insufficient chemical capability of some substrates results in uncoupling.

Thanks to the combination of highly sophisticated spectroscopic techniques, crystal structure analysis, and site-directed mutagenesis methods we are now just on the way to make plausible quantitative correlations to structural parameters of the protein which might be relevant for the leakage phenomenon. The accessibility of the protein structure for water molecules seems to be one important issue. It is increasingly accepted that water molecules play a significant role in the structural dynamics of proteins and P450 is not an exception.

The application of new techniques or the new combination of different approaches such as rapid-mixing and freeze-quenching of intermediates and high-frequency EPR and Mössbauer analysis revealed a new uncoupling side reaction which may open a new way to look at the P450 reaction mechanism.

ACKNOWLEDGMENTS

The author thanks all former co-workers and students, in particular H. Schulze, C. Mouro, J. Contzen, N. Legrand, E. Deprez, R. Schwarzer, A. Kariakin, M. Richter, and M. Gerhardt, for contributing to projects related to the subject of this chapter. Special gratitude goes to A. X. Trautwein, V. Schünemann, and F. Lendzian for the fruitful collaboration in studying P450 intermediates. The German Research Foundation, the European Commission, the Institut National de la Santé et de la Recherche Médicale, the German Academic Exchange Service, and the Volkswagen Foundation are acknowledged for financial support.

ABBREVIATIONS

CD	circular dichroism
CNDO	complete neglect of differential overlap
CPO	chloroperoxidase from *Caldariomyces fumago*
DFT	density functional theory
ENDOR	electron-nuclear double resonance
EPR	electron spin paramagnetic resonance
ESEEM	electron spin-echo envelope modulation
EXAFS	extended absorption fine structure spectroscopy
FAD	flavin adenine dinucleotide
FMN	flavin mononucleotide
FTIR	Fourier transform infrared
IEHT	iterative extended Hückel theory
INDO	intermediate neglect of differential overlap
INDO-CI	INDO with configuration interaction
iNOSox	oxygenase domain of inducible nitric oxide synthase
mCPBA	meta-chloro-perbenzoic acid
MDC	magnetic circular dichroism
NAD(P)H	nicotinamide adenine (di)nucleotide phosphate, reduced form
NHE	normal hydrogen electrode
NMR	nuclear magnetic resonance
nNOSox	oxygenase domain of neuronal nitric oxide synthase
P420	inactive form of P450
P450	cytochrome P450
P450cam	P450 from *Pseudomonas putida* (CYP101)

P450BMP	the heme protein domain of P450BM-3 from *Bacillus megaterium*, (CYP102)
P450eryF	P450 from *Saccaropolyspora ertherea* (CYP107A)
P450LM2	microsomal P450 from rabbit (CYP2B4)
P450LM4	microsomal P450
P450scc	cholesterol side chain cleavage P450 (CYP11A1)
P450terp	P450 from *Pseudomonas* strain hydroxylating α-terpineol (CYP108)
Pdx	putidaredoxin
QM/MM	quantum chemical/molecular mechanics
SCE	saturated calomel electrode
TMCH	3,3,5,5-tetramethylcyclohexanone

REFERENCES

1. C. Jung, V. Schünemann and F. Lendzian, *Biochem. Biophys. Res. Commun.*, *338*, 355–364 (2005).
2. C. Jung and O. Ristau, *Die Pharmazie*, *33*, 329–331 (1978).
3. I. Schlichting, J. Berendzen, K. Chu, A. M. Stock, S. A. Maves, D. E. Benson, R. M. Sweet, D. Ringe, G. A. Petzko, and S. G. Sligar, *Science*, *287*, 1615–1622 (2000).
4. S. Nagano and T. L. Poulos, *J. Biol. Chem.*, *280*, 31659–31663 (2005).
5. T. L. Poulos and E. F. Johnson, in *Cytochrome P450 – Structure, Mechanism, and Biochemistry* (P. R. Ortiz de Montellano, ed.), 3rd edn, Kluwer Academic/Plenum Publishers, New York, 2005, pp. 87–114.
6. S. Shaik and S. P. De Visser, in *Cytochrome P450, Structure, Mechansim, and Biochemistry*, 3rd edn (P. R. Ortiz de Montellano, ed.), Kluwer Academic/Plenum Publishers, New York, 2005, pp. 45–85.
7. B. O. Nölting, C. Jung, and G. Snatzke, *Biochim. Biophys. Acta*, *1100*, 171–176 (1992).
8. C. Mouro, C. Jung, A. Bondon, and G. Simonneaux, *Biochemistry*, *36*, 8125–8134 (1997).
9. C. Jung, *J. Mol. Recognition*, *13*, 325–351 (2000).
10. A. Kariakin, D. Davydov, J. A. Peterson, and C. Jung, *Biochemistry*, *41*, 13514–13525 (2002).
11. S. S. Pochapsky, T. C. Pochapsky, and J. W. Wei, *Biochemistry*, *42*, 5649–5656 (2003).
12. T. Tosha, S. Yoshioka, K. Ishimori, and I. Morishima, *J. Biol. Chem.*, *279*, 42836–42843 (2004)
13. W. M. Atkins and S. G. Sligar, *Biochemistry*, *29*, 1271–1275 (1990).
14. R. Lange and C. Balny, *Biochim. Biophys. Acta*, *1595*, 80–93 (2002).
15. V. Schünemann, F. Lendzian, C. Jung, J. Contzen, A.-L. Barra, S. G. Sligar, and A. X. Trautwein, *J. Biol. Chem.*, *279*, 10919–10930 (2004).
16. H. Rein, C. Jung, O. Ristau, and J. Friedrich, in *Cytochrome P-450* (K. Ruckpaul and H. Rein, eds), Akademie-Verlag Berlin, 1984, pp. 163–249.
17. J. A. Peterson, *Arch. Biochem. Biophys.* *144*, 678–693 (1971).

18. B. K. Hawkins and J. H. Dawson, in *Relationships between Structure and Function of Cytochrome P-450 – Experiments, Calculations, Models* (K. Ruckpaul and H. Rein, eds), *Frontiers in Biotransformation*, Vol. 7, Akademie Verlag GmbH, Berlin, 1992, pp. 218–278.
19. M. Sono, R. Perera, S. Jin, T. M. Makris, S. G. Sligar, T. A. Bryson, and J. H. Dawson, *Arch. Biochem. Biophys., 436*, 40–49 (2005).
20. C. Mouro, A. Bondon, C. Jung, J. D. De Certaines, and G. Simonneaux, *Eur. J. Biochem., 267*, 216–221 (2000).
21. G. B. Crull, J. W. Kennington, A. R. Garber, P. D. Ellis, and J. H. Dawson, *J. Biol. Chem., 264*, 2649–2655 (1989).
22. N. Legrand, A. Bondon, G. Simonneaux, C. Jung, and E. Gill, *FEBS Lett., 364*, 152–156 (1995).
23. P. Hildebrandt, in *Relationships between Structure and Function of Cytochrome P-450 – Experiments, Calculations, Models* (K. Ruckpaul and H. Rein, eds), *Frontiers in Biotransformation*, Vol. 7, Akademie Verlag GmbH, Berlin, 1992, pp. 166–215.
24. P. M. Champion, I. C. Gunsalus, and G. C. Wagner, *J. Am. Chem. Soc., 100*, 3743–3751 (1978).
25. T. Sjodin, J. F. Christian, I. D. G. Macdonald, R. Davydov, M. Unno, S. G. Sligar, B. M. Hoffman, and P. M. Champion, *Biochemistry, 40*, 6852–6859 (2001).
26. A. X. Trautwein, E. Bill, E. L. Bominaar, and H. Winkler, *Structure and Bonding, 78*, 1–95 (1991).
27. V. Schünemann, C. Jung, A. X. Trautwein, D. Mandon, and R. Weiss, *FEBS Lett., 479*, 149–154 (2000).
28. H. Thomann, M. Bernardo, D. Goldfarb, P. M. H. Kroneck, and V. Ullrich, *J. Am. Chem. Soc., 117*, 8243–8251 (1995).
29. B. M. Hoffman, *Proc. Natl. Am. Soc., 100*, 3575–3578 (2003).
30. L. A. Andersson and J. H. Dawson, *Structure and Bonding, 74*, 1–39 (1990).
31. M. T. Green, J. H. Dawson, and H. B. Gray, *Science, 304*, 1653–1656 (2004).
32. J. H. Dawson and K. S. Eble, in *Advances in Inorganic and Bioinorganic Mechanisms*, Academic Press Inc. Ltd, London, 1986, Vol. 4, pp. 1–64.
33. D. V. Lewis, *Cytochromes P450: Structure, Function and Mechanism*, Taylor & Francis, Ltd, London, 1996.
34. P. R. Ortiz de Montellano, *Cytochrome P450, Structure, Mechanism, and Biochemistry*, 3rd edn, Kluwer Academic/Plenum Publishers, New York, 2005.
35. H. Frauenfelder, S. G. Sligar, and P. G. Wolynes, *Science, 254*, 1598–1603 (1991).
36. C. Jung, *Biochim. Biophys. Acta, 1595*, 309–328 (2002).
37. C. Jung, G. Hui Bon Hoa, D. Davydov, E. Gill, and K. Heremans, *Eur. J. Biochemistry, 233*, 600–606 (1995).
38. C. Jung, in *High Pressure Research in BioScience and BioTechnology, Proceedings of the XXXIVth Meeting of the European High Pressure Research Group*, Leuven, Belgium, Sept. 1–5, 1996 (K. Heremans, ed.), Leuven University Press, 1997, pp 35–38.
39. J. A. Kornblatt and G. Hui Bon Hoa, *Biochemistry, 29*, 9370–9376.
40. H. Schulze H, G. Hui Bon Hoa, and C. Jung, *Biochim. Biophys. Acta, 1338*, 77–92 (1997).
41. C. C. Page, C. C. Moser, X. Chen, and P. L. Dutton, *Nature, 402*, 47–52 (1999).
42. O. Ristau, *Periodicum Biologorium, 83*, 39–49 (1981).

43. B. W. Griffin and J. A. Peterson, *J. Biol. Chem.*, *250*, 6445–6451 (1975).
44. S. Maričič, in *Cytochrome P-450* (K. Ruckpaul and H. Rein, eds), Akademie-Verlag Berlin, 1984, pp. 250–276.
45. T. L. Poulos, B. C. Finzel, and A. J. Howard, *Biochemistry*, *25*, 5314–5322 (1986).
46. J. D. Lipscomb, S. G. Sligar, M. J. Namtvedt, and I. C. Gunsalus, *J. Biol. Chem.*, *251*, 1116–1124 (1976).
47. V. Schünemann, C. Jung, J. Terner, A. X. Trautwein, and R. Weiss, *J. Inorg. Biochem.*, *91*, 586–596 (2002).
48. T. L. Poulos, B. C. Finzel, and A. J. Howard, *J. Mol. Biol.*, *195*, 687–700 (1987).
49. H. Schulze, O. Ristau, and C. Jung, in *Cytochrome P450: Biochemistry, Biophysics and Molecular Biology* (M. C. Lechner, ed.), John Libbey Eurotext, Paris, 1994, pp. 595–598.
50. S. G. Sligar, *Biochemistry*, *15*, 5399–5406 (1976).
51. M. T. Fisher and S. G. Sligar, *Biochemistry*, *24*, 6696–6701 (1985).
52. J. Blanck, H. Rein, M. Sommer, O. Ristau, G. Smettan, and K. Ruckpaul, *Biochem. Pharmacology*, *32*, 1683–1688 (1983).
53. R. J. Lawson, D. Leys, M. J. Sutcliffe, C. A. Kemp, M. R. Cheesman, S. J. Smith, J. Clarkson, W. E. Smith, I. Haq, J. B. Perkins, and A. W. Munro, *Biochemistry*, *43*, 12410–12426 (2004).
54. B. Simgen, J. Contzen, R. Schwarzer, R. Bernhardt, and C. Jung, *Biochem. Biophys. Res. Commun.*, *269*, 737–742 (2000).
55. R. Raag and T. L. Poulos, in *Relationships between Structure and Function of Cytochrome P-450 – Experiments, Calculations, Models* (K. Ruckpaul and H. Rein, eds), *Frontiers in Biotransformation*, Vol. 7, Akademie Verlag GmbH, Berlin, 1992, pp. 1–43.
56. C. Jung, H. Schulze, and E. Deprez, *Biochemistry*, *35*, 15088–15094 (1996).
57. G. Hui Bon Hoa, M. A. McLean, and S. G. Sligar, *Biochim. Biophys. Acta*, *1595*, 297–308 (2002).
58. S. G. Sligar and I. C. Gunsalus, *Biochemistry*, *11*, 2290–2295 (1979).
59. G. Niaura, V. Reipa, M. P. Mayhew, M. Holden, and V. L. Vilker, *Arch. Biochem. Biophys.*, *409*, 102–112 (2003).
60. J. Tanabe and S. Sugano, *J. Phys. Soc. (Japan)*, *9*, 766–779 (1954).
61. D. L. Harries and G. H. Loew, *J. Am. Chem. Soc.*, *115*, 8775–8779 (1993).
62. V. Guallar and R. A. Friesner, *J. Am. Chem. Soc.*, *126*, 8501–8508 (2004).
63. K. Ruckpaul, H. Rein, and J. Blanck, in *Basis and Mechanisms of Regulation of Cytochrome P-450* (K. Ruckpaul and H. Rein, eds), Vol. 1 of *Frontiers in Biotransformation*, Akademie Verlag GmbH, Berlin, 1989, pp. 1–65.
64. D. L. Harris, J.-Y. Park, L. Gruenke, and L. Waskell, *Proteins: Structure, Function, and Bioinformatics*, *55*, 895–914 (2004).
65. C. Di Primo, E. Deprez, G. Hui Bon Hoa, and P. Douzou, *Biophys. J.*, *68*, 2056–2061 (1995).
66. M. T. Fisher and S. G. Sligar, *Biochemistry*, *26*, 4797–4803 (1987).
67. S. K. Lüdemann, V. Lounnas, and R. C. Wade, *J. Mol. Biol.*, *303*, 797–811 (2000).
68. R. C. Wade, D. Motiejunas, K. Schleinkofer, S. Sudarko, P. J. Winn, A. Banerjee, A. Kariakin, and C. Jung, *Biochim. Biophys. Acta*, *1754*, 239–244 (2005).
69. H. Shimada, R. Makino, M. Unno, T. Horiuchi, and Y. Ishimura, in *Cyochrome P450-Biochemistry, Biophysics and Molecular Biology* (M. C. Lechner, ed.) John

Libbey Eurotext, Paris, 1994, pp. 299–306.
70. J. N. Onuchic, D. N. Beratan, J. R. Winkler, and H. B. Gray, *Annu. Rev. Biophys. Biomol. Struct., 21*, 349–377 (1992).
71. P. M. Champion, E. Münk, P. G. Debrunner, T. H. Moss, J. D. Lipscomb, and I. C. Gunsalus, *Biochem. Biophys. Acta, 376*, 579–582 (1975).
72. M. Sharrock, P. G. Debrunner, C. Schulz, J. D. Lipscomb, V. Marshall, and I. C. Gunsalus, *Biochim. Biophys. Acta, 420*, 8–26 (1976).
73. M. T. Fisher and S. G. Sligar, *J. Am. Chem. Soc., 107*, 5018–5019 (1985).
74. F. P. Guengerich and W. W. Johnson, *Biochemistry, 36*, 14741–14750 (1997).
75. R. Raag and T. L. Poulos, *Biochemistry, 28*, 7586–7592 (1989).
76. S. Nagano, T. Tosha, K. Ishimori, I. Morishima, and T. L. Poulos, *J. Biol. Chem., 279*, 42844–42849 (2004).
77. J. Contzen and C. Jung, *Biochemistry, 37*, 4317–4324 (1998).
78. C. Jung, *Chem. Phys. Letters, 113*, 589–596 (1985).
79. L. K. Hanson, W. A. Eaton, S. G. Sligar, I. C. Gunsalus, M. Gouterman, and C. R. Connel, *J. Am. Chem. Soc., 98*, 2672–2674.
80. G. H. Loew and M. M. Rohmer, *J. Am. Chem. Soc., 102*, 3655–3657 (1980).
81. N. Suzuki, T. Higuchi, Y. Urano, K. Kikuchi, Y. Uekusa, T. Uchida, T. Kitagawa, and T. Nagano, *J. Am. Chem. Soc. 121*, 11571–11572 (1999).
82. F. Ogliaro, S. Cohen, S. P. de Visser and S. Shaik, *J. Am. Chem. Soc., 122*, 12892–12893 (2000).
83. S. Yoshioka, S. Takahashi, K. Ishimori, and I. Morishima, *J. Inorg. Biochem., 81*, 141–151 (2000).
84. S. Yoshioka, T. Tosha, S. Takahashi, K. Ishimori, H. Hori, and I. Morishima, *J. Am. Chem. Soc., 124*, 14571–14579 (2002).
85. C.-A. Yu and I. C. Gunsalus, *J. Biol. Chem., 249*, 102–106 (1974).
86. C. Jung, *Studia Biophysica, 93*, 225–230 (1983).
87. T. G. Spiro and I. H. Wasbotten, *J. Inorg. Biochem., 99*, 34–44 (2005).
88. C. Jung, S. A. Kozin, B. Canny, J.-C. Chervin, and G. Hui Bon Hoa, *Biochem. Biophys. Res. Commun., 312*, 197–203 (2003).
89. C. Jung, B. Canny, J. C. Chervin, and G. Hui Bon Hoa, in *Proceedings of the 13th Int. Conference on Cytochrome P450*, Prague, 2003, Monduzzi Editore – International Division, Bologna, Italy, pp. 195–199.
90. A. Walch and G. H. Loew, *J. Am. Chem. Soc., 104*, 2352–2356 (1982).
91. J. Blanck, G. Smettan, and S. Greschner, in *Cytochrome P-450* (K. Ruckpaul and H. Rein, eds), Akademie-Verlag Berlin, 1984, pp. 111–162.
92. R. W. Estabrook, *Drug Metabolism and Disposition, 31*, 1461–1473 (2003).
93. C. Trétreau, C. Di Primo, R. Lange, H. Tourbez, and D. Lavalette, *Biochemistry, 36*, 10262–10275 (1997).
94. C. Jung, N. Bec, and R. Lange, *Eur. J. Biochem., 269*, 2989–2996 (2002).
95. M. Unno, K. Ishimori, Y. Ishimura, and I. Morishima, *Biochemistry, 33*, 9762–9768 (1994).
96. M. Kato, R. Makino, and T. Iizuka, *Biochim. Biophys. Acta, 1246*, 179–184 (1995).
97. J. H. Dawson and S. P. Cramers, *FEBS Lett., 88*, 127–130 (1978).
98. D. Harris, G. Loew, and L. Waskell, *J. Am. Chem. Soc., 120*, 4308–4318 (1998).
99. Z. S. Herman and G. H. Loew, *J. Am. Chem. Soc., 102*, 1815–1821 (1980).

100. O. Bangcharoenpaurpong, A. K. Rizos, P. M. Champion, D. Jollie, and S. G. Sligar, *J. Biol. Chem.*, *261*, 8089–8092 (1986).
101. M. Cerdonio, A. Congiu-Castellano, L. Calabrese, S. Morante, B. Pispisa, and S. Vitale, *Proc. Natl. Acad. Sci. USA*, *75*, 4916–4919 (1978).
102. G. M. Rosen, E. Finkelstein, and E. J. Rauckman, *Arch. Biochem. Biophys.*, *215*, 367–378 (1982).
103. S. Nagano, J. R. Cupp-Vickery, and T. L. Poulos, *J. Biol. Chem.*, *280*, 22102–22107 (2005).
104. B. J. Brazeau, B. J. Wallar, and J. D. Lipscomb, *Biochem. Biophys. Res. Comm.*, *312*, 143–148 (2003).
105. T. Tosha, S. Yoshioka, H. Hori, S. Takahashi, K. Ishimori, and I. Morishima, *Biochemistry*, *41*, 13883–13893 (2002).
106. M. C. Glascock, D. P. Ballou, and J. H. Dawson, *J. Biol. Chem. 280*, 42134–42141 (2005).
107. D. L. Harris and G. H. Loew, *J. Am. Chem. Soc.*, *120*, 8941–8948 (1998).
108. R. Davydov, R. Kappl, J. Hüttermann, and J. A. Peterson, *FEBS Lett.*, *295*, 113–115 (1991).
109. R. Davydov, T. M. Makris, V. Kofman, D. E. Werst, S. G. Sligar, and B. M. Hoffman, *J. Am. Chem. Soc.*, *123*, 1403–1415(2001).
110. P. D. Nyman and P. G. Debrunner, *J. Inorg. Biochem.*, *30*, 346 (1991).
111. A. I. Archakov and A. A. Zhukov, in *Basis and Mechanisms of Regulation of Cytochrome P-450* (K. Ruckpaul and H. Rein, eds), *Frontiers in Biotransformation*, Vol. 1, Akademie-Verlag Berlin, 1989, pp. 151–175.
112. A. A. Zhukov and A. I. Archakov, *Biochem. Biophys. Res. Commun.*, *109*, 813–818 (1982).
113. J. Blanck, O. Ristau, A. A. Zhukov, A. I. Archakov, H. Rein, and K. Ruckpaul, *Xenobiotica*, *21*, 121–135 (1991).
114. I. A. Pikuleva, A. Puchkaev, and I. Björkhem, *Biochemistry*, *40*, 7621–7629 (2001).
115. D. Kim and F. P. Guengerich, *Biochemistry*, *43*, 981–988 (2004).
116. T. M. Makris, I. L. Denisov, I. Schlichting, and S. G. Sligar, in *Cytochrome P450, Structure, Mechanism, and Biochemistry* (P. R. Ortiz de Montellano, ed.), 3rd edn, Kluwer Academic/Plenum Publishers, New York, 2005, pp. 149–182.
117. S. G. Sligar, T. M. Makris, and I. G. Denisov, *Biochem. Biophys. Res. Commun.*, *338*, 346–354 (2005).
118. M. Newcomb and R. E. P. Chandrasena, *Biochem. Biophys. Res. Commun.*, *338*, 394–403 (2005).
119. T. L. Poulos, *Biochem. Biophys. Res. Commun.*, *338*, 337–345 (2005).
120. H. Yeom, S. G. Sligar, H. Li, T. L. Poulos, and A. J. Fulco, *Biochemistry*, *34*, 14733–14740 (1995).
121. M. Vidakovic, S. G. Sligar, H. Li, and T. L. Poulos, *Biochemistry*, *37*, 9211–9219 (1998).
122. A. D. N. Vaz, S. J. Pernecky, G. M. Raner, and M. J. Coon, *Proc. Natl. Acad. Sci. USA*, *93*, 4644–4648 (1996).
123. A. D. N. Vaz, D. F. McGinnity, and M. J. Coon, *Proc. Natl. Acad. Sci. USA*, *95*, 3555–3560 (1998).
124. S. Jin, T. M. Makris, T. A. Bryson, S. G. Sligar, and J. H. Dawson, *J. Am. Chem. Soc.*, *125*, 3406–3407 (2003).

125. P. R. Ortiz de Montellano and J. J. De Voss, in *Cytochrome P450 – Structure, Mechanism, and Biochemistry* (P. R. Ortiz de Montellano, ed.), 3rd edn, Kluwer Academic/Plenum Publishers, New York, 2005, pp. 183–245.
126. J. T. Groves, in *Cytochrome P450, Structure, Mechanism, and Biochemistry*, (P. R. Ortiz de Montellano, ed.), 3rd edn, Kluwer Academic/Plenum Publishers, New York, 2005, pp. 1–43.
127. H. Fujii, *Coord. Chem. Rev.*, *226*, 51–60 (2002).
128. F. P. Guengerich, *Chem. Res. Toxicol.*, *14*, 611–650 (2001).
129. J. T. Groves, K. Shalyaev, and J. Lee, in *Biochemistry and Binding: Activation of Small Molecules, The Porphyrin Handbook* (K. M. Kadish, K. M. Smith, and R. Guillard, eds) Vol. 4, Academic Press, 2000, pp. 17–40.
130. K. L. Stone, R. K. Behan, and M. T. Green, *Proc. Natl. Am. Soc. USA*, *102*, 16563–16565 (2005).
131. J. C. Schöneboom, F. Neese, and W. Thiel, *J. Am. Chem. Soc.*, *127*, 5840–5853 (2005).
132. L. D. Gorsky, D. R. Koop, and M. J. Coon, *J. Biol. Chem.*, *259*, 6812–6817 (1984).
133. R. Rapoport, I. Hanukoglu, and D. Sklan, *Analyt. Biochem.*, *218*, 309–313 (1994).
134. J.-L. Primus, K. Teunis, D. Mandon, C. Veeger, and I. M. C. M. Rietjens, *Biochem. Biophys. Res. Commun.*, *272*, 551–556 (2000).
135. S. Hashimoto, R. Nakajiama, I. Yamazaki, Y. Tatsuno, and T. Kitagawa, *FEBS Lett.*, *208*, 305–307 (1986).
136. H. Staudt, F. Lichtenberger, and V. Ullrich, *Eur. J. Biochem.*, *46*, 99–106 (1974).
137. C. Larroque, R. Lange, L. Maurin, A. Bienvenue, and J. E. van Lier, *Arch. Biochem. Biophys.*, *282*, 198–201 (1990).
138. V. Ullrich, H. J. Staudinger, in *Handbuch der experimentellen Pharmakologie* (B. B. Brodie, J. R. Gillette, and H. S. Ackermann, eds), Vol. 28/2, Springer Verlag, 1971, pp. 251–263.
139. M. M. Palcic, R. Rutter, T. Araiso, L. P. Hager, and H. B. Dunford, *Biochem. Biophys. Res. Commun.*, *94*, 1123–1127 (1980).
140. T. Egawa, D. A. Proshlyakov, H. Miki, R. Makino, T. Ogura, T. Kitagawa, and Y. Ishimura, *J. Biol. Inorg. Chem*, *6*, 46–54 (2001).
141. R. Rutter, L. P. Hager, H. Dhonau, M. Hendrich, M. Valentine, and P. Debrunner, *Biochemistry*, *23*, 6809–6816 (1984).
142. T. Egawa, H. Shimada, and Y. Ishimura, *Biochem. Biophys. Res. Commun.*, *201*, 1464–1469 (1994).
143. D. G. Kellner, S.-C. Hung, K. E. Weiss, and S. G. Sligar, *J. Biol. Chem.*, *277*, 9641–9644 (2002).
144. T. C. Pederson, R. H. Austin, and I. C. Gunsalus, in *Microsomes and Drug Oxidations* (V. Ullrich, I. Roots, A. Hildebrandt, and R. W. Estabrook, eds), Pergamon Press Oxford, 1977, pp. 275–283.
145. G. C. Wagner, M. M. Palcic, and H. B. Dunford, *FEBS Lett.*, *156*, 244–248 (1983).
146. S. G. Sligar, B. S. Shastry, and I. C. Gunsalus, in *Microsomes and Drug Oxidations* (V. Ullrich, I. Roots, A. Hildebrandt, and R. W. Estabrook, eds), Pergamon Press Oxford, 1977, pp. 202–209.
147. S. Prasad and S. Mitra, *Biochem. Biophys. Res. Commun.*, *314*, 610–614 (2004).
148. T. Spolitak, J. H. Dawson, and D. P. Ballou, *J. Biol. Chem.*, *280*, 20300–20309 (2005).

149. C. Jung, V. Schünemann, F. Lendzian, A. X. Trautwein, J. Contzen, M. Galander, L. H. Böttger, M. Richter, and A.-L. Barra, *Biol. Chem.*, *386*, 1043–1053 (2005).
150. C. Jung, F. Lendzian, V. Schünemann, M. Richter, L. H. Böttger, A. X. Trautwein, J. Contzen, M. Galander, D. K. Ghosh, and A.-L. Barra, *Magn. Reson. Chem.*, *43*, S84–S95 (2005).
151. Y. Zhang and E. Oldfried, *J. Am. Chem. Soc.*, *126*, 4470–4471 (2004).
152. V. Schünemann, C. Jung, F. Lendzian, A.-L. Barra, Th. Teschner, and A. X. Trautwein, *Hyperfine Interactions*, *156/157*, 247–256 (2004).
153. L. A. LeBrun, U. Hoch, and P. R. Ortiz de Montellano, *J. Biol. Chem.*, *277*, 12755–12761 (2002).
154. K. R. Henne, K. L. Kunze, Y.-M. Zheng, P. Christmas, R. J. Soberman, and A. E. Retti, *Biochemistry*, *40*, 12925–12931 (2001).
155. C.-L. Kuo, G. M. Raner, A. D. N. Vaz, and M. J. Coon, *Biochemistry*, *38*, 10511–10518 (1999).
156. A. I. Archakov, I. I. Karuzina, N. A. Petushkova, A. V. Lisitsa, and V. G. Zgoda, *Toxicology in Vitro*, *16*, 1–10 (2002).
157. R. C. Zanger, D. R. Davydov, and S. Verma, *Toxicol. Appl. Pharmacol.*, *199*, 316–331 (2004).
158. T. L. Macdonald, W. G. Gutheim, R. B. Martin, and F. P. Guengerich, *Biochemistry*, *28*, 2071–2077 (1989).
159. M. R. DeFelippis, C. P. Murthy, M. Faraggi, and M. H. Klapper, *Biochemistry*, *28*, 4847–4853 (1989).
160. J. A. Stubbe and W. A. van der Donk, *Chem. Rev.*, *98*, 705–762 (1998).
161. A. V. Wells, P. Li, and P. M. Champion, *Biochemistry*, *31*, 4384–4393 (1992).
162. X.-C. Yu and H. W. Strobel, *Biochemistry*, *34*, 5511–5517 (1995).
163. Y. Imai and R. Sato, *Can. J. Biochem.*, *1*, 419–426 (1967).
164. J. Wang, D. J. Stuer, and D. L. Rousseau, *Biochemistry*, *34*, 7080–7087 (1995).
165. R. Perera, M. Sono, J. A. Sigman, T. D. Pfister, Y. Lu, and J. H. Dawson, *Proc. Natl. Acad. Sci. USA*, *100*, 3641–3646 (2003).
166. M. J. Cheesman, B. R. Baer, Y. M. Zheng, E. M. Gillam, and A. E. Rettie, *Chem. Biol. Interact.*, *146*, 157–64 (2003).
167. L. Eisenstein, P. Debey, and P. Douzou, *Biochem. Biophys. Res. Commun.*, *77*, 1377–1383 (1977).
168. C. Bonfils, C. Balny, P. Douzou, and P. Maurel, in *Biochemistry, Biophysics and Regulation of Cytochrome P-450* (J. A. Gustafsson, J. Carlstedt-Duke, A. Mode, and J. Rafter, eds), Elsevier/Northholland, Amsterdam, 1980, pp. 559–564.
169. H. Zhang, L. Gruenke, D. Arscott, A. Shen, C. Kasper, D. L. Harris, M. Glavanovich, R. Johnson, and L. Waskell, *Biochemistry*, *42*, 11594–11603 (2003).
170. I. F. Sevrioukova and J. A. Peterson, *Arch. Biochem. Biophys.*, *317*, 397–404 (1995).
171. D. D. Oprian and M. J. Coon, *J. Biol. Chem.*, *257*, 8935–8944 (1982).
172. I. G. Denisov, S.-C. Hung, K. A. Weiss, M. A. McLean, Y. Shiro, S.-Y. Park, P. M. Champion, and S. G. Sligar, *J. Inorg. Biochem.*, *87*, 215–226 (2001).
173. K. J. French, A. A. Rock, D. A. Rock, J. I. Manchester, B. M. Goldstein, and J. P. Jones, *Arch. Biochem. Biophys.*, *398*, 188–197 (2002).
174. P. J. Loida and S. G. Sligar, *Protein Engineering*, *6*, 207–212 (1993).

175. S. Kadkhodayan, E. D. Coulter, D. M. Maryniak, T. A. Bryson, and J. H. Dawson, *J. Biol. Chem., 270*, 28042–28048 (1995).
176. M. R. Lefever and L. P. Wackett, *Biochem. Biophys. Res. Commun., 201*, 373–378 (1994).
177. C. Kim, H. Kim, and O. Han, *Biosci. Biotechnol. Biochem., 65*, 752–757 (2001).
178. R. Rapoport, D. Sklan, and I. Hanugloku, *Arch. Biochem. Biophys., 317*, 412–416 (1995).
179. G. M. Raner, J. A. Hatchell, M. U. Dixon, T. L. Joy, A. E. Haddy, and E. R. Johnston, *Biochemistry, 41*, 9601–9610(2002).
180. J. Limburg, L. A. LeBrun, and P. R. Ortiz de Montellano, *Biochemistry, 44*, 4091–4099 (2005).
181. R. C. Tuckey and H. Kamin, *J. Biol. Chem., 258*, 4232–4237 (1983).
182. A. M. Lambeir, C. A. Appleby, and H. B. Dunford, *Biochim. Biophys. Acta, 828*, 144–150 (1985).
183. B. Zhao, F. P. Guengerich, M. Voehler, and M. R. Waterman, *J. Biol. Chem., 280*, 42188–42197 (2005).

8

Cytochromes P450 – Structural Basis for Binding and Catalysis

Konstanze von König and Ilme Schlichting

Department of Biomolecular Mechanisms, Max Planck Institute for Medical Research,
Jahnstrasse 29, D-69120 Heidelberg, Germany
<ilme.schlichting@mpimf-heidelberg.mpg.de>

1. INTRODUCTION	236
2. LIGAND BINDING: SUBSTRATE RECOGNITION AND ACCESS TO THE DISTAL POCKET	237
3. ARCHITECTURE OF THE ACTIVE SITE OF CYP101	239
4. THE DISTAL ACID-ALCOHOL PAIR	242
4.1. CYP101	242
4.1.1. The Conserved Alcohol Functionality: Thr252	242
4.1.2. The Acidic Group: Asp251	244
4.2. CYP102	246
4.3. CYP107A	248
4.4. CYP55A1	249
5. EXPERIMENTAL CHARACTERIZATION OF REACTION INTERMEDIATES. RADIOLYSIS AS A TOOL TO STUDY REDOX REACTIONS	250
6. CRYSTAL STRUCTURES OF OXY-FERROUS COMPLEXES	253
6.1. Wild-type and Mutant CYP101	253
6.2. Wild-type and Mutant CYP107A	256
6.3. CYP158A2	257
7. MECHANISM: SUMMARY, CONCLUSIONS, SPECULATIONS	258
ACKNOWLEDGMENTS	260
ABBREVIATIONS	260
REFERENCES	261

1. INTRODUCTION

Cytochrome P450 enzymes form a super family of extremely versatile heme containing monooxygenases that catalyze a wide variety of reactions such as hydroxylations, epoxidations, heteroatom oxygenation, or dealkylations [1–3]. P450s are ubiquitous enzymes that have been found in all organisms. Currently, there are more than 6000 identified P450 sequences available that are collected and annotated in a variety of web sites, such as the one maintained by David Nelson (http://drnelson.utmem.edu/CytochromeP450.html).

As detailed in Chapters 11, 12, 15, and 16 of this volume, P450 enzymes have essentially two main functions: they play important roles in the biosynthesis of, e.g., steroids and prostaglandins, as well as in the catabolism of xenobiotics. It is evident that these enzymes have to meet different criteria concerning specificity. In general, biosynthetically or metabolically active P450s have well-defined substrates that are turned over with high regioselectivity or stereospecificity, implying an excellent if not rigid fit between substrate and active site. In contrast, P450s which act on substances that are exogenous to the organism need to have flexible active sites that can accommodate and adapt to the respective compound(s). In line with the diverse nature of substrates, their sizes vary significantly. Nitric oxide for example, is reduced by P450nor (CYP55), bulky heptapeptides attached to an acyl carrier protein are cyclized via phenol coupling reactions by the oxy proteins (CYP165) to yield the peptide antibiotic vancomycin. In contrast to these highly specific P450s, others are rather promiscuous; the human liver enzyme P4503A4, for example, metabolizes over 50% of the currently marketed drugs, implying significant plasticity of the active site. Although the basic mechanism for substrate recognition is understood in the framework of the conserved P450 structure, there are still a lot of open questions, particularly in respect to quantitative predictions. Complicating the picture of substrate-induced conformational changes in the P450 systems are complex homo- and heterotropic cooperativities that are present in many mammalian metabolic P450s. Such interactions are of critical importance in determining the metabolic profile and drug-drug interactions present in human P450 isozymes. Solving these issues is important not only for a functional annotation of sequence data, but also for a prediction of the pharmacokinetics of drugs.

P450 enzymes have been dubbed molecular blow torches due to their amazing capability to catalyze aliphatic hydroxylation reactions at ambient pressure and temperature. Since there are no direct synthetic methods for specific hydroxylations of unactivated hydrocarbons, there is significant interest in exploiting P450s, including the related chloroperoxidases (CPO) synthetically, either directly or by small molecule mimicry for the development of new, selective catalysts.

Tremendous progress has been made in the understanding of the workings of the enzymes since their discovery some 40 years ago by the combination of information obtained by the analysis of directed mutants, from high resolution crystal structures of both bacterial and eukaryotic (including) mammalian P450s, characterization of intermediates by different spectroscopic techniques, including the analysis of reaction intermediates using cryogenic or fast kinetic techniques, and of quantum chemical and molecular dynamics computational studies. Despite this wealth of information, there are still a lot of open questions, e.g., concerning the enzymatic mechanism. P450s activate molecular oxygen by the sequential two-electron reduction to the formal oxidation state of hydrogen peroxide. Although a great deal is known about this process, one of the unresolved mysteries concerns the nature of the catalytically active oxygen species. In this chapter, we will review structural aspects of ligand binding and catalysis by cytochrome P450 enzymes. The emphasis will be on the role model of P450s, the bacterial P450cam (CYP101) that catalyzes the stereo- and regiospecific hydroxylation of camphor to 5-exo-hydroxycamphor, thus allowing *Pseudomonas putida* to use camphor as a carbon source.

2. LIGAND BINDING: SUBSTRATE RECOGNITION AND ACCESS TO THE DISTAL POCKET

As described in Chapter 3 of this volume, the P450 fold is highly conserved, yet there is enough structural diversity to allow for binding of substrates of significantly different sizes to different P450s. Using bioinformatic analysis, Gotoh identified six 'substrate recognition sites' (SRS) [4] that consist of secondary structural elements lining the active site, namely the B'-helix region (SRS1), parts of the F- and G-helices (SRS2 and SRS3), a part of the I-helix (SRS4), the β4 hairpin (SRS5) and the K-helix β2 connecting region (SRS6) (See Figure 1 of Chapter 3 for the nomenclature of the secondary elements). The SRS determine P450 substrate specificity, point mutations within SRSs significantly affect substrate specificities [4]. The SRSs are mobile protein regions that move upon substrate binding in an induced fit mechanism, thereby closing off the active site [5]. The structural dynamics of these elements differs in P450 enzymes resulting in equilibria that favor either open or closed forms of the substrate-free enzyme. Substrate access is not obvious in the latter case, as exemplified by P450cam. Although this problem does not exist in the open forms (e.g., P450 BM3, Oxy proteins), they do present the challenge of predicting the correct binding mode of the substrate and its interactions with the protein.

A great deal of insight into ligand access to the active site has come from the comparison of crystal structures of different P450s in the presence or absence of ligands. As described in detail in Chapter 3 of this volume, the information obtained from free and ligand-bound P450cam [6,7], P450 BM3 (CYP102) [8–10] and of different ligand complexes of CYP119 [11,12] has shown (i) a high plasticity of the active site and (ii) a structurally conserved channel allowing access of the substrate to the active site. This channel is provided by movement of the F/G helices, and the B'helix.

Ligand binding to cytochromes P450 has not only been studied experimentally but also by computation using random expulsion molecular dynamics (REMD) and steered molecular dynamics simulations in which the ligand is pulled along a predetermined route. These studies identified three main classes of ligand egress routes with classes defined according to the secondary structure elements surrounding clusters of trajectories at the protein surface [13]. These pathways agree with the ones found by a thermal motion pathway analysis of the crystallographic temperature factors of the coordinates where putative ligand escape routes are identified as connected chains of protein atoms with elevated temperature factors leading from the active site to the exterior. One common pathway, termed pw2a, was identified in CYP101, CYP102A1, and CYP107A1. This pathway runs between the F/G loop region, the B'helix/BC loop region and the β1 sheet. These regions are highly variable in sequence and structure and therefore allow the enzymes to adapt the extent of channel opening and the mechanism of ligand passage specific to a certain substrate [14]. For instance, the exit of the product from P450cam, 5-hydroxy-camphor, requires only relatively small main chain motions, whereas a large product like the macrocyclic product synthesized by P450eryF demands extensive opening motions.

With the growing number of P450 structures solved, experimental evidence became available in support of the other ligand pathways identified by REMD simulations [15]. The recent observation of an opening at the pathway termed pw1, that runs from the heme via the C/C' and H or L helices to the G/H loop and β2 in the crystal structure of CYP51, and its location close to a region of high flexibility in cytochrome P450cam suggest a functional role. The passage of a molecule the size of camphor along pw1 in cytochrome P450cam is, however, energetically unfavorable [16]. The observation of xenon binding sites (PDB code: 1 uyu) close to this pathway in P450cam [15] suggests that it may serve as an access route for oxygen to the active site. Xenon binds in a rather hydrophobic cavity (separated from the active site by Leu245, Leu246, and Phe163) that might act as a vestibule for storing and controlling the supply of oxygen to the active site. Such docking sites for gaseous ligands have been shown to exist for nitric oxide and dioxygen in *Ascaris suum* myoglobin, for example [17].

Different routes for substrate access and product exit may exist in membrane bound P450s. Generally, the substrates of P450s tend to be rather hydrophobic and, in the case of membrane-bound P450s, often come from the lipid bilayer. P450s associate with the membrane by insertion of the N-terminal helix into the membrane and by dipping of the rather hydrophobic G' helix/FG loop region into the membrane. This would presumably facilitate the opening of pathway pw2a to permit access of hydrophobic substrates to the active site and egress of hydrophobic products. However, if the product is more hydrophilic than the substrate, it may be expelled into the aqueous environment via a different route [15].

3. ARCHITECTURE OF THE ACTIVE SITE OF CYP101

CYP101 catalyzes the hydroxylation of camphor to 5-exo-hydroxy-camphor. Stereo- and regio specificity is obtained by a snug fit of the substrate (1R)-camphor into the heme pocket. This is provided by numerous van der Waals interactions between the camphor methyl groups and the protein as well as a hydrogen bond between the keto group of camphor and the hydroxyl group of Tyr96. The I-helix forms a wall of the heme pocket and contains the conserved signature amino acid sequence ^{248}**GGLDT**^{252} which is centered at a kink in the middle of the helix. The kink is caused by the absence of the α-helix defining hydrogen bond (i — i+4) between the carbonyl oxygen atom of Gly248 and the amide of Thr252. Instead of the backbone, the side chain hydroxyl of Thr252 takes part in this hydrogen bond. This is caused by a widening of the I helix due to an interaction of the backbone carbonyl oxygen atom of Leu250 and the terminal amino group of Lys178 (which also hydrogen bonds to Asp251) and a two pronged salt bridge between the side chains of Asp251 and Arg186 (see Figure 1). The importance of this interaction is evidenced by the fact that mutations of Lys178 and Arg186 result in uncoupling [18] (see also Table 1). It is interesting to note that although Asp251 is highly conserved, these interactions are not. Thus, the structural basis for the frequently observed widening of the I-helix varies among the different cytochromes P450.

Since its detection in the first crystal structure of a cytochrome P450 [19], the ferric form of CYP101, the groove in the I-helix has been implicated in the binding of oxygen and catalytic water molecules. Therefore, it is surprising that CYP176A1 (P450cin), the closest homologue (26% sequence identity) to CYP101 with respect to size, shape, and chemical composition of the substrates, 1,8-cineole versus camphor, has a 'regular' straight I-helix [20]. This has been traced to the occurrence of Asn242 instead of the active site Thr, which forms a hydrogen bond with the substrate ether oxygen atom. However, although the corresponding active site Asn240 in CYP165B1 (OxyB) superimposes perfectly, in this case the I-helix displays the typical kink [21] (see Figure 2).

There is one important difference in the I-helix conformations of the ferric forms of CYP101, and other cytochromes P450: in CYP101, the backbone amide of Thr252 points towards the axis of the I-helix whereas it points towards the

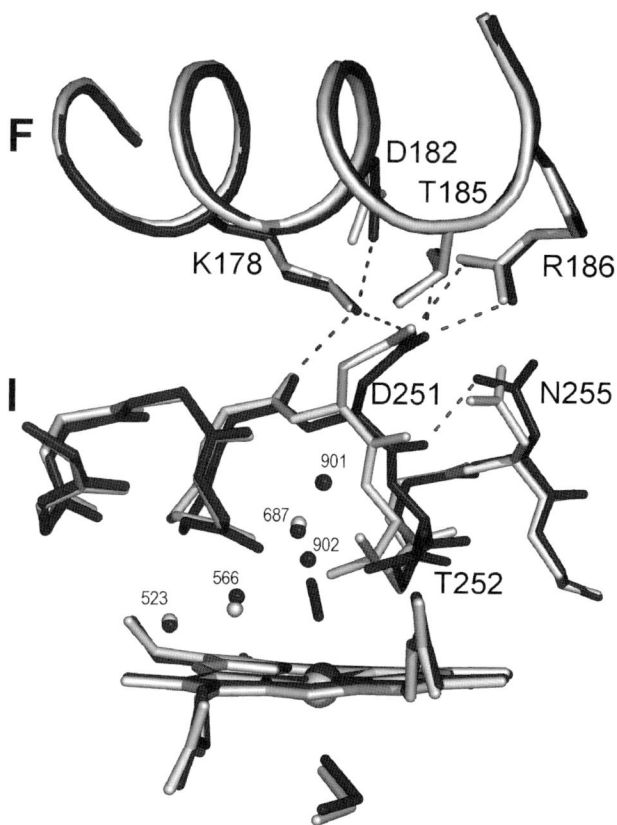

Figure 1. The active site of wild-type P450cam in the ferric (gray) and the oxyferrous state (black). Shown is the functionally important kink in the I-helix and the interactions of the conserved acid (D251)-alcohol (T252) pair. The I-helix is 'pulled open' by interactions of the carbonyl oxygen of Leu250 and the carboxylate of Asp251 with Lys178 and Arg186 located in the adjoining F helix. Upon O_2 binding, Thr252 rotates to form a hydrogen bond with the distal oxygen atom. The backbone carbonyl and amide of Asp251 and Thr252, respectively, flip and two new water molecules, Wat901 and Wat902, bind. The flipped conformation is stabilized by a hydrogen bond between the carbonyl of Asp251 and the side chain of Asn255.

active site in the other CYPs. Mutation of Thr252 to Ala ([22], PDB code 2cp4) or Ile ([23], PDB code 1geb), or the replacement of camphor by 2-phenylimidazole ([24], PDB code 1phe) induces a flip in the 251–252 backbone such that it overlays with the other P450 conformations. Importantly, in wild-type CYP101 the backbone flips upon oxygen binding thereby creating a new water binding site (see Figure 1).

Table 1. Kinetic properties of CYP101 mutants in which residues thought to be directly or indirectly involved in proton transfer were replaced.

CYP101 Mutant	Rate of O_2 Consumption [μmol min^{-1}/ μmol P450]	5-OH-camphor Formation [%]	H_2O_2 Formation [%]	Reference
wild-type	1350	97	3	[18]
Lys178Ala	880	96	7	[18]
Arg186Ala	490	60	13	[18]
Asp251Gly	21	99	2	[18]
Asp251Ala	3	89	12	[18]
Asp251Asn	6[a]	92	5	[42]
Thr252Ala	ND	5	51	[30]
Thr252Ala	1100	6	83	[29]
Thr252Val	420	22	45	[29]
Thr252Ser	1100	81	15	[29]
Thr252OMe	410	100	ND	[34]
Thr252Ile	277	100	3	[23]
Asp251Asn/ Thr252Ala	5	52	64	[23]
Glu366Met	940	85	12	[32]

[a] Calculated from the reported values in [42].
ND, not determined.

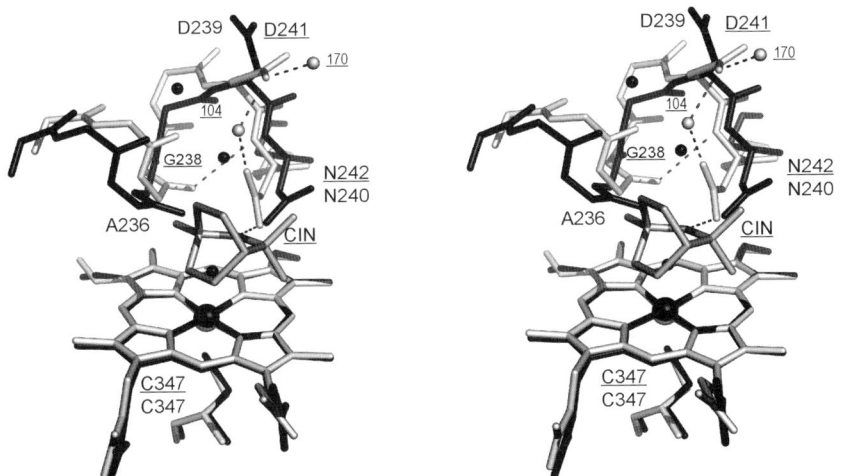

Figure 2. Stereoview of the superposition of P450cin (gray) and OxyB (black). In contrast to P450cam, the closely homologous P450cin has a 'regular' straight I-helix, where the backbone amide of Asn242 forms a hydrogen bond with the carbonyl oxygen of Gly238. This interaction is not caused by the occurrence of an asparagine (Asn242) at the position of the active site threonine because OxyB, which also has an asparagine at this position, displays the typical kink in the I-helix.

4. THE DISTAL ACID-ALCOHOL PAIR

Despite the wide range of reactions catalyzed by P450s, the enzymes do not only share a common overall fold but also several highly conserved residues that have been implicated in the catalytic mechanism [25]. Of particular importance is an acidic residue in the I-helix followed by a threonine. Examination of aligned sequences of P450 isozymes shows that this acid-alcohol side chain pair is conserved in a large majority, but not all cytochrome P450s. These residues are typically aspartate or glutamate and either threonine, serine or asparagine [26,27]. P450s that do not contain the conserved alcohol functionality include CYP107A and CYP158A2, which have an alanine at the position of the conserved threonine (Ala245 in both enzymes). In CYP176A as well as in CYP165B1, the conserved threonine is replaced by an asparagine (Asn242 and Asn240, respectively). The alignment of the 58 human P450 sequences shows that 80% of these sequences retain this acid-alcohol pair. Site-directed mutagenesis in combination with a wide variety of other techniques was used to assess the roles of both residues. So far, no clear picture has evolved (see Table 2). On the contrary, it has been suggested that despite its conservation, the threonine may play different roles or has a differing degree of importance in the various P450 enzymes. Some important findings in bacterial and fungal P450s are summarized in the following paragraphs.

4.1. CYP101

4.1.1. The Conserved Alcohol Functionality: Thr252

In P450 CYP101, the acid alcohol pair appears as the already mentioned residues Asp251 and Thr252. Given their proximity to the heme dioxygen binding site [28], numerous mutagenesis experiments were performed to probe their role in stabilizing the oxygen complex [28], distal pocket hydrogen bonding networks, and in contributing to proton delivery to the iron bound peroxoanion. Mutant enzymes that have the threonine residue replaced by alanine [29,30], valine, and serine [29] display normal kinetic parameters of pyridine nucleotide oxidation and dioxygen consumption, yet the enzyme is uncoupled, resulting almost entirely in hydrogen peroxide production rather than the oxidation product 5-hydroxycamphor (see Table 1).

The crystal structure of the Thr252Ala mutant [22] seemed to be able to explain the uncoupling effect since a new water molecule (Wat720) was observed in the active site which forms a hydrogen bond with the backbone amide of Ala252 and which might destabilize the dioxygen-heme complex (see Figure 4B in Section 6.1). Ishimura and colleagues showed that the extent of uncoupling reaction resulting in peroxide formation was smaller if the Thr residue was replaced by an amino acid that may form a hydrogen bond, irrespective of its size

Table 2. The conserved acidic and alcohol residues have been studied in great detail in various CYP enzymes. Some isozymes with non-conserved residues at the corresponding position are also listed.

CYP Enzyme	Species	Residue	Suggested Role
		Acidic Residue	
101 (cam)	*Pseudomonas putida*	Asp251	second electron transfer [42], proton donation to Thr252 or O_2 [43,49,50]
1A2	mammalian	Glu318	spin character of resting state [97], active site architecture [97], substrate binding [98]
2D6	mammalian	Val308	
55A1 (nor)	*Fusarium oxysporum*	Ala242	
aromatase	mammalian	Asp309	active site architecture [99]
		Alcohol Residue	
101 (cam)	*Pseudomonas putida*	Thr252	oxygen activation [29,30] and stabilization [28], proton delivery [42], facilitation of (hydro)peroxo species formation [38], H-bond acceptor for hydroperoxo species [40]
102 (BM3)	*Bacillus megaterium*	Thr268	oxygen activation [54], substrate recognition [53], positioning of active site water [8]
107A (EryF)	*Saccaropolyspora erythraea*	Ala245	
158A2	*Streptomyces coelicolor*	Ala245	
165B1 (OxyB)	*Amycolatopsis orientalis*	Asn240	
176A (cin)	*Citrobacter braakii*	Asn242	
1A2 (former P450$_d$)	mammalian	Thr319	proton transfer [91]
2B4	mammalian	Thr302	equilibrium of oxygenating species [92], proton delivery [93]
2D6	mammalian	Thr309	equilibrium of oxygenating species, substrate-, regioselectivity [94]
2E1	mammalian	Thr303	proton delivery [95]
55A1 (nor)	*Fusarium oxysporum*	Thr243	electron transfer, spin character of resting state [65]
Laurate (ω-1)-hydroxylase	mammalian	Thr301	substrate specificity [96]

or pK_a [31–33]. In the Thr252Ser [29] and Thr252Asn [32] mutants more than half of the enzyme's activity with regards to both pyridine nucleotide oxidation and hydroxylation turnover are retained (see Table 1). This has been interpreted as the threonine serving either as a direct proton source to the peroxoanion species, or being part of critical hydrogen bond networks and placement of water molecules which in turn serve in proton delivery and dioxygen bond scission. To distinguish between these possibilities, Kimata et al. performed an unnatural amino acid mutagenesis experiment and replaced Thr252 with methoxythreonine (Thr-OCH$_3$), removing proton donating ability [34]. This mutant had almost unchanged kinetic parameters and retained nearly full coupling of reducing equivalents into product. From this result it was derived that a 'free' hydroxyl group at position 252 is not directly donating protons to a peroxoanion intermediate but is involved in a hydrogen bonding network that stabilizes active site water molecules that are the immediate source of catalytic protons [35]. One needs to bear in mind, however, that the mutation was not verified on the level of the protein. This would have been important because cytochrome P450s catalyze oxidative demethylation [36,37] and a single turnover could have resulted in restoration of a functional hydroxyl at the enzyme active site.

The crystal structure of the oxy complex of the wild type enzyme [38] showed that Thr252 interacts with both the dioxygen ligand and with two catalytic water molecules that populate the active site upon O$_2$ binding (see Figure 4B in Section 6.1). Hence it was deduced that Thr252 appears to aid in the formation of the ferric hydroperoxy and subsequently the oxy ferryl species by supporting a dioxygen-water hydrogen bond network required for the timely addition of the two protons [38]. However, the observation of the hydroperoxo species in the Thr252Ala mutant in cryoradiolysis studies suggests that the threonine is not required for the first proton transfer [39]. Recently, the role of Thr252 has been extended to include the stabilization of the hydroperoxy intermediate [40] and the placement of the water molecule donating the second proton required for the transient formation of a dihydroperoxo species ([5c] in Figure 3 in Section 5) [40,41].

4.1.2. The Acidic Group: Asp251

Mutagenesis of Asp251 showed that the conserved acid residue plays a different role in catalysis since the mutant's phenotype differs from the one obtained by replacing the succeeding threonine. Instead of the normally high but decoupled rate of pyridine nucleotide consumption seen in the Thr252Ala mutant, the replacement of Asp251 with asparagine causes a strong decrease in the NADH consumption rate [42–44] while the ability to form product is retained (see Table 1). While these phenotypical consequences of Asp251 mutation are similar throughout many P450 enzymes, the kinetic efficiency of the mutant enzymes often depends on the identity of the substrate being metabolized [45–48]. Unlike the conserved hydroxyl moiety in the immediate vicinity of the heme iron that fulfills an essential hydrogen bonding task, a number of mutations of the acidic residue have revealed that the observed phenotypical variance is poorly explained by a discrete physicochemical property of the inserted amino acid side chain. For example, both the Ala251 and Gly251 substitutions behave similarly to that of the isosteric Asp251Asn mutation [32].

Gerber and Sligar suggested that the reason for the decrease of the catalytic turnover of the Asp251Asn mutant by over two orders of magnitude is the second electron transfer and linked protonation events following formation of the ferrous-oxy complex [42]. The observation of a new spectral intermediate under steady-state turnover conditions with an absorbance maximum at 421 nm indicated a new rate-limiting step in the mutant enzyme. A model was proposed according to which the carboxylate side chain of Asp251 represents the proton donor to Thr252 or bound oxygen and is part of a proton relay system together with the solvent accessible residues Lys178, Asp182, and Arg186. This relay would allow proton delivery to the buried active site while still restricting water access [42,43]. Benson et al. [49] further analyzed the oxidation and oxygenation state of the unknown intermediate by combining UV-Vis spectroscopy, EPR, and gas chromatography to concurrently monitor the relative concentrations of the spectral intermediate, reduced putidaredoxin and product formation. The new intermediate was characterized as being one electron reduced from oxy-P450 with an intact dioxygen bond, hence corresponding to the peroxoanion species. Thus, these experiments showed that the first proton transfer is the rate-limiting step and not the second electron transfer [49].

These results were confirmed by radiolytic reduction of the oxy complex of the camphor-bound Asp251Asn mutant at cryogenic temperatures. Using this approach, the second electron was injected into the catalytic cycle, and EPR and ^{1}H-ENDOR spectra were measured after the annealing of the complex at higher temperatures. The mutation results in the build up of the peroxoanion species rather than immediate protonation to form the hydroperoxo species as shown in the wild-type enzyme upon reduction at 77 K [39].

In order to distinguish between the effect of the solvent and the role of solvent exchangeable moieties in the proton transfer events, solvent isotope effect measurement were performed [50]. The turnover rates of the Asp251Asn mutant in H_2O/D_2O mixtures showed a larger kinetic solvent isotope effect than wild-type (ca. 5-fold), i.e., the turnover dropped as a function of solvent D_2O content, suggesting that proton transfer becomes rate-limiting in the mutant. A pH profile demonstrated that the catalytic rates increased linearly with the proton concentration, suggesting that internal solvent molecules could provide an alternative source of protons. Correspondingly, a proton inventory yielded that a much larger number of protons is involved in the rate-limiting step in the mutant enzyme. This would be consistent with the presence of extra solvent molecules to compensate for the disrupted proton delivery pathway via Asp251 [50].

The crystallographic analysis of the Asp251Asn mutant has provided additional evidence for Asp251's involvement in proton delivery [50]. As described in Section 6.1, the Asp251 side chain points away from the distal pocket in the ferric wild-type complex and forms salt bridges with Arg186 and Lys178. Mutagenesis to Asn results in the simultaneous rotation of the Asn251 and Lys178 side chains to the protein surface, resulting in loss of their interaction (see Figure 4A in Section 6.1). This is compensated by a new interaction of the Asn251's side-chain amide group with Asp182 and Thr181, and of Asn251's side chain and backbone carbonyl with Asn255 (2.8 Å and 3.4 Å, respectively). Structurally, it is difficult to rationalize the observed kinetic solvent isotope effect and pH dependence data of the Asn251 single mutant, in part due to the difficulty in visualizing mobile active site water molecules. However, even well-ordered water molecules observed in the wild-type are missing in the oxy complex of the Asp251Asn mutant, which is consistent with the kinetic finding of the first proton transfer being rate-limiting.

4.2. CYP102

P450 BM3 catalyzes the hydroxylation of fatty acids at the ω-1, ω-2, and ω-3 positions, as well as epoxidations of double bonds. It is a catalytically self-sufficient enzyme, containing both the heme containing oxygenase and the reductase domain on the same polypeptide chain. Crystal structures have been determined of the ligand free [10] and complexed heme domain of CYP102 [8,9]. Studies on the interactions of CYP102 with the high affinity substrate N-palmitoylglycine showed how ligand binding induces the closing of a solvent accessible channel and triggers a low spin–high spin transition [8]. The structural changes shield the active site from solvent and reduce the heme iron reduction potential ($\Delta Em[\text{wt} + \text{arachidonate}(120\,\mu\text{M})] = +138\,\text{mV}$) [51], moreover, they free the 6th coordination site for the binding of oxygen. Upon binding of N-palmitoylglycine to ferric CYP102, a water molecule is displaced from its binding site at the heme iron and the hydrogen bond with the Ala264 carbonyl (corresponding to Gly248 in CYP101) is broken due to a 1.7 Å shift of the latter away from the heme iron. The ligand induced rearrangement of the I-helix results in the formation of two new water binding sites, one of which hosts the 'ferric water' that got displaced parallel to the heme plane and off to the side of the iron (Wat500). In the new position, it is stabilized by hydrogen bonds with Ala264 and Thr268. This water binding site is associated with the high spin complex (H-site) and so close to the one typical for the low spin complex (L-site) that the two sites are mutually exclusive.

Resonance Raman studies of CYP102 have shown that an equilibrium exists between the five-coordinate high-spin and the six-coordinate low-spin states in the presence of saturating concentrations of substrate [52]. The structural data by Haines et al. [8] explain this by an equilibrium between the L- and H-sites such that in the wild-type enzyme the H-site is occupied predominately, although not exclusively. In the Thr268Ala mutant, the stabilizing hydrogen bond is not available to keep the water molecule at the H-site, thus the equilibrium is shifted toward the L-site. In fact, about 40% of the enzyme remains in the low spin state when complexed to arachidonic acid [53]. The effects of the Thr mutation on catalysis are similar in CYP102 and CYP101 in that the catalytic activity is affected but not abolished. However, the effects are less dramatic in CYP102 suggesting that the threonine may fulfill a similar role in all P450s although with a differing degree in importance. In CYP102, the Thr268Ala mutation leads to smaller rates of O_2 and NADPH consumption when using lauric [54] and arachidonic acid [53] as substrate. Concomitant with the decrease in substrate hydroxylation rate is a loss of coupling efficiency (16% coupling for lauric acid oxidation [54], 9% for arachidonic acid oxidation [53]).

Mutation of Thr268 to Ala and Asn reduces the degree of substrate-induced heme reduction potential (ΔEm [T268A + arachidonate($120 \mu M$)] = $+73 mV$, ΔEm [T268N + arachidonate($120 \mu M$)] = $+9 mV$), resulting in a slowed rate for the first electron transfer and a reduced stability of the oxy-ferrous complex [51].

These investigations shed light on the question whether the conserved threonine takes part in dioxygen stabilization. The rate constants for the decay of the oxy-ferrous species demonstrated that the dioxygen complex was neither destabilized by the Thr268Ala mutation nor stabilized by Thr268Asn substitution [51]. This result suggests that the stabilizing effect is not exerted by hydrogen bonding interactions of the threonine but can be mainly attributed to the heme reduction potential, as suggested by Ost et al. [55].

The overall topology of the Thr268Ala [54] and Thr268Asn mutants [51] is the same as the wild-type, there are only local differences around the site of mutation. Interestingly, no additional water molecules were observed in the active site of the mutant that could be credited for uncoupling [54]. Unexpectedly, the Thr268 mutation to Ala (which removes both the hydrogen bonding and proton donating capability of this residue) and to Asn (which maintains the hydrogen bonding capacity) results in the same phenotype. Both mutants show smaller turnover rates and are highly uncoupled, and display lower rate constants for the first electron transfer and a less stable oxy-ferrous species. This was unexpected for the T268Asn mutant since these results contradicted the findings of Kimata et al. who had shown that replacing the conserved threonine by an unnatural amino acid (methoxy threonine) with intact hydrogen bonding capacity did not disturb the function of CYP101 [34].

Both the Thr268Asn and the Thr268Ala mutations significantly diminish the substrate-induced spin-state shift from low- to high-spin and subsequent heme potential changes. Since the initiation of catalysis is usually accompanied by spin conversion, it was proposed that the mutants are impaired in substrate recognition [51,53]. One should bear in mind, however, that these conclusions were based on crystal structures of the ferric, substrate free forms of the enzyme. Substrate binding induces a significant change of the active site and its associated water molecules as evidenced by the crystal structure of the wild-type complexed with N-palmitoylglycine [8]. This structure suggested a further conformational change upon oxygen binding. Similar, but so far unidentified changes are expected for the Thr268 mutants. Their knowledge is essential for deriving a meaningful structure-activity relationship.

In CYP102, substrate binding serves as a switch that transforms the enzyme from a form of low activity to one of high catalytic competence. It is therefore interesting to compare the structures of CYP102 and oxy CYP101, focusing on the active site. In ferric CYP102, the I-helix takes up the flipped conformation observed in the oxy form of CYP101. Interestingly, one of the two ligand induced water molecules, Wat501, occupies roughly the same position as Wat902 in CYP101. The other water molecule, Wat500, binds at a similar site as the distal oxygen atom of O_2 in oxy CYP101. It is most likely that Wat500 is displaced upon oxygen binding to the site corresponding to Wat901 in oxy CYP101. The structural similarities in the water structure suggest that the same mechanism for protonation is used in CYP102 and CYP101, with Wat500 and Wat901, respectively, donating the first proton to the distal oxygen.

4.3. CYP107A

Even though the presence of an active-site alcohol functionality is highly conserved throughout the landscape of P450 isozymes, one important exception has further established the importance of an active-site hydrogen bonding network in dioxygen mediated catalysis. CYP107 (P450eryF), which catalyzes the hydroxylation of 6-deoxyerythronolide B in the erythromycin biosynthetic pathway, has an alanine (Ala245) instead of a threonine at the position of the alcohol functionality [56,57]. Similar to the Thr252Ala mutant of CYP101, the kink in the I-helix is retained and the backbone takes up the flipped conformation observed in the oxy complex of CYP101 (see Figure 5 in Section 6.2).

The 5-hydroxyl group of the 6-deoxyerythronolide B (6-DEB) macrolide substrate forms a hydrogen bond with an active site water molecule (Wat564) and positions it such that is almost superimposable with the typical placement of the threonine hydroxyl [58–60]. Both removal of the substrate hydroxyl and reintroduction of the threonine residue (Ala245Thr) have effects consistent with substrate assisted catalysis. This mechanism is thought to be the reason why P450eryF is not as versatile in substrate metabolism as other P450s [61]. However, upon mutation of Ala245 to Ser or Thr, testosterone is metabolized at different sites [62] and hydroxylation of 6-DEB is decreased due to the interruption of the hydrogen bonding interactions [58].

4.4. CYP55A1

CYP55 (P450nor) is a nitric oxide (NO) reductase that is involved in fungal denitrification. The enzyme catalyzes the reduction of NO to N_2O, a non-standard P450 reaction: $2NO + NAD(P)H + H^+ \rightarrow N_2O + NAD(P)^+ + H_2O$. Interestingly, this reaction does not require protein redox partners. In contrast to CYP101, the active site of CYP55 is rather hydrophilic, a cluster of arginines and lysines has been shown to be crucial for binding of NAD(P)H [63]. While CYP55 contains the conserved threonine (Thr243), it has an alanine (Ala242) at the position of the conserved acidic residue [64]. The side chain of Thr243 points away from the 6th ligand position and forms a hydrogen bond with Wat173 that interacts with the I-helix main chain peptide carbonyl of the *i-4* residue Ala239. The Ala242-Thr243 backbone conformation corresponds to the 'flipped' conformation of Asp251 observed in the oxy-ferrous complex of CYP101. Analogously, the carbonyl of Ala242 forms a hydrogen bond with the side chain of Asn246 (corresponding to Asn255 in CYP101) and the amide interacts with Wat175, which is positioned similarly as Wat901 in CYP101.

In order to investigate the function of Thr243, it was mutated to 18 different amino acids [65]. No pronounced effects on optical absorption and CD spectral properties were observed, consistent with the structural data [66] that show only subtle changes in the heme environment. The catalytic activity, however, was drastically altered depending on the properties of the introduced amino acid. A decrease in NO reduction was attributed to a diminished reduction by NADH, which transfers electrons directly to the enzyme. This result led to the proposal that Thr243 is essential for NO reduction and may be involved in electron transfer from NADH. The crystal structure of CYP55 complexed with the NADH analogue nicotinic acid adenine dinucleotide (NAAD) [67] shows that the hydroxyl group of Thr243 interacts with the carboxyl group of NAAD. Because of the distance of Thr243 from the catalytic site it is likely that the functional role of Thr243 is stereo-selective binding of the nicotinic ring but not proton delivery. The role of the conserved P450 hydroxyl may be fulfilled by Ser286 instead. [68,69]. Its hydroxyl group is close to the C4 atom of the nicotinic ring and the oxygen atom of heme bound NO, respectively, and NO reductase activity is abolished upon mutation. In contrast to the Thr243 mutants, mutation of Ser286 results in drastic changes in hydrogen bonding networks through destabilization of water networks [68,69]. The water networks involving Ser286 appear critical for activity, as even the relatively conservative Ser286Thr mutation results in a rotameric configuration of the side chain that destabilizes the water network and a loss of activity.

It is interesting to note that the highly conserved sequence A/G-G-X-D-T in the I-helix is used for a different purpose in CYP55 and CYP101. In the latter, this motif serves in binding of water molecules, whereas in CYP55 it binds NAD(P)H [67].

5. EXPERIMENTAL CHARACTERIZATION OF REACTION INTERMEDIATES. RADIOLYSIS AS A TOOL TO STUDY REDOX REACTIONS

The great diversity of reactions catalyzed by P450 enzymes appears to be based on two unique properties: The ability of their heme iron to exist under several oxidation states with different reactivities and active sites that provide a great variety of substrates access to the catalytic iron. Thus, within the framework of the conserved P450 structure (Chapter 3 of this volume), the active sites must achieve enough structural diversity to provide the required specificity, yet conserve a mechanism of catalysis and solvent exclusion. In particular, bulk water access to the active site is restricted to allow for control of the proton transfer required for catalysis and to minimize side reactions, such as the conversion of activated dioxygen to reactive oxygen species (superoxide or peroxide). The main features of the common catalytic cycle of cytochromes P450 shown in Figure 3 are two one-electron reduction steps of the heme iron, interspersed by oxygen binding, followed by two proton transfer events yielding an oxoferryl intermediate. Each step involves the input of a specific 'ingredient' in the form of electrons, substrates, and protons. This allows for a high degree of regulation, minimizing (i) unproductive and thus wasteful use of reducing equivalents via a linkage between substrate binding and redox potential changes and (ii) the harmful production of reactive oxygen species due to uncoupling. For a detailed description of the steps occurring along the reaction cycle see Chapters 7 and 11 of this volume.

A detailed chemical mechanism includes characterization of all steps occurring along the reaction coordinate, in particular of the intermediate species. Their direct experimental analysis can be quite challenging, or impossible if they are formed more slowly than they decay and therefore do not accumulate. Another experimental difficulty can result from the finite lifetime of reaction intermediates, in particular if their decay times are shorter than the data collection times. Typically, this problem is circumvented by either speeding up data collection or by slowing the reaction, e.g., by reducing the temperature. In addition to slowing kinetics, temperature can be used to separate different reaction steps if the respective rate coefficients have different temperature dependencies (for example, different activation enthalpies). One way of exploiting this is to trap a reaction intermediate by freeze-quenching, analyze it, increase the temperature, and to then let the reaction proceed to the next 'temperature-limited' intermediate. A further temperature raise restarts the reaction which can thus be analyzed in a stop-and-go fashion. Very often intermediates accumulate because a barrier of the reaction cannot be overcome at temperatures below the glass transition temperature (around 180–200 K), at which non-harmonic collective motions are 'frozen out'. Thus, kinetic studies at cryogenic temperatures have a number of

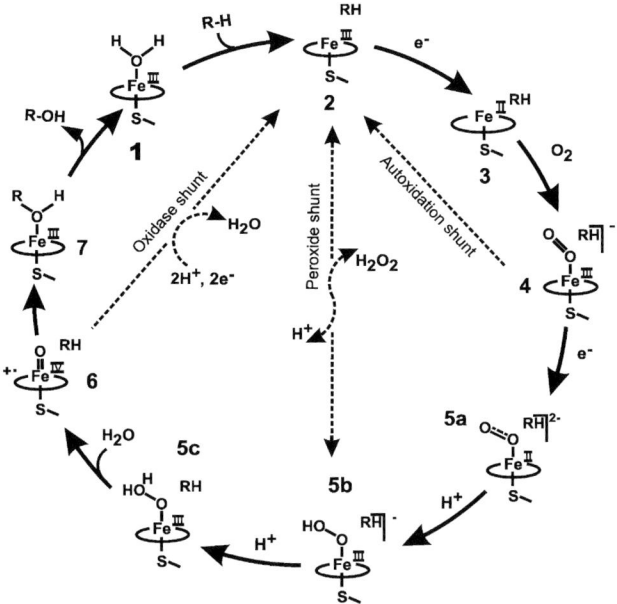

Figure 3. The catalytic cycle of P450 enzymes consists of several steps after substrate displaces [2] the iron bound water molecule [1]: Oxygen binding to the reduced heme iron [3] and formation of an oxygenated heme Fe^{2+}-OO or Fe^{3+}-OO$^-$ [4], the one-electron reduction of this complex yields a ferric peroxo Fe^{3+}-OO^{2-} complex [5a], which is protonated to form the hydroperoxo Fe^{3+}-OOH$^-$ species [5b], the second protonation of the distal oxygen results in an unstable transient Fe-OOH$_2$ [5c] that decays by heterolytic scission of the O-O bond and formation of a water molecule, producing the fleeting high valent porphyrin metal-oxo complex, often described as a ferryl-oxo π-cation porphyrin radical [6] and referred to as 'compound I'.

advantages, but suffer from the lack of easy mixing of reactants when studying frozen solutions or cryo-cooled crystals.

This is particularly noticeable when the reactants are large, for example redox partners that are needed to provide the second electron ([4] → [5a]). Fortunately, redox reactions can be started relatively easily by irradiation of the sample with high-energy photons (X-rays or γ-rays from a ^{60}Co source). Radiolysis of solvent molecules (water, cryoprotectant, e.g., glycerol) results in the generation of free electrons that can reduce for example ferric or oxyferrous heme complexes. Keeping the sample at cryogenic temperatures has not only the benefit of slowing the decay of the intermediates generated (e.g., a peroxo complex) but has the additional advantage that the radiolysis side products, the organic radicals, are immobilized and cannot react with the species under investigation. Raising the temperature allows the reaction to proceed and to 'anneal' the reactive intermediate, lowering the temperature again allows one to follow the reaction on a time-scale compatible with the chosen data collection method. Depending on the barriers, chemical transformations e.g. by protonation, and conformational relaxations can be analyzed. Cryoradiolysis has been used to generate and to subsequently spectroscopically analyze the reaction intermediates of P450 [39,70], NOS [71], CPO [72], hemeoxygenase [73], and HRP [74].

In the case of P450, the oxy complex was cryotrapped and irradiated with an external ^{60}Co source [70] or internally with ^{32}P [75]. Subsequently, the temperature was raised transiently, then the sample was cooled again and analyzed by UV-VIS, ^{1}H-ENDOR, and EPR spectroscopy. This protocol was followed to temperatures above the glass transition temperature. When keeping the sample at 6 K, a hydrogen bonded peroxo species [5a] can be observed. In the wild-type enzyme, it converts largely, but not completely, to the hydroperoxo complex [5b] at ~70 K, and does so completely at temperatures near ~147 K. No further intermediates are observed until product formation [7] sets in upon annealing for 1 min at ~190–210 K, i.e., at temperatures above the glass transition. Initially, the product 5-exo-hydroxycamphor, is coordinated to the ferric heme iron in a non-equilibrium configuration. Although there is no direct spectroscopic indication for the generation of an oxoferryl/π cation radical intermediate [6] (Fig. 3), ENDOR spectroscopy of the initial non-equilibrium product state in D_2O and H_2O buffers shows that the trapped hydroxyl proton originates from camphor C5 which is consistent with product formation via [6]. Annealing at temperatures ≥ 220 K results in the formation of the equilibrium product state in which the product hydroxyl is no longer trapped.

The same methodology was used to study active site mutants suspected to be involved in proton transfer. The kinetics of the hydroperoxo [5b] formation differ in the Thr252Ala and Asp251Asn mutants. In contrast to the wild-type enzyme, the former converts completely at 77 K whereas the Asp251Asn mutant does not convert [5a] to [5b] significantly until temperatures above 130 K are reached. Upon further annealing at ~190–210 K, the uncoupled Thr252Ala mutant reverts to the ferric heme state [2] without product formation, presumably via hydrogen peroxide formation. This suggests that the hydroperoxo species [5b] is a key intermediate at or near the branch-point that leads either to product formation or to non-productive uncoupling and that Thr252Ala is important for the second proton transfer. In contrast, the Asp251Asn mutation influences the first proton transfer, once [5b] is formed it qualitatively turns over to product, in contrast to the wild-type the non-equilibrium product state accumulates appreciably at 215 K before relaxing to the equilibrium product state.

Recently, the photo-reductive powers of X-rays have been recognized as both a problem and a means to 'titrate' redox-steps when using appropriate data collection schemes [38,76–78]. The major difference to the 'chemical' cryoradiolysis experiments is that the X-rays serve both as a pump and a probe, complicating the experimental design and data evaluation. An additional, independent analysis method, such as UV-VIS microspectrophotometry of crystals during diffraction data collection is therefore highly advisable, in particular, if the high-intensity radiation of third generation synchrotron sources is used [100]. If photo-reduction is observed, full datasets need to be sliced together using partial datasets of crystals that were exposed to X-rays briefly enough before significant reduction occurs [76,78,79]. An alternative route is the data collection at different wavelength [38].

6. CRYSTAL STRUCTURES OF OXY-FERROUS COMPLEXES

In contrast to the ferric complex, the ferrous and oxy-ferrous complexes of cytochrome P450s are unstable. Their life times depend strongly on the specific enzyme, with CYP101 being one of the long lived forms. The ferrous form can be generated by soaking crystals in dithionate solutions under anaerobic conditions, followed by freeze quenching. So far, the ferrous form of CYP101 is the only chemically reduced form available. (It is unknown how many of the crystals of ferric complexes were reduced (unwantedly and unknowingly) by the X-ray beam to the ferrous form.) As expected from spectroscopic studies, the ferric [**2**] (Fig. 3) and ferrous forms [**3**] of CYP101 are very similar [38]. Oxygen binds to the ferrous P450, yielding the Fe^{2+}-OO (ferrous dioxygen), or Fe^{3+}-OO$^-$ (ferric superoxide) complex [80] [**4**].

The oxy complex [**4**] can be generated from the ferrous form by washing the crystals in dithionate free solution with subsequent transfer to an oxygenated solution. The experimental problem of the conflicting times for the generation of the oxy complex and its decay can be addressed by either speeding up the generation step through use of a pressure cell to increase the oxygen concentration [38] or by slowing the decay of the oxy complex by doing the soak at subzero (-10 to $-5\,°C$) temperature [40,81,82]. So far, crystal structures of oxy complexes have been determined of CYP101 [38,40], CYP107A [81], and CYP158A2 [82]. In all cases, oxygen binds end on ($\eta 1$), with a geometry similar to the one observed in myoglobin [83], however, the effects of oxygen binding on the conformation of the active site and the water structure differ in the three P450s.

6.1. Wild-type and Mutant CYP101

In CYP101, oxygen binding induces a rotation of Thr252 to form a hydrogen bond with the bound O_2^- of the ferric superoxo complex, thereby loosening the hydrogen bond with the backbone carbonyl of Thr248 (see Figure 1 in Section 3). Due to salt bridges between the side chains of Asp251 and Lys178, Thr185, and Arg186, the backbone cannot follow suit; the resulting strain is released by a backbone flip of the Asp251 carbonyl towards the heme cavity [38,40]. This new conformation is stabilized by a hydrogen bond with Asn255 and positions the Asp251 amide towards the heme pocket thereby creating a binding site for a new water molecule, Wat901, which also forms a hydrogen bond with the hydroxyl group of Thr252. A second new water molecule, Wat902, binds within hydrogen bonding distance of Wat901, the Thr252 hydroxyl, and Wat687.

Due to the stabilizing interactions they are involved in, Thr252 and Asp251 represent 'knots' in the I-helix that prevent dissipation of the O_2-induced strain further along the backbone which would eliminate the need of the Thr252-Asp251 backbone rearrangement. This is the case for the Asp251Asn mutant. Asn251 interacts with Thr181, Asp182, Arg186, the backbone carbonyl of Val247, and Asn255 [50]. Based on the structure of the oxy complex of wild-type CYP101, we had predicted [38] that the latter interaction would render an oxygen-induced backbone flip less favorable since Asn255 is not available for stabilization. Indeed, double (unflipped and flipped) conformations of the Asn251-Thr252 backbone are observed in the oxy complex of Asp251Asn [41]. The other consequence of the Asn251 interactions for structural changes induced by oxygen binding is that the rotation of the Thr248 side chain induces subtle changes in a longer stretch of the backbone so that no large widening of the I-helix is required [40] to dissipate the strain. This, together with the Asn251 conformation such that it 'leans' into the active site cavity, sterically prevents binding of Wat901. Even in the flipped conformation of the backbone, Wat901 would be too close to the β-carbon of Asn251 (2.3 Å). This renders binding of Wat901 unfavorable. Since the first proton transfer is rate-limiting in the Asp251Asn mutant [39,43], it is suggestive that Wat901 is responsible for this step [38].

The highly uncoupled Thr252Ala mutant shows a contrasting behavior. Whereas the Asp251Asn mutant does not display the full opening (see Figure 4A), presumably due to dissipation of the strain past residue 251, the Thr252Ala mutant is missing the 'spring' provided by the Thr252OH – Gly248 carbonyl interaction, the mutant is fully relaxed to the open form even in the absence of oxygen (see Figure 4B). Thus, the oxygen binding hole is fully open, the backbone has rearranged, a water molecule (Wat720) is present close to the position taken up by Wat901 in the oxy complex of the wild-type. The structure of the oxy complex of the Thr252Ala mutant has been published recently by Nagano and Poulos [40]. The active site water corresponding to Wat901 moves by ~0.6 Å and is 3.1 Å (wild-type: 3.4 Å) apart from the distal oxygen atom of O_2. The water might be better positioned for proton transfer, which would be in line with the observation that hydroperoxo complex formation occurs more readily in the mutant than in the wild-type enzyme. A careful analysis of the water structure shows that the displacement of Wat901 is transmitted in displacements of the other water molecules resulting in the net loss of one water molecule; Wat567 moved to take up the position of Wat902. The resulting void is filled by an alternative conformation of Val253. It is suggestive to link the change in water structure of the mutant with the functional finding that the second proton

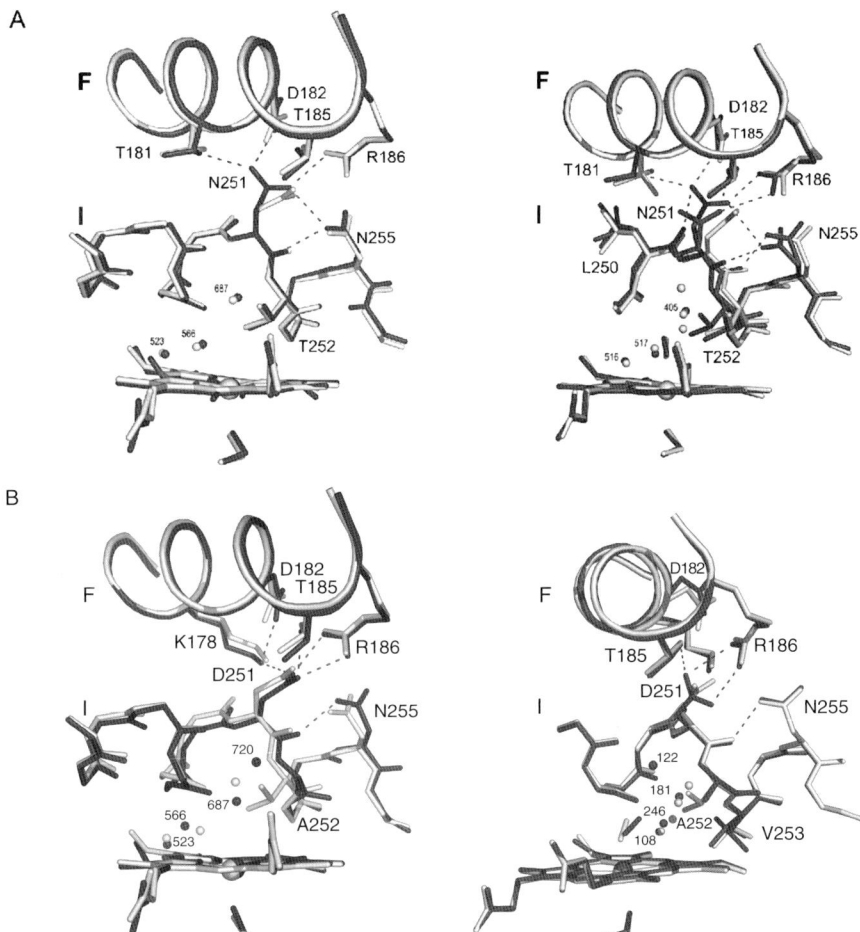

Figure 4. Structural comparison of ferric (left) and oxy-ferrous (right) states of wild-type (gray) and mutant (black) enzymes of P450cam. In the ferric state of the D251N mutant (A), Asn251 forms a hydrogen bond with Asn255 which reduces the probability of backbone flipping in the oxy ferrous state. The opening of the I-helix kink is reduced and binding of active site water molecules is hindered due to steric clashes. — The T252A mutant (B) shows a contrasting behavior, the backbone already takes up the flipped conformation in the ferric state. Therefore, a water molecule can enter the opened active site and bind in close proximity to the position of Wat901 observed in the oxy ferrous complex of the wild-type enzyme.

transfer is affected. This would imply that the water chain involving Wat902 takes part in the second protonation step. This interpretation differs from the one by Nagano and Poulos [40] who reported very minor changes of the water structure. They concluded that the role of Thr252 is to stabilize the hydroperoxo intermediate which favors addition of the second proton over peroxide release. The uncoupling by the Thr252Ala mutant was explained by a lack of stabilization of the hydroperoxo intermediate by Thr252, favoring peroxide release over O-O bond cleavage. Another argument against this interpretation is the observation of the high stability of the hydroperoxo complex in the Thr252Ala mutant [39].

6.2. Wild-type and Mutant CYP107A

CYP107 (P450eryF) has an alanine at the position of the conserved threonine residue. Based on the crystal structure of the ferric substrate bound complex of P450eryF [58], it was suggested that the substrate 5-OH replaces the missing Thr-OH group and positions an active site water molecule (Wat519) such that it can donate protons to the iron-linked dioxygen. It was therefore a surprise to see that oxygen binding displaces Wat519 observed in the ferric complex, and that the structure of the oxy complex of P450eryF (PDB code 1z8o) strongly resembles that of the ferric complex. In particular, there are no new water molecules at the active site that could serve as proton donors to the bound oxygen molecule. Based on the hydrogen bond between O_2 and the substrate 5-OH, Nagano, Cupp-Vickery, and Poulos suggested that the substrate donates one proton while the second comes from the hydrogen bond network involving Wat63-Glu360-Ser246-Wat53-Ala241 carbonyl in the I-helix cleft [81].

The structural basis for the observation of Wat63 in both the ferric and the oxy complex of CYP107A1 is that the I-helix residues always take up the flipped conformation that is characterized by the amide of the conserved acidic residue pointing towards the oxygen binding site. Indeed, an overlay of the oxy complexes of CYP107/CYP101 based on the superposition of the heme iron and pyrrole nitrogen atoms and the sulphur atom of the heme ligating cysteine shows (see Figure 5) that the positions of the backbone atoms of Asp244/Asp251 and Ala245/Thr252 superimpose perfectly, whereas the rest of the I-helix does not. Wat63 corresponds to Wat902 ($\Delta 0.4 \text{ Å}$).

It is interesting to note that both water molecules are connected to a conserved glutamate (Glu360/Glu366), although by a different hydrogen bond network. In CYP101, the network is water based: Glu366–Wat523–Wat566–Wat687–Wat902, whereas in CYP107 the water molecules have no space and the network involves a side chain: Glu360–Ser246-hydroxyl–Wat63. Replacement of Ala245 by Thr or Ser (inversely corresponding to the Thr252Ala mutation in CYP101) results in loss of Wat63 due to steric clashes with the side chain and an increase in decoupling [81]. This supports the proposal that the water molecule corresponding to Wat902 provides the second proton. It would therefore seem that the source of the first proton is enzyme specific in CYP101 (Wat901 positioning via the backbone flip) and CYP107 (substrate assisted), whereas the second proton transfer is conserved involving a conserved water molecule (Wat902/Wat63). This is an extension and modification of the proposal by Nagano, Cupp-Vickery,

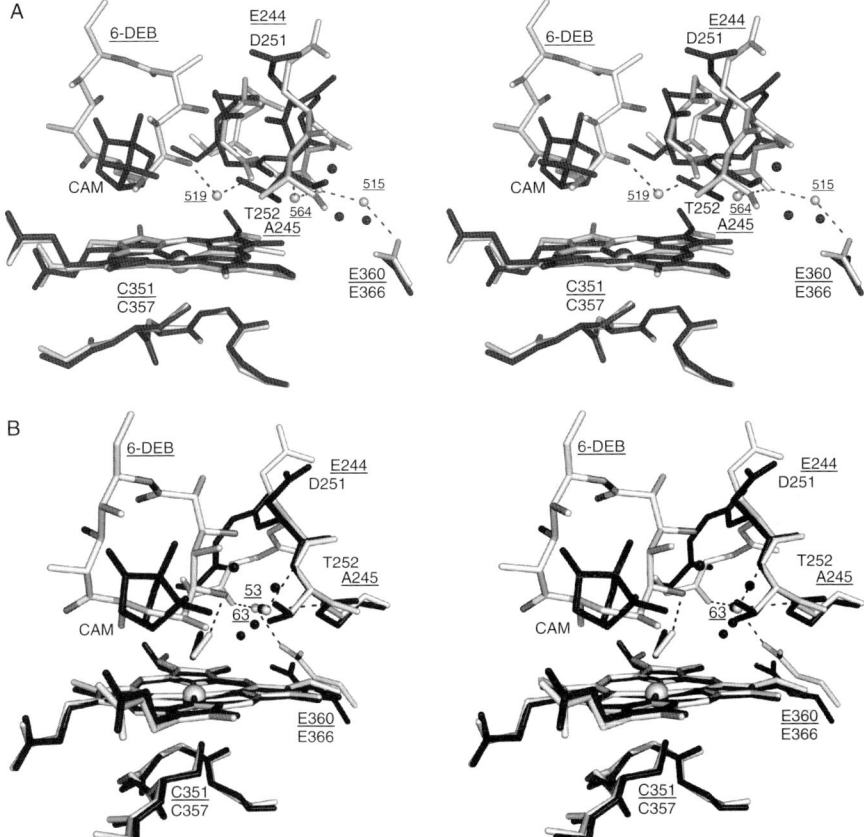

Figure 5. Stereoview of the structural superposition of P450cam (black) and P450eryF (gray) in the ferric (A) and oxy ferrous (B) states. In P450eryF, the backbone of the I-helix takes up the flipped conformation in both functional states. Interestingly, in the oxy ferrous state, the backbones of the acid-alcohol residues (P450cam:Asp251/Thr252, P450eryF: Glu244/Ala245) superimpose perfectly, and Wat63 corresponds to Wat902. The observation of a hydrogen bond between the 6-DEB substrate hydroxyl and the distal oxygen atom of O_2 suggests involvement of the substrate in proton transfer because no water molecules are observed.

and Poulos [81] who stated that it appears that the details of the proton shuttle machinery differ substantially in CYP107 and CYP101.

6.3. CYP158A2

CYP158A2 from *Streptomyces coelicolor* catalyzes the phenolic oxidative C-C coupling of two to three molecules of flaviolin to biflaviolin and triflaviolin, respectively. Like CYP107A1, CYP158A2 contains an Ala at the position of the conserved Thr residue. The crystal structure of CYP158A2 complexed with two molecules of flaviolin [82] suggested that flaviolin's 2-OH group is involved in binding and that its 5-OH and/or 7-OH groups may stabilize catalytically important water molecules. Three water molecules in the active site (Wat505, Wat600, and Wat640) were discussed to have structural and functional roles in dioxygen binding, and/or proton transfer for oxygen activation.

Upon oxygen binding, the three water molecules remain at the same positions; two of them interact with the distal oxygen atom of O_2 [84]. Their disappearance in the structure of CYP158A2 complexed with a substrate analogue lacking the 5- and 7-OH groups correlates with a decrease of catalytic activity. Therefore, these two water molecules are believed to be the direct proton donors to the iron-linked distal oxygen atom and that the substrate's 5- and 7-hydroxyl groups position these active site water molecules. Thus, in CYP158A2 substrate specificity is achieved by a direct involvement of the substrate in positioning of the catalytic water molecules. A continuous water chain to the bulk solvent exists that is predicted to serve as direct proton relay pathway. This is in contrast to CYP101 and CYP107, where no continuous hydrogen bond connection between the bound dioxygen molecule and the protein surface has been observed.

7. MECHANISM: SUMMARY, CONCLUSIONS, SPECULATIONS

The structural, spectroscopic, and computational characterization of reaction intermediates, together with mutagenesis of critical residues have helped elucidate the mechanism of P450 monooxygenase catalysis (recently reviewed in [85,86]). A great deal of insight has been obtained in the role of the highly conserved acid and alcohol functionalities, including the function and stabilization of active site water molecules. Of particular importance is P450cam, which has become the paradigm for understanding the structure-function analysis of cytochromes P450 due to the wealth of data available, which includes the first crystal structure of the super family and the very first site-directed mutation of a protein. This success was greatly helped by the high stability of the enzyme and its functional complexes. It has therefore been tempting to extrapolate the knowledge gained on CYP101 to other systems. However, it seems to turn out that CYP101 is not as generic as believed initially. For example, the conformation of the I-helix observed in the ferric state of P450cam is not observed in other CYPs, the backbone flip induced by oxygen binding is therefore the exception and not the rule. Likewise, the oxygen induced binding of water molecules may not be as general as assumed. What is needed to clarify this point are more high resolution structures of oxy complexes of different P450s.

Met. Ions Life Sci. **3**, 235–265 (2007)

P450cam has Thr252 at the position of the conserved alcohol functionality. It seems to have at least three functions. The first role is to act as a spring that controls the width and the backbone conformation of the kink in the I-helix [28] via the hydrogen bond between its hydroxyl side chain and the carbonyl of the conserved Gly248. Upon oxygen binding, Thr252 rotates to interact with the proximal oxygen atom, which induces a widening of the I-helix and a flip of the Asp251-Thr252 backbone to reduce strain since Asp251 acts as an anchor point [38]. The conformational switch stabilizes two new water molecules, Wat901 and Wat902 that are oriented by Thr252 in the active site. The lack of the observation of Wat901 in the oxy complex of the Asp251Asn mutant – which is rate-limited in the first proton transfer step – suggests that Wat901 is involved in the first proton transfer to the nearby oxygen molecule, resulting in the hydroperoxo complex [5b] (Fig. 3). Its stabilization is the second role of Thr252 [38]. The third role of Thr252 is the stabilization and orientation of a water molecule that serves as direct source or as a stepping stone for the second proton. The observation that both CYP107 and CYP101 have a water molecule (Wat63 and Wat902, respectively) at the same position and the fact that this water molecule disappears and the chain that it is part of loses a member in the uncoupling Ala254Thr and Thr252Ala mutants, respectively, suggests that this water molecule is involved in the second protonation step. Uncoupling could be explained by the lack of a timely second proton transfer (via Wat902) and the subsequent dissociation of the weakly bound hydroperoxo moiety which may pick up a proton when diffusing out of the protein, thereby generating hydrogen peroxide.

For productive oxygen bond scission, the second proton attacks the distal oxygen atom, water dissociates and the oxyferryl complex [6] (Fig. 3) is formed. Recent calculations on CYP101 [87] showed that [5b] and [6] are separated by significant barriers and a dihydroperoxo intermediate [5c]. We propose that a water molecule is pulled off from [5c] by the backbone back flip yielding [6]. This is supported by the fact that while the hydroperoxo species [5b] accumulates below the glass transition temperature, subsequent steps occur only beyond the glass transition [39,75], implying that large unharmonic motions may be involved. Moreover, the backbone is back to the normal unflipped conformation in the product complex [38,88].

Proton transfer is believed to occur via a proton relay system comprising water molecules and active site residues. Three water channels have been suggested, involving: (i) a flipping arginine channel [89], (ii) Asp251 [42,87,101] and (iii) Glu366 [87,90]. The first one is unlikely to be of functional relevance because it does not connect up to the active site water molecules. The second one requires a conformational change of the Asp251 side chain such that it rotates into the active site thereby breaking a number of salt bridges. The conformational change might be triggered by a protonation of Asp251. MD calculations show that this might be possible, in particular, if a new predicted, but experimentally

Met. Ions Life Sci. **3**, 235–265 (2007)

not observed, water molecule moves along into the active site, bridging the protonated Asp251 and Wat901 [101]. This model provides for a direct proton transfer from the solvent to the bound oxygen molecule.

The extent of the conformational change of Asp251 suggests that the side chain rotation should occur only above the glass transition temperature. Thus, this model would explain why product formation takes place only above 190 K but fails to explain why the Asp251Asn mutant is active (i.e., where does the second proton come from?). The third water/proton channel that has been proposed involves Glu366 which is the origin of a continuous chain of water molecules that connect to the oxygen molecule. QM/MM calculations show that this pathway is also possible [87]. What argues against Glu366's role as the ultimate proton donor is that first, it does not connect to solvent, and second, its substitution to Met has only very minor consequences on turnover (Table 1) [33]. What is intriguing, however, is that this water channel is disrupted in the Thr252Ala mutant due to the one missing water molecule. Since this mutant decouples it is suggestive that the integrity of this water chain is important for productive protonation. Moreover, an analogous H-bonding network between Glu and a water molecule corresponding to Wat902 (Wat63) is observed in P450eryF. Thus, a third role of Thr252 becomes apparent, namely the organization of the water chain which serves as source of protons (see Sections 1 and 2).

ACKNOWLEDGMENTS

The stimulating discussions with our long-time collaborators Drs Christiane Jung, Ilia G. Denisov, Thomas M. Makris, and Stephen G. Sligar are gratefully acknowledged.

ABBREVIATIONS

CPO	chloroperoxidase
6-DEB	6-deoxyerythronolide
HRP	horseradish peroxidase
HS	high spin
LS	low spin
MD	molecular dynamics
NAAD	nicotinic acid adenine dinucleotide
NO	nitric oxide
NOS	nitric oxide synthase
QM/MM	quantum mechanics/molecular mechanics
REMD	random expulsion molecular dynamics
SRS	substrate recognition site

REFERENCES

1. F. P. Guengerich, *J. Biol. Chem.*, 266, 10019–10022 (1991).
2. F. P. Guengerich, *Chem. Res. Toxicol.*, 14, 611–650 (2001).
3. D. Mansuy, *Comp. Biochem. Physiol., Part C.: Pharmacol. Toxicol. Endocrinol.*, 121, 5–14 (1998).
4. O. Gotoh, *J. Biol. Chem.*, 267, 83–90 (1992).
5. O. Pylypenko and I. Schlichting, *Annu. Rev. Biochem.*, 73, 991–1018 (2004).
6. I. J. Dmochowski, B. R. Crane, J. J. Wilker, J. R. Winkler, and H. B. Gray, *Proc. Natl. Acad. Sci. USA*, 96, 12987–12990 (1999).
7. R. Raag and T. L. Poulos, *Biochemistry*, 30, 2674–2684 (1991).
8. D. C. Haines, D. R. Tomchick, M. Machius, and J. A. Peterson, *Biochemistry*, 40, 13456–13465 (2001).
9. H. Li and T. L. Poulos, *Nature Struct. Biol.*, 4, 140–146 (1997).
10. K. G. Ravichandran, S. S. Boddupalli, C. A. Hasermann, J. A. Peterson, and J. Deisenhofer, *Science*, 261, 731–736 (1993).
11. S. Y. Park, K. Yamane, S. I. Adachi, Y. Shiro, K. E. Weiss, S. A. Maves, and S. G. Sligar, *J. Inorg. Biochem.*, 91, 491–501 (2002).
12. J. K. Yano, L. S. Koo, D. J. Schuller, H. Li, P. R. Ortiz de Montellano, and T. L. Poulos, *J. Biol. Chem.*, 275, 31086–31092 (2000).
13. S. K. Ludemann, O. Carugo, and R. C. Wade, *J. Mol. Model.*, 3, 1–5 (1997).
14. P. J. Winn, S. K. Ludemann, R. Gauges, V. Lounnas, and R. C. Wade, *Proc. Natl. Acad. Sci. USA.*, 99, 5361–5366 (2002).
15. R. C. Wade, P. J. Winn, I. Schlichting, and S. Sudarko, *J. Inorg. Biochem.*, 98, 1175–1182 (2004).
16. S. K. Ludemann, V. Lounnas, and R. C. Wade, *J. Mol. Biol.*, 303, 813–830 (2000).
17. M. Brunori, F. Cutruzzola, C. Savino, C. Travaglini-Allocatelli, B. Vallone, and Q. H. Gibson, *Trends Biochem. Sci.*, 24, 253–255 (1999).
18. H. Shimada, R. Makino, M. Imai, T. Horiuchi, and Y. Ishimura, in *Dioxygen Activation and Homogeneous Catalytic Oxidation* (L. I. Shimandi, ed.), Elsevier, Amsterdam, 1991, pp. 313–319.
19. T. L. Poulos, B. C. Finzel, I. C. Gunsalus, G. C. Wagner, and J. Kraut, *J. Biol. Chem.*, 260, 16122–16130 (1985).
20. Y. T. Meharenna, H. Li, D. B. Hawkes, A. G. Pearson, J. De Voss, and T. L. Poulos, *Biochemistry*, 43, 9487–9494 (2004).
21. K. Zerbe, O. Pylypenko, F. Vitali, W. Zhang, S. Rouset, M. Heck, J. W. Vrijbloed, D. Bischoff, B. Bister, R. D. Sussmuth, S. Pelzer, W. Wohlleben, J. A. Robinson, and I. Schlichting, *J. Biol. Chem.*, 277, 47476–47485 (2002).
22. R. Raag, S. A. Martinis, S. G. Sligar, and T. L. Poulos, *Biochemistry*, 30, 11420–11429 (1991).
23. T. Hishiki, H. Shimada, S. Nagano, T. Egawa, Y. Kanamori, R. Makino, S. Park, Y, S. Adachi, Y. Shiro, and Y. Ishimura, *J. Biochem. (Tokyo)*, 128, 965–974 (2000).
24. T. L. Poulos and A. J. Howard, *Biochemistry*, 26, 8165–8174 (1987).
25. D. R. Nelson and H. W. Strobel, *J. Biol. Chem.*, 263, 6038–6050 (1988).
26. D. W. Nebert, D. R. Nelson, M. Adesnik, M. J. Coon, R. W. Estabrook, F. J. Gonzalez, F. P. Guengerich, I. C. Gunsalus, E. F. Johnson, B. Kemper,

W. Levin, I. R. Phillips, R. Sato, and M. R. Waterman, *DNA Cell Biol.*, *8*, 1–13 (1989).

27. D. W. Nebert, D. R. Nelson, M. J. Coon, R. W. Estabrook, R. Feyereisen, Y. Fujiikuriyama, F. J. Gonzalez, F. P. Guengerich, I. C. Gunsalus, E. F. Johnson, J. C. Loper, R. Sato, M. R. Waterman, and D. J. Waxman, *DNA Cell Biol.*, *10*, 1–14 (1991).
28. T. L. Poulos, B. C. Finzel, and A. J. Howard, *J. Mol. Biol.*, *195*, 687–700 (1987).
29. M. Imai, H. Shimada, Y. Watanabe, Y. Matsushima-Hibiya, R. Makino, H. Koga, T. Horiuchi, and Y. Ishimura, *Proc. Natl. Acad. Sci. USA*, *86*, 7823–7827 (1989).
30. S. A. Martinis, W. M. Atkins, P. S. Stayton, and S. G. Sligar, *J. Am. Chem. Soc.*, *111*, 9252–9253 (1989).
31. Y. Ishimura, H. Shimada, and M. Suematsu, *Oxygen Homeostasis and its Dynamics*. (International Symposium held at Minato-ku, Tokyo, December 8–13, 1996), *Keio Univ. Symp. Life Sci. Med.*, *1*, 1998, 619 (1998).
32. H. Shimada, R. Makino, M. Unno, T. Horiuchi, and Y. Ishimura, *Cytochrome P450*. 8th International Conference (M. C. Lechner and J. Libbey, eds), Eurotext, Paris, 1994, pp. 299–306.
33. H. Shimada, Y. Watanabe, M. Imai, R. Makino, H. Koga, T. Horiuchi, and Y. Ishimura, in *Dioxygen Activation and Homogeneous Catalytic Oxidation* (L. I. Shimandi, ed.), Elsevier, Amsterdam, 1991.
34. Y. Kimata, H. Shimada, T. Hirose, and Y. Ishimura, *Biochem. Biophys. Res. Commun.*, *208*, 96–102 (1995).
35. H. Shimada, S. G. Sligar, H. Yeom, and Y. Ishimura, in *Catalysis by Metal Complexes*, Keio University, Tokyo, 1997, pp. 195–221.
36. D. W. Nebert, J. R. Robinson, and H. Kon, *J. Biol. Chem.*, *248*, 7637–7647 (1973).
37. F. Scheller, R. Renneberg, P. Mohr, G. R. Janig, and K. Ruckpaul, *FEBS Lett.*, *71*, 309–312 (1976).
38. I. Schlichting, J. Berendzen, K. Chu, A. M. Stock, S. A. Maves, D. E. Benson, R. M. Sweet, D. Ringe, G. A. Petsko, and S. G. Sligar, *Science*, *287*, 1615–1622 (2000).
39. R. Davydov, T. M. Makris, V. Kofman, D. E. Werst, S. G. Sligar, and B. M. Hoffman, *J. Am. Chem. Soc.*, *123*, 1403–1415 (2001).
40. S. Nagano and T. L. Poulos, *J. Biol. Chem.*, *280*, 31659–31663 (2005).
41. T. M. Makris, I. G. Denisov, S. G. Sligar, K. Chu, R. M. Sweet, and I. Schlichting, submitted (2006).
42. N. C. Gerber and S. G. Sligar, *J. Am. Chem. Soc.*, *114*, 8742–8743 (1992).
43. N. C. Gerber and S. G. Sligar, *J. Biol. Chem.*, *269*, 4260–4266. (1994).
44. H. Shimada, R. Makino, M. Imai, T. Horiuchi, and Y. Ishimura, *International Symposium on Oxygenases and Oxygen Activation* (S. Yamamoto, M. Nozaki, and Y. Ishimura, eds), Osaka, 1991, pp. 133–136.
45. S. W. Ellis, G. P. Hayhurst, G. Smith, T. Lightfoot, M. M. S. Wong, A. P. Simula, M. J. Ackland, M. J. E. Sternberg, M. S. Lennard, G. T. Tucker, and C. R. Wolf, *J. Biol. Chem.*, *270*, 29055–29058 (1995).
46. K. Hiroya, M. Ishigooka, T. Shimizu, and M. Hatano, *Faseb J.*, *6*, 749–751 (1992).
47. H. Y. Yeom and S. G. Sligar, *Arch. Biochem. Biophys.*, *337*, 209–216 (1997).

48. D. Zhou, K. R. Korzekwa, T. L. Poulos, and S. Chen, *J. Biol. Chem.*, 267, 762–768 (1992).
49. D. E. Benson, K. S. Suslick, and S. G. Sligar, *Biochemistry*, 36, 5104–5107 (1997).
50. M. Vidakovic, S. G. Sligar, H. Li, and T. L. Poulos, *Biochemistry*, 37, 9211–9219. (1998).
51. J. P. Clark, C. S. Miles, C. G. Mowat, M. D. Walkinshaw, G. A. Reid, S. N. Daff, and S. K. Chapman, *J. Inorg. Biochem.*, 100, 1075–19090 (2006).
52. T.-j. Deng, L. M. Proniewicz, J. R. Kincaid, H. Yeom, I. D. Macdonald, and S. G. Sligar, *Biochemistry*, 38, 13699–13706 (1999).
53. G. Truan and J. A. Peterson, *Arch. Biochem. Biophys.*, 349, 53–64 (1998).
54. H. Yeom, S. G. Sligar, H. Y. Li, T. L. Poulos, and A. J. Fulco, *Biochemistry*, 34, 14733–14740 (1995).
55. T. W. Ost, J. Clark, C. G. Mowat, C. S. Miles, M. D. Walkinshaw, G. A. Reid, S. K. Chapman, and S. Daff, *J. Am. Chem. Soc.*, 125, 15010–15020 (2003).
56. J. F. Andersen and C. R. Hutchinson, *J. Bacteriol.*, 174, 725–735 (1992).
57. S. Donadio and C. R. Hutchinson, *Gene*, 100, 231–235 (1991).
58. J. R. Cupp-Vickery, O. Han, C. R. Hutchinson, and T. L. Poulos, *Nature Struct. Biol.*, 3, 632–637 (1996).
59. J. R. Cupp-Vickery, H. Li, and T. L. Poulos, *Proteins*, 20, 197–201 (1994).
60. J. R. Cupp-Vickery and T. L. Poulos, *Nature Struct. Biol.*, 2, 144–153. (1995).
61. J. F. Andersen, K. Tatsuta, H. Gunji, T. Ishiyama, and C. R. Hutchinson, *Biochemistry*, 32, 1905–1913 (1993).
62. H. Xiang, R. A. Tschirret-Guth, and P. R. Ortiz De Montellano, *J. Biol. Chem.*, 275, 35999–36006 (2000).
63. T. Kudo, N. Takaya, S. Y. Park, Y. Shiro, and H. Shoun, *J. Biol. Chem.*, 276, 5020–5026 (2001).
64. S. Y. Park, H. Shimizu, S. I. Adachi, Y. Shiro, T. Iizuka, A. Nakagawa, I. Tanaka, H. Shoun, and H. Hori, *FEBS Lett.*, 412, 346–350 (1997).
65. N. Okamoto, Y. Imai, H. Shoun, and Y. Shiro, *Biochemistry*, 37, 8839–8847 (1998).
66. E. Obayashi, H. Shimizu, S. Y. Park, H. Shoun, and Y. Shiro, *J. Inorg. Biochem.*, 82, 103–111 (2000).
67. R. Oshima, S. Fushinobu, F. Su, L. Zhang, N. Takaya, and H. Shoun, *J. Mol. Biol.*, 342, 207–217 (2004).
68. H. Shimizu, E. Obayashi, Y. Gomi, H. Arakawa, S. Y. Park, H. Nakamura, S. I. Adachi, H. Shoun, and Y. Shiro, *J. Biol. Chem.*, 275, 4816–4826 (2000).
69. H. Shimizu, S. Y. Park, D. Lee, H. Shoun, and Y. Shiro, *J. Inorg. Biochem.*, 81, 191–205 (2000).
70. R. Davydov, I. D. G. Macdonald, T. M. Makris, S. G. Sligar, and B. M. Hoffman, *J. Am. Chem. Soc.*, 121, 10654–10655 (1999).
71. R. Davydov, A. Ledbetter–Rogers, P. Martasek, M. Larukhin, M. Sono, J. H. Dawson, B. S. Masters, and B. M. Hoffman, *Biochemistry*, 41, 10375–10381 (2002).
72. D. Metodiewa and H. B. Dunford, *Biochem. Biophys. Res. Commun.*, 168, 1311–1317 (1990).

73. R. M. Davydov, T. Yoshida, M. Ikeda-Saito, and B. M. Hoffman, *J. Am. Chem. Soc.*, *121*, 10656–10657 (1999).
74. I. G. Denisov, T. M. Makris, and S. G. Sligar, *J. Biol. Chem.*, *277*, 42706–42710 (2002).
75. I. G. Denisov, T. M. Makris, and S. G. Sligar, *J. Biol. Chem.*, *276*, 11648–11652 (2001).
76. G. I. Berglund, G. H. Carlsson, A. T. Smith, H. Szoke, A. Henriksen, and J. Hajdu, *Nature*, *417*, 463–468 (2002).
77. D. Bourgeois and A. Royant, *Curr. Opin. Struct. Biol.*, *15*, 538–547 (2005).
78. O. Carugo and K. Djinovic Carugo, *Trends Biochem. Sci.*, *30*, 213–219 (2005).
79. R. Fedorov, I. Schlichting, E. Hartmann, T. Domratcheva, M. Fuhrmann, and P. Hegemann, *Biophys. J.*, *84*, 2474–2482 (2003).
80. J. P. Collman, J. I. Brauman, T. R. Halbert, and K. S. Suslick, *Proc. Natl. Acad. Sci. USA.*, *73*, 3333–3337 (1976).
81. S. Nagano, J. R. Cupp-Vickery, and T. L. Poulos, *J. Biol. Chem.*, *280*, 22102–22107 (2005).
82. B. Zhao, F. P. Guengerich, M. Voehler, and M. R. Waterman, *J. Biol. Chem.*, *280*, 42188–42197 (2005).
83. J. Vojtechovsky, K. Chu, J. Berendzen, R. M. Sweet, and I. Schlichting, *Biophys. J.*, *77*, 2153–2174 (1999).
84. B. Zhao, F. P. Guengerich, A. Bellamine, D. C. Lamb, M. Izumikawa, L. Lei, L. M. Podust, M. Sundaramoorthy, J. A. Kalaitzis, L. M. Reddy, S. L. Kelly, B. S. Moore, D. Stec, M. Voehler, J. R. Falck, T. Shimada, and M. R. Waterman, *J. Biol. Chem.*, *280*, 11599–11607 (2005).
85. I. G. Denisov, T. M. Makris, S. G. Sligar, and I. Schlichting, *Chem. Rev.*, *105*, 2253–2277 (2005).
86. S. Shaik, D. Kumar, S. P. De Visser, A. Altun, and W. Thiel, *Chem. Rev.*, *105*, 2279–2328 (2005).
87. D. Kumar, H. Hirao, S. P. De Visser, J. Zheng, D. Wang, W. Thiel, and S. Shaik, *J. Phys. Chem. B*, *109*, 19946–19951 (2005).
88. H. Li, S. Narasimhulu, L. M. Havran, J. D. Winkler, and T. L. Poulos, *J. Am. Chem. Soc.*, *117*, 6297–6299 (1995).
89. T. I. Oprea, G. Hummer, and A. E. Garcia, *Proc. Natl. Acad. Sci. USA.*, *94*, 2133–2138 (1997).
90. D. L. Harris and G. H. Loew, *J. Am. Chem. Soc.*, *116*, 11671–11674 (1994).
91. T. Shimizu, Y. Murakami, and M. Hatano, *J. Biol. Chem.*, *269*, 13296–13304 (1994).
92. A. D. N. Vaz, S. J. Pernecky, G. M. Raner, and M. J. Coon, *Proc. Natl. Acad. Sci. U. S. A.*, *93*, 4644–4648 (1996).
93. A. L. Blobaum, D. L. Harris, and P. F. Hollenberg, *Biochemistry*, *44*, 3831–3844 (2005).
94. P. H. Keizers, L. H. Schraven, C. de Graaf, M. Hidestrand, M. Ingelman–Sundberg, B. R. van Dijk, N. P. Vermeulen, and J. N. Commandeur, *Biochem. Biophys. Res. Commun.*, *338*, 1065–1074 (2005).
95. A. L. Blobaum, U. M. Kent, W. L. Alworth, and P. F. Hollenberg, *J. Pharmacol. Exp. Ther.*, *310*, 281–290 (2004).
96. Y. Imai and M. Nakamura, *FEBS Lett.*, *234*, 313–315 (1988).

97. A. G. Krainev, T. Shimizu, M. Ishigooka, K. Hiroya, M. Hatano, and Y. Fujii-Kuriyama, *Biochemistry, 30*, 11206–11211 (1991).
98. K. Hiroya, Y. Murakami, T. Shimizu, M. Hatano, and P. R. Ortiz de Montellano, *Arch. Biochem. Biophys., 310*, 397–401 (1994).
99. N. Kadohama, C. Yarborough, D. Zhou, S. Chen, and Y. Osawa, *J. Steroid Biochem. Mol. Biol., 43*, 693–701 (1992).
100. T. Beitlich, K. Kühnel, C. Schulze-Briese, R. L. Shoeman, and I. Schlichting, J. *Synchrotron Rad.*, in press (2007).
101. S. Taraphder and G. Hummer, *J. Am. Chem. Soc., 125*, 3931–3940 (2003).

9

Beyond Heme-Thiolate Interactions: Roles of the Secondary Coordination Sphere in Cytochrome P450 Systems

Yi Lu[1,2] and Thomas D. Pfister[1]

[1]Department of Biochemistry and
[2]Department of Chemistry, University of Illinois at Urbana-Champaign, A322 Chemical and Life Sciences Building, Box 8-6, 600 S. Matthews Avenue, Urbana, IL 61801, USA
<yi-lu@uiuc.edu>

1. OVERVIEW OF CYTOCHROME P450 ACTIVE SITE STRUCTURE	268
2. SECONDARY COORDINATION SPHERE ON THE PROXIMAL SIDE	269
2.1. Structural Features	269
2.2. Functional Significance	270
2.3. Insights from Synthetic Models	272
2.4. Insights from Designed Protein Models	273
3. SECONDARY COORDINATION SPHERE ON THE DISTAL SIDE	275
3.1. Structural Features	275
3.2. Functional Significance	277
3.3. Insights from Synthetic Models	279
3.4. Insights from Designed Protein Models	280
4. SUMMARY AND OUTLOOK	280
ACKNOWLEDGMENTS	281
ABBREVIATIONS	281
REFERENCES	281

1. OVERVIEW OF CYTOCHROME P450 ACTIVE SITE STRUCTURE

The cytochrome P450 (P450) active site contains several conserved structural features that are catalytically important. Perhaps the most notable feature of the P450 active site is the heme-thiolate bond formed by proximal cysteine ligation to the heme iron (Figure 1). Detailed descriptions and discussions of P450 structures can be found in earlier chapters of this book, including Chapters 3 and 8, and books and reviews published previously (see [1] and references therein). Interestingly, P450s are not the only heme-thiolate containing proteins, as other heme proteins such as nitric oxide synthase (NOS), chloroperoxidase (CPO) and CO-sensing transcriptional activator (CooA) also contain heme-thiolate ligation. Therefore, the function of P450s goes well beyond the presence of a heme-thiolate ligation.

The subtle roles that certain amino acids play in the formation of a secondary coordination sphere around the heme-thiolate on the proximal side and around the substrate-binding site in the distal pocket are just as important; they allow the enzymes to function efficiently as monooxygenases and help fine-tune their selectivity. Here, we examine the secondary coordination spheres on both the

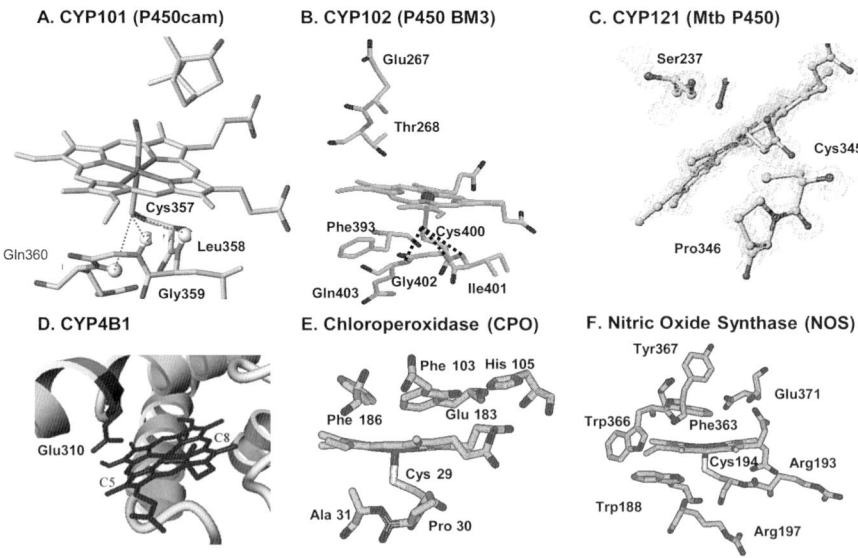

Figure 1. Active site structures of P450, CPO, and NOS showing proximal side secondary coordination observed in (A) CYP101 (P450cam), (B) CYP102 (P450 BM3), (C) CYP121 (Mtb P450), (D) CYP4B1 (E) CPO, and (F) NOS. (Compiled from information given in [7,12,74] and the PDB files 1FAG, 1CPO and 2NOS).

proximal and distal sides of P450, provide structural and functional evidence of their importance, and summarize the insights gained from studies of the enzyme system as well as synthetic and protein models.

2. SECONDARY COORDINATION SPHERE ON THE PROXIMAL SIDE

The secondary sphere interactions on the proximal side of P450s are primarily involved in fine-tuning the properties of the heme and coordination of the conserved Cys axial ligand to heme iron. The availability of high-resolution X-ray structures has provided much insight into some of the structural features present in the secondary coordination sphere. The functional significance of these structural features have been revealed through site-directed mutagenesis, comparisons with other P450s containing residues that deviate from consensus residues, and comparisons with other monooxygenase enzymes. Additional insights have also been obtained from the design and study of both synthetic small molecule models and biosynthetic protein models.

2.1. Structural Features

Sequence comparisons of members of the P450 superfamily have revealed several highly conserved residues on the proximal side of the heme, in addition to the cysteine ligand. As evident from the X-ray crystal structures available for P450s, many of these conserved residues are located close to the conserved cysteine ligand and the heme, forming a secondary coordination sphere on the proximal side of the heme. For example, the X-ray crystal structure of CYP101A1 (P450cam – a prototypical P450) shows that the axial Cys357 forms three H-bonds with the amide protons of Leu358 (3.34 Å), Gly359 (2.47 Å) and Gln360 (2.82 Å) (Figure 1A) [2]. These interactions have also been observed in the X-ray crystal structures of other P450s, including CYP102A1 (P450 BM3) [3], CYP107A1 (P450eryF) [4], CYP108 (P450terp) [5], and CYP152A1 (P450$_{BSβ}$) [6]. Structural data also shows that another highly conserved residue, Phe350 in CYP101A1 (P450cam) or Phe393 in CYP102A1 (P450 BM3), located seven residues from the axial Cys, stacks with the heme in addition to packing with the axial Cys ligand (see Figure 1B) [3]. Furthermore, the side chains of Pro392 and Gln403 in CYP102A1 (P450 BM3) sandwich the highly conserved Phe into place.

Several other P450s, including those from *Mycobacterium tuberculosis* (Mtb), have a Pro (346 in CYP121) adjacent to the axial Cys (345 in CYP121), which distorts one of the four heme pyrrole rings as shown in Figure 1C [7]. Because it is immediately adjacent to the axial Cys, this interaction is proposed to have even greater functional implications on the electronic properties of the heme than on the highly conserved Phe seven amino acids (one helical turn) from the axial Cys (see Section 2.2) [7]. One distinct feature of Pro is that it lacks an amide proton interaction such that these P450s have one less H-bond to the axial thiolate. In this respect, these P450s are more like CPO, which also has a Pro at this position. Another highly conserved residue, Trp96 in CYP102A1 (P450 BM3), interacts with the heme propionate groups, thus forming part of the secondary coordination sphere of the reactive heme iron. In CYP101A1 (P450cam), the heme propionate groups also interact with Arg112, Arg 299, and His355 [2]. Finally, covalent heme attachment via a conserved Glu to a heme methyl group is observed in the CYP4 family (see Figure 1D). This can also be considered as part of the secondary coordination sphere of the heme iron.

2.2. Functional Significance

Perhaps the most prominent structural feature in the secondary coordination sphere of the promixal side is the presence of a hydrogen bonding network around the Cys, such as those provided by Leu358, Glu359 and Gln360 to the axial Cys357 in CYP101A1 (P450cam). One function of these H-bond interactions is to reduce the negative charge on the cysteine sulfur [8], resulting in destabilization of the reduced Fe^{2+} form [9] and thus an increase of the redox potential of the heme [10,11]. Site-directed mutagenesis has been a powerful tool used to gain insight into the roles of specific amino acids in enzyme function. However, since these hydrogen bonds are provided by the backbone amide groups, mutation of these residues to amino acids other than Pro will probably leave these amides intact. To perturb such a hydrogen bonding pattern, mutagenesis of residues to Pro at positions 358 [12] and 360 [11] has been successful in removing the amide proton. The Leu358Pro mutant CYP101A1 (P450cam) exhibited 30-fold tighter binding to putidaredoxin (Pdx) [12]. NMR data shows that the Leu358Pro mutant of CYP101A1 (P450cam) mimics the Pdx bound form [12]. These results provided insight into structural changes that may be occurring upon binding of CYP101A1 (P450cam) to its redox partner Pdx.

A hydrogen bonding network around the Cys axial ligand has been also observed in another monooxygenase, nitric oxide synthase [13] and in chloroperoxidase [14], which may explain the ability of this heme-thiolate peroxidase to carry out some of the P450-type reactions. However, CPO has only two of the three H-bonds to the axial Cys29 due to the presence of Pro30 (which lacks an amide proton) at the position analogous to Leu358 (Figure 1E) [14]. NOS, which also has a proline at this position, has an additional H-bond from the imidazole N-H of Trp (Figure 1F) [13]. Functional differences in certain P450s that also have a Pro similar to the one found in CPO can also offer additional insight. In all of these monooxygenases it is apparent that at least a certain degree of hydrogen bonding to the axial Cys is required and that the exact nature likely depends on both the presence of other structural factors and on the functional requirements of the enzyme.

Site-directed mutagenesis has provided the most conclusive functional evidence for the conserved Phe393 (CYP102A1 (P450 BM3)) which contacts both the heme and the axial Cys ligand. To probe its involvement in P450 chemistry, the proximal Phe393 in CYP102A1 (P450 BM3) was mutated to His [15]. The crystal structure of this variant revealed that the five-membered imidazole ring occupies the same space as the six-membered ring of Phe393 and no obvious conformational change of the heme was detected. The mutation also had no effect on product distribution or regioselectivity. It was, however, shown that the mutation exerted a large effect on the electronic properties of the heme in the CYP102A1 (P450 BM3) system [15]. It was thus proposed that the role of the Phe was to provide hydrophobicity and electrochemical neutrality, as well as to protect the axial Cys from solvent exposure. A small group of P450s, including CYP3, CYP8, and CYP74, lack the proximal Phe (393 in CYP1021A (P450 BM3)). These members of the P450 superfamily lack the ability to activate molecular oxygen and generally perform isomerization or rearrangement reactions instead. Interesting insight into the role of the proximal Phe has also been obtained from a related heme-thiolate protein, NOS. NOS has a Trp at the position homologous to Phe393 in CYP102A1 (P450 BM3), indicating that the presence of other aromatic residues at this position are capable of monooxygenase activity. Another highly conserved tryptophan residue (Trp96 in CYP102A1 (P450 BM3)) interacts with the heme propionate groups, and has been shown to have a strong influence on the spin state of the heme [16].

Certain CYP4 family members (including CYP4A, CYP4B and CYP4F) contain a covalently attached heme [17]. Elimination of the covalent ester linkage between the conserved Glu and the heme methyl group resulted in a 40–100% decrease in activity depending on the variant used [18]. Mutation of Glu310 to Ala in CYP4B1 resulted in a 40% decrease in activity for the hydroxylation of laurate, while activities of the Glu310Gly and Glu310Asp mutants were less than 3% of wild-type. Interestingly, most P450s function perfectly well without this type of covalent attachment.

This type of covalent attachment has also been observed in mammalian peroxidases, i.e., lactoperoxidase (LPO) and myeloperoxidase (MPO) [19,20]. Lactoperoxidase shows increased activity as a result of this covalent attachment [20]. It has been suggested that covalent attachment of the heme in mammalian peroxidases (LPO and MPO) serves to protect the heme from modification and confers higher activity than mutants lacking the covalent attachment [21]. The crystal structure of CYP101A1 (P450cam) for example, shows that the heme is completely surrounded by amino acids such that there are no exposed edges [2]. Thus, it is likely that most P450s use other means to 'protect' their heme (i.e., the heme binding pocket or secondary coordination of the heme cofactor to prevent undesired modification of the heme).

2.3. Insights from Synthetic Models

Numerous synthetic model complexes have been generated to mimic the active site of P450. Chapter 2 in this volume gives a more detailed account of their contribution to the understanding of P450 chemistry in general. This chapter will focus more specifically on models that offer insight into the secondary coordination sphere in P450s.

Early model complexes mimicked thiolate ligation in the ferric and ferrous-CO bound forms [22,23]. P450 model compounds even allowed for the detection of reactive iron-oxygen (ferryl) species prior to their detection in the native enzymes [24]. Later models sought to incorporate the secondary coordination found in P450s. An alkyl-thiolate coordinated heme complex (Figure 2A) was synthesized in which the axial thiolate was protected by bulky pivoloyl groups, thus giving it remarkable stability [25]. This model mimics a key function of the secondary sphere around the axial ligand by providing a hydrophobic environment and demonstrates its importance in maintaining stable thiolate ligation. While this complex was capable of performing P450 chemistry with hydrogen peroxide and other peroxy acids, it was unable to form a stable dioxygen

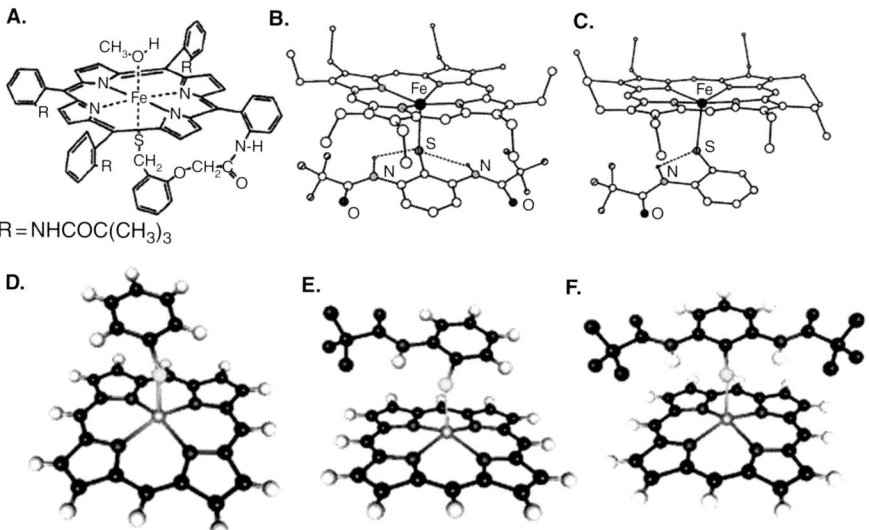

Figure 2. Synthetic models with proximal secondary coordination: (A) P450 model with pivoloyl groups to provide a hydrophobic pocket, models with H bonding to proximal Cys; (B) FeIII(OEP) (S-2,6-(CF$_3$CONH)$_2$C$_6$H$_3$), (C) FeIII(OEP)(S-2-CF$_3$CONHC$_6$H$_4$) (3), (D) Fe(OEP)SPh, (E) Fe(OEP)L1 and (F) Fe(OEP)L2. Ethyl groups in the original molecules (D-F) are truncated to hydrogens for simplicity. (Compiled from data provided in [25,27,28]).

adduct [25]. In another study it was demonstrated that a thiolate-iron porphyrin complex could cleave O_2 and insert an oxygen atom into non-activated C-H bonds [26]. This complex and others of its type, however, have lower redox potentials ($E_o = -600$ to -700 mV) than P450 enzymes (e.g., -290 mV for CYP101A1 (P450cam)). A possible reason for this large difference was inferred from the X-ray structures of P450s which show H-bonding to the axial Cys, thereby decreasing the charge density on the sulfur and raising the redox potential.

In more recent model complexes, secondary sphere H-bonding interactions have been introduced. A series of thiolate ligated porphyrins (Figure 2B-C) have been synthesized with increasing H-bonding strength to the thiolate ligand, which demonstrated that H-bonding increased the redox potential of the site [27]. XAS studies and DFT calculations on similar model complexes (Figure 2D-F) show that H-bonding to the thiolate ligand decreases the covalency of the metal-sulfur bond [28]. Models have also been made using the α-helical peptide fragments corresponding to P450cam (as well as the α-helical peptide fragments in CPO) containing the Cys ligand to a modified iron porphyrin, to clarify the effect of cooperativity between the α-helix and H-bonds to the axial Cys sulfur [10]. Indeed, they found significant contribution from the helical dipole.

As an alternative to using secondary sphere interactions, P450 model complexes with SO_3^- ligands instead of thiolates have been synthesized, which were designed to reduce the charge on the axial ligand [29]. While these complexes have an oxy anion as the proximal ligand instead of a thiolate, they show adequate reactivity for reactions catalyzed by P450s (e.g., epoxidation, hydroxylation, N-dealkylation reactivity, and diol cleavage). This study also supports the role of secondary sphere H-bonding as a means of reducing charge density on the sulfur.

2.4. Insights from Designed Protein Models

Since heme-thiolate ligation is the main feature in the primary coordination sphere of P450s (shown in Figure 3A), the axial His ligand of myoglobin (Mb) (shown in Figure 3B) was mutated to Cys to mimic the heme-thiolate ligation of P450 [30]. This mutation resulted in a variant with a similar UV-vis spectrum to that of P450 in the resting ferric state of the protein. The variants also exhibit enhanced oxygenation activity. The His93Gly variant of Mb has also been used to study thiolate ligation by binding exogenous thiol ligands in the cavity created by the mutation [31].

Based on the same principle, the axial His175 of C*c*P (shown in Figure 3C) was also replaced with a Cys. In contrast to Mb, the heme-thiolate ligation was not observed. Instead, the Cys was rapidly oxidized to cysteic acid [32]. It was then recognized that the secondary coordination spheres around the axial ligand coordinated to the heme are different among Mb, cytochrome *c* peroxidase (C*c*P), and P450s. While a conserved Phe (350 in CYP101A1 (P450cam)) is packed against the heme-thiolate in P450 (shown in Figure 3A), a conserved

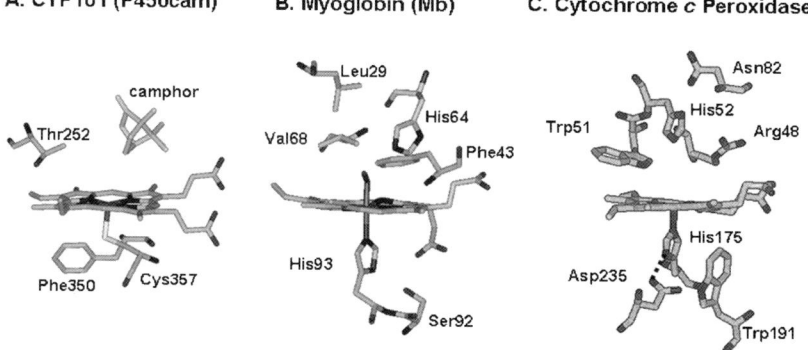

Figure 3. Active sites of (A) P450 and protein scaffolds used for P450 models (B) myoglobin and (C) cytochrome *c* peroxidase. (Compiled from PDB files 1AKD, 1MBO and 2CYP).

Asp235 is present at a similar position in C*c*P. Since the presence of a negatively charged Asp235 will probably destabilize the negatively charged Cys, the Asp235 (shown in Figure 3C) was replaced with Leu, which is close in steric structure and yet provides a similar hydrophobic environment as Phe350 in CYP101A1 (P450cam)) [33]. This Asp235Leu mutation potentially mimics features of the secondary coordination provided by the highly conserved proximal Phe in P450s, such as increased hydrophobicity and exclusion of water from the vicinity of the axial Cys. This secondary coordination sphere mutation turned out to be critical in creating a stable heme-thiolate ligation in not only the resting ferric state, but also the reduced ferrous heme state [33–35].

It is interesting to compare how the secondary coordination sphere of P450, Mb, and C*c*P influence the stability of heme-thiolate ligation (Figure 3). In P450, the Cys is next to a hydrophobic Phe (350 in CYP101A1 (P450cam)), which can stabilize the neutral Cys-ferric ligation. On the other hand, Mb has a peptide bond carbonyl group in the corresponding position. Even though the peptide carbonyl group is more polar, it is not enough to destabilize the heme-thiolate ligation. Therefore, the axial His mutation to Cys was in itself, enough to result in a stable heme-thiolate ligation in Mb. The presence of a conserved (amongst heme peroxidases) Asp235 in C*c*P, however, provides a negative environment next to the negatively charged Cys and thus, can destabilize the heme-thiolate ligation. The only way for a stable heme-thiolate ligation to form in C*c*P is to provide a similar hydrophobic environment with the D235L mutation. This is an excellent example illustrating the importance of the secondary coordination sphere in protein design.

Most of these protein models are designed to mimic the ferric and or ligand bound states of P450. A cytochrome *c* peroxidase variant, C*c*P(His175Cys/Sap235Leu), and Mb(His93Gly) reconstituted with exogenous thiols were studied in the reduced ferrous state and it was found that the Cys became protonated, forming a neutral thiol proximal ligand [35]. A protein model of the ferrous CO complex was later successfully engineered in cytochrome b_{562} [36]. The native Met7 axial ligand was replaced with Gly to create a ligand binding pocket while the His102 heme ligand was replaced with Cys. In addition, Glu4 and Glu8 were mutated to Ser. The Soret band of the ferrous-CO bound form of this quadruple mutant was at 442 nm [36]. Resonance Raman studies also support thiolate coordination to the ferrous CO bound enzyme [36]. The presence of positively charged Arg98 and Arg102 which flank the axial His have been proposed to stabilize the anionic thiolate, decreasing the charge in the ferrous state [36]. This would mimic the role of H-bonding, to the axial Cys of P450 from the backbone amides, in reducing the charge on the sulfur and stabilizing the Fe^{2+} form.

3. SECONDARY COORDINATION SPHERE ON THE DISTAL SIDE

Secondary sphere interactions on the distal side of P450s are primarily involved in forming a structured H-bonding network and providing a proton source for the activation of dioxygen as well as substrate binding. Again, the availability of high-resolution X-ray structures has provided much insight into the structural features required to form a distal secondary coordination sphere (Section 3.1). Their functional significance has been obtained from biochemical studies of native enzymes and their variants (Section 3.2) as well as synthetic model complexes (Section 3.3) and biosynthetic protein models (Section 3.4).

3.1. Structural Features

The available P450 X-ray crystal structures have revealed a number of features in the distal pocket. First, much information has been obtained from the resting state (ferric) structures of CYP101A1 (P450cam) with and without the substrate camphor bound. Furthermore, the crystal structure of a short lived oxy intermediate (Figure 4A) revealed the role of water in contributing to P450 catalysis [37]. A new water molecule that is absent in the ferric structures was shown to bridge the dioxygen and Thr252. This structure also shed new light onto the structural dynamics of how the secondary coordination sphere changes in the P450 catalytic cycle. In the ferric resting state structure, the backbone oxygen of Gly248 in CYP101A1 (P450cam) is H-bonded to the Thr252 hydroxyl side chain. Upon reduction, this H-bond is broken as the backbone rotates. In the oxy structure, Thr252 coordinates a water molecule, which is bound to the terminal oxygen of O_2 [37]. Besides allowing for H-bonding of water to Thr252, the conformation of the oxy structure prevents unfavorable interactions between Asp251 and the water bound to dioxygen [37].

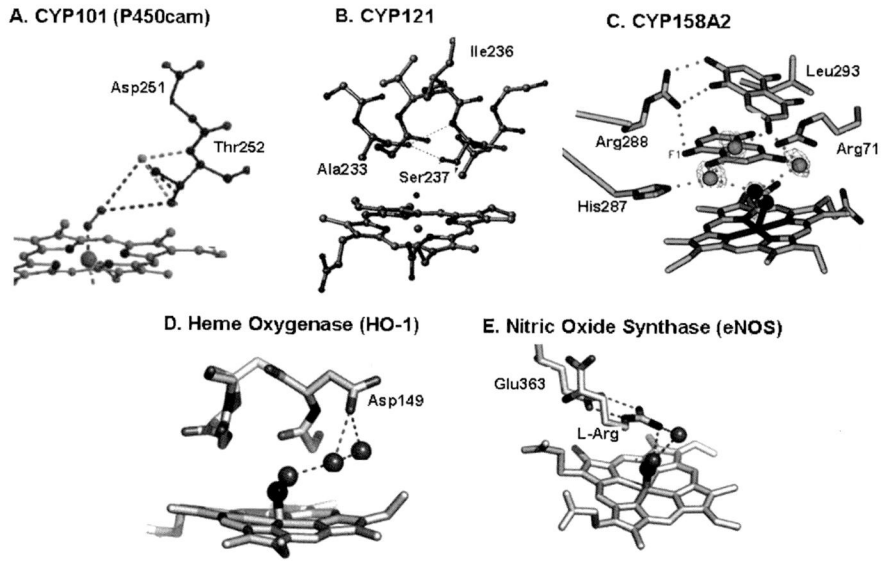

Figure 4. Distal side secondary coordination showing distal H-bonding networks observed in P450, heme oxygenase-1 (HO-1), and NOS structures. (A) CYP101 (P450cam) with conserved Thr252, (B) CYP121 with Ser237, (C) H-bonding of flaviolin substrate to water in CYP158A2 (Ala245 was omitted for clarity), (D) HO-1 with NO bound, (E) eNOS with NO bound. (Compiled from data provided in [7,37,39,41]).

While Thr is the most conserved residue at position 252 (CYP101A1 (P450cam) numbering), some P450s have Ser (Ser237 in CYP121, shown in Figure 4B) H-bonded to the water ligand bound to the heme [7]. Even though the Thr is not conserved in this case, the positioning of water or more specifically, the protons required for oxygen cleavage, is. Again, as in the case of Thr252 (CYP101A1 (P450cam)), a secondary sphere coordination around the Ser237 in CYP121 is apparent. Ser237 is H-bonded to the backbone carbonyl of Ala 232 and side chain of Arg386. Furthermore, in addition to forming an additional H-bond with Gln285, Arg386 is wedged between Ile236 and Phe280.

Other P450s lack a residue capable of H-bonding to the bound dioxygen via water (such as Thr or Ser) altogether. CYP107A1 (P450eryF) and CYP158A2 both have Ala245 at an analogous position to Thr252 in CYP101A1 (P450cam). In CYP107A1 (P450eryF), the water does not directly bind to the dioxygen. Thus, it is not surprising to find that no analogs of Thr252 in CYP107A1 (P450eryF) exist. In this enzyme, the 5-OH group of the substrate (6-deoxyerythronolide B) binds directly to the terminal oxygen [38]. It has been suggested that in this case, the substrate would serve as the proton source for the cleavage of dioxygen instead of water [38].

The ferrous dioxygen structure of CYP158A2 (Figure 4C) shows an intricate H-bonding network between flaviolin (substrate) and a water molecule that is coordinated to dioxygen [39]. A second substrate molecule is also involved in this network of H-bonds between substrate and protein and active site water molecules [39]. Whereas CYP107A1 (P450eryF) uses the substrate-OH as a proton source, the substrate coordinated water molecule is likely the source of protons for dioxygen cleavage in CYP158A2, which is supported by the functional studies discussed below (see Section 3.2).

In addition to positioning protons for catalysis, secondary sphere interactions can also be involved in substrate binding. An Arg-Tyr substrate binding motif has been identified in CYP102A1 (P450 BM3) [40]. This motif is also conserved in several non-P450 enzymes. While these residues coordinate substrate, there have so far been no reports on secondary coordination of these residues, but it would be interesting to see future investigations into this area.

The crystal structures of other monooxygenases also show extensive networks of secondary sphere interactions, similar to those found in P450s [41,42]. Heme oxygenases (HO) and NOS also have highly ordered water molecules that have been proposed to be involved as proton sources for the cleavage of dioxygen [41,42]. In human HO-1, Asp140 coordinates two water molecules (Figure 4D), including one bound to the terminal oxygen of O_2 [43]. In contrast, in endothelial NOS (eNOS), the bound arginine substrate forms H-bonds to a water bound to the terminal oxygen (Figure 4E) [13]. The Arg itself also forms hydrogen bonds with both oxygens of the Glu363 side chain.

3.2. Functional Significance

The functional significance of correctly positioning protons for dioxygen cleavage using a hydrogen-bonding network can be inferred from the numerous strategies employed by P450s. While the Thr (252 in CYP101A1 (P450cam)) is not strictly conserved throughout the P450 superfamily, the positioning of water or more specifically, the protons required for oxygen cleavage, is. The requirement of a secondary coordination sphere to ensure that these water/protons are precisely positioned also appears to be conserved, although it can come in a number of different forms. Even P450s that lack a residue capable of forming an H-bond bridge (e.g., CYP107A1 (P450eryF)) via water to the heme oxygen and substrate molecules bind directly to the terminal oxygen [38]. The requirement of a proton source for the cleavage of dioxygen remains invariable [38].

Mutation of the highly conserved Thr252 in CYP101A1 (P450cam) has offered insight into its role in positioning the water that H-bonds to the dioxygen in many P450s [44]. Not surprisingly, mutation to Ser results in a mutant that retains 85% of its activity [45]. The Thr252Asn mutation resulted in 57% of wild-type (WT) O_2 consumption for camphor hydroxylation [46]. Although the Thr252Ala mutant is only 10% as active as the WT in camphor hydroxylation, O_2 consumption was shown to be near that of WT [45]. Resonance Raman (RR) studies on Thr252Ala are consistent with the presence of extra water molecules in these mutants [44]. The presence of these water molecules are confirmed in the X-ray crystal structure of Thr252Ala [47]. Interestingly, mutation of Ala245 in CYP101A1 (P450eryF) (which occupies an analogous position to Thr252 in CYP101A1 (P450cam)) to Ser or Thr resulted in a dramatic loss of activity (down to 16% and < 1%, respectively) [48]. The crystal structure showed that the newly introduced Thr side chain disrupted H-bonding of the substrate to the dioxygen. On the other hand, the Asp251Asn mutation in CYP101A1 (P450cam) resulted in a decrease in activity of two orders of magnitude [49]. As with Thr252Ala, RR studies on Asp251Asn are consistent with the presence of extra water molecules [44].

The binding of multiple substrates can also affect the coordination of the primary substrate and/or provide a secondary coordination sphere for key active site residues and/or waters. Evidence for homotropic cooperativity (enhancement of activity by a second substrate molecule) has been observed in the sigmoidal (non Michaelis-Menten kinetics) plots for reactions of progesterone and testosterone with CYP3A4 [50,51]. Evidence has also been obtained for heterotropic cooperativity (enhancement of activity by a second molecule other than the substate) [52]. Recent cooperativity data suggests that CYP3A4 may even be able to bind three molecules [53]. One of the proposed effects of stacking two (or three) molecules into the binding pocket is to reduce freedom of motion [54]. The emergence of more mutagenesis studies and structures with multiple compounds bound will likely soon shed more light onto the nature of these interactions and the coordination of residues contributing to increased activity.

CYP3A4 has a Phe (304) in its distal site that has been implicated in substrate binding. Mutation of this residue to Ala resulted in increased progesterone oxidation (at the 6β position), implicating it in substrate binding [55]. Mutation of this residue to Trp yielded a mutant with hyperbolic kinetics instead of the typical sigmoidal kinetics associated with homotropic cooperativity [56]. The double mutant L211F/D214E also lost homotropic cooperativity for testosterone and progesterone [57]. These results show the importance of the amino acid side chains in binding substrate molecules and in regulating their interaction with a second substrate in the active site.

Other functional insights have been obtained from selectivity studies. Experiments with CYP101A1 (P450cam) indicated that either the 5-endo or 5-exo proton was extracted from d-camphor [58]. This is in comparison to recent isotope labeling studies with 6β-testosterone (a prototypic substrate for CYP3A4) which showed absolute preference for the 6β proton [59]. In both cases, rebound was very selective. Slight differences in selectivity may reflect differences in the degree of coordination of the substrates amongst various P450s.

CYP158A2 is another example of a P450 which lacks this conserved Thr, yet it is catalytically active in oxidizing flaviolin. Flaviolin has hydroxyl groups that could be involved in H-bonding. To verify the role of the substrate hydroxyl groups, studies with a substrate analogue, 2-hydroxy-1,4-naphthoquinone, which lacks two hydroxyl groups were carried out [39]. Activity towards this substrate analogue was decreased 70-fold, indicating that the substrate hydroxyl groups function by coordinating active site water molecules [39]. Even though kinetic isotope effects (KIE) in P450 reactions have been studied for about 30 years [60], they have not really been used to directly study the contribution of the secondary coordination sphere. Kinetic solvent isotope effect (KSIE) experiments with CYP158A2 support proton transfer from water molecules in the dioxygen cleavage step [39].

The participation of substrates (or effector molecules) in the secondary coordination of amino acid residues, water, dioxygen, and other substrate molecules has been established from a number of functional studies (discussed in Section 3.2). As discussed above, the structure for CYP158A2 clearly shows two molecules of flaviolin in the active site. While kinetic studies of CYP3A4, the major liver P450 in humans, suggest homo and heterotropic cooperativity of substrate, there are still no available structures with substrate bound in catalytically productive conformations [61]. Structures do, however, show that the active site is large enough to accommodate multiple substrates [61].

Highly ordered water molecules in HOs (Figure 4D) and NOSs (Figure 4E) have also been proposed to be involved in providing protons for oxygen cleavage in these systems [41,42]. While HO and NOS are also both monooxygenases, they have very specific substrates; heme for HO and arginine for NOS. NOS is similar to P450 in that it is proposed to heterolytically cleave dioxygen, whereas HO cleaves dioxygen homolytically and incorporates the terminal oxygen into the substrate (heme). One factor dictating this reactivity in HO is a low 'ceiling' in the HO cavity which forces dioxygen to assume a more bent conformation. Another factor is the presence of different hydrogen bonds. For example, while two hydrogen bonds are present in P450 to promote heterolytic cleavage of O_2, only one hydrogen bond is present in HO, which is sufficient for homolytic cleavage of O_2 [62].

3.3. Insights from Synthetic Models

Early models (discussed in Section 2.3) concentrated on modeling the heme-thiolate ligation. Although the model contained an axial thiolate protected by bulky pivoloyl groups to give it remarkable stability, it was unable to form a stable dioxygen adduct [25]. More complex models with oxygen binding pockets and hydrogen bonding were required to more closely mimic the native P450 system. Synthetic models have been made to mimic the H-bonding provided by Thr252 [63,64]. For example, twin coronet porphyrin (TCP) complexes contain hydroxylated bi-naphthyl groups capable of H-bonding to the oxygen bound to the heme iron [63,64]. The resulting RR spectra assigned to the (O-O) stretching mode ($\sim 1140\,\text{cm}^{-1}$) are very similar to those of oxy-CYP101A1 (P450cam).

3.4. Insights from Designed Protein Models

While many of the designed protein models have placed a greater emphasis on modeling the thiolate ligand (discussed above in Section 2.4), certain distal side features have been explored. Notably, many of the models in CcP and Mb contain a distal His to provide a hydrogen bonding network [30,32,33,65,66]. This eliminates the need for an elaborate network of water molecules, which seems so heavily dependent on the presence of a secondary coordination sphere to provide rigidity and order to this network. Myoglobin and heme peroxidase mutants with the native distal His ligand have also been used with the native proximal His ligand to catalyze a number of reactions normally catalyzed by P450s [67–72]. A covalently attached Trp has also been engineered in the active site of Mb, mimicking a bound indole substrate to demonstrate P450 hydroxylation chemistry in the absence of thiolate ligation [73]. This variant contains additional mutations to the distal pocket, including the mutation of the distal His64 to Glu. Again, the Glu can provide protons and function as an acid-base catalyst similar to the native His.

The distal His has been replaced with a Gly to provide a more hydrophobic pocket to model the active site of P450 [65]. Trans effects of His64Ile and His64Val on the His93Cys ligand were also studied [66]. These mutants produced P450-like UV-vis, NMR, and EPR spectra in the ferric state but failed to maintain thiolate ligation when reduced [66]. While many of the protein models eliminate the need for an elaborate network of water molecules, and hence the need for a secondary coordination sphere to provide rigidity to this network, these models are typically unable to use molecular oxygen and only function via the peroxide shunt pathway.

4. SUMMARY AND OUTLOOK

Numerous structural and functional studies of P450s and P450 model systems have already contributed to the elucidation of the role of the secondary coordination sphere in P450 chemistry. This understanding will no doubt be expanded as more structures and results from additional studies become available. From these studies, it has become abundantly clear that the primary coordination sphere in itself is not enough to confer the functional properties of enzymes such as P450. The secondary coordination sphere must also play an important role in defining the characteristics of each metalloenzyme.

Understanding these secondary coordination sphere interactions will allow for an improved understanding of how P450s function, as well as the synthesis of better P450 models. This will in turn lead to advancements in the areas of drug discovery, new compound synthesis and bioremediation.

ACKNOWLEDGMENTS

We are grateful for contributions made by Lu group members, whose works are cited in the references. The Lu group work on metalloprotein design and engineering has been supported by the US National Science Foundation (CHE 05-52008) and the National Institute of Health, National Institute of General Medical Sciences (GM062211).

ABBREVIATIONS

CcP	yeast cytochrome c peroxidase
CooA	CO-sensing transcriptional activator
CPO	chloroperoxidase
6-DEB	deoxyerythronolide B
DFT	density functional theory
eNOS	endothelial nitric oxide synthase
HO	heme oxygenase
KIE	kinetic isotope effect
KSIE	kinetic solvent isotope effect
LPO	lactoperoxidase
Mb	myoglobin
MPO	myeloperoxidase
Mtb	*Mycobacterium tuberculosis*
NOS	nitric oxide synthase
OEP	octaethylporphyrin
Pdx	putidaredoxin
RR	Resonance Raman
TCP	twin coronet porphyrin
WT	wild-type
XAS	X-ray absorption spectroscopy

REFERENCES

1. P. R. Ortiz de Montellano, *Cytochrome P450: Structure, Mechanism, and Biochemistry*, 3rd edn, Plenum Press, New York, 2005, pp 685.
2. T. L. Poulos, B. C. Finzel, I. C. Gunsalus, G. C. Wagner, and J. Kraut, *J. Biol.Chem.*, 260, 16122–16130 (1985).
3. K. G. Ravichandran, S. S. Boddupalli, C. A. Hasemann, J. A. Peterson, and J. Deisenhofer, *Science*, 261, 731–736 (1993).
4. J. R. Cupp-Vickery and T. L. Poulos, *Nat. Struct. Biol.*, 2, 144–153 (1995).
5. C. A. Hasemann, K. G. Ravichandran, J. A. Peterson, and J. Deisenhofer, *J.Mol.Biol.* 236, 1169–1185 (1994).

6. D.-S. Lee, A. Yamada, H. Sugimoto, I. Matsunaga, H. Ogura, K. Ichihara, S.-i. Adachi, S.-Y. Park, and Y. Shiro, *J. Biol. Chem., 278*, 9761–9767 (2003).
7. D. Leys, C. G. Mowat, K. J. McLean, A. Richmond, S. K. Chapman, M. D. Walkinshaw, and A. W. Munro, *J. Biol. Chem., 278*, 5141–5147 (2003).
8. T. L. Poulos, *J. Biol. Inorg. Chem., 1*, 356–359 (1996).
9. T. L. Poulos, J. Cupp-Vickery, and H. Li, in *Cytochrome P-450: Structure, Mechanism, and Biochemistry* (P. R. Ortiz de Montellano, ed.), Plenum Press, New York, 1995, pp. 125–150.
10. T. Ueno, Y. Kousumi, K. Yoshizawa-Kumagaye, K. Nakajima, N. Ueyama, T.-a. Okamura, and A. Nakamura, *J. Am. Chem. Soc., 120*, 12264–12273 (1998).
11. S. Yoshioka, T. Tosha, S. Takahashi, K. Ishimori, H. Hori, and I. Morishima, *J. Am. Chem. Soc., 124*, 14571–14579 (2002).
12. T. Tosha, S. Yoshioka, K. Ishimori, and I. Morishima, *J. Biol. Chem., 279*, 42836–42843 (2004).
13. C. S. Raman, H. Li, P. Martasek, V. Kral, B. S. S. Masters, and T. L. Poulos, *Cell, 95*, 939–950 (1998).
14. M. Sundaramoorthy, J. Terner, and T. L. Poulos, *Structure, 3*, 1367–1377 (1995).
15. T. W. Ost, C. S. Miles, A. W. Munro, J. Murdoch, G. A. Reid, and S. K. Chapman, *Biochemistry, 40*, 13421–13429 (2001).
16. T. W. Ost, A. W. Munro, C. G. Mowat, P. R. Taylor, A. Pesseguiero, A. J. Fulco, A. K. Cho, M. A. Cheesman, M. D. Walkinshaw, and S. K. Chapman, *Biochemistry, 40*, 13430–13438 (2001).
17. K. R. Henne, K. L. Kunze, Y.-M. Zheng, P. Christmas, R. J. Soberman, and A. E. Rettie, *Biochemistry, 40*, 12925–12931 (2001).
18. Y.-M. Zheng, B. R. Baer, M. B. Kneller, K. R. Henne, K. L. Kunze, and A. E. Rettie, *Biochemistry, 42*, 4601–4606 (2003).
19. I. M. Kooter, A. J. Pierik, M. Merkx, B. A. Averill, N. Moguilevsky, A. Bollen, and R. Wever, *J. Am. Chem. Soc., 119*, 11542–11543 (1997).
20. G. D. DePillis, S.-I. Ozaki, J. M. Kuo, D. A. Maltby, and P. R. Ortiz De Montellano, *J. Biol.Chem., 272*, 8857–8860 (1997).
21. L. Huang, and P. R. Ortiz de Montellano, *Arch. Biochem. Biophys., 446*, 77–83 (2006).
22. J. P. Collman, and T. N. Sorrell, *J. Am. Chem. Soc., 97*, 4133–4134 (1975).
23. J. P. Collman, T. N. Sorrell, and B. M. Hoffman, *J. Am. Chem. Soc., 97*, 913–914 (1975).
24. J. T. Groves, R. C. Haushalter, M. Nakamura, T. E. Nemo, and B. J. Evans, *J. Am. Chem. Soc., 103*, 2884–2886 (1981).
25. T. Higuchi, S. Uzu, and M. Hirobe, *J. Am. Chem. Soc., 112*, 7051–7053 (1990).
26. H. Patzelt, and W. D. Woggon, *Helv. Chim. Acta, 75*, 523–530 (1992).
27. N. Ueyama, N. Nishikawa, Y. Yamada, T.-a. Okamura, S. Oka, H. Sakurai, and A. Nakamura, *Inorg.Chem., 37*, 2415–2421 (1998).
28. A. Dey, T. Okamura, N. Ueyama, B. Hedman, K. O. Hodgson, and E. I. Solomon, *J. Am. Chem. Soc., 127*, 12046–12053 (2005).
29. D. Meyer, T. Leifels, L. Sbaragli, and W.-D. Woggon, *Biochem. Biophys. Res. Commun., 338*, 372–377 (2005).
30. S. Adachi, S. Nagano, K. Ishimori, Y. Watanabe, I. Morishima, T. Egawa, T. Kitagawa, and R. Makino, *Biochemistry, 32*, 241–252 (1993).

31. M. P. Roach, S. Franzen, P. S. H. Pang, S. G. Boxer, W. H. Woodruff, and J. H. Dawson, *J. Inorg. Biochem., 67*, 134 (1997).
32. K. Choudhury, M. Sundaramoorthy, A. Hickman, T. Yonetani, E. Woehl, M. F. Dunn, and T. L. Poulos, *J. Biol. Chem., 269*, 20239–20249 (1994).
33. J. A. Sigman, A. E. Pond, J. H. Dawson, and Y. Lu, *Biochemistry, 38*, 11122–11129 (1999).
34. J. A. Sigman, H. K. Kim, X. Zhao, J. R. Carey, and Y. Lu, *Proc. Nat. Acad. Sci., 100*, 3629–3634 (2003).
35. R. Perera, M. Sono, J. A. Sigman, T. D. Pfister, Y. Lu, and J. H. Dawson, *Proc. Natl. Acad. Sci. USA, 100*, 3641–3646 (2003).
36. T. Uno, A. Yukinari, Y. Tomisugi, Y. Ishikawa, R. Makino, J. A. Brannigan, and A. J. Wilkinson, *J. Am. Chem. Soc., 123*, 2458–2459 (2001).
37. I. Schlichting, J. Berendzen, K. Chu, A. M. Stock, S. A. Maves, D. E. Benson, R. M. Sweet, D. Ringe, G. A. Petsko, and S. G. Sligar, *Science, 287*, 1615–1622 (2000).
38. S. Nagano, J. R. Cupp-Vickery, and T. L. Poulos, *J. Biol. Chem., 280*, 22102–22107 (2005).
39. B. Zhao, F. P. Guengerich, M. Voehler, and M. R. Waterman, *J. Biol. Chem., 280*, 42188–42197 (2005).
40. M. A. Noble, C. S. Miles, S. K. Chapman, D. A. Lysek, A. C. Mackay, G. A. Reid, R. P. Hanzlik, and A. W. Munro, *Biochem. J., 339*, 371–379 (1999).
41. T. L. Poulos, *Biochem. Biophys. Res. Commun., 338*, 337–345 (2005).
42. T. L. Poulos, *Drug Metab. Dispos., 33*, 10–18 (2005).
43. D. J. Schuller, A. Wilks, P. R. Ortiz de Montellano, and T. L. Poulos, *Nat. Struct. Biol., 6*, 860–867 (1999).
44. T.-j. Deng, I. D. G. Macdonald, M. C. Simianu, M. Sykora, J. R. Kincaid, and S. G. Sligar, *J. Am. Chem. Soc., 123*, 269–278 (2001).
45. M. Imai, H. Shimada, Y. Watanabe, Y. Matsushima-Hibiya, R. Makino, H. Koga, T. Horiuchi, and Y. Ishimura, *Proc. Natl. Acad. Sci. USA, 86*, 7823–7787 (1989).
46. H. Shimada, Y. Watanabe, M. Imai, R. Makino, H. Koga, T. Horiuchi, and Y. Ishimura, *Stud. Surf. Sci. Catal., 66*, 313–319 (1991).
47. R. Raag, S. A. Martinis, S. G. Sligar, and T. L. Poulos, *Biochemistry, 30*, 11420–11429 (1991).
48. J. R. Cupp-Vickery, O. Han, C. R. Hutchinson, and T. L. Poulos, *Nat. Stuct. Biol., 3*, 632–637 (1996).
49. M. Vidakovic, S. G. Sligar, H. Li, and T. L. Poulos, *Biochemistry, 37*, 9211–9219 (1998).
50. G. E. Schwab, J. L. Raucy, and E. F. Johnson, *Mol. Pharmacol., 33*, 493–499 (1988).
51. Y.-F. Ueng, T. Kuwabara, Y.-J. Chun, and F. P. Guengerich, *Biochemistry, 36*, 370–381 (1997).
52. M. Shou, J. Grogan, J. A. Mancewicz, K. W. Krausz, F. J. Gonzalez, H. V. Gelboin, and K. R. Korzekwa, *Biochemistry, 33*, 6450–6455 (1994).
53. N. A. Hosea, G. P. Miller, and F. P. Guengerich, *Biochemistry, 39*, 5929–5939 (2000).
54. M. Shou, Q. Mei, M. W. Ettore, Jr., R. Dai, T. A. Baillie, and T. H. Rushmore, *Biochem. J., 340*, 845–853 (1999).

55. T. L. Domanski, J. Liu, G. R. Harlow, and J. R. Halpert, *Arch. Biochem. Biophys.*, *350*, 223–232 (1998).
56. T. L. Domanski, Y.-A. He, G. R. Harlow, and J. R. Halpert, *J. Pharmacol. Exp. Ther.*, *293*, 585–591 (2000).
57. G. R. Harlow, and J. R. Halpert, *Proc. Nat. Acad. Sci. USA*, *95*, 6636–6641 (1998).
58. M. H. Gelb, D. C. Heimbrook, P. Malkonen, and S. G. Sligar, *Biochemistry*, *21*, 370–377 (1982).
59. J. A. Krauser, and F. P. Guengerich, *J. Biol. Chem.*, *280*, 19496–19506 (2005).
60. C. Mitoma, D. Yasuda, J. Tagg, and M. Tanabe, *Biochim. Biophys. Acta*, *136*, 566–567 (1967).
61. E. E. Scott, and J. R. Halpert, *Trends Biochem. Sci.*, *30*, 5–7 (2005).
62. H. Fujii, X. Zhang, T. Tomita, M. Ikeda-Saito, and T. Yoshida, *J. Am. Chem. Soc.*, *123*, 6475–6484 (2001).
63. M. Matsu-Ura, F. Tani, S. Nakayama, N. Nakamura, and Y. Naruta, *Angew. Chem., Int. Ed. Engl.*, *39*, 1989–1991 (2000).
64. F. Tani, M. Matsu-ura, S. Nakayama, M. Ichimura, N. Nakamura, and Y. Naruta, *J. Am. Chem. Soc.*, *123*, 1133–1142 (2001).
65. T. Matsui, S. Nagano, K. Ishimori, Y. Watanabe, and I. Morishima, *Biochemistry*, *35*, 13118–13124 (1996).
66. D. P. Hildebrand, J. C. Ferrer, H.-L. Tang, M. Smith, and A. G. Mauk, *Biochemistry*, *34*, 11598–11605 (1995).
67. S.-i. Ozaki, and P. R. Ortiz de Montellano, *J. Am. Chem. Soc.*, *117*, 7056–7064 (1995).
68. S. I. Rao, A. Wilks, and P. R. Ortiz de Montellano, *J. Biol. Chem.*, *268*, 803–809 (1993).
69. D. C. Levinger, J. A. Stevenson, and L. K. Wong, *J. Chem. Soc. Chem. Comm.*, 2305–2306 (1995).
70. S. Ozaki, I. Hara, T. Matsui, and Y. Watanabe, *Biochemistry*, *40*, 1044–1052 (2001).
71. V. P. Miller, G. D. DePillis, J. C. Ferrer, A. G. Mauk, and P. R. Ortiz de Montellano, *J. Biol. Chem.*, *267*, 8936–8942 (1992).
72. R. Z. Harris, S. L. Newmyer, and P. R. Ortiz de Montellano, *J. Biol. Chem.*, *268*, 1637–1645 (1993).
73. T. D. Pfister, T. Ohki, T. Ueno, I. Hara, S. Adachi, Y. Makino, N. Ueyama, Y. Lu, and Y. Watanabe, *J. Biol. Chem.*, *280*, 12858–12866 (2005).
74. B. R. Baer, J. T. Schuman, A. P. Campbell, M. J. Cheesman, M. Nakano, N. Moguilevsky, K. L. Kunze, and A. E. Rettie, *Biochemistry*, *44*, 13914–13920 (2005).

10

Interactions of Cytochrome P450 with Nitric Oxide and Related Ligands

Andrew W. Munro, Kirsty J. McLean, and Hazel M. Girvan

Manchester Interdisciplinary Biocentre, School of Chemical Engineering and Analytical Science University of Manchester, 131 Princess Street, Manchester M1 7DN, UK
<andrew.munro@manchester.ac.uk>

1. INTRODUCTION. INTERACTIONS OF LIGANDS AND SUBSTRATES WITH P450 ENZYMES: GENERAL FEATURES	286
2. NITRIC OXIDE AND ITS INTERACTIONS WITH P450s	289
2.1. Synthesis and Biology of Nitric Oxide. Comparisons between Nitric Oxide Synthase and P450/CPR Enzymes	290
2.2. Binding of Nitric Oxide to P450s, Spectroscopic Analysis, and Relevance to Catalysis	293
2.3. Cytochrome P450 nor. A P450 Designed to Metabolize Nitric Oxide	295
2.4. Nitration and Nitrosylation of Cytochrome P450 Redox Systems by Reactive Nitrogen Oxides and NO-Dependent Inhibition of P450s	298
3. INTERACTIONS OF IMIDAZOLES AND SUBSTITUTED IMIDAZOLES WITH P450s	300
3.1. Azoles as Inhibitors of P450s in Laboratory and Clinical Uses	300
3.2. Spectroscopic Analysis of Azole Drug Binding to P450s	302
3.3. Structural Studies of Azole Drug Binding to P450s	303
4. OTHER LIGANDS AND INHIBITORS OF P450 FUNCTION	305
4.1. Carbon Monoxide and the Characterization of P450s	305
4.2. Pyridines, Cyanides, and Other Reversible Inhibitors of P450s	308
4.3. Irreversible Binding of P450 Inhibitors	309

5. CONCLUSIONS AND FUTURE PROSPECTS	310
ACKNOWLEDGMENTS	310
ABBREVIATIONS	311
REFERENCES	311

1. INTRODUCTION. INTERACTIONS OF LIGANDS AND SUBSTRATES WITH P450 ENZYMES: GENERAL FEATURES

The cytochromes P450 are a superfamily of heme *b*-containing oxygenases responsible for a vast array of chemical reactions in organisms from all of the domains of life [1]. Their activities include dehydrogenations, epoxidations, and reductions, but they are most famous for hydroxylations on unactivated carbon centers [2]. Their capacity to perform regio- and stereoselective oxygenations on a wide range of metabolically and biotechnologically important molecules has encouraged scientists from different disciplines to utilize various P450s for transformations relevant to, e.g., synthesis of antibiotics or chiral synthons [3,4]. Moreover, several P450s have been extensively engineered to facilitate the enhancements of desired activities, or the alteration of their activity profiles in favor of novel substrates [5]. Atomic structures of a number of P450s are now available, and both rational and forced evolution approaches have been taken to alter the activities of a number of P450s to enable their biotechnological exploitation [6,7].

The P450s' activities in oxygen activation rely critically on the nature of the heme ligation in the active site. The P450s contain non-covalently bound heme, except in the cases of some CYP4 family P450s, where turnover-dependent linkage of the 5-methyl group of the heme macrocycle to a conserved glutamate residue has been shown to occur (e.g., [8]). In the resting state, the heme iron is in the ferric (Fe^{3+}) form, and is coordinated axially by a conserved cysteine residue (proximal position) and by a water molecule in the distal position. Removal of the water is known to favour conversion of the heme iron from a low spin ($S = 1/2$) to a high-spin ($S = 5/2$) form in several P450s, and to cause an elevation of the reduction potential of the heme iron by up to $\sim 140\,\text{mV}$ (e.g., [9,10]). This occurs on binding of substrate in the active site of many P450s and may act as a 'trigger' for electron transfer from the redox partner (at least in some of the 'better regulated' P450s) and serve to prevent non-specific reactions such as production of superoxide and hydrogen peroxide [11]. The departure of the axial water also creates a binding site for the substrate dioxygen, and this ligand will bind to the ferrous heme iron following delivery of the first electron from the redox partner, to form a ferric superoxy species [12]. Subsequent steps in the catalytic cycle leading to substrate oxygenation are (a) the delivery of a second electron from the redox partner (which is typically a flavoprotein or iron sulfur protein) to form a ferric peroxy intermediate, (b) protonation of the peroxy iron species formed, leading to formation of a ferric hydroperoxy intermediate, (c) further proton delivery leading to formation of a ferryl-oxo species and ejection of a molecule of water, (d) attack of the ferryl-oxo species on the closely bound substrate and formation of oxygenated product, and (e) dissociation of the oxygenated (frequently hydroxylated) product and restoration of the ferric resting state of the enzyme [13].

Central features of P450-mediated catalysis are the presence of the conserved cysteine in the proximal position on the heme iron, and the exchangeable water molecule on the distal side, which is replaced by dioxygen during the catalytic cycle. The relatively weak binding of dioxygen to the P450 heme iron means that it is readily replaceable by stronger ligands – and this has proven important in the context of inhibition of certain P450s by heme coordinating molecules. This has been exploited both biomedically (in the context of inhibitors of key P450s in human therapy and antifungal antibiotic treatment) and for exploration of P450 structure and function [14]. Indeed, a number of the P450s for which we currently have atomic structural data have been crystallized in the presence of azole-based drugs, which are well known inhibitors of P450 function (see Section 3) (e.g., [15]). Thus, a key difference between the mode of binding of substrates and inhibitors to the P450s is that the former occupy positions close to the heme iron in readiness for oxidative attack by the ferryl-oxo form of the P450 heme iron; but do not physically bind to the heme iron. The majority of the latter coordinate the heme iron to inactivate redox function of the P450, and competitively inhibit binding of the substrate dioxygen to the 6th (axial) position on the iron [14]. However, it should be noted that inhibitors of P450 function that do not coordinate the heme iron have also been developed. These include the acetylene-based suicide inhibitors, such as terminal acetylenes that alkylate the heme in, e.g., fatty acid hydroxylase P450s (see Section 4.3), and C-19 methyl substituted steroids that act as competitive inhibitors of the aromatase P450 involved in estrogen synthesis [14].

Among the small molecule heme-coordinating inhibitors of P450s, imidazole, carbon monoxide, cyanide, and nitrogen monoxide (nitric oxide) are some of the best known. A convenient marker of the binding of these ligands to the P450s is their effect on the UV-visible absorption spectrum of the P450. Many characterized prokaryotic P450s are predominantly low-spin in the absence of substrate, whereas several human and other eukaryotic P450s have significant contents of high-spin heme iron, even in the absence of substrate. The P450 heme has its major (Soret) band located at $\sim 418\,nm$ for the low-spin form, and at $\sim 390\,nm$ for the high-spin form. For species with mixed spin-state, the Soret band is broadened and contains both features. The ligation of the

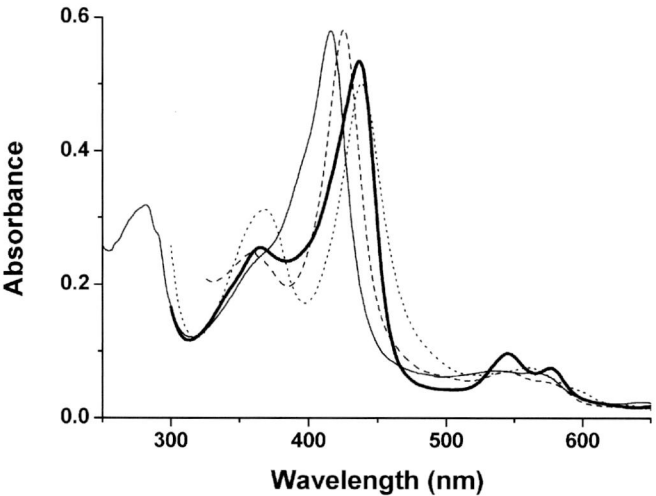

Figure 1. Spectral changes occurring on coordination of P450 heme iron by NO and other ligands. Spectral characteristics are shown for the ferric form of *Mycobacterium tuberculosis* CYP121 (approx. 5.5 μM) in its oxidized, ligand-free form (thin solid line) and on binding to NO (Fe^{3+}-NO complex, thick solid line), cyanide (500 mM, dotted line) and imidazole (1 M, dashed line). Ligand-bound spectra are corrected for any contributions from the ligands themselves. Soret maxima are at 416.5, 437, 438, and 425 nm, respectively [82].

heme-coordinating inhibitors converts the spectrum to single states at saturation, with red shifts in the absorption maximum of the Soret band (Figure 1).

The mentioned convenient switches in absorption features are frequently used for determination of the binding constant (K_d, often referred to as K_s to indicate spectral binding constant) for the ligand. Imidazole itself (which coordinates the ferric iron via a nitrogen atom) is usually a rather weak ligand to the P450 heme iron, with K_d values of the order of several hundred μM or more. Coordination of the heme iron by imidazole shifts the P450 Soret maximum to ∼425 nm. Nonetheless, the affinity of imidazole for the heme iron is an important consideration in the development of purification schemes for expressed P450s, since use of imidazole to elute His-tagged enzymes will generate an inactivated form, and separation methods (e.g., extensive dialysis) will be required to remove the ligand. As discussed in Section 3.1, substituted imidazoles and triazoles can be substantially stronger inhibitors with binding constants in the very low μM or nM range. This enhanced binding is usually explicable in terms of increased affinity for the P450 active site brought about by the presence of bulky hydrophobic and other substituents attached to the coordinating azole group (e.g., [16]). Cyanide is also a relatively weak binder of P450 heme iron (millimolar K_d values are typical), and the affinity of cyanide for cytochrome oxidase (and other enzymes) is a major reason that cyanide and substituted cyanides are not seriously considered as therapeutics. The cyanide complex of P450s typically has its Soret maximum shifted to beyond 430 nm. As explained in Section 4.1, the binding of carbon monoxide exclusively to the ferrous form of P450 was important for the identification of this enzyme class and for its nomenclature. The Soret band of this complex is shifted to ~ 450 nm for the native form of the enzyme.

In recent years, there has been an explosion of interest in the diatomic gas nitric oxide (NO) due to the recognition of its importance in eukaryotic physiology as a mediator of neural transmission, immune function and vasodilation [17]. As discussed in Section 2.1 below, the class of heme enzymes that generates NO (nitric oxide synthase or NOS) is P450-like in its chemistry and heme iron coordination. The coordination of NO to P450 heme iron can occur to either the ferric or ferrous form, with Soret shifts to ~ 430–435 nm. The increasing understanding of the regulatory role of NO (the reaction product) on nitric oxide synthase and the realization that higher oxides of NO have the ability to post-translationally modify proteins and to alter their function has led to enhanced interest into how NO might interact with P450 enzymes (both covalently and non-covalently) to modify their activities (e.g., [18]).

The overall aim of this chapter is to examine the mechanism of interaction of P450 enzymes with NO and with other heme-coordinating or modifying inhibitor molecules. In so doing, we will describe how these interactions have become important in the characterization of the P450s, for the development of inhibitors (both for metabolic and antibiotic treatments), and for the determination of cytochrome P450 structure. In addition, we will compare and contrast the P450s with nitric oxide synthases – the P450-like enzymes designed to generate the heme coordinating inhibitor NO. Finally, we will look at the convenient spectroscopic methods for examining the interaction of the P450s with inhibitor molecules and for the determination of the affinities of these inhibitors for the P450 hemes. Collectively, this chapter will provide an overview of the most important features associated with the interactions of P450s with NO and other heme-coordinating ligands, and the relevance to physiology and understanding of P450 structure and mechanism.

2. NITRIC OXIDE AND ITS INTERACTIONS WITH P450s

As explained in the introduction section, interest in the interaction of nitric oxide with P450s has increased dramatically in recent years with the discovery and characterization of nitric oxide synthases as P450-like enzymes, and also with the more general realization of the importance of NO as a key regulator of cellular function (e.g., [19]). Moreover, P450s whose physiological role is in NO reduction have recently been characterized biochemically and structurally [20].

In this chapter we will examine what is currently known about NO's biochemical synthesis, physiological roles and interactions with hemoproteins and P450s.

2.1. Synthesis and Biology of Nitric Oxide. Comparisons between Nitric Oxide Synthase and P450/CPR Enzymes

In the cytochromes P450, the heme iron is coordinated proximally by the anionic form of a conserved cysteine (i.e., the thiolate form). This form of heme iron coordination appears to be that compatible with the monooxygenase catalytic activity of the P450s, and protonation of the thiolate (which has been shown to be reversible in a number of P450s) leads to inactivation of P450 function [21,22] (see also Section 4.1). Until the discovery of nitric oxide synthases, the P450s were considered to be the only oxygenase enzymes to bind heme iron via a proximal cysteinate [23]. The characterization of the nitric oxide synthases in the first half of the 1990's demonstrated conclusively (using both structural and spectroscopic methods) that these enzymes were not only P450-like in their oxygenase activity, but also bound heme b via a conserved cysteinate [24].

Until the mid to late 1980s, NO was considered only to be a product of bacterial metabolism and a pollutant chemical [23]. However, production of NO in eukaryotic cells was reported in the late 1980s and the realization of its global importance in physiology was soon made. In particular, NO is now known to bind to the heme iron of the cell signalling enzyme guanylate cyclase, inducing the breakage of its proximal His-Fe bond and activating the enzyme for the production of cyclic guanosine monophosphate (cGMP) from the substrate GTP [25]. The cGMP then initiates a cascade of signalling events through its activation of calcium fluxes in the cell.

The enzymes predominantly responsible for the production of NO in eukaryotes are the NOSs, a class of flavocytochromes constructed from the fusion of a diflavin reductase domain with a cysteinate-ligated heme b-containing oxygenase domain [26]. The reductase domain is evolutionarily related to cytochrome P450 reductase and to other known diflavin reductases (e.g., methionine synthase reductase and the cancer-related novel reductase 1) and catalyzes the dehydrogenation of NADPH and the transfer of electrons through FAD and FMN cofactors to the NOS heme domain [27]. Electron transfer within the reductase and from FMN-to-heme is controlled by the binding of calmodulin (CaM), which appears to release a conformational lock in the reductase [28]. Two major classes of mammalian NOS are recognized: the constitutive and inducible isoforms. The neuronal (nNOS) and endothelial (eNOS) enzymes are constitutive, and are regulated by changes in intracellular calcium levels, which affect the binding of the Ca^{2+}/CaM activator complex. The inducible class (iNOS) is characterized by the macrophage enzyme, which binds Ca^{2+}/CaM very tightly and is not subject to regulation by calcium levels, but is regulated by cytokines [17]. The heme domains of the NOS isoforms catalyze two P450-like monooxygenation reactions using electrons delivered from the reductase.

The substrate L-arginine is hydroxylated to N-hydroxy-L-arginine, and this product is used as the substrate for a second reaction, which generates L-citrulline and NO as products [17] (Figure 2). While a ferryl-oxo species (as in P450 catalysis) is considered to be that which hydroxylates L-arginine, the possibility that preceding intermediates (i.e., the ferric hydroperoxy or even the ferric superoxy) could participate in the second catalytic step (oxygenation of N-hydroxy-L-arginine) remains to be resolved [29]. Eukaryotic NOS reactions differ considerably from those of the P450s in that the second electron can be derived either from the reductase FMN or from the bound cofactor tetrahydrobiopterin (H_4B) [17]. The latter is not bound by P450s, but the nature of the H_4B-dependent activity is reminiscent of that of cytochrome b_5, which can act as an allosteric effector and may deliver the second electron to certain eukaryotic P450s [30].

Related NOS enzymes also exist in bacteria (e.g., *Bacillus subtilis*), but those bacterial NOS enzymes characterized to date do not have a fused reductase system. Electrons are delivered instead from separate redox partners – with flavodoxins/ferredoxins and their cognate NAD(P)H-dependent reductases being obvious candidates [31]. This type of bacterial redox system is shared by several so-called 'class I' P450s from bacteria and from mammalian adrenal mitochondria [32]. However, even the complex bi-domain construction of the eukaryotic NOS enzymes has a forerunner in the P450 world – with the well characterized P450 BM3 (CYP102A1) enzyme from *Bacillus megaterium* being the first enzyme shown to contain FAD, FMN and heme in a single polypeptide, and to be formed from the fusion of a diflavin reductase to a heme-thiolate protein [33].

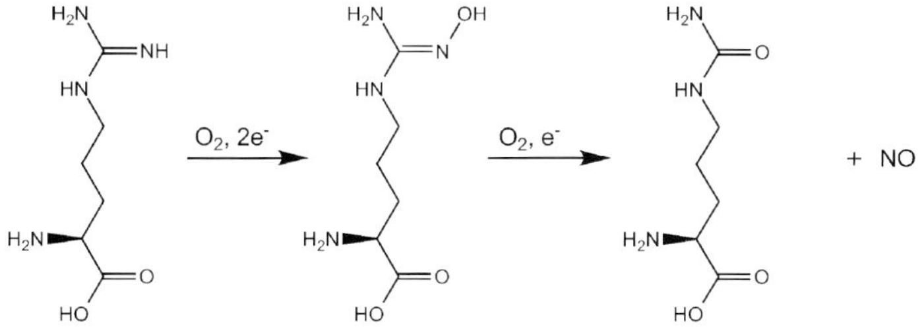

Figure 2. Catalytic reactions of nitric oxide synthase. Nitric oxide synthases catalyze the oxidation of L-arginine using NADPH as an electron donor and molecular oxygen for oxidation of the substrate [17]. The reaction takes place in two steps, involving first the N-hydroxylation of a guanidino nitrogen of L-arginine to form N-hydroxy-L-arginine, and then the further NOS-dependent oxidation of the product to form L-citrulline and NO. A net two electrons (one molecule of NADPH) are used for the first step, and one electron ($0.5 \times$ NADPH) for the second step. A molecule of dioxygen is used and a molecule of water produced at each step.

A number of P450 BM3-like enzymes have subsequently been identified in other microbes (e.g., [34]). Eukaryotic NOS enzymes have been shown to be functional in NO synthesis only in the dimeric form [35], and recent studies on P450 BM3 also indicate that dimeric status is essential for the fatty acid hydroxylase activity of this P450 [36].

Structures of various heme domains of NOS enzymes have been solved, and reveal a completely different fold to that seen for the P450s (e.g., [24,37]) (Figure 3). There is a more extensive content of β sheet in the NOS oxygenase domains, and they are dimeric. The structure of bacterial NOS isoforms are similar to those for their eukaryotic counterparts. Thus, convergent evolution appears to have resulted in similar solutions to oxygenase catalysis (using a thiolate-ligated heme center), but within a different structural framework. Indeed, the eukaryotic NOS enzymes and P450 BM3 (and its homologues) have taken a step further and fused diflavin reductases in order to enable efficient and controlled electron delivery to the heme iron [33]. Readers are directed to excellent recent reviews on the NOS family of enzymes for further details of their structure and mechanism (e.g., [17,38]).

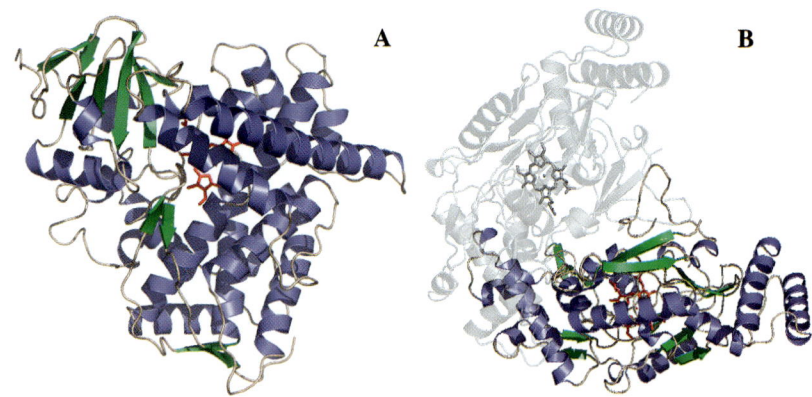

Figure 3. Structural comparison of cytochrome P450 and nitric oxide synthase. The P450s and NOS enzymes are heme thiolate proteins that catalyze similar oxygenation chemistry. However, their common reactivity appears to be the product of convergent evolution, since the three dimensional structures of P450s and the oxygenase domains of NOS are markedly different. Panel A shows the atomic structure of the heme domain of flavocytochrome P450 BM3. The protein structure resembles a triagonal prism with mainly helical structure (PDB code 2HPD) [137]. Panel B shows the dimeric structure of the oxygenase domain of rat neuronal NOS (PDB code 1LZX) [138]. For clarity, only one of the molecules in the dimer is colored. NOS adopts a more extended structure than seen in the P450s, with a less deeply buried heme. Helical and sheet segments are shown in blue and green, respectively, with the heme macrocycle presented in red.

The fact that the NOS enzymes generate a potent inhibitory ligand for the heme iron as their reaction product raises intriguing questions with respect to the regulation of their catalysis. The first product observed in NOS catalysis is a ferric heme-NO adduct (rather than free NO) [17]. This is clearly an inactive species with respect to further *de novo* NO synthesis, and most NO formed by NOS catalysis may first form a Fe^{3+}-NO adduct prior to exiting from the enzyme and returning the enzyme to the resting state. This is a productive pathway. However, an alternative (futile) pathway exists by which the Fe^{3+}-NO form is reduced by a single electron from the NOS reductase, producing a stable Fe^{2+}-NO form. NO release from this species is slow, and ultimately the species will likely collapse (re-forming the starting Fe^{3+} form of NOS) following reaction with dioxygen and non-productive release of nitrate [39]. The NOS enzymes appear to have evolved to maximize NO production at the expense of the futile pathway. A major feature in this adaptation is having a slow rate of heme reduction from the reductase, to enable partitioning of the reaction pathways in favor of NO production. In the productive cycle, the first electron is delivered to the NOS heme from the reductase FMN, and the 2nd electron is derived from H_4B at up to 30-fold the rate of the reductase-dependent electron transfer [40]. This ensures that collapse of the ferric superoxy form is minimized in the productive cycle and that NO formation is efficient. However, the slow FMN-dependent reduction of the Fe^{3+}-NO complex retards the futile cycle and maximizes NO product release [17]. The intricate interplay between NO as a product and as an inhibitor for NOS enzymes also has interesting connotations for its interactions with P450 enzymes.

2.2. Binding of Nitric Oxide to P450s, Spectroscopic Analysis, and Relevance to Catalysis

In similar fashion to the NOS enzymes, P450s are able to bind NO to both the ferric and ferrous forms of the heme iron. The ferrous complexes of P450s are relatively unstable (at least in some P450 isoforms, see below) and may be subject to attack by molecular oxygen [18]. The ferric complexes are rather more stable in the presence of oxygen, but can decay (typically over periods of minutes to hours) due to oxidation of free NO to nitrite in aerobic solution [41]. Analysis of the UV-visible spectral properties of the Fe^{3+}-NO complexes of various P450s indicates that their Soret maxima are shifted to $\sim$ 430–435 nm. The spectra of these complexes are also characterized by increased intensity of absorption in the visible (α and β band) region – with development of features at $\sim$ 575 and 543 nm in the case of the P450 BM3 enzyme [18] (Table 1). EPR spectroscopy indicates that the Fe^{3+}-NO complexes of both P450 BM3 (CYP102A1) and the *Pseudomonas putida* camphor hydroxylase P450 cam (CYP101A1) are essentially EPR-silent, suggesting that a spin-paired system exists, i.e., that the

Table 1. Electronic absorption properties for P450 and NOS enzymes in their complexes with nitric oxide. The spectral absorption maxima (Soret and visible region bands) for the Fe^{3+}-NO and Fe^{2+}-NO complexes of various P450 enzymes and NOS isoforms are given. Data are compiled from [18,42,60,97,135,136].

Heme-thiolate enzyme	Fe^{3+}-NO		Fe^{2+}-NO	
	Soret	Visible	Soret	Visible
CYP102A1	435	575/543	437	559
CYP101A1	430	571/541	438	557
CYP2B4	433	575/543	—	—
CYP1A2	431	571/541	400	552
CYP55A1	431	571/539	434	558
nNOS	440	580/549	436	567
iNOS	443	585/549	436	560

unpaired electron of the heme iron is coupled to that of the bound NO [42]. However, the Fe^{2+}-NO complexes of the same P450s are spectrally isolatable under anaerobic conditions, and have Soret bands at 438 nm (P450 cam) and 437 nm (P450 BM3), with single absorption maxima (at 559 nm for P450 BM3 and 557 nm for P450 cam) in the visible region [18,42]. The Soret absorption bands for the Fe^{2+}-NO complexes of P450s are of lower intensity than those of the Fe^{3+}-NO complexes, or of the resting ferric species (e.g., coefficients of $\sim 95\,mM^{-1}\,cm^{-1}$ for ferric P450 BM3 heme domain, $86\,mM^{-1}\,cm^{-1}$ for the Fe^{3+}-NO complex and $63\,mM^{-1}\,cm^{-1}$ for the Fe^{2+}-NO complex) [18].

The EPR active Fe^{2+}-NO complex of P450 cam was obtained both in presence and absence of camphor substrate, and showed a rhombic spectrum centered at $g = \sim 2.01$, with a three-fold splitting of the high-field g-value [42]. Such triple hyperfine structure has also been observed in other cytochromes, such as chloroperoxidase and lactoperoxidase [43,44]. The EPR spectrum for the Fe^{2+}-NO complex of P450 BM3 heme domain showed similar features, but with less well-defined hyperfine splitting [18]. This might reflect heterogeneity in the conformation of the Fe^{2+}-NO species. Both the P450 BM3 and P450 cam Fe^{2+}-NO complexes gave EPR spectra indicative of the integrity of the cysteinate linkage *trans* to the NO (i.e., the P450 form of the enzymes). However, the P450 BM3 Fe^{2+}-NO complex was substantially more susceptible to oxidation than the respective P450 cam complex and had to be prepared under strictly anaerobic conditions [18]. The Fe^{2+}-NO complexes of certain eukaryotic P450s may be more prone to breakage of the cysteinate linkage. For instance, the Fe^{2+}-NO complex of rat liver microsomal P450 samples showed time-dependent conversion to the P420 form (see Section 4.1 below for discussion of the P450/P420 equilibrium) and freeze quenching of the complex immediately on preparation was required to observe the EPR signal from the true hexacoordinated form [42].

By contrast, the Fe^{2+}-NO complex of bovine P450scc (CYP11A1) was found to be stable for several hours on ice, and was further stabilized on binding the substrate cholesterol [45].

In P450 BM3, the Fe^{2+}-NO complex is substantially stabilized in a F393H mutant form [18]. The phylogenetically conserved residue phenylalanine 393 interacts with the Fe-Cys bond on the proximal face of the heme and F393H (and other mutants) also stabilize the ferrous oxy species, and have profound effects on the heme iron reduction potential [46]. In neuronal NOS, tryptophan 409 forms a hydrogen bond with the cysteine thiolate ligand, and W409F/Y mutants have increased catalytic activity that is apparently due to faster aerobic decay of the Fe^{2+}-NO complexes [47]. The UV-visible absorption spectra of the W409F/Y mutant Fe^{2+}-NO complexes show Soret bands shifted to 406.5 nm from 436 nm for the wild-type nNOS [48]. This occurs due to loss of cysteinate coordination and formation of a 5-coordinate NO complex, as evidenced also from resonance Raman (RR) analysis [48,49]. Evidently, conserved aromatic residues that interact with the cysteine thiolate in both NOS and P450 enzymes have important roles to play in stabilization of the Fe-Cys bond and of the oxy and nitrosyl complexes. On the distal side of the CYP1A2 P450 heme, the acidic residue glutamate 318 (conserved as a Glu or Asp residue in most P450s) also impacts on NO coordination of the heme iron. In wild-type CYP1A2 the Soret band for the Fe^{3+}-NO complex is located at 431 nm. However, in the E318A mutant this Fe^{3+}-NO complex collapses rapidly to give a 5-coordinate NO-bound ferrous species with a maximum at 400 nm [50].

2.3. Cytochrome P450 nor. A P450 Designed to Metabolize Nitric Oxide

In the early 1990s a novel type of P450 was characterized by Shoun and Tanimoto [51]. P450 nor (CYP55A1) from *Fusarium oxysporum* was shown to catalyze reduction of nitrogen monoxide to dinitrogen oxide according to the following scheme:

$$2NO + NADH + H^+ \rightarrow N_2O + NAD^+ + H_2O$$

While reductive reactions have been recognized to be feasible in certain other oxygenase P450s (e.g., reductive dehalogenation and reduction of hydrazines under low oxygen tension conditions), P450 nor is clearly an enzyme that has evolved to perform reductive rather than oxygenase catalysis [20,52]. P450 nor was shown to interact directly with NADH, bypassing the requirement for redox partners to facilitate electron delivery. Spectroscopic analysis (including EPR) demonstrated that the resting enzyme had a cysteinate-coordinated ferric heme iron that was a mixture of low-spin and high-spin forms at ambient temperature, and that the Fe^{2+}-CO complex had its Soret maximum at 448 nm, consistent with a *bona fide* P450 enzyme [53]. NADPH was shown to be far less efficient than NADH as a coenzyme for NO reduction, or heme reduction in the presence of CO [20]. P450 nor-like enzymes have now been cloned and expressed from other fungi, suggesting that this activity is widespread in these eukaryotes, and that the mechanism of NO reduction is distinct from that observed for the bacterial membranous heme *b*- and *c*-containing NO reductases (e.g., [54]). Spectroscopic analysis (IR, resonance Raman and EXAFS) indicates tighter binding of NO to ferric P450 nor than to, e.g., P450 cam, with a substantially shorter Fe-N bond distance [55].

Mitochondrial and cytosolic forms of the *F. oxysporum* P450 nor appear to occur, distinguished by different modifications at the N-terminus [56]. The longer isoform was co-isolated with mitochondria (P450 norA), while the other isoform was cytosolic. The *F. oxysporum* P450 nor is induced by nitrite/nitrate and participates in an energy generating pathway for reduction of these molecules to N_2O. The pathway may be important under conditions of oxygen limitation, which explained its presence in mitochondria and helped to clarify how toxic effects of oxidative decay of NO and the NO complexes of P450 nor can be minimized. P450 nor's reductive removal of NO from mitochondria also prevents its action as an inhibitor of the fungal respiratory chain [57]. In the fungus *Cylindrocarpon tonkinese*, two separate genes (CYP55A2 and A3) were found to code for distinct mitochondrial and cytosolic isoforms. In this case, the cytosolic form had higher affinity for NADPH, suggesting that detoxification of NO was also a requirement in the cytoplasm, and possibly that this isoform could also participate in $NADP^+$ formation to fuel the pentose phosphate pathway [58]. In a further member of the CYP55 family (CYP55A4 from *Trichosporon cutaneum*), the gene was found to be induced by dioxygen, suggesting a further adaptive response to avoiding toxicity from NO oxidation [59].

In deconvoluting mechanistic aspects of the P450 nor reaction, identification of distinctive spectral intermediates has proved important. The Fe^{3+}-NO complex has its Soret maximum at 431 nm, but on stopped-flow mixing of the complex with NADH an intermediate with Soret peak at 444 nm appears [60]. The intermediate is considered to be a Fe^{2+}-NOH species, which accumulates following hydride transfer from NADH. Evidence in favor of this conclusion comes also from the fact that the spectral intermediate is formed when sodium borohydride is used as a surrogate hydride donor, and when the ferric P450 nor is exposed to hydroxylamine radicals [61]. The ligand-free structure of *F. oxysporum* P450 nor showed that the distal heme pocket is open to solvent and is an obvious binding site for NADH. Residues Ser 286, Thr 243 and Asp 393 and an active site water define a hydrogen-bonding network to the surface of the protein, and could be involved in proton relay to the heme iron (Figure 4). In the structure of the Fe^{2+}-CO complex of P450 nor (where the Fe^{2+}-CO system is isoelectronic and isosteric with the catalytically relevant Fe^{3+}-NO complex), a water molecule is located 3 Å from the CO ligand and forms a hydrogen bond with the hydroxyl

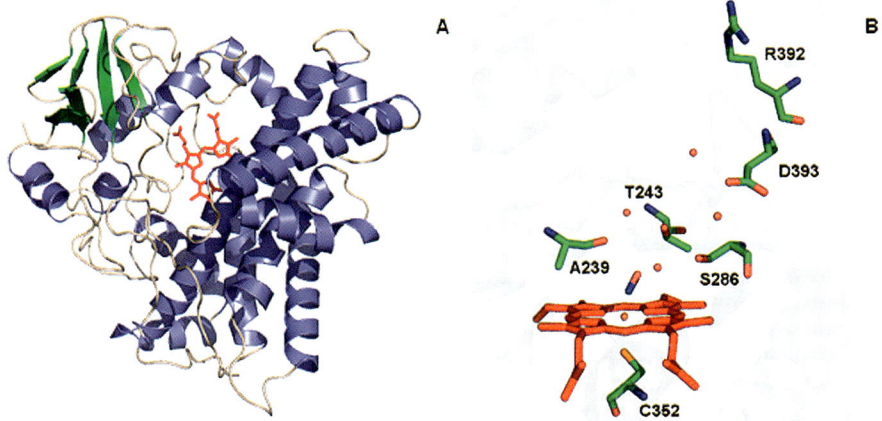

Figure 4. Atomic structural features of P450 nor. Panel A shows the typical P450 fold adopted by the *Fusarium oxysporum* nitric oxide reductase P450 nor (CYP55A1) [62]. The P450 is predominantly α helical (blue) with a smaller β sheet-rich domain (green). The heme (red) is sandwiched between the α and β domains. Panel B shows the active site structure of the NO complex of P450 nor, with Cys 352 being the proximal ligand to the heme iron (PDB code 1CL6). On the distal side of the iron, the heme pocket is relatively open to solvent and residues Ser 286, Thr 243, and Asp 393 (together with active site water) create a hydrogen bonded network to the protein surface and are proposed to participate in a proton delivery pathway during NO reduction. Arg 392 is part of a positively charged cluster that likely defines a NADH-binding site [20].

of Ser 286, which is in turn linked to another water and on to Asp 393 [62]. A scheme for the catalytic process involves protonations after hydride transfer from NADH, followed by attack of a second NO molecule to facilitate breakage of the Fe-N bond and generation of N_2O and water as products, as follows:

$$Fe^{3+} + NO \xrightarrow{NADH+H^+} Fe^{3+}[NOH^-] + H^+ \xrightarrow{NO} Fe^{3+} + N_2O + H_2O$$

Kinetic analysis of the P450 nor reaction indicate that hydride transfer is rate-limiting at high NO concentration, but that reaction with the second molecule of NO is likely rate-limiting under physiological conditions, and that the affinity of NO in this step is rather low ($K_d \sim 600\,\mu M$) [20].

Structural comparison of P450 nor with other well characterized prokaryotic P450s emphasizes the more exposed nature of the heme pocket in P450 nor and indicates that interactions between the conserved F/G and I helix regions are key to this structural perturbation [62]. In the heme distal pocket of P450 nor, a number of positive residues replace hydrophobic amino acids found in other P450s (notably, lysines 62, 64, and 291, and arginine 392). Structural and mutagenesis studies are consistent with this positively charged region forming a NADH binding site above the plane of the heme [63,64]. Hydride transfer concomitant with proton relay involving (at least) Ser 286 and water molecules is thought to facilitate catalysis, and the apparent distortion from linearity of the Fe-N-O axis observed from spectroscopic analysis may be an important factor in increasing the electrophilic character of the nitrogen to enable effective NO reduction [20].

2.4. Nitration and Nitrosylation of Cytochrome P450 Redox Systems by Reactive Nitrogen Oxides and NO-Dependent Inhibition of P450s

The sphere of interest in NO and its biology has extended in recent years with the realization that important modifications of proteins can be produced by NO and products derived from it *in vivo*. NO has an unpaired electron and will react with other radicals. Its reaction with O_2 leads to production of nitrogen dioxide (NO_2), which can then react with another molecule of NO to form dinitrogen trioxide (N_2O_3). Both NO_2 and N_2O_3 are stronger oxidants than NO itself, and N_2O_3 can react with water to form nitrite (NO_2^-), another powerful oxidant [65]. The reaction of NO with superoxide ($O_2^{\bullet-}$) produces peroxynitrite ($ONOO^-$) at a faster rate than superoxide dismutase can remove superoxide from solution. Peroxynitrite is another powerful oxidant that is recognized to modify several proteins and which has received much attention as the mediator of many of the pathological effects of NO [66]. A number of other oxides of NO may be relevant to cellular physiology and pathology [67].

The NO-dependent modification of proteins by tyrosine nitration and by sulfhydryl (cysteine) nitrosylation has been of particular interest. S-nitrosylation effected by NO is a widely recognized protein modification and may present the major route for NO's control over cellular signalling pathways, as well as a basis for control mediated by cellular redox status [68]. Studies of the glycolytic enzyme glyceraldehyde-3-phosphate dehydrogenase (GAPDH) have shown that NO-dependent nitrosylation (at Cys 149) results in covalent attachment of NAD(H) substrate – probably facilitated by transnitrosation of the active site S-nitrosothiol to the nicotinamide ring, followed by attack of the protein thiolate on the modified bound coenzyme [69]. More recently, the nitrosylation of GAPDH has been shown to be of importance in apoptosis. S-nitrosylated GAPDH binds to and stabilizes an E3 ubiquitin ligase (Siah1), and the complex translocates to the nucleus to degrade nuclear proteins [70]. While little information is yet available relating to regulation of specific P450s by nitrosylation, a proteomics study of nitrated proteins in NO donor-treated murine cells showed NADPH-cytochrome P450 reductase to be a major target of S-nitrosylation [71].

As discussed above, NO plays an important auto-regulatory role through binding the heme iron in NOS isoforms [18], and further important physiological and regulatory roles are known to be effected through its coordination of heme iron in other hemoproteins. For instance, NO competes with oxygen to reversibly inhibit mitochondrial cytochrome *c* oxidase [66] and binds to the ferrous heme iron of soluble guanylate cyclase to release the axial histidine ligand and thus activate cyclic GMP production [72]. NO binding to heme iron of P450s facilitates transient inhibition of the P450 BM3 enzyme, and similar reversible inhibition of mammalian CYP1A1 and CYP2B1 enzymes is also seen [18]. However, some irreversible inactivation of CYP2B1 was also observed [73]. It has now been established that nitration of Tyr 190 in CYP2B1 (readily achieved by treatment with peroxynitrite) leads to enzyme inactivation, and that this conserved residue interacts with Glu 149 to play an important role in enzyme stabilization [74]. Prostacyclin synthase (CYP8A1) is also inhibited by peroxynitrite-mediated tyrosine nitration [75]. In rat CYP4A1 and CYP4A3, peroxynitrite also nitrates tyrosine residues and inhibits production of 20-hydroxyeicosatetraenoic acid (20-HETE), a key mediator of renal vasoconstriction [76]. Similar phenomena have also been observed in the P450 BM3 and P450 cam systems [77,78]. In P450 BM3, peroxynitrite treatment leads to enzyme inactivation and modifications of the active site residues Tyr 334 and (in an F87Y mutant) Y87. However, loss of fatty acid hydroxylase function in this P450/P450 reductase fusion enzyme actually appears to result from oxidation of a thiol in the reductase domain [77]. For both P450 BM3 and mammalian CYP2B4, the aerobic reaction of the P450s with NO resulted in progressive nitration of one or more tyrosine residues (including Tyr 51 in P450 BM3) that was readily detected using resonance Raman spectroscopy and by following the development of prominent signals associated with nitro stretching and aromatic ring breathing vibrations of the nitrotyrosine(s) formed [18,79].

While several mammalian P450s are now recognized to be post-translationally modified by glycosylation or phosphorylation, the nitration field is still relatively young and it is likely that several more P450 isoforms will be found to be modified by NO and its oxides in the coming years [80]. In particular, the importance of NO and its control over energy generating pathways in the mitochondrion is becoming increasingly recognized [81]. Peroxynitrite is able to inhibit each of the four complexes of the mitochondrial respiratory chain (as well as other mitochondrial proteins), but recent studies suggest that mitochondrial protein nitration may be reversible through the action of an as yet uncharacterized denitratase enzyme [66,81]. The mitochondrial systems of nitration and denitration are dependent on NO production and responsive to oxygen levels. Thus, tyrosine nitration appears not to be a dead end and/or pathological modification for at least selected proteins, and may instead be a newly recognized and reversible means of enzyme regulation [81]. In the context of mitochondrial enzymes, this could clearly have major ramifications for the regulation of the steroid transforming P450 isoforms in adrenal mitochondria.

3. INTERACTIONS OF IMIDAZOLES AND SUBSTITUTED IMIDAZOLES WITH P450s

The azoles (typically imidazole- and triazole-based compounds) are the best known of the inhibitory ligands for P450 enzymes. They are frequently employed as ligands for P450s in studies of enzyme inhibition, for crystallography of P450s, and clinically as inhibitors of fungal growth and replication through inhibition of the sterol demethylase (CYP51) isoform. Here we contextualize the features associated with azole binding to P450 heme iron and the importance of these molecules for biomedical applications, as well as for structural and biophysical characterization of P450s.

3.1. Azoles as Inhibitors of P450s in Laboratory and Clinical Uses

The azole drugs coordinate predominantly to the ferric heme iron of P450s in the distal position and prevent reactivity with dioxygen. The displacement of the sixth axial ligand (i.e., water) results in a type II spectral change, typically shifting the Soret maximum to ~ 425 nm [14]. Imidazole itself is a hydrophilic molecule and is usually a rather weak inhibitor of P450s (K_d values often $> 100\,\mu M$). However, more hydrophobic derivatives of imidazole and triazole can be substantially tighter binding P450 inhibitors. Such molecules typically have higher affinity for the lipophilic regions of the P450 active site and may also benefit from stabilizing interactions between functional groups on the drug and specific amino acid sidechains [14]. Even the attachment of a relatively small apolar group can have huge effects on the binding to P450s. For instance, 4-phenylimidazole binds to *M. tuberculosis* CYP121 with a K_d of $\sim 30\,\mu M$, compared to $\sim 50\,mM$ for imidazole itself [82]. Various substituted imidazoles and triazoles have been developed as inhibitors of aromatase, the key P450 involved in transformation of androgen steroids to estrogens. Some of these have great potential in breast cancer treatment. For example, the aromatase inhibitor drugs letrozole and its forerunner fadrozole (Figure 5) have shown great promise in treatment of post-menopausal women suffering from advanced breast cancer (e.g., [14,83]). However, the best known and most widely clinically used azole-based P450 inhibitors are those which inactivate the fungal lanosterol 14α-demethylase (CYP51) enzymes.

Several azole- and triazole-based drugs have been used clinically as antifungals (Figure 5). Their effects are considered to be mediated through inhibition of fungal CYP51, which catalyzes the demethylation of lanosterol to ergosterol. CYP51 inhibition leads to accumulation of methylated sterols and alterations in membrane permeability that affect fungal cell viability [84]. The market for this type of drug is buoyant and a large number of different molecules have been

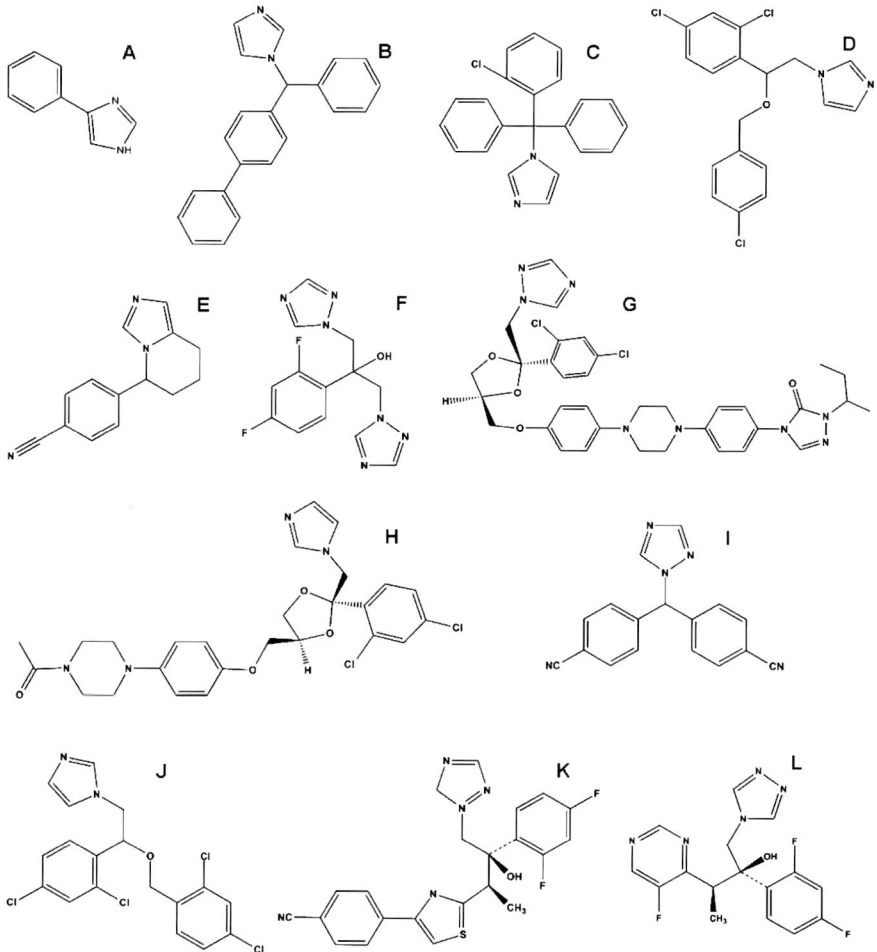

Figure 5. Imidazole- and triazole-based ligands used as P450 inhibitors and in structural analysis of P450 enzymes. Structures are depicted for 4-phenylimidazole (A), bifonazole (B), clotrimazole (C), econazole (D), fadrozole (E), fluconazole (F), itraconazole (G), ketoconazole (H), letrozole (I), miconazole (J), ravuconazole (K), and voriconazole (L). The molecules C, D, F, G and J-L are widely used as antifungals and/or have also been shown to bind tightly to *M. tuberculosis* P450s [90]. E and I are aromatase inhibitors developed as drugs for treatment of breast cancer [14]. Atomic structures of P450 enzymes have been solved in complex with A, B, F and H (e.g., [15,105]).

employed clinically. Many are for topical application to fungal skin infections, due to adverse reactions with human P450s. For instance, the leading antifungal ketoconazole inhibits hepatic CYP3A4 [85] and the C17-C20 lyase (CYP17) which catalyzes the 17α-hydroxylation of pregnenolone and progesterone, and then the cleavage of the C 20,21-acetyl group to produce the corresponding androgens [86]. Miconazole is also a potent inhibitor of CYP3A4 and a number of CYP2 family members (e.g., [87]). Both azole drugs inactivate sterol demethylase activity at nanomolar concentrations, but cross-reactivity with host P450s prevents their systemic use. However, later generation azoles that evade severe cross-reactivity with human P450 isoforms are now available. Among the best known of the systemically tolerated azoles is fluconazole, but others such as voriconazole and ravuconazole have also reached the clinic in recent years (e.g., [88]) (Figure 5). As with most other antibiotics, resistance to azoles is encountered widely – with typical mechanisms including increased efflux and mutations of the CYP51 to diminish azole binding [87].

However, an encouraging new avenue of application of the antifungal azoles has come with the realization that selected drugs bind avidly to P450s from other parasites and pathogens. For instance, disubstituted imidazoles have been shown to dramatically decrease parasitemia in *Trypanosoma cruzi*-infected mice, likely through inhibition of the parasite CYP51 [89]. In addition, tight binding of azoles to *M. tuberculosis* P450s (particularly CYP121) has raised hopes that azole-based drugs might be useful antibiotics to prevent the further spread of strains that have developed resistance to virtually all leading antitubercular agents [90].

3.2. Spectroscopic Analysis of Azole Drug Binding to P450s

The direct coordination of the ferric P450 heme iron by nitrogen atoms from azole/triazole drugs results in type II spectral shifts, as indicated above. The binding is reversible (but often very tight) and usually involves the interaction of a single molecule of the inhibitor with the P450. This binding phenomenon lends itself to determination of binding constants (K_d values), usually by plotting absorption changes (as the Soret red shifts on binding the azole) versus the drug concentration and fitting the data to a hyperbolic function (e.g., [91]). The binding of azole drugs and other ligands to P450 enzymes can also be characterized by other spectroscopic methods, including electron paramagnetic resonance (EPR), magnetic circular dichroism (MCD) and resonance Raman (RR) (e.g., [91–93]). EPR is perhaps the most readily applicable among these methods, and demonstrates that the azole complexes of P450s are low-spin ferric complexes with rhombic character. The g-values for several P450-azole complexes fall within the ranges of $g_z = 2.44–2.62$, $g_y = 2.25–2.28$ and $g_x = 1.83–1.93$ and may overlap those for the aqua-ligated ferric forms of the P450s. For comparison, the g-values for the aqua-ligated forms of P450 BM3 and P450 cam are 2.42, 2.26, and 1.92, and 2.45, 2.26, and 1.92, respectively. The g-values depend on both the nature of the axial ligands and their orientation to the heme iron, thus EPR spectroscopy does sometimes not provide a definitive indication of azole coordination to a ferric P450 heme iron, although a g_z value of > 2.5 might indicate a nitrogenous axial ligand. However, binding of azoles to P450s does convert them to a low-spin form, and this form is clearly distinguished from any high-spin azole-free enzyme (e.g., g-values of 8.0, 4.0 and 1.8 for high-spin P450 cam) [94].

Magnetic circular dichroism spectroscopy can provide a more diagnostic tool for establishing azole coordination to a P450, based on the absolute position of a porphyrin-to-ferric iron charge-transfer (CT) transition in the near infra red (NIR) region. While the contributions from the cysteinate axial ligand dominate the MCD spectrum (as with EPR), the wavelength of the NIR CT transition is dependent on the nature of the ligand, and alterations in ligand orientation affect mainly the intensity (and not position) of the transition. That said, in the case of P450 BM3 there is a relatively minor shift of the NIR CT band from ~ 1075 nm to ~ 1085 nm for wild-type, substrate-free (Cys-aqua coordinated) and the fatty acid-bound form of the A264E mutant, in which glutamate 264 replaces water as the distal axial ligand [91].

Resonance Raman is another commonly used tool in the characterization of P450s and their substrate/ligand complexes. The real strength of this method comes from its ability to assign P450 heme iron spin-state and oxidation state (e.g., [95]). Useful information is also forthcoming relating to, e.g., the geometry of the substituent groups (i.e., methyls, vinyls, and propionates). The technique is also useful with respect to diagnosis of the chemical nature of the proximal ligand in cytochromes [96]. In the case of P450s (and NOS enzymes), the Fe-Cys bond stretching modes are detected at ~ 350 cm^{-1} (distinct from those cytochromes with, e.g., histidine or imidazolate ligation at ~ 200–260 cm^{-1}). A distal ligand may also be identifiable by its stretching and/or bending modes. For instance, the binding of NO *trans* to the thiolate to both Fe^{3+} and Fe^{2+} forms of rat nNOS was demonstrated by detection of the Fe-NO stretching mode (ν_{Fe-NO}) at 540 cm^{-1} and 536 cm^{-1}, respectively. Addition of the substrate L-arginine to the Fe^{2+}-NO complex induced a further shift of the ν_{Fe-NO} band to 549 cm^{-1} [97]. In the P450 cam system, the ν_{Fe-NO} band for the Fe^{3+}-NO complex shifts from 528 cm^{-1} to 522 cm^{-1} in the presence of camphor, and to 524 cm^{-1} in the presence of norcamphor [98]. Confirmation of the identities of these and other modes is possible through isotopic substitution of the heme iron (^{54}Fe rather than ^{56}Fe) or of the distal or proximal ligands. For instance, ^{34}S or ^{15}N labelling for cysteine/histidine proximal heme ligands, or use of e.g., ^{15}N^{18}O or ^{13}C^{18}O for distal inhibitory ligands to a P450 heme iron [99].

3.3. Structural Studies of Azole Drug Binding to P450s

Recent years have seen the determination of several cytochrome P450 atomic structures, often in the presence of substrates and ligands. A number of P450 structures have been determined in complex with imidazole, phenylimidazoles, and (more recently) complex azole antifungals (Figure 6). These have provided important information relating to, e.g., the modes of binding of these drugs,

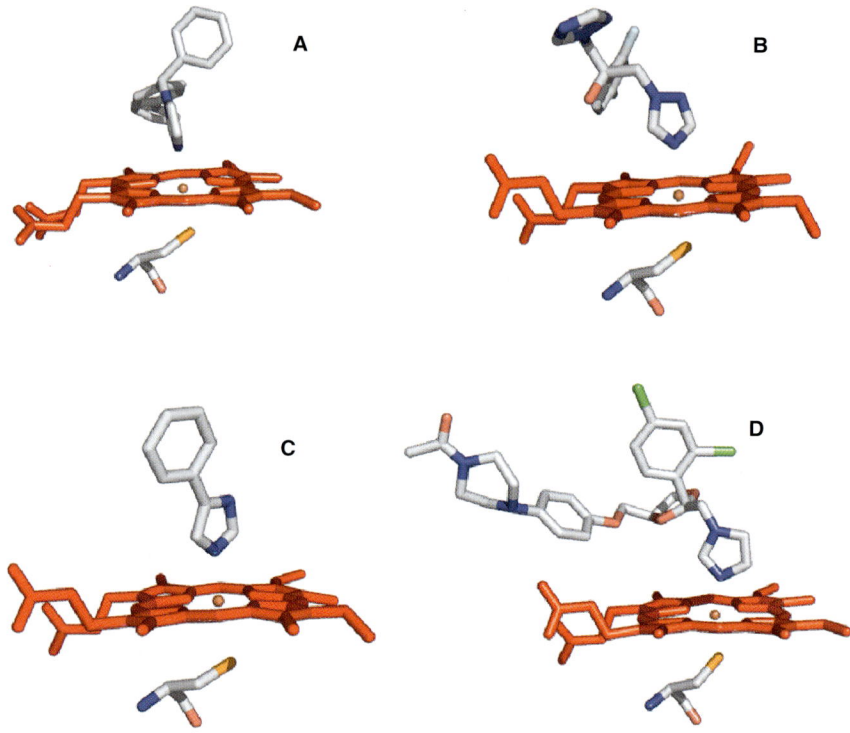

Figure 6. Structures of P450 heme-azole complexes. The heme macrocycles, cysteinate proximal ligands and azole distal ligands are shown from structurally resolved cytochrome P450 complexes. A: CYP2B4 bound to bifonazole (PDB code 2BDM [102]); B: CYP51 with fluconazole (1EA1 [15]); C: CYP119 with 4-phenylimidazole (1F4T [100]); D: P450 eryF with ketoconazole (1JIN [105]).

the geometry of the N-Fe-S system created, and (in the case of the bulkier azoles) active site interactions that enable tight binding interactions. The *Sulfolobus solfataricus* CYP119 structure has been resolved only in the presence of 4-phenylimidazole and imidazole [100]. Similarly, mammalian P450 CYP2B4 was solved with bifonazole and 4-chlorophenylimidazole coordinated to the heme iron, and CYP2C5 with the pyrazole derivative 4-methyl-N-methyl-N-(2-phenyl-2H-pyrazol-3-yl)benzenesulfonamide (DMZ) bound (although DMZ is a substrate and does not ligate the heme iron) [101–103]. *Saccharopolyspora erythraea* P450 eryF has been resolved in the ketoconazole-ligated state, and *Mycobacterium tuberculosis* CYP51 bound to both fluconazole and 4-phenylimidazole [15].

P450s are known to undergo considerable conformational alterations in solution, and the successes achieved in crystallization of several azole-P450 complexes may be in large part due to their occupancy of active site space and the stabilization of the P450. Restriction of conformational freedom in the proteins on binding the ligands may result in a limited set of conformers and a higher chance of nucleation. In the case of the CYP51 fluconazole-bound structure, fluconazole was exchanged into a 4-phenylimidazole-bound P450 crystal by soaking [15]. Distortions of the I helix (and adjacent G and H helices) and a helix-coil transition (in the C helix) occurred in response to binding of the bulkier azole antifungal. In CYP119, substantial differences were also observed between the imidazole and 4-phenylimidazole-bound structures. In the imidazole-bound form there is unwinding of the C-terminal end of the F helix, thus increasing the size of the F/G loop region and enabling stabilizing interactions between the loop and the imidazole ligand [100,104]. In the ketoconazole-bound P450 eryF structure, several protein side chain azole hydrophobic interactions and hydrogen bonds stabilize the binding of the drug. Ketoconazole is considerably larger than the natural substrate (the erythromycin precursor 6-deoxyerythronolide B, 6-DEB), and extends from the region occupied by 6-DEB into a water-filled region seen in the 6-DEB-bound structure, displacing several waters from the active site. As with CYP51 and fluconazole, P450 eryF protein backbone changes occur to accommodate the ketoconazole in the B and B' helices, and in the I helix [105].

Some deviations from linearity are noted in the geometry of coordination of the azole nitrogens with the heme iron in the P450-azole structures. However, due to differing levels of resolution in the atomic structures and refinement procedures used, it is not certain in some cases whether these deviations are actually caused by constraints imposed by the P450 active sites.

4. OTHER LIGANDS AND INHIBITORS OF P450 FUNCTION

4.1. Carbon Monoxide and the Characterization of P450s

The binding of carbon monoxide to P450s has a pivotal position in the history of the characterization of these enzymes and in their nomenclature. The spectroscopic properties of the hepatic isoforms, their hemoprotein nature, and their importance to drug metabolism were established through the formation of the Fe^{2+}-CO complex and the demonstration that loss of oxidative function accompanied complex formation. The dead end complex has its Soret maximum close to 450 nm for the native form of the enzymes [106–108]. CO binds exclusively to the ferrous form of the heme iron, and the shift of the Soret band to ~ 450 nm has become accepted as a characteristic of the cysteinate-ligated P450 enzymes.

Solubilization of liver microsomes with detergents resulted in a Soret shift to ~ 420 nm and the formation of the inactive P420 complex [107,108]. Several subsequent studies have demonstrated that similar spectral transitions and enzyme inactivation can be induced by e.g., addition of chaotropes or sulfhydryl reagents, or by application of high hydrostatic pressures (e.g., [109,110]). The P420 transition results from the loss of cysteine thiolate linkage to the heme iron, with simple protonation to the thiol able to produce the P420 form [21] (Figure 7).

The P450s were the first cysteinate-coordinated cytochromes characterized, but other types of so-called heme thiolate proteins have subsequently been discovered, most of which contain b-type heme [23]. The P450s and nitric oxide synthases catalyze oxygen activation, but the Fe^{2+}-CO complex of NOS enzymes is typically at $\sim$ 443–444 nm (rather than 450 nm). The extracellular

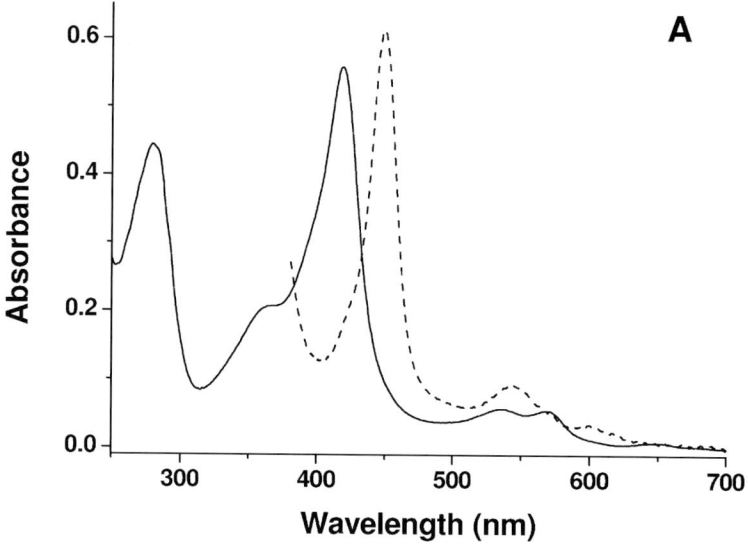

Figure 7. The cytochrome P450 and P420 Fe^{2+}-CO complexes in selected cytochrome P450 enzymes. Panel A shows the near complete spectral conversion from the oxidized (Fe^{3+}, solid line, Soret maximum at 418 nm) to the ferrous-CO (Fe^{2+}-CO, dashed line, Soret at 449 nm) form for the heme domain of *Bacillus megaterium* flavocytochrome P450 BM3 (CYP102A1, 5.8 μM). There is virtually no trace of the inactive P420 form of the Fe^{2+}-CO complex with Soret maximum close to 420 nm [91]. By contrast, the Fe^{2+}-CO P450 complex of the *Mycobacterium tuberculosis* CYP51 is highly unstable and collapses quickly to the P420 species (Panel B). The first spectrum (thin solid line) in panel B is collected immediately following formation of the Fe^{2+}-complex of CYP51 ($\sim$ 3 μM) and shows the presence of both the P450 form (Soret maximum at approx. 448.5 nm) and the P420 form (Soret maximum at 421.5 nm) in roughly equal proportions. Spectra shown as dashed lines are collected over the following 10 minutes and show the progressive formation of P420 at the expense of P450. The spectrum shown as a thick solid line is collected at 15 minutes and shows the complete formation of the P420 form [22]. Arrows indicate directions of absorption change during the P450/P420 transition.

Met. Ions Life Sci. **3**, 285–317 (2007)

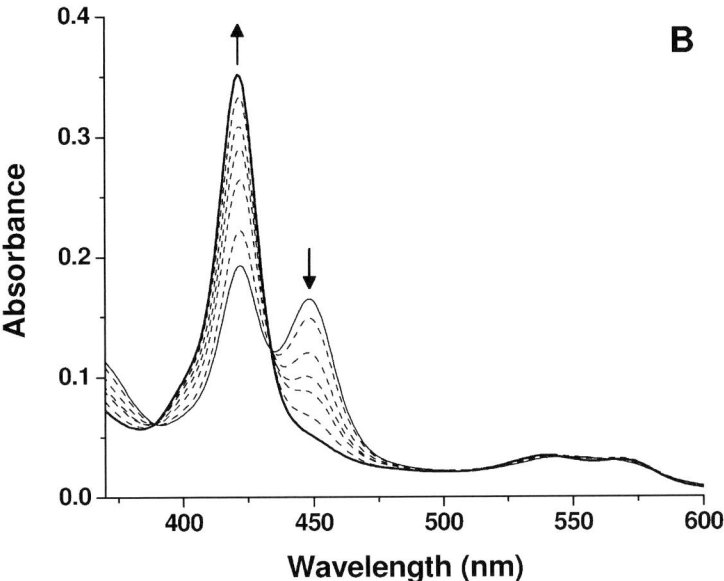

Figure 7. (Continued).

chloroperoxidase from the mould *Caldariomyces fumago* also has the Soret for its Fe^{2+}-CO complex at 443 nm [111,112]. Cystathionine β-synthase has its Soret maximum at ∼ 448 nm in absence of CO. Structural details for the human enzyme confirmed conclusions made from spectroscopic studies of the rat form, showing that the heme is hexacoordinated with both cysteinate and histidine axial ligands [113,114]. Binding of CO to this enzyme results in a Soret shift to 420 nm, probably due to replacement of the thiolate ligand by CO [114]. The role of heme in the PLP-dependent cystathionine β-synthase reaction is unclear. No redox changes are required in the condensation reaction catalyzed; heme-depleted human enzyme retains activity and the yeast form does not bind heme [115,116]. However, redox regulation may be important for the human form, since the enzyme is less active under reducing conditions [117].

The transcriptional regulator CooA from *Rhodospirillum rubrum* and mammalian heme-regulated eIF2-kinase are further heme thiolate proteins that are activated by binding of CO and NO, respectively [118,119]. These events lead to expression of enzymes involved in CO oxidation and to protein phosphorylation resulting in regulation of protein synthesis, respectively. In the case of CooA, structural effects mediated by binding of CO are transmitted to a DNA binding domain that controls the expression of relevant genes. The CooA heme, unusually, is coordinated by a proline *trans* to either a cysteine (Fe^{3+}) or a histidine (Fe^{2+}), dependent on the redox state of the iron. The role and mechanism of the redox-dependent ligand switch seems to involve replacement of the proline ligand by CO in the ferrous enzyme form, and structural changes have been spectroscopically observed on binding of CO. These involve heme domain residues that contact the DNA-binding domain and provide a likely means of signal transmission [120,121]. In the case of CooA, heme reduction induces the dissociation of the cysteinate ligand. In the case of eIF2 kinase, NO ligand binding results in dissociation of the thiolate ligand and the formation of a 5-coordinate NO-bound state (and Soret shift to ~ 398 nm) [23]. Thus, the retention of the heme thiolate on reduction and/or binding of CO (and NO) is not a common feature of the heme thiolate proteins, and the P450s give the most spectacular red shift of the Soret band on binding CO among the heme thiolate enzyme family.

Recently, two thiolate-ligated c-type hemes have been demonstrated spectroscopically in the SoxA component of SoxAX – a redox protein from the bacterium *Rhodovulum sulfidophilum* that plays an important role in photosynthetic oxidation of sulfite and thiosulfate. The first heme (heme 1) is hexacoordinated with histidine and cysteinate as ligands, whereas the second (heme 2) has a modified cysteine (cysteine persulfide) and histidine as ligands. On addition of sodium dithionite, heme 1 remains oxidized whereas heme 2 is reduced concomitant with what seems to be (from a combination of spectroscopic and structural studies on the reduced form) protonation of the thiolate to the thiol [122,123]. This is the first example of thiolate-ligated c-type hemes in a protein.

4.2. Pyridines, Cyanides, and Other Reversible Inhibitors of P450s

A number of other types of ligand have been shown to coordinate the heme iron in P450s. Cyanide is an ionic ligand that typically binds relatively weakly and preferably to the ferric heme iron in P450s. Substituted cyanides (e.g. alkyl cyanides) are also often weak P450 inhibitors. The weak binding of the negatively charged cyanide ion is a direct consequence of the high electron density that is placed on the iron as a result of proximal coordination by the cysteinate ligand [14,124]. Various agents that inhibit P450s by coordinating the heme iron via an oxygen atom or by stabilizing the binding of the distal water are also known. These include various alcohols, ethers, and lactones. These are typically weak inhibitors, and usually induce a Soret shift to ~ 415 nm [14].

In addition to the imidazoles and triazoles discussed above, pyridine derivatives have also been investigated extensively as P450 inhibitors. One of the first molecules to be used clinically as a P450 inhibitor was the pyridine derivative metyrapone, which was used as an inhibitor of the human steroid 11β-hydroxylase enzyme (CYP11B1) that catalyzes the final step in synthesis of cortisol [125]. Metyrapone was used as a treatment for hormonal disorders, including Cushing's disease (hypercortisolism). Further information on the interactions of reversible inhibitors with P450 enzymes is available in the excellent review of Correia and Ortiz de Montellano [14].

4.3. Irreversible Binding of P450 Inhibitors

Various types of P450 suicide inhibitors (i.e., molecules that undergo activation by their P450 target prior to irreversible modification of the P450) have been reported. Selected sulfur-containing and halogenated compounds fall into this class. For instance, the substituted thiophene tienilic acid is oxidatively metabolized by mammalian CYP2C9 and CYP2C10 to a product (likely a thiophene sulfoxide rather than an epoxide) that covalently modifies the P450 by interaction with a nucleophilic group on the protein [14,126]. Ticlopidine is a related thiophene that is in clinical use for prevention of platelet aggregation, and is likely to be activated in similar fashion to tienilic acid and to act as a mechanism based inhibitor of both CYP2C19 (the target molecule) and CYP2B6 [14,127].

Other sulfur-containing P450 inhibitor molecules that are in clinical use include the estrogen receptor-modulating raloxifene, which is suicidally activated by CYP3A4 (and to a lower extent *in vivo* by CYP2D6), possibly by oxidative attack on thiophene or phenolic functionalities [14,128]. In addition, the HIV-1 protease inhibitor ritonavir is suicidally activated by CYP3A enzymes, likely producing fragments with reactive thiazolyl groups that modify and inactivate the P450s [129]. Perhaps the best known example of a chlorinated inhibitor of P450 is chloramphenicol. This mechanism-based inhibitor has been shown to inactivate CYP2B1. The chloramphenicol is converted to an intermediate (an oxamyl chloride) that is either hydrolyzed to the acid, or modifies an important lysine residue. The key lysine is likely on the protein surface and the modification appears to prevent electron transfer from the redox partner (since use of cumene hydroperoxide and other surrogate electron donors restores ethoxycoumarin O-deethylase activity) [130]. Chlorinated steroids (with the methyl group that is normally oxidized replaced by a dichloromethyl functionality) have also been investigated as possible selective mechanism-based inhibitors of various eukaryotic P450s (e.g., [14,131]).

Another major class of P450 suicide inhibitors are molecules containing terminal olefinic groups, or with acetylenic groups either terminal or internal to the inhibitor. There is a large literature describing such inhibitors and the covalent modifications of heme and/or protein that occur when the P450 target attempts to oxidize across the relevant carbon-carbon double or triple bond. N-alkylation of the P450 porphyrin (with destruction of the chromophore, i.e., loss of its UV-visible spectral characteristics) is a common result of the reaction with the inhibitor. In the case of the barbiturate derivative and terminal olefin secobarbital, complete inhibition of the P450 CYP2B1 is observed without total loss of the chromophore, and this results from a combination of heme alkylation, protein modification and the formation of an epoxide product [14,132]. Terminal acetylenes and terminal olefins alkylate the P450 heme group, but certain terminal acetylenes appear to exert their major inhibitory action through protein modification. In at least the cases of the inhibitors 10-undecynoic acid (an inhibitor of CYP4B1) and 1-ethynylpyrene (an inhibitor of CYP1A enzymes), this appears to result from the oxidative attack from the P450 on the carbon triple bond leading to migration of the terminal hydrogen to the next carbon atom. A reactive ketene intermediate is formed that will either acylate the protein or become hydrolyzed to a carboxylic acid product [133]. Several suicide-type inhibitors have been designed and synthesized as mechanistic probes for specific P450s and their reactions, and as potential therapeutic agents (e.g., [134]). A recent excellent review provides detailed insights into these molecules and their applications [14].

5. CONCLUSIONS AND FUTURE PROSPECTS

Inhibition of P450 enzymes through coordination of the heme iron or by irreversible modification of the protein or heme has been the topic of intensive investigation since the discovery of these enzymes almost 50 years ago, and due to the increasing recognition of P450s as critical for eukaryotic metabolism and of prime interest for biotechnological applications.

A recent addition to the database of knowledge on P450 inhibition has come with the realization that NO is a critical player in the control of physiological processes in both prokaryotes and eukaryotes, and with the demonstrations that NO can act both as a reversible inhibitor of P450 function, and that its oxides can also give rise to covalent modifications of P450s and other proteins that are likely to be important in regulation of P450 catalysis. In particular, the study of protein nitration is a rapidly developing field and the next few years are likely to provide novel information on reactions between key P450 isoforms and nitric oxides that have fundamental impacts on, e.g., cell physiology, drug metabolism, and adaptation to environment.

ACKNOWLEDGMENTS

The authors wish to acknowledge the Biotechnology and Biological Sciences (BBSRC, UK) and Engineering and Physical Sciences Research Councils (EPSRC, UK) and the EU (network grants X-TB and NM4TB) for the support of research in the AWM laboratory. AWM and KJM also acknowledge the Royal Society (UK) and Leverhulme Trust for the award of a Royal Society Leverhulme Trust Senior Research Fellowship. The authors are grateful to Dr David Leys (University of Manchester) and to Dr Myles Cheesman (University of East Anglia) for useful discussions.

ABBREVIATIONS

CaM	calmodulin
cGMP	cyclic guanosine monophosphate
CPR	cytochrome P450 reductase
CT	charge transfer
CYP	cytochrome P450
6-DEB	6-deoxyerythronolide B
DMZ	4-methyl-N-methyl-N-(2-phenyl-2H-pyrazol-3-yl)benzenesulfonamide
eNOS	endothelial nitric oxide synthase
EPR	electron paramagnetic resonance
EXAFS	X-ray absorption fine structure
FAD	flavin adenine dinucleotide
FMN	flavin mononucleotide
GAPDH	glyceraldehyde-3-phosphate dehydrogenase
GTP	guanosine triphosphate
H_4B	tetrahydrobiopterin
20-HETE	20-hydroxyeicosatetraenoic acid
iNOS	inducible nitric oxide synthase
IR	infrared
MCD	magnetic circular dichroism
NADH	nicotinamide adenine dinucleotide (reduced form)
NADPH	nicotinamide adenine dinucleotide phosphate (reduced form)
NIR	near infrared
nNOS	neuronal nitric oxide synthase
NOS	nitric oxide synthase
PLP	pyridoxal phosphate
RR	resonance Raman

REFERENCES

1. S. G. Sligar, *Essays Biochem.*, **34**, 71–83 (1999).
2. I. G. Denisov, T. M. Makris, S. G. Sligar, and I. Schlichting, *Chem. Rev. 105*, 2253–2277 (2005).
3. J. F. Andersen, K. Tatsuka, H. Gunji, T. Ishiyama, and C. R. Hutchinson, *Biochemistry, 32*, 1905–1913 (1993).
4. T. Kubo, P. Peters, M. W. Meinhold, and F. H. Arnold, *Chemistry, 12*, 1216–1220 (2006).
5. T. W. B. Ost, C. S. Miles, J. Murdoch, Y. Cheung, G. A. Reid, S. K. Chapman, and A. W. Munro, *FEBS Lett. 486*, 173–177 (2000).
6. C. F. Harford-Cross, A. B. Carmichael, F. K. Allan, P. A. England, D. A. Rouch, and L. L. Wong, *Protein Eng. 13*, 121–128 (2000).

7. P. Peters, M. W. Meinhold, A. Glieder, and F. H. Arnold, *J. Am. Chem. Soc. 125*, 13442–13450 (2003).
8. Y. M. Zheng, B. R. Baer, M. B. Kneller, K. R. Henne, K. L. Kunze, and A. E. Rettie, *Biochemistry 42*, 4601–4606 (2003).
9. S. G. Sligar and I. C. Gunsalus, *Proc. Natl. Acad. Sci. USA 73*, 1078–1082 (1976).
10. S. N. Daff, S. K. Chapman, K. L. Turner, R. A. Holt, S. Govindaraj, T. L. Poulos, and A. W. Munro, *Biochemistry 36*, 13816–13823 (1997).
11. J. T. Groves, in *Cytochrome P450: Structure, Mechanism and Biochemistry* (P. R. Ortiz de Montellano, ed.), 3rd edn, Kluwer Academic/Plenum Publishers, New York, 2005, pp. 1–43.
12. T. M. Makris, R. Davydov, I. G. Denisov, B. M. Hoffman, and S. G. Sligar, *Drug Met. Rev. 34*, 691–708 (2002).
13. M. J. Coon, *Annu. Rev. Pharmacol. Toxicol. 45*, 1–25 (2005).
14. M. A. Correia and P. R. Ortiz de Montellano, in *Cytochrome P450: Structure, Mechanism and Biochemistry* (P. R. Ortiz de Montellano, ed.), 3rd edn, Kluwer Academic/Plenum Publishers, New York, 2005, pp. 247–322.
15. L. M. Podust, T. L. Poulos, and M. R. Waterman, *Proc. Natl. Acad. Sci. USA 98*, 3068–3073 (2001).
16. A. Lupetti, R. Danesi, M. Campa, M. Del Tacca, and S. L. Kelly, *Trends Mol. Med. 8*, 76–81 (2002).
17. D. J. Stuehr, J. Santolini, Z. Q. Wang, C. C. Wei, and S. Adak, *J. Biol. Chem. 279*, 36167–36170 (2004).
18. L. G. Quaroni, H. E. Seward, K. J. McLean, H. M. Girvan, T. W. B. Ost, M. A. Noble, S. M. Kelly, N. C. Price, M. R. Cheesman, W. E. Smith, and A. W. Munro, *Biochemistry 43*, 16416–16431 (2004).
19. W. Xu, I. G. Charles, and S. Moncada, *Cell. Res. 2005*, 63–65 (2005).
20. A. Daiber, H. Shoun, and V. Ullrich, V. *J. Inorg. Biochem. 99*, 185–193 (2005).
21. R. Perera, M. Sono, J. A. Sigman, T. D. Pfister, Y. Lu, and J. H. Dawson, *Proc. Natl. Acad. Sci. USA 100*, 3641–3646 (2003).
22. K. J. McLean, A. J. Warman, H. E. Seward, K. R. Marshall, M. R. Cheesman, M. R. Waterman, and A. W. Munro, *Biochemistry*, in press (2006).
23. T. Omura, *Biochem. Biophys. Res. Commun. 338*, 404–409 (2005).
24. B. R. Crane, A. S. Arvai, R. Gachhui, C. Wu, D. K. Ghosh, E. D. Getzoff, and D. J. Stuehr, *Structure 278*, 425–431 (1997).
25. M. Russwurm and D. Koesling, *Biochem. Soc. Symp. 71*, 51–63 (2004).
26. D. J. Stuehr, *Biochim. Biophys. Acta 1411*, 217–230 (1999).
27. E. D. Garcin, C. M. Bruns, S. J. Lloyd, D. J. Hosfield, M. Tiso, R. Gachhui, D. J. Stuehr, J. A. Tainer, and E. D. Getzoff, *J. Biol. Chem. 279*, 37918–37927 (2004).
28. D. H. Craig, S. K. Chapman, and S. Daff, *J. Biol. Chem. 277*, 33987–33994 (2002).
29. C. C. Wei, Z. Q. Wang, A. S. Arvai, C. Hemann, R. Hille, E. D. Getzoff, and D. J. Stuehr, *Biochemistry 42*, 1969–1977 (2003).
30. T. Shimada, R. L. Mernaugh, and F. P. Guengerich, *Arch. Biochem. Biophys. 435*, 207–216 (2005).
31. J. Santolini, M. Roman, D. J. Stuehr, and T. A. Mattioli, *Biochemistry 45*, 1480–1489 (2006).
32. A. W. Munro and J. G. Lindsay, *Mol. Microbiol. 20*, 1115–1125 (1996).

33. A. W. Munro, D. G. Leys, K. J. McLean, K. R. Marshall, T. W. B. Ost, S. Daff, C. S. Miles, S. K. Chapman, D. A. Lysek, C. C. Moser, C. C. Page, and P. L. Dutton, *Trends Biochem. Sci. 27*, 250–257 (2002).
34. M. C. Gustafsson, O. Roitel, K. R. Marshall, M. A. Noble, S. K. Chapman, A. Pessegueiro, A. J. Fulco, M. R. Cheesman, C. Von Wachenfeldt, and A. W. Munro, *Biochemistry 43*, 5474–5487 (2004).
35. U. Siddhanta, A. Presta, B. Fan, D. Wolan, D. L. Rousseau, and D. J. Stuehr, *J. Biol. Chem. 273*, 18950–18958 (1998).
36. R. Neeli, H. M. Girvan, A. Lawrence, M. J. Warren, D. Leys, N. S. Scrutton, and A. W. Munro, *FEBS Lett. 579*, 5582–5588 (2005).
37. H. Li, C. S. Raman, C. B. Glaser, E. Blasko, T. A. Young, J. F. Parkinson, M. Whitlow, and T. L. Poulos, *J. Biol. Chem. 274*, 21276–21284 (1999).
38. D. J. Stuehr, C. C. Wei, Z. Wang, and R. Hille, *Dalton Trans. 2005*, 3427–3435 (2005).
39. J. Santolini, A. L. Meade, and D. J. Stuehr, *J. Biol. Chem. 276*, 48887–48898 (2001).
40. C. C. Wei, B. R. Crane, and D. J. Stuehr, *Chem. Rev. 103*, 2365–2383 (2003).
41. F. T. Bonner and G. Stedman, in *Methods in Nitric Oxide Research* (M. Feelisch and J. S. Stamler, eds), John Wiley and Sons, Inc., New York, 1996, pp. 3–18.
42. D. H. O'Keefe, R. E. Ebel, and J. A. Peterson, *J. Biol. Chem. 253*, 3509–3516 (1978).
43. R. Chiang, R. Makino, W. E. Spomer, and L. P. Hager, *Biochemistry 14*, 4166–4171 (1975).
44. Y. Yonetani, H. Yamamoto, J. E. Erman, J. S. Leigh Jr., and G. H. Reed, *J. Biol. Chem. 247*, 2447–2455 (1972).
45. M. Tsubaki, A. Hiwatashi, Y. Ichikawa, and H. Hori, *Biochemistry 26*, 4527–4534 (1987).
46. T. W. B. Ost, C. S. Miles, A. W. Munro. J. Murdoch, G. A. Reid, and S. K. Chapman, *Biochemistry 40*, 13421–13429 (2001).
47. S. Adak, Q. Wang, and D. J. Stuehr, *J. Biol. Chem. 275*, 17434–17439 (2000).
48. M. Couture, S. Adak, D. J. Stuehr, and D. L. Rousseau, *J. Biol. Chem. 276*, 38280–38288 (2001).
49. H. L. Voegtle, M. Sono, S. Adak, A. E. Pond, T. Tomita, R. Perera, D. B. Goodin, M. Ikeda-Saito, D. J. Stuehr, and J. H. Dawson, *Biochemistry 42*, 2475–2484 (2003).
50. R. Nakano, H. Sato, A. Watanabe, O. Ito, and T. Shimizu, *J. Biol. Chem. 271*, 8570–8574 (1996).
51. T. Shoun and T. Tanimoto, T., *J. Biol. Chem. 266*, 11078–11082 (1991).
52. F. P. Guengerich, *Curr. Drug Metab. 2*, 93–115 (2001).
53. M. Umemara, F. Su, N. Takaya, Y. Shiro, and H. Shoun, *Eur. J. Biochem. 271*, 2887–2894 (2004).
54. U. Flock, J. Reimann, and P. A. Adelroth, *Biochemical Soc. Trans. 34*, 188–190 (2004).
55. E. Obayashi, K. Tsukamoto, S. Adachi, S. Takahashi, M. Nomura, T. Iizuka, H. Shoun, and H. Shiro, *J. Am. Chem. Soc. 119*, (1997).
56. K. Nakahara and H. Shoun, *J. Biochem. 120*, 1082–1087 (1996).

57. H. Shoun and T. Tanimoto, *J. Biol. Chem.* **266**, 11078–11082 (1991).
58. T. O. Watsuji, N. Takaya, A. Nakamura, and H. Shoun, *Biotechnol. Biochem.* **67**, 1109–1114 (2003).
59. L. Zhang, N. Takaya, T. Kitazume, T. Kondo, and H. Shoun, *Eur. J. Biochem.* **268**, 3198–3204 (2001).
60. Y. Shiro, M. Fujii, T. Iizuka, S.-I. Adachi, K. Tsukamoto, K. Nakahara, and H. Shoun, *J. Biol. Chem.* **270**, 1617–1623 (1995).
61. A. Daiber, T. Nauser, N. Takaya, T. Kudo, P. Weber, C. Hultschig, H. Shoun, and V. Ullrich, V., *J. Inorg. Biochem.* **88**, 343–352.
62. S.-Y. Park, H. Shimizu, S.-I. Adachi, A. Nakagawa, I. Tanaka, K. Nakahara, H. Shoun, E. Obayashi, H. Nakamura, T. Iizuka, and Y. Shiro, *Nature Struct. Biol.* **4**, 827–832 (1997).
63. L. Zhang, T. Kudo, N. Takaya, and H. Shoun, H., *J. Biol. Chem.* **277**, 33842–33847 (2002).
64. T. Kudo, N. Takaya, S. Y. Park, Y. Shiro, and H. Shoun, *J. Biol. Chem.* **276**, 5020–5026 (2001).
65. J. S. Beckman, D. A. Wink, and J. P. Crow, in *Methods in Nitric Oxide Research* (M. Feelisch and J. S. Stamler, eds), John Wiley and Sons, Inc., New York, 1996, pp. 61–70.
66. G. C. Brown, *Biochim. Biophys. Acta* **1504**, 45–67 (2001).
67. M. Kelm and K. Yoshida, in *Methods in Nitric Oxide Research* (M. Feelisch and J. S. Stamler, eds), John Wiley and Sons, Inc., New York, 1996, pp. 46–58.
68. D. T. Hess, A. Matsumoto, S.-O. Kim, H. E. Marshall, and J. S. Stamler, *Nat. Rev. Mol. Cell. Biol.* **6**, 150–166 (2005).
69. S. Mohr, J. S. Stamler, and B. Brune, *J. Biol. Chem.* **271**, 4209–4214 (1996).
70. M. R. Hara, N. Agrawal, S. F. Kim, M. B. Cascio, M. Fujimoro, Y. Ozeki, M. Takahashi, J. H. Cheah, S. K. Tankou, L. D. Hester, C. D. Ferris, S. D. Hayward, S. H. Snyder, and A. Sawa, *Nat. Cell. Biol.* **7**, 665–674 (2005).
71. T. Kuncewicz, E. A. Sheta, I. L. Goldknopf, and B. C. Kone, *Mol. Cell. Proteomics* **2**, 156–163 (2003).
72. L. J. Ignarro, J. B. Adams, P. M. Horwitz, and K. S. Wood, *J. Biol. Chem.* **261**, 4997–5002 (1986).
73. D. A. Wink, Y. Osawa, J. F. Darbyshire, C. R. Jones, S. C. Eshenaur, and R. W. Nims, *Arch. Biochem. Biophys.* **300**, 115–123 (1993).
74. H. L. Lin, H. Zhang, L. Waskell, and P. F. Hollenberg, *Chem. Res. Toxicol.* **18**, 1203–1210. (2005).
75. M. Bachschmid, S. Thurau, M. H. Zou, and V. Ullrich, *FASEB J.* **17**, 914–916.
76. M. H. Wang, J. Wang, H. H. Chang, B. A. Zand, M. Jiang, A. Nasjletti, and M. Laniado-Schwartzmann, *Am. J. Pysiol. Renal Physiol.* **285**, F295–302 (2003).
77. A. Daiber, S. Herold, C. Schoneich, D. Namgaladze, J. A. Peterson, and V. Ullrich, *Eur. J. Biochem.* **267**, 6729–6739 (2000).
78. A. Daiber, C. Schoneich, P. Schmidt, C. A. Jung, and V. Ullrich, *J. Inorg. Biochem.* **81**, 213–220 (2000).
79. L. Quaroni, J. Reglinski, C. R. Wolf, and W. E. Smith, *Biochim. Biophys. Acta* **1296**, 5–8 (1996).
80. M. Aguiar, R. Masse, and B. F. Gibbs, *Drug Metab. Rev.* **37**, 379–404 (2005).

81. T. Koeck, D. J. Stuehr, and K. S. Aulak, *Biochem. Soc. Trans. 33*, 1399–1403 (2005).
82. K. J. McLean, M. R. Cheesman, S. L. Rivers, A. Richmond, D. Leys, S. K. Chapman, G. A. Reid, N. C. Price, S. M. Kelly, J. Clarkson, W. E. Smith, and A. W. Munro, *J. Inorg. Biochem. 91*, 527–541 (2002).
83. T. Tominaga, I. Adachi, Y. Sasaki, T. Tabei, T. Ikeda, Y. Takatsuka, M. Toi, T. Suwa, and Y. Ohashi, Y., *Annal. Oncol. 14*, 62–70 (2003).
84. E. Rodriguez-Fernandez, J. L. Manzano, J. J. Benito, R. Hermosa, E. Monte, and J. J. Criado, *J. Inorg. Biochem. 99*, 1558–1572 (2005).
85. D. E. Moody, S. L. Walsh, D. E. Rolline, J. A. Neff, and W. Huang, *Clin. Pharmacol. Ther. 76*, 154–166 (2004).
86. S. Haidar, P. B. Ehmer, S. Barassin, C. Batzl-Hartmann, and R. W. Hartmann, *J. Steroid Biochem. Mol. Biol. 84*, 555–562 (2003).
87. W. Zhang, Y. Ramamoorthy, T. Kilicarslan, H. Nolte, R. F. Tyndale, and E. M. Sellers, *Drug Met. Dispos. 30*, 314–318 (2002).
88. P. Kale and L. B. Johnson, *Drugs Today 41*, 91–105 (2005).
89. F. Buckner, K. Yokoyama, J. Lockman, K. Aikenhead, J. Ohkanda, M. Sadilek, S. Sebti, W. Van Voorhis, A. Hamilton, and M. H. Gelb, *Proc. Natl. Acad. Sci. USA 100*, 15149–15153 (2003).
90. K. J. McLean, K. R. Marshall, A. Richmond, I. S. Hunter, K. Fowler, T. Kieser, S. S. Gurcha, G. S. Besra, and A. W. Munro, *Microbiology 148*, 2937–2949 (2002).
91. H. M. Girvan, K. R. Marshall, R. J. Lawson, D. Leys, M. G. Joyce, J. Clarkson, W. E. Smith, M. R. Cheesman, and A. W. Munro, *J. Biol. Chem. 279*, 23274–23286 (2004).
92. J. McKnight, M. R. Cheesman, A. J. Thomson, J. S. Miles, and A. W. Munro, *Eur. J. Biochem. 213*, 683–687 (1993).
93. P. Anzenbacher and J. Hudecek, *J. Inorg. Biochem. 87*, 209–213 (2001).
94. J. D. Lipscomb, *Biochemistry 19*, 3590–3599 (1980).
95. J. S. Miles, A. W. Munro, B. N. Rospendowski, W. E. Smith, J. McKnight, and A. J. Thomson, *Biochem. J. 288*, 503–509 (1992).
96. J. Hudecek, E. Anzenbacherova, P. Anzenbacher, A. W. Munro, and P. Hildebrandt, *Arch. Biochem. Biophys. 383*, 70–78 (2000).
97. J. Wang, D. L. Rousseau, H. M. Abu-Soud, and D. J. Stuehr, *Proc. Natl. Acad. Sci. USA 91*, 10512–10516 (1994).
98. S. Hu and J. R. Kincaid, *J. Am. Chem. Soc. 113*, 2843–2850 (1991).
99. J. Wang, W. S. Caughey, and D. L. Rousseau, in *Methods in Nitric Oxide Research* (M. Feelisch and J. S. Stamler, eds), John Wiley and Sons, Inc., New York, 1996, pp. 427–454.
100. J. K. Yano, L. S. Koo, D. J. Schuller, H. Li, P. R. Ortiz de Montellano, and T. L. Poulos, *J. Biol. Chem. 275*, 31086–31092 (2000).
101. E. E. Scott, M. A. White, Y. A. He, E. F. Johnson, C. D. Stout, and J. R. Halpert, *J. Biol. Chem. 279*, 27294–27301 (2004).
102. Y. Zhao, M. A. White, B. K. Muralidhara, L. Sun, J. R. Halpert, and C. D. Stout, *J. Biol. Chem. 281*, 5973–5981 (2006).
103. M. R. Wester, E. F. Johnson, C. Marques-Soares, P. M. Dansette, D. Mansuy, and C. D. Stout, *Biochemistry 42*, 6370–6379 (2003).

104. T. L. Poulos and E. F. Johnson, in *Cytochrome P450: Structure, Mechanism and Biochemistry* (P. R. Ortiz de Montellano, ed.), 3rd edn, Kluwer Academic/Plenum Publishers, New York, 2005, pp. 87–114.
105. J. R. Cupp-Vickery, C. Garcia, A. Hofacre, and K. McGee-Estrada, *J. Mol. Biol. 311*, 101–110 (2001).
106. M. Klingenberg, *Arch. Biochem. Biophys. 75*, 376–386 (1958).
107. T. Omura and R. Sato, *J. Biol. Chem. 237*, 1375–1376 (1962).
108. T. Omura and R. Sato, *J. Biol. Chem. 239*, 2370–2378 (1964).
109. A. W. Munro, J. G. Lindsay, J. R. Coggins, S. M. Kelly, and N. C. Price, *Biochim. Biophys. Acta. 1296*, 127–137 (1996).
110. C. Di Primo, G. Hui Bon Hoa, P. Douzou, and S. G. Sligar, *Eur. J. Biochem. 209*, 583–588 (1992).
111. P. F. Hollenberg and L. P. Hager, *J. Biol. Chem. 248*, 2630–2633 (1973).
112. J. H. Dawson, J. R. Trudell, G. Barth, R. E. Linder, E. Bunnenberg, C. Dierassi, R. Chiang, and L. P. Hager, *J. Am. Chem. Soc. 98*, 3709–3710 (1976).
113. M. Meiyer, M. Janosik, V. Kery, J. P. Kraus, and P. Burkhard, *EMBO J. 20*, 3910–3916 (2001).
114. T. Omura, H. Sadano, T. Hasegawa, Y. Yoshida, and S. Kominami, *J. Biochem. 96*, 1491–1500 (1984).
115. K. H. Jhee, P. McPhie, and E. W. Miles, *Biochemistry 39*, 10548–10556 (2000).
116. S. Bruno, P. Schiarett, P. Burkhardt, J. P. Kraus, M. Janosik, and A. Mozzarelli, *J. Biol. Chem 276*, 16–19 (2001).
117. S. Taoka, S. Ohja, X. Shan, W. D. Kruger, and R. Banerjee, *J. Biol. Chem. 273*, 25179–25184 (1998).
118. S. Aono, K. Ohkubo, T. Matsuo, and H. Nakajima, *J. Biol. Chem. 273*, 25757–25764 (1998).
119. J. Igarashi, A. Sato, T. Kitagawa, T. Yoshimura, S. Yamauchi, I. Sagami, and T. Shimizu, *J. Biol. Chem. 279*, 15752–15762 (2004).
120. T. Yamashita, Y. Hoashi, K. Watanabe, Y. Tomisugi, Y. Ishikawa, and T. Uno, *J. Biol. Chem. 279*, 21394–21400 (2004).
121. T. Yamashita, Y. Hoashi, Y. Tomisugi, Y. Ishikawa, and T. Uno, *J. Biol. Chem. 279*, 47320–47325 (2004).
122. M. R. Cheesman, P. J. Little, and B. C. Berks, *Biochemistry 40*, 10562–10569 (2001).
123. V. A. Bamford, S. Bruno, T. Rasmussen, C. Appia-Ayme, M. R. Cheesman, B. C. Berks, and A. M. Hemmings, *EMBO J. 21*, 5599–5610 (2002).
124. M. Sono and J. H. Dawson, *J. Biol. Chem. 257*, 5496–5502 (1982).
125. O. V. Dominguez and L. T. Samuels, *Endocrinology 73*, 304–309 (1963).
126. L. L. Koenigs, R. M. Peter, A. P. Hunter, A. E. Haining, A. E. Rettie, T. Friedberg, M. P. Pritchard, M. Shou, T. H. Rushmore, and W. F. Trager, *Biochemistry 38*, 2312–2319 (1999).
127. N. T. Ha-Duong, S. Dijols, A. C. Macherey, J. A. Goldstein, P. M. Dansette, and D. Mansuy, *Biochemistry 40*, 12112–12122 (2001).
128. Q. Chen, J. S. Ngui, G. A. Doss, R. W. Wang, X. Cai, F. P. DiNinno, T. A. Blizzard, M. L. Hammond, R. A. Stearns, D. C. Evans, T. A. Baillie, and W. Tang, *Chem. Res. Toxicol. 15*, 907–914 (2002).

129. T. Koudriakova, E. Iatsimirskaia, I. Utkin, E. Gangl, P. Vouros, E. Storozhuk, D. Orza, J. Marinina, and N. Gerber, *Drug Metab. Dispos. 26*, 552–561 (1998).
130. J. R. Halpert, *Biochem. Pharmacol. 30*, 875–881 (1981).
131. J. R. Halpert, J. Y. Jaw, C. Balfour, E. A. Mash, and E. F. Johnson, *Arch. Biochem. Biophys. 264*, 462–471 (1988).
132. K. He, Y. A. He, G. D. Sklarz, J. R. Halpert, and M. A. Correia, *J. Biol. Chem. 271*, 25864–25872 (1996).
133. W. K. Chan, Z. Sui, and P. R. Ortiz de Montellano, *Chem. Res. Toxicol. 6*, 38–45 (1993).
134. A. Nagahisa, R. W. Spencer, and W. H. Orme-Johnson, *J. Biol. Chem. 258*, 6721–6723 (1983).
135. R. E. White and M. J. Coon, *J. Biol. Chem. 257*, 3073–3078 (1982).
136. A. R. Hurshman and M. A. Marletta, *Biochemistry 34*, 5627–5634 (1995).
137. K. G. Ravichandran, S. S. Boddupalli, C. A. Hasemann, J. A. Peterson, and J. Deisenhofer, *Science 261*, 731–736 (1993).
138. H. Li, H. Shimizu, M. Flinspach, J. Jamal, W. Yang, M. Xian, T. Cai, E. Z. Wen, Q. Jia, P. G. Wang, and T. L. Poulos, *Biochemistry 41*, 13868–13875 (2002).

11

Cytochrome P450-Catalyzed Hydroxylations and Epoxidations

Roshan Perera,[1] *Shengxi Jin,*[1] *Masanori Sono,*[1] *and John H. Dawson*[1,2]

[1]Department of Chemistry and Biochemistry, and [2]School of Medicine, University of South Carolina, Columbia, SC 29208, USA
<perera@scripps.edu>
<jinshengxi@gmail.com>
<sono@mail.chem.sc.edu>
<dawson@sc.edu>

1. INTRODUCTION	320
2. THE CYTOCHROME P450 ENZYMES	322
2.1. The Catalytic Reaction Cycle of Cytochrome P450	323
2.2. Properties of the P450 Dioxygen Complex	325
3. THREE-DIMENSIONAL STRUCTURES OF THE ACTIVE SITES OF CYTOCHROME P450 ENZYMES	327
3.1. Substrate Binding Specificity of P450	330
3.2. Role of the Distal Pocket Threonine in P450 Catalysis	331
4. ROLE OF THE CYS LIGAND: THE PROXIMAL THIOLATE 'PUSH' AND DISTAL PROTON-DELIVERY	333
5. MULTIPLE MECHANISMS OF P450 CATALYSIS	337
5.1. Mechanism of Hydroxylation	337
5.2. Mechanism of Epoxidation	342
6. MULTIPLE OXIDANTS IN P450 CATALYSIS	343
7. TWO STATES THEORY	347

Metal Ions in Life Sciences, Volume 3   Edited by Astrid Sigel, Helmut Sigel and Roland K. O. Sigel
© 2007 John Wiley & Sons, Ltd

8. INFLUENCE OF SUBSTRATE ON THE SPECTRAL
   PROPERTIES AND REACTIVITY OF P450 INTERMEDIATES   348
9. FORMATION AND REACTIVITY OF TRANSIENT P450
   OXYGEN INTERMEDIATES   351
10. SUMMARY AND FUTURE PROSPECTIVE   352
    ACKNOWLEDGMENTS   353
    ABBREVIATIONS   353
    REFERENCES   354

1. INTRODUCTION

Heme iron-containing enzymes are remarkably versatile given the obvious restriction that all contain a heme iron center in a protein environment. The most common heme prosthetic group, protoporphyrin IX, consists of a highly conjugated and symmetrical macrocycle (Figure 1) coordinated to an iron atom. The wide variety of chemical reactions that heme enzymes catalyze include oxygen activation and insertion (P450 cytochromes) [1–8a], substrate oxidation (peroxidases) and disproportionation (catalases) [8], nitric oxide biosynthesis (nitric oxide synthase) [8a,9] as well as non-enzymatic functions of oxygen transport (hemoglobin and myoglobin) [10] and electron transfer (cytochromes) [11]. Factors known to contribute to the range of reactions that heme-containing enzymes catalyze include the identity of the proximal heme iron ligand and the shape of the distal substrate binding pocket including the positioning of catalytically important amino acids. A key intermediate in the enzymatic reaction cycle of oxidative heme enzymes is a transient high-valent heme oxygen atom species known as compound I (Cpd I). However, as will be discussed in this review, some of the versatility of heme enzymes may derive from the involvement of additional oxidant states other than Cpd I and from the influence that the substrate can have on the reactivity of heme iron intermediates. Understanding the structure, function and redox reactions of these enzymes is essential to the full elucidation of their mechanisms of action.

Dioxygen and Peroxide Activation. Aerobic organisms utilize dioxygen for two main purposes. The first major pathway involves the four-electron reduction of dioxygen to give two molecules of water coupled to the oxidation of electron rich molecules, such as glucose, to produce ATP as a way to store source of energy for the organism (Equation 1).

$$O_2 + 4H^+ + 4e^- \rightarrow 2H_2O \qquad E^{0'} = +0.815\,V \qquad (1)$$

The second use of dioxygen by aerobic organisms is to carry-out the incorporation of one (monooxygenase) or both (dioxygenase) oxygen atoms into

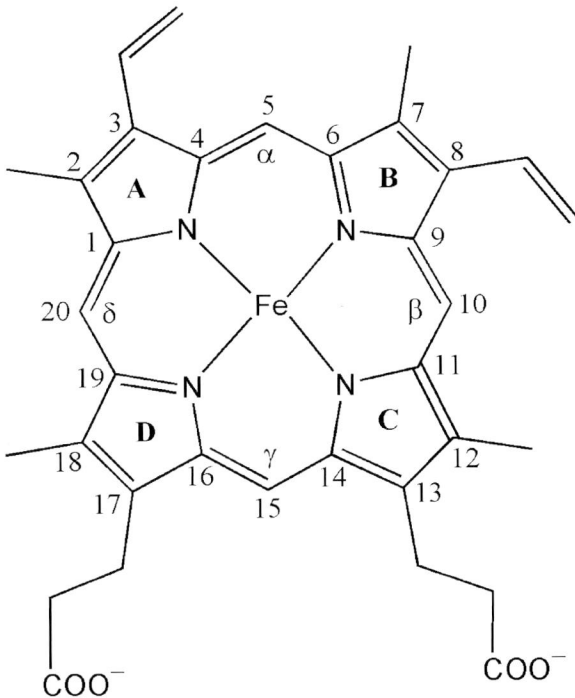

Figure 1. Protoheme (iron protoporphyrin IX, heme *b*). The Greek letters α to δ identify the site of each methine bridge and the Roman letters A to D indicate the location of each pyrrole ring. Note that an IUPAC numbering system is used here and thus the carbons at 2, 3, 7, 8, 12, 13, 17, and 18 positions correspond to those at 1, 2, 3, 4, 5, 6, 7, and 8 positions in the classic Fischer numbering system.

organic molecules in the biosynthesis of important biomolecules or in the solubilization of xenobiotic components for excretion purposes [4,12–14]. These enzymes fall into two categories: Oxygenases incorporate O-atoms from O_2 into organic substrates while oxidases oxidize substrates. Oxygenases further divide into monooxygenases, which incorporate one O atom of O_2 and reduce the other to H_2O (Equation 2), and dioxygenases, which incorporate both O atoms (Equation 3). Peroxidases [8], like oxidases, oxidize substrates, but do so while reducing H_2O_2 to H_2O (Equation 4).

$$\text{Monooxygenase} \quad R_3\text{C-H} + O_2 + 2H^+ + 2e^- \rightarrow R_3\text{C-OH} + H_2O \quad (2)$$

$$\text{Dioxygenase} \quad AH + O_2 \rightarrow A(O_2)H \quad (3)$$

$$\text{Peroxidase} \quad AH_2 + H_2O_2 \rightarrow A + 2H_2O \quad (4)$$

2. THE CYTOCHROME P450 ENZYMES

The cytochrome P450 (P450, CYP) superfamily of cysteinate-ligated heme-containing enzymes has extensive distribution throughout the plant, animal, and microorganism kingdoms. These monooxygenase enzymes metabolize a large number of hydrophobic endogenous and xenobiotic substrates [15–19]. Their reactivity covers a vast number of reactions including hydroxylation of unactivated alkanes, conversion of alkenes to epoxides, aromatics to phenols, and sulfides to sulfoxides or sulfones [15–19]. The classification of a heme protein as a P450 is derived from the absorption spectrum of its Fe^{II}-CO complex, which has a characteristic maximum near 450 nm [20]. The common feature of all known P450 enzyme isoforms is that they utilize cysteine-ligated heme protein active sites to catalyze monooxygenase reactions. Though P450 enzymes are capable of an incredibly diverse range of reactions, the most thoroughly studied reaction has been the hydroxylation of unactivated alkane C-H bonds. The activation of molecular oxygen for monooxygenase activity requires two reduction equivalents, as shown in equation (2). P450 enzymes are critical to many biological processes including steroid hormone biosynthesis (Figures 2 and 3) (see also Chapter 12 of this volume), drug metabolism (see also Chapters 16 and 17 of this volume), and the detoxification of xenobiotics [15–19] (see also Chapter 15 of this volume).

P450-CAM (CYP 101) is a water-soluble P450 enzyme from a strain of *Pseudomonas putida* grown on 1R-(+)-camphor as the sole source of carbon. P450-CAM has been widely studied over the past four decades because it is one

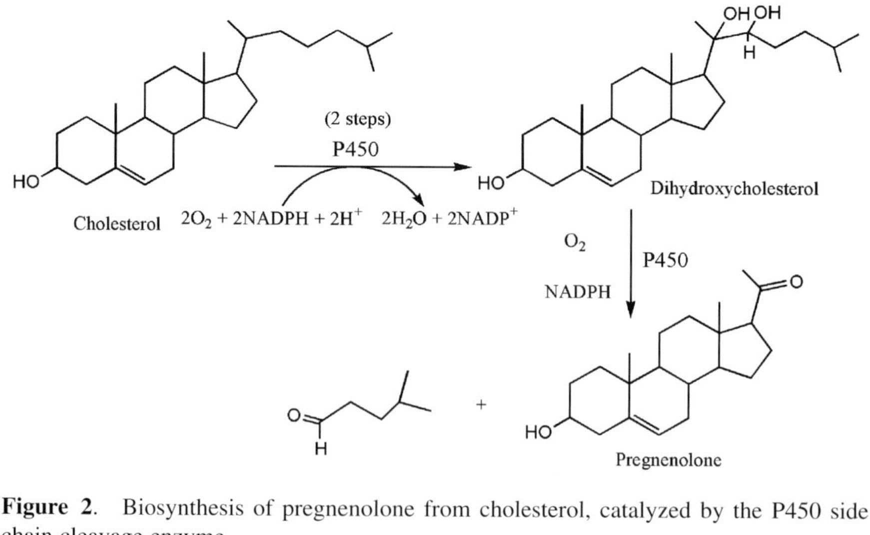

Figure 2. Biosynthesis of pregnenolone from cholesterol, catalyzed by the P450 side chain cleavage enzyme.

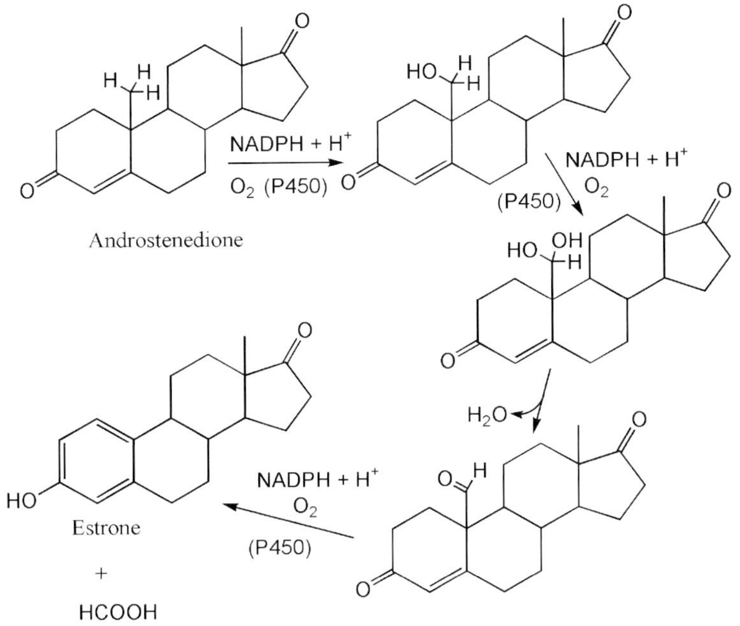

Figure 3. Biosynthesis of estrone from androstenedione, catalyzed by P450-aromatase.

of the few non-membrane-bound P450s. Because it is water-soluble, it is also relatively easy to purify. The specifics of the electron transport systems vary among P450s, but in all cases electrons flow from NADH or NADPH to the P450 heme iron active site [3,4,21]. For P450-CAM, the two electron transport proteins are putidaredoxin reductase (PdR, a flavoprotein) and putidaredoxin (Pdx, an iron-sulfur protein). For most mammalian P450 enzymes, a single electron transfer flavoprotein containing both FMN and FAD mediates the delivery of electrons to the P450 heme iron.

2.1. The Catalytic Reaction Cycle of Cytochrome P450

The enzymatic catalytic cycle carries substrate (RH) to its oxidized product (ROH) (Equation 5) through a series of steps. Structurally and biochemically, the best-characterized P450 is P450-CAM, which catalyzes the regio- and stereospecific hydroxylation of 1R-camphor. The reaction cycle for P450-CAM (Figure 4) is likely to be shared by most members of the P450 family and involves four well-characterized, isolable states (**A–D**). The reaction pathway of P450-CAM consists of reversible substrate binding, which converts the six coordinate, low spin, water-bound Fe^{III} resting P450 form (**A**) to a camphor-bound five coordinate, high spin Fe^{III} state (**B**). The addition of substrate is accompanied by a

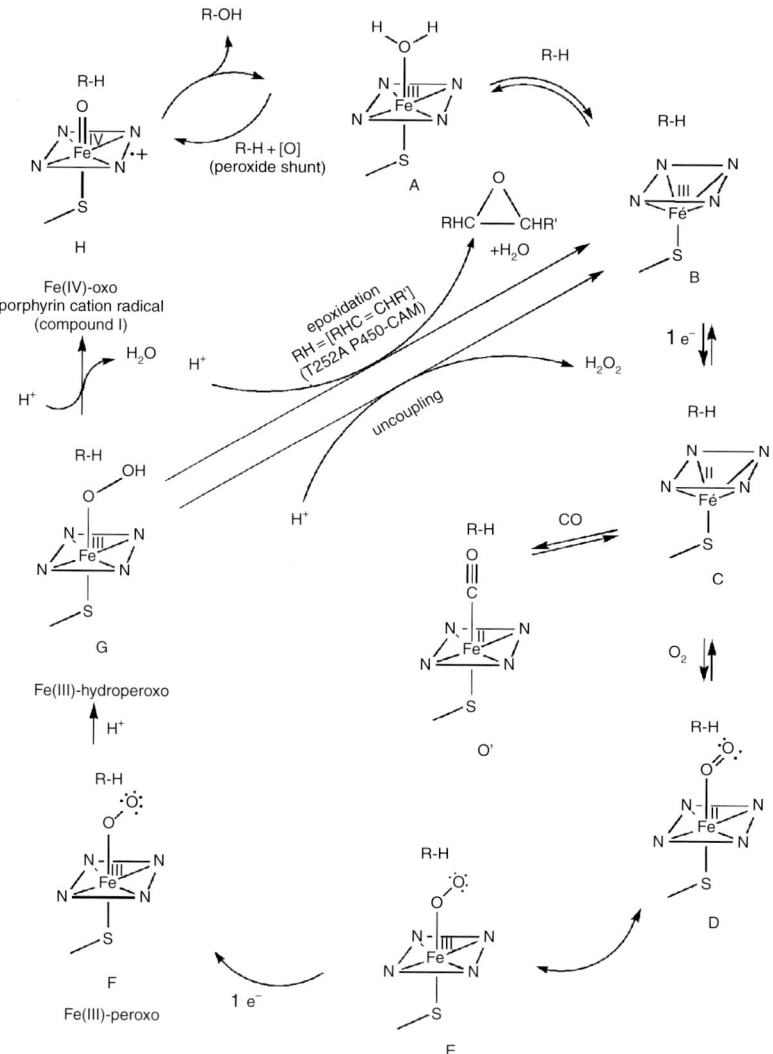

Figure 4. Catalytic cycle of cytochrome P450-CAM. Parallelograms with nitrogen atoms at the four corners represent a porphyrin ring. A possible alternative cycle for the mutant T252A P450-CAM-catalyzed olefin epoxidation is also shown.

considerable increase in the reduction potential ($-330\,\text{mV} \rightarrow -173\,\text{mV}$) of the heme iron [7]. Thus, substrate binding to the Fe^{III} enzyme initiates catalysis by facilitating the first electron transfer to reduce the heme center to the five coordinate Fe^{II} complex (**C**). The camphor-bound Fe^{II} enzyme (**C**) then binds dioxygen to give the six coordinate oxyferrous intermediate (**D**). The second order rate constant for dioxygen binding is $1.7 \times 10^6\,M^{-1}s^{-1}$ at $0\,°C$ for P450-CAM [22]. The dioxygen complex can be thought of as a resonance hybrid of Fe^{II}-O_2 (**D**) and Fe^{III}-O_2^- (**E**). Although much less stable than the oxyferrous complexes of dioxygen transport and storage proteins (hemoglobin and myoglobin) [10], oxyferrous P450-CAM is moderately stable in the presence of camphor; it autoxidizes back to the Fe^{III} state at a rate constant of $0.01\,s^{-1}$ ($t_{1/2} = 69\,s$) at room temperature [22,23]. The superoxide generated in this uncoupling reaction then disproportionates and generates hydrogen peroxide, a source of destructive hydroxyl radicals.

$$R_3C\text{-}H + O_2 + NAD(P)H + H^+ \rightarrow R_3C\text{-}OH + H_2O + NADP^+ \quad (5)$$

The iron-dioxygen intermediate **E** has a η^1-superoxide ion coordinated to the Fe^{III} center with an unpaired electron on the terminal oxygen atom [3]. Addition of CO to **C** yields the Fe^{II}-CO inhibitor adduct (**D′**) with its characteristic Soret absorption maximum at $\sim 450\,nm$. Addition of the second electron to **E**, which is the rate-limiting step of the P450-CAM catalytic cycle, gives a Fe^{III}-peroxo complex (**F**). Addition of the first proton yields a Fe^{III}-hydroperoxide species (**G**). Coon and co-workers provided initial evidence that the Fe^{III}-hydroperoxo P450 state (**G**) can serve as a second electrophilic oxygen donor in olefin epoxidations [24]. In 2003, Jin et al. reported even stronger evidence that state **G** can serve as a second electrophilic oxidant in olefin epoxidations [25]. Finally, addition of a second proton to the outer oxygen of the Fe^{III}-hydroperoxide species (**G**) leads to heterolytic cleavage of the O-O bond to produce a molecule of water and a strongly oxidizing species, the high valent ferryl Fe^{IV}=O porphyrin π-cation radical P450 Cpd I intermediate **H**. Intermediate **H** is analogous to the high-valent Cpd I species of the peroxidases and is generally thought to be the 'active oxygen' form of P450 that hydroxylates unactivated hydrocarbon substrates [8].

The heterolytic cleavage of the O-O bond in (**G**) is the critical step for the Cpd I species formation. If the bond is broken homolytically, the resulting species is called Compound II (Cpd II, one-electron reduced Cpd I). Cpd II is not reactive enough to catalyze oxygen transfer reactions on nonactivated alkanes. The cycle can be turned over by 'shunt' pathways as well. In the 'shunt' pathway, the Cpd I species (**H**) can be attained from intermediate **A** using oxygen atom donors such as peroxyacids (e.g., *meta*-chloroperoxybenzoic acid) [26–28]. Similarly, with some P450s, addition of H_2O_2 to **A** or **B** is assumed to yield the Cpd I species via the Fe^{III}-hydroperoxo species (**G**) [5].

2.2. Properties of the P450 Dioxygen Complex

The dioxygen complex is the last well-characterized intermediate in the P450 reaction cycle. Oxyferrous P450-CAM has an O-O stretching vibration of

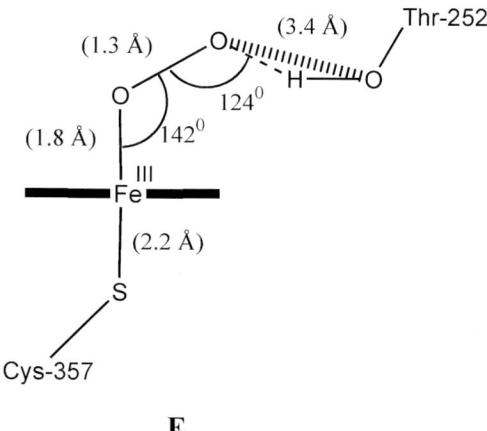

Figure 5. The measured geometries for the Fe^{II}-O_2 complex of P450-CAM [31], determined from X-ray coordinates and prepared using Protein Data Bank file (1DZ8).

$1141\,cm^{-1}$ and a Fe-O-O bending mode at $401\,cm^{-1}$ with a Fe-O-O angle of 125–130° as determined by resonance Raman spectroscopy [29,30]. The Fe-O-O angle derived from the published oxy-P450-CAM crystal structure is 142° (Figure 5) [31].

Shaik and coworkers [32–34] have used density functional theory (DFT) to calculate that the five coordinate Fe^{III} ground state is a sextet ($S = 5/2$) which in turn will be reduced by filling the $d_{x^2-y^2}$ orbital to generate the quintet ($S = 2$) (^5C in Figure 6 [27]) five coordinate Fe^{II} complex. The binding of dioxygen to this state (**C**) produces a singlet ($S = 0$)(^1D) Fe^{II}-O_2 adduct (Figure 6). One electron from the Fe^{II} center and one from the triplet oxygen pair create a Fe^{III}-O_2^- (**E**) complex. This Fe^{II}-O_2/Fe^{III}-O_2^- resonance hybrid forms mainly as a result of the empty π^* (O-O) orbital that can mix with the doubly occupied d_{yz} (Fe) orbital of the singlet ground state Fe^{II}-O_2 form (π^*_{yz} orbital has a higher mixing capability with the π and π^* of the dioxygen moiety) [34,35].

Molecular dynamics (MD) simulations of P450-CAM and P450eryF (CYP107A) carried out using MM/MD calculations show that the Fe^{II}-dioxygen species is stabilized by a hydrogen bond formed between the distal (outer) oxygen and the hydroxyl group of Thr-252 in P450-CAM. But in P450eryF, the distal oxygen is linked through Ser-246 and Ala-241 amino acids and several water molecule hydrogen-bonding network [36,37]. Transfer of the second electron to **E** would result in a nucleophilic, doublet ground state Fe^{III}-peroxo species (**F**). One implication of these calculations is the critical role of the proton delivery system (in the form of hydrogen-bonding network) in controlling the double protonation of the distal O-atom of bound dioxygen in order to generate the electrophilic Cpd I species (**H**).

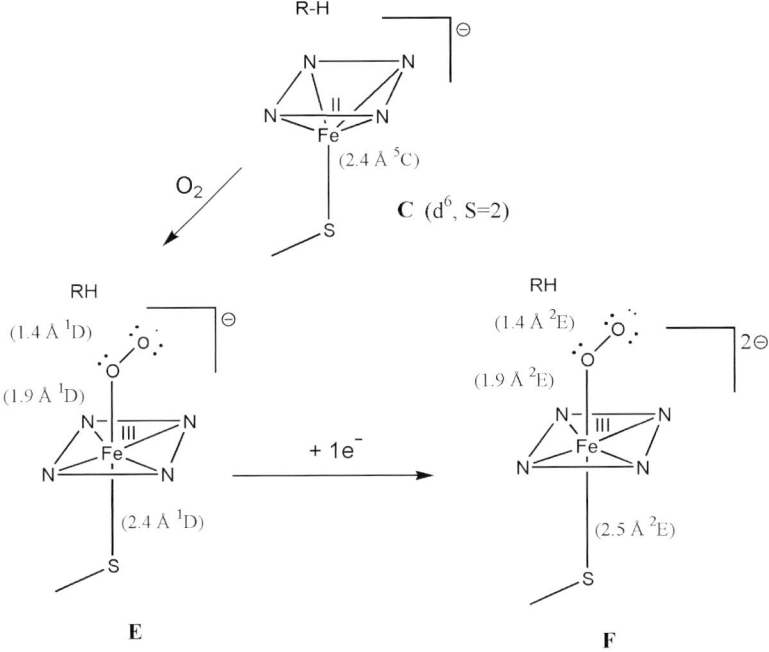

Figure 6. DFT geometries (in parentheses; bond lengths are in Å; spin state in superscript) **C**: Fe^{II} complex in a quintet ground state; **E**: Singlet-state Fe^{III}-O_2^- complex; **F**: Fe^{III}-OO^{2-} (reduced form) in doublet ground state (S = 1/2). RH = substrate [33]. Parallelograms with nitrogen atoms at the four corners represent a porphyrin ring.

3. THREE-DIMENSIONAL STRUCTURES OF THE ACTIVE SITES OF CYTOCHROME P450 ENZYMES

Even though the discovery of P450 took place over forty-five years ago, the exact nature of the active species responsible for the oxygen insertion step has been the subject of intensive debates. Researchers continue to puzzle over fundamental questions such as: How does the enzyme activate dioxygen? Why are the catalytic activities of P450 and peroxidases so different even though both systems form quite similar active oxygen species during their catalytic cycles? What factors allow over 4000 P450 isoforms (http://drnelson.utmem.edu/cytochromep450.html) to catalyze reactions on a million or more substrates with very broad substrate and stereochemical specificity? The answers to many of these questions likely lie in the three-dimensional structures of the different P450 enzymes.

Based on the crystal structures (atomic coordinates) of camphor-bound oxy-P450-CAM [31], the active site conformation of the dioxygen complex of this enzyme is schematically shown in Figure 7. The residues shown include the

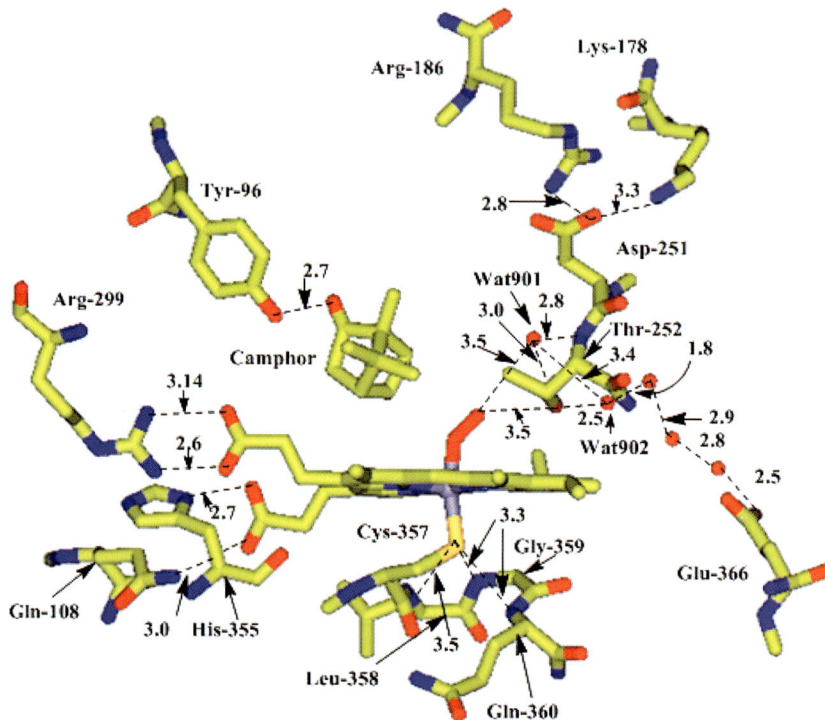

Figure 7. Schematic representation of the active site structure of the Fe^{II}-O_2 complex of P450-CAM and a possible proton delivery network involving a series of hydrogen bonded water molecules (the middle part of the right hand side) [31], determined from X-ray coordinates and drawn using Protein Data Bank file (1DZ8).

dioxygen-heme unit, and several adjacent amino acid groups (Tyr96, Cys357, Asp251, Thr252) and water molecules that are located near the heme, as well as the bound camphor substrate. It is known from a crystallographic characterization of wild-type oxy-P450-CAM that two new hydrogen bond-stabilized water molecules (WAT901 and WAT902), not seen in the Fe^{III}, Fe^{II}, and Fe^{II}-CO enzyme crystals, exist on the distal side of the heme. These water molecules likely serve with other aligned water molecules as the source of the protons required for formation of the reactive oxygen intermediate [31]. The hydroxyl group of T252 and WAT901 serve as hydrogen bond donors to the distal oxygen atom of the heme iron-bound dioxygen for wild-type P450-CAM. Although the Thr252-O···O-O-Fe distance (3.5 Å) is somewhat longer than a usual hydrogen bond distance (≤ 3.0 Å), it is considered to be within hydrogen bonding distance [31]. In the camphor-bound Fe^{III}-cyanide complex of P450-CAM, which Schlichting and coworkers [38] have proposed as a good structural model for oxyferrous P450-CAM [31], the heme iron-coordinated cyanide indeed forms a hydrogen bond to the oxygen atom of the Thr252 hydroxyl group with the Fe-C-N···O-Thr252 distances of 2.7 and 3.0 Å (for the two differently bent Fe-C-N conformers).

Phenobarbital-induced rabbit liver microsomal P450 (CYP 2B4) is a very important mammalian P450 that has been extensively studied over the years. The crystal structure of the Fe^{III} state of this enzyme with an imidazole inhibitor bound [39] has enabled researchers to compare and contrast the heme environment of a mammalian drug metabolizing P450 with the bacterial camphor-hydroxylating P450-CAM. The distal pocket of CYP 2B4 shows a network for proton delivery based on the hydrogen bonding involving Thr302, Glu301 and two adjacent water molecules (Figure 8). It is likely that the Thr302 hydroxyl

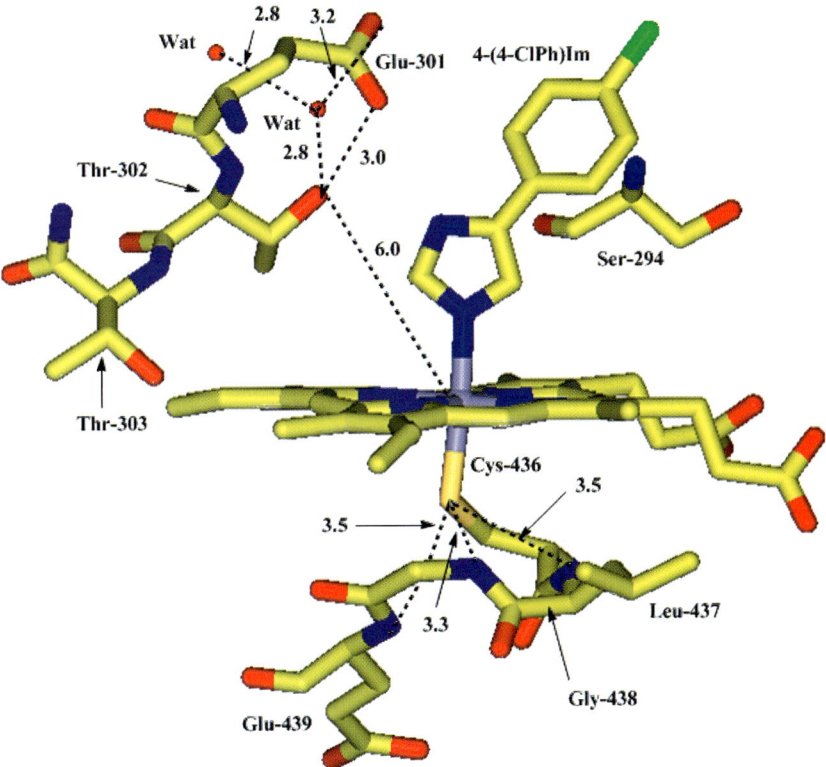

Figure 8. Schematic representation of the active site structures of the ferric wild-type microsomal CYP 2B4 with the specific inhibitor of 4-(4-chlorophenyl)imidazole bound to the heme iron [39], drawn based on the X-ray coordinates using Protein Data Bank file (1SUO).

group forms a hydrogen bond to the outer-oxygen atom in the dioxygen complex of CYP 2B4. Glu301 presumably plays a similar role to Asp251 in bacterial P450-CAM [40]. Asp251 plays a major role in the kinetics of the proton transfer to Fe^{III}-peroxo complex in wild type P450-CAM [41]. This proton transfer is the rate-determining step in turnover of the Asp251Asn mutant unlike second electron transfer for wild-type P450-CAM. Consequently, the kinetic isotope solvent effect for the Asp252Asn mutant is 10 compared to 1.8 for the wild-type enzyme [40]. Furthermore, the turnover rate for camphor hydroxylation of this mutant is ~ 2 orders of magnitude slower than for native (wild-type) P450-CAM. Another prominent feature is that both the Glu301 carboxylate group oxygen (3.0 Å) as well as an adjacent water (2.8 Å) (not shown) molecule are hydrogen bonded to the Thr hydroxyl group (Figure 8).

3.1. Substrate Binding Specificity of P450

Generally, the substrate binding pockets of P450 enzymes are hydrophobic. This is consistent with the main role of P450 enzymes in the activation and detoxication of hydrophobic xenobiotic chemicals, including many carcinogens. But each P450 isoform has a unique binding pocket for its natural substrate. In P450-CAM, the hydroxyl group of Tyr-96 (Figure 7) forms a hydrogen bond with the keto-group of camphor to orient the substrate close to the iron. The rest of the camphor molecule is hydrophobic and is held in position by a number of hydrophobic side chains that line the substrate binding pocket.

The active pocket of CYP 2B4 contains an iron protoporphyrin IX, heme *b*, which is axially coordinated by a cysteine thiolate (Cys-436) [39,42]. A loosely coordinated water molecule is displaced when the substrate binds to create a vacant coordination site where molecular oxygen binds after the iron is reduced to the Fe^{II} state. Figure 8 shows the specific bulky inhibitor of 4-(4-chlorophenyl)imidazole bound at the active site [39]. But its natural substrate, benzphetamine (BP) (Figure 9), is a tertiary amine with a pK_a value of 8.6 [43]. The CYP 2B4 substrate binding site is hydrophobic with a low dielectric constant

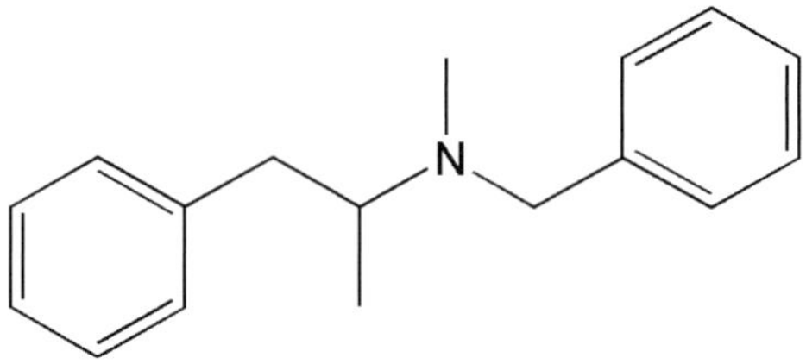

Benzphetamine (BP)

Figure 9. Structure of benzphetamine.

and therefore only neutral BP will bind. At pH 7.5, there will be a six-fold excess of the cationic form over the neutral form of benzphetamine [44].

The redox potentials reported for Fe^{III}/Fe^{III} P450-CAM in the presence and absence of camphor are $-170\,mV$ and $-300\,mV$, respectively [7,44]. The calculated redox potential of wild type Fe^{III} CYP 2B4 in the presence of 1mM BP is $-245\,mV$ [45]; in the absence of substrate, it is $\sim -330\,mV$ [45,46]. In the substrate-free structure of CYP 2B4, the distal pocket is occupied by a water molecule that coordinates to the Fe^{III} heme iron (low spin, six coordinate). However, hydrophobic, bulky BP displaces this water molecule to form the high spin, five coordinate heme and increases the heme iron redox potential considerably by 85 mV.

The fatty acid hydroxylase from *Bacillus megaterium* P450 BM3 (CYP 102) [47,48] is composed of heme- and FMN/FAD-containing reductase domains linked together on a single polypeptide (i.e., a self-sufficient P450 enzyme). The protein is a soluble, multidomain electron-transfer enzyme, and represents an excellent model system for studying structure/function relationships in P450s and the mechanism of electron transfer. Figure 10 (*top*) shows the crystal structure of a complex between the heme- and FMN-binding domains of P450 BM3 (but without the FAD domain) was found to be the simplest model to follow the FMN to heme intramolecular electron transfer [49]. Figure 10 (*bottom*) shows the structure of the cytochrome P450 BM3 heme domain complexed with the fatty acid substrate, palmitoleic acid. The X-ray structure of the Fe^{III} heme iron in substrate-free heme domain reveals an iron-cysteinate Fe-S bond of 2.25 Å [50,51]. The figure also shows the substrate binding pocket where fatty acids bind. Careful analysis reveals that there are no hydrophilic amino acids in the substrate binding channel, which facilitates fatty acid hydroxylation by bringing elongated, highly hydrophobic fatty acids close to the reactive center of the protein. Homogeneous dioxygen complexes of both the heme domain and holo enzyme of P450 BM3 have recently been prepared and spectroscopically characterized [52].

3.2. Role of the Distal Pocket Threonine in P450 Catalysis

The highly conserved threonine residue (Thr252 in P450-CAM, Thr302 in CYP 2B4, and Thr268 in P450 BM3) has been shown to play a vital role in a proton transfer cascade from bulk solvent to the outer oxygen. Previous investigations on P450-CAM T252A mutant have suggested that Thr252 might operate as a general acid catalyst by donating a proton to oxygen during substrate hydroxylation [53–55]. The Thr has a hydrophilic side chain that could accept external protons to operate in a proton shuttle without becoming negatively ionized. The Thr252 side chain hydroxyl group in P450-CAM could participate in hydrogen bonding with distal (outer) oxygen of the oxy-P450 complex [31]. As shown in Figure 8, in the crystal structure of Fe^{III} CYP 2B4, the Thr hydroxyl group oxygen is only

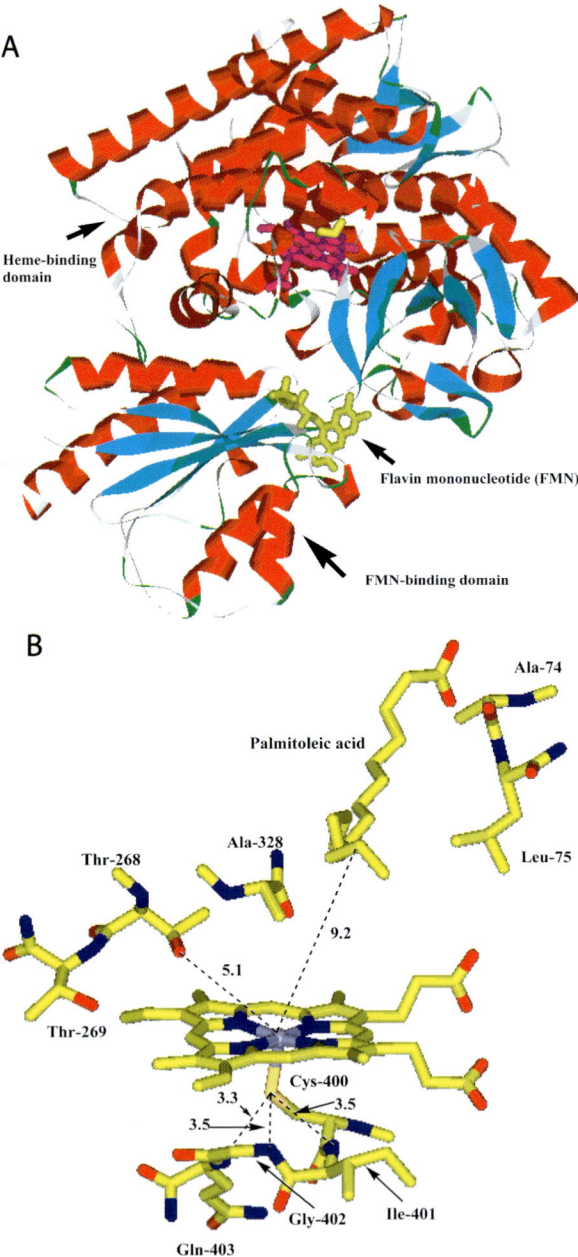

Figure 10. A: The crystal structure of the complex between the heme- and FMN-binding domains of bacterial cytochrome P450 BM3 (CYP 102) from *Bacillus megaterium* as a prototype for the complex between eukaryotic microsomal P450s and P450 reductase. B: The active site structure of the ferric P450 BM3 heme domain. Both structures were drawn based on the X-ray coordinates [51] using Protein Data Bank files (1BVY and 1FAG).

6 Å away from the heme iron. When the oxyferrous intermediate is formed, this should be within the hydrogen bond distance of outer oxygen of the oxyferrous CYP 2B4 complex similar to the case of P450-CAM. Furthermore, replacement of Thr to an Ala residue yields a more hydrophobic environment to the distal heme pocket that contributes the significant change in the properties of the oxyferrous complex as demonstrated for P450-CAM [56,57]. This interruption of the proton delivery may be also attributed to the prominent structural changes, which are also reflected in spectral changes [56]. Previous catalytic work done on the T302A CYP 2B4 mutant also shows a high accumulation of the Fe^{III}-hydroperoxo intermediate **G** (Figure 4), which could, therefore, become the active oxidant [24].

The crystal structure of 4-(4-chlorophenyl)imidazole inhibitor-bound Fe^{III} microsomal CYP 2B4 has revealed that a water molecule is present inside the active site near Thr302 even in the Fe^{III} state as mentioned above [39]. The recent crystal structure of the dioxygen complex of the T252A P450-CAM mutant has shown that a corresponding water molecule near Ala252 and one close to the Glu251-Ala252 peptide NH group are still present in the dioxygen complex of this mutant. It has also been shown that the absence of the Thr hydroxyl group hydrogen bond to the distal oxygen does not affect the orientation of the Fe^{II}-O-O unit in the T252A P450-CAM mutant [57].

The highly conserved Thr268 in P450 BM3 has been shown to play a vital role in a proton transfer cascade from bulk solvent to the outer oxygen via the side chain of the Thr268 hydroxyl group in P450 BM3. Mutation of Thr268 to alanine influences the effects of substrate on the spin state and redox potential changes of the heme iron, decreases the coupling efficiency as well as reduces the rate of substrate oxidation (16% coupling for lauric acid oxidation and 9% coupling of arachidonic acid oxidation by T268A P450 BM3) [58,59].

4. ROLE OF THE CYS LIGAND: THE PROXIMAL THIOLATE 'PUSH' AND DISTAL PROTON-DELIVERY

Heme-containing peroxidases generate high-valent transient Cpd I intermediates by reaction of the Fe^{III} resting enzyme with hydrogen peroxide [8,60–62]. What is unique to the P450 enzymes is the ability to form this same type of high-valent intermediate via a reaction cycle (Figure 4) involving an *oxyferrous* heme iron intermediate. One-electron reduction of this latter state followed by protonation of the distal oxygen gives the Fe^{III}-hydroperoxo state that is set up for heterolytic cleavage of the oxygen-oxygen bond after addition of a second proton to the distal oxygen [41]. The ability of P450 enzymes to retain the anionic thiolate ligand upon reduction to the Fe^{II} state is critical to this process. Stabilization of the axial thiolate ligand by hydrogen bonding appears to be of considerable importance for the retention of the anionic ligand following reduction [63]. In

P450-CAM, the axial cysteinate sulfur atom is stabilized by NH···S hydrogen bonds from adjacent peptide backbone amide protons of Leu 358 (3.5 Å), Gly 359 (3.3 Å), Gln 360 (3.3 Å) (Figure 7). Morishima and coworkers studied the role of the axial thiolate in P450-CAM through the mutation of the Gln360 to Pro and Leu358 to Pro amino acid residue [64,65]. In each of these mutants, one of the conserved amide protons that are proposed to stabilize the negative charge of the thiolate sulfur by NH···S hydrogen bonding has been removed. This increases the electron donation of the axial thiolate as evaluated by a change in reduction potential of -71 mV (Q360L) and -36 mV (L358P). The L358P mutant showed an increased push effect (see below) of the axial cysteine when compared with the wild-type enzyme, which led to increased protonation of the inner oxygen and decreased protonation of the outer oxygen atom in the putative Fe^{III}-hydroperoxo intermediate of P450-CAM.

In 1976, Dawson et al. proposed that heterolytic cleavage of the O-O bond in the P450 intermediate **G** is facilitated by the proximal thiolate ligand (Figure 11) [66]. It was posited that cysteine, being the most polarizable of all the

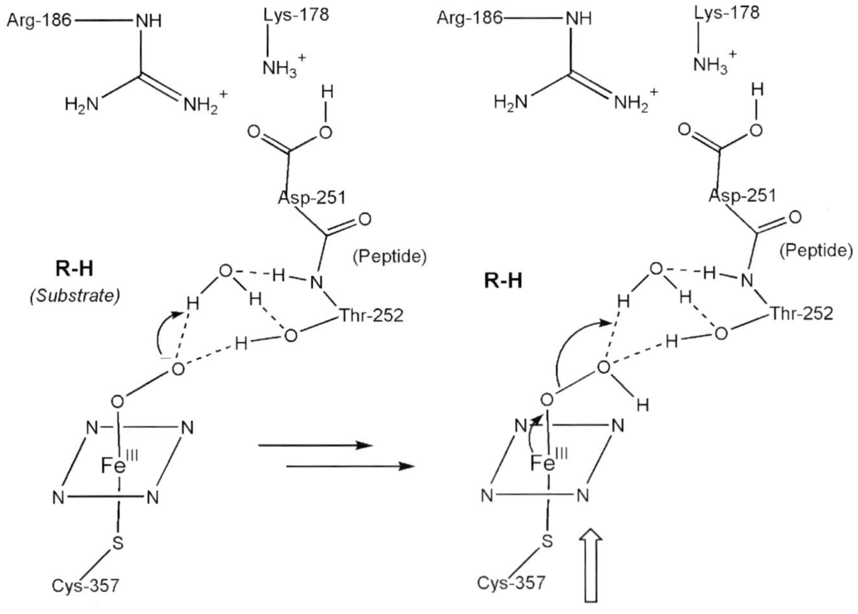

Figure 11. Schematic representation of the initially proposed push-pull mechanism for oxygen-oxygen bond cleavage to generate Cpd I species of P450-CAM [4,8a]. Left: Protonation of the Fe^{III}-peroxo intermediate (**F**). Right: Protonation of the Fe^{III}-hydroperoxo state (**G**) to generate P450 Cpd I (**H**). The lettering (**F**, **G**, and **H**) correspond to those for the intermediates shown in Figure 4. Parallelograms with nitrogen atoms at the four corners represent a prophyrin ring.

amino acid ligands, provides a strong proximal 'push' of electron density through the heme and into the O-O bond of the *trans* peroxide promoting heterolytic cleavage. Several lines of evidence have been reported by the Dawson laboratory in support of this proposal [4,67]. Once the O-O bond is heterolytically cleaved, the electron-donating character of the axial thiolate ligand also helps to stabilize the high valent nature of the resulting Cpd I species. A similar, although weaker, 'push' from a proximal histidine imidazolate ligand was proposed to play a comparable role in the generation of cytochrome *c* peroxidase Cpd I by Poulos and Kraut in 1980 [68]; the imidazolate 'push' was suggested to work in concert with a 'pull' from conserved distal His and Arg residues to heterolytically cleave the O-O bond of iron bound peroxide en route to Cpd I [68,69].

For P450, Gerber and Sligar have proposed a distal 'charge relay' involving highly conserved amino acids to play the role of the 'pull' analogous to the 'push-pull' mechanism for peroxidases [40]. The distal charge relay involves two conserved amino acids, Thr252 and Asp-251, plus water to carry out the proton delivery (Figure 11) [40]. The role of Thr252 has been difficult to assess because of the uncoupling reactions [53,54]. Recently, it has been proposed that the Thr252 residue stabilizes the Fe^{II} P450-O_2 intermediate through hydrogen bond formation between its hydroxyl group and the distal oxygen of the bound O_2. Substitution of Asp-251 with Asn or Gly yields mutants in which the rate of electron transfer from Pdx to oxyferrous P450-CAM is considerably reduced and it has been suggested that this is due to the important role of the Asp-251 residue in the proton delivery conduit [40,57,70].

The Fe^{III}-hydroperoxo species (**G**), sometimes called Compound 0 (Cpd 0), is the precursor of Cpd I (**H**). The latter is generated by protonation of Cpd 0, likely via proton transfer from Thr-252 to the distal (outer) oxygen atom in Cpd 0 (Figure 11) [40,53,54,71]. If the proximal (inner) oxygen is protonated, the resulting complex will release hydrogen peroxide from the heme returning it to the Fe^{III} resting state, **B** or **A** depending on the presence or absence of substrate, respectively. This is called uncoupling and is a serious competitor for the formation of Cpd I species. We know relatively little about the factors that control this uncoupling process and promote formation of the Cpd I species, although it is clear that Thr-252 is very important to tight coupling of electron transfer and hydroxylation in P450-CAM. A theoretical study has indicated that uncoupling of Cpd 0 ($\sim$ 6 kcal mol^{-1}) is less exothermic than that of formation of Cpd I from Cpd 0 [72].

Examination of the Fe(II)-dioxygen crystal structure of P450-CAM [31] creates some questions regarding the proton delivery system which is responsible for the so called 'pull' effect (Figure 11) [40,53,54,71]. The main problem is the Asp-251 residue and an unfavorable orientation of its carboxylate group in the crystal structure as shown in Figure 7. However, it is now believed that Asp-251 is responsible for the stabilization of the water molecule (WAT-901) via a 'carbonyl switch' (i.e., rotation of Asp-251 into a position to interact with WAT-901) [31]. Furthermore, the crystal structure in Figure 7 reveals an

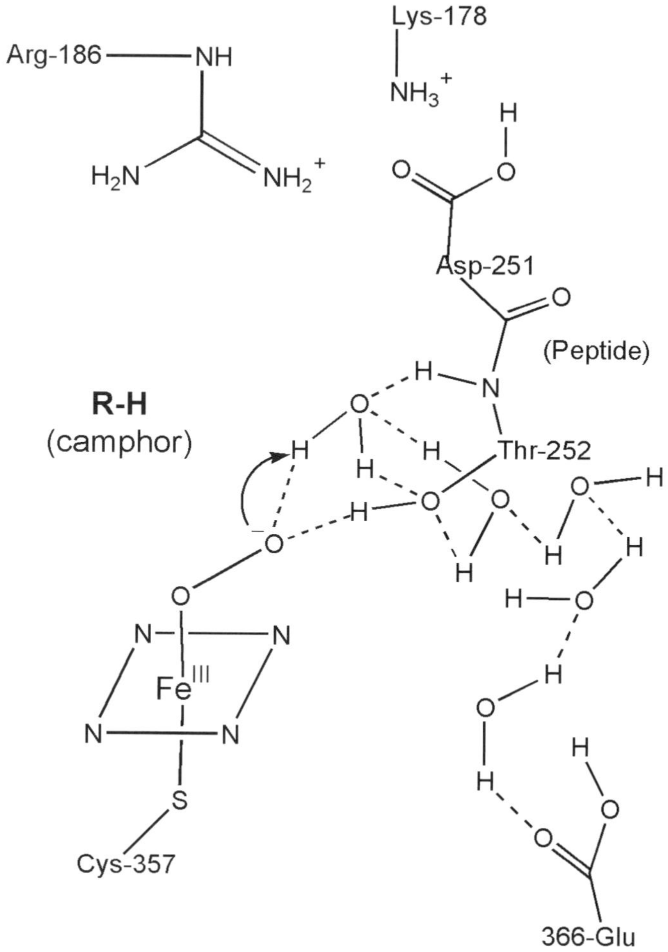

Figure 12. Schematic view of the water channel in the Fe^{II}-O_2 complex of the P450-CAM [31]. This pathway would serve as a conduit for protons needed later in the reaction cycle (see also Figure 7). Parallelograms with nitrogen atoms at the four corners represent a prophyrin ring.

extensive hydrogen-bonding network in the Fe^{II}-dioxygen moiety leading from the outer oxygen to the carboxylate group of the highly conserved amino acid residue, Glu-366. A water-accessible channel together with Thr-252 (Figures 7 and 12) can act as the proton source for the heterolytic cleavage of the O-O bond of P450-CAM. This is also the case for P450eryF [37,73].

The key role of the cysteinate ligand has also been confirmed by site-directed mutagenesis. The mutation of the Cys357 to histidine in P450-CAM generates a mutant that is incapable of camphor hydroxylation but retains the catalytic properties of a proximal histidine ligated protein [74]. This is seen in the ability of the C357H mutant to perform hydrogen peroxide–dependent guaiacol peroxidation at a rate that is two orders of magnitude higher than wild-type P450-CAM. This loss of monooxygenase activity of C357H is attributed to the change (apparent collapse) in active site structure [74]. The C436H/A298E double mutant of CYP 2B4 has also been shown to be incapable of oxygenating substrates, but exhibits an NADPH oxidase activity (yielding an H_2O_2 as product) as well as an H_2O_2-dependent pyrogallol peroxidation activity that is considerably greater than that of the wild-type CYP 2B4 [75]. The results of these studies provide support for the proposed role of the thiolate ligand in strongly facilitating O-O bond cleavage by P450 enzymes as well as aiding in other important roles such as substrate binding and electron transfer [74,75].

Induced coordination structure alterations of heme iron are thus intrinsically linked to enzymatic activity in the catalytic cycle. Triplet dioxygen binds to Fe^{II} P450 to form the dioxygen adduct by pairing an electron from triplet oxygen and one electron from Fe^{II} to form the Fe^{III}-O_2^- adduct. This intermediate dissociates to Fe^{III} and O_2^- via an 'uncoupling' (autoxidation) reaction. Investigations of the dioxygen-bound complex of microsomal P450s have been hampered due to their high rate of autoxidation [46]. The electronic structure and the chemical properties of the P450 Fe^{II}-O_2 species are different from those of the analogous oxyferrous complexes of hemoglobin or myoglobin. The stability of the oxyferrous states of these oxygen carriers relies on the less electron rich histidine proximal ligand in comparison to the electron donating cysteine in P450s. The anionic cysteinate proximal ligand increases the autoxidation reaction that leads to the breakdown of the ferrous-O_2 complex [2,30].

5. MULTIPLE MECHANISMS OF P450 CATALYSIS

This section mainly focuses on two main reactions of P450 enzymes, hydroxylation and epoxidation. There are no universally accepted mechanisms for these P450 reactions. The main reason for this could be the range and complexity of the reactions carried out by P450 enzymes in the face of its versatility as a catalyst. Indeed, monooxygenase enzymes perform one of the most intriguing as well as difficult reactions, the insertion of an oxygen atom from molecular oxygen into unactivated C-H bonds under mild conditions.

5.1. Mechanism of Hydroxylation

Two main mechanisms have been proposed for the hydroxylation of the sp^3-hybridized C-H bonds of saturated hydrocarbons by P450, a concerted

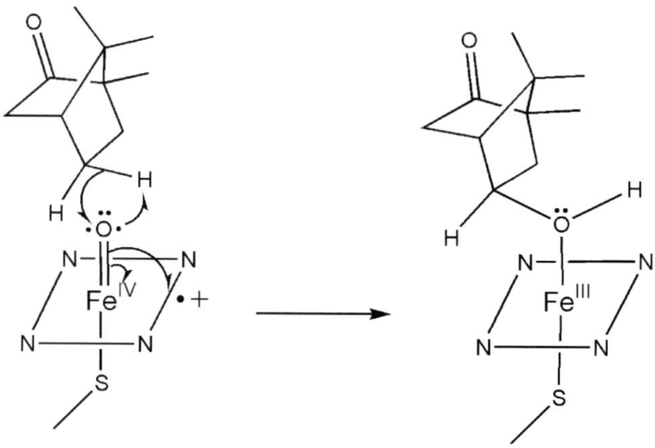

Figure 13. The proposed concerted mechanism in hydroxylation reactions catalyzed by P450 enzymes. Parallelograms with nitrogen atoms at the four corners represent a prophyrin ring.

mechanism (Figure 13) and a radical abstraction/recombination mechanism (Figure 14). The concerted mechanism for unactivated alkane hydroxylation was initially proposed based on the retention of stereochemistry and small apparent kinetic isotope effects (KIE) ($k_H/k_D < 2$) that had been observed [76–78]. A concerted 'oxene' insertion mechanism starts from the formation of highly reactive compound I species with two unpaired electrons in p-π : d-π-antibonding orbitals of the ferryl. As shown in Figure 13, the proposed concerted mechanism proceeds without formation of a radical intermediate and would be expected to display relatively low apparent intermolecular isotope effects ($k_H/k_D < 2$) since the C-H bond is not fully broken during O-atom insertion.

However, in the late 1970s, Hjelmeland et al. [79] and Groves et al. [80] determined the apparent isotope effects for benzylic hydroxylation and aliphatic hydroxylations, respectively, and found them to be much larger ($k_H/k_D > 11$). The reason for this difference is that the new studies determined the KIE for an *intramolecular* reaction, i.e., competition for two equivalent C-H bonds within an individual molecule, whereas the former approach was for the *intermolecular* competition between C-H or C-D bonds on separate molecules. Examination of the relative rate of C-H versus C-D within an individual molecule provides a more accurate measure of the KIE when C-H bond breakage is *not* the rate-limiting step in the overall reaction. Since the second electron transfer is the rate-limiting step in many P450 catalytic reactions (Figure 4), this approach provides a better gauge of the KIE in P450-catalyzed reactions.

The larger KIE observed by Hjelmeland and Groves suggested a complete breakage of C-H bond during hydroxylation. Figure 14 (top) indicates three possible resonance forms of the compound I species involved in P450 hydroxylations, where the radical might be located on (i) the conjugated porphyrin ring, (ii) the sulfur atom or (iii) the oxygen atom. The lack of evidence for the involvement of carbocations (no rearrangement products) and carboanions (extremely low acidity of C-H bonds) laid the foundation for a two step radical mechanism, generally known as the 'oxygen rebound' mechanism (Figure 14) and attributed to Groves [81]. The basis for this proposal came from the loss of stereochemistry during hydrogenation of a norbornane derivative [80,82]. Therefore, hydroxylation reactions catalyzed by P450s are not stereospecific. For example, stereochemical scrambling was observed for the stereospecifically tetradeuterated norbornane [17] using purified liver microsomal P450 (CYP 2B4) (Figure 15(i)). The observation of an *endo* alcohol (b) with only three deuterium atoms and an *exo* alcohol with all four deuterium atoms (a') required the stereochemical scrambling during catalysis [80]. This led to the proposal that a carbon radical intermediate was a critical component of the oxygen activation process (Figure 14).

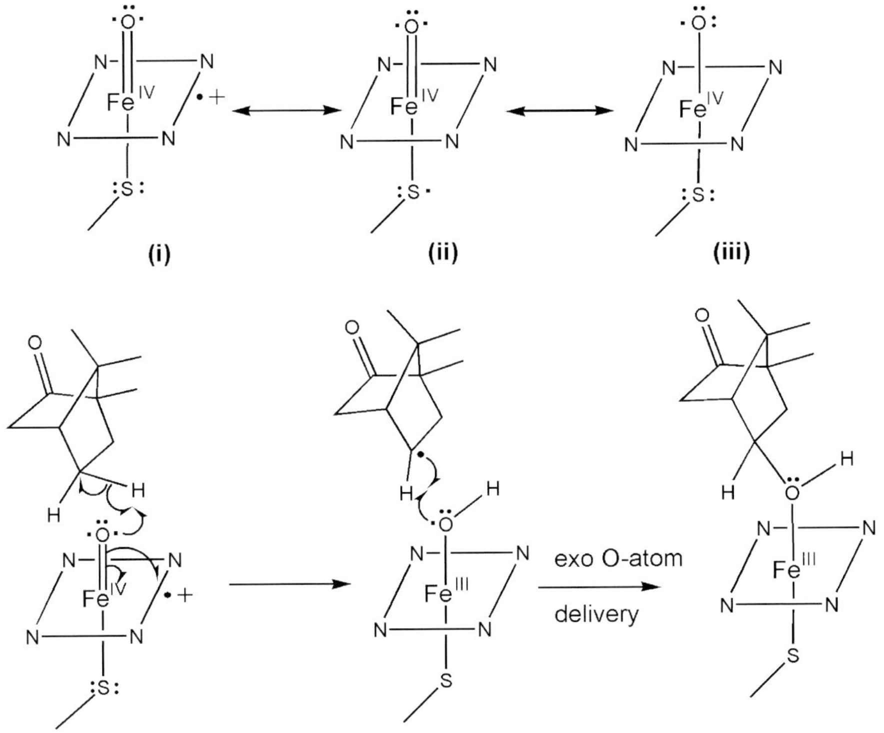

Figure 14. The oxygen rebound mechanism (a cage controlled radical pathway) for hydroxylation reactions catalyzed by P450 enzymes. Parallelograms with nitrogen atoms at the four corners represent a porphyrin ring.

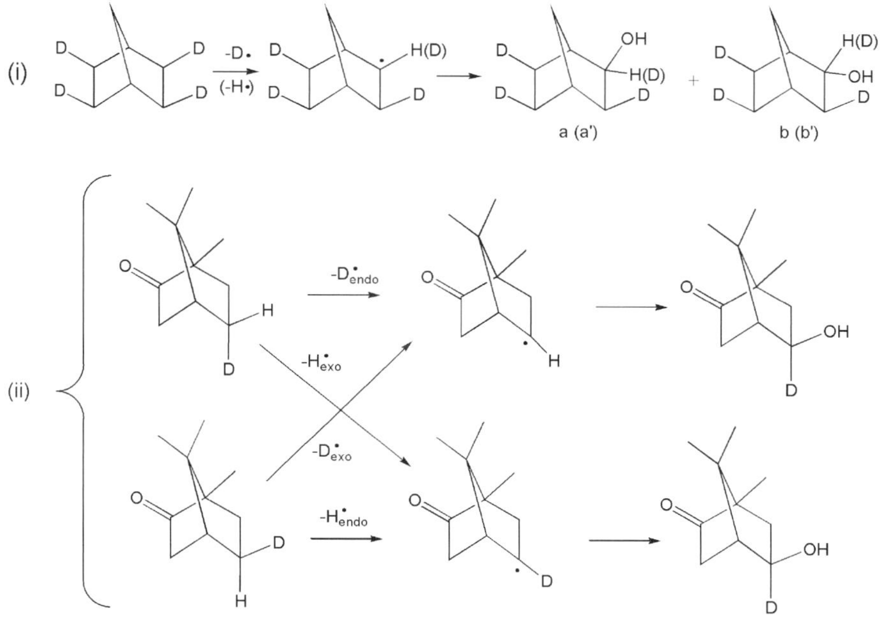

Figure 15. Hydroxylation reactions with stereochemical scrambling: (i) (2,3,5,6-*exo*-d_4)-norbornane with CYP 2B4 [80] and (ii) (5-*exo*- and 5-*endo*-d_1)-camphor with P450-CAM [33].

Furthermore, Sligar and coworkers reported that hydroxylation of either 5-*endo* or 5-*exo*-d_1 camphor led to loss of either *endo* or *exo* deuterium atom, but always led to the formation of the 5-*exo* hydroxycamphor product (Figure 15(ii)) [83]. This second example of stereochemical scrambling provided strong additional evidence for the two step oxygen rebound mechanism. Additional experiments by Groves and Adhyam [84] and White [85] further confirmed the involvement of stereochemical scrambling during hydroxylation. Furthermore, more thorough theoretical and spectroscopic investigations have supported the radical hydroxylation mechanism by P450s. The oxygen rebound (cage-radical) mechanism via the iron-oxo triplet state as shown in Figure 14 is the generally accepted P450 hydroxylation mechanism.

The cage controlled radical, or oxygen rebound mechanism (Figure 14), involves hydrogen atom abstraction to generate a carbon radical intermediate. The observed intrinsic isotope effects ($k_H/k_D < 5$–12) in hydroxylation reactions is often taken to be evidence for a nonconcerted mechanism involving an intermediate radical generated after abstraction of a H-atom by the high-valent iron-oxo species. The use of ultrafast 'radical clocks' [86–88] for the P450-catalyzed hydroxylation of hydrocarbons supports the existence of a radical pathway. In the first study of this kind, the oxidation of bicyclo[2.1.0]pentane

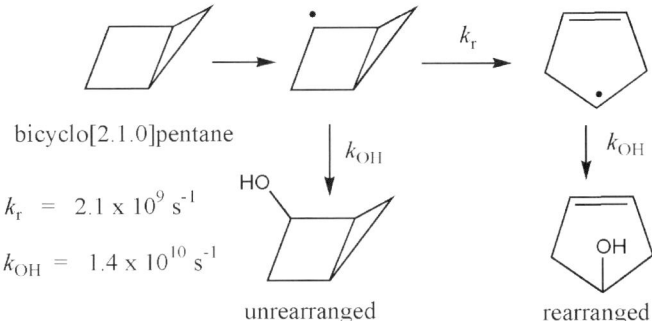

Figure 16. Bicyclo[2.1.0]pentane acts as a radical clock during hydroxylation with P450 2B1. Examination of the rate of the hydroxyl radical transfer reaction (k_r) (a ratio of rearranged to unrearranged products) produces an estimate of the rate of the radical rebound reaction between the substrate carbon radical and the enzyme iron-bound hydroxyl radical [89,90].

(Figure 16) was found to give a ratio of rearranged and unrearranged products that implicated a recombination rate of $\sim 10^{10}\,\mathrm{s}^{-1}$, and thus a lifetime of 50 ps, for the carbon radical intermediate [89,90]. Subsequent experiments suggested a somewhat faster rate for primary carbon radicals [91].

However, when more elegant and faster radical clocks were examined, the results obtained were confusing and conflicting. The calculated rates for the recombination reaction required by the data were so rapid (in the order of $10^{12} - 10^{13}\,\mathrm{s}^{-1}$) that the existence of a true carbon radical intermediate became questionable [86,92,93]. Rates of this magnitude are more consistent with a transition state rather than an actual intermediate. Furthermore, the subsequent development of probes that can distinguish between radical and cation intermediates suggested that the intermediate might be predominantly a cation rather than a radical and therefore that the oxidizing species might be something altogether different from the Cpd I species. One alternative hydroxylating species that has been proposed based on these studies is the Fe^{III}-hydroperoxide complex preceding the Cpd I species in the catalytic activation of dioxygen by P450 [17,76,82,83]. However, this notion was challenged by density functional theory (DFT) calculations which predict that the hydroperoxide complex should be a poor hydroxylating agent, and only inferential evidence has been advanced for its involvement [33].

Recent radical clock studies with norcarane, spiro[2.5]octane [94] and cyclopropyl fatty acids [95] seem to have led to a consensus that agrees with the initial radical clock studies implicating radical lifetimes that range from 16 to 50 ps. Trace amounts of cationic rearrangement products were detected in the norcarane but not in the cyclopropyl fatty acid studies. When put together these results suggest that differences in the measured radical lifetimes might reflect steric or electronic differences, because secondary radicals generally give slower apparent recombination rates than primary radicals [96].

Computational investigations by Shaik and coworkers have provided additional insight into the possible origin of the discrepancies among the experimental results [97–101]. In particular, DFT calculations suggest a two-state reactivity paradigm in which hydrogen abstraction yields either a doublet or quartet radical species, depending on which of the two possible spin states of Cpd I actually abstracts the hydrogen (more below).

5.2. Mechanism of Epoxidation

The insertion of an oxygen atom to C=C π-bonding network of an alkene generates an epoxide product. Enantioselective epoxidation still remains a synthetic challenge. But most often P450 carries out an epoxidation reaction with retention of stereochemistry. Even though this suggests a concerted path for the oxygen atom transfer (Figure 17(b)), the mechanism of olefin epoxidation by P450 is a

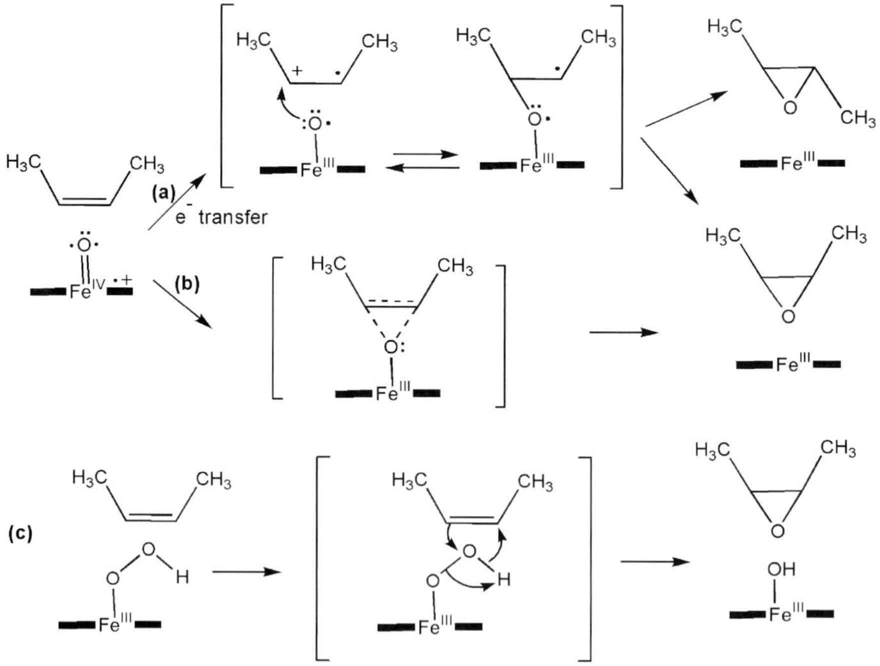

Figure 17. Mechanisms of olefin epoxidation (a) by P450 Cpd I involving a charge-transfer complex, step-wise insertion of oxygen with epimerization, (b) by P450 Cpd I leading to a concerted insertion of oxygen, and (c) by Fe^{III}-hydroperoxo P450 in a concerted reaction. The scheme is drawn based on information provided in [24,114]. The horizontal thick bars on both sides of the iron atom represent the porphyrin plane.

stepwise process. The earlier suggested pathway involving Cpd I is shown in Figure 17(a). This stepwise process is initiated by formation of a charge transfer complex followed by electron transfer to an olefin cation radical. This now can proceed to epoxide product via radical or cationic paths.

However, investigations by Coon and co-workers into alkene epoxidation have suggested that Fe^{III}-hydroperoxo species in P450 can serve as a second electrophilic oxygen donor (Figure 17(c)) [24]. In 2003, Jin et al. provided especially strong evidence in support of this proposal [25]. They reported that T252A P450-CAM can efficiently carry out the epoxidation reaction, although it hardly produces any hydroxylated product of camphor. The inability of this mutant to hydroxylate camphor is presumed to be due to almost complete lack of Cpd I formation species. The epoxidation of olefin carried out by the T252A mutant therefore provides compelling evidence that Fe^{III}-hydroperoxo P450 is capable of functioning as a second P450 oxidant.

6. MULTIPLE OXIDANTS IN P450 CATALYSIS

The ferryl Fe^{IV}=O porphyrin π-radical cation radical Cpd I species (**H**, Figure 4) is widely considered to be the active oxidant in P450-catalyzed reactions, especially for the hydroxylation reaction [6,102]. This proposal developed from the well-known chemical properties of peroxidases and porphyrin model compounds [103] as well as from the need for a highly active species capable of catalyzing oxygen atom insertion into unactivated carbon-hydrogen bonds [4,80]. However, other intermediates in the P450 catalytic cycle (Figure 4) such as the Fe^{III}-peroxo and Fe^{III}-hydroperoxo species (**F** and **G**, respectively) have been identified as potential oxidants that could insert an oxygen atom into substrates in deformylation [4] and aromatization [105] reactions.

In particular, the most widely accepted mechanism for the final oxidative deformylation step catalyzed by P450-aromatase in the demethylation of androgen steroids to give estrogens was initially proposed by Akhtar et al. [106] to involve Fe^{III}-peroxo P450. The aromatase reaction proceeds via three stages whereby the methyl group is hydroxylated twice to give, after loss of water, an aldehyde (formyl) group. Finally, in the P450-catalyzed lyase reaction, the formyl group is oxidatively removed as formic acid with concomitant formation of the C_1-C_{10} double bond (Figure 18) [17]. Akhtar et al. [106] proposed that the final step is initiated by nucleophilic attack of the Fe^{III}-peroxo P450 on the electrophilic carbonyl carbon to generate a hemiacetal type intermediate (Figure 18). This is followed by a concerted cyclic rearrangement of the hemiacetal intermediate (**2**) to give the olefin product (**3**), formic acid, and regenerate the Fe^{III} P450 resting state. This type of oxidative deformylation which leads to a double bond upon removal of a methyl group has been shown to occur by several researchers. Cole and Robinson demonstrated that a 2,4-dien-3-ol analog of the 19-aldehyde

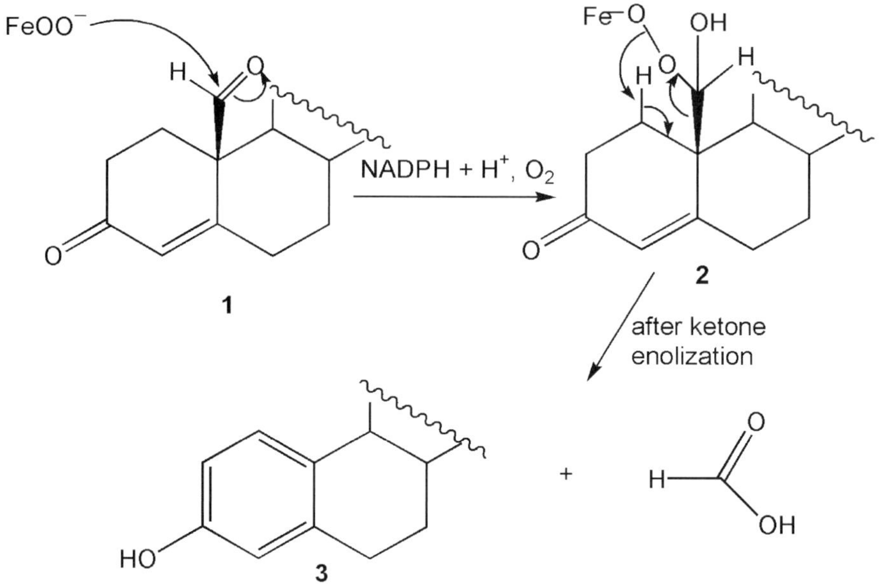

Figure 18. Proposed mechanism for deformylation by the ferric-peroxo species [106].

intermediate non-enzymatically reacts with H_2O_2 to produce the corresponding estrogen derivative [107].

Coon and coworkers found that CYP 2B4 can bring about a comparable reaction, the oxidative deformylation of cyclohexane carboxaldehyde and of various xenobiotic aldehydes to olefins and formate [108–110]. They also found that CYP 2B4 can convert 3-oxodecalin-4-ene-10-carboxaldehyde, a bicyclic analog of the steroidal 19-aldehyde intermediate (**1**, Figure 18), to an aromatic analog of estrone, thus mimicking the oxidative deformylation reaction of P450-aromatase [109]. In addition, the deformylation of cyclohexane carboxaldehyde to cyclohexene could be catalyzed by CYP 2B4 using H_2O_2 in place of dioxygen plus two electrons, presumably due to the ability of H_2O_2 to generate the Fe^{III}-peroxo species (**F**) to serve as oxidant. In contrast, artificial oxidants such as cumyl hydroperoxide, iodosobenzene, and m-chloroperbenzoic acid, which are presumed to directly form P450 Cpd I (**H**), were incapable of catalyzing the deformylation reactions [110].

In a related iron porphyrin model study, Watanabe et al. [111] reported that the Cpd I model complex exclusively oxidizes aldehyde compounds to the corresponding carboxylic acids, rather than to deformylated compounds. Although these results seem to argue persuasively for the key involvement of the Fe^{III}-peroxo P450 intermediate in the deformylation step, Ahmed and Davis have argued that deformylation is done exclusively by Cpd I [112]. Therefore, the identity of the second electrophilic oxidant remains a subject of speculation. Clearly, additional experimental work will be needed to further resolve the mechanism of oxidative deformylation by P450-aromatase.

Furthermore, more recent investigations with the threonine to alanine mutants of CYP 2B4 and 2E1 reveal changes in reaction rates for the epoxidation of olefins and hydroxylation of various substrates were observed. It is believed that this mutation disrupts the proton delivery system to the active site heme iron center [113] leading to the accumulation of alternate oxidants other than Cpd I and thus supporting the concept that there are three distinct oxidants in the P450 reaction cycle (Figure 4), namely: the Fe^{III}-peroxo state (**F**), the Fe^{III}-hydroperoxo intermediate (**G**) and the Cpd I (**H**) adduct [24].

The multiple oxidants model for P450 was initially proposed by Coon and coworkers [24,114] primarily based on intermolecular and intramolecular kinetic isotope effects obtained for hydroxylation of the *trans*-2(p-trifluoromethyl)cyclopropylmethane. These results indicate that in addition to Cpd I, the Fe^{III}-hydroperoxo species is a plausible electrophilic oxidant for hydroxylation reaction [115]. The kinetic isotope effect on the product distribution support the concept that three distinct oxidants are participating in P450 catalysis: nucleophilic Fe^{III}-peroxo, nucleophilic or electrophilic Fe^{III}-hydroperoxo, and electrophilic Cpd I [115] (Figure 19). The latter is shown as two spin states in equilibrium for reasons to be discussed in the next section.

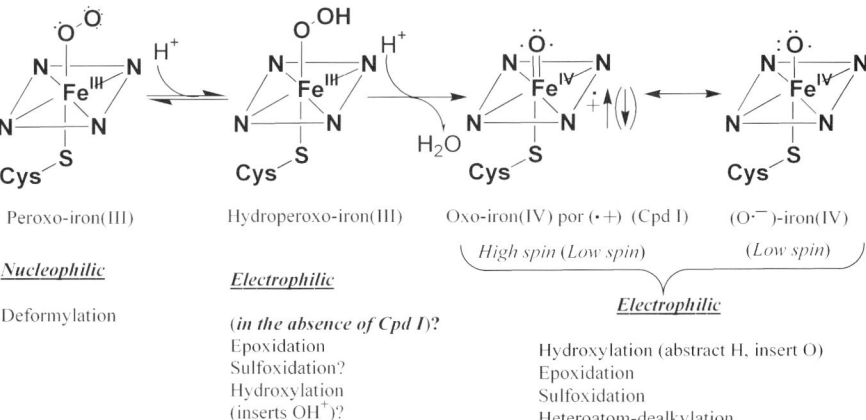

Figure 19. Multiple P450 oxidants and their potential roles in various reactions [24,115–117]. Parallelograms with nitrogen atoms at the four corners represent a prophyrin ring. In the high spin and low spin states of the Cpd I, the porphyrin (por) radical spin ($S = 1/2$) is parallel and antiparallel to the spin of the Fe = O triplet electrons ($S = 1$), respectively.

Single active oxidant

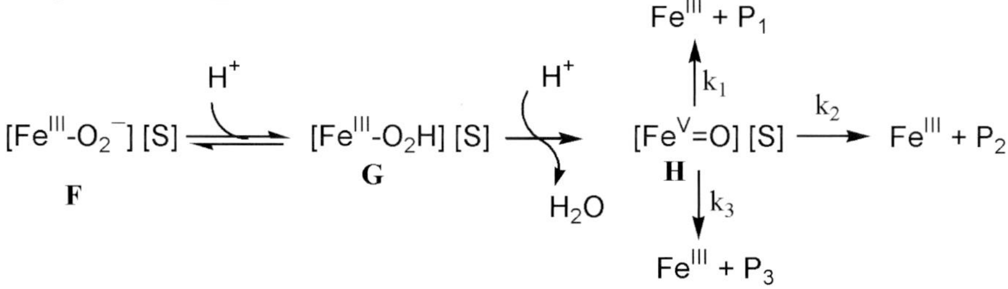

$$[Fe^{III}\text{-}O_2^-][S] \underset{}{\overset{H^+}{\rightleftharpoons}} [Fe^{III}\text{-}O_2H][S] \overset{H^+}{\underset{H_2O}{\rightarrow}} [Fe^V=O][S] \overset{k_2}{\rightarrow} Fe^{III} + P_2$$

with $[Fe^V=O][S]$ also going via k_1 to $Fe^{III} + P_1$ and via k_3 to $Fe^{III} + P_3$.

F = $[Fe^{III}\text{-}O_2^-][S]$, G = $[Fe^{III}\text{-}O_2H][S]$, H = $[Fe^V=O][S]$

$$\delta P_1 : \delta P_2 : \delta P_3 = k_1[Fe^V=O][S] : k_2[Fe^V=O][S] : k_3[Fe^V=O][S] = k_1 : k_2 : k_3$$

Multiple active oxidants

$$[Fe^{III}\text{-}O_2^-][S] \overset{H^+}{\rightleftharpoons} [Fe^{III}\text{-}O_2H][S] \overset{H^+}{\underset{H_2O}{\rightarrow}} [Fe^V=O][S]$$

F $\xrightarrow{k_1}$ $Fe^{III} + P_1$; G $\xrightarrow{k_2}$ $Fe^{III} + P_2$; H $\xrightarrow{k_3}$ $Fe^{III} + P_3$

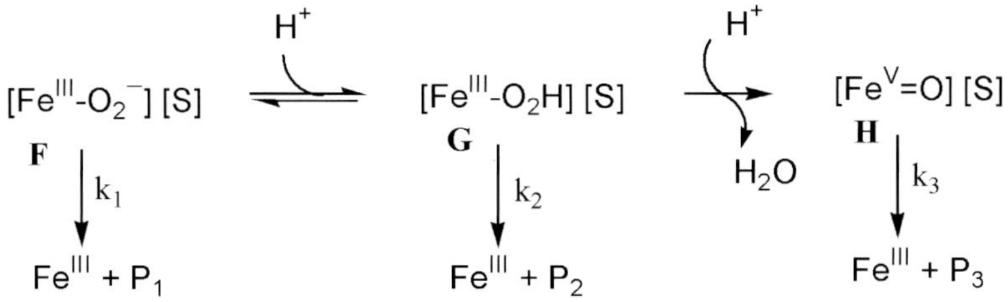

$$\delta P_1 : \delta P_2 : \delta P_3 = k_1[Fe^{III}\text{-}O_2^-][S] : k_2[Fe^{III}\text{-}O_2H][S] : k_3[Fe^V=O][S]$$

$P_1; P_2; P_3$: multiple products
$k_1; k_2; k_3$: rate constants
S: substrate

Figure 20. Model representing steady-state kinetics for product(s) formation from single or multiple active oxygen species (see also [117]). The letterings (**F, G,** and **H**) correspond to those for the intermediates shown in Figure 4. The Cpd I species is represented by $Fe^V=O$ using a formal oxidation state (V) of the heme iron.

As shown in Figure 20, the ratio of product formation ($\delta P_1/\delta P_2$) can be used to identify the single active oxidant or multiple active oxidants participation in catalysis [116,117]. For example, if product(s) is (are) formed during catalysis by the action of single oxidant, then the $\delta P_1/\delta P_2$ and ratio of rate constants (k_1/k_2) would remain constant (such as a mutation to disturb the proton delivery; T252A mutant in P450-CAM), but multiple oxidants show $\delta P_1/\delta P_2$ depends on the

Table 1. Investigation of multiple oxidants hypothesis using threonine to alanine mutants on rates of cyclohexene oxidation (nmol/min/nmol P450) by NH_2-terminal-trancated cytochromes P450 $\triangle$2B4 and $\triangle$2E1.

Products	Cyclohexene oxide (δP_1)	1-Cyclohexene-3-ol (δP_2)	$\dfrac{\delta P_1}{\delta P_2}$
$\triangle$2B4	48.7 ± 4.4	29.3 ± 1.7	~1.6
$\triangle$2B4T302A	13.0 ± 0.5	7.8 ± 1.1	~1.6
$\triangle$2E1	6.7 ± 0.1	6.6 ± 0.7	~1.0
$\triangle$2E1T303A	11.4 ± 0.2	4.8 ± 0.1	~2.4

Compiled from [116].

concentration of the each oxidant (Figure 20). Table 1 summarizes the products ratio ($\delta P_1/\delta P_2$) [cyclohexene oxide to 1-cyclohexene-3-ol] for cyclohexene oxidation by CYP 2B4 and CYP 2E1 and their threonine to alanine mutants. Examination of the data in the table indicates that for the 2B4 and its mutant T302A products ratio remains the same (~1.6), where as for the 2E1 and its T303A mutant the ratio is increased from 1.0 to 2.4. Therefore, this latter result suggests the involvement of additional oxidants (multiple active oxidants) in catalysis during formation of two products.

Substantial evidence for the involvement of second electrophilic oxidants in the P450 reaction cycle has been obtained in studies of substrate oxidation by P450 mutants. Figure 19 summarizes the different P450 oxidant species and their functions. P450 Cpd I will certainly do most of the reactions efficiently. If the Fe^{III}-hydroperoxo intermediate serves as an electrophilic oxidant, then this oxidant could be an epoxidizing catalyst since that reaction does not need as strong an oxidant as is needed for hydroxylation.

7. TWO STATES THEORY

In the late 1990s, Shaik and coworkers [97,118] presented a theoretical model that predicts multiple reaction pathways for the ferryl porphyrin π-cation radical P450 Cpd I species. The two states model involves two different spin states for reactions of Cpd I with substrate (Figure 19) and is an extension of the model for multistate gas phase reactions of small iron-oxygen cations. The two spin states arise from two different coupling modes between the spin of the porphyrin radical (·+ in Figure 19) (S = 1/2) and the Fe=O triplet electrons (S = 1) in Cpd I. More recently, the model has been further refined in a number of publications by Shaik and coworkers (recently reviewed in [34,35]) and by Yoshizawa et al. [119]. Essentially, the model focuses on the conclusions that different spin states of the Cpd I species can catalyze reactions via different pathways and therefore produce different outcomes. This theory has been used to explain almost every aspect of P450 reactions [33,120–123]. They suggest that all P450 reactivity can be explained by the competition between two different spin states of P450 Cpd I: high spin (quartet, $S = 3/2$) and low spin (doublet, $S = 1/2$) [97,118]. The 'two states' hypothesis explains many puzzling results previously reported in mechanistic studies of P450 enzymes. The calculations show that the low spin species recombines without a measurable barrier to give the alcohol, whereas the quartet species must overcome an energy barrier and is thus susceptible to competing reactions, such as rearrangements, prior to recombination to give the alcohol. The proportion of the two spin states is sensitive to the environment, so that different extents of doublet and quartet reactions are possible with different enzymes and different substrates.

In the case of radical clock substrates, the apparent radical recombination rate will be artificially elevated by the extent of the doublet reaction, because the unrearranged product formed by this pathway will be added to that formed by the quartet reaction, the only one of the two reactions that is in fact relevant to determination of the recombination rate [101]. The measured recombination rates thus represent the maximum possible rather than necessarily the actual rate, because it assumes that all of the reaction proceeds via the quartet state. Very recently, the two states theory has been expanded to both the C-H hydroxylation and C=C epoxidation reactions [124].

8. INFLUENCE OF SUBSTRATE ON THE SPECTRAL PROPERTIES AND REACTIVITY OF P450 INTERMEDIATES

There are many factors that affect the electronic nature of the active heme center. Among such factors, vinyl substituent conjugation with the porphyrin π-system, propionate group hydrogen bonding, the proximal thiolate push and the distal pocket hydrogen bonding network are prominent. Substrate binding can also perturb the heme environment as sterically diverse substrates bind to the hydrophobic substrate binding pocket of individual P450 enzymes. For example, substrate binding has been studied with CYP 2B4 using surface enhanced resonance Raman scattering (SERRS) [125]. Two asymmetric vinyl modes were identified that were assigned to two structurally different vinyl alignments and interactions with the heme porphyrin π-system. The lower energy ($\sim 1622\,cm^{-1}$) band appears for a coplanar vinyl substituent while a higher frequency ($\sim 1629\,cm^{-1}$) band arises from a vinyl group that is tilted or out of plane with the heme and therefore in poor conjugation with the heme porphyrin π-system. The authors found that the energy and intensity of the vinyl modes were influenced by substrate binding and suggested that such interactions might be the result of heme distortions that could affect the catalytic activity of the enzyme.

Sono et al. have recently examined the issue of whether the nature of the substrate can directly influence the spectral properties of oxyferrous P450, the last stable intermediate in the P450 reaction cycle [56]. Using low temperatures ($-40\,°C$) and cryosolvents [60% (v/v) ethylene glycol] to stabilize this short-lived state with P450-CAM, it was investigated in the presence of the natural substrate, camphor, and of several camphor analogs as well as in the substrate-free form. The oxyferrous derivative of T252A P450-CAM, a mutant lacking the conserved distal Thr hydroxyl group that provides a stabilizing hydrogen bond to the heme iron-coordinated dioxygen, was also studied to monitor the influence of this hydrogen bond. The UV-visible absorption and magnetic circular dichroism (MCD) spectra of these various oxyferrous adducts were found to be broadly similar. However, while the MCD spectrum of camphor-bound oxyferrous P450-CAM was similar to that of the substrate-free oxyferrous enzyme, the spectrum of the oxyferrous enzyme was seen to differ detectably in the presence of the various non-natural substrate analogs. The spectra of the oxyferrous T252A mutant and wild-type enzyme were similar overall except for 2–6 nm blue shifts in the position of the Soret absorption band for the mutant. 5-Methylenylcamphor (a camphor analogue used to study epoxidation reactions) had an anomalous binding mode for the mutant compared with that for the wild-type enzyme. This study indicated that the structures of the camphor analogs can sensitively influence the physical (spectroscopic) properties of the P450 dioxygen complex and could therefore also affect its reactivity. The ability of substrate to modulate the reactivity of P450 intermediates could be an important factor in explaining the extensive range of reactions catalyzed by the enzyme [56].

Although P450 researchers have hypothesized on the nature of the reactive oxygen intermediates that follow the oxyferrous state (Figure 4) for over two decades, it is only recently that direct studies of these intermediates have been possible. An important breakthrough has been the development of a cryoradiolytic reduction technique initially developed by Davydov and Hoffman that has made it feasible to directly study the Fe^{III}-peroxo and -hydroperoxo P450 states [31,41,126,127]. Hoffman and coworkers have used γ-radiation by ^{60}Co and, more recently, Sligar and coworkers have used γ-radiation by ^{32}P to generate hydrated electrons to serve as the electron donor to reduce the substrate-bound oxyferrous complex at a very low temperature (77 K). The initial one electron-reduced dioxygen species has been characterized by EPR and ENDOR spectroscopy as a low spin Fe^{III}-peroxo complex [128]. Controlled annealing of this transient state led to generation of the Fe^{III}-hydroperoxo state in selected cases. The use of ENDOR spectroscopy was crucial to clearly monitor the transition from the Fe^{III}-peroxo species to the Fe^{III}-hydroperoxo complex. Further raising of the temperature to 190 K led to decay of the Fe^{III}-peroxo complex and formation of the natural product, 5-*exo*-hydroxycamphor, bound to the Fe^{III} heme iron. The radiolytic technique also allowed the Sligar group to use UV-Vis absorption spectroscopy to characterize the Fe^{III}-hydroperoxo species. Unfortunately, in no case was Cpd I detected.

More recently, Hoffman, Dawson, and their coworkers have used EPR and ENDOR spectroscopy to examine the physical properties and reactivity of the one-electron cryoreduced P450-CAM intermediates formed in the presence of a variety of substrates including the native substrate, camphor, as well as several alternate substrates such as 5-methylenylcamphor [56,128]. Parallel studies were performed on the camphor- and 5-methylenylcamphor-bound derivatives of the T252A mutant. The properties and reactivity of the initially formed one-electron reduced oxyheme and of both the Fe^{III}-peroxo and -hydroperoxo P450-CAM intermediates were found to be considerably modulated by bound substrate. This included alterations in the heme center magnetic properties as well as substantial changes in reactivity. In particular, the presence of any substrate increased the lifetime of Fe^{III}-hydroperoxo P450-CAM by a minimum of twenty-fold. Among the substrates, 5-methylenylcamphor stood out as having an exceptionally strong influence on the properties and reactivity of the intermediates, especially in

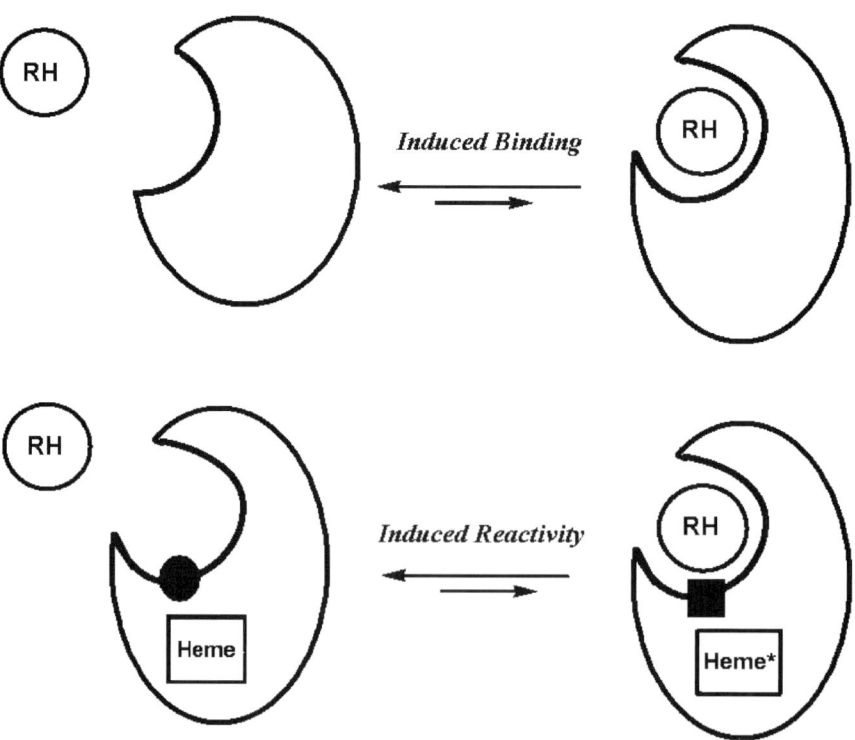

Figure 21. 'Induced reactivity' brought about by substrate binding (see also [128]).

the T252A mutant. For example, Fe^{III}-hydroperoxo P450-CAM (T252A) with 5-methylenyl-camphor bound did not lose H_2O_2, as occurred with camphor-bound Fe^{III}-hydroperoxo P450-CAM (T252A), but instead decayed with formation of the epoxide of 5-methylenylcamphor. These data clearly showed that substrate can significantly modulate the properties of both the monoxygenase active-oxygen intermediates and the proton-delivery network that encompasses them [56,128].

The spectral studies described above that have revealed the influence of substrate on the properties of the oxyferrous, Fe^{III}-peroxo and Fe^{III}-hydroperoxo P450-CAM intermediates [52,118] has led to the proposal that the substrate can modulate the reactivity of P450, 'induced reactivity'. This could be a contributing factor in explaining the extensive range of reactions catalyzed by the enzyme [56]. Hoffman et al. summarized this concept by comparing it to the classical idea of induced fit binding of substrates to promote enzyme catalysis (Figure 21) [56,128]. Induced reactivity involves the influence of the substrate on the reactivity of transient catalytic intermediates to produce a new mechanistic pathway not available in the absence of substrate.

9. FORMATION AND REACTIVITY OF TRANSIENT P450 OXYGEN INTERMEDIATES

Two recent studies have investigated the formation and reactivity of transient P450 oxygen intermediates. Both studies involved the use of rapid scanning stopped-flow absorption spectroscopy and followed up on important earlier discoveries. The first involved the generation of the short-lived oxyferrous state of P450-CAM in the first mix of a double-mixing stopped-flow instrument followed by reaction with reduced putidaredoxin (Pdx), the natural electron donor in the P450-CAM system [129]. The second study reported optimum conditions for formation of P450 Cpd I in a single mix of the stopped-flow instrument.

The single turnover of camphor-bound oxyferrous cytochrome P450-CAM with one equivalent of dithionite-reduced Pdx was examined at 3 °C for the appearance of transient intermediates [129]. After generating oxyferrous P450-CAM in the presence of excess camphor in the first mix of a double mixing rapid scanning stopped-flow instrument, the resulting oxyferrous enzyme was reacted with reduced Pdx. Three consecutive enzyme derivatives were seen: a perturbed oxyferrous species, an intermediate that consisted of a mixture of high and low spin Fe(III) states, and, finally, high spin camphor-bound Fe^{III} P450-CAM. The rates of the first two steps, ~ 140 and $\sim 85\,s^{-1}$, were assigned to formation of the perturbed oxyferrous state and to electron transfer from reduced Pdx to oxyferrous P450-CAM, respectively. The latter rate had previously been reported by Brewer and Peterson [130] from data collected using single wavelength single-mix stopped-flow instrumentation. In the reaction of reduced Pdx with the oxyferrous enzyme in the presence of stoichiometric substrate, three phases with similar rates were seen despite the fact that the reaction generates low spin Fe^{III} P450-CAM, presumably due to substrate hydroxylation during the reaction. The single turnover reaction following reaction of dioxygen with a preformed complex of reduced P450-CAM and reduced Pdx in the presence of excess camphor also led to phases with similar rates. The researchers proposed that generation of the perturbed oxyferrous state resulted from alteration of hydrogen bonding to the proximal cysteine ligand, increasing the reduction potential of the oxyferrous state and triggering electron transfer from reduced Pdx. The perturbed oxyferrous species was suggested to be a direct spectral signature of the effector role of Pdx on P450-CAM reactivity (i.e., during catalysis) [131].

In 1994, Egawa et al. reported that addition of m-chloroperoxybenzoic acid to substrate-free Fe^{III} P450-CAM led to generation of P450 Cpd I, a $Fe^{IV}=O$ porphyrin π-cation radical (Figure 4, intermediate **H**) [26]. By contrast, Schunemann and coworkers [27] found that reaction of substrate-free Fe^{III} P450-CAM with peracetic acid led to formation of a different high-valent species consisting of a $Fe^{IV}=O$ heme center plus a tyrosyl radical as monitored by freeze-quench Mössbauer and EPR spectroscopy. To clarify these disparate observations, Spolitak et al. [28] re-examined the reactions of substrate-free Fe^{III} P450-CAM with both peracids as a function of pH and temperature. At pH > 7, a significant fraction of Cpd I was formed transiently and evidence for acylperoxo complex formation *en route* to Cpd I was described. Alternatively, at low pH, only a species with a Soret absorption band at 406 nm, presumably the same $Fe^{IV}=O$ + tyrosyl radical species seen by Schunemann and coworkers was observed. Because of rapid heme destruction resulting in the bleaching of the heme chromophore, it was not possible to fully characterize the steps subsequent to formation of the $Fe^{IV}=O$ derivatives. Interestingly, heme bleaching was avoided by including peroxidase substrates (e.g., guaiacol), which underwent typical peroxidase reactions with concomitant regeneration of Fe^{III} P450. Addition of ascorbate to either of the $Fe^{IV}=O$ species also reformed the Fe^{III} state with minimal loss of heme absorbance. These results [28] indicated that typical peroxidase chemistry occurs with P450-CAM and offered an explanation for the contrasting results reported earlier by Egawa et al. and by Schunemann and coworkers [26,132]. The delineation of improved conditions (pH, temperature, choice of peracid) for generating highly oxidized species with P450-CAM should be valuable for their further characterization.

10. SUMMARY AND FUTURE PROSPECTIVE

Unfortunately, neither the optical nor EPR studies have been able to detect the Cpd I species. ENDOR data shows that the P450 wild-type and mutant

Fe^{III}-hydroperoxo iron species convert to hydroxycamphor-bound Fe^{III} and resting Fe^{III} states, respectively, without a detectable Cpd I species. In summary, cytochrome P450 has been subject to intense research for the past 45 years. The gathered knowledge in the field, both structural and functional, will pave the way for researchers to generate next generation bio-catalysts. In the future, engineered synthetic porphyrin combined with natural and unnatural amino acids incorporated model catalysts may carry out the many organic reactions, including hydroxylation of alkane and epoxidation of olefins, in mild reaction conditions.

ACKNOWLEDGMENTS

Work in the Dawson laboratory on cytochrome P450-CAM has been supported by the NIH (GM-26730). We thank Prof. John T. Groves for helpful discussions. We also wish to thank Profs Thomas A. Bryson, David P. Ballou, and Brian M. Hoffman for their contributions to our collaborative studies on the mechanism of action of P450.

ABBREVIATIONS

ATP	adenosine 5′-triphosphate
BP	benzphetamine
Cpd I or II	compound I or II of a heme enzyme, an $Fe^{IV}=O$ π-cation radical species or an $Fe^{IV}=O$ species, respectively
CYP	cytochrome P450
CYP 2B4	phenobarbital-inducible rabbit liver microsomal cytochrome P450 (P450LM2)
CYP 2E1	alcohol-inducible rabbit liver microsomal cytochrome P450 (P450LM3A)
DFT	density functional theory
ENDOR	electron-nuclear double resonance spectroscopy
EPR	electron paramagnetic resonance
EXAFS	extended X-ray absorption fine structure
FAD	flavin adenine dinucleotide
FMN	flavin mononucleotide
heme	iron protoporphyrin IX (heme *b*)
HO	heme oxygenase
KIE	kinetic isootope effect
MCD	magnetic circular dichroism
MD	molecular dynamics simulations
MM	molecular mechanics simulations

P450	cytochrome P450
P450 BM3	fatty acid-hydroxylating cytochrome P450 from *Bacillus megaterium*
P450-CAM	camphor-hydroxylating cytochrome P450 from *Pseudomonas putida*
P450eryF	macrolide-hydroxylating cytochrome P450 from *Saccaropolyspora erythraea*
PdR	putidaredoxin reductase
Pdx	putidaredoxin
PPIX	protoporphyrin-IX
RH	substrate
ROH	oxidized product

REFERENCES

1. P. R. Ortiz de Montellano, ed., *Cytochrome P450: Structure, Mechanism and Biochemistry*, 3rd edn, Kluwer Academic/Plenum Publishers, New York, 2005.
2. I. G. Denisov, T. M. Makris, S. G. Sligar, and I. Schlichting, *Chem. Rev., 105*, 2253–2277 (2005).
3. B. Meunier, S. P. de Visser, and S. Shaik, *Chem. Rev., 104*, 3947–3980 (2004).
4. M. Sono, M. P. Roach, E. D. Coulter, and J. H. Dawson, *Chem. Rev., 96*, 2841–2887 (1996).
5. Y. Watanabe and J. T. Groves, *Enzymes, 20*, 405–452 (1992).
6. R. E. White and M. J. Coon, *Annu. Rev. Biochem., 49*, 315–356 (1980).
7. I. C. Gunsalus, T. C. Pederson, and S. G. Sligar, *Annu. Rev. Biochem., 44*, 377–407 (1975).
8. (a) A. E. Pond, A. P. Ledbetter, M. Sono, D. B. Goodin, and J. H. Dawson, in *Handbook on Electron Transfer* (V. Balzani, ed.), Volume 3, 2001, and *Part 1, Biological Systems* (H. B. Gray and J. R. Winkler, eds), Wiley-VCH, Weinheim, pp. 56–104. (b) H. B. Dunford, *Heme Peroxidases*, John Wiley and Sons, Inc., New York, 1999. (c) J. Everse, K. E. Everse, and M. B. Grisham, eds, *Peroxidases in Chemistry and Biology*, CRC Press, Boca Raton, FL., 1991.
9. (a) P. L. Feldman, O. W. Griffith, and D. J. Stuehr, *Chem. Eng. News, 71* (51), 26–38 (1993). (b) M. A. Marletta, *J. Biol. Chem., 268*, 12231–12234 (1993). (c) D. S. Bredt and S. H. Snyder, *Annu. Rev. Biochem., 63*, 175–195 (1994). (d) S. Moncada and E. A. Higgs, *FASEB J., 10*, 1319–1330 (1995). (e) D. J. Stuehr, *Biochim. Biophys. Acta, 1411*, 217–230 (1999). (f) C. S. Raman, P. Martasek, and B. S. S. Masters, in *Porphyrin Handbook* (K. M. Kadish, K. M. Smith and R. Guilard, eds), Vol. 4, Academic Press, London, UK, 2000, pp. 293–339.
10. E. Antonini and M. Brunori, in *Hemoglobin and Myoglobin in Their Reactions with Ligands*, North-Holland Publishing Co., Amsterdam, 1971.
11. D. Dolphin, ed., *The Porphyrins*, Vol. VII, Academic Press, New York, 1979.
12. O. Hayaishi, ed., *Molecular Mechanisms of Oxygen Activation*, Academic Press, New York, 1974.

13. J. S. Valentine, in *Bioinorganic Chemistry* (I. Bertini, H. B. Gray, S. J. Lippard, and J. S. Valentine, eds), University Science Books, Mill Valley, CA, USA, 1994, pp. 253–313.
14. M. Costas, M. P. Mehn, M. P. Jensen, and L. Que, Jr., *Chem. Rev., 104*, 939–986 (2004).
15. F. P. Guengerich, *Chem. Res. Toxicol., 14*, 611–650 (2001).
16. F. P. Guengerich, in *Cytochrome P450: Structure, Mechanisms and Biochemistry* (P. R. Ortiz de Montellano, ed.), 3rd edn, Kluwer Academic/Plenum Publishers, New York, 2005, pp. 377–530.
17. P. R. Ortiz de Montellano and J. J. De Voss, in *Cytochrome P450: Structure, Mechanisms and Biochemistry* (P. R. Ortiz de Montellano, ed.), 3rd edn, Kluwer Academic/Plenum Publishers, New York, 2005, pp. 183–245.
18. J. H. Capdevila, V. Y. Holla, and J. R. Falck, in *Cytochrome P450: Structure, Mechanism and Biochemistry* (P. R. Ortiz de Montellano, ed.), 3rd edn, Kluwer Academic/Plenum Publishers, New York, 2005, pp. 531–545.
19. I. A. Pikuleva, *Drug. Metab. Disp., 34*, 513–520 (2006).
20. T. Omura and R. Sato, *J. Biol. Chem., 239*, 2370–2378 (1964).
21. M. J. I. Paine, N. S. Scrutton, A. W. Munro, A. Gutierrez, G. C. K. Roberts, and C. R. Wolf, in *Cytochrome P450: Structure, Mechanisms and Biochemistry* (P. R. Ortiz de Montellano, ed.), 3rd edn, Kluwer Academic/Plenum Publishers, New York, 2005, pp. 115–148.
22. P. J. Loida and S. G. Sligar, *Biochemistry, 32*, 11530–11538 (1993).
23. L. Eisenstein, P. Debey, and P. Douzou, *Biochem. Biophys. Res. Commun., 77*, 1377–1383 (1977).
24. A. D. Vaz, D. F. McGinnity, and M. J. Coon, *Proc. Natl. Acad. Sci. USA, 95*, 3555–3560 (1998).
25. S. Jin, T. M. Makris, T. A. Bryson, S. G. Sligar, and J. H. Dawson, *J. Am. Chem. Soc., 125*, 3406–3407 (2003).
26. T. Egawa, H. Shimada, and Y. Ishimura, *Biochem. Biophys. Res. Commun., 201*, 1464–1468 (1994).
27. V. L. F. Schunemann, C. Jung, J. Contzen, A. L. Barra, S. G. Sligar, and A. X. Trautwein, *J. Biol. Chem., 279*, 10919–10930 (2004).
28. T. Spolitak, J. H. Dawson, and D. P. Ballou, *J. Biol. Chem., 280*, 20300–20309 (2005).
29. T. Egawa, T. Ogura, R. Makino, Y. Ishimura, and T. Kitagawa, *J. Biol. Chem., 266*, 10246–10248 (1991).
30. I. D. G. MacDonald, S. G. Sligar, J. F. Christian, M. Unno, and P. M. Champion, *J. Am. Chem. Soc., 121*, 376–380 (1999).
31. I. Schlichting, J. Berendzen, K. Chu, A. M. Stock, S. A. Maves, D. E. Benson, R. M. Sweet, D. Ringe, G. A. Petsko, and S. G. Sligar, *Science, 287*, 1615–1622 (2000).
32. F. Ogliaro, S. P. de Visser, and S. Shaik, *J. Inorg. Biochem., 91*, 554–567 (2002).
33. F. Ogliaro, S. P. de Visser, S. Cohen, P. K. Sharma, and S. Shaik, *J. Am. Chem. Soc., 124*, 2806–2817 (2002).
34. S. Shaik and S. P. de Visser, in *Cytochrome P450: Structure, Mechanisms and Biochemistry* (P. R. Ortiz de Montellano, ed.), 3rd edn, Kluwer Academic/Plenum Publishers, New York, 2005, pp. 45–86.

35. S. Shaik, D. Kumar, S. P. de Visser, A. Altun, and W. Thiel, *Chem. Rev., 105*, 2279–2328 (2005).
36. D. L. Harris and G. H. Loew, *J. Am. Chem. Soc., 116*, 11671–11674 (1994).
37. D. L. Harris and G. H. Loew, *J. Am. Chem. Soc., 118*, 6377–6387 (1996).
38. R. Fedorov, D. K. Ghosh, and I. Schlichting, *Arch. Biochem. Biophys. 409*, 25–31(2003).
39. E. E. Scott, M. A. White, Y. A. He, E. F. Johnson, C. D. Stout, and J. R. Halpert, *J. Biol. Chem., 279*, 27294–27301 (2004).
40. N. C. Gerber and S. G. Sligar, *J. Biol. Chem. 269*, 4260–4266 (1994).
41. R. Davydov, T. M. Makris, V. Kofman, D. E. Werst, S. G. Sligar, and B. M. Hoffman, *J. Am. Chem. Soc., 123*, 1403–1415 (2001).
42. P. A. Williams, J. Cosme, V. Sridhar, E. F. Johnson, and D. E. McRee, *Mol. Cell., 5*, 121–131 (2000).
43. D. L. Harris, J.-Y. Park, L. Gruenke, and L. Waskell, *Proteins, 55*, 895–914 (2004).
44. S. G. Sligar and I. C. Gunsalus, *Proc. Natl. Acad. Sci. USA, 73*, 1078–1082 (1976).
45. H. Zhang, L. Gruenke, D. Arscott, A. Shen, C. Kasper, D. L. Harris, M. Glavanovich, R. Johnson, and L. Waskell, *Biochemistry, 42*, 11594–11603 (2003).
46. F. P. Guengerich, D. P. Ballou, and M. J. Coon, *J. Biol. Chem., 250*, 7405–7414 (1975).
47. L. O. Nahri and A. J. Fulco, *J. Biol. Chem., 261*, 7160–7169 (1986).
48. R. T. Ruettinger, L. P. Wen, and A. J. Fulco, *J. Biol. Chem., 264*, 10987–10995 (1989).
49. I. Sevrioukova, G. Truan, and J. A. Peterson, *Biochemistry, 35*, 7528–7535 (1996).
50. J. K. Yano, F. Blasco, H. Li, R. D. Schmid, A. Henne, and T. J. Poulos, *J. Biol. Chem., 278*, 608–616 (2003).
51. H. Li and T. L. Poulos, *Acta Crystallogr., D51*, 21–32 (1995).
52. R. Perera, M. Sono, G. M. Raner, and J. H. Dawson, *Biochem. Biophys. Res. Commun., 338*, 365–371 (2005).
53. M. Imai, H. Shimada, Y. Watanabe, Y. Matsushima-Hibiya, R. Makino, H. Koga, T. Horiuchi, and Y. Ishimura, *Proc. Natl. Acad. Sci. USA, 86*, 7823–7827 (1989).
54. S. A. Martinis, W. M. Atkins, P. S. Stayton, and S. G. Sligar, *J. Am. Chem. Soc., 111*, 9252–9253 (1989).
55. R. Raag, S. A. Martinis, S. G. Sligar, and T. L. Poulos, *Biochemistry, 30*, 11420–11429(1991).
56. M. Sono, R. Perera, S. Jin, T. M. Makris, S. G. Sligar, T. A. Bryson, and J. H. Dawson, *Arch. Biochem. Biophys., 436*, 40–49 (2005).
57. S. Nagano and T. L. Poulos, *J. Biol. Chem., 280*, 31659–31663 (2005).
58. G. Truan and J. A. Peterson, *Arch. Biochem. Biophys., 349*, 53–64 (1998).
59. H. Yeom, S. G. Sligar, H. Li, T. L. Poulos, and A. J. Fulco, *Biochemistry, 34*, 14733–14740 (1995).
60. D. Dolman, G. A. Newell, and M. D. Thurlow, *Can. J. Biochem., 53*, 495–501(1975).
61. P. Jones and H. B. Dunford, *J. Theor. Biol., 69*, 457–470 (1977).
62. M. B. McCarthy and R. E. White, *J. Biol. Chem. 258*, 9153–9158 (1983).
63. R. Perera, M. Sono, J. A. Sigman, T. D. Pfister, Y. Lu, and J. H. Dawson, *Proc. Natl. Acad. Sci., USA, 100*, 3641–3646 (2003).

64. S. Yoshioka, S. Takahashi, K. Ishimori, and I. Morishima, *J. Inorg. Biochem., 81*, 141–151 (2000).
65. S. Yoshioka, T. Tosha, S. Takahashi, K. Ishimori, H. Hori, and I. Morishima, *J. Am. Chem. Soc., 124*, 14571–14579 (2002).
66. J. H. Dawson, R. H. Holm, J. R. Trudell, G. Barth, R. E. Linder, E. Bunnenberg, C. Djerassi, and S. C. Tang, *J. Am. Chem. Soc., 98*, 3707–3709 (1976).
67. J. H. Dawson, *Science, 240*, 433–439 (1988).
68. T. L. Poulos and J. Kraut, *J. Biol. Chem., 255*, 8199–8205 (1980).
69. T. L. Poulos, *Adv. Inorg. Biochem.*, 7, 1–36 (1988).
70. H. Shimada, S. Nagano, Y. Ariga, M. Unno, T. Egawa, T. Hishiki, Y. Ishimura, F. Masuya, T. Obata, and H. Hori, *J. Biol. Chem., 274*, 9363–9369 (1999).
71. J. Aikens and S. Sligar, *J. Am. Chem. Soc., 116*, 1143–1144 (1994).
72. D. L. Harris and G. H. Loew, *J. Am. Chem. Soc., 120*, 8941–8948 (1998).
73. G. H. Loew and D. L. Harris, *Chem. Rev., 100*, 407–420 (2000).
74. (a) K. Auclair, P. Moenne-Loccoz, and P. R. Ortiz de Montellano, *J. Am. Chem. Soc., 123*, 4877–4885 (2001). (b) S. Yoshikawa, S. Takahashi, H. Hori and I. Morishima, *Eur. J. Biochem., 268*, 252–259 (2001).
75. K. P. Vatsis, H. Peng, and M. J. Coon, *Arch. Biochem. Biophys., 434*, 1128–1138 (2005).
76. R. E. McMahon, H. R. Sullivan, J. C. Craig, and W. E. Pereira, *Arch. Biochem. Biophys., 132*, 575–577 (1969).
77. M. Hamberg and I. Bjorkhem, *J. Biol. Chem., 246*, 7411–7416 (1971).
78. S. Shapiro, J. U. Piper, and E. Caspi, *J. Am. Chem. Soc., 104*, 2301–2305 (1982).
79. L. M. Hjelmeland, L. Aronow, and J. R. Trudell, *Biochem. Biophys. Res. Commun., 76*, 541–549 (1977).
80. J. T. Groves, G. A. McClusky, R. E. White, and M. J. Coon, *Biochem. Biophys. Res. Commun., 81*, 154–160 (1978).
81. J. T. Groves, *J. Chem. Ed., 62*, 928–932 (1985).
82. J. T. Groves and Y. Z. Han, in *Cytochrome P450: Structure, Mechanism and Biochemistry* (P. R. Ortiz de Montellano, ed.), 2nd edn, Plenum, New York, 1995, pp. 3–48.
83. M. H. Gelb, D. C. Heimbrook, P. Malkonen, and S. G. Sligar, *Biochemistry, 21*, 370–377 (1982).
84. J. T. Groves and D. V. Adhyam, *J. Am. Chem. Soc., 106*, 2177–2181 (1984).
85. R. E. White, J. P. Miller, L. V. Favreau, and A. Bhattacharyya, *J. Am. Chem. Soc., 108*, 6024–6031 (1986).
86. M. Newcomb and P. H. Toy, *Acc. Chem. Res., 33*, 449–455 (2000).
87. J. F. Atkinson and K. U. Ingold, *Biochemistry, 32*, 9209–9214 (1993).
88. P. H. Toy, N. Newcomb, and P. F. Hollenberg, *J. Am. Chem. Soc., 120*, 7719–7729 (1998).
89. P. R. Ortiz de Montellano and R. A. Stearns, *J. Am. Chem. Soc., 109*, 3415–3420 (1987).
90. V. W. Bowry and K. U. Ingold, *J. Am. Chem. Soc., 113*, 5699–5707 (1991).
91. J. K. Atkinson and K. U. Ingold, *Biochemistry, 32*, 9209–9214 (1993).
92. M. Newcomb, R. Shen, S. Y. Choi, P. H. Toy, P. F. Hollenberg, A. D. N. Vaz, and M. J. Coon, *J. Am. Chem. Soc., 122*, 2677–2686 (2000).

93. R. E. Chandrasena, K. P. Vatsis, M. J. Coon, P. F. Hollenberg, and M. Newcomb, *J. Am. Chem. Soc.*, *126*, 115–126 (2004).
94. M. Newcomb, R. Shen, Y. Lu, M. J. Coon, P. F. Hollenberg, D. A. Kopp, and S. J. Lippard, *J. Am. Chem. Soc.*, *124*, 6879–6886 (2002).
95. M. J. Cryle, P. R. Ortiz de Montellano, and J. J. De Voss, *J. Org. Chem.*, *70*, 2455–2469 (2005).
96. K. Auclair, Z. Hu, D. M. Little, P. R. Ortiz de Montellano, and J. T. Groves, *J. Am. Chem. Soc.*, *124*, 6020–6027 (2002).
97. S. Shaik, M. Filatov, D. Schroder, and H. Schwarz, *Chem. Eur. J.*, *4*, 193–199 (1998).
98. N. Harris, S. Cohen, M. Filatov, F. Ogliaro, and S. Shaik, *Angew. Chem., Int. Ed.*, *39*, 2003–2207 (2000).
99. D. Schroder, S. Shaik, and H. Schwarz, *Acc. Chem. Res.*, *33*, 139–145 (2000).
100. D. Kumar, S. P. de Visser, and S. Shaik, *J. Am. Chem. Soc.*, *125*, 13024–13025 (2003).
101. D. Kumar, S. P. de Visser, P. K. Sharma, S. Cohen, and S. Shaik, *J. Am. Chem. Soc.*, *126*, 1907–1920 (2004).
102. F. P. Guengerich and T. L. Macdonald, *FASEB J.*, *4*, 2453–2459 (1990).
103. J. T. Groves, *Ann. N. Y. Acad. Sci.*, *471*, 99–107 (1986).
104. A. D. Vaz, S. J. Pernecky, G. M. Raner, and M. J. Coon, *Proc. Natl. Acad. Sci. USA*, *93*, 4644–4648 (1996).
105. Y. Watanabe and Y. Ishimura, *J. Am. Chem. Soc.*, *111*, 8047–8049 (1989).
106. M. Akhtar, M. R. Calder, D. L. Corina, and J. N. Wright, *Biochem. J.*, *201*, 569–580 (1982).
107. P. A. Cole and C. H. Robinson, *J. Am. Chem. Soc.*, *110*, 1284–1285 (1988).
108. A. D. N. Vaz, E. S. Roberts, and M. J. Coon, *J. Am. Chem. Soc.*, *113*, 5886–5887 (1991).
109. A. D. N. Vaz, K. J. Kessell, and M. J. Coon, *Biochemistry*, *33*, 13651–13661 (1994).
110. E. S. Roberts, A. D. N. Vaz, and M. J. Coon, *Proc. Natl. Acad. Sci. USA*, *88*, 8963–8966 (1991).
111. Y. Watanabe, K. Takehira, M. Shimizu, T. Harakawa, and H. Orita, *J. Chem. Soc. Chem. Commun.*, 927–928 (1990).
112. S. Ahmed and P. Davis, *Bioorg. Med. Chem. Lett.*, *5*, 2789–2794 (1995).
113. M. Newcomb, D. Aebisher, R. Shen, R. E. P. Chandrasena, P. F. Hollenberg, and M. J. Coon, *J. Am. Chem. Soc.*, *125*, 6064–6065 (2003).
114. M. Newcomb, P. F. Hollenberg, and M. J. Coon, *Arch. Biochem. Biophys.*, *409*, 72–79 (2003).
115. S. Jin, T. A. Bryson, and J. H. Dawson, *J. Biol. Inorg. Chem.*, *9*, 644–653 (2004).
116. F. P. Guengerich, A. D. N. Vaz, G. N. Raner, S. J. Pernecky, and M. J. Coon, *Mol. Pharmacol.*, *51*, 147–151 (1997).
117. M. J. Coon, A. D. Vaz, D. F. McGinnity, and H. M. Peng, *Drug Metab. Dispos.*, *12*, 1190–1193 (1998).
118. M. Filatov and S. Shaik, *J. Phys. Chem. A.*, *102*, 3835–3846 (1998).
119. K. Yoshizawa, T. Kamachi, and Y. Shiota, *J. Am. Chem. Soc.*, *123*, 9806–9816 (2001).

120. F. Ogliaro, N. Harris, S. Cohen, M. Filatov, S. P. de Visser, and S. Shaik, *J. Am. Chem. Soc.*, *122*, 8977–8989 (2000).
121. F. Ogliaro, S. Cohen, S. P. de Visser, and S. Shaik, *J. Am. Chem. Soc.*, *122*, 12892–12893 (2000).
122. S. P. de Visser, F. Ogliaro, P. K. Sharma, and S. Shaik, *Angew. Chem. Int. Ed.*, *41*, 1947–1951 (2002).
123. P. K. Sharma, S. P. de Visser, and S. Shaik, *J. Am. Chem. Soc.*, *125*, 8698–8699 (2003).
124. H. Hirao, D. Kumar, W. Thiel, and S. Shaik, *J. Am. Chem. Soc.*, *127*, 13007–13018 (2005).
125. I. D. G. Macdonald, G. C. M. Smith, C. R. Wolf, and W. E. Smith, *Biochem. Biophys. Res. Commun.* *226*, 51–58 (1996).
126. R. Davydov, I. D. Macdonald, T. M. Makris, S. G. Sligar, and B. M. Hoffman, *J. Am. Chem. Soc.*, *121*, 10654–10655 (1999).
127. I. G. Denisov, S. Hung, K. E. Weiss, M. A. McLean, Y. Shiro, S. Park, P. M. Champion, and S. G. Sligar, *J. Inorg. Biochem.*, *87*, 215–226 (2001).
128. R. Davydov, R. Perera, S. Jin, T. C. Yang, T. A. Bryson, M. Sono, J. H. Dawson, and B. M. Hoffman, *J. Am. Chem. Soc.*, *127*, 1403–1413 (2005).
129. M. C. Glascock, D. P. Ballou, and J. H. Dawson, *J. Biol. Chem.*, *280*, 42134–42141 (2005).
130. C. B. Brewer and J. A. Peterson, *J. Biol. Chem.*, *263*, 791–798 (1988).
131. J. D. Lipscomb, S. G. Sligar, M. J. Namtvedt, and I. C. Gunsalus, *J. Biol. Chem.*, *251*, 1116–1124 (1976).
132. V. Schunemann, C. Jung, A. X. Trautwein, D. Mandon, and R. Weiss, *FEBS Lett.*, *479*, 149–154 (2000).

12

Cytochrome P450 and Steroid Hormone Biosynthesis

Rita Bernhardt[1] and Michael R. Waterman[2]

[1]Institut für Biochemie, Universität des Saarlandes, P. O. Box 151150,
D-66041 Saarbrücken, Germany
<ritabern@mx.uni-saarland.de>

[2]Vanderbilt University School of Medicine, Nashville, TN 37232-0146, USA
<michael.waterman@vanderbilt.edu>

1. INTRODUCTION	362
2. STEROIDOGENIC P450s	363
2.1. Mitochondrial P450 Systems	365
2.1.1. CYP11A1	365
2.1.2. CYP11B1 and CYP11B2	366
2.1.3. Adrenodoxin	367
2.1.4. Adrenodoxin Reductase	368
2.2. Microsomal P450 Systems	369
2.2.1. CYP17	369
2.2.2. CYP21	370
2.2.3. CYP19	370
2.2.4. NADPH-Cytochrome P450 Reductase	371
2.3. Electron Transfer and Its Regulation	371
2.3.1. Mitochondrial Systems	371
2.3.2. Microsomal Systems	373
2.3.3. Steroidogenic P450s as Drug Targets	373
3. STEROID HORMONE BIOSYNTHESIS IN THE ADRENAL CORTEX	374
3.1. Acute Regulation	375
3.2. Chronic Regulation	376
3.3. Conclusion	379

4. STEROID HORMONE BIOSYNTHESIS IN THE GONADS   379
   4.1. Testis   379
      4.1.1. Conclusion   381
   4.2. Ovary   381
      4.2.1. Extragonadal Aromatase   382
5. EXTRAADRENAL AND EXTRAGONADAL
   STEROIDOGENESIS   382
   5.1. Other Steroidogenic Organs: Placenta and Brain   382
   5.2. Endothelial Cells and Heart   383
   5.3. Skin   384
   5.4. Other Tissues   384
   5.5. Conclusion   385
6. OUTLOOK FOR THE FUTURE   385
   ACKNOWLEDGMENTS   386
   ABBREVIATIONS   387
   REFERENCES   387

1. INTRODUCTION

Steroids comprise a large class of important chemical and biological compounds. Sterols, bile acids, steroid hormones, vitamin D as well as many secondary metabolites and poisons from plants and animals such as digitalis glycosides and saponines belong to this class of compounds. Biosynthesis of sterols (cholesterol in animals, ergosterol in fungi and phytosterol in plants (Figure 1)) is dependent on P450-catalyzed reactions, of which the oxidative removal of the 14α-methyl

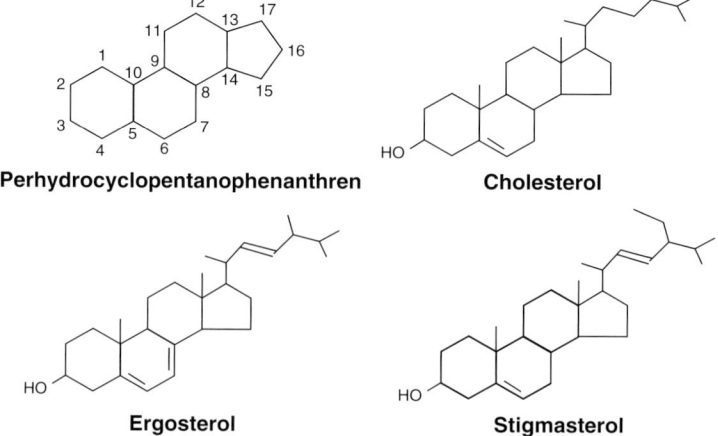

Figure 1. Important sterols in mammals, fungi and plants. The basic ring system from which the steroids derived is perhydrocyclopentanophenathrene.

group from the intermediate compounds (lanosterol in the synthesis of cholesterol and ergosterol and obtusiferiol in the synthesis of phytosterols) by sterol 14α-demethylase (CYP51) is the common essential step for the formation of all functional sterols.

Sterols are essential constituents of the plasma membrane of all eukaryotic cells, and CYP51 seems to be the most ancient form of all eukaryotic P450s [1]. In mammals, cholesterol is the common precursor of all steroid hormones, and its conversion to pregnenolone is the initial and rate-limiting step in hormone biosynthesis in steroidogenic tissues such as gonads and adrenal glands. Cytochrome P450 systems are involved in the biosynthesis of a broad variety of steroids and their contribution to the biosynthesis of steroid hormones is the topic of this chapter.

2. STEROIDOGENIC P450s

Steroid hormones are produced in multi-step pathways that involve the participation of up to six P450s (Figure 2): CYP11A1 (cholesterol side chain cleavage cytochrome P450 or P450scc), CYP17 (17α-hydroxylase/17,20-lyase cytochrome P450 or P450c17), CYP21 (21-hydroxylase cytochrome P450

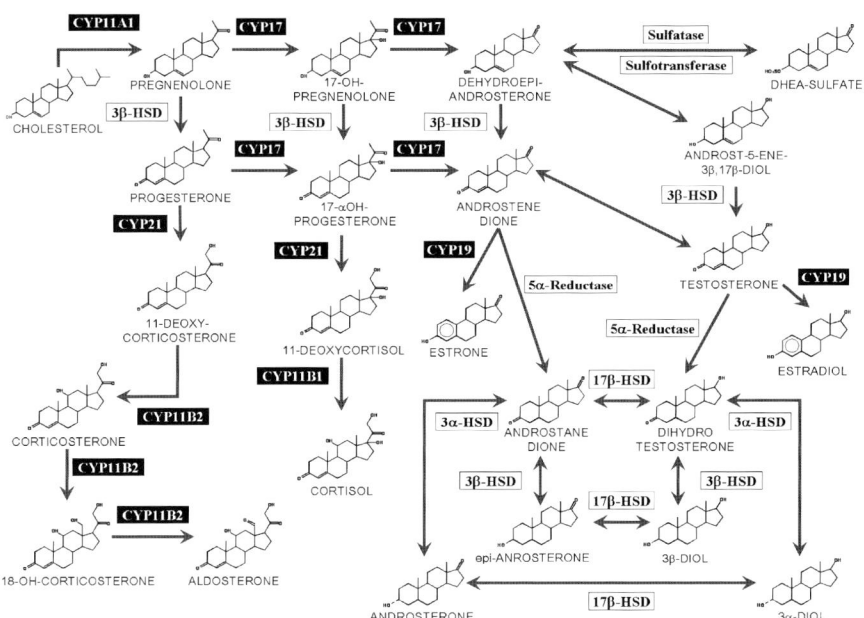

Figure 2. Biosynthesis of steroid hormones from cholesterol.

or P450c21), CYP11B1 (11β-hydroxylase cytochrome P450 or P45011β), CYP11B2 (aldosterone synthase cytochrome P450 or P450aldo) and CYP19 (aromatase cytochrome P450 or P450arom) [2]. While the adrenal gland produces primarily glucocorticoids, mineralocorticoids and androgens, the main site for sex hormone (testosterone and estrogens) production is located in the gonads. The final steps in cortisol (main human glucocorticoid) and aldosterone (main human mineralocortiocid) synthesis are performed by two very closely related enzymes, namely CYP11B1 and CYP11B2 [3–5].

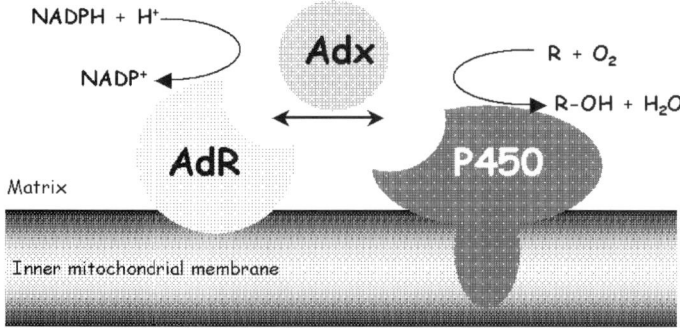

Figure 3. Scheme of mitochondrial and microsomal electron transfer chains.

Mammalian cytochrome P450 systems involved in the biosynthesis of steroid hormones can be subdivided into two major groups according to their subcellular localization: microsomal and mitochondrial P450s, and both utilize NADPH as the electron donor of the monooxygenation reactions, whereas most bacterial P450s receive electrons from NADH. Two components, the flavoprotein adrenodoxin reductase (AdR), being associated with the mitochondrial membrane, and a soluble iron-sulfur protein named adrenodoxin (Adx), catalyze the electron transfer from NADPH to mitochondrial P450s (Figure 3) [6–8]. On the other hand, microsomal P450 systems consist of two membrane-bound components, cytochrome P450 and NADPH-cytochrome P450 reductase. There is no general expression pattern of mammalian P450s, and control mechanisms range from drug-inducible liver enzymes that function in the metabolism of drugs and xenobiotics to hormonally controlled expression of steroid hydroxylases [9]. Detailed overviews on the cytochrome P450 dependent steroid hormone biosynthesis have been previously published [9,10].

2.1. Mitochondrial P450 Systems

There are three reactions in steroid hormone biosynthesis, which are catalyzed by mitochondrial P450 systems (Figure 2) the side-chain cleavage reaction, performed by CYP11A1, the 11β-hydroxylation of 11-deoxycortisol to produce cortisol, the main human glucocorticoid, catalyzed by CYP11B1 and the formation of aldosterone, the main human mineralocorticoid, realized by three consecutive hydroxylations and a water extraction from deoxycorticosterone via corticosterone and 18-hydroxycorticosterone, catalyzed by CYP11B2. All three cytochromes get the necessary electrons (six in case of CYP11A1 and CYP11B2 and two in case of CYP11B1) from NADPH via AdR and Adx. The structure and function of the five proteins is discussed in more detail in the following subchapters.

2.1.1. CYP11A1

CYP11A1 catalyzes the first and (after uptake of cholesterol into mitochondria) rate-limiting step of biosynthesis of all steroid hormones, the side-chain cleavage of cholesterol to produce pregnenolone (Figure 2). The cDNA of bovine CYP11A1 was first cloned by Morohashi et al. [11]. Subsequently, human CYP11A1 was cloned from adrenals and placenta [12]. The gene is mapped on chromosome 15q23-q24 [13]. The reaction catalyzed includes 3 consecutive hydroxylation steps: The first hydroxylation leading to 22R hydroxycholesterol, the second to 20α, 22R dihydroxycholesterol and the third leads to cleavage of the side chain of cholesterol yielding pregnenolone. Under normal circumstances the intermediates do not dissociate from the enzyme. The bovine enzyme was expressed in *E. coli* in 1991 [14], the human form a few years later [15].

Throughout this chapter it will become clear that development of bacterial expression systems for steroidogenic P450s in the Waterman laboratory has allowed detailed studies of these enzymes both biochemically and biophysically. So far several attempts to obtain crystals suitable for X-ray analysis and to resolve the 3D structure of CYP11A1 have been without success. The most straightforward strategies include solubilization of the protein by introducing charged instead of hydrophobic amino acid residues onto the surface of the protein molecule [16] and cross-linking it with the soluble adrenodoxin redox partner [17,18].

CYP11A1 is localized in all three zones of the adrenal cortex as well as in gonads and other tissues as discussed below. It was suspected that defects in this gene would lead to spontaneous abortion because placental steroidogenesis and placental progesterone is required to suppress uterine contractility and maintain pregnancy. However, patients have been described very recently as carrying either a *de novo* heterozygous mutation in the CYP11A1 gene or being compound heterozygotes [19,20]. Both patients were suffering from the so-called congenital lipoid adrenal hyperplasia, a severe disorder in which cholesterol accumulates within steroidogenic cells and synthesis of all adrenal and gonadal steroids is impaired, but the mutations are obviously not lethal during fetal life.

2.1.2. CYP11B1 and CYP11B2

In humans, the CYP11B family contains two members, CYP11B1 and CYP11B2, producing cortisol and aldosterone respectively. In the course of cloning and analyzing the CYP11B1 gene, White and coworkers isolated a cross-hybridizing gene CYP11B2 [3] whose sequence was about 95% identical to the well known CYP11B1 gene in coding regions and 90% identical in introns. The 5' upstream region had diverged considerably from that of CYP11B1, suggesting that this second gene, if expressed, may be regulated differently. Mornet et al. [3] determined that the CYP11B1 and CYP11B2 genes both contain nine exons. The eight introns are identical in location. Both genes have been located on chromosome 8q21. CYP11B enzymes of other species have also been studied, and in bovine [21], porcine [22], and frog [23] adrenal cortex, synthesis of gluco- and mineralocorticoids is catalyzed by a single enzyme, while besides humans, rat [24], mouse [25], and guinea pig [26,27] contain two distinct isoforms specialized in the formation of either minerallo- or glucocorticoids. The reason for these interspecies differences is unknown. Enzymes with 11β-hydroxylase activity have also been found in several fungi [28]; however, none of the genes for these enzymes has been cloned to date and their relation to the CYP11B family remains unclear.

Cortisol, the major human glucocorticoid, is synthesized in the *zona fasciculata/reticularis* of the adrenal cortex, and expression of human CYP11B1 has been demonstrated to be limited to this part of the adrenal cortex [29,30].

By contrast, aldosterone secretion and CYP11B2 expression [31] takes place in the *zona glomerulosa*.

The genes encoding the two human enzymes are tandemly arranged approximately 45 kb apart on chromosome 8 [32,33], and chimeric CYP11B1/CYP11B2 genes that result from unequal crossing over between these two genes have been found in patients suffering from familial hyperaldosteronism type I (also called glucocorticoid remediable hyperaldosteronism; see below) and congenital adrenal hyperplasia, CAH [34–37]. Defects in the CYP11B1 gene are the second most frequent cause for CAH comprising about 5–10% of cases of this disease [38]. Defects in the CYP11B2 gene can occur at two levels, either the 18-hydroxylation or the 18-oxidation steps. The first case (called CMO I-defect) leads to overproduction of corticosterone, whereas the latter case (called CMO II-defect) is characterized by an increase of 18OH cortisosterone. Recent reviews summarize the identified mutations in both genes [38–41]. The transcriptional regulation of both genes has been intensively studied and is also covered in recent reviews [42–44].

2.1.3. *Adrenodoxin*

Adrenodoxin (Adx) is an iron-sulfur protein (ferredoxin) of the [2Fe-2S] type and in adrenals supplies electrons from a NADPH-dependent AdR to cytochrome P450scc (CYP11A1), which catalyses the side-chain cleavage of cholesterol, and to cytochromes of the CYP11B family (Figure 2) that are involved in the formation of cortisol and aldosterone [2,45,46]. The same Adx participates in electron transfer from AdR to different mitochondrial cytochromes P450 of liver and kidney [47]. Adrenodoxin deficiencies have not been described so far and it has been discussed that a loss of progesterone production upon Adx deficiency in a fetus would lead to spontaneous abortion at about 6–7 weeks [48]. However, recently mutations in the CYP11A1 gene have been described in two patients (see above), and it is conceivable that mutations in the Adx gene (those leading to only partial loss in the electron transfer ability) are not lethal.

The gene for human adrenodoxin is located on chromosome 11q22, while its pseudogenes are on 20q11-q12 [49]. Its redox partners are mapped to 17q24-q25 (AdR), 15q23-q24 (CYP11A1), and to 8q21 (CYP11B1 and CYP11B2). Adrenodoxin mRNA is translated in the cytoplasm as a higher molecular weight preprotein including 58 extra-amino acids at the N-terminus of bovine Adx [50] and is processed upon mitochondrial import [51–54], for a recent review see [55]. The mitochondrial P450s are also synthesized as preproteins and proteolytically processed in a similar fashion [55].

The zonal localization of human Adx has not yet been studied. Bovine Adx was shown by immunocytochemical methods to be present at high concentration in the *zona fasciculata* and *zona reticularis*, while low concentrations seem to be present in the *zona glomerulosa* [56,57]. A recent informative overview on the light and electron microscopic immunohistochemical localization of adrenal steroidogenic enzymes of different species has been published by Ishimura and Fujita [58].

The resolution of the bovine adrenodoxin three dimensional structure has been a matter of intense scientific work for about 20 years [59,60], but it was not until the end of the 1990s that it was resolved. Obviously, proteolytic digestion of amino acids from the C-terminal part of the protein as well as high flexibility of the C-terminal region hindered the formation of diffracting crystals so that only after truncation of the molecule the crystallization of this ferredoxin and resolution of its structure to 1.85 Å was feasible [18,61]. Later on, applying special methods to protect the protein from proteolysis, it was also possible to obtain diffracting crystals of the wild-type full length protein with 2.5 Å resolution [62]. It turned out that the C-terminal portion is highly flexible also in these crystals. The structures of the truncated form and the wild-type protein are essentially identical. Data on the structure of bovine Adx in solution has also been obtained by NMR. It was shown that the NMR and X-ray structures are very close to each other [63]. Structural differences have, however, been observed between the oxidized and the reduced forms of the protein with the main differences focused at the regions being involved in redox partner binding [63,64]. Reviews on the structure and physiological importance of Adx in steroidogenesis have recently been published [8,65].

2.1.4. Adrenodoxin Reductase

Adrenodoxin reductase (AdR) (EC 1.18.1.2) is a flavoprotein containing FAD as prosthetic group. The protein was purified 40 years ago by Omura [66]. The cDNA of AdR has first been isolated from bovine adrenals [67,68]. Cloning of human cDNA has been performed by Miller and colleagues [69,70]. The gene has been mapped to chromosome 17q24-q25 [13,71]. The human protein is synthesised as a precursor protein including an extra-peptide of 32 amino acids at the N-terminal sequence which is hydrophobic and rich in arginine. Two forms of human AdR mRNA, differing by the presence or absence of 18 bases in the middle of the sequence, arise from alternate splicing at the 5′ end of exon 7. The protein is localized at the surface of the inner mitochondrial membrane. The FAD and NADPH binding domains are in the N-terminal part of the protein.

Bovine and human AdR have been expressed in *E. coli* [72,73]. The three-dimensional structure of bovine AdR, which has been resolved with 2.8 Å resolution [6], shows a highly asymmetric charge distribution rendering the cleft between the FAD- and the NADPH-domains almost completely basic, while the opposite side of the protein is predominantly acidic. This cleft has been proposed to bind to negatively charged residues of Adx.

Defects in AdR also have not been found so far in patients just as noted above for Adx. It has, however, to be mentioned that AdR was suggested to be involved in apoptosis by a p53 mediated pathway [74]. The authors proposed that this might be due to the production of reactive oxygen species (ROS) by AdR. However, previous studies in Hanukoglu's group demonstrated that there is little or no 'leakiness' connected with AdR [75]. Very recently it was clearly demonstrated that over-expression of human Adx leads to ROS production, to a disruption of the mitochondrial transmembrane potential, to cytochrome c release and to caspase activation and thus to apoptosis [76] which is consistent with the studies of Hanukoglu's group [75].

2.2. Microsomal P450 Systems

There are three important reactions in steroid hormone biosynthesis which are catalyzed by cytochromes P450 in the endoplasmic reticulum: the 17α-hydroxylation of pregnenolone or progesterone and the 17,20 lyase reaction performed by CYP17 to produce dehydroepiandrosterone and androstenedione, respectively, the hydroxylation of progesterone and 17α-hydroxyprogesterone at the 21 position by CYP21 and the aromatization of the A ring of the steroid molecule catalyzed by CYP19. While all three reactions take place in gonads, only CYP17 and CYP21 are expressed in the adrenal cortex.

2.2.1. CYP17

The CYP17-dependent reactions are of special interest since on the one hand they occur at the branch point leading to cortisol or corticosterone and on the other hand they lead to precursors of sex hormones (Figure 2). CYP17 converts pregnenolone to 17-OH pregnenolone and further on to dehydroepiandrosterone and it converts progesterone to 17-OH–progesterone and androstenedione. There were long-standing disputes as to whether there are two enzymes for these reactions or not and it was not until the early 1980s that homogeneous proteins catalyzing both activities were purified from testis and adrenals [77–80].

The cDNAs for bovine adrenal and human CYP17 were cloned in the late 1980s [81,82]. The gene of the human enzyme has been mapped to chromosome 10q24.3 [13]. Heterologous expression of this cDNA and measurement of the corresponding activities demonstrated that CYP17 is capable of performing both, 17α-hydroxylation and 17,20 lyase reaction [81,83,84]. Surprisingly, 17-OH progesterone is a poor substrate for human CYP17 so that most androstenedione is produced by the action of 3β-hydroxysteroid dehydrogenase on dehydroepiandrosterone, and only a minimal amount is formed from 17-OH progesterone meaning that most human sex steroid production is from dehydroepiandrosterone (see Figure 2). There are severe defects connected with CYP17 deficiencies (for reviews see [85,86]). Of special interest is the observation that cytochrome b_5 can interact with CYP17 and specifically stimulate lyase reaction [84] as does phosphorylation of the protein (see below).

2.2.2. CYP21

The 21-hydroxylation of steroids in microsomes of different tissues is catalysed by CYP21. This P450 was the first for which involvement in steroid hormone biosynthesis was proved and the first P450 used to prove its activity by photochemical action spectrum [87]. It converts progesterone to 11-deoxycorticosterone and 17-OH-progesterone to 11-deoxycortisol (Fig. 2) and thus promotes formation of both, glucocorticoids and mineralocorticoids. The cDNA of human CYP21 has been cloned [88,89]. Today, more than 6,300 P450s are known throughout the biological kingdoms [1]. For those tested, the majority of these have been easily expressed in bacterial systems which permits relatively large scale production for biochemical and biophysical studies. The development of expression systems for microsomal P450s using the pCWori$^+$ vector [90] and for mitochondrial P450s using the TRC99A vector [14] have made it easy to study eukaryotic forms. In very few cases, however, expression of functional forms has been very difficult and CYP21 has been perhaps the most difficult. Only this year, 2006, has a viable expression system for bovine CYP21 been developed [91].

The gene for CYP21 is localized on chromosome 6p21.3, very close to a pseudogene. This is the reason for rather frequent cross-over events between both DNA segments leading to formation of inactive proteins. Defects in the CYP21 gene are the primary cause of CAH, accounting for about 90% of analyzed cases of this disease [92]. Thus, congenital adrenal hyperplasia is mainly caused by defects of CYP21 [93,94] although in 8–9% of the patients with CAH, CYP11B1 mutations are the cause [92–94]. Patients suffering from these defects lack formation of significant amounts of glucocortioids and mineralocorticoids, while the precursors, 17-OH steroids, accumulate.

2.2.3. CYP19

CYP19, also called aromatase, is responsible for converting androgens into the corresponding estrogens. This enzyme catalyzes an aromatization of the A ring as indicated in Figure 2. The main sites for estrogen synthesis are granulosa cells of the ovary, the placenta, and adipose tissue, but this sex hormone is also produced in testis and brain [95].

The cDNA of human CYP19 was cloned in 1986 [96]. The gene is localized on chromosome 15q21.1. The crystal structure is not known, but a three-dimensional model has been published [97,98]. Structure-function relationships have been analyzed using site-directed mutagenesis. Thus, it was demonstrated that Glu302 plays an important role in the catalytic activity of this enzyme [97]. Furthermore, the participation of several aromatic amino acids in substrate recognition has been postulated [98]. Estrogens play an important role in the development of certain hormone-dependent cancers such as breast cancer [99] and aromatase inhibitors are in many cases important in treatment of breast cancer. Deficiency of the CYP19 gene leads to sexual infantilism, polycystic ovaries, and pseudo-hermaphroditism [100–102].

2.2.4. NADPH-Cytochrome P450 Reductase

NADPH-cytochrome P450 reductase (CPR) plays a very important role in microsomal P450 systems, since it provides them with electrons from NADPH for oxygen activation and steroid hydroxylation (Figure 3). This reductase contains besides FAD, which is also a cofactor of the mitochondrial NADPH-dependent adrenodoxin reductase, FMN at 1:1 ratio [103]. The cDNA of this enzyme has been cloned from rat liver in 1985 [104] and has been expressed in *Escherichia coli* [105]. The human reductase was cloned in 1989 [71,106]. The gene has been localized on chromosome 7q11.2. CPR provides electrons not only to the P450s involved in steroidogeniesis, CYP17, CYP21 and CYP19, but also to all other P450s contained in the endoplasmic reticulum such as those involved in the metabolism of drugs and xenobiotics.

Very recently, it has been shown that some patients with Antley-Bixler Syndrome suffering from disordered steroidogenesis are deficient in CPR [107–109]. This observation was rather unexpected, since CPR knockout mice do not survive past the middle of fetal development [110,111]. It is still not very well understood how this defect affects other vital functions of mammals like cholesterol biosynthesis, and metabolism of drugs, xenobiotics etc., where CPR is thought to be essential.

2.3. Electron Transfer and Its Regulation

Protein–protein interaction and electron transfer is a necessary prerequisite for oxygen activation and product formation in cytochrome P450 systems. As shown in Figure 3, two different electron transport chains are involved in steroid hormone biosynthesis. Thus, protein–protein interaction may be an important regulatory factor in steroidogenesis.

2.3.1. Mitochondrial Systems

As mentioned above, mitochondrial cytochrome P450-dependent steroid hydroxylases require a NADPH-dependent redox system consisting of adrenodoxin reductase and adrenodoxin [2,65]. The rate-limiting step in the reaction cycle of mitochondrial steroid hydroxylation has not yet been unambiguously defined. In the similarly organized camphor hydroxylase system the second electron transfer from the ferredoxin (putidaredoxin) to CYP101 has been identified as the slowest step [112] in substrate hydroxylation suggesting that the importance of ferredoxin-mediated electron transfer could be similarly high in mitochondrial adrenal steroid hydroxylation. Although the rate-limiting step in the mitochondrial system is not clear, there is considerable data describing the electron transfer and the mechanism of protein–protein interaction in mitochondrial steroid hydroxylase systems as being limiting.

It has been shown that positively charged residues on the proximal surface of the cytochrome P450 on the one hand [113–115] and both negatively charged amino acids [116,117] as well as hydrophobic properties on the surface of Adx on the other hand [118–120] are important for the correct function of the ferredoxin as an efficient electron carrier. Moreover, the C-terminal domain of Adx has been demonstrated to modulate interaction with the redox partners [61]. Furthermore, the loop region of Adx has been demonstrated to be of fundamental importance for the electron transfer from AdR to Adx and from Adx to the different P450s [122,123]. For a more detailed overview of the redox partner interactions in mitochondrial steroid hydroxylase systems the reader is referred to recent reviews [8,39,120,124].

So far only little attention has been paid to the role of redox partner interaction in regulating steroid hormone production in adrenals [65]. It should, however, be mentioned that results with various Adx concentrations and Adx mutants that display different electron transfer properties support the proposal that an increase in the amount of Adx or in the electron transfer rate leads to an enhancement of the 'intermediate flow' towards aldosterone. This observation could be of physiological importance [125]. Further modulation of steroid hormone production is realized via interaction among the different P450s. Competition between rat CYP11A1 and CYP11B1 for electron transfer is observed *in vitro* [126] and with both the human and bovine enzymes in COS-1 cells at low availability of reducing equivalents [127] resulting in a decrease of cortisol and aldosterone synthesis, respectively. In addition, with the bovine enzymes a specific interaction was observed indicating allosteric effects of bovine CYP11A1 on bovine CYP11B1 in reconstituted liposomal membranes [128,129] and in COS-1 cells at high electron supply due to cotransfection of Adx [127]. In contrast, human CYP11A1 does not show such a specific interaction with human CYP11B2 protein.

Regulation of protein interaction and thereby steroid hormone production has also been demonstrated by phosphorylation of Adx [121,130]. A stimulatory effect has been demonstrated for cAMP-dependent phosphorylation of Adx on both bovine CYP11A1 and CYP11B1 [130]. Interestingly, CK2-kinase-dependent phosphorylation revealed a differential effect on these two enzymes. While CYP11A1-catalyzed substrate conversion was increased, the activity of CYP11B1-dependent product formation was not altered [121]. Although so far no data on the *in vivo* relevance of this observation are available, the *in vitro* data show the potential of phosphorylation on differentially regulating CYP11A1 and CYP11B1 reactions [121].

2.3.2. Microsomal Systems

Protein–protein interactions in microsomal steroidogenic P450 systems have not yet been extensively studied. Early investigations in the drug metabolizing liver systems, especially with CYP2B4 and CPR, suggested the participation of charge–pair recognition in this process [131–135]. This has been supported by studying other members of the P450 family (for review see [136]). Interestingly, microsomal P450s are shown to have a common charge distribution on the proximal surface of the P450 molecule supposed to be involved in redox partner binding although individual members of this family have slightly different but nonetheless unique charge distributions [137,138]. This charge distribution was shown to be different from that of mitochondrial P450s [137] reflecting the various sizes of the redox partners, CPR and Adx. X-ray structures of rat and yeast CPRs are now available to assist in understanding the structure/function of this reductase [139,140].

Very recently, it was shown that P450/CPR interaction can be affected in patients suffering from defects in steroidogenesis. Impairment of P450/CPR interaction has been found to be the cause of selective disruption of 17,20-lyase activity of CYP17 in patients having R347H and R358Q replacements in their corresponding proteins [141]. Molecular modelling of CYP17 showed that the two mutated residues altered the distribution of the electrostatic charge on the proximal surface in the region predicted to interact with CPR [142].

Moreover, the electron transfer protein cytochrome b_5 selectively augments 17,20-lyase activity [84,143–145] whereby the apo-protein is as effective as the holo cytochrome b_5. This indicates that cytochrome b_5 functions as an allosteric facilitator and not as electron donor. The mechanism of this augmentation is not well understood yet.

Following up on CYP17 being a phosphoprotein [146] phosphorylation increases the lyase activity, whereas dephosphorylation ablates lyase activity without reducing 17α-hydroxylase activity. The mechanism by which serine phosphorylation promotes 17,20 lyase activity is not known. Since the phosphorylated residue is located near the redox partner binding site it may increase the affinity of CYP17 for CPR or cytochrome b_5 or both [147].

2.3.3. Steroidogenic P450s as Drug Targets

Overproduction of steroid hormones can be involved in breast or prostate cancer, in hypertension and in heart fibrosis (Figure 4). Besides inhibiting the action

Steroid hydroxylases as targets

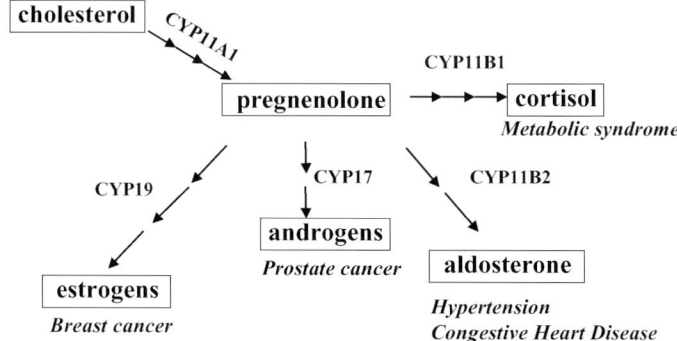

Figure 4. Steroid hydroxylases as potential drug targets.

of the steroid hormones on steroid hormone receptors by using antihormones, which often is connected with severe side-effects, more recently the steroid hydroxylases themselves have turned out to be promising new targets for drug development (for recent reviews see [95,148]). This is especially true of inhibitors of aromatase which have been used in clinical practice for several years to treat breast cancer (for review see [149]).

Since the three-dimensional structures of steroid hydroxylases are not yet available, computer models of the corresponding CYPs may help to develop new inhibitors of these enzymes. During the past years the necessary test systems have been developed and new compounds have been synthesized which display selective and specific inhibition of CYP19, CYP17, CYP11B2, and CYP11B1 [148]. With some of these potential new drugs clinical trials are under way. Although for some of the steroidogenic P450s a role as drug target is very new, it can be expected that in the near future some of the compounds inhibiting steroidogenic P450s will contribute to our arsenal of important new and selective drugs.

3. STEROID HORMONE BIOSYNTHESIS IN THE ADRENAL CORTEX

The adrenal cortex produces a complex array of steroid hormones including glucocorticoids, mineralocorticoids, and androgens. This gland surrounds the adrenal medulla and consists of different regions which produce different steroids. Just below the capsule of the adrenal lies the region of the adrenal cortex named the glomerulosa in which the major mineralocorticoid, aldosterone is produced. Between the glomerulosa and the adrenal medulla lie two regions of the cortex called the fasiculata and the reticularis which are required for the synthesis of glucocorticoids and the adrenal androgens. As shown above in Section 2, different steroid hydroxylases are located in mitochondria and in the endoplasmic reticulum. Mitochondria in the adrenocortex serve the same general roles of mitochondria in other tissues to produce energy, while at the same time they synthesize steroid hormones, making them very specialized mitochondria. The initial step of steroid hormone biosynthesis in the glomerulosa and the fasiculata/reticularis occurs in the mitochondrion where CYP11A1 converts cholesterol to pregnenolone (Figure 2). The presence or absence of other steroidogenic P450s is what separates the steroid hormone biosynthetic function of the glomerulosa from that of the fasiculata/reticularis.

The absence of CYP17 in the glomerulosa assures that no androgens or glucocorticoids will be produced. Secondly, CYP11B2 is present in the glomerulosa and not in the fasiculata/reticularis assuring that aldosterone is produced. The reverse, CYP17 present and CYP11B2 absent in the fasiculata/reticularis, assures glucocorticoid biosynthesis as well as that of adrenal androgens. While in several animal species, including humans, both CYP11B1 and 11B2 are present in mitochondria of the glomerulosa, some glomerulosa (i.e., bovine) contain a single peptide chain, CYP11B1, which catalyzes both 11β-hydroxylation and aldosterone biosynthesis. When success in crystallization and structure determination of human and bovine CYP11B1 is achieved, it will be interesting to compare their structures to learn why a single bovine enzyme can catalyze the activities of two different human enzymes.

Steroidogenesis in the fasiculata/reticularis of the adrenal cortex is regulated by the peptide hormone adrenocorticotrophin (ACTH) or tropin produced in the anterior pituitary [150,151]. ACTH binds to a cell surface receptor of these cells and activates the synthesis of cAMP leading to both acute and chronic regulation of steroidogenesis. Acute regulation activates the mobilization of the substrate for steroid hormone biosynthesis, cholesterol, to the inner mitochondrial membrane where CYP11A1 is located. Chronic regulation activates the biosynthesis of steroidogenic P450s, their essential redox systems and other steroidogenic enzymes assuring that optimal steroidogenic capacity is maintained in the adrenal cortex.

3.1. Acute Regulation

The acute regulation of steroidogenesis in the adrenal cortex has been known for more than 40 years, over which time a number of different mechanisms have been proposed as the basis of this regulation. Currently considerable detail is known about how ACTH via cAMP regulates the mobilization of cholesterol into the CYP11A1 active site. First, stimulation of cholesterol mobilization involves *de novo* protein biosynthesis [152]. The rapid acute regulatory response to ACTH is found to be mediated through the cAMP second messenger signaling pathway. This leads to mobilization of cholesterol from lipid stores within steroidogenic cells to the mitochondrion [153,154]. cAMP activates cholesterol ester hydrolase which cleaves the ester function from cholesterol esters [155]. Specifically the cAMP-response activates synthesis of the steroidogenic acute regulatory protein (StAR) [156]. There is strong evidence that StAR is essential for steroidogenesis [157], particularly from data in patients suffering from a form of congenital adrenal hyperplasia where both adrenal and gonadal steroid hormone biosynthesis is greatly reduced due to mutations in the StAR gene [156–162].

StAR is found to play a key or even essential role in mediating transport of cholesterol from outside the outer mitochondrial membrane into the inner mitochondrial membrane from which it can enter the active site of CYP11A1 and be converted to pregnenolone [157,163]. A cyclic AMP response-element binding protein, CREB, or perhaps its close relative CREM, is responsible for the protein kinase A (PKA) mediated response between ACTH and elevated StAR levels [156]. The biophysical basis by which StAR stimulates cholesterol transport remains unclear, however, two important findings provide insight into this process. Mutagenesis studies of StAR suggest that StAR activity requires a pH-dependent protein globule transition on the outer mitochondrial membrane [163]. Further the peripheral-type benzodiazepine receptor (PBR) interacts with StAR on the outer mitochondrial membrane to facilitate cholesterol transfer across this membrane to the inner mitochondrial membrane, and then to CYP11A1 [164].

When thinking about cholesterol in the inner mitochondrial membrane, we realize that there are two pools of cholesterol present. First, as with the inner membrane of all mitochondria there is membrane cholesterol that maintains the fluidity of the membrane. Second, in response to ACTH through cAMP, StAR and PBR, a pool of steroidogenic cholesterol accumulates in the inner membrane.

3.2. Chronic Regulation

Using primary bovine adrenocortical cells in culture, Waterman, Simpson and their colleagues characterized a second cAMP-dependent pathway in steroidogenesis, the regulation of transcription of the genes encoding steroid hydroxylases, their reductases and related enzymes such as 3β-hydroxysteroid dehydrogenase (3β-HSD) [150,151]. The key to uncovering this level of regulation was the ability to use a very stable primary cell culture system, bovine adrenocortical cells. They grow into confluent monolayers in which the steroidogenic pathways can be readily manipulated with ACTH or cAMP. Developmental and tissue specific regulation of these steroidogenic genes underlies this chronic regulation. Studies carried out by Morohashi, Omura et al. [165] and by Parker and his colleagues [166] led to the discovery of a zinc-finger orphan receptor now known as steroidogenic factor 1 (SF-1). SF-1 is expressed in a number of cell types, predominantly steroidogenic cells and the steroidogenic genes contain one or more SF-1 binding site in their 5'-flanking region, (AGGTCA). Developmental studies have shown that SF-1 is expressed prior to expression of steroidogenic enzymes [166]. It is thought to play an essential role in the development of steroidogenic organs as well as in the developmental and tissue-specific expression of steroidogenic enzymes.

The chronic regulation of transcription of steroidogenic genes via cAMP in response to ACTH in the adrenal cortex assures that optimal steroidogenic capacity is maintained throughout life [150,151]. Both acute and chronic responses to ACTH are mediated by cAMP through the action of protein kinase A. ACTH binds to its cell surface receptor on adrenocortical cells and through a coupled G-protein and elevated levels of cAMP activate PKA which activates the transcription process. To date, the details of this transcription process have been established in detail in one steroidogenic gene, human CYP17, as will be described below [167].

Promoter bashing studies of the 5'-flanking region of steroidogenic genes were expected to uncover a common cAMP-response system associated with each gene, not unlike the common SF-1-dependent system found in these genes which controls developmental and tissue-specific regulation. However, surprisingly it has been found that each gene seems to have its own complex biochemical mechanism responsible for cAMP-dependent transcription [168]. Further, although SF-1 is found in all mammalian species studied, different species apparently utilize different cAMP-dependent mechanisms for the same gene. The biological reason for this rather broad diversity in the mechanism by which optimal steroidogenic capacity is maintained compared to the common mechanism for development/tissue specific regulation is not understood. Perhaps the explanation lies in the broad variation in steroidogenesis found between species. It certainly is true that the physiological requirement for steroid hormones varies dramatically between species, particularly that for adrenal steroids. Thus, complexity is the reason that the biochemical details of only one cAMP-dependent transcription system has been elucidated so far [168].

A dual-specificity phosphatase has been found to be essential for ACTH/cAMP-dependent regulation of several steroidogenic genes in the human adrenal cortex. For human CYP17, mitogen-activated protein kinase phosphatase-1 (MKP-1) which is a nuclear dual-specificity phosphatase, has been found to be a key player in cAMP-dependent transcription [167]. cAMP induces MKP-1 mRNA and protein expression within 30 minutes of exposure. Furthermore, increased phosphorylation of MKP-1 occurs and silencing MKP-1 expression with antisense oligonucleotides attenuates cAMP-stimulated human CYP17 expression [167].

The 5'-flanking region of human CYP17 contains a cAMP response sequence which overlaps with the SF-1 binding site. In human adrenal cells, SF-1 is phosphorylated by the ERK1/2 phosphatase leading to complex formation with several nuclear proteins including mSin3A, PSF and p54nrb [169]. When MKP-1 is phosphorylated by PKA in response to ACTH binding to its receptor, it is activated and it dephosphorylates SF-1. As a result, mSin3A is released from the above complex and CYP17 transcription is activated [167]. Therefore, phosphorylation/dephosphorylation of MKP-1 and SF-1 are the key events in cAMP-dependent regulation of the human CYP17 gene.

For a long time after discovery of SF-1, the question of whether or not this orphan receptor bound a ligand remained unresolved. Recently, phospholipids have been identified as ligands for SF-1 [170]. More recently in human adrenocortical cells, SF-1 has been found to bind sphingosine and cAMP stimulation which leads to dephosphorylation of SF-1, decreases the amount of sphingosine bound to SF-1 [171]. Thus it seems that sphingosine as a ligand for SF-1 is inhibitory of transcription. It may very well be that displacement of spinogosine from SF-1 by phospholipids activates CYP17 transcription [171].

It is evident that there is a need for extensive and detailed analysis of the biochemical details of cAMP-dependent transcription of different steroidogenic genes. It can be expected that there will be certain features in common among the different genes, but overall considerable variation from one gene to the next will be found [172].

While peptide hormones from the anterior pituitary play very important roles in regulating steroid hormone biosynthesis, it is important to look at regulation of steroidogenesis from a broader view. In addition to cAMP-dependent regulation, it is clearly established that cAMP-independent regulation is present in different steroidogenic organs in different species throughout life. This regulatory path involves different growth factors and is generally quite modest. However, it provides a basal level of transcription upon which the cAMP-dependent mechanisms build and therefore is an important piece of the regulatory processes leading to optimal steroidogenesis that exists in steroidogenic cells throughout life [150,151]. A third level transcriptional regulation of steroidogenic P450 gene expression is tissue specific. That is, only a limited subset of cells produce steroid hormones. In the steroidogenic factories, adrenal cortex and gonads, SF-1 is thought to be the key tissue/cell-specific transcription factor. However, several other cell types are also known to express different profiles of steroid hormones, cell types which have not been found to express SF-1. Therefore, SF-1 can not be the only mechanism by which cell-specific regulation is controlled. Finally, there is a fourth level of transcriptional regulation of steroidogenic genes, developmental regulation. The best example of this is the upregulation of the testosterone biosynthetic pathway in the fetal testis at precisely the correct time to assure development of the male phenotype (secondary sex characteristics).

The above paragraph presents an overview of a complex multifactorial transcriptional process which seems to vary in details in different steroidogenic sites in the same species and also in the same steroidogenic site between different species. There is not a great deal of biochemical information which allows us to understand the basis of these variations. However, steroid hormone production and function is different amongst species providing one explanation for the variability. However, the story is even more complex when considering the function of SF-1. Disruption of the mouse SF-1 gene (SF-1 knockout) generates a phenotype which does not have steroidogenic organs. Adrenal glands, testes or ovaries are absent from these knockout animals [173]. This clearly demonstrates a higher level of function of SF-1 than simply regulation of steroidogenic gene transcription.

3.3. Conclusion

The precise biochemical details of both acute regulation and, in almost all cases, of chronic regulation of steroidogenesis in the adrenal cortex remain to be established. This will require the efforts of a number of laboratories over the next several years. In the end, perhaps the most detailed understanding of regulation of a metabolic pathway will be revealed and as a result, not only will regulation of steroidogenesis become clear but a paradigm for studying other metabolic pathways will emerge.

4. STEROID HORMONE BIOSYNTHESIS IN THE GONADS

Steroid hormone production in the gonads begins similarly to steroid hormone biosynthesis in the adrenal cortex, but then shifts to pathways that focus strictly on the biosynthesis of sex hormones. In the testis this is primarily biosynthesis of testosterone in Leydig cells which cluster between the testicular tubules. In the ovary the primary steroid hormone products are progesterone and estrogen involving two cell types in the ovarian follicle, the thecal cells and the granulosa cells. A schematic view of these biosynthetic pathways is seen in Figure 5. Like all biosynthetic pathways for steroid hormones, the biosynthetic pathways in the gonads begin with conversion of cholesterol to pregnenolone in mitochondria. The remainder of the pathways occurs in the endoplasmic reticulum. Acute regulation of the mitochondrial pathway in steroidogenic cells in the testis and ovary is the same as that described previously for the adrenal cortex.

4.1. Testis

Chronic regulation in the testis requires the action of luteinizing hormone (LH) in place of ACTH. However, because the timely synthesis of testosterone in fetal Leydig cells is necessary for development of the male phenotype in all vertebrate species, close association between SF-1 and LH-dependent transcription of CYP17 and CYP11A1 occurs in fetal life and the action of LH to regulate transcription of these and other steroidogenic genes (3β-HSD) continues uninterrupted throughout life [174,175]. It appears that while LH regulates CYP11A1 and CYP17 gene expression differently both within the same species and between

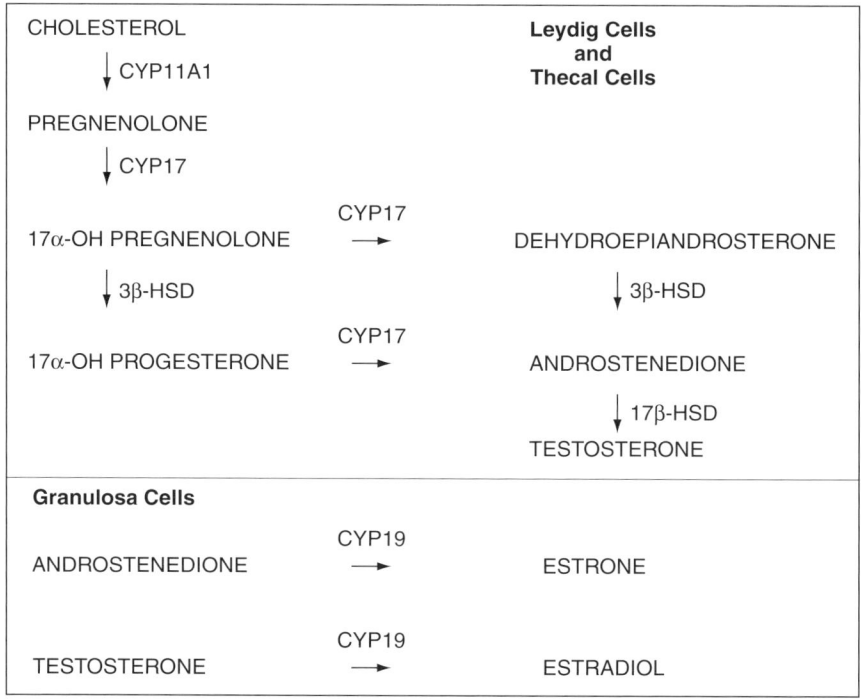

Figure 5. Sex hormone biosynthesis in testis and ovary.

species, the final result of continuing testosterone biosynthesis throughout fetal and adult life in all vertebrates is the same [176]. Diversity of testosterone biosynthesis in Leydig cells from different species may be important in establishing the magnitude of this production. Just as suggested for ACTH in the adrenal cortex the chronic role of LH in regulation of testosterone production is to maintain optimal steroidogenic capacity in Leydig cells.

In species such as human and cattle, CYP17 has a very poor Δ^4 17,20-lyase activity [176]. Cytochrome b_5, a ubiquitously distributed electron transfer protein in endoplasmic reticulum has been shown to enhance 17α-hydroxylase and/or 17,20-lyase activities of CYP17 from different species. *In vitro* studies have shown that cytochrome b_5 only modestly enhances Δ^5 17α-hydroxylase of human CYP17, has no effect on its Δ^4 17-hydroxylase activity, but has a significant positive effect on both Δ^4 and Δ^5–17,20-lyase activities [84,143–145]. As a result of a favorable cytochrome b_5/P450 ratio (primarily CYP17) in human testis [177], both Δ^4 and Δ^5 pathways participate in testosterone biosynthesis. Studies of the action of cytochrome b_5 on the function of CYP17 have been seminal in establishing the role of protein-protein interaction in P450 activities, opening the door to a new view of monooxygenase activity.

4.1.1. Conclusion

Testosterone biosynthesis in the Leydig cells is regulated in very complex patterns. Developmental regulation of steroidogenic enzyme synthesis is very important to assure 'maleness' which is essential for reproduction in vertebrates. Building on this developmental regulation, regulation of a peptide hormone (LH) assures that sufficient testosterone is produced throughout life assuring reproductive capacity is maintained. While steroidogenesis in the testis is not essential for maintaining life of the individual, it is essential for continuation of vertebrate species.

4.2. Ovary

The ovary contains two types of steroidogenic cells, both located in the follicles. The walls of these structures are lined with thecal cells and the ovum floating in the follicular fluid is surrounded by granulosa cells. The thecal cells contain both CYP11A1 and CYP17 and produce the androgens, androstenedione and testosterone. Steroidogenesis in thecal cells is regulated by LH via cAMP. The granulosa cells do not contain CYP17 and therefore produce progesterone as a major steroid hormone. Also, the androgens from the thecal cells migrate to the granulose cells where CYP19 (aromatase cytochrome P450) is present. This P450 is responsible for the conversion of androgens to estrogens, estrone being produced from androstenedione and estrogen from testosterone. Progesterone biosynthesis in the granulose cells is regulated by LH or by follicle stimulating hormone (FSH) both via cAMP. Aromatase activity leading to estrogen production is regulated by FSH.

The ovary presents an interesting example of tissue-specific gene expression under physiological control. Using bovine tissue it has been found that ovarian follicles contain relatively low levels of CYP11A1, CYP17, and adrenodoxin [178]. Following ovulation the follicle undergoes transformation to form the corpus luteum which is a very active steroidogenic tissue producing progesterone. The transition from follicle to corpus luteum is accompanied by a surge in LH which causes, through cAMP, a dramatic increase in CYP11A1 and adrenodoxin levels and a sharp fall in CYP17 level [179,180]. This is quite different from the adrenal where cAMP enhances transcription of all three genes coordinately. The corpus luteum contains two cell types, small luteal cells derived from the theca interna and large luteal cells derived from the granulosa cells. As noted above, CYP11A1 and CYP17 exist in the thecal cells and CYP11A1 is found in low levels in the granulosa cells [178]. In the corpus luteum both cell types contain high levels of CYP11A1 and undetectable CYP17 [181]. This is an excellent example of tissue or cell specific regulation of expression of steroidogenic genes.

4.2.1. Extragonadal Aromatase

Aromatase cytochrome P450 expressed in the ovary is regulated by SF-1 and cAMP in patterns similar to those outlined herein for other steroidogenic P450s. However, CYP19 is widely expressed in other tissues including adipose tissue, bone, specific sites in the brain, vascular endothelial cells, and smooth muscle cells where in humans it plays important roles in health and disease [182,183]. The expression of this P450 in breast tissue has been clearly found to be associated with certain types of breast cancer [184] and different CYP19 azole inhibitors are used in treatment of these cancers [148].

The human aromatase gene has been found to have a very complex 5'-regulatory region [185], the most complex discovered with any CYP gene to date. The regulation of CYP19 in different cell types remains to be clearly established but it will involve growth factors and other regulatory systems in addition to cAMP. In very few cases, aromatase deficiency has been found leading to complex phenotypes in both males and females [186].

5. EXTRAADRENAL AND EXTRAGONADAL STEROIDOGENESIS

It has long been anticipated that steroidogenesis is limited to the primary steroidogenic tissues such as gonads and adrenals. Expression of steroidogenic enzymes is, however, also found in other steroidogenic organs like placenta and brain [187–190]. Profiling transcripts for steroidogenic enzymes in human fetal tissues reveals expression of some of these enzymes (CYP11A1, CYP21, CYP11B1, CYP19) in non-endocrine tissues such as aorta, liver, kidney, heart, lung, pancreas, prostate, stomach, and thymus. CYP17 transcripts have also been found in all these tissues except lung. Interestingly, CYP11B2 was found only in adrenals in these studies [191]. The role of steroidogenesis in non-endocrine tissues is not yet clear, but it can be postulated that the steroids may be involved in important paracrine and autocrine actions.

5.1. Other Steroidogenic Organs: Placenta and Brain

It was shown that in porcine placenta CYP11A1 and CYP17 were detected although with lower expression than in the adrenal gland and testis [187]. Studies using human placenta revealed that CYP11A1, the CYP11B enzymes, and CYP19 genes, proteins and activities can be detected throughout pregnancy [189].

Dehydroepiandrosterone was found in the brain of adult male rats 25 years ago [192]. The presence of immunoreactive CYP11A1 in rat and human brain has been established in the white matter and in primary cultures of rat forebrain glial cells [193]. Detection of mRNAs for Adx, CYP11A1 and CYP11B1, but not for CYP11B2 and CYP17 in rat brains and in cultured rat glial cells has been described [194]. Subsequently it was found that CYP11A1 and CYP17 mRNAs can be detected in all brain regions, whereas CYP21 was detected only in the brain stem [195]. These authors were also able to detect CYP11B2, besides CYP11B1, in rat but not in mouse brain. Expression of CYP11B1 was primarily in the cerebrum, whereas CYP11B2 was detected in all brain regions. Corticoid and aldosterone biosynthesis in brain has been demonstrated by Gomez-Sanchez and coworkers [196,197]. CYP11B1 and CYP11B2 have been found in fetal rat hippocampal neurons [198]. In addition, hippocampal neurons express CYP11A1, CYP17 and CYP19 (for review see [190]). Besides the cytochromes P450, expression of StAR, 3β- and 17β-hydroxysteroid dehydrogenases, and 5α-reductase has been found in the brain and thus the brain is able to synthesize neurosteroids as paracrine neuromodulators [199] (for reviews see [190,200]).

The mRNA of CYP11B1 was clearly localized in layers II-IV of the neocortex and in layer II of the piriform cortex of rat using *in situ* hybridization. No significant differences in the expression level of CYP11B1 have been observed in different brain regions of normal and hypertensive transgenic rats [201]. Interestingly, it was found that CYP11B2 expression in the brain is not increased by angiotensin II stimulation or decreased by high sodium diet as it is in the adrenal gland indicating regulatory implications [202]. While expression of cytochromes P450 in the brain of rodents is well established, only limited information is available on expression of these enzymes in human brain. Sex- and age-dependent differences in CYP11A1 mRNA expression have been established. It was demonstrated that CYP11A1 mRNA is expressed approximately 200 times lower in the temporal lobe, frontal lobe, and hippocampus than in adrenal tissue. Its concentration in the temporal lobe increased markedly during childhood and reached adult levels in puberty. Moreover, CYP11A1 mRNA is significantly higher in the temporal and frontal lobe cortex of women than in that of men [203].

It has to be mentioned that expression of cytochromes P450 in brain can be very low, so that in some cases contradicting results for their expression (e.g., CYP11B2, CYP21) have been described (for review see [200]).

5.2. Endothelial Cells and Heart

Mineralocorticoids have been suggested for a long time to act on blood vessels leading to increased vasoreactivity and peripherial resistance. However, only recently has expression of CYP11B2 and synthesis of aldosterone in blood vessels been demonstrated [204–207]. The authors clearly demonstrated that production of aldosterone in the rat mesenteric artery was not changed by adrenalectomy. Levels of CYP11B1 and CYP11B2 expression have been determined using competitive PCR and it was found that both angiotensin II and potassium increased CYP11B2 but not CYP11B1 mRNA. This data showed for the first time the potential role of locally produced aldosterone in blood pressure regulation and cardiovascular disease.

While expression of CYP11B1 and CYP11B2 in blood vessels was well accepted, the expression of these enzymes in the heart was a matter of contradiction. Silvestre et al. [208] described CYP11B2 expression in rodent heart, where it was regulated by angiotensin II and potassium or sodium diet. More recently, and in contrast to findings in the normal human heart, the expression of CYP11B2 was demonstrated in the failing left ventricle on the basis of coronary-sinus levels of aldosterone exceeding those found in the aorta [209–212]. Chronic elevation of aldosterone has been diagnosed in congestive heart failure or myocardial fibrosis [213]. Especially in congestive heart failure, elevated aldosterone levels lead to an increase in blood volume and may stimulate cardiac fibroblasts resulting in cardiac hypertrophy, myocardial fibrosis, ventricular arrhythmia, and other adverse effects [214,215]. Thus, the RALES (Randomized Aldactone Evaluation Study) and EPHESUS (Eplerenone Post Acute Myocardial Infarction Efficacy and Survival Study) trials were done to determine whether aldosterone antagonists (spironolactone, eplerenone) reduce mortality in patients with severe heart failure [216,217]. It was clearly demonstrated that the group treated with the aldosterone antagonists revealed a decreased risk of mortality and an improvement of the heart disease. It can be suggested that local production of aldosterone in the failing heart is also involved in the induction of myocardial fibrosis. The mechanism of the pathophysiological changes and the possibilities of a knowledge-based treatment of this disease are, however, not yet well understood.

5.3. Skin

Already in 1996 evidence was provided that mRNAs for CYP11A1, CYP17, and CYP21 were expressed in the skin [218]. Since the skin is the organ where synthesis of vitamin D takes place, it can be anticipated that the electron donors for P450 reactions are also expressed in this organ. Very recently, the presence of CYP11A1, AdR, CYP17, and the transcription factor SF-1 was documented in human facial skin, human sebocytes and in the human sebaceous gland cell line SEB-1. In addition, biochemical activity could be demonstrated for both P450s demonstrating that human skin has steroidogenic capacity [219]. The impact of these observations is not clear nor is the interplay between steroid hormone and vitamin D synthesis in this organ.

5.4. Other Tissues

There are only a few reports besides that of Pezzi et al. [191] showing the presence of steroidogenic P450s in other tissues. *In situ* hybridization with specific oligonucleotide probes as well as immunocytochemistry with specific antibodies showed that the parietal cells of the gastric mucosa, contrary to other cell types, strongly express CYP17 [220]. Production of aldosterone and expression of CYP11B2 has been shown using RT-PCR and Southern blots in kidney, liver and lung of rats [221]. RT-PCR was also used to demonstrate the expression of CYP11A1, CYP17, CYP11B1, CYP11B2 as well as 3β-hydroxysteroid dehydrogenase in the rat cochlea, although in varying amounts depending on the region studied [222].

It was suggested that steroid hormones, particularly mineralocorticoids, are candidates for controlling the homeostasis of endolymph as steroid receptors are widely expressed in the cochlea. The expression pattern suggests that enzymes leading to mineralocortioids and sex steroids but not to glucocorticoids may be expressed in the rat cochlea. This is again in line with a paracrine role of these steroids which was also suggested for the steroids produced in brain, heart or skin.

CYP11A1 gene expression is observed in the primitive gut of mouse embryo and when examined in SF-1 knockout mouse it is evident that expression at this site does not require this nuclear transcription factor [223]. Also CYP17 is expressed in rat liver very early after partuition reaching a very high level which rapidly declines to undetectable levels. The biological role of this pattern of expression is unknown [224].

5.5. Conclusion

It has to be stated that we are still far from understanding the function of steroids in these extra-adrenal tissues and it can be anticipated that at least in some cases independent regulatory pathways have been developed for these organs so that a complex interplay between the different levels of regulation may occur. Detailed understanding of the purpose of these steroidogenic systems and their regulation will be important in providing a complete understanding of steroid hormones.

6. OUTLOOK FOR THE FUTURE

It is very clear that new developments in molecular biology and biochemistry have led to many new discoveries in the study of cytochromes P450 involved in steroid hormone biosynthesis. These new approaches open the door for additional important discoveries in the coming years. The authors use this final section of this chapter to emphasize our priorities for these new discoveries:

- At present there are no high resolution X-ray structures determined for any of the steroidogenic P450s. Structures have been determined for several microsomal drug metabolizing P450s which should point the way for structure determination of CYP17, CYP21, and CYP19. It is reasonable to imagine that such structures will appear within the next year or two. There are multiple reasons why such structures would be important: First, of course, is the intrinsic understanding of how these monooxygenases work that would emerge from the structures. Second, both CYP17 and P450 aromatase catalyze multistep reactions and the details of the CYP structure in assuring these very specific steps will emerge from such studies. Third, a variety of mutations are known in human CYP21, CYP17, and P450 aromatase. Knowing the structure of these enzymes will reveal how these mutations lead to their established phenotypes. Finally, P450 aromatase and CYP17 are drug targets and their structure determination can help drive the discovery of new drugs targeted to these P450s. – No structures have been determined for mitochondrial P450s. Once such information exists for mitochondrial steroidogenic P450s (CYP11A1, CYP11B1, CYP11B2), similar information to that outlined above for microsomal P450s will become available.

- For steroidogenic P450s that are drug targets we can expect that high-throughput screening using large libraries of potential drugs should reveal new types of molecules which will be the basis for further drug design.
- The role of phosphorylation of proteins involved in steroidogenesis and perhaps other types of posttranslational modifications as well, is an important topic for detailed investigation.
- The role of intermediates and products from the steroidogenic pathways as ligands for nuclear transcription factors and how binding of these molecules regulates transcription is beginning to be understood. We can expect much more detailed information on the biochemical and physiological roles that steroid hormones play to emerge within the next few years.
- Likewise, the details of multifactoral transcriptional regulation of genes encoding steroidogenic enzymes will be unravelled in the next few years.
- The biochemical and physiological roles of steroidogenic P450s expressed in a large number of nonsteroidogenic sites will continue to emerge. Clearly these enzymes play a number of different jobs in biology, the majority of which remain unknown.

Understanding these many important features of steroidogenic P450s will require considerable effort from a large number of laboratories. Very interesting experiments leading to generation of important new knowledge await these laboratories and the authors look forward to watching this information unfold.

ACKNOWLEDGMENTS

R. B. was supported by grants from the Deutsche Forschungsgemeinschaft (Be 1343/12) and the Fonds der Chemischen Industrie. M. R. W. has been supported by NIH grants GM37942 and NIH T30ES000267-40 and the Natalie Overall Warren Chair in Biochemistry.

ABBREVIATIONS

ACTH	adrenocorticotrophin hormone
AdR	adrenodoxin reductase
Adx	adrenodoxin
3β-HSD	hydroxysteroid dehydrogenase
CAH	congenital adrenal hyperplasia
CMO	corticosterone methyl oxidase
CPR	cytochrome P450 reductase
CYP11A1	cholesterol side chain cleavage cytochrome P450 or P450scc
CYP11B1	11β-hydroxylase cytochrome P450 or P45011β
CYP11B2	aldosterone synthase cytochrome P450 or P450aldo
CYP17	17α-hydroxyplase/17,20-lyase cytochrome P450 or P450c17
CYP19	aromatase cytochrome P450 or P450arom
CYP21	21-hydroxylase cytochrome P450 or P450c21
CYP51	sterol 14α-demethylase
DHEA	dehydroepiandrosterone
EPHESUS	Eplerenone Post Acute Myocardial Infarction Efficacy and Survival Study
FAD	flavin adenine dinucleotide
FMN	flavin mononucleotide
FSH	follicle stimulating hormone
LH	luteinizing hormone
MKP-1	mitogen-activated protein kinase phosphatase-1
NADH	nicotinamide adenine dinucleotide, reduced
NADPH	nicotinamide adenine dinucleotide phosphate, reduced
PBR	peripheral-type benzodiazapine receptor
PCR	polymerase chain reaction
PKA	protein kinase A
RALES	Randomized Aldosterone Evaluation Study
ROS	reactive oxygen species
SF-1	steroidogenic factor-1
StAR	steroidogenic acute regulatory protein

REFERENCES

1. D. R. Nelson, *Comp. Biochem. Physiol. C Pharmacol. Toxicol. Endocrinol.*, *121*, 15–22 (1998).
2. R. Bernhardt, *Rev. Physiol. Biochem. Pharmacol.*, *127*, 137–221 (1996).

3. E. Mornet, J. Dupont, A. Vitek, and P. C. White, *J. Biol. Chem.*, 264, 20961–20967 (1989).
4. T. Kawainoto, Y. Mitsuuchi, T. Ohnishi, Y. Ichikawa, Y. Yokoyama, H. Sumimoto, K. Toda, K. Miyahara, I. Kuribayashi, and K. Nakao, *Biochem. Biophys. Res. Commun.*, 173, 309–316 (1990).
5. T. Kawamoto, Y. Mitsuuchi, K. Toda, K. Miyahara, Y. Yokoyama, K. Nakao, K. Hosoda, Y. Yamamoto, H. Imura, Y. Shizuta, *FEBS Lett.*, 269, 345–349 (1990).
6. G. A. Ziegler, C. Vonrhein, I. Hanukoglu, and G. E. Schulz, *J. Mol. Biol.*, 289, 981–990 (1999).
7. R. Bernhardt, C. Pei-rang, A. Grinberg, F. Hanneman, P. Weissgerber, J. Muller, and U. Heinemann, *Molecular Steroidgenesis*, (M. Okamoto, Y. Ishimura, and H. Nawata, eds), Universal Academy Press, Inc., Tokyo, Japan, 2000, pp. 53–56.
8. A. V. Grinberg, F. Hannemann, B. Schittler, J. Muller, U. Heinemann, and R. Bernhardt, *Proteins*, 40, 590–612 (2000).
9. T. Omura, Y. Ishimura, and Y. Fujii-Kuriyama, *Cytochrome P-450*, VCH, Weinheim, Germany, 1993.
10. K. Ruckpaul and H. Rein, *Molecular Mechanism of Adrenal Steroidogenesis and Aspects of Regulation and Application*, Vol. 3, Akademie Verlag, Berlin, 1990.
11. K. Morohashi, H. Yoshioka, O. Gotoh, Y. Okada, K. Yamamoto, T. Miyata, K. Sogawa, Y. Fujii-Kuriyama, and T. Omura, *J. Biochem. Tokyo*, 102, 559–568 (1987).
12. B. C. Chung, K. J. Matteson, R. Voutilainen, T. K. Mohandas, and W. L. Miller, *Proc. Natl. Acad. Sci. USA*, 83, 8962–8966 (1986).
13. R. S. Sparkes, I. Klisak, and W. L. Miller, *DNA Cell Biol.*, 10, 359–365 (1991).
14. A. Wada, P. A. Mathew, H. J. Barnes, D. Sanders, R. W. Estabrook, and M. R. Waterman, *Arch. Biochem. Biophys.*, 290, 376–380 (1991).
15. S. T. Woods, J. Sadleir, T. Downs, T. Triantopoulos, M. J. Headlam, and R. C. Tuckey, *Arch. Biochem. Biophys.*, 353, 109–115 (1998).
16. I. A. Pikuleva, *Mol. Cell Endocrinol.*, 215, 161–164 (2004).
17. E. C. Muller, A. Lapko, A. Otto, J. J. Muller, K. Ruckpaul, and U. Heinemann, *Eur. J. Biochem.*, 268, 1837–1843 (2001).
18. A. Muller, J. J. Muller, Y. A. Muller, H. Uhlman, R. Bernhardt, and U. Heinemann, *Structure*, 6, 269–280 (1998).
19. T. Tajima, K. Fujieda, N. Kouda, J. Nakae, and W. L. Miller, *J. Clin. Endocrinol. Metab.*, 86, 3820–3825 (2001).
20. N. Katsumata, M. Ohtake, T. Hojo, E. Ogawa, T. Hara, N. Sato, and T. Tanaka, *J. Clin. Endocrinol. Metab.*, 87, 3808–3813 (2002).
21. A. Wada, T. Ohnishi, Y. Nonaka, M. Okamoto, and T. Yamano, *J. Biochem. (Tokyo)*, 98, 245–256 (1985).
22. K. Yanagibashi, M. Haniu, J. E. Shively, W. H. Shen, and P. Hall, *J. Biol.Chem.*, 261, 3556–3562 (1986).
23. Y. Nonaka, H. Takemori, S. K. Halder, T. Sun, M. Ohta, O. Hatano, A. Takakusu, and M. Okamoto, *Eur. J. Biochem.*, 229, 249–256 (1995).
24. N. Matsukawa, Y. Nonaka, Z. Ying, J. Higaki, T. Ogihara, and M. Okamoto, *Biochem. Biophys Res. Commun.*, 169, 245–252 (1990).
25. L. J. Domalik, D. D. Chaplin, M. S. Kirkman, R. C. Wu, W. W. Liu, T. A. Howard, M. F. Seldin, and K. L. Parker, *Mol. Endocrinol.*, 5, 1853–1861 (1991).

26. H. E. Bulow, K. Mobius, V. Bahr, and R. Bernhardt, *Endocr. Res.* 22 479–484 (1996).
27. H. E. Bulow and R. Bernhardt, *Eur. J. Biochem.*, 269, 3838–3846 (2002).
28. R. Megges, M. Müller-Frohne, D. Pfeil, and K. Ruckpaul, *Frontiers in Biotransformation*, (K. Ruckpaul and H. Rein, eds), Akademie Verlag, Berlin, 1990, pp. 204–243.
29. B. Erdmann, K. Denner, H. Gerst, D. Lenz, and R. Bernhardt, *Endocr. Res.*, 21, 425–435 (1995).
30. B. Erdmann, H. Gerst, H. Bulow, D. Lenz, V. Bähr, and R. Bernhardt, *Histochemistry Cell Biol.*, 104, 301–307 (1995).
31. L. Pascoe, X. Jeunemaitre, M. C. Lebrethon, K. M. Curnow, C. E. Gomez-Sanchez, J. M. Gasc, J. M. Saez, and P. Corvol, *J. Clin. Invest.*, 96, 2236–2246 (1995).
32. S. C. Chua, P. Szabo, A. Vitek, K. H. Grzeschik, M. John, and P. C. White, *Proc. Natl. Acad. Sci. USA*, 84, 7193–7197 (1987).
33. M. J. Wagner, Y. Ge, M. Siciliano, and D. E. Wells, *Genomica*, 10, 114–125 (1991).
34. R. P. Lifton, R. G. Dluhy, M. Powers, G. M. Rich, S. Cook, S. Ulick, and J. M. Lalouc, *Nature*, 355, 262–265 (1992).
35. L. Pascoe, K. M. Curnow, L. Slutsker, J. M. Connell, P. W. Speiser, M. I. New and P. C. White, *Proc. Natl. Acad. Sci. USA*, 89, 8327–8331 (1992).
36. A. A. MacConnachie, K. F. Kelly, A. McNamara, S. Loughlin, L. J. Gates, G. C. Inglis, A. Jamieson, J. M. Connell, and N. E. Haites, *J. Clin. Endocrinol. Metab.*, 83, 4328–4331 (1998).
37. M. Hampf, N. T. Dao, N. T. Hoan, and R. Bernhardt, *J. Clin. Endocrinol. Metab.*, 86, 4445–4452 (2001).
38. J. Peters, M. Hampf, B. Peters, and R. Bernhardt, *Handbuch der Molekularen Medizin* (D. Ganten and K. Ruckpaul, eds), Springer-Verlag, Berlin/Heidelberg/New York, 1998, pp. 413–452.
39. M. Lisurek and R. Bernhardt, *Mol. Cell. Endo.*, 215, 149–159 (2004).
40. M. Peter, *Semin Reprod. Med.*, 20, 249–254 (2002).
41. P. C. White, *Mol. Cell. Endocrinol.*, 217, 81–87 (2004).
42. P. Ferrari, *Curr. Hypertens. Rep.*, 4, 18–24 (2002).
43. R. P. Lifton, A. G. Gharavi, and D. S. Geller, *Cell*, 104, 545–556 (2001).
44. P. C. White, *Endocrinol. Metab. Clin. North Am.*, 30, 61–79 (2001).
45. H. Naganuma, M. Ojima, and N. Sasano, *Tohoku J. Exp. Med.*, 155, 81–96 (1988).
46. W. L. Miller, *J. Steroid Biochem. Mol. Biol.*, 55, 607–616 (1995).
47. S. A. Usanov, V. L. Chashchin, and A. A. Akhrem, *Frontiers in Biotransformation*, Vol. 3 (K. Ruckpaul and H. Rein, eds), Akademie Verlag, Berlin, Germany, 1990, pp. 1–57.
48. W. L. Miller, *J. Clin. Endocrinol. Metab.*, 83, 1399–1400 (1998).
49. Y. Morel, J. Picado-Leonard, D. A. Wu, C. Y. Chang, T. K. Mohandas, B. C. Chung, and W. L. Miller, *Am. J. Hum. Genet.*, 43, 52–59 (1988).
50. T. Okamura, M. E. John, M. X. Zuber, E. R. Simpson, and M. R. Waterman, *Proc. Natl. Acad. Sci. U.S.A.*, 82, 5705–5709 (1985).
51. N. Nabi and T. Omura, *Biochem. Biophys. Res. Commun.*, 97, 680–686 (1980).
52. N. Nabi and T. Omura, *J. Biochem. (Tokyo)*, 94, 1529–1538 (1983).

53. N. Nabi, T. Ishikawa, M. Ohashi, and T. Omura, *J. Biochem. (Tokyo)*, 94, 1505–1515 (1983).
54. M. F. Matocha and M. R. Waterman, *J. Biol. Chem.*, 260, 12259–12265 (1985).
55. T. Omura, *J. Biochem. (Tokyo)*, 123, 1010–1016 (1998).
56. Y. Ichikawa, M. Ichikawa, S. Ishigami, A. Hiwatashi, and T. Okuno, *Biochim. Biophys. Acta*, 529, 263–269 (1978).
57. F. Mitani, Y. Ishimura, S. Izumi, and K. Watanabe, *Acta Endocrinol. (Copenh.)*, 90, 317–327 (1979).
58. K. Ishimura and H. Fujita, *Microsc. Res. Tech.*, 36, 445–453 (1997).
59. K. Suhara, S. Takemori, and M. Katagiri, *Biochim. Biophys. Acta (Protein Structure)*, 263, 272–278 (1972).
60. R. W. Estabrook, K. Suzuki, J. I. Mason, J. Baron, W. E. Taylor, E. R. Simpson, J. Purvis, and J. McCarthy, *Adrenodoxin: An Iron-Sulfur Protein of Adrenal Cortex Mitochondria* (W. Lovenberg, ed.) Academic Press, New York, 1973, pp.193–223.
61. H. Uhlmann, R. Kraft, and R. Bernhardt, *J. Biol. Chem.*, 269, 22557–22564 (1994).
62. I. A. Pikuleva, K. Tesh, M. R. Waterman, and Y. Kim, *Arch. Biochem. Biophys.*, 373, 44–55 (2000).
63. D. Beilke, R. Weiss, F. Lohr, P. Pristovsek, F. Hannemann, R. Bernhardt, and H. Rutherjans, *Biochemistry*, 41, 7969–7978 (2002).
64. M. Kostic, R. Bernhardt, and T. C. Pochapsky, *Biochemistry*, 42, 8171–8182 (2003).
65. R. Bernhardt, Current Opinion in *Endocrinol. & Diabetes*, 7, 109–115 (2000).
66. T. Omura, E. Sanders, R. W. Estabrook, D. Y. Cooper, and O. Rosenthal, *Arch. Biochem. Biophys.*, 117, 660–673 (1966).
67. Y. Nonaka, H. Murakami, Y. Yabusaki, S. Kuramitsu, H. Kagamiyama, T. Yamano, and M. Okamoto, *Biochem. Biophys. Res. Commun.*, 145, 1239–1247 (1987).
68. Y. Sagara, Y. Takata, T. Miyata, T. Hara, and T. Horiuchi, *J. Biochem. (Tokyo)*, 102, 1333–1336 (1987).
69. S. B. Solish, J. Picado-Leonard, Y. Morel, R. W. Kubn, T. K. Mohandas, I. Hanukoglu, and W. L. Miller, *Proc. Natl. Acad. Sci. USA*, 85, 7104–7108 (1988).
70. D. Lin, Y. F. Shi, and W. L. Miller, *Proc. Natl. Acad. Sci. USA*, 87, 8516–8520 (1990).
71. E. A. Shephard, I. R. Phillips, I. Santisteban, L. F. West, C. N. Palmer, A. Ashworth, and S. Povey, *Ann. Hum. Genet.*, 53, 291–301 (1989).
72. M. E. Brandt and L. E. Vickery, *Arch. Biochem. Biophys.*, 294, 735–740 (1992).
73. Y. Sagara, A. Wada, Y. Takata, M. R. Waterman, K. Sekimizu, and T. Moriuchi, *Biol. Pharm. Bull.*, 16, 627–630 (1993).
74. P. M. Hwang, F. Bunz, J. Yu, C. Rago, T. A. Chan, M. P. Murphy, G. F. Kelso, R. A. Smith, K. W. Kinzler, and B. Vogelstein, *Nat. Med.*, 7, 1111–1117 (2001).
75. R. Rapoport, D. Sklan, and I. Hanukoglu, *Arch. Biochem. Biophys.*, 317, 412–416 (1995).
76. E. Derouet-Humbert, K. Roemer, and M. Bureik, *Biol. Chem.*, 386, 453–461 (2005).
77. S. Nakajin, P. F. Hall, and M. Onoda, *J. Biol. Chem.*, 256, 6134–6139 (1981).
78. S. Nakajin, M. Shinoda, M. Haniu, J. E. Shively, and P. F. Hall, *J. Biol. Chem.*, 259, 3971–3976 (1984).

79. S. Kominami, K. Shinzawa, and S. Takemori, *Biochem. Biophys. Res. Commun.*, *109*, 916–921 (1982).
80. K. Suhara, Y. Fujimura, M. Shiroo, and M. Katagiri, *J. Biol. Chem.*, *259*, 8729–8736 (1984).
81. M. X. Zuber, E. R. Simpson, and M. R. Waterman, *Science*, *234*, 1258–1261 (1986).
82. B. C. Chung, J. Picado-Leonard, M. Haniu, M. Bienkowski, P. F. Hall, J. E. Shively, and W. L. Miller, *Proc. Natl. Acad. Sci. USA*, *84*, 407–411 (1987).
83. P. Swart, A. C. Swart, M. R. Waterman, R. W. Estabrook, and J. I. Mason, *J. Clin. Endocrinol. Metab.*, *77*, 98–102 (1993).
84. R. J. Auchus, T. C. Lee, and W. L. Miller, *J. Biol. Chem.*, *273*, 3158–3165 (1998).
85. T. Yanase, *J. Steroid Biochem. Mol. Biol.*, *53*, 153–157 (1995).
86. W. L. Miller, *Endocrinology*, *146*, 2544–2550 (2005).
87. R. W. Estabrook, D. Y. Cooper, and O. Rosenthal, *Biochem.* *338*, 741–755 (1963).
88. Y. Higashi, H. Yoshioka, M. Yamane, O. Gotoh, and Y. Fujii-Kuriyama, *Proc. Natl. Acad. Sci. USA*, *83*, 2841–2845 (1986).
89. P. C. White, D. Grossberger, B. J. Onufer, D. D. Chaplin, M. I. New, B. Dupont, and J. L. Strominger, *Proc. Natl. Acad. Sci. USA*, *82*, 1089–1093 (1985).
90. H. J. Barnes, M. P. Arlotto, and M. R. Waterman, *Proc. Natl. Acad. Sci. USA*, *88*, 5597–5601 (1991).
91. M. Arase, M. R. Waterman, and N. Kagawa, *Biochem. Biophys. Res. Commun.*, *281*, 6290–6295 (2006).
92. P. W. Speiser and P. C. White, *N. Eng. J. Med.*, *349*, 776–788 (2003).
93. C. J. Migeon and P. A. Donohoue, *Endocrinol. Metab. Clin. North. Am.*, *20*, 277–296 (1991).
94. M. I. New, *Horm. Res.*, *37*, 22–33 (1992).
95. E. R. Simpson, C. Clyne, G. Rubin, W. C. Boon, K. Robertson, K. Britt, C. Speed, and M. Jones, *Annu. Rev. Physiol.*, *64*, 93–127 (2002).
96. C. T. Evans, D. B. Ledesma, T. Z. Schulz, E. R. Simpson, and C. R. Mendelson, *Proc. Natl. Acad. Sci. USA*, *83*, 6387–6391 (1986).
97. S. Graham-Lorence, M. W. Khalil, M. C. Lorence, C. R. Mendelson, and E. R. Simpson, *J. Biol. Chem.* *266*, 11939–11946 (1991).
98. S. Graham-Lorence, B. Amarneh, R. E. White, J. A. Peterson, and E. R. Simpson, *Protein Sci.*, *4*, 1065–1080 (1995).
99. N. Siegelman-Danieli and K. H. Buetow, *Br. J. Cancer*, *79*, 456–463 (1999).
100. M. Shozu, K. Akasofu, T. Harada, and Y. Kubota, *J. Clin. Endocrinol. Metab.*, *72*, 560–566 (1991).
101. N. Harada, H. Ogawa, M. Shozu, K. Yamada, K. Suhara, E. Nishida, and Y. Takagi, *J. Biol. Chem.*, *267*, 4781–4785 (1992).
102. Y. Ito, C. R. Fisher, F. A. Conte, M. M. Grumbach, and E. R. Simpson, *Proc. Natl. Acad. Sci. USA*, *90*, 11673–11677 (1993).
103. T. Iyanagi and H. S. Mason, *Biochemistry*, *12*, 2297–2308 (1973).
104. T. D. Porter and C. B. Kasper, *Proc. Natl. Acad. Sci. USA*, *82*, 973–977 (1985).
105. T. D. Porter, T. E. Wilson, and C. B. Kasper, *Arch. Biochem. Biophys.*, *254*, 353–367 (1987).
106. S. Yamano, T. Aoyama, O. W. McBride, J. P. Hardwick, H. V. Gelboin, and F. J. Gonzalez, *Mol. Pharmacol.*, *36*, 83–88 (1989).

107. C. E. Flück, T. Tajima, A. V. Pandey, W. Arlt, K. Okuhara, C. F. Verge, E. W. Jabs, B. B. Mendonca, K. Fujieda, and W. L. Miller, *Nature Genetics, 36*, 228–230 (2004).
108. W. Arlt, E. A. Walker, N. Draper, H. E. Ivison, J. P. Ride, F. Hammer, S. M. Chalder, M. Borucka-Mankiewicz, B. P. Hauffa, E. M. Malunowicz, P. M. Stewart, and C. H. Shackleton, *Lancet, 363*, 2128–2135 (2004).
109. N. Huang, A. V. Pandey, V. Agrawal, W. Reardon, P. D. Lapunzina, D. Mowat, E. W. Jabs, G. Van Vliet, T. Sack, C. E. Flück, and W. L. Miller, *Am. J. Hum. Genet., 76*, 729–749 (2005).
110. A. L. Shen, K. A. O'Leary, and C. B. Kasper, *Biol. Chem., 277*, 6536–6541 (2002).
111. D. M. Otto, C. J. Henderson, D. Carrie, M. Davey, T. E. Gunderson, R. Blomhoff, R. H. Adams, C. Tickle, and C. R. Wolf, *Mol. Cell. Biol., 23*, 6103–6116 (2003).
112. S. G. Sligar and I. C. Gunsalus, *Biochemistry, 18*, 2290–2295 (1979).
113. A. Wada and M. R. Waterman, *J. Biol. Chem., 267*, 22877–22882 (1992).
114. G. I. Lepesheva, T. N. Azeva, N. V. Strushkevich, A. A. Gilep, and S. A. Usanov, *Biochemistry (Mosc.), 65*, 1409–1418 (2000).
115. T. N. Azeva, A. A. Gilep, G. I. Lepesheva, N. V. Strushkevich, and S. A. Usanov, *Biochemistry (Mosc.), 66*, 564–575 (2001).
116. V. M. Coghlan and L. E. Vickery, *J. Biol. Chem., 266*, 18606–18612 (1991).
117. V. M. Coghlan and L. E. Vickery, *J. Biol. Chem., 267*, 8932–8935 (1992).
118. V. Beckert, R. Dettmer, and R. Bernhardt, *J. Biol. Chem., 269*, 2568–2573 (1994).
119. V. Beckert and R. Bernhardt, *J. Biol. Chem., 272*, 4883–4888 (1997).
120. M. Bureik, M. Lisurek, and R. Bernhardt, *Biol. Chem., 383*, 1537–1551 (2002).
121. M. Bureik, A. Zollner, N. Schuster, M. Montenarh, and R. Bernhardt, *Biochemistry, 44*, 3821–3830 (2005).
122. F. Hannemann, M. Rottmann, B. Schiffler, J. Zapp, and R. Bernhardt, *J. Biol. Chem., 276*, 1369–1375. (2001).
123. A. Zöllner, F. Hannemann, M. Lisurek, and R. Bernhardt, *J. Inorg. Biochem., 91*, 644–654 (2002).
124. R. Bernhardt, *Biophysics of Electron Transfer and Molecular Bioelectronics* (C. Nicolini, ed.), Plenum Press, New York, 1998, pp. 51–66.
125. P. R. Cao and R. Bernhardt, *Eur. J. Biochem., 265*, 152–159 (1999).
126. T. Yamazaki, B. C. McNamara, and C. R. Jefcoate, *Mol. Cell Endocrinol,. 95*, 1–11 (1993).
127. P. R. Cao and R. Bernhardt, *Eur. J. Biochem., 262*, 720–726 (1999)
128. S. Ikushiro, S. Kominami, and S. Takemori, *J. Biol. Chem., 267*, 1464–1469 (1992).
129. S. Kominami, D. Harada, and S. Takemori, *Biochim. Biophys. Acta, 1192*, 234–240 (1994).
130. N. Monnier, G. Defaye, and E. M. Chambaz, *Eur. J. Biochem., 169*, 147–153 (1987).
131. R. Bernhardt, A. Makower, G. R. Jänig, and K. Ruckpaul, *Biochim. Biophys. Acta, 785*, 186–190 (1984).
132. R. Bernhardt, K. Pommerening, and K. Ruckpaul, *Biochem. Int., 4*, 823–832 (1987).

133. R. Bernhardt, R. Kraft, A. Otto, and K. Ruckpaul, *Biomed. Biochim. Acta, 47*, 581–592 (1988).
134. R. Bernhardt, R. Kraft, A. Otto, and K. Ruckpaul, *Cytochrome P450: Biochemistry and Biophysics* (I. Schuster, ed.) Taylor & Francis, London, New York, Philadelphia, 1989, pp. 320–323.
135. S. Shen and H. W. Strobel, *Arch. Biochem. Biophys., 294*, 83–90 (1992).
136. R. Bernhardt, *Handbook of Experimental Pharmacology*, Vol. 105, *Cytochrome P450* (J. Schenkman and H. Greim, eds) Springer-Verlag, Berlin, 1993, pp. 547–560.
137. J. A. Peterson and S. E. Graham, *Structure, 6*, 1079–1085 (1998).
138. S. E. Graham and J. A. Peterson, *Arch. Biochem. Biophys., 369*, 24–29 (1999).
139. M. Wang, D. L. Roberts, R. Paschke, T. M. Shea, B. S. Masters, and J. J. Kim, *Proc. Natl. Acad. Sci. USA, 94*, 8411–8416 (1997).
140. D. C. Lamb, Y. Kim, L. V. Yermalitskaya, V. N. Yermalitsky, G. I. Lepesheva, S. L. Kelly, M. R. Waterman, and L. M. Podust, *Structure, 14*, 51–61 (2006).
141. D. H. Geller, R. J. Auchus, B. B. Mendonca, and W. L. Miller, *Mol. Endocrinol., 13*, 167–175 (1999).
142. R. J. Auchus and W. L. Miller, *Mol. Endocrinol., 13*, 1169–1182 (1999).
143. M. Onoda and P. F. Hall, *Biochem. Biophys. Res. Commun., 108*, 454–460 (1982).
144. M. Katagiri, N. Kagawa, and M. R. Waterman, *Arch. Biochem. Biophys. 317*, 343–347 (1995).
145. S. Kominami, N. Ogawa, R. Morimune, H. De-Ying, and S. Takemori, *J. Steroid Biochem. Mol. Biol., 42*, 57–64 (1992).
146. L. Zhang, H. Rodriguez, S. Ohno, and W. L. Miller, *Proc. Natl. Acad. Sci. USA, 92*, 10619–10623 (1995).
147. W. L. Miller, *Med. Prince. Pract., 14*, 58–68 (2005).
148. T. Hakki and R. Bernhardt, *Pharmacology & Therapeutics* (2006), in press.
149. Y. Tamaki, Y. Miyoshi, S. J. Kim, Y. Tanji, T. Taguchi and S. Hoguchi, *Expert Rev. Anticancer Ther., 3*, 193–201 (2003).
150. E. R. Simpson and M. R. Waterman, *Ann. Rev. Physiol., 30*, 427–440 (1988).
151. M. R. Waterman and E. R. Simpson *Recent Progress in Hormone Research* (J. H. Clark, ed.), Vol. 45, Academic Press, New York, 1989, pp. 533–566.
152. L. D. Garren, R. L. Ney, and W. W. Davis, *Proc. Natl. Acad. Sci. USA*, 53, 1443–1450 (1965).
153. C. T. Privalle, J. F. Crivello, and C. R. Jefcoate, *Proc. Natl. Acad. Sci. USA, 80*, 702–706 (1983).
154. C. R. Jefcoate, B. C. McNamara, I. Artenenko, and T. Yamazaki, *J. Steroid Biochem. Mol. Biol., 43*, 751–767 (1982).
155. W. H. Trzeciak and G. S. Boyd, *Eur. J. Biochem., 46*, 201–207 (1974).
156. P. R. Manna, M. T. Dyson, D. W. Eubank, B. J. Clark, E. Lalli, P. Sassone-Corsi, A. J. Zellezink, and D. M. Stocco, *Mol. Endocrinol., 16*, 184–199 (2002).
157. W. L. Miller and J. F. Strauss III, *J. Steroid Biochem. Mol. Biol., 69*, 131–141 (1999).
158. H. S. Bose, T. Sugawara, J. F. Strauss III, and W. L. Miller, *New Eng. J. Med., 335*, 1870–1878 (1996).

159. D. Lin, T. Sugawara, J. F. Strauss III, B. J. Clark, D. M. Stocco, P. Saengler, A. Rogol, and W. L. Miller, *Science*, 267, 1828–1831 (1995).
160. M. K. Tee, D. Lin, T. Sugwara, J. A. Holt, Y. Guiguen, B. Buckingham, J. L. Strauss III, and W. L. Miller, *Hum. Mol. Genet*, 4, 2299–2305 (1995).
161. W. L. Miller, *J. Mol. Endocrinol.* 19, 227–240 (1997).
162. X. Chen, B. Y. Baker, M. A. Abulajabber, and W. L. Miller, *J. Clin. Endocrinol. Metab.*, 90, 835–840 (2005).
163. B. Y. Baker, D. C. Yaworsky, and W. L. Miller, *J. Biol. Chem.*, 280, 41753–41760 (2005).
164. T. Hanet, Z. X. Yao, H. S. Bose, C. T. Wall, Z. Han, W. Li, D. B. Hales, W. L. Miller, M. Culty, and V. Papadopoulos, *Mol. Endocrinol.*, 19, 540–554 (2005).
165. K. Morohashi, U. M. Zanger, S. Honda, M. Hara, M. R. Waterman, and T. Omura, *Mol. Endocrinol.*, 7, 1196–1204 (1993).
166. Y. Ikeda, D. S. Lala, X. Luo, E. Kim, M. P. Moisan, and K. L. Parker, *Mol. Endocrinol.* 7, 852–860 (1993).
167. M. B. Sewer and M. R. Waterman, *Endocrinol.* 143, 1769–1777 (2002).
168. M. B. Sewer and M. R. Waterman, *Rev. Endocr. Metab. Disord.*, 2, 269–274 (2001).
169. M. B. Sewer, V. Q. Nguyen, C-J. Huant, P. W. Tucker, N. Kagawa, and M. R. Waterman, *Endocrinol.*, 143, 1280–1290 (2002).
170. I. N. Krylova, E. P. Sablin, J. Moore, R. X. Xu, G. M. Waitt, J. A. McKay, D. Juzuniene, J. M. Bynum, K. Madauss, V. Montana, L. Lebedva, M. Suzawa, J. D. Williams, S. P. Williams, R. K. Guy, J. W. Thornton, R. J. Fletterick, T. M. Willson, and H. A. Ingraham, *Cell*, 120, 343–355 (2005).
171. A. N. Urs, E. Dammer, and M. B. Sewer, *Endocrinol.*, 147, 5249–5258 (2006).
172. M. B. Sewer and M. R. Waterman, *J. Mol. Endo.* 29, 163–174 (2002).
173. Y. Ikeda, W. H. Shen, H. A. Ingraham, and K. L. Parker, *Mol. Endocrinol*, 8, 654–662 (1994).
174. B. A. Cooke, D. R. E. Abayasekara, S. K. Choi, M. P. Lopez-Ruiz and A. P. West, *Progress in Endocrinology* (R. Mornex, C. Jaffiol, and J. Leclere, eds), The Parthenon Publishing Group, UK, 1993, pp. 615–618.
175. M. R. Waterman, *Sex Differentiation: Clinical and Biological Aspects* (I. A. Hughes, ed.), Vol. 20, *Frontiers in Endocrinology*, Christengraf, Via Anagnina, Rome, 1996, pp. 61–69.
176. T. Imai, H. Globerman, J. Gertner, N. Kagawa, and M. R. Waterman, *J. Biol. Chem.*, 268, 19681–19689 (1993).
177. J. I. Mason, R. W. Estabrook, and J. L. Purvis, *Ann. N. Y. Acad. Sci.*, 212, 406–419 (1973).
178. R. J. Rodgers, H. F. Rodgers, P. F. Hall, M. R. Waterman, and E. R. Simpson, *J. Reprod. Fertil.*, 78, 627–638 (1986).
179. R. J. Rodgers, M. R. Waterman, and E. R. Simpson, *Endocrinol.*, 118, 1366–1374 (1986).
180. R. J. Rodgers, M. R. Waterman, and E. R. Simpson, *Mol. Endocrinol.* 1, 274–279 (1987).
181. R. J. Rodgers, H. F. Rogers, M. R. Waterman, and E. R. Simpson, *J. Reprod. Fertil.*, 78, 639–659 (1986).

182. E. R. Simpson, Y. Zhao, V. R. Agarwal, M. D. Michael, S. E. Bulun, M. M. Hinshelwood, S. Graham-Lorence, T. Sun, C. R. Fisher, K. Qin, and C. R. Mendelson, *Rec. Prog. Horm. Res.*, 52, 185–213 (1997).
183. E. Simpson, M. Jones, S. Davis, and G. Rubin in *Molecular Steroidogenesis* (M. Okamoto, Y. I. Shimara, and N. Nawata, eds), University Academy Press, Tokyo, 2000, pp. 125–132.
184. H. Sasano and N. Harada, *Endocrine Rev.*, 19, 593–607 (1998).
185. G. D. Means, M. Mahendroo, C. J. Corbin, J. M. Mathis, F. E. Powell, C. R. Mendelson, and E. R. Simpson, *J. Biol. Chem.*, 264, 19385–19391 (1989).
186. A. Morishima, M. M. Grumbach, E. R. Simpson, C. Fisher, and K. Qin, *J. Clin. Endocrinol. Metab.*, 80, 3689–3698 (1995).
187. A. J. Conley, W. E. Rainey, and J. I. Mason, *J. Mol. Endocrinol.* 12, 155–165 (1994).
188. S. J. Chou, K. N. Lai, and B. Chung, *J. Biol. Chem.* 271, 22125–22129 (1996).
189. M. Pasanen, N. Takata, Y. Nojo, A. Furukawa, H. Yaumatsu, T. Kimoto, T. Enami, K. Suzuk, N. Tanabe, H. Ishii, H. Mukai, T. Takahashi, T. A. Hattori, and S. Kawato, *Adv. Drug Deliv. Rev.*, 38, 81–97 (1999).
190. K. Shibuya, *Biochim. Biophys. Acta. 1619*, 301–316 (2003).
191. V. Pezzi, J. M. Mathis, W. E. Rainey, and B. R. Carr, *J. Steroid Biochem. Mol. Biol.*, 87, 181–189 (2003).
192. C. Corpechot, P. Robel, M. Axelson, J. Sjovall, and E. Baulieu, *Proc. Natl. Acad. Sci. USA*, 78, 4704–4707 (1981).
193. C. LeGoascogne, P. Robel, M. Gaueazou, N. Sananes, E. R. Boy, and M. R. Waterman, *Science*, 237, 1212–1215 (1987).
194. S. H. Mellon and C. F. Deschepper, *Brain Res.*, 629, 283–292 (1993).
195. M. Stromstedt and M. R. Waterman, *Brain Res. Mol. Brain Res.*, 34, 75–88 (1995).
196. C. E. Gomez-Sanchez, M. Zhou, E. H. Cozza, H. Morita, F. C. Eddleman, and E. P. Gomez-Sanchez, *Endocr. Res.* 22, 463–470 (1996).
197. C. E. Gomez-Sanchez, M. Y. Zhou, H. Morita, E. N. Cozza, M. F. Foecking, and E. P. Gomez-Sanchez, *Endocrinology*, 138, 3369–3373 (1997).
198. S. M. MacKenzie, M. Lai, C. J. Clark, R. Fraser, C. E. Gromez-Sanchez, J. R. Jeckl, J. M. Connell, and E. Davies, *J. Mol. Endocrinol.* 29, 319–325 (2002).
199. S. R. King, S. D. Ginsberg, T. Ishii, R. C. Smith, K. L. Parker, and D. J. Lamb, *Eur. J. Endocrinol.*, 145, 4775–4780 (2004).
200. B. Stoffel-Wagner, *Eur. J. Endocrinol,.* 145, 669–679 (2001).
201. B. Erdmann, H. Gerst, A. Lippoldt, H. Bulow, D. Ganten, K. Fuxe, and R. Bernhardt, *Brain Res.*, 733, 73–82 (1996).
202. E. P. Gomez-Sanchez and C. E. Gomez-Sanchez, *Trends Endocrinol. Metab.*, 14, 444–446 (2003).
203. M. Watzka, F. Bidlingmaier, J. Schramm, D. Klingmuller, and B. Stoffel-Wagner, *J. Neuroendocrinol.*, 11, 901–905 (1999).
204. R. Takeda, H. Hatakeyama, Y. Takeda, H. Iki, I. Miyamori, W. P. Sheng, M. Yamamoto, and I. A. Blair, *Steroids*, 60, 120–124 (1995).
205. Y. Takeda, I. Miyamori, T. Yoneda, H. Hatakeyama, S. Inaba, K. Furukawa, H. Mabuchi, and R. Takeda, *J. Clin. Endocrinol. Metab.*, 81, 2797–2800 (1996).
206. Y. Takeda, T. Yoneda, M. Demura, I. Miyamori, and H. Mabuchi, *Endocrinology*, 141, 1901–1904 (2000).

207. P. Wu, Z. Guo, Y. Zhang, W. Lui, X. Liang, R. Zhang, W. Lai, Y. Takeda, M. Isamu, and R. Takeda, *Horm. Res. 50*, 28–31 (1998).
208. J. S. Silvestre, V. Robert, C. Heymes, B. Aupetit-Faisant, C. Mouas, J. M. Moalic, B. Swynghedauw, and C. Delcayre, *J. Biol. Chem., 273*, 4883–4891 (1998).
209. K. M. Kayes-Wandover and P. C. White, *J. Clin. Endocrinol. Metab., 85*, 2519–2525 (2000).
210. Y. Mizuno, M. Yoshimura, H. Yasue, T. Sakamoto, H. Ogawa, K. Kugiyama, E. Harada, M. Nakayama, S. Namakura, T. Ito,, Y. Shimasaki, Y. Saito, and K. Nakao, *Circulation, 103*, 72–77 (2001).
211. M. Yoshimura, S. Nakamura, T. Ito, M. Nakayama, E. Harada, Y. Mizuno, T. Sakamoto, M. Yamamura, Y. Saito, K. Nakao, H. Yasue, and H. Ogawa, *J. Clin. Endocrinol. Metab., 87*, 3936–3940 (2002).
212. M. Satoh, M. Nakamura, H. Saitoh, H. Satoh, T. Akatsu, J. Iwasaka, T. Masuda, and K. Hiramori, *Clin. Sci. (Lond.), 102*, 381–386 (2002).
213. C. G. Brilla, *Herz, 25*, 299–306 (2000).
214. F. J. Ramires, Y. Sun, and K. T. Weber, *J. Mol. Cell. Cardiol., 30*, 475–483 (1998).
215. M. Young and J. W. Funder, *Trends Endocrinol. Metab., 11*, 224–226 (2000).
216. B. Pitt, F. Zannad, W. J. Remme, R. Cody, A. Castaigne, A. Perez, J. Palensky, and J. Wittes, *N. Eng. J. Med., 341*, 709–717 (1999).
217. B. Pitt, G. Williams, W. Remme, F. Martinez, J. Lopez Sendon, F. Zannad, J. Neaton, B. Roniker, S. Hurley, D. Burns, R. Bittman, and J. Kleiman, *Cardiovasc. Drugs. Ther., 15*, 79–87 (2001).
218. A. Slominski, G. Ermak, and M. Mihm, *J. Clin. Endocrinol. Metab., 81*, 2746–2749 (1996).
219. D. Thiboutot, S. Jabara, J. M. McAllister, A. Sivarajah, K. Gilliland, Z. Cong, and G. Dawson, *J. Invest. Dermatol. 120*, 905–914 (2003).
220. C. LeGoascogne, N. Sananes, B. Eychenne, M. Gouezou, E. E. Baulien, and P. Robel, *Endocrinology, 136*, 1744–1752 (1995).
221. P. Wu, X. Liang, Y. Dai, H. Liu, Y. Zang, Z. Guo, R. Zhang, W. Lai, Y. Zhang, and Y. Liu, *Chin. Med. J. (Engl.), 112*, 414–418 (1999).
222. E. Lecain, T. H. Yang, and P. Tran Ba Huy, *Acta Otolaryngol., 123*, 187–191 (2003).
223. D. S. Keeney, Y. Ikeda, M. R. Waterman, and K. Parker, *Mol. Endocrinol., 19*, 1091–1098 (1995).
224. S. Vianello, M. R. Waterman, L. D. Valle, and L. Colombo, *Endocrinol., 138*, 3166–3174 (1997).

13

Carbon-Carbon Bond Cleavage by P450 Systems

James J. De Voss and Max J. Cryle

School of Molecular and Microbial Sciences, University of Queensland, Brisbane,
Queensland 4072, Australia
<j.devoss@uq.edu.au>

1. INTRODUCTION	398
2. CLEAVAGE BETWEEN OXYGENATED CARBONS	398
2.1. Diols	398
2.2. Keto Alcohols	403
3. CLEAVAGE ALPHA TO OXYGENATED CARBONS	408
3.1. Ketones	408
3.2. Aldehydes	408
3.3. Alcohols	416
3.4. Acids	419
3.5. Ethers	421
4. CLEAVAGE ALPHA TO CARBON BEARING NITROGEN	422
5. CARBON-CARBON BOND CLEAVAGE INVOLVING PEROXIDES	423
5.1. Hydroperoxy Fatty Acids	423
5.2. Prostaglandins	426
6. GENERAL CONCLUSIONS	429
ACKNOWLEDGMENTS	430
ABBREVIATIONS	430
REFERENCES	430

1. INTRODUCTION

The cytochromes P450 have long been recognised as powerful oxidants capable of catalysing a wide variety of substrate transformations. These features are perhaps nowhere more evident than in their catalysis of C-C bond cleavage reactions. Some of these reactions are relatively simple, occurring in one mechanistic step with loss of a single, activated carbon whilst others are much more dramatic, involving multiple oxidative transformations and skeletal rearrangement and/or fragmentation of the substrate.

Almost counter intuitively, these reactions are usually associated with elaborate biosynthetic pathways that build up complex secondary metabolites with significant biological activities. They are classified here according to the functional group(s) that is (are) adjacent to the site of C-C bond cleavage, even if these moieties are introduced by the P450 during the course of a multi-step oxidative transformation. Excluded from this discussion are substrates specifically designed to undergo oxidative rearrangement as mechanistic probes, such as cyclopropyl containing compounds.

2. CLEAVAGE BETWEEN OXYGENATED CARBONS

2.1. Diols

One widely studied steroidal P450 that has been implicated in a C-C bond cleavage reaction is $P450_{scc}$ (CYP11A1). This mammalian P450 is involved in the biosynthesis of pregnenolone from cholesterol, which proceeds via the oxidative scission of the cholesterol side chain [1]. $P450_{scc}$ catalyses a three-step oxidation process, whereby sequential hydroxylation reactions at C22 and C20 lead to the formation of a vicinal diol, which then undergoes oxidative cleavage to produce pregnenolone and isocaproic aldehyde (Figure 1) [2–5]. The overall cleavage process consumes 3 moles of NADPH and 3 moles of $O_{2(g)}$, with labelling studies indicating that the added atoms of oxygen present in 22R-hydroxycholesterol, 20S,22R-dihydroxycholesterol, and pregnenolone arise from molecular oxygen [5].

The oxidation process is efficiently controlled: both intermediates, 22R-hydroxycholesterol and 20S,22R-dihydroxycholesterol, bind more tightly to $P450_{scc}$ than cholesterol (≥ 700 times) [6] and there is increased stabilisation of oxygen binding in the presence of cholesterol and oxygenated intermediates [7,8]. Turnover of labelled 22S-$^3H_1$-cholesterol has shown that the label is retained in isocaproic aldehyde following diol cleavage, demonstrating direct cleavage of the diol and discounting the possibility of further oxidation at C22 prior to C-C bond cleavage [9].

The mechanism of cleavage of the diol in 20S,22R-dihydroxycholesterol is yet undefined, with two possible pathways postulated (Figure 2). The first involves

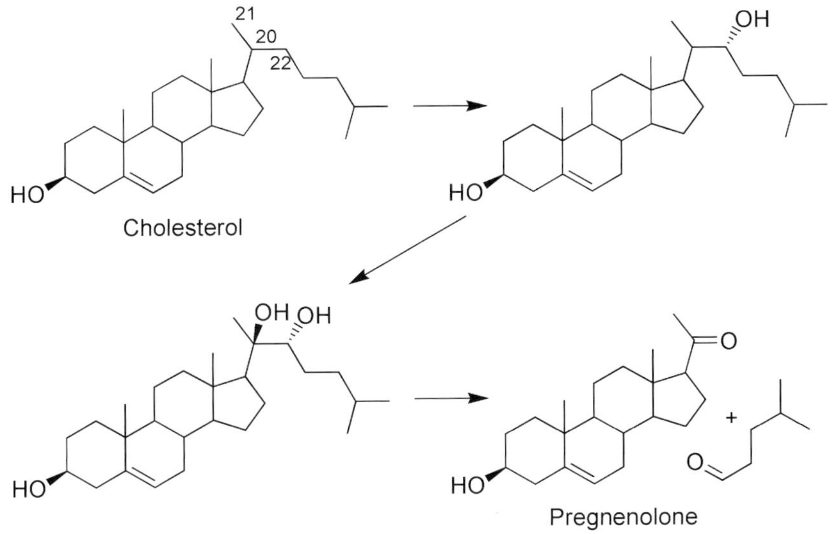

CARBON-CARBON BOND CLEAVAGE BY P450 SYSTEMS

Figure 1. Proposed pathway for the P450$_{scc}$ catalysed conversion of cholesterol to pregnenolone and 4-methylpentanal.

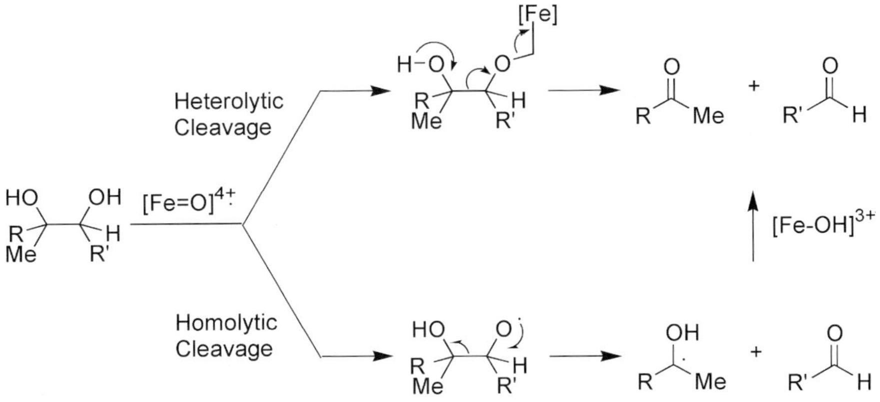

Figure 2. Possible mechanisms of diol cleavage catalysed by P450$_{scc}$, where R = the sterol nucleus and R' = $CH_2CH_2CH(CH_3)_2$.

the formation of a hydroperoxide from one of the hydroxyl groups of the diol with the species produced then undergoing cleavage to produce two carbonyl compounds (upper pathway in Figure 2). The alternative mechanism is initiated by hydrogen radical abstraction from one of the hydroxyl groups. The resultant species then undergoes homolytic cleavage of the diol and recombination of the radical produced with the iron-oxygen complex affording the observed carbonyl

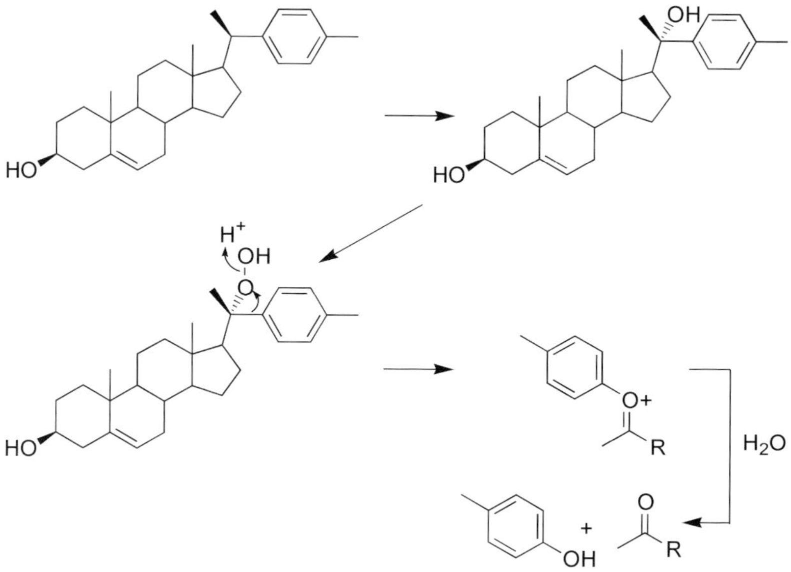

Figure 3. A possible mechanism for the P450$_{scc}$ catalysed conversion of (20S)-20-(p-tolyl)-5-pregnen-3β-ol to pregnenolone and phenol, where R = sterol nucleus.

products (lower pathway in Figure 2). Formation of a hydroperoxide may be seen as precedented by the exchange of oxygen between hydrogen peroxide and water via a putative ferryl species in model systems [10].

The chemistry of such a hydroperoxide would also explain nicely the intriguing early observation that (20S)-20-(p-tolyl)-5-pregnen-3β-ol is cleaved to pregnenolone and presumably phenol by P450$_{scc}$ (Figure 3) [11]. This remarkable transformation would be analogous to the well-known formation of acetone and phenol from cumene hydroperoxide under acid catalysis.

Recent studies conducted with P450$_{scc}$ have identified potential new and physiologically relevant substrates for this enzyme: 7-dehydrocholesterol and vitamin D$_3$, both of which are linked to cholesterol biosynthesis. 7-Dehydrocholesterol is the precursor of cholesterol and is derived from lathosterol, whilst vitamin D$_3$ is produced upon photochemical rearrangement of 7-dehydrocholesterol. P450$_{scc}$ catalyses the oxidative cleavage of 7-dehydrocholesterol in the same manner as for cholesterol [12], whilst the oxidation of vitamin D$_3$ yields products of C22 and C20 oxidation (the C22-alcohol and C20,22-diol) as well as two unidentified products, which have been tentatively identified as a diol and a triol [13]. However, oxidative cleavage of these oxygenated vitamin D$_3$ intermediates has not been reported [13].

Another biosynthetic example of C-C bond cleavage utilising a diol intermediate has been recently identified with P450$_{Biol}$ (CYP107H1). This enzyme has been shown to cleave an aliphatic chain via a diol intermediate (Figure 4). Encoded by

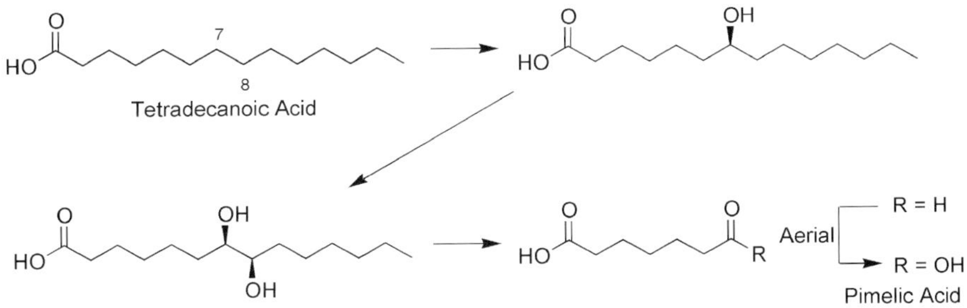

Figure 4. Proposed mechanism of P450$_{Biol}$ catalysed C-C bond cleavage that produces pimelic acid from tetradecanoic acid.

a gene in the biotin biosynthetic operon of *Bacillus subtilis*, P450$_{Biol}$ was implicated in the formation of a biological equivalent of pimelic acid (heptanedioic acid) through analysis of bacterial mutants [14]. Cloning and over-expression in *E. coli* resulted in isolation of a complex between the P450 and acylated acyl carrier protein (ACP) as well as the P450 alone [15]. P450$_{Biol}$ was shown to cleave the acyl moiety of the complex to produce a pimeloyl ACP and could also catalyse the oxidation of free fatty acids to produce pimelic acid as well as a suite of hydroxylated fatty acids [15,16]. Careful analysis of these hydroxylated fatty acid products indicated that whilst the hydroxy fatty acids production was not regiospecific, no oxidation of the terminal methyl group was observed [17]. This result discounted the production of long chain diacids as pimeloate precursors as the biological function of P450$_{Biol}$. Subsequently, a series of potential oxygenated intermediates in the C-C bond cleavage reaction were synthesised and incubated with the enzyme [18]. Pimelic acid production was found to increase as the substrate was changed from the C_{14} fatty acid to the 7-hydroxy derivative, with the *threo*-7,8-diol the best substrate (Fig. 4). Other derivatives, such as the 7- or 8-oxo, 8-hydroxy or *erythro*-7,8 diol, did not result in an increase in pimelic acid formation. As was shown with P450$_{scc}$, the oxygenated intermediates bound much more tightly to the enzyme than the parent substrate [18].

These results suggest a C-C bond cleavage mechanism in which the P450 operates on one face of the fatty acid chain in an extended conformation to produce a *threo*-diol, which then undergoes cleavage to produce two aldehyde fragments. The oxo-acid initially formed is highly susceptible to aerial oxidation and it is seen, along with pimelic acid, in the cleavage of the *threo*-7,8-diol [18]. Interestingly, only a slight preference was found for the 7-(S) enantiomer of the intermediate alcohol and the derived *threo* diol. This may be caused by the fact that the true substrate is an ACP bound fatty acid and, if true, this would make P450$_{Biol}$ one of a growing number of P450s found to act on carrier protein bound substrates [19].

Investigations into the oxidation of the antimicrobial agent olanexidine (1–3(3,4-dichlorobenzyl)-5-octylbiguanide) with mammalian systems have

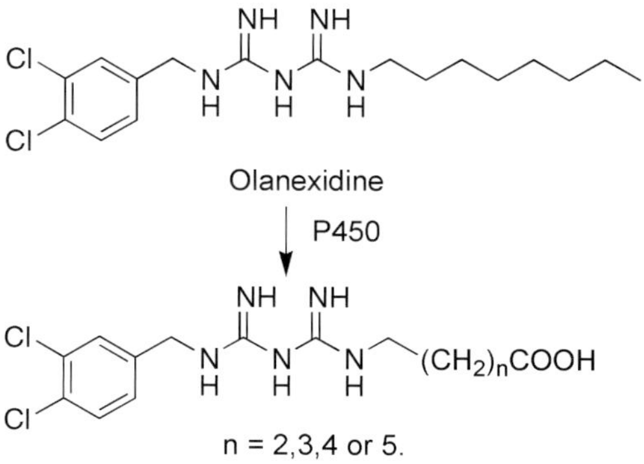

Figure 5. Urinary metabolites from the P450 mediated oxidation of olanexidine.

shown that the oxidation of the octyl side chain is P450 catalysed [20–22]. The major *in vivo* products of olanexidine metabolism in dogs have been identified as three chain-shortened carboxylic acids (Figure 5) [21]. Studies on the *in vitro* oxidation of olanexidine by dog liver microsomes have revealed a complex array of products, all of which are caused by oxidation of the octyl side chain (Figure 6). Three mono-hydroxylated products have been identified, along with

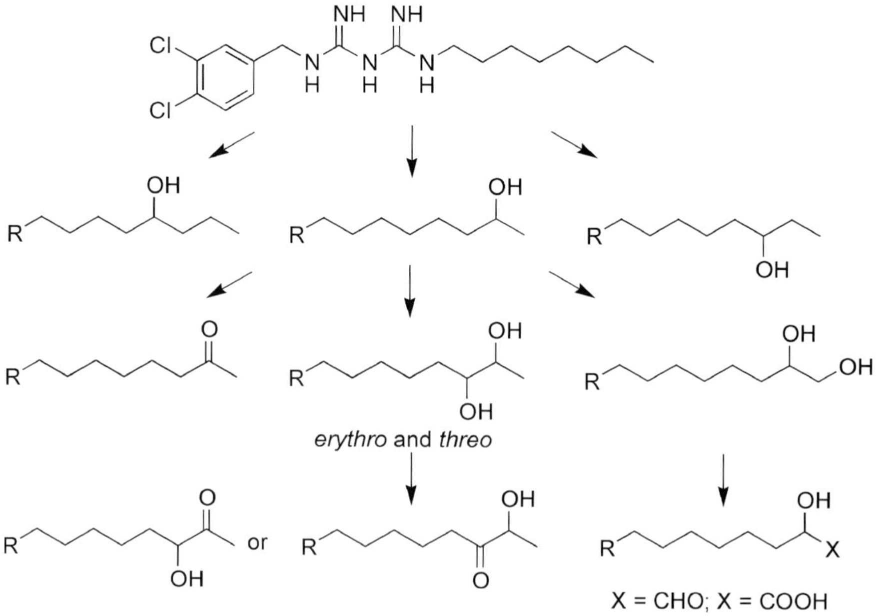

Figure 6. Oxygenated products from the P450 mediated oxidation of olanexidine (R = 1-(3,4-dichlorobenzyl)biguanidino moiety).

Met. Ions Life Sci. **3**, 397–435 (2007)

products resulting from further oxidation including a ketone and internal and terminal diols (Figure 6) [20].

Incubation of the internal *erythro* and *threo* diols with dog liver microsomes yielded further oxidation, with both a hydroxy-ketone and a chain shortened carboxylic acid produced (Figure 6). Both diastereomers yielded the same ratio of products and were oxidised at identical rates, indicating there was no stereochemical preference in this oxidation [20]. Investigation into the processing of terminal diol metabolite resulted in the identification of three further products including a hydroxy-aldehyde, a hydroxy-acid and a chain shortened carboxylic acid (Figure 6) [22]. This work indicated that dog liver microsomes and, more specifically dog CYP2D enzymes [20], are capable of oxidative C-C bond cleavage of unactivated hydrocarbon chains via a diol intermediate: the α-hydroxy ketone is a possible substrate for cleavage, but the different effects of specific P450 inhibitors on α-hydroxy ketone formation and C-C bond cleavage argue against a precursor–product relationship [20].

2.2. Keto Alcohols

C-C bond cleavage via an α-hydroxy ketone intermediate has been shown to be catalysed by the mammalian P450 CYP17A. This P450, like P450$_{scc}$, is involved in steroid biosynthesis and acts upon both pregnenolone and progesterone [23,24]. Unlike P450$_{scc}$, however, CYP17A has been demonstrated to utilise a range of mechanistic pathways in order to catalyse C-C bond cleavage. The major pathway through which CYP17A operates involves the cleavage of an α-hydroxy-ketone (Figure 7) [25]. This species arises from hydroxylation

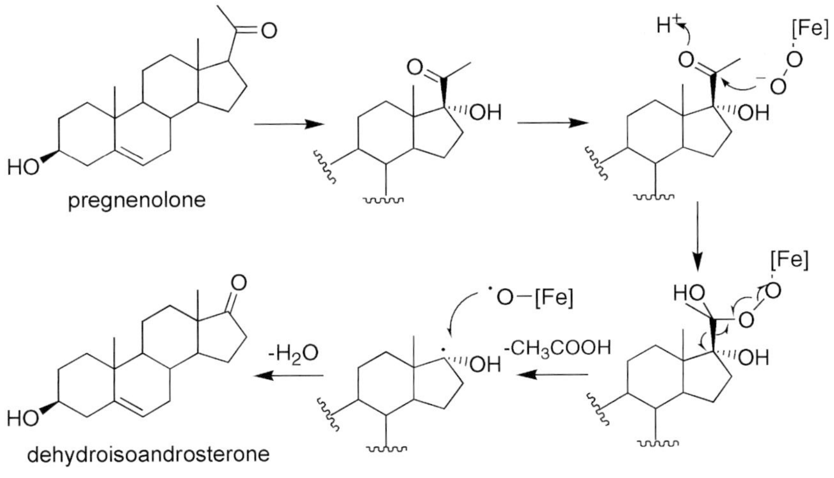

Figure 7. Proposed pathway for CYP17A mediated oxidation of pregnenolone to dehydroisoandrosterone.

at C17, with cleavage affected by attack by of the iron dioxygen species upon the carbonyl. This is proposed to result in the formation of an alkoxy radical species through homolytic cleavage of the peroxy species, with concurrent loss of acetic acid. The alkoxy radical produced then undergoes rebound onto the ferryl oxygen and the geminal diol produced dehydrates to a ketone (Figure 7). Such a mechanism has been supported by labelling studies showing that one oxygen atom in the acetic acid molecule cleaved from the steroid nucleus comes from molecular oxygen [26–28]. The oxygenation of the steroid at C17 has also been shown to result in a molecular oxygen derived oxygen atom in the ketone or alcohol formed at this position [25].

Other minor pathways have also been identified by which CYP17A catalyses C-C bond cleavage. The first involves initial attack of the ferric peroxy species upon the ketone of progesterone [29]. The resulting intermediate undergoes bond migration of carbon onto the distal oxygen of the peroxide species, with regeneration of the carbonyl functionality following loss of the oxyferryl species (Figure 8). This results in the formation of 17-O-acetyltestosterone, with the reaction mechanism recognisable as a Baeyer-Villiger type rearrangement (Figure 8) [29]. The second minor pathway that CYP17A utilises to effect a carbon-carbon bond cleavage involves iron peroxy attack on the carbonyl functionality of pregnenolone (Figure 9) [28,30,31]. This pathway affords two products, both of which result from C20-C17 cleavage and as in the Baeyer-Villiger pathway, the selectivity of the enzyme for hydroxylation chemistry relegates this reaction pathway to a smaller role. The formation of an alkene or alcohol is determined by the pathway taken after the initial oxidation, with disproportionation resulting in formation of the alkene and oxygen rebound giving rise to the alcohol (Figure 9) [25]. Results of CYP17A catalysed oxidation of substrates bearing deuterium labels at the C16, C17, and the methyl group α to the ketone are in agreement with this mechanism [25].

The role of the ferric peroxo moiety in the mechanism has been supported by mutagenesis studies in which Thr^{306} has been replaced by an alanine [27].

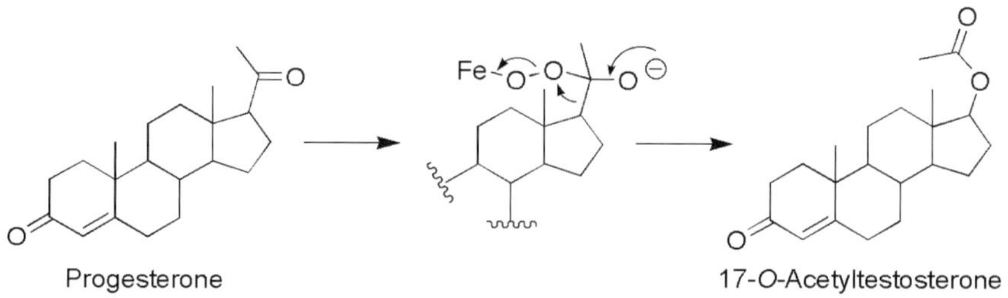

Figure 8. Baeyer-Villiger type oxidation of progesterone to 17-O-acetyltestosterone catalysed by CYP17A.

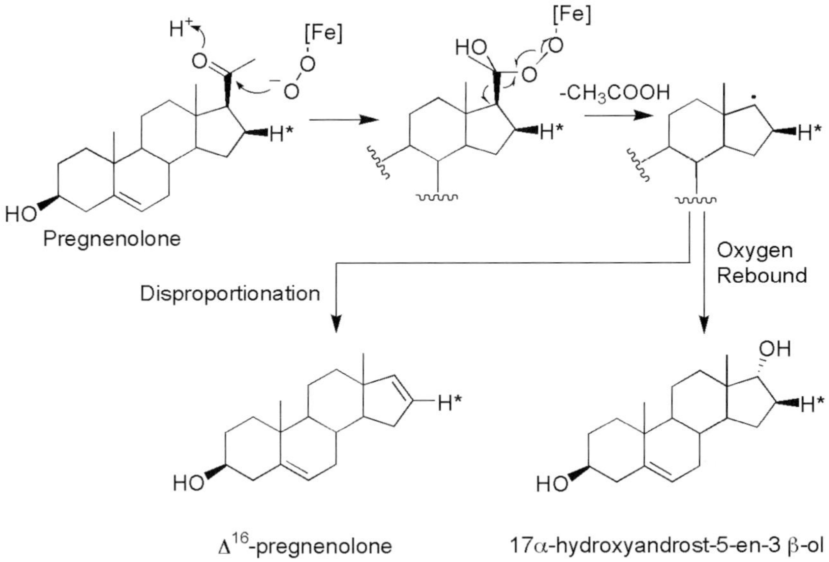

Figure 9. Mechanistic proposal for formation of minor products in the CYP17A catalysed oxidation of pregnenolone via a radical fragmentation of the peroxyhemiacetal intermediate.

This threonine residue is believed to control the correct delivery of the protons required to cleave the O-O bond and form the ferryl species. As expected, its loss results in a ~20 fold reduction in C17 hydroxylation (ferryl dependent) activity but a smaller decrease in C17-C20 lyase (ferric peroxo) activity [27]. Protonation of the ferric peroxide intermediate has been proposed to govern whether the reaction proceeds via a ferryl dependent (C17α hydroxylation) or a peroxy adduct (C17-C20 lyase) pathway [32].

The aldehyde corresponding to pregnenolone has also been used as a mechanistic probe with CYP17A. It is reported to undergo exclusive cleavage of the C17-C20 bond, producing formic acid and either the 17α-hydroxylated compound or the $\Delta^{16,17}$-olefin via analogous pathways to those proposed for formation of the corresponding minor cleavage products formed from pregnenolone [33]. It is believed that the more electrophilic carbonyl of the aldehyde favours the direct bond scission pathway by more effectively trapping the ferric peroxide moiety. The aldehyde is not oxidised to the corresponding acid, which suggests that the Baeyer-Villiger pathway is not a major one [33]. These experiments also provide evidence for the bifurcation of a single pathway leading to the two minor products of pregnenolone oxidation: deuteration of the C16α position leads to an enrichment in deuteration of the 17α-hydroxy product, which suggests that an isotope effect is inducing partitioning away from the

$\Delta^{16,17}$-olefin [33]. Cleavage of a C-C bond α to an aldehyde is discussed further in Section 3.2.

A further example of P450 involvement in a biosynthetic C-C bond cleavage reaction, potentially through a hydroxy ketone intermediate, can be found in the CYP24 catalysed conversion of 25-hydroxy-vitamin D_3 into a chain shortened alcohol and $1\alpha,25$-dihydroxy-vitamin D_3 into calcitroic acid (Figure 10) [34–39]. This reaction is reported to involve a six-step oxidation pathway, beginning with ketone formation via a two-step oxidation, followed by production of a hydroxy-ketone that undergoes C-C bond cleavage [35]. The product of this reaction is stated to be an alcohol [38,39], which, in the case of $1\alpha,25$-dihydroxyvitamin D_3, is then proposed to undergo two further oxidations to yield the acid (Figure 10) [35]. The product of cleavage of the hydroxy-ketone is unlikely to be a primary alcohol, however, as the oxidation state of the product would be expected to at least correspond to that of an aldehyde. Direct observation of the

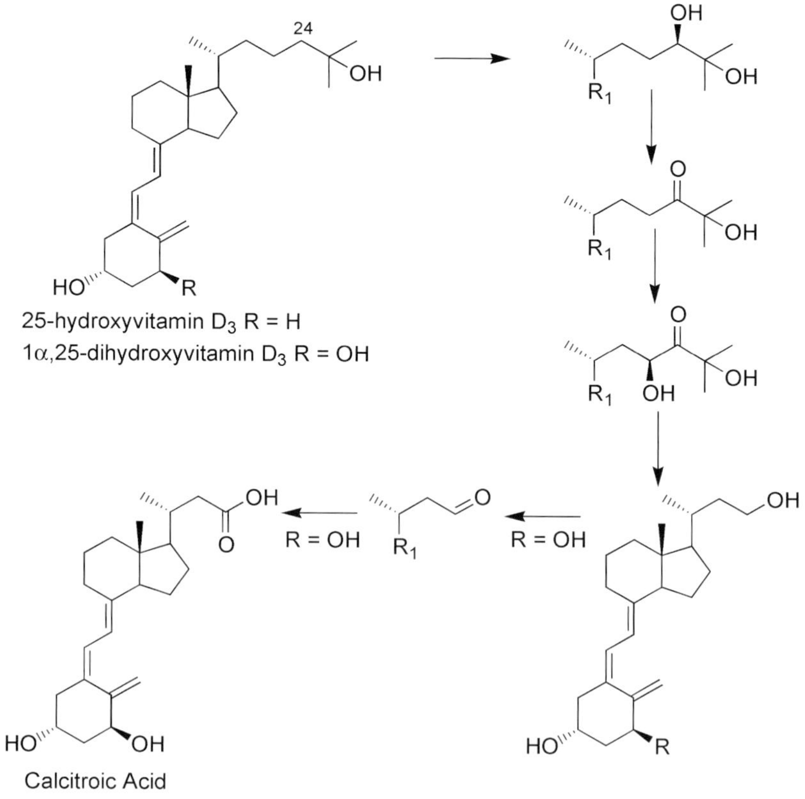

25-hydroxyvitamin D_3 R = H
$1\alpha,25$-dihydroxyvitamin D_3 R = OH

Calcitroic Acid

Figure 10. Oxidative transformations of 25-hydroxyvitamin D_3 and $1\alpha,25$-dihydroxyvitamin D_3 catalysed by CYP24 that lead to C-C cleavage.

chain cleaved aldehyde product of turnover of $1\alpha,25$-dihydroxyvitamin D_3 was not possible as it co-eluted with the 24-keto oxidation product, so it is unclear if this reaction also occurs *in vitro* [35]. Possible explanations could involve other unidentified intermediates or multiple pathways operating in parallel with further clarification of this pathway warranted.

CYP24 also metabolises $1\alpha,25$-dihydroxyvitamin D_3 via a different oxidation sequence, which is initiated by hydroxylation at C23 (Figure 11) [34,39]. This pathway does not culminate in a C-C bond cleavage but rather consists of sequential hydroxylation reactions [34,39]. However, a C-C bond cleavage reaction similar to that seen in psoralen biosynthesis (*vide infra*) can occur from the initial radical produced by oxidation at C23 leading to formation of an alkene (Figure 12) [36]. This is thought to arise from hydrogen abstraction β to the

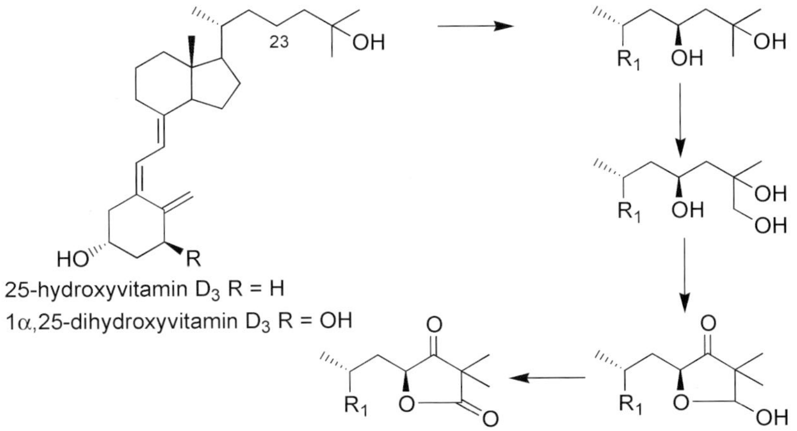

Figure 11. Alternate oxidative transformations at C23 of 25-hydroxyvitamin D_3 and $1\alpha,25$-dihydroxyvitamin D_3 catalysed by CYP24.

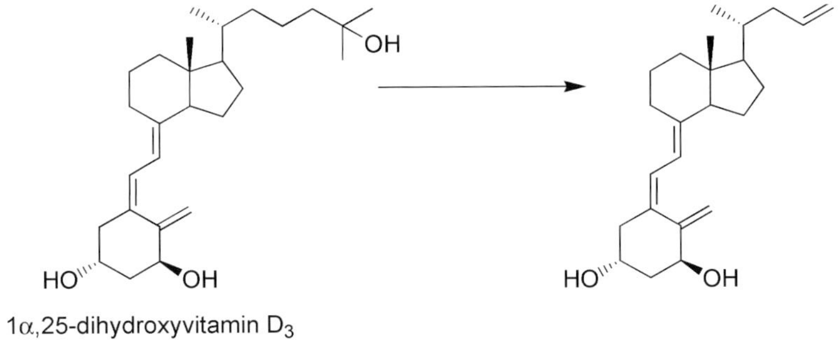

Figure 12. Alkene formed from $1\alpha,25$-dihydroxyvitamin D_3 via CYP24 catalysed C-C cleavage.

tertiary alcohol with subsequent loss of acetone affording the alkene. The rate of formation of this alkene correlates with the rate of C23 hydroxylation, which supports differing fates for the C23 radical intermediate [36].

Further complexity in the CYP24 cleavage pathway has been identified by investigation of vitamin D_2 metabolism. There is an apparent switch in activity of CYP24 between C24 hydroxylation of vitamin D_2 for *in vitro* experiments and C25 hydroxylation for *in vivo* experiments [40]. Whilst vitamin D_2 is hydroxylated by CYP24, there is no evidence of further oxidation or C-C bond cleavage occurring [41]. This may be due to the C24 methyl group preventing ketone formation or a function of the C22-C23 double bond. However, the fact remains that the major metabolic endpoint for vitamin D_2 is the same as for vitamin D_3 (calcitroic acid) and thus a C-C cleavage reaction must be catalysed by some unknown enzymic process.

3. CLEAVAGE ALPHA TO OXYGENATED CARBONS

3.1. Ketones

The CYP17A mediated oxidative cleavage of pregnenolone/progesterone without prior C17 hydroxylation provides one of the two clearly documented examples of cleavage of a C-C bond α to a ketone. The manifold of products formed by CYP17A indicates some of the variety of mechanistic pathways that might be envisioned. Addition of the ferric peroxy species to the carbonyl forms an intermediate that can subsequently decompose by one of two pathways [25]. A radical mechanism leads to the formation of an alkoxy radical, which affords an alcohol or olefinic product following C-C bond cleavage (Figure 9) [25]. An ionic mechanism (Baeyer-Villiger) leads to an ester product in which insertion of oxygen has occurred with retention of configuration (Figure 8) [29].

A more defined P450 catalysed Baeyer-Villiger type oxidation than that noted for CYP17A is performed by members of the CYP85A family [42,43]. The oxidation of castasterone to brassinolide, a potent plant steroid responsible for plant development, requires the oxidation of a cyclohexanone ring to the corresponding lactone (Figure 13). This reaction is catalysed by CYP85A2 in *Arabidopsis thaliana* and by CYP85A3 in tomato (*Lycopersicon esculentum*) [42,43]. Castasterone itself is formed via a sequential oxidation from deoxocastasterone and is catalysed by members of the CYP85A family, including CYP85A2 [43]. Mechanistic studies for the reactions catalysed by the CYP85A family have not been reported to date but are clearly of significant interest.

3.2. Aldehydes

Cleavage of a C-C bond α to an aldehyde has already been discussed in the context of the CYP17A catalysed oxidation of an aldehyde analogue of pregnenolone

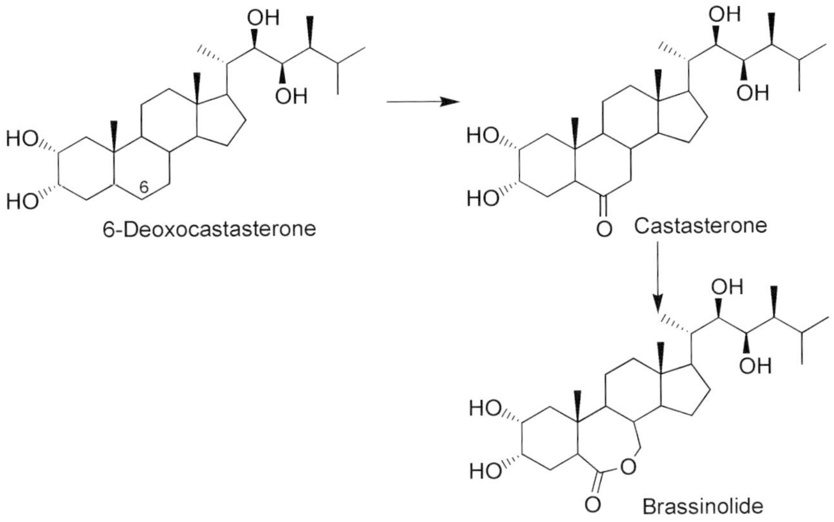

Figure 13. Baeyer-Villiger like conversion of castasterone to brassinolide catalysed by CYP85.

(Section 2.2). Such reactions are also believed to play a central role in the activities of several other P450s including the important steroid biosynthetic enzymes aromatase (CYP19) and 14α-demethylase (CYP51). It should be noted, however, that P450 catalysed aldehyde oxidation does not necessarily result in C-C bond cleavage and often oxidation leads to the corresponding carboxylic acid [44]. The factors that govern partitioning between these different modes of oxidation are not well understood at present [44].

Aromatase (CYP19), like P450$_{scc}$ plays an essential role in the biosynthesis of steroid hormones. It catalyses the aromatisation of the C_{19} androgen, androstenedione to the C_{18} oestrogen oestrone (Figure 14), as well as similar aromatisations of testosterone and 16α-hydroxyandrostendione [45,46]. These reactions proceed through three sequential oxidations at the angular C19 methyl group that result in its eventual loss with concomitant aromatisation of the A-ring of the substrate. Each oxidative step has been shown to require a molecule of NADPH and oxygen, with the first two steps unexceptional P450 catalysed hydroxylation steps [47]. The initial reaction which produces the C19 primary alcohol proceeds with retention of configuration [48,49], whilst the second oxidation yields a *gem*-diol intermediate via initial abstraction of the 19-pro-*R* hydrogen [50,51]. This *gem*-diol is then believed to dehydrate to yield the more stable, observed C19 aldehyde. There is an observable tritium isotope effect on the first hydroxylation step [52], but not on the subsequent one with (19-^{3}H)androst-4-ene-3,17-dione or comparable compounds [53,54], which is explicable by lack of the kind of *inter*-molecular effect commonly suppressed in P450 reactions.

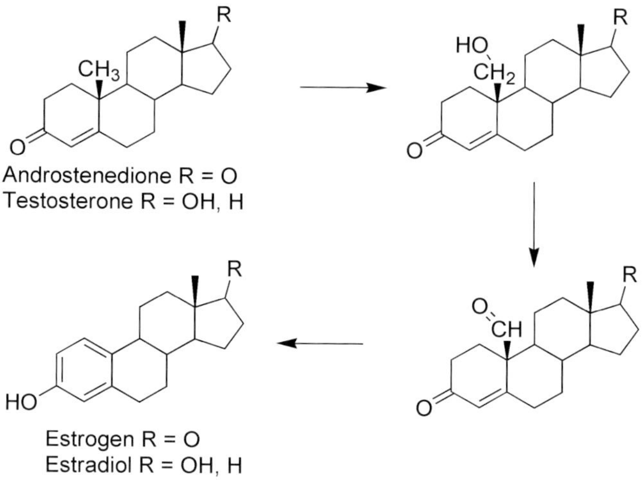

Androstenedione R = O
Testosterone R = OH, H

Estrogen R = O
Estradiol R = OH, H

Figure 14. Proposed pathway for the aromatisation reaction catalysed by CYP19.

The mechanism of the third oxidative transformation, involving C-C bond cleavage and aromatisation has attracted the most attention. In this reaction the 1β and 2β hydrogens are lost into water [55–60] and the C_{19} carbon as formate, containing an oxygen atom from the first and third oxidation steps [61,62]. A large number of different mechanisms have been proposed to account for this transformation involving the intermediacy of a steroid containing a C19 formyl group and variously, a 4,5-epoxide [63], a 1β-hydroxyl [60], a 2β-hydroxyl [64,65] or a C19 peroxide [61,66] as well as a possible enzymic Schiff base formed from the 3-keto moiety [67]. Several of these intermediates are known to be converted spontaneously [64] or by aromatase [68] into oestrone but none of them are currently accepted as lying upon the major pathway for aromatisation. This is primarily due to ^{18}O labelling studies, which indicate that an oxygen atom from the third oxidative step is incorporated into formate [61,63].

The currently accepted mechanism[25,69–71] explains all experimental observations and is supported by model studies [72–74] and analogy with the mechanisms of other P450s such as CYP17A and CYP2B4 (*vide infra*) (Figure 15). In this mechanism, the ferric peroxide intermediate is believed to add to the electrophilic aldehyde carbonyl, which yields a peroxyhemiacetal intermediate. Fragmentation of this intermediate affords an alkoxy radical that, through loss of formic acid, produces a C10 radical. Removal of the 1β hydrogen and enolisation of the carbonyl group is also required to produce the aromatised A-ring contained in the product. Recent model studies by Valentine and coworkers provide strong support for the involvement of the ferric peroxo in the mechanism [72]. These experiments demonstrate that a model peroxo ferric porphyrin complex will quantitatively convert an enolised androstenedione model compound into

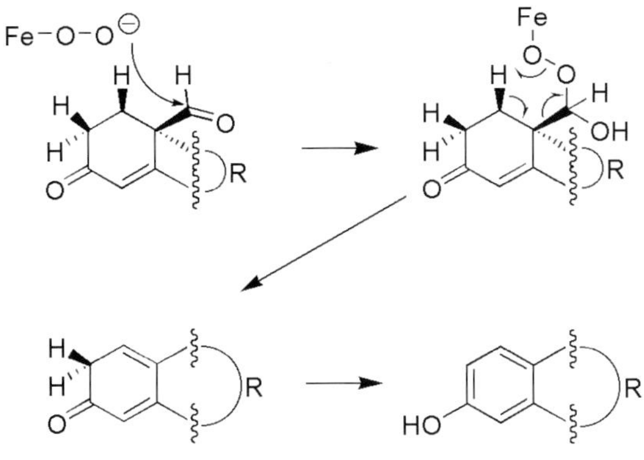

Figure 15. A currently accepted mechanism for the C-C cleavage/aromatization reaction catalysed by aromatase (CYP19), where R = sterol nucleus. The timing enolisation of the carbonyl with respect to the addition of the ferric peroxide to the aldehyde and to C-C and O-O bond fission is still uncertain.

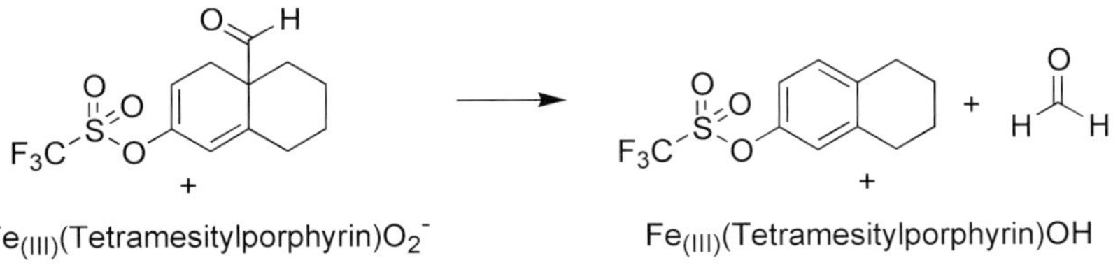

Figure 16. Conversion of an enolized analog of the aromatase (CYP19) substrate to the corresponding aromatized compound by a model peroxo ferric porphyrin complex.

the corresponding aromatic compound, with the loss of formate (Figure 16). This model is unable to catalyse this reaction with androstenedione due to the susceptibility of the electron deficient C4-C5 double bond to epoxidation. This follows suggestions in the literature that enolisation of the C3 carbonyl occurs prior to C19-C10 bond cleavage and additionally activates the 1β hydrogen towards loss [74,75]. However, whether the chemoselectivity required (C-C bond cleavage versus epoxidation) can be achieved enzymatically via prior enolisation or by selective positioning of the substrate within the active site remains to be established.

Another well-studied P450 involved in steroid biosynthesis is CYP51, also known as $P450_{14DM}$. This enzyme has been shown to catalyse the oxidative cleavage of a methyl group from the steroids lanosterol and 24,25-dihydrolanosterol (DHL), amongst others [76,77]. This oxidative process has been shown to proceed via a three-step oxidation process, with an initial conversion of the methyl group to a primary alcohol (Figure 17) [76–78]. The initially formed alcohol has been shown to be a much better substrate for CYP51 than the unoxidised

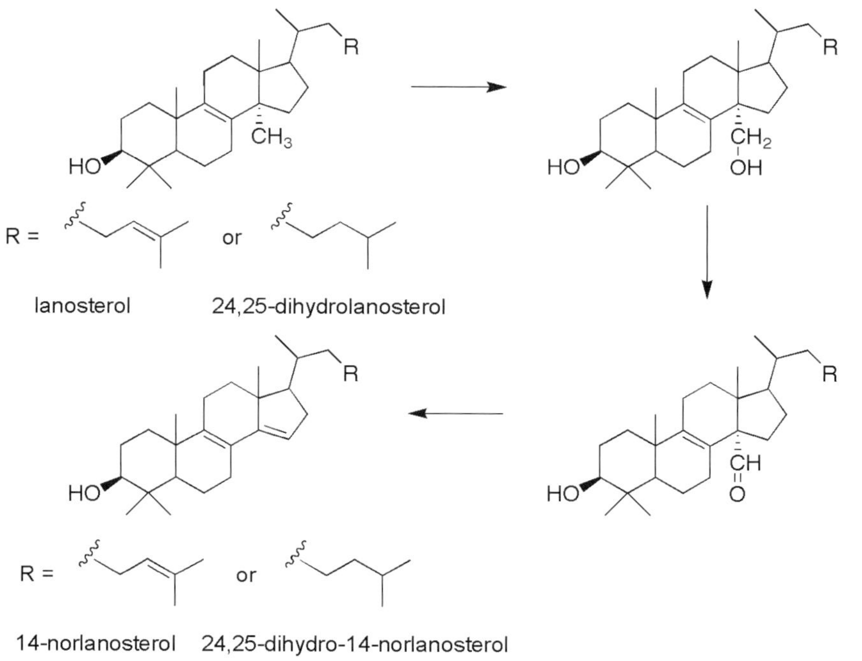

Figure 17. The proposed, stable intermediates in the oxidative removal of the 14α-methyl group of lanosterol.

substrate, supporting a multiple-step oxidation process in a manner analogous to that seen for other P450s that catalyse multi-step transformations. The alcohol is then further oxidised to the corresponding aldehyde, a reaction which proceeds through formation of a *gem*-diol [78,79]. The binding of such an aldehyde to CYP51 is comparable to that of the alcohol. A further oxidative step results in C-C cleavage with release of aldehyde carbon as formic acid and with concomitant incorporation of a double bond into the ring skeleton of both nor-14-DHL and nor-14-lanosterol (Figures 17 and 18) [79].

The oxidative transformation of the aldehyde intermediate catalysed by CYP51 is remarkably similar to that catalysed by aromatase with its analogous aldehyde. As with the aromatase catalysed reaction, the formate group contains one dioxygen derived oxygen atom and incorporates one oxygen and one hydrogen atom from the aldehyde precursor [80], with the formation of the $\Delta^{14,15}$ double bond initiated by a stereospecific loss of the *syn* 15α hydrogen [81–83]. Aldehyde oxidation is proposed to occur via initial peroxyhemiacetal formation through addition of the ferric peroxo intermediate to the electrophilic aldehyde (Figure 18). Due to the isolation and spectroscopic identification of the 14α-formyloxy compound, a Baeyer-Villiger like ionic mechanism was proposed for

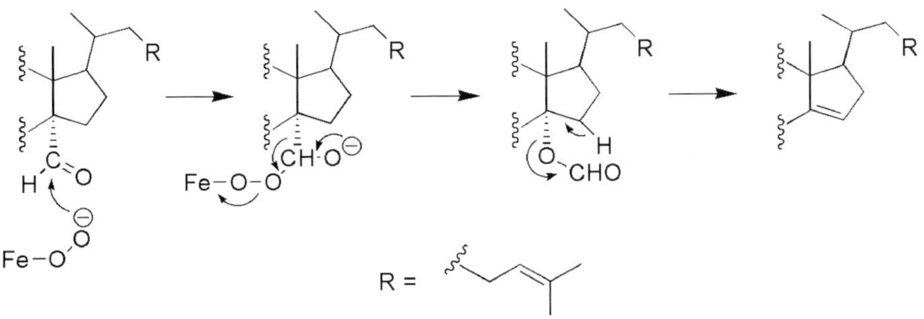

Figure 18. A possible mechanism for the final step in the oxidative cleavage of the 14α-methyl moiety of lanosterol that employs the isolated Baeyer-Villiger rearrangement product. An alternate pathway proceeding via a radical decomposition of the peroxyhemiacetal intermediate may also lead to the observed product.

the decomposition of the peroxyhemiacetal (Figure 18) [84]. A radical decomposition mechanism may still predominate, however, with the Baeyer-Villiger product a minor pathway occurring alongside the major pathway proceeding via simultaneous elimination of C32 and the 15α hydrogen. Support for the ionic pathway was gained by the conversion of the 14α-formyloxy intermediate into the final product by the enzyme [84], although this reaction would be expected to be a facile reaction, subject to standard acid-base catalysis. Further elucidation of the mechanism of this enzyme is clearly of interest so that the relative importance of the ionic and the radical pathways of peroxyhemiacetal decomposition in C-C cleaving P450s may be determined.

The mechanism of demethylation of steroidal biosynthetic P450s appears to be uniform, proceeding via sequential oxidations to alcohol and aldehyde intermediates. Interception of the carbonyl by the ferric peroxy anion produces an intermediate species, which then undergoes a rearrangement with resultant C-C bond cleavage. This results in the incorporation of a double bond into the ring structure and loss of formic acid. Interestingly, a vicinal dioxygenated intermediate cannot be used to facilitate these C-C cleavages as in the case of $P450_{scc}$ and the major pathway of CYP17 due to the presence of a quaternary centre adjacent to the methyl group to be removed in lanosterol and DHL. However, as seen in the following examples, a quaternary carbon adjacent to the carbonyl is not a requirement for C-C cleavage to occur.

Some xenobiotic metabolising enzymes, particularly CYP2B4, are also reported to be capable of catalysing the conversion of some aldehydes with α or β branching into chain-shortened alkenes with the loss of formate [85,86]. Although most work has been carried out with CYP2B4, other isoforms such as CYP1A2, 2E1, 2C3, and 3A6 are all reported to catalyse this type of transformation [85]. The relative frequency of this reaction compared to aldehyde oxidation is low, for example, only 2% of the overall reaction in the CYP2B4 catalysed oxidation of 2-phenylpropionaldehyde [87]. This reaction is useful, however, as a model for understanding the mechanism of the C-C bond cleavage reactions of CYP17, 19, and 51 and has significantly assisted in the formulation of the proposals for their mechanisms. This CYP2B4-catalysed deformylation reaction was enabled by P450 reductase/NADPH or H_2O_2 but not by cumene hydroperoxide or iodosyl benzene [86], a result that indicates that it is the ferric peroxo species that adds to the aldehyde and not the ferryl species. Fragmentation of the intermediate hydroperoxyhemiacetal then occurs to produce formate and the alkene (Figure 19).

The formation of a one carbon reduced alcohol has also been reported in a process apparently analogous to the 17α-hydroxy C_{19} products reported from CYP17A catalysed oxidation of pregnenolone and its analogues [87]. It is important to note at this point that deformylation is thought to involve the ferric peroxo species, whilst oxidation to the acid is believed to proceed via the ferryl species [87].

The relevance of the CYP2B4 catalysed deformylation reaction as a model for CYP51 is clearly demonstrated by the aromatisation of the androstenedione analogue 3-oxodecalin-4-ene-10-carboxaldehyde to the corresponding tetrahydronaphthalene (cf. Fig. 16) with concomitant formate production [88,89]. Deuterium isotope studies showed identical results for the specificity of loss of the 1β and C2 hydrogens as well as for the fate of the formyl hydrogen when compared with similar studies of the reaction catalysed by aromatase. Further support for the role of ferric peroxo species in the CYP2B4 catalysed deformylation has come from mutagenesis studies replacing the residue thought to facilitate O-O bond cleavage in CYP2B4 (Thr^{302}) with an alanine residue [89]. This mutation was expected to favour the peroxo pathway and decrease the availability of the

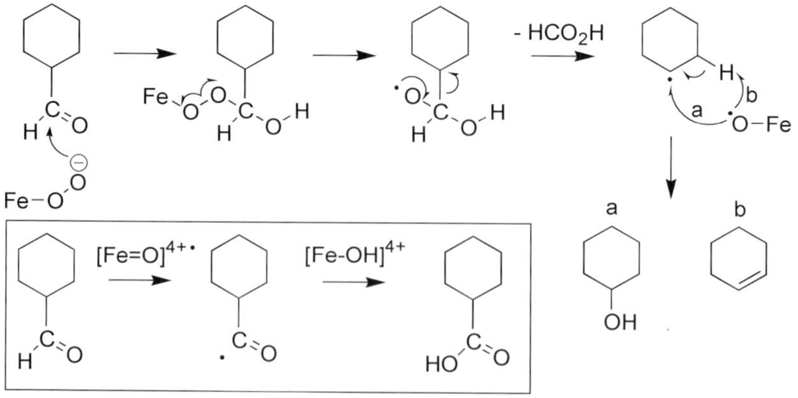

Figure 19. The CYP2B4 catalysed oxidation of cyclohexanecarboxaldehyde is believed to proceed via the ferryl species to yield the carboxylic acid (boxed) and via the ferric peroxo species to yield cyclohexanol (a) and cyclohexene (b) via C-C cleavage.

ferryl species. As predicted, normal P450 catalysed reactions such as aldehyde to carboxylic acid oxidation were suppressed but deformylation to the alkene and alcohol products was significantly enhanced [89].

Cytochrome P450s can also interact with aldehydes to generate the corresponding hydrocarbon and CO_2 [90]. Hydrocarbons are abundant components of cuticular lipids in most insects and are implicated in their chemical communication. It has been demonstrated that in microsomes derived from the housefly, *Musca domestica*, hydrocarbons are formed from the corresponding aldehyde with concomitant generation of CO_2 and with all the characteristics expected of a P450-mediated reaction (Figure 20). These include a requirement for NADPH and oxygen, with the reaction inhibited by both CO and an antibody to house fly P450 reductase [90]. Labelling studies showed that deuterium atoms at the C1, C2 and C3 positions of the substrate aldehyde were all retained [91]. In addition, active oxygen donors such as hydrogen peroxide, cumene hydroperoxide and iodosylbenzene all support hydrocarbon production to some extent. As iodosylbenzene supports oxidation, this clearly indicates that the ferric peroxide species is not the active oxidant in this case.

On the basis of the above results, an unusual mechanism has been proposed [91], with a more conventional version portrayed here (Figure 20). In this mechanism, oxidation of the aldehyde to a dioxirane or its resonance form, a carbonyl oxide occurs first. Dioxiranes are known to decompose with release of CO_2 and formation of two radicals that can recombine as shown to form a hydrocarbon [92]. Recombination of these radicals would be favoured by retention of the fragments within the active site. Complete elucidation of the reaction mechanism awaits identification and purification of the P450 responsible, with recent studies indicating that this is an important, general reaction in insects for hydrocarbon formation [93].

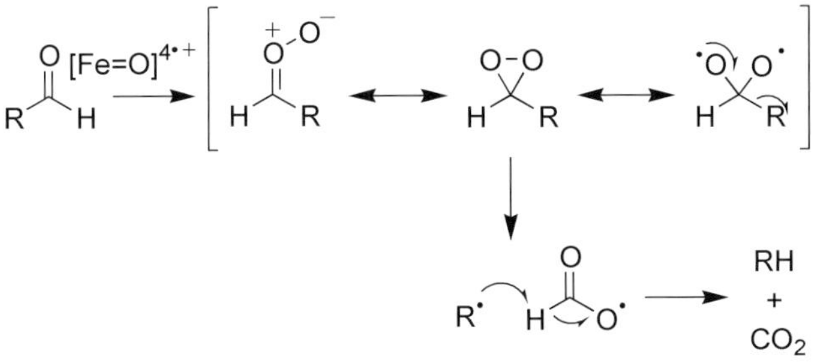

Figure 20. A possible mechanism for the P450 catalysed conversion of an aldehyde into the corresponding hydrocarbon and CO_2. This transformation is important in the biosynthesis of insect derived hydrocarbons (R = alkyl chain).

3.3. Alcohols

Another mechanistic possibility that has been demonstrated for P450-catalysed C-C bond cleaving reactions is found in the biosynthesis of psoralen, a phototoxic furocoumarin, which is produced in plants in response to infection or insect injury (Figure 21). Based upon the presence in plants of oxygenated precursors of psoralen, a biosynthetic pathway involving a P450-mediated sequential oxidation mechanism followed by C-C bond cleavage was initially envisaged. However, labelling studies relying upon deuterated marmesin precursors showed that cleavage of the three carbon side chain produced acetone, which retains all hydrogens from marmesin [94].

Labelling of the ring system suggests that the mechanism of cleavage involves the decomposition of an initially formed carbon radical β to the C-C bond to be cleaved. Through hydrogen abstraction *syn* to this hydroxyl group and β-cleavage of the carbon radical formed, a double bond is incorporated into the substrate. Oxygen rebound from the Fe(IV)OH species onto the carbon radical then results in the formation of a *gem* diol, which dehydrates to the observed carbonyl fragment that is lost from the substrate (Figure 21) [94].

Oxidation of a methyl ketone analogue of marmesin has also been reported to yield some psoralen, but the majority of product observed results from formation of the double bond α, β to the ketone. This result suggests that the different fates of these substrates depend upon the stabilisation of the intermediate radical

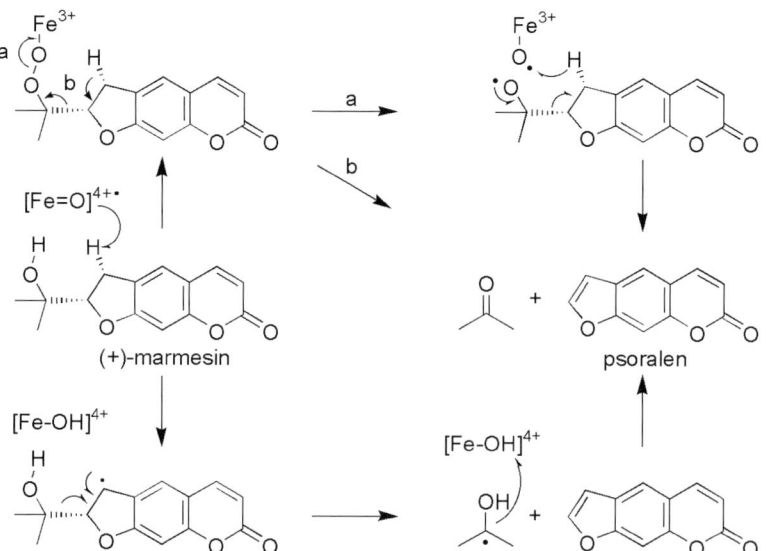

Figure 21. Mechanistic possibilities for the P450 catalysed conversion of (+)-marmesin into psoralen and acetone.

species generated, with the hydroxyl moiety present in psoralen capable of greater stabilisation than the ketone present in the marmesin derivative. This lack of stabilisation may lead to oxygen rebound onto the initial product radical formed rather than C-C cleavage. This would generate a β-hydroxy ketone that would yield the conjugated the α, β unsaturated ketone via dehydration.

The significance of this type of C-C bond cleaving reaction in plants is highlighted by the formation of other metabolites through oxidative bond cleavage with concomitant double bond formation [95–98]. In particular, secologanin synthase from *Catharanthus roseus* (CYP72A1) is believed to catalyse the C-C bond cleavage reaction that transforms loganin into secologanin, the final common non-nitrogenous precursor of many plant indole alkaloids (Figure 22) [98,99]. Due to the structure in this case, the reaction involves cleavage of a carbocyclic ring rather than fragmentation of the substrate. This activity was demonstrated *in vitro* with CYP72A1 heterologously expressed in *E. coli* as a fusion with its homologous P450 reductase [99].

The CYP24 oxidation of vitamin D_3 also results in C-C bond cleavage reaction similar to that observed in psoralen biosynthesis, in which the product formed is a chain shortened alkene (see Section 2.2) [36].

A reaction that involves cleavage of the C-C bond α to a phenol occurs in aflatoxin biosynthesis [100]. Aflatoxins are mycotoxins produced by strains in the fungal genus *Aspergillus* and exhibit significant complexity in their biogenesis. Genetic evidence suggested that a single P450 was responsible for the transformation of *O*-methylsterigmatocystin to aflatoxin B_1 [101]. A P450 from *A. parasiticus* was subsequently demonstrated to be capable of catalysing this reaction and the conversion of 11-hydroxy *O*-methylsterigmatocystin *in vivo*. These experiments were combined with copious results from *in vivo* isotope labelling studies and led to the mechanistic hypothesis shown in Figure 23 [100]. After C-C bond cleavage and formation of the proposed hydrolytically unstable lactone, spontaneous decarboxylation, dehydration, rearrangement and *O*-demethylation reactions are presumed to proceed (Figure 23). Two plausible mechanisms for the C-C bond cleavage process have been proposed, the first of which involves a Baeyer-Villiger type oxidation of the keto

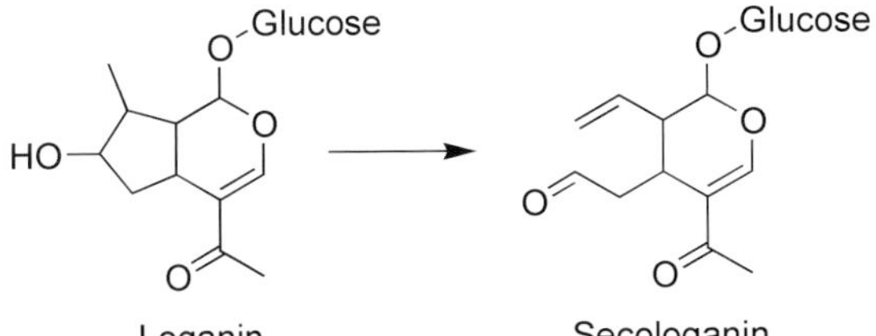

Figure 22. Loganin is converted into secologanin via a P450 catalysed C-C bond cleavage reaction analogous to the one seen in psoralen biosynthesis.

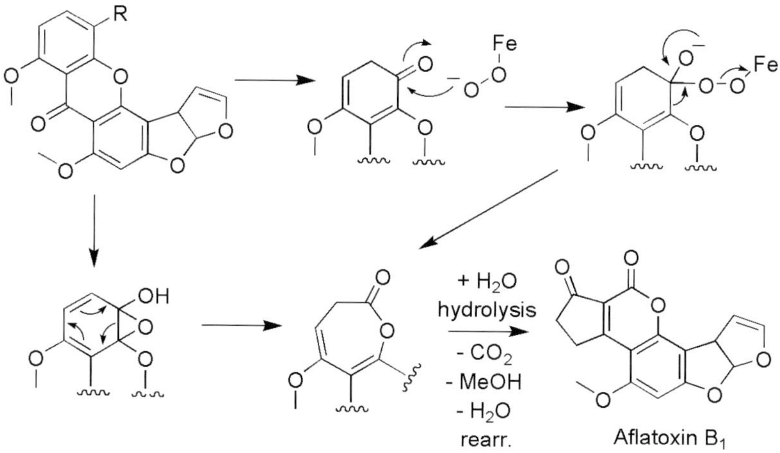

Figure 23. Conversion of O-methylsterigmaticystin (R = H) to aflatoxin B_1 is catalysed by a P450 via 11-hydroxy O-methylsterigmaticystin (R = OH). Mechanistic proposals for the P450 catalysed C-C bond cleavage include a Baeyer-Villiger type addition of the ferric peroxo species to the keto tautomer of the phenolic substrate or oxidation to an epoxide intermediate typical of ferryl catalysed aromatic oxidations.

tautomer of the phenol, whilst the second involves a rearrangement of the epoxide intermediate in aromatic oxidation [100]. Delineation of the mechanism will require further experimentation with purified enzyme, mutants and substrate analogues.

The gibberellins (GAs) are important plant hormones, which possess strikingly complex structures. Several similarly remarkable multi-functional P450s have been implicated in their biosynthesis in both plants and fungi [102,103]. CYP88A (from *A. thaliana* and barley (*Hordeum vulgare* cv Himalaya)) is capable of the three-step oxidative sequence required to convert *ent*-kaurenoic acid to GA_{12} (Figure 24) [104]. The key step in the proposed mechanism involves cleavage of a C-C bond to yield an aldehyde, via cleavage α to an alcohol during oxidative ring contraction. The mechanism has not been investigated but the process follows the pathway predicted for a α-hydroxy carbocation, as would be seen during a pinacol rearrangement of a diol (Figure 25). Such a cation could arise from a diol formed by CYP88A catalysed C6 hydroxylation under the influence of the Lewis acidic heme iron, or directly, via a single electron transfer process from the radical intermediate formed during hydroxylation.

A further P450, from the fungus *Gibberella fujikuroi*, has been shown to catalyse a similar pinacol-like transformation [105]. A 6,7-diol was isolated during these studies, but was not further transformed by the enzyme: this suggests that such a compound is not an intermediate in this reaction but rather the result of an alternative pathway in which the intermediate cation is trapped by

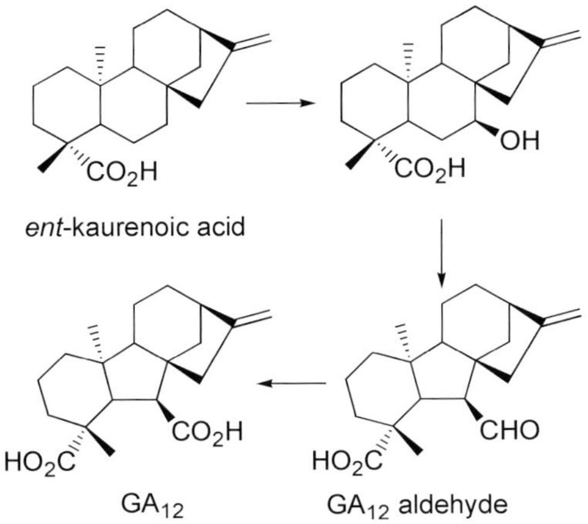

Figure 24. CYP88A catalyses the three oxidative transformations required to convert *ent*-kaurenoic acid into GA_{12}.

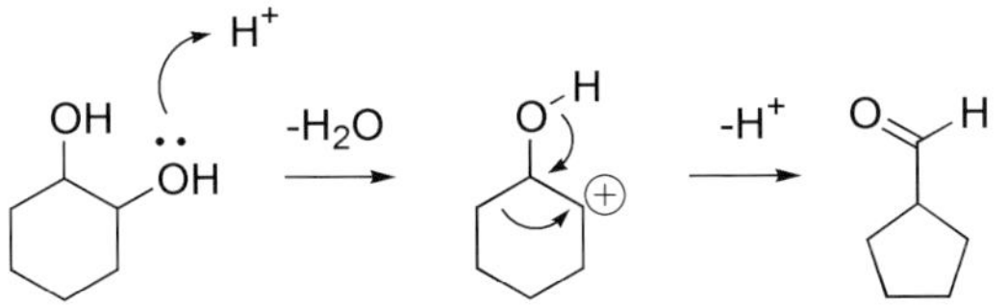

Figure 25. Pincol rearrangement of a cyclic diol results in ring contraction.

water rather than undergoing rearrangement. This P450 was also proposed to be capable of catalysing at least *seven* other biosynthetically significant oxidative transformations as well as the three assigned to CYP88A, explaining the various GA metabolites found in *G. fujikuroi*. One of these transformations is a proposed oxidative cleavage of the vicinal 6,7-diol analogous to those discussed in Section 2.1 above. Clearly, the results of *in vitro* characterisation of the catalytic capabilities of this enzyme will be of great interest.

3.4. Acids

P450 catalysed oxidative decarboxylation has been reported twice in the literature. The first example relates to a P450 purified from the yeast *Rhodotorula minuta* which is responsible for the formation of isobutene from isovalerate (Fig. 26) [106]. A large isotope effect upon isobutene formation was found when the β-hydrogen was substituted with deuterium ($k_H/k_D = 14$), which implies that the cleavage of this bond is the rate-determining step. β-Branching within the substrate also appeared to be necessary for alkene formation. Direct

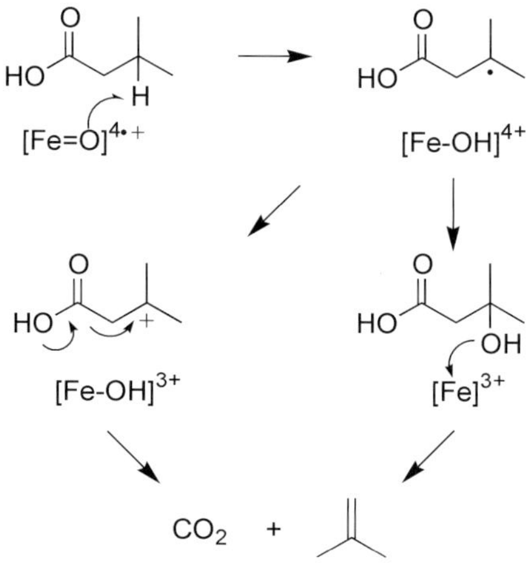

Figure 26. Mechanistic alternatives for the P450 catalysed conversion of valerate into isobutene and CO_2. A less probable direct hydride abstraction mechanism has also been proposed.

hydride abstraction followed by decarboxylation of the resultant cation was initially proposed [106]. Other mechanistic possibilities include: hydrogen atom abstraction to give a carbon radical, in turn followed by electron transfer to generate the corresponding carbocation; or tertiary alcohol formation followed by ionisation to the carbocation via interaction with the Lewis acidic heme iron. The detection method utilised to identify isobutene as a product of oxidation (headspace gas chromatography) is unable to rule out the formation of non-volatile products of isovalerate oxidation, which in turn make identification of the exact mechanism of this decarboxylation reaction difficult [106]. This P450 has subsequently been shown to be involved in phenylalanine catabolism, with its primary metabolic function the oxidation of benzoate to 4-hydroxybenzoate [107,108].

Hirobe et al. have reported the P450 catalysed decarboxylation of certain α-substituted carboxylic acids to one-carbon chain shortened alcohols, with the α-carbon attached to a phenyl group or three other substituents (Figure 27) [109]. This transformation, originally observed in iron-porphyrin model systems, was reproduced *in vivo* in rats and in rat liver microsomes for two therapeutically important carboxylic acids [109]. Mechanistic details are sparse, but the proposed decomposition of a carboxyl radical appears feasible, as α-substitution is known to influence the decomposition of such radicals. Possible sources for the radical intermediate include direct interaction of the acid and the ferryl species, or via the decomposition of an intermediate peroxyacid. This latter mechanism has precedent in the CYP2B4 catalysed conversion of 2-phenylperacetic acid to CO_2 and benzyl alcohol via homolysis of the O-O bond [110].

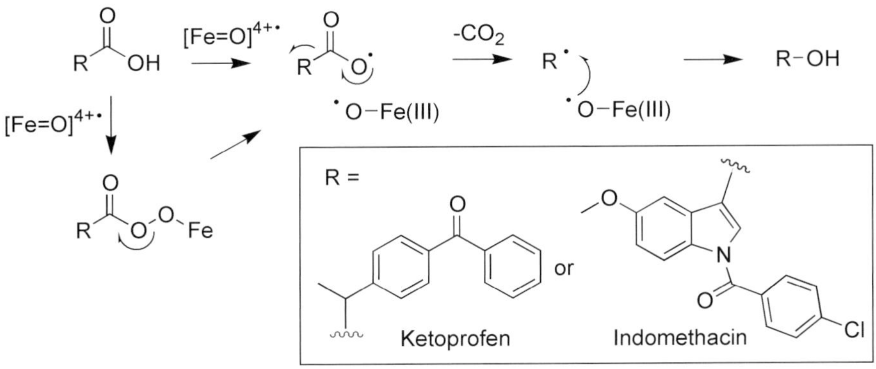

Figure 27. Mechanistic possibilities for the oxidative decarboxylation of some carboxylic acids catalysed by some P450s and iron-porphyrin model systems.

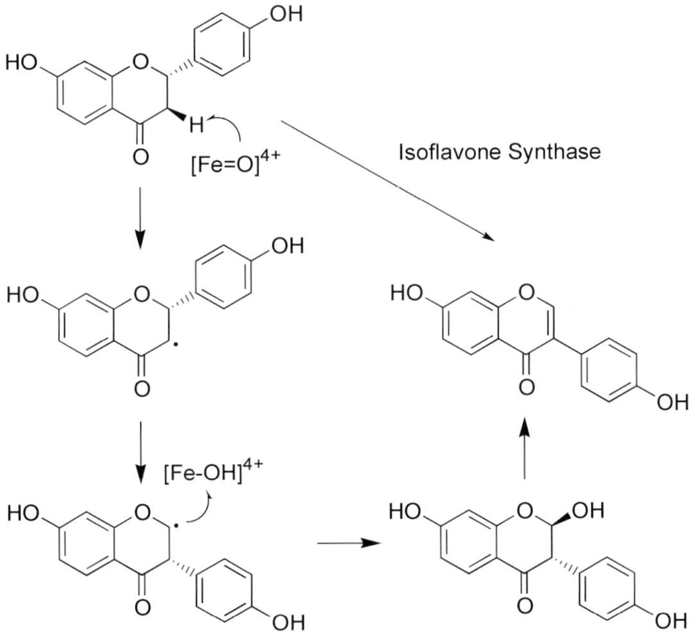

Figure 28. A possible mechanism for the formation of isoflavone from 2S-flavone catalysed by isoflavone synthase (CYP93C) via an oxidative aryl migration.

3.5. Ethers

Isoflavone synthase (CYP93C) catalyses the formation of isoflavone from 2S-flavone via an unusual oxidative aryl migration (Figure 28) [111–113]. (This C-C bond cleavage occurs α to an ether and is classified as such here, but it is unclear whether this is a mechanistically significant feature.) Little is known about the reaction, but it is postulated to proceed via 3β hydrogen abstraction to give a carbon radical anchimerically stabilized by the adjacent phenol [112,114]. Oxygen rebound onto the C2 position would give the unstable 2-hydroxyisoflavone, which would undergo dehydration to afford the isoflavone (Figure 28). Support for this mechanism is provided by the isolation of the 3β-hydroxyisoflavone as a side product of the reaction [114]. Further investigations with purified protein would doubtless provide more information about this unusual transformation [114].

4. CLEAVAGE ALPHA TO CARBON BEARING NITROGEN

An example of C-C bond cleavage with rearrangement β to a nitrogen atom has been reported to occur with human liver P450s (predominantly the CYP3A4 isoform), with several tetramethylpiperidine containing compounds transformed into the corresponding dimethylpyrrolidine derivatives (Figure 29).

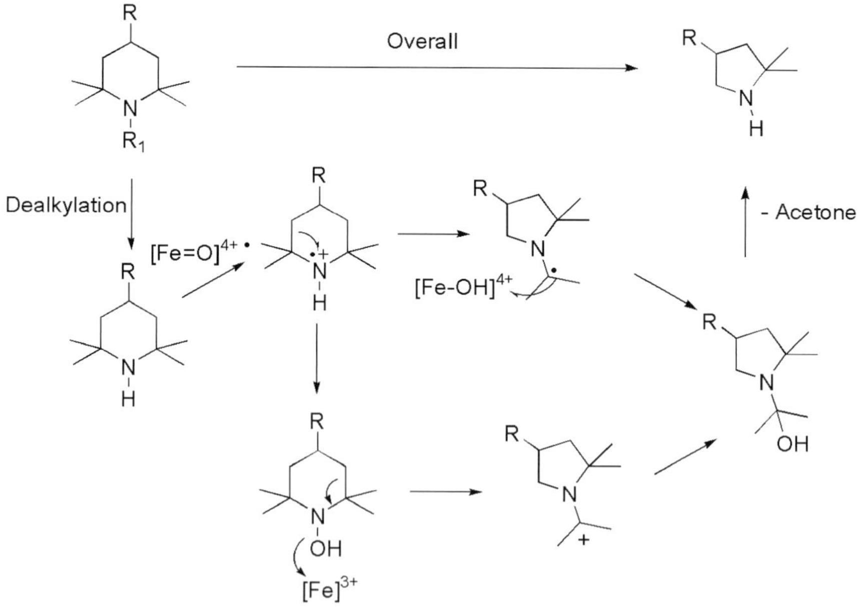

Figure 29. Several tetramethylpiperidines are converted into the corresponding dimethylpyrrolidines by a number of P450s that metabolise xenobiotics (particularly CYP3A4). Two mechanistic possibilities for this process are shown. (R = p-nitophenyl-NH-, R_1 = H, or R = $(C_6H_5)_2$HCO-), R_1 = CH_3.

This transformation has been reported to occur in other mammalian species and it is therefore speculated that it is a general metabolic pathway.[115] The mechanism of the reaction is not well understood, but the nature of the pyrrolidines produced suggests that the cleavage reaction begins from a secondary amine, formed via N-dealkylation in the case of tertiary amines (Figure 29) [115].

Several mechanistic possibilities have been suggested, including the interaction of the iron-oxo species with the nitrogen atom to afford a hydroxylamine or a nitrogen radical cation intermediate. Ionisation of the hydroxylamine to the corresponding nitrogen cation radical is postulated for the first mechanistic alternative, a pathway comparable to the P450 catalysed dehydration of oximes to nitriles [116]. As this cation can no longer be stabilised by α-hydrogen elimination rearrangement then occurs. In the second path, the steric congestion of the surrounding methyl groups may possibly slow oxygen rebound and the nitrogen radical cation rearranges (Figure 29). The piperidine/pyrrolidine rearrangement has precedent in the chemistry of N-fluoroamines, which undergo this reaction in the presence of a Lewis acid [117]. This latter reaction, however, presumably involves the equivalent of a nitrogen cation, rather than a cation radical species favouring the first pathway suggested. As the majority of the mechanistic detail presented here is speculation, further investigations are clearly required to determine the mechanism of this unusual and interesting transformation.

5. CARBON-CARBON BOND CLEAVAGE INVOLVING PEROXIDES

As well as including enzymes capable of catalysing *oxidative* bond cleavage (discussed in the previous sections), the P450 family includes members that do not oxidise the substrate but still catalyse C-C bond cleavage. These P450s possess unusual characteristics, which are explained by the nature of the substrates for such enzymes: these substrates contain peroxide moieties, the rearrangement of which these P450s have evolved to catalyse. This section will deal with these enzymes in two parts, differentiated by the substrates they utilise, either prostaglandins or hydroperoxy fatty acids.

5.1. Hydroperoxy Fatty Acids

Several subfamilies of CYP74, a family of non-classical P450s found in a wide variety of plant species (including *A. thaliana*, flaxseed (*Linum usitatissimum*), tulip bulbs (*Tulipa gesneriana cv. Red Apeldoorn*), tomato (*L. esculentum*), and potato (*Solanum tuberosum*)) have been shown to carry out C-C bond cleavage of hydroperoxide containing fatty acids. All members of the CYP74 family act upon unsaturated hydroperoxide fatty acids (most commonly

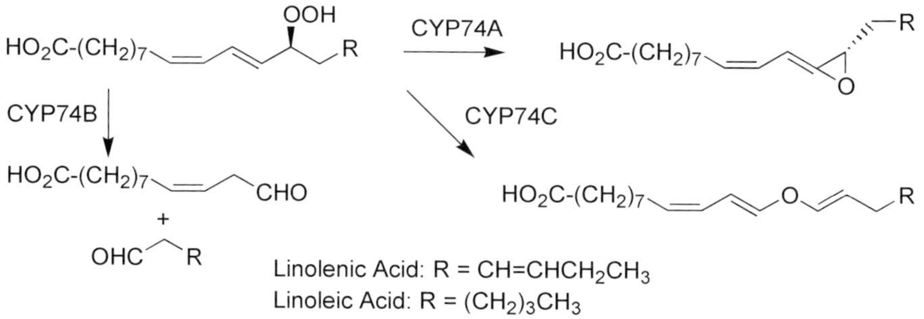

Figure 30. Transformations of unsaturated fatty acids performed by members of the CYP74 subfamily include the formation of allene oxides (CYP74A), reactive aldehydes (CYP74B) and divinyl ethers (CYP74C).

linolenic and linoleic acids) and catalyse their conversion into one of three different categories of product (Figure 30). These products are allene oxides (catalysed by CYP74A) [118], short-chain aldehydes/ω-oxoacids (catalysed by CYP74B) [119,120], and divinyl ethers (catalysed by CYP74D) [120,121], the latter two of which involve C-C cleavage (Figure 30). A further subfamily, CYP74C, has also been identified with similar activity to that of CYP74B but which exhibit less substrate specificity [122].

CYP74A was the first identified subfamily of CYP74 and displayed a series of non-classical P450 characteristics, which were later found to be universal across the CYP74 family. These include the absence of several features of proto-typical P450s, including difficulty in obtaining a reduced, ferrous form of the enzyme. The affinity of the ferrous form of CYP74 for carbon monoxide was also greatly reduced over typical P450s and made obtaining 450nm difference absorption spectra difficult. There were alterations to the highly conserved I-helix [123], including the absence of the highly conserved threonine residue, which is responsible for facilitating O-O cleavage in the normal P450 catalytic cycle. Indeed, the majority of enzymes in the CYP74 family replace this threonine residue with an isoleucine or alanine residue [123]. Further differences exist in that these enzymes are capable of sustaining turnover without dioxygen or NAD(P)H [119,124–126]. These changes to 'normal' P450 structure are dictated by the unusual nature of the substrates, which, as indicated in the introduction to this section, do not require oxidation. Instead, the members of the CYP74 family catalyse the rearrangement of the unsaturated hydroperoxide functionality with control of the rearrangement leading to the different product classes for these enzymes. The nature of this rearrangement as opposed to a more normal oxidation reaction allows the members of CYP74 very rapid rates of turnover, some as high as $2000\,s^{-1}$.

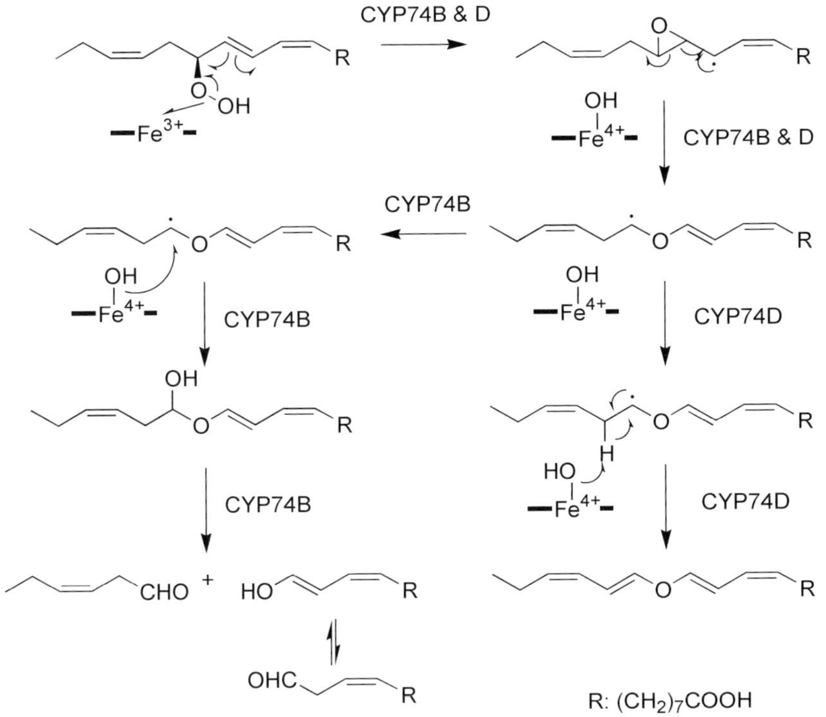

Figure 31. Postulated mechanisms for the C-C cleavage reactions of unsaturated fatty acids catalysed by CYP74B and CYP74D.

From a C-C bond cleavage perspective, the interest in CYP74 mechanism lies with CYP74B and CYP74D [120]. The postulated mechanism that these enzymes utilise is remarkably similar, with the fate of an intermediate allylic radical determining the nature of the products (Figure 31). Following initial homolytic cleavage of the substrate hydroperoxide, an intermediate epoxide is formed, with concomitant formation of an allylic radical. Ring opening of the epoxide then forms a radical/vinyl ether intermediate.

At this point, the mechanisms of the two subfamilies diverge: in the case of CYP74D, abstraction of a hydrogen atom α to the carbon-centred radical by the Fe(IV)-OH species is proposed to lead to alkene formation with concomitant release of water and regeneration of the resting state Fe(III) species (Figure 31) [120]. Alternatively, this may be represented as an electron transfer to produce a stabilised cation which then undergoes loss of a proton to yield the product. The mechanism of CYP74B differs in that the Fe(IV)-OH species undergoes rebound onto the radical/vinyl ether intermediate forming a hemiacetal [120,127]. Decomposition of this hemiacetal leads directly to the formation of the aldehyde species, whilst the ω-oxoacid arises from tautomerisation of the initially produced enol (Figure 31) [120,127]. The experimental evidence to support this mechanism is somewhat sparse, however, with the only intermediate identified to date being the TMS-derivative of the intermediate enol formed by CYP74B [127]. Clearly further research would be invaluable into this family of non-classical P450s.

5.2. Prostaglandins

The involvement of two non-classical P450s in the biosynthesis of the prostaglandins has been reported [125,126,128,129]. The prostaglandins are important cell signalling molecules and are ultimately derived from arachidonic acid. The two P450s involved are prostacyclin and thromboxane synthases (CYP8 [130] and CYP5 [131–133]), which are involved in the biosynthesis of prostacyclin and thromboxane, respectively (Figure 32). These compounds both play important but opposite roles in vasoconstriction and platelet aggregation: thromboxane causes vasoconstriction and platelet aggregation [134], whilst prostacyclin is a vasodilator and potent inhibitor of platelet aggregation [135]. Both CYP5 [136] and CYP8 [137] utilise prostaglandin endoperoxide PGH_2 as a substrate and both catalyse the rearrangement of the peroxide moiety along with, in the case of CYP5, the cleavage of a C-C bond [125,126,138]. Curiously, although both enzymes accept the same substrate and catalyse hydroperoxide

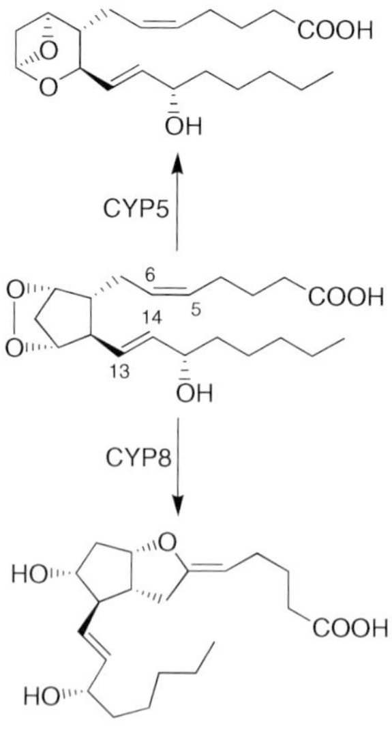

Figure 32. CYP5 and CYP8 are involved in the biosynthesis of the prostaglandins thromboxane and prostacyclin, respectively.

lysis, the sequence similarity of the two enzymes is only 16% and thus they are placed in different families.

The initial step in the mechanism of both enzymes involves the homolytic cleavage of the peroxide oxygen-oxygen bond via interaction with the haem-bound Fe(III) (Figure 33). This generates an Fe(IV)-OR intermediate along with an oxygen radical, with CYP5 and CYP8 differing in which of the oxygen atoms of the peroxide interacts with the iron. In the case of CYP8 the C-5,6

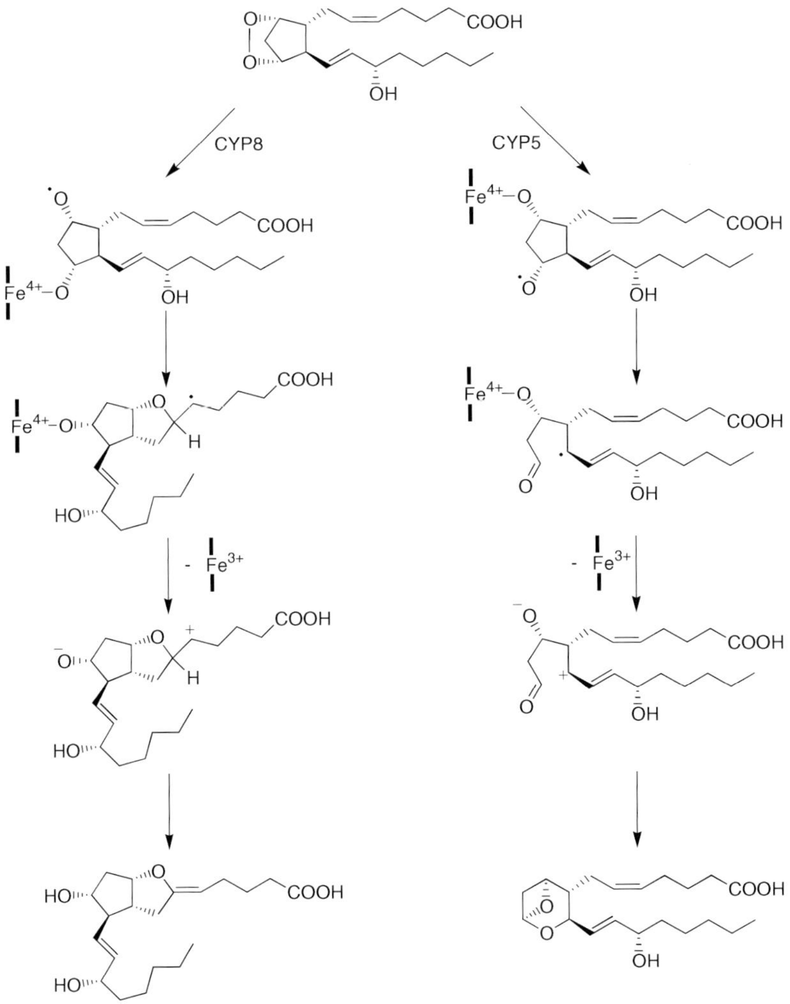

Figure 33. The mechanisms by which prostaglandin endoperoxide PGH_2 is believed to be transformed by CYP5 into thromboxane and by CYP8 into prostacyclin.

double bond is attacked by the oxygen radical, which results in the formation of a furan ring with an adjacent carbinyl radical. Electron transfer from the carbon radical to the haem then occurs, regenerating the initial haem-bound Fe(III) and forming a zwitterionic intermediate. Proton transfer and double bond formation then proceeds, leading to the formation and release of prostacyclin (Figure 33).

In the biosynthesis of thromboxane, the oxygen radical generated by CYP5 rearranges via C-C bond cleavage within the pentane ring to afford an aldehyde/allylic radical intermediate. Electron transfer to the haem, with concomitant generation of a zwitterionic intermediate and release of the haem-bound Fe(III) then occurs in a similar manner to CYP5. Attack of the carbonyl on the carbocation then initiates formation of thromboxane (Figure 33). CYP5 in contrast to CYP8, however, invariably yields products from competing reaction pathways. These products, hydroxyhepatrienoic acid and malondialdehyde, are always produced in at least a (1:1):1 ratio with thromboxane (Figure 34) [125,136,139]. Early studies suggested CYP5 employed two active sites to catalyse these two different reactions [140], but more recent studies of the structure of the enzyme do not appear to support this [139].

An alternative explanation is that the rearrangement of the carbon-based radical intermediate produced by CYP5 can occur either to generate a zwitterionic intermediate, or via further homolytic C-C bond cleavage with resultant regen-

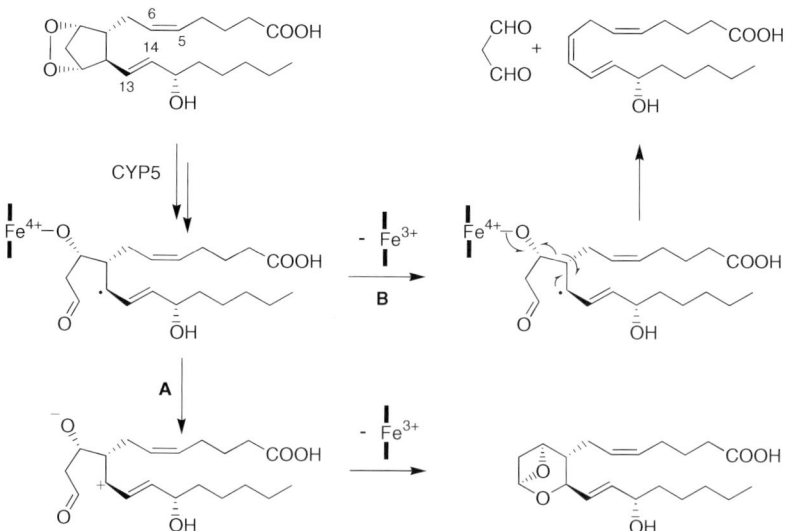

Figure 34. A mechanistic rationale for why the biosynthesis of thromboxane by CYP5 from prostaglandin endoperoxide PGH$_2$ (pathway A) is always accompanied by the formation of hydroxyhepatrienoic acid and malondialdehyde (pathway B).

eration of the resting state haem and the observed alternate products (Figure 34) [125,126,138]. This type of rearrangement is also seen when other P450s, such as P450$_{cam}$ [141], act upon prostaglandin endoperoxide PGH$_2$ [142]. The reaction rate for these enzymes is always less than that seen with CYP5, whilst the generation of thromboxane was never observed [141,142]. The stabilising effect of the C-13,14 double bond is important in mediating the generation of the zwitterionic intermediate, as if this is absent only hydroxyhepatrienoic acid and malondialdehyde are formed from homolytic C-C bond cleavage [126]. The exact role of these latter two products of CYP5 catalysis is unsure, but evidence has been found that links malondialdehyde to DNA aggregation and cancer [142]. In a mechanistically comparable process, CYP8 produces comparable products to hydroxyhepatrienoic acid and malondialdehyde when 5,6-dihydroprostaglandin endoperoxide PGH$_2$ is used as a mechanistic probe [126]. This stems from the inability of the oxygen radical to rearrange via addition to the C-5,6-double bond to a more stable carbon based radical, which in turn leads to the fragmentation of the oxygen radical intermediate [126].

CYP5 and CYP8 represent important members of the P450 superfamily in both their biological function and mechanism of action. The ability of CYP5 to partition between two different reaction pathways is made more interesting by the selectivity of CYP8 when acting upon the same substrate. Further research into the biological significance of hydroxyhepatrienoic acid and malondialdehyde is also warranted due to their concomitant production with thromboxane.

6. GENERAL CONCLUSIONS

Whilst the type and mechanism of the P450-mediated C-C bond cleavage reactions presented in this chapter are diverse, it is important to recognise that they will probably constitute just a selection of the actual array of such reactions catalysed in Nature. To highlight this point, recent studies have identified another multi-functional biosynthetic P450 capable of catalysing C-C bond cleavage (Figure 35). This fungal P450 is involved in gibberellin formation and is capable of catalysing the demethylation of an angular carbon along the biosynthetic pathway (Figure 35) [143]. The formation of the lactone in the product is believed to be concomitant with concomitant demethylation, and this suggests that a different pathway is operating from that seen with aromatase and 14α-demethylase. Mechanistically, this may represent a biosynthetically novel oxidative decarbonylation or decarboxylation reaction involving an alcohol or the corresponding cation as the initial product. Interception of this species by the adjacent carboxylate could then occur, resulting in lactone formation (Figure 35). As this example demonstrates, the variety of mechanistic alternatives

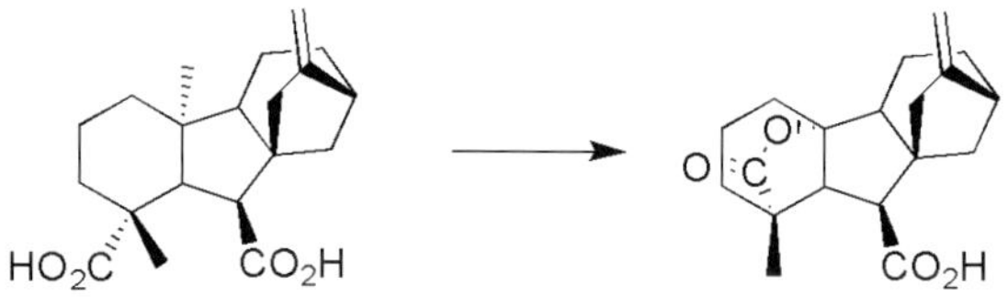

Figure 35. Gibberellin 20-oxidase, a multi-functional P450 isolated from the fungus *Gibberella fujikuroi* catalyses the angular demethylation of GA_{12} to produce the lactone GA_9.

utilised by P450s to carry out C-C bond cleavage are yet to be fully explored and defined.

ACKNOWLEDGMENTS

This work was funded in part by the Australian Research Council through the provision of grants *LP0560595* and *DP0451923*.

ABBREVIATIONS

ACP acyl carrier protein
DHL 24,25-dihydrolanosterol
GA   gibberellin
PG   prostaglandin
TMS trimethylsilyl

REFERENCES

1. R. J. Kraaipoel, H. J. Degenhart, and J. G. Leferink, *FEBS Lett.*, 57, 294–300 (1975).
2. M. Shikita and P. F. Hall, *Proc. Nat. Acad. Sci. USA*, 71, 1441–1445 (1974).
3. E. R. Simpson and G. S. Boyd, *Biochem. Biophys. Res Commun.*, 24, 10–17 (1966).
4. E. R. Simpson and G. S. Boyd, *Eur. J. Biochem.*, 2, 275–285 (1967).
5. S. Burstein, B. S. Middleditch, and M. Gut, *J. Biol. Chem.*, 250, 9028–9037 (1975).
6. J. D. Lambeth, S. E. Kitchen, A. A. Farooqui, R. Tuckey, and H. Kamin, *J. Biol. Chem.*, 257, 1876–1884 (1982).
7. R. C. Tuckey and H. Kamin, *J. Biol. Chem.*, 257, 9309–9314 (1982).
8. R. C. Tuckey and H. Kamin, *J. Biol. Chem.*, 258, 4232–4237 (1983).
9. C.-Y. Byon and M. Gut, *Biochem. Biophys. Res. Commun.*, 94, 549–552 (1980).
10. J.-L. Primus, K. Teunis, D. Mandon, C. Veeger, and I. M. C. M. Rietjens, *Biochem. Biophys. Res. Commun.*, 272, 551–556 (2000).

11. R. B. Hochberg, P. D. McDonald, M. Feldman, and S. Lieberman, *J. Biol. Chem.*, *249*, 1274–1285 (1974).
12. A. Slominski, J. Zjawiony, J. Wortsman, I. Semak, J. Stewart, A. Pisarchik, T. Sweatman, J. Marcos, C. Dunbar, and R. C. Tuckey, *Eur. J. Biochem.*, *271*, 4178–4188 (2004).
13. O. Guryev, R. A. Carvalho, S. Usanov, A. Gilep, and R. W. Estabrook, *Proc. Nat. Acad. Sci. USA*, *100*, 14754–14759 (2003).
14. S. Bower, J. B. Perkins, R. R. Yocum, C. L. Howitt, P. Rahaim, and J. Pero, *J. Bacteriol.*, *178*, 4122–4130 (1996).
15. J. E. Stok and J. J. De Voss, *Arch. Biochem. Biophys.*, *384*, 351–360 (2000).
16. A. J. Green, S. L. Rivers, M. Cheesman, G. A. Reid, L. G. Quaroni, I. D. G. Macdonald, S. K. Chapman, and A. W. Munro, *J. Biol. Inorg. Chem.*, *6*, 523–533 (2001).
17. M. J. Cryle, N. J. Matovic, and J. J. De Voss, *Org. Lett.*, *5*, 3341–3344 (2003).
18. M. J. Cryle and J. J. De Voss, *Chem. Commun.*, 86–87 (2004).
19. M. J. Cryle, J. E. Stok, and J. J. De Voss, *Aus. J. Chem.*, *56*, 749–762 (2003).
20. K. Umehara, S. Kudo, Y. Hirao, S. Morita, T. Ohtani, M. Uchida, and G. Miyamoto, *Drug Metab. Disp.*, *28*, 1417–1424 (2000).
21. K. Umehara, S. Kudo, Y. Hirao, S. Morita, M. Uchida, M. Odomi, and G. Miyamoto, *Drug Metab. Disp.*, *28*, 887–894 (2000).
22. K. Umehara, Y. Shimokawa, T. Koga, T. Ohtani, and G. Miyamoto, *Xenobiotica*, *34*, 61–71 (2004).
23. H. J. Barnes, M. P. Arlotto, and M. R. Waterman, *Proc. Nat. Acad. Sci. USA*, *88*, 5597–5601 (1991).
24. M. X. Zuber, E. R. Simpson, and M. R. Waterman, *Science*, *234*, 1258–1261 (1986).
25. M. Akhtar, D. Corina, S. Miller, A. Z. Shyadehi, and J. N. Wright, *Biochemistry*, *33*, 4410–4418 (1994).
26. M. Akhtar, D. L. Corina, S. L. Miller, A. Z. Shyadehi, and J. N. Wright, *J. Chem. Soc., Perkin Trans. 1*, 263–267 (1994).
27. P. Lee-Robichaud, M. E. Akhtar, and M. Akhtar, *Biochem. J.*, *330*, 967–974 (1998).
28. K. Shimizu, *J. Biol. Chem.*, *253*, 4237–4241 (1978).
29. A. Y. Mak and D. C. Swinney, *J. Am. Chem. Soc.*, *114*, 8309–8310 (1992).
30. S. Nakajin, M. Takahashi, M. Shinoda, and P. F. Hall, *Biochem. Biophys. Res. Commun.*, *132*, 708–713 (1985).
31. D. L. Corina, S. L. Miller, J. N. Wright, and M. Akhtar, *Chem. Commun.*, 782–783 (1991).
32. D. C. Swinney and A. Y. Mak, *Biochemistry*, *33*, 2185–2190 (1994).
33. P. Lee-Robichaud, A. Z. Shyadehi, J. N. Wright, M. Akhtar, and M. Akhtar, *Biochemistry*, *34*, 14104–14113 (1995).
34. T. Sakaki, N. Sawada, K. Komai, S. Shiozawa, S. Yamada, K. Yamamoto, Y. Ohyama, and K. Inouye, *Eur. J. Biochem.*, *267*, 6158–6165 (2000).
35. T. Sakaki, N. Sawada, Y. Nonaka, Y. Ohyama, and K. Inouye, *Eur. J. Biochem.*, *262*, 43–48 (1999).
36. N. Sawada, T. Kusudo, T. Sakaki, S. Hatakeyama, M. Hanada, D. Abe, M. Kamao, T. Okano, M. Ohta, and K. Inouye, *Biochemistry*, *43*, 4530–4537, (2004).
37. E. A. Weinstein, D. S. Rao, M. L. Siu–Caldera, K. Y. Tserng, M. R. Uskokovic, S. Ishizuka, and G. S. Reddy, *Biochem. Pharm.*, *58*, 1965–1973 (1999).

38. M. Akiyoshi-Shibata, T. Sakaki, Y. Ohyama, M. Noshiro, K. Okuda, and Y. Yabusaki, *Eur. J. Biochem.*, 224, 335–343 (1994).
39. M. J. Beckman, P. Tadikonda, E. Werner, J. Prahl, S. Yamada, and H. F. DeLuca, *Biochemistry*, 35, 8465–8472 (1996).
40. S. Masuda, M. Kaufmann, V. Byford, M. Gao, R. St-Arnaud, A. Arabian, H. L. J. Makin, J. C. Knutson, S. Strugnell, and G. Jones, *J. Steroid Biochem. Mol. Biol.*, 89–90, 149–153 (2004).
41. R. L. Horst, J. A. Omdahl, and S. Reddy, *J. Cell. Biol.*, 88, 282–285 (2003).
42. T. Nomura, T. Kushiro, T. Yokota, Y. Kamiya, G. J. Bishop, and S. Yamaguchi, *J. Biol. Chem.*, 280, 17873–17879 (2005).
43. T.-W. Kim, J.-Y. Hwang, Y.-S. Kim, S.-H. Joo, S. C. Chang, J. S. Lee, S. Takatsuto, and S.-K. Kim, *Plant Cell*, 17, 2397–2412 (2005).
44. S. C. Davis, Z. Sui, J. A. Peterson, and P. R. Ortiz de Montellano, *Arch. Biochem. Biophys.*, 328, 35–42 (1996).
45. E. A. Thompson, Jr. and P. K. Siiteri, *J. Biol. Chem.*, 249, 5373–5378 (1974).
46. J. T. Kellis, Jr. and L. E. Vickery, *J. Biol. Chem.*, 262, 4413–4420 (1987).
47. E. A. Thompson, Jr. and P. K. Siiteri, *J. Biol. Chem.*, 249, 5364–5372 (1974).
48. E. Caspi, T. Arunachalam, and P. A. Nelson, *J. Am. Chem. Soc.*, 105, 6987–6989 (1983).
49. E. Caspi, T. Arunachalam, and P. A. Nelson, *J. Am. Chem. Soc.*, 108, 1847–1852 (1986).
50. D. Arigoni, R. Battaglia, M. Akhtar, and T. Smith, *Chem. Commun.*, 185–186 (1975).
51. Y. Osawa, K. Shibata, D. Rohrer, C. Weeks, and W. L. Duax, *J. Am. Chem. Soc.*, 97, 4400–4402 (1975).
52. S. Miyairi and J. Fishman, *Biochem. Biophys. Res. Commun.*, 117, 392–398 (1983).
53. S. Miyairi and J. Fishman, *J. Biol. Chem.*, 260, 320–325 (1985).
54. M. Numazawa, K. Midzuhashi, and M. Nagaoka, *Biochem. Pharm.*, 47, 717–726 (1994).
55. H. J. Brodie, K. J. Kripalani, and G. Possanza, *J. Am. Chem. Soc.*, 91, 1241–1242 (1969).
56. J. Fishman and H. Guzik, *J. Am. Chem. Soc.*, 91, 2805–2806 (1969).
57. J. Fishman, H. Guzik, and D. Dixon, *Biochemistry*, 8, 4304–4309 (1969).
58. J. Fishman and M. S. Raju, *J. Biol. Chem.*, 256, 4472–4477 (1981).
59. Y. Osawa, N. Yoshida, M. Fronckowiak, and J. Kitawaki, *Steroids*, 50, 11–28 (1987).
60. J. D. Townsley and H. J. Brodie, *Biochemistry*, 7, 33–40 (1968).
61. M. Akhtar, M. R. Calder, D. L. Corina, and J. N. Wright, *Biochem. J.*, 201, 569–580 (1982).
62. E. Caspi, J. Wicha, T. Arunachalam, P. Nelson, and G. Spiteller, *J. Am. Chem. Soc.*, 106, 7282–7283 (1984).
63. P. Morand, D. G. Williamson, D. S. Layne, L. Lompa-Krzymien and J. Salvador, *Biochemistry*, 14, 635–638 (1975).
64. H. Hosoda and J. Fishman, *J. Am. Chem. Soc.*, 96, 7325–7329 (1974).
65. J. Goto and J. Fishman, *Science*, 195, 80–81 (1977).
66. D. F. Covey and W. F. Hood, *Cancer Res.*, 42, 3327–3333 (1982).
67. D. D. Beusen and D. F. Covey, *J. Steroid Biochem.*, 20, 931–934 (1984).

68. M. Numazawa, A. Yoshimura, M. Tachibana, M. Shelangouski, and M. Ishikawa, *Steroids, 67*, 185–193 (2002).
69. M. Akhtar, D. Corina, J. Pratt, and T. Smith, *Chem. Commun.*, 854–856 (1976).
70. M. Akhtar, V. C. O. Njar, and J. N. Wright, *J. Steroid Biochem. Mol. Biol., 44*, 375–387 (1993).
71. D. E. Stevenson, J. N. Wright, and M. Akhtar, *J. Chem. Soc., Perkin Trans. 1*, 2043–2052 (1988).
72. D. L. Wertz, M. F. Sisemore, M. Selke, J. Driscoll, and J. S. Valentine, *J. Am. Chem. Soc., 120*, 5331–5332 (1998).
73. Y. Goto, S. Wada, I. Morishima, and Y. Watanabe, *J. Inorg. Biochem., 69*, 241–247 (1998).
74. P. A. Cole and C. H. Robinson, *J. Am. Chem. Soc., 110*, 1284–1285 (1988).
75. S. Graham-Lorence, B. Amarneh, R. E. White, J. A. Peterson, and E. R. Simpson, *Protein Sci., 4*, 1065–1080 (1995).
76. Y. Aoyama, Y. Yoshida, and R. Sato, *J. Biol. Chem., 259*, 1661–1666 (1984).
77. Y. Yoshida and Y. Aoyama, *J. Biol. Chem., 259*, 1655–1660 (1984).
78. Y. Aoyama, Y. Yoshida, Y. Sonoda, and Y. Sato, *J. Biol. Chem., 262*, 1239–1243 (1987).
79. Y. Aoyama, Y. Yoshida, Y. Sonoda, and Y. Sato, *J. Biol. Chem., 264*, 18502–18505 (1989).
80. A. Z. Shyadehi, D. C. Lamb, S. L. Kelly, D. E. Kelly, W.-H. Schunck, J. N. Wright, D. Corina, and M. Akhtar, *J. Biol. Chem., 271*, 12445–12450 (1996).
81. P. J. Ramm and E. Caspi, *J. Biol. Chem., 244*, 6064–6073 (1969).
82. G. F. Gibbons, L. J. Goad, and T. W. Goodwin, *Chem. Commun.*, 1458–1460 (1968).
83. M. Akhtar, A. D. Rahimtula, I. A. Watkinson, D. C. Wilton, and K. A. Munday, *Eur. J. Biochem., 9*, 107–111 (1969).
84. R. T. Fischer, J. M. Trzaskos, R. L. Magolda, S. S. Ko, C. S. Brosz, and B. Larsen, *J. Biol. Chem., 266*, 6124–6132 (1991).
85. E. S. Roberts, A. D. N. Vaz, and M. J. Coon, *Proc. Nat. Acad. Sci. USA, 88*, 8963–8966 (1991).
86. A. D. N. Vaz, E. A. Roberts, and M. J. Coon, *J. Am. Chem. Soc., 113*, 5886–5887 (1991).
87. G. M. Raner, E. W. Chiang, A. D. N. Vaz, and M. J. Coon, *Biochemistry, 36*, 4895–4902 (1997).
88. A. D. N. Vaz, K. J. Kessell, and M. J. Coon, *Biochemistry, 33*, 13651–13661 (1994).
89. A. D. Vaz, S. J. Pernecky, G. M. Raner, and M. J. Coon, *Proc. Nat. Acad. Sci. USA, 93*, 4644–4648 (1996).
90. J. R. Reed, D. Vanderwel, S. Choi, G. Pomonis, R. C. Reitz, and G. J. Blomquist, *Proc. Nat. Acad. Sci. USA, 91*, 10000–10004 (1994).
91. J. R. Reed, D. R. Quilici, G. J. Blomquist, and R. C. Reitz, *Biochemistry, 34*, 16221–16227 (1995).
92. W. Adam, R. Curci, M. E. Gonzalez Nunez, and R. Mello, *J. Am. Chem. Soc., 113*, 7654–7658 (1991).
93. S. Mpuru, J. R. Reed, R. C. Reitz, and G. J. Blomquist, *Insect Biochem. Mol. Biol., 26*, 203–208 (1996).

94. V. Stanjek, M. Miksch, P. Lueer, U. Matern, and W. Boland, *Angew. Chem. Int. Ed.*, *38*, 400–402 (1999).
95. W. Boland, *Pure Appl. Chem.*, *65*, 1133–1142 (1993).
96. U. Matern, *Rec. Adv. Phytochem.*, *33*, 161–183 (1999).
97. D. Spiteller, A. Jux, J. Piel, and W. Boland, *Phytochem.*, *61*, 827–834 (2002).
98. H. Yamamoto, N. Katano, A. Ooi, and K. Inoue, *Phytochem.*, *53*, 7–12 (2000).
99. S. Irmler, G. Schroder, B. St-Pierre, N. P. Crouch, M. Hotze, J. Schmidt, D. Strack, U. Matern, and J. Schroder, *Plant J.*, *24*, 797–804 (2000).
100. D. W. Udwary, L. K. Casillas, and C. A. Townsend, *J. Am. Chem. Soc.*, *124*, 5294–5303 (2002).
101. R. Prieto and C. P. Woloshuk, *Appl. Environ. Microbiol.*, *63*, 1661–1666 (1997).
102. C. A. Helliwell, W. J. Peacock, and E. S. Dennis, *Methods Enzymol.*, *357*, 381–388 (2002).
103. R. C. Coolbaugh, *Proc. Plant Growth Reg. Soc.*, *24*, 10–14 (1997).
104. C. A. Helliwell, P. M. Chandler, A. Poole, E. S. Dennis, and W. J. Peacock, *Proc. Nat. Acad. Sci. USA*, *98*, 2065–2070 (2001).
105. M. C. Rojas, P. Hedden, P. Gaskin, and B. Tudzynki, *Proc. Nat. Acad. Sci. USA*, *98*, 5838–5843 (2001).
106. H. Fukuda, T. Fujii, E. Sukita, M. Tazaki, S. Nagahama, and T. Ogawa, *Biochem. Biophys. Res. Commun.*, *201*, 516–522 (1994).
107. H. Fukuda, K. Nakamura, E. Sukita, T. Ogawa, and T. Fujii, *J. Biochem.*, *119*, 314–318 (1996).
108. C. Shimaya and T. Fujii, *J. Biosci. Bioeng.*, *90*, 465–467 (2000).
109. M. Komuro, T. Higuchi, and M. Hirobe, *Bioorg. Med. Chem.*, *3*, 55–65 (1995).
110. R. E. White, S. G. Sligar, and M. J. Coon, *J. Biol. Chem.*, *255*, 11108–11111 (1980).
111. T. Hakamatsuka, M. F. Hashim, Y. Ebizuka, and U. Sankawa, *Tetrahedron*, *47*, 5969–5978 (1991).
112. M. F. Hashim, T. Hakamatsuka, Y. Ebizuka, and U. Sankawa, *FEBS Lett.*, *271*, 219–222 (1990).
113. T. Akashi, T. Aoki, and S.-I. Ayabe, *Plant Physiol.*, *121*, 821–828 (1999).
114. Y. Sawada, K. Kinoshita, T. Akashi, T. Aoki, and S.-i. Ayabe, *Plant J.*, *31*, 555–564 (2002).
115. W. Yin, G. A. Doss, R. A. Stearns, A. G. Chaudhary, C. E. Hop, R. B. Franklin, and S. Kumar, *Drug Metab. Disp.*, *31*, 215–223 (2003).
116. J.-L. Boucher, M. Delaforge, and D. Mansuy, *Biochemistry*, *33*, 7811–7818, (1994).
117. O. D. Gupta, R. L. Kirchmeier, and J. N. M. Shreeve, *J. Am. Chem. Soc.*, *112*, 2383–2386 (1990).
118. N. Tijet and A. R. Brash, *Prostaglandins Lipid Med.*, *68–69*, 423–431 (2002).
119. K. Matsui, M. Shibutani, T. Hase, and T. Kajiwara, *FEBS Lett.*, *394*, 21–24 (1996).
120. A. N. Grechkin, *Prostaglandins Lipid Med.*, *68–69*, 457–470 (2002).
121. M. Stumpe, R. Kandzia, C. Goebel, S. Rosahl, and I. Feussner, *FEBS Lett.*, *507*, 371–376, (2001).
122. N. Tijet, C. Schneider, B. L. Muller, and A. R. Brash, *Arch. Biochem. Biophys.*, *386*, 281–289 (2001).
123. W.-C. Song, C. D. Funk, and A. R. Brash, *Proc. Nat. Acad. Sci. USA*, *90*, 8519–8523 (1993).

124. E. Psylinakis, E. M. Davoras, N. Ioannidis, M. Trikeriotis, V. Petrouleas, and D. F. Ghanotakis, *Biochim. Biophys. Acta, 1533*, 119–127 (2001).
125. V. Ullrich and R. Brugger, *Angew. Chem. Int. Ed., 33*, 1911–1919 (1994).
126. T. Tanabe and V. Ullrich, *J. Lipid Mediat. Cell Signal., 12*, 243–255 (1995).
127. A. N. Grechkin, L. S. Mukhtarova, and M. Hamberg, *FEBS Lett., 549*, 31–34 (2003).
128. V. Ullrich, L. Castle, and P. Weber, *Biochem. Pharm., 30*, 2033–2036 (1981).
129. M. Haurand and V. Ullrich, *J. Biol. Chem., 260*, 15059–15067 (1985).
130. A. Miyata, S. Hara, C. Yokoyama, H. Inoue, V. Ullrich, and T. Tanabe, *Biochem. Biophys. Res. Commun., 200*, 1728–1734 (1994).
131. T. Tanabe, C. Yokoyama, A. Miyata, H. Ihara, T. Kosaka, K. Suzuki, Y. Nishikawa, T. Yoshimoto, S. Yamamoto, R. Nusing, and V. Ullrich, *J. Lipid Med. Cell Signal., 6*, 139–144 (1993).
132. L. Zhang, M. B. Chase, and R. F. Shen, *Biochem. Biophys. Res. Commun., 194*, 741–748 (1993).
133. L. H. Wang, K. Ohashi, and K. K. Wu, *Biochem. Biophys. Res. Commun., 177*, 286–291 (1991).
134. M. L. Ogletree, *Fed. Proc., 46*, 133–138 (1987).
135. S. Moncada and J. R. Vane, *Pharmacol. Rev., 30*, 293–331 (1978).
136. M. Hecker, M. Haurand, V. Ullrich, U. Diczfalusy, and S. Hammarstroem, *Arch. Biochem. Biophys., 254*, 124–135 (1987).
137. J. A. Salmon and R. J. Flower, *Methods Enzymol., 86*, 93–99 (1982).
138. V. Ullrich, *Arch. Biochem. Biophys., 409*, 45–51 (2003).
139. L.-H. Wang and R. J. Kulmacz, *Prostaglandins Lipid Med., 68–69*, 409–422 (2002).
140. H. John, K. Cammann, and W. Schlegel, *Prostaglandins Lipid Med., 56*, 53 (1998).
141. M. Hecker and V. Ullrich, *J. Biol. Chem., 264*, 141–150 (1989).
142. J. P. Plastaras, F. P. Guengerich, D. W. Nebert, and L. J. Marnett, *J. Biol. Chem., 275*, 11784–11790 (2000).
143. B. Tudzynski, M. C. Rojas, P. Gaskin, and P. Hedden, *J. Biol. Chem., 277*, 21246–21253 (2002).

14

Design and Engineering of Cytochrome P450 Systems

Stephen G. Bell, Nicola Hoskins, Christopher J. C. Whitehouse, and Luet L. Wong

Department of Chemistry, Inorganic Chemistry Laboratory,
Oxford University, South Parks Road, Oxford, OX1 3QR, UK
<stephen.bell@chem.ox.ac.uk>
<luet.wong@chem.ox.ac.uk>

1. INTRODUCTION	438
1.1. The Activity of P450 Enzymes	438
1.2. The Role of the Substrate	439
1.3. Engineering P450 Enzymes	441
1.4. The Structure of P450 Enzymes	441
2. ENGINEERING BACTERIAL CYTOCHROME P450 SYSTEMS	444
2.1. Site-Directed Mutagenesis	444
2.1.1. Site-Directed Mutagenesis of *P450*cam (CYP101A1)	444
2.1.2. Site-Directed Mutagenesis of $P450_{BM3}$ (CYP101A2)	452
2.1.3. Other Bacterial P450 Enzymes	456
2.2. Directed Evolution	457
2.2.1. Approaches for the Directed Evolution of Enzymes	457
2.2.2. Directed Evolution of *P*450cam (CYP101A1)	458
2.2.3. Directed Evolution of $P450_{BM3}$ (CYP101A2)	458
3. ENGINEERING MAMMALIAN CYTOCHROME P450 ENZYMES	462
3.1. Rational Engineering	463

    3.2. Random Mutagenesis 465
        3.2.1. CYP1A2 465
        3.2.2. CYP2A6 466
        3.2.3. CYP2B1 466
        3.2.4. CYP2C Family 467
        3.2.5. CYP3A4 467
4. ENGINEERING PLANT P450 ENZYMES 467
5. CONCLUSIONS AND OUTLOOK 468
    ABBREVIATIONS 469
    REFERENCES 469

1. INTRODUCTION

Cytochrome P450 enzymes possess the unique catalytic activity of carbon-hydrogen bond oxidation using atmospheric dioxygen as the oxidant. This monooxygenase reactivity has no equivalent in synthetic methodologies where high oxidation state heavy metal compounds (e.g. RuO_2, $Pb(OAc)_4$, chromic acid), high energy compounds (e.g. H_2O_2, organic hydroperoxides) together with a heavy metal catalyst, or heterogeneous metal oxide catalysts that require high temperatures and often high pressures, are used. With concerns for the environmental impact of human activities growing and energy costs rising, the application of enzymes in synthesis is increasingly coming into focus. The P450 monooxygenase reaction has potential applications in areas such as the manufacture of synthetic intermediates, novel fine chemicals and pharmaceuticals, benign synthesis and bioremediation (see also Chapter 15 of this volume). However, whilst some P450 enzymes, especially those involved in xenobiotic oxidation, oxidize a wide range of molecules, most P450s are substrate specific and show high product selectivity (see also Chapter 1 of this volume). In order to apply these enzymes in synthesis, the substrate range and selectivity have to be expanded. One method is to screen many different organisms for new activity patterns. The other is to engineer the necessary specificity into existing enzymes – the subject of this chapter.

1.1. The Activity of P450 Enzymes

The P450 reaction involves the activation of dioxygen with two electrons and two protons, leading to the formation of a highly reactive intermediate that attacks the substrate.

$$RH + O_2 + 2e^- + 2H^+ \rightarrow R-OH + H_2O$$

Mechanistically the dioxygen activation reaction can be considered as the heterolytic cleavage of the O=O bond, forming one molecule of water and an oxygen atom which is bound to the heme iron. Formally the oxidation state of the iron increases by two, from ferric in the resting state of the enzyme to Fe(V) in the active intermediate, but a more plausible structure is an oxyferryl Fe^{IV} porphyrin cation (compound I) species:

$$O_2 + 2e^- + 2H^+ \rightarrow H_2O + "O"$$
$$(Por)Fe^{III} + "O" \rightarrow [(Por)Fe^V=O \rightarrow (Por)^+Fe^{IV}=O]$$

Compound I will attack almost all functional groups (except C–F and C–Cl bonds) located close to it. In addition to C–H bond oxidation, P450 enzymes therefore show other activities such as olefin epoxidation and heteroatom oxidation (see also Chapter 11 of this volume) [1,2].

Complications in P450 catalysis arise from the non-complementarity of the electron transfer and chemical steps. Physiologically the two electrons are typically derived from NAD(P)H but these cannot be transferred directly to the heme prosthetic group which can only accept one electron at a time (cf flavin-dependent monooxygenases) [3–5]. Instead they are initially donated in the form of hydride, typically to a flavin-dependent electron transfer reductase protein, and then transferred one at a time to the heme via another prosthetic group in the same reductase protein or a separate electron transfer protein (see also Chapter 6 of this volume). The first electron generates the ferrous heme which binds dioxygen. Delivery of the second electron gives an intermediate that can be considered as a peroxide bound to the ferric iron in an end-on fashion, i.e., both electrons from NAD(P)H are transferred to the dioxygen. The bound peroxo group is rapidly protonated on the distal oxygen to give the hydroperoxo species. A second protonation at the distal oxygen generates the incipient water leaving-group and commits the enzyme to compound I formation:

$$(Por)Fe^{III} \rightarrow (Por)Fe^{II} \rightarrow [(Por)Fe^{II}(O_2) \rightarrow (Por)Fe^{III}(O_2^-)]$$
$$(Por)Fe^{III}-O-O^- \rightarrow (Por)Fe^{III}-O-OH \rightarrow (Por)Fe^{III}-O-{}^+OH_2$$
$$\rightarrow (Por)Fe^V=O + H_2O$$

1.2. The Role of the Substrate

Once formed, the highly reactive compound I species should in principle attack the substrate to form the product. Selectivity is determined by substrate binding. If the substrate is mobile then the most activated functional group in the molecule will be selectively attacked. Hence C–H bond oxidation rates will decrease in the order benzylic > allylic > tertiary > secondary > primary, and olefin epoxidation will also compete. A less activated C–H bond will be selectively attacked only if it is significantly closer to the heme iron.

If the substrate is not bound close to the heme iron, the reaction may be slowed down sufficiently for the electron transfer proteins to reduce the highly oxidizing Compound I species by two electrons, returning the enzyme to the ferric resting state (see also Chapter 7 of this volume). This side reaction, in which the P450 enzyme is acting as an oxidase, causes uncoupling of the NAD(P)H consumed from product formation. In some cases oxidase activity dominates and little to no product is formed even though fast NAD(P)H turnover is observed. The other uncoupling pathway is the dissociation of hydroperoxide from the iron centre before it is protonated on the distal oxygen. The two electrons from NAD(P)H are then used to form hydrogen peroxide. This can occur if the proton delivery pathway or the Fe–OOH bond is disturbed by substrate binding, especially if the substrate is too large for the substrate pocket, and if a water molecule can approach the Fe–O bond and protonate the iron-bound oxygen because the substrate is too small to exclude all water molecules from the vicinity of the heme [6–8]. The pK_a of an isolated water molecule is too high to protonate the heme-bound oxygen and an acidic side chain is probably involved.

Substrate binding is also important in that it opens the kinetic 'gate' that minimizes the redox cycling of cellular NAD(P)H with dioxygen to form hydrogen peroxide and water. In most P450 enzymes, the first electron transfer to form the ferrous heme is slow (k_{et} 1–10 min^{-1}) in the absence of substrate but can be accelerated by orders of magnitude when a substrate is bound. By Marcus theory the activation energy of electron transfer must be reduced when the substrate becomes bound [9].

$$k_{et} \propto H_{AB}^2 \exp\left[-\frac{(\Delta G^o + \lambda)^2}{4\lambda k_B T}\right]$$

Without a bound substrate the ferric heme is typically hexa-coordinate with water as the sixth ligand. The reduction potential is low (hence ΔG^o is less favorable) and the reorganization energy (λ) for the electron transfer is high because the Fe–OH$_2$ bond has to be broken to form the penta-coordinate ferrous heme. If, in binding, the substrate displaces all waters from the substrate pocket, the ferric heme becomes penta-coordinate, the reduction potential is increased and the reorganization energy lowered, leading to fast electron transfer [10]. The absolute rate of electron transfer and the steady state rate of the catalytic reaction depend on the driving force (ΔG^o) and the electronic coupling (H_{AB}) between the electron transfer protein donor and the ferric heme acceptor.

The maximal catalytic turnover rates of different P450 enzymes depend on the reduction potential of the electron transfer protein, the delocalized nature of the prosthetic group and the distance of closest approach of this group to the heme. For any given P450 enzyme, the NAD(P)H oxidation rate may vary by substrate because some compounds are not bound sufficiently close to the heme to expel the water ligand, meaning electron transfer is slower. However, the catalytic turnover rate can still be high for certain substrates even if a water molecule remains in situ - provided that the Fe–OH$_2$ interaction is weakened, making both the thermodynamic driving force and the kinetic reorganization energy barrier more favorable.

1.3. Engineering P450 Enzymes

In order to prepare a specific target compound from a starting substrate, the P450 enzyme has to be modified so that the substrate can enter the active site and bind close to the heme, displacing the water ligand (or at least weakening the Fe–OH$_2$ interaction) and initiating the catalytic cycle. The target product will only be formed with high selectivity if the correct C–H bond or functional group is held closest to the heme iron. The product yield will be low unless water is displaced from the substrate pocket and the substrate is neither mobile nor too far from the oxygen atom of the compound I intermediate. This poses a significant challenge to our chemical understanding of enzyme/substrate recognition even if the starting substrate is structurally related to the enzyme's natural substrate (if this is known). It is therefore not surprising that P450 engineering is an iterative process, be it by active site redesign or random mutagenesis and directed evolution.

1.4. The Structure of P450 Enzymes

Although only a small number of the 3500 or so P450 cytochromes currently known have been structurally characterized, the crystal structures that are available show a highly conserved tertiary structure (see also Chapter 3 of this volume). The enzyme resembles a trigonal prism, typically measuring $\sim 60 \times 60 \times 30$ Å, with the heme plane parallel to one of the triangular faces. Electron transfer proteins are believed to dock on the proximal side of the heme. On the distal side, where the substrate binding pocket is located, large structural variations occur, and it is these that give rise to the huge diversity of substrate specificity exhibited by the P450 superfamily. The distal I-helix, which runs virtually the full length of the protein, contains a pronounced and characteristic kink caused by hydrogen bonding. This kink forms an oxygen binding groove that is part of the proton delivery pathway for O–O bond activation (see also Chapter 8 of this volume).

The recently reported structure of CYP158A2 from *Streptomyces coelicolor* is typical in several regards, but it also displays a number of interesting and unusual features (Figure 1) [11,12]. In the absence of substrate, solvent can penetrate the active site and the binding cleft is filled with water molecules. Such open conformations are rare in P450 enzymes, many of which have enclosed active sites with

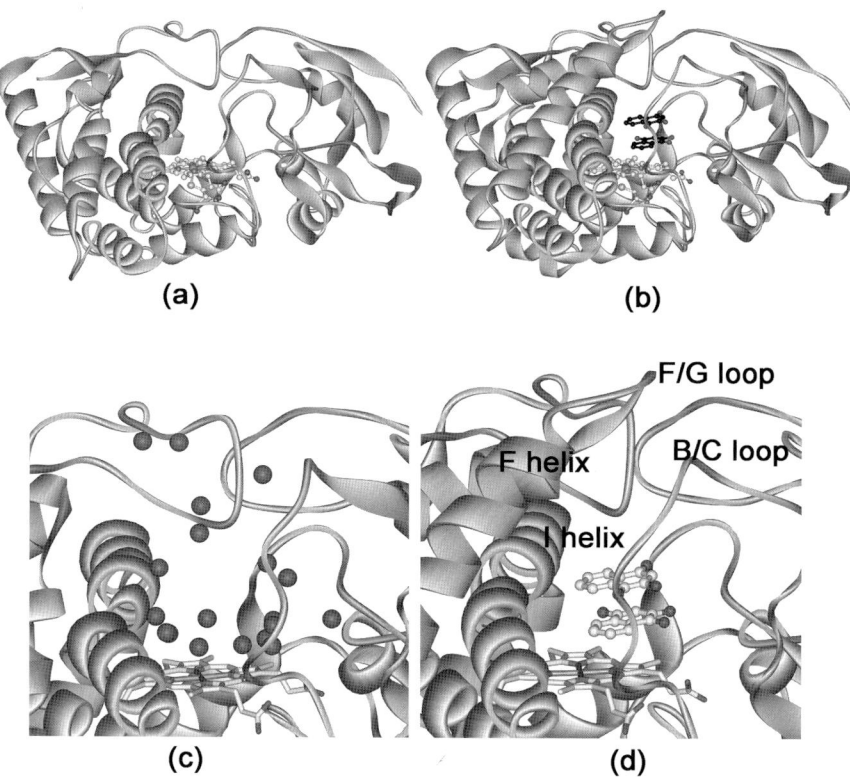

Figure 1. The structure of CYP158A2, as viewed along the distal I helix. Dramatic changes take place when flaviolin binds. In the substrate-free form the active site is exposed to solvent (a), with the F and G helices oriented away from the I helix, and the B/C and F/G loops away from the substrate pocket, which is filled with ordered water molecules (c). When flaviolin binds to the enzyme, the F helix moves towards the I helix and the F/G and B/C loops (b) both close over the active site to enclose the two molecules of flaviolin that are bound within the pocket (d).

no obvious substrate access channel (notable exceptions include CYP2B5 and the fatty acid-oxidizing CYP102A1 from *Bacillus megaterium*). The I-helix kink in CYP158A2 is relatively muted as the enzyme lacks a conserved threonine residue (Thr252 in CYP101A1, Thr268 in CYP102A1, etc). The F and G helices, which are packed against the I-helix in most P450 structures, are rotated out of the active site and away from the β_4 strand, taking with them the B/C and F/G loops, which normally form part of the substrate pocket. The low-spin heme iron is coordinated by a water molecule hydrogen-bonded to two other active site water molecules, one of which forms a hydrogen bond with the carbonyl of Gly242.

The natural substrate of CYP158A2 appears to be flaviolin, which causes dramatic structural rearrangements when it binds. Two molecules occupy the active site, where they are in van der Waals contact, a configuration consistent with dimerisation (the dimer and trimer have been characterized as reaction products). The simultaneous binding of more than one molecule within the active site has been proposed for certain mammalian P450 enzymes, but this is the first P450 structure to show two substrate molecules in an apparently functionally relevant arrangement. The protein fold changes from the open form to a closed form on substrate binding. This involves the movement of around 35% of the amino acid residues – and in particular those in the six substrate-recognition regions (SRS). The F helix moves towards the I helix and into the active site. The G helix is disrupted into a shorter helix and a 3_{10} helix. The F/G loop moves towards the B/C loop to close the active site at the top, bringing the latter into contact with the substrate, and the heme water ligand is displaced, as would be expected from the spin state shift observed when flaviolin binds. In the majority of P450 enzymes, most if not all solvent is displaced from the active site when a substrate binds. Here, however, very unusually, no less than nine ordered water molecules are present, all intimately involved in flaviolin binding. The active site remains fairly open even in this altered form – to the extent that there may be room for a third flaviolin molecule to bind, resulting in trimerization.

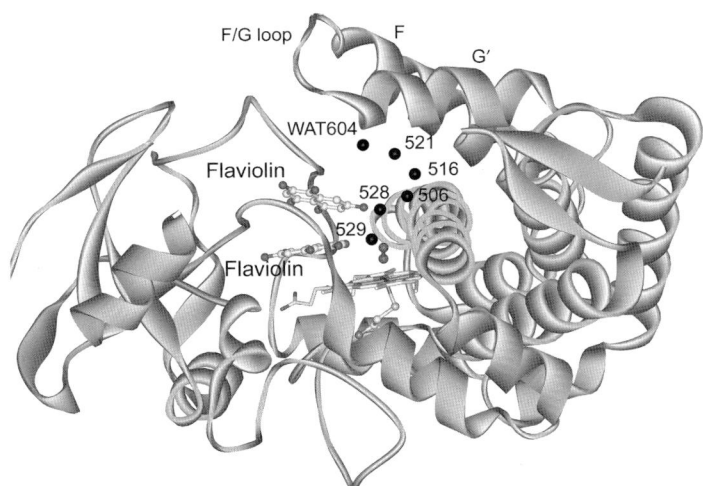

Figure 2. The structure of the ferrous-oxy form of CYP158A2 showing the heme-bound dioxygen molecule and the chain of hydrogen-bonded water molecules that connect the distal oxygen atom to the surface of the enzyme. This hydrogen-bonding network is probably the pathway for delivering the two protons necessary for O–O bond activation. In contrast to CYP101A1 and most other P450 enzymes the side chain of a threonine residue in the I helix is not involved.

In CYP101A1, oxygen binding triggers the enrolment of two extra water molecules to serve as conduits for proton delivery in the O–O bond activation step, while the highly conserved Thr252 serves mainly to stabilize dioxygen binding [13–15]. The structure of the ferrous-oxy flaviolin-bound form of CYP158A2 shows contrastingly little change in the arrangement of the water molecules around the oxygen binding groove. Those present in the ferric resting state of the enzyme are retained and interact with dioxygen when it binds. But protons are delivered by a separate chain of continuous ordered water molecules linking the distal oxygen atom to the external solvent milieu (Figure 2).

2. ENGINEERING BACTERIAL CYTOCHROME P450 SYSTEMS

Two distinct approaches to engineering bacterial systems exist [16–23]: site-directed mutagenesis and directed evolution. Site-directed mutagenesis has been applied extensively to CYP101 and CYP102 systems alike, while the majority of directed evolution work has been carried out on CYP102 systems.

2.1. Site-Directed Mutagenesis

2.1.1. Site-Directed Mutagenesis of P450cam (CYP101A1)

Cytochrome P450cam (CYP101A1) catalyses the stereospecific oxidation of camphor to 5-*exo*-hydroxycamphor in *Pseudomonas putida* – the first step in its metabolism [24]. The protein has been thoroughly studied and was the first P450 enzyme to be structurally characterized [25] – an important consideration in enzyme engineering, given that effective active site redesign depends on the amino acid side chains defining the active site and contacting the substrate being known. Camphor is bound in a specific orientation with C5 directly above the heme iron (Figure 3). There are numerous non-covalent contacts with active site residues and a hydrogen bond between the camphor carbonyl and the phenol side chain of Tyr96 [26].

The first investigation into the effects of active site changes on CYP101A1 activity and substrate specificity was carried out by Atkins and Sligar who introduced the Y96F substitution to remove the hydrogen bond [27]. Camphor binding was slightly weakened and other oxidation products (total ~10%) were observed in addition to 5-*exo*-hydroxycamphor. Hydrogen bonding evidently plays a key role in orientating the substrate correctly within the binding site and determining the regioselectivity of oxidation. Wild-type CYP101A1 was shown to oxidize ethylbenzene at the benzylic C–H bond, with some stereo selectivity [28]. Engineering larger hydrophobic residues into the upper region of

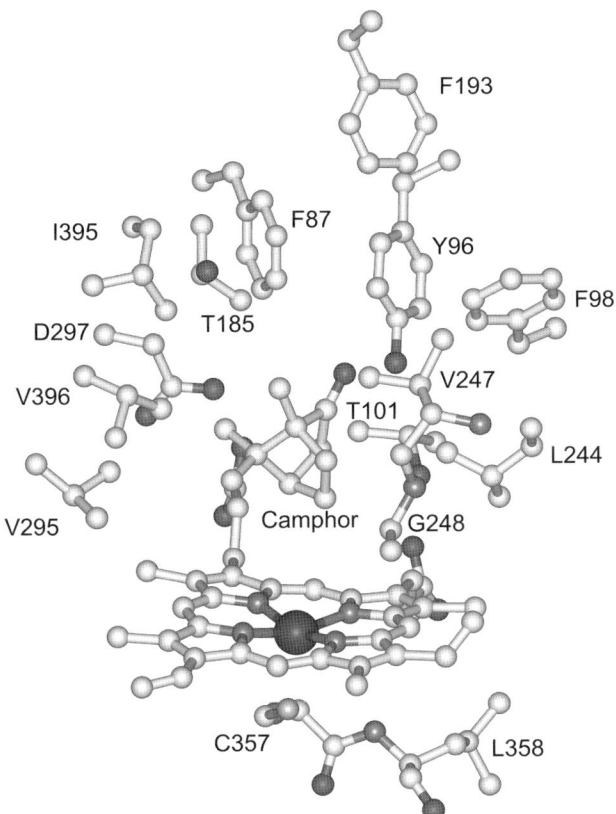

Figure 3. The active site structure of P450cam (CYP101A1) with bound camphor, as viewed from the I-helix threonine residue believed to be involved in oxygen binding and activation. The substrate pocket is capped by Thr185. Phe193 is proposed to be responsible for substrate recognition. The camphor carbonyl group forms a hydrogen bond to Tyr96 that anchors the molecule and defines the C5 regioselectivity of camphor oxidation. Phe98, Tyr96, Phe87 and Ile395 are in close contact and, together with Val247 and Thr185, probably constitute the region of the enzyme that opens up to allow substrate entry. The side chains of Val295 and Val396, which lies directly above it, both contact camphor.

the substrate pocket increased the ratio of hydroxylated product to water production by up to four-fold relative to the wild-type, while comparable substitutions at residues near the porphyrin plane resulted in diminished product yield. Bulky side chains high in the pocket force substrates to bind closer to the heme and retard the oxidase uncoupling pathway. However, when such substitutions are introduced close to the heme they tend to increase uncoupling because substrates must bind further away.

Relatively high NAD(P)H turnover rates can generally be achieved for P450 enzymes engineered to oxidize non-natural substrates, but obtaining tight coupling and high product selectivity presents more of a challenge. The most dramatic improvements have been reported for hydrophobic substrates – specifically alkanes [29], alicyclic compounds, substituted benzenes and polyaromatic hydrocarbons [30]. Mutants with increased activity can often be identified by screening for colored indole oxidation products (vide infra) following site-saturation mutagenesis at Tyr96 and Phe98 [31]. Ortiz de Montellano and co-workers projected the activity of the wild-type enzyme and its L244A mutant towards non-natural substrates using a computer-based substrate screening protocol. When the turnover rates of these substrates were subsequently determined by experiment they showed significant correlation with the predicted data [32–34]. It will be interesting to see if the underlying docking algorithms can be developed to predict oxidation products with similar levels of accuracy.

Other engineering applications include substituting Ala at Cys334 to prevent protein dimerisation via disulfide bond formation [35], and altering the potassium ion binding site by introducing aspartate at Glu84 to promote camphor and cation binding [36]. The oxidation of 2-ethylhexanol to hexanoic acid by CYP101A1 mutants has also been reported [37,38].

2.1.1.1. Alkanes. Engineering CYP101A1 into a more versatile alkane hydroxylase is an area of ongoing interest. The size of the substrate and its shape in relation to the active site are crucial considerations. Hydrophobic substitutions at Tyr96 accelerated linear and branched alkane oxidation compared to the wild-type enzyme. Coupling was higher for the Y96F mutant than for Y96A on account of the latter having a larger substrate pocket which allowed increased substrate mobility [29]. Varying the size of the active site can also bring about specificity changes. The Y96F/V247L mutant showed 4-fold higher activity towards linear hexane than the branched analogue, 3-methylpentane, while the Y96A/V247L mutant, with a larger active site, showed 2-fold higher activity for the more sterically demanding 3-methylpentane [39].

Increases in CYP101A1 activity towards smaller alkanes were achieved by introducing bulky side chains at residues high up in the active site to oblige the substrate to bind closer to the heme iron. The F87W/Y96F/T101L/V247L mutant oxidized butane with a turnover rate of $750\,min^{-1}$ compared with $0.4\,min^{-1}$ for the wild-type enzyme [40]. Scrutinizing the active site for cavities into which small molecules might be able to retreat when forced nearer to the heme identified further target sites for mutagenesis. The L358P mutation first used by Morishima and co-workers [41] was also added. This tightened the active site, pushed the heme towards the distally bound substrate, and eliminated a hydrogen bond between Leu358 and Cys357 (the proximal heme ligand). This in turn increased the σ-donor strength of the thiolate ligand, rendering O–O bond cleavage more facile and reducing uncoupling rates [41,42].

The F87W/Y96F/T101L/T185M/L244M/V247L/G248A/L294M/L358P mutant oxidized propane at 500 min^{-1} with 86% coupling. The NADH turnover rate for ethane was even higher ($\sim$ 800 min^{-1}) but the coupling for ethanol formation was low (10.5%) [43]. With fast and efficient ethane oxidation now within reach, methane oxidation has become the next tantalizing frontier.

2.1.1.2. Aromatic Hydrocarbons. Styrene and naphthalene were oxidized by the Y96F mutant at rates 25 and 140 times those of the wild-type enzyme respectively, with much improved coupling, demonstrating the value of employing hydrophobic residues at Tyr96 when oxidizing aromatic compounds [30,44]. The Y96A mutant gave less dramatic increases as its larger active site pocket allows readier access by water molecules, leading to looser substrate binding and higher oxidase activity. Styrene and ethylbenzene oxidation activity was further increased by incorporating the V247L mutation, especially in combination with Y96F [45]. Sibbesen et al. found increases in the oxidation rates and coupling of ethylbenzene and its derivatives by wild-type CYP101A1 and the mutants T185L and T185F, but there was little correlation between mutation type and activity or product selectivity [46].

The size and planarity of polycyclic aromatic hydrocarbons relative to camphor impinges significantly on coupling efficiencies when they act as substrates in CYP101A1 turnovers. For example, the NADH turnover rates of pyrene when oxidized by mutants Y96A and Y96F were in the region of three times those of the wild-type enzyme, yet 60-fold (Y96A) and 80-fold (Y96F) increases in coupling efficiency resulted in substantial increases in substrate oxidation rates. Substitutions at Phe87 by residues with smaller side chains greatly enhanced the rates of phenanthrene, fluroanthene (Figure 4), pyrene and benzo[a]pyrene oxidation compared to the equivalent substitutions at Tyr96 [47]. One interesting observation was that fast NADH turnover could be observed without an appreciable (<5%) shift of the heme spin state to high spin. It appeared that, although a water molecule remained bound, the Fe–OH$_2$ interaction was insufficiently strong for its cleavage to present a material barrier to electron transfer.

2.1.1.3. Polychlorinated Benzenes. Polychlorinated aromatics constitute hazardous environmental contaminants as many are resistant to biodegradation, the level of recalcitrance generally increasing with the degree of chlorination. Chlorinated phenols, being more reactive, are readily degraded by microorganisms, and novel biodegradation systems can therefore be engineered by genetically augmenting such organisms with P450 enzymes that have the capacity to oxidize polychlorinated benzenes to phenols [48]. As with other compound types, the hydrophobic Y96F mutation has been used to promote substrate binding, while bulky substitutions have been introduced higher up in the active site at Phe87, Phe98, and Val247 to force the planar benzene rings closer to the heme.

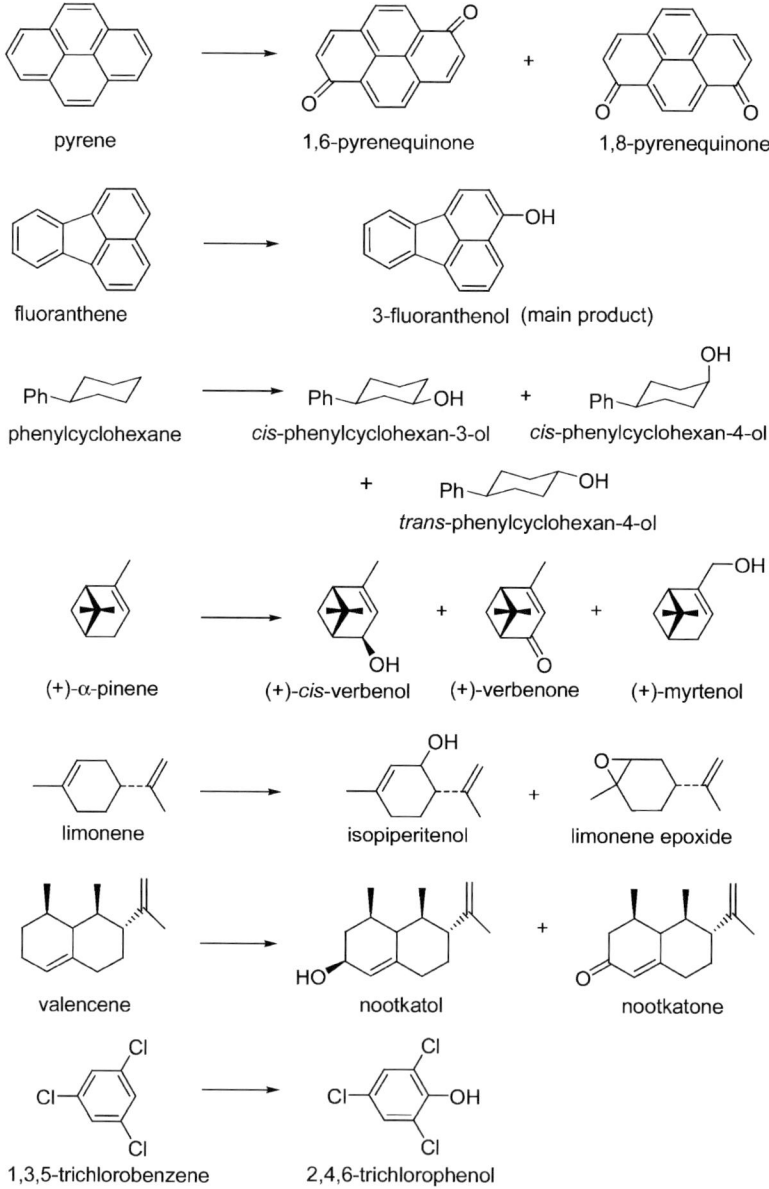

Figure 4. CYP101A1 substrates and oxidation products.

The various mutants showed levels of activity two to three orders of magnitude higher than those of the wild-type enzyme for the oxidation of di-, tri- and tetra-chlorobenzenes, with coupling efficiencies as high as 95%. Oxidation activity and coupling for pentachlorobenzene (PeCB) were low [48–50].

The crystal structure of the F87W/Y96F/V247L mutant complexed with 1,3,5-trichlorobenzene (TCB), once determined, served as a platform for structure-based active site redesign. Multiple substrate binding orientations exist, with the most productive having TCB parallel to, and in van der Waals contact with, the heme (Figure 5) [50]. The F87W and V247L side chains, which are also in van der Waals contact, point towards each other across the top of the active site and act as a lid that closes off the substrate pocket. It appeared that PeCB could not bind in the productive parallel orientation on account of steric interference with Leu244 (Figure 5b). The L244A mutation was therefore introduced to create extra space in the binding pocket. F87W/Y96F/L244A/V247L oxidized PeCB 45 times faster than the parent mutant, while hexachlorobenzene oxidation activity was increased 200-fold. Both compounds are oxidized to pentachlorophenol which is known to be readily biodegradable.

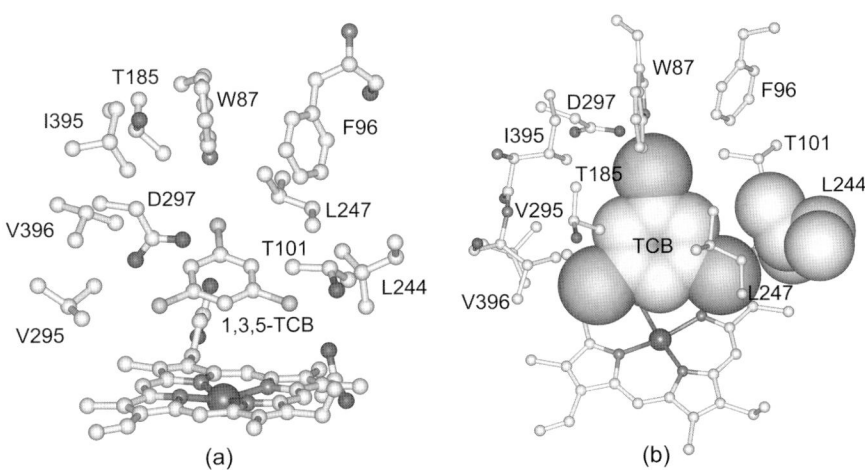

Figure 5. The structure of CYP101A1 with 1,3,5-trichlorobenzene bound. (a) The parallel binding mode in the F87W/Y96F/V247L mutant. There is a hydrogen bond between the indole NH of Trp87 and a substrate chlorine atom. A C–H bond is positioned over the heme iron, indicating that 2,4,6-trichlorophenol should be the product, as is observed. (b) View of the substrate binding mode from the top of the active site showing the close contact between a substrate C–H group and the Leu244 side chain that precludes the binding of more highly chlorinated compounds such as pentachlorobenzene in the parallel binding mode. The Leu244A mutation enhances pentachlorobenzene oxidation activity.

2.1.1.4. Alicyclics.

CYP101A1 was engineered to oxidize phenylcyclohexane selectively at positions C2, C3 or C4 of the cyclohexane ring (chemical systems, by contrast, would be expected to oxidize the benzylic position) [51–53]. The Y96F mutant gave 81% cis-3-phenylcyclohexanol (34% ee) and the Y96F/V247A mutant 97% (42% ee), while the Y96F/V247L mutant gave 83% trans-4-phenylcyclohexanol.

Terpenoid compounds constitute one of the largest groups of biological molecules. Terpenes, the parent hydrocarbons, make interesting substrates for monooxidation studies because the resulting products include fragrance and flavoring compounds, insect pheromones, and precursors to pharmaceuticals. (+)-α-Pinene is structurally related to camphor, S-limonene less so, though the carbonyl group that forms a hydrogen bond with Tyr96 in camphor is absent from both compounds. As before, the Y96F mutation strengthened substrate binding and increased oxidation rates and coupling [54]. Pinene was largely oxidized to verbenol and the pinene oxides while limonene gave mainly isopiperitenol and the limonene oxides. Mutations that increased the active site volume gave higher proportions of the oxides, indicating that substrate mobility increased when the substrate pocket was enlarged, causing the reactive olefinic bond to be preferentially attacked.

Trp was introduced at Phe87 and Leu at Val247 as a means of reducing substrate lability. Despite weakening substrate binding, both mutations substantially increased pinene oxidation rates and improved coupling efficiency while enhancing the selectivity for verbenol formation (70% for the F87W/Y96F/V247L mutant). The S-limonene oxidation rate of the F87W/Y96F/V247L mutant was lower than those for the F87W/Y96F and Y96F/V247L mutants but the triple mutant showed substantially higher selectivity($\sim$ 90%) for isopiperitenol formation. The crystal structure of this triple mutant complexed with (+)-α-pinene revealed two substrate binding orientations (Figure 6). In one orientation the C3 atom was positioned over the heme iron (Figure 6a), consistent with verbenol formation. In the other the heme iron was closest to the allylic C10 methyl group, a configuration that should give myrtenol (Figure 6b). Since no myrtenol was detected in practice, it was proposed that the two orientations interconvert, leading predominantly to products formed from oxidation of the allylic CH_2 group at C3, which is significantly more reactive than the methyl group at C10. It was also apparent that the side chains of residues Trp87 and Leu247 were in competition: Leu247 contacted the C10 methyl and pulled it away from Trp87, favoring the orientation expected to give myrtenol. Analysis of enzyme/substrate interactions and dynamics led to introduction of the L244A mutation, which increased the selectivity for (+)-cis-verbenol to 86% for the F87W/Y96F/L244A mutant [55].

(+)-valencene, a sesquiterpene found in orange zest, is the likely biological precursor of (+)-nootkatone, its C2 ketone derivative and a prized fragrance found in grapefruit juice. Valencene is larger and more hydrophobic than

DESIGN AND ENGINEERING OF CYTOCHROME P450 SYSTEMS

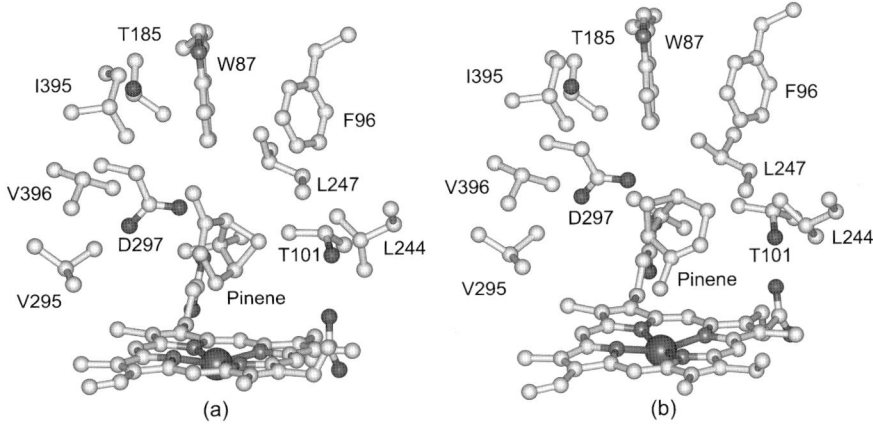

Figure 6. The two binding modes of (+)-α-pinene in the F87W/Y96F/V247L mutant of CYP101A1. Orientation (a) resembles that of camphor binding in the wild-type enzyme, but only one geminal methyl group contacts the Thr101 side chain whereas both methyls contact Val295 in the wild-type. The allylic C10 methyl is closer to Leu247 than Trp87. This configuration suggests that (+)-cis-verbenol will be formed, as is observed in practice. Orientation (b), which can be generated from (a) by rotation of the substrate, places the C10 methyl closest to the heme iron, suggesting that (+)-myrtenol will be formed, but this product is not observed in practice. Rapid interconversion between the two orientations leads to the more reactive allylic CH_2 group on C3 being preferentially attacked.

camphor, and predictably neither wild-type CYP101A1 nor the Y96F mutant is active towards it [56]. The Y96A and F87A/Y96F mutations were introduced to free up space in which to accommodate the isopropenyl group of valencene. Ideally this would place C2 closest to the heme iron and in the correct orientation for oxidation to nootkatol and thence to nootkatone. The Y96A mutant did not oxidize valencene but the F87A/Y96F mutant yielded a mixture of nootkatol, nootkatone and further oxidation products.

Additional substitutions were introduced with a view to improving selectivity. The V247L mutation retarded the further oxidation of nootkatol while the L244A mutation had the opposite effect, promoting the further oxidation both of nootkatol and nootkatone. The effect of substitutions at Phe87 was more subtle, the F87V mutation favoring the oxidation of nootkatol but not of nootkatone while the F87A and F87L mutations both retarded the oxidation of nootkatol to nootkatone. In combination studies it was found that the F87A/Y96F/L244A/V247L mutant gave 86% (+)-*trans*-nootkatol, 4% nootkatone, and 10% of products from nootkatone oxidation (mainly 9-hydroxynootkatone) while the F87V/Y96F/L244A mutant gave 38% nootkatol and 47% nootkatone.

2.1.2. Site-Directed Mutagenesis of $P450_{BM3}$ (CYP101A2)

The crystal structure of the heme domain of CYP102A1 from *B. megaterium* was published in 1993 [57], prompting mutagenesis studies of residues of obvious strategic importance, both close to the active site and along the hydrophobic substrate access channel connecting it to the protein surface (Figure 7). One of the earliest involved Thr268, a conserved residue that together with Glu267 is part of the kink in the I helix where dioxygen binding is stabilized [58]. The T268A and E267Q mutations both caused severe uncoupling in substrate turnovers [59]. It was suggested that the water molecule hydrogen-bonded to these two residues could be part of a proton delivery pathway. A later investigation into the behaviour of T268A towards a range of substrates implied that the threonine side chain also played a role in substrate orientation [60].

Phe87 is a crucial residue for substrate binding in CYP102A1 (Figure 7). Its aromatic side chain is known to flip from a horizontal configuration to a perpendicular one when palmitic acid (one of the enzyme's natural substrates)

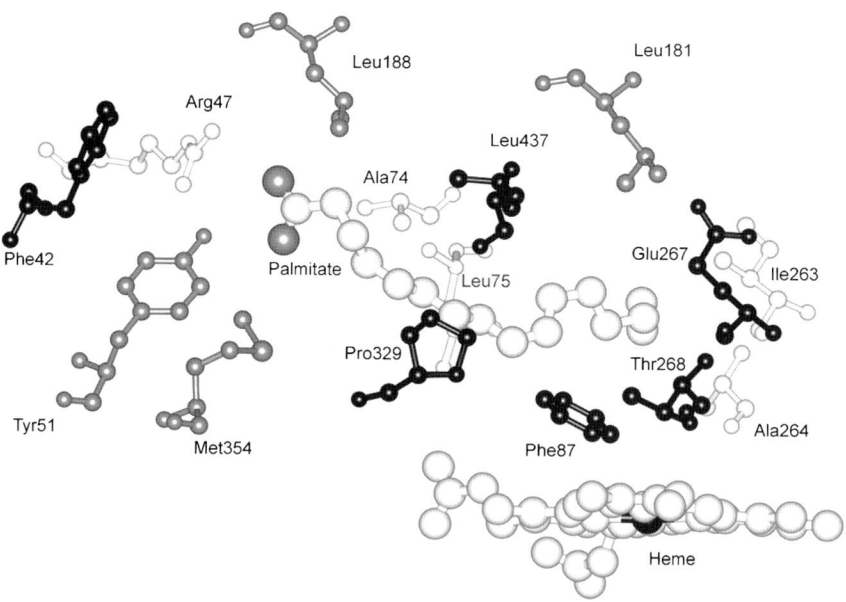

Figure 7. The active site and substrate channel structure of CYP102A1. The aromatic ring of Phe87 is positioned directly between the heme iron and the natural substrate, palmitate, which is oxidized at or around the ω-2 position. Arg47 and Tyr51 are located opposite one another in the mouth of the access channel, where they play a crucial role in attracting and binding substrate. Phe42 is thought to 'cap' the channel once substrate is bound. Glu267 and Thr268 jointly bind a water molecule and assist with dioxygen binding.

binds [57,61], and clearly has a profound influence on substrate access. The F87V mutant was found to have different reaction profiles to those of the wild-type, forming epoxides rather than hydroxylation products, while the F87Y mutant proved inactive [62]. The F87A mutation significantly increased the rate of shorter-chain fatty acid oxidation, with hydroxylation apparently taking place largely at the ω position [63], an outcome common in the CYP4A1 family but unknown at that time for CYP102A1 which typically hydroxylates at around the ω-2 position. However, this finding has been questioned [64]. The F87Y and F87G mutants showed some activity towards shorter-chain substrates, though less than the wild-type enzyme [65].

As it became clear that CYP102A1 could be adapted to oxidize non-natural substrates, mutations were routinely carried out at Phe87 (principally F87A, F87G, and F87V) as a means of improving activity and opening up alternative regiochemical pathways. Targets included polycyclic aromatic hydrocarbons [66], propylbenzene and 3-chlorostyrene [67], (+)-valencene [56], indole [68], phenols [69], β-ionone [70] and the epoxidation of terminal alkenes (Figure 8) [71]. The F87V mutation was used in the recent development of a fluorescence assay for CYP102A1, along with L188Q (first reported by Schmid and co-workers [105], see Section 2.2.3) and R47L [72].

Arg47 is positioned close to the opening of the substrate access channel opposite Tyr51. Together, these two residues are thought to stabilize the binding of fatty acid substrates by tethering their carboxylate groups. Early work showed that enzymic activity could be nullified by substituting Glu for Arg at residue 47 to reverse the polarity of the side chain, while the R47A mutation reduced fatty acid oxidation activity but accelerated the epoxidation reaction [62]. The process was taken a stage further by testing the mutant R47E on C12-C16 trimethylammonium compounds, whose ionic character diametrically opposed that of the enzyme's natural substrates [73]. Not only did the mutant hydroxylate these cationic compounds efficiently. It also displayed activity towards fatty acids with shorter chain lengths than hitherto investigated, lending credence to the idea that Tyr51 also played a role in substrate anchoring. Subsequent reports suggested that Arg47 made a more significant contribution to binding than Tyr51 [65], but that the consequences of introducing Ala at Tyr51 were generally more pronounced than those of introducing it at Arg47, and that it was when acting in concert that the two residues were at their most influential [64].

An effective catalyst for shorter-chain fatty acid oxidation was designed by installing an R47/Y51-like duo further down the substrate channel [74]. Various mutations of channel residues Leu75, Leu181, Ile263, and Leu437 were tested, with L75T/L181K proving the most efficacious. At the same time, the R47A/Y51F mutant was observed to have lower activity than the wild-type enzyme towards fatty acids irrespective of chain length, implying that the carboxylate binding site also played a part in attracting substrate into the channel.

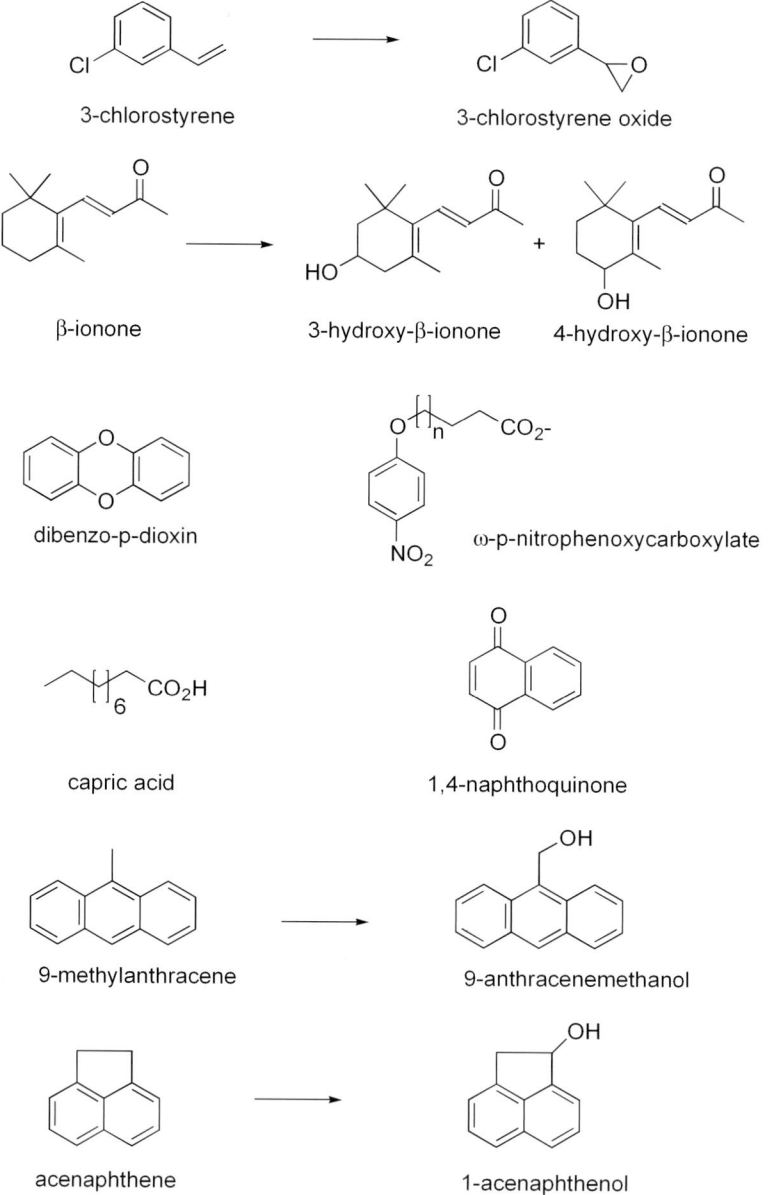

Figure 8. CYP102A1 substrates, oxidation products and assays.

Another residue believed to participate in substrate enticement and capture is Phe42, which appears to close across the mouth of the channel in the crystal structure of the substrate-bound enzyme [57,61]. The k_{cat}/K_m ratios for the F42A mutant proved to be significantly lower than those for the wild-type, partly because solvent water was excluded from the carboxylate binding site when the hydrophobic lid was in place, and partly because Phe42 formed part of a hydrophobic patch on the surface of the enzyme where substrate was thought to dock [65].

It became common in due course for Arg47 and/or Tyr51 to be disenabled when engineering CYP102A1 to oxidize non-natural substrates, with R47L/Y51F often being the variant of choice [66,70]. The rates of polyaromatic hydrocarbon oxidation were enhanced by replacing Ala264 with Gly to free up space near the heme and facilitate substrate access, though it emerged that the effects of downsizing Ala264 and Phe87 were not always additive. Changes to a number of product mix profiles were achieved by using the I263A mutation, but it worked poorly in combination with R47L/Y51F [56,75].

CYP102A1 is unusual among prokaryotic P450 cytochromes in that its heme domain is coded onto the same polypeptide chain as the reductase domain responsible for delivering electrons from NADPH via FMN and FAD. The exceptional catalytic efficiency conferred by the high inter-domain encounter frequency that can be achieved as a result of this configuration has attracted much attention. Site-directed mutagenesis has been extensively used to probe inter-domain electron transfer pathways. Three of the earliest CYP102A1 point mutations were in the reductase domain [76]. In the heme domain, Phe96, a highly conserved residue, was mutated in the hope of substantiating its implied function as an electron transfer mediator, but it was concluded that its role was instead concerned with heme binding [77]. Introducing cysteine at Leu104 and attaching a bulky substituent to the modified residue led to a dramatic drop-off in electron transfer rates, confirming that FMN docks on the proximal side of the heme [78].

The highly conserved Phe393 did not affect electron transfer or heme-stability, but rather played a part in the activation of oxygen, possibly controlling the equilibrium between the rate of heme reduction and the oxygen binding rate [79]. Of particular interest was the observation that the stability of the oxy-ferrous complexes of these mutants increased with reduction potential, suggesting that the second FMN-to-heme electron transfer might be becoming rate-limiting [80]. On a related theme, CYP102A1 was altered at Ala264 to mimic an equivalent glutamate residue that is highly conserved in eukaryotic CYP4 systems where it esterifies autocatalytically with the heme macrocycle [81]. Rather than showing the same behaviour, the introduced glutamate displayed the capacity to coordinate directly to the heme iron in place of solvent water, particularly at high substrate concentrations, without seriously compromising enzymic activity. According to the crystal structure, the substrate-free A264E mutant appeared to adopt the conformation characteristic of the substrate-bound wild-type enzyme, but with iron in the low-spin state [82]. The A264H mutant was subsequently used to incapacitate the heme domain as part of an experiment demonstrating that the tendency of CYP102A1 to lose activity at low concentrations was due to the separation of catalytically competent dimeric units into inactive monomers [83]. Activity was recovered by mixing A264H and an FMN-depleted mutant, demonstrating that inter-monomer (FMN-to-heme) electron transfer could support oxygenase activity in the dimer.

2.1.3. Other Bacterial P450 Enzymes

P450eryF (CYP107A1), like CYP158A2 from *Streptomyces coelicolor (vide supra)*, lacks the conserved I helix threonine residue, rendering it unable to oxidize compounds other than the natural substrate, deoxyerythronolide. However, Xiang, Tschirret-Guth, and Ortiz de Montellano showed that by replacing Ala245 with threonine the oxidation of non-natural substrates (in this study testosterone) was possible [84]. Building on this work, Khan and coworkers went on to show that further mutations in the Ser171, Ile174, Leu175, and Gly91 positions affected both the binding and the oxidation rate of testosterone and 7-benzyloxyquinoline (7-BQ). In particular, the G91A/A245T mutant showed a 4-fold higher k_{cat} value with 7-BQ than the parent A245T mutant [85]

Mutations on the surface and at the active site of CYP119 from the thermophilic *Sulfolobus solfataricus* have been shown to increase activity when deployed in conjunction with electron transfer proteins from the CYP101A1 system [86]. Mutagenesis has also been used to investigate the stability of this enzyme [87,88].

In summary, the compact binding pocket of CYP101A1 limits the number of possible substrate orientations, meaning that altering the size and/or hydrophobicity of active site residues often has a relatively predictable outcome. Tyr96 is almost invariably mutated to improve activity when non-natural substrates are oxidized, while other residues have been targeted on a case by case basis. Rational mutagenesis in CYP102A1 has followed a somewhat different trajectory, partly because the binding mode of non-natural substrates is less well understood and partly because there is a substrate access channel with which to work as well as an active site. Although engineering key residues such as Phe87, Arg47, and Tyr51 has typically brought results that fulfil expectations, the outcomes of combining mutations have proved less straightforward to forecast. Similarly, whilst targeting highly conserved residues for mutation has in general been a fruitful exercise, the insights into the workings of the enzyme that have emerged have not always been in line with anticipation.

2.2. Directed Evolution

2.2.1. Approaches for the Directed Evolution of Enzymes

2.2.1.1. Random or Semi-Random Methods. Approaches to directed evolution have been reviewed [89–91]. Random mutations can be introduced by error-prone PCR [92], saturation mutagenesis [93] or random priming [94]. The latter two methods can be used to target specific regions of the enzyme while error-prone PCR randomly generates nucleotide changes throughout the entire sequence. Because purely random mutations may cause the protein to become unstable or fail to fold properly rather than changing the substrate binding or catalytic properties, only a few mutations are typically introduced in each round of mutagenesis. Screening methods need to be sensitive as the improvements in activity that result may be relatively small. The evolution of the enzyme is linear and many rounds of mutation and selection may be required.

2.2.1.2. Recombination Methods. Recombining structurally similar proteins with high sequence identity can lead to significant improvements in activity without the need for detailed structural information. Several methods have been developed to recombine genes.

DNA family shuffling [95–97]. The parental DNA is subjected to fragmentation before being reassembled at random in a primerless PCR step. The recombined fragments are then amplified to full-length sequences which are screened for improvements. As the parental genes usually differ widely, new mutants with a significant number of mutations commonly emerge in each round of shuffling. The number of recombinations is high, resulting in a diverse shuffled library, but important structural elements in the parent proteins are often conserved. The mutations introduced will usually have been present in at least one of the parents, which minimizes the number of destabilizing mutations compared to random mutagenesis.

The parent proteins are required to have significant sequence identity ($\geq 70\%$). The libraries generated are typically very large so further screening is required to detect mutants with significantly improved activity.

Staggered extension process (StEP) [98]: This technique requires no fragmentation. The PCR amplification of sequences uses very short extension times, leading to switching of primers between templates and hence recombination between multiple sequences. Like the family shuffling method, the two parents must have significant sequence identity.

Combinatorial libraries enhanced by recombination in yeast (CLERY) [99]: This combines conventional DNA shuffling with *in vivo* recombination in yeast. The *in vivo* recombination improves the shuffling and reduces the proportion of parental DNA in the library. The protein expression in yeast may be low compared to other organisms.

Sequence homology independent protein recombination (SHIPREC) [100]: A technique developed to allow crossover between distantly related genes. A dimer of the two parent genes is digested to form random sized fragments. Fragments with lengths corresponding to either parent sequence are isolated and treated to produce blunt ends. The remaining ends are fused by circularization and then linearized by a restriction digest. This restriction enzyme site has been engineered in the linker region of the fusion in order to reverse the position of the genes in the fusion. The fragment size selection step ensures that the amino acids that meet at the crossover are in similar positions in the parent proteins. The technique allows crossover along the entire length of the gene. A pre-selection technique is required to remove frame-shifted mutants from the library. Functional mutants are biased towards recombination at the ends of the sequences, possibly because recombination at the core leads to significant perturbation of structure.

Incremental truncation for the creation of hybrid enzymes (ITCHY) and the *combination of DNA shuffling and ITCHY (SCRATCHY)* have been developed and applied to other enzyme systems [101].

2.2.2. Directed Evolution of P450cam (CYP101A1)

Directed evolution has been applied to naphthalene oxidation by CYP101A1 using the peroxide shunt pathway as a screening method. Mutants with turnover rates up to 20 times faster than the wild-type enzyme were found [102]. However, problems associated with the assay were subsequently shown to have led to activity levels being overestimated [103].

2.2.3. Directed Evolution of P450$_{BM3}$ (CYP101A2)

An early random mutagenesis study revealed that the affinity of the P25Q mutant for palmitate was some two orders of magnitude lower than that of the wild-type enzyme [104]. Pro25 would not ordinarily have been targeted for alteration, since it occupies a peripheral position in the protein structure. The powerful secondary effects of the mutation were attributed to the proximity of Arg47.

In the course of developing a CYP102A1 mutant capable of medium-chain fatty acid hydroxylation, Schmid and co-workers [105] encountered indigo formation and concluded that this must be the product of indole oxidation (the indole present having arisen from the degradation of tryptophan in the growth medium). Not only did this breakthrough result in the discovery of the A74G/F87V/L188Q mutant, still one of the most potent CYP102A1 variants currently known for non-natural substrates. It further confirmed that CYP102A1 could be engineered to hydroxylate compounds that differed considerably in character from its natural substrates [105]. The 'Schmid' mutant was shown to hydroxylate a range of polycyclic aromatic hydrocarbons, as well as various alkanes, cycloalkanes, arenes,

Table 1. Oxidation rates (nmol min^{-1} (nmol P450)$^{-1}$) of the Schmid triple mutant and related CYP102A1 variants.

Mutant	Naphthalene	Fluorene	Acenaphthene	9-Methylanthracene
		CYP102A1		
Wild-type	0.76	0.014	0.12	0.002
F87V	111	2.8	2.2	0.71
F87V/L188Q	115	22.7	76.2	9.2
A74G/F87V/L188Q	160	52.8	109	21.8
		CYP102A3 (vide infra)		
(G75G)/F88V/S189Q	143	–	–	–

and heteroarenes [106,107], and could render polychlorinated dibenzo-p-dioxins 90% less toxic [108] (Table 1).

Taking the F87A mutant as a starting point for rational evolution, sublibraries with secondary mutations at eight other sites of known or likely importance were randomly generated. These were spectroscopically screened for activity using an ω-p-nitrophenoxycarboxylic acid assay [109,110]. The most promising were then combined in a stepwise fashion to give the mutant F87A/L188K/A74G/R47F/V26T, which also oxidized capric acid [111]. Random and site-directed mutagenesis experiments were run side by side on β-ionone, ultimately generating a pair of triple mutants with roughly equal activity – R47L/Y51F/F87V (rational) and A74E/F87V/P386S (random) [70]. When the two were combined, however, the coupling was halved, underlining the measure of the challenge that engineering the CYP102A1 system can pose.

Screening for activity with a p-nitrophenoxyoctane assay, Arnold and co-workers took CYP102A1 through two rounds of directed evolution and identified an octane hydroxylase five times as potent as the wild-type enzyme [112]. A further three rounds produced mutant '139-3', which was capable of oxidizing alkanes as short as butane and propane [113]. Whereas 2-octanol had accounted for just 17% of the product mix when the Schmid mutant had been used to oxidize octane [107], the 139-3 mutant gave 66% (Table 2). Only one of the mutations in 139-3 involves a residue with substrate contact (Val78), the remainder being located in positions that are of no obvious strategic importance (though Arg255 is part of a salt bridge). Such obscurity of mechanism is typical of directed evolution and makes it an invaluable complementary tool to site-directed mutagenesis.

By applying a combination of rational mutation techniques to constrict the active site of 139-3 and directed evolution to improve its propane affinity, not only were further rate improvements achieved, but also significant regio- and stereochemical shifts in product distribution profiles [114]. A mutant with

Table 2. Engineering alkane hydroxylation by CYP102A1 by directed evolution.

	\multicolumn{5}{c}{CYP102A1 enzyme}				
	Wild-type	139-3	9-10A	1-12G	35-E11
Number of mutations	0	10	13	15	17
			Octane		
Product formation rate[a]	30	480	540	150	420
% 1-octanol	0	1	1	5	ND
% 2-octanol	17	61	53	82	89
% other products	83	38	46	13	11
% ee/favored product	ND	58%S	50%S	39%R	65%S
			Propane		
Product formation rate[a]	0	12	23	160	210
% 1-propanol	0	3	8	11	ND
% 2-propanol	0	97	92	89	ND

[a] (nmol min^{-1} (nmol P450)$^{-1}$)
ND: not determined

the designed A82L mutation incorporated, gave 22% 2-octanol, while mutant '1-12G' yielded 82% (Table 2). More crucially, several variants yielded small quantities of terminal hydroxylation products for a range of alkanes. A wide variance in enantioselectivities was also observed. Another mutant ('9-10A') was now developed into an ethane hydroxylase, again by employing directed evolution while deliberately constricting the active site [115]. Contrasting outcomes were achieved in the hydroxylation of an achiral cyclopentanecarboxylic acid derivative, with 139-3 giving principally the S,S product while 1-12G formed the R,R diastereoisomer [116]. High levels of selectivity were observed in the epoxidation of terminal alkenes using a rationally engineered variant of mutant 9-10A involving active site substitutions such as F87V and I263A [71].

With an eye to harnessing the peroxide shunt as an electron source for P450 hydroxylations in place of NAD(P)H, it had earlier been shown that mutations F87A and F87G (though not F87V) were significantly more active in peroxide conditions than the wild-type enzyme, whose turnover rates are negligible [117]. Sequential rounds of directed evolution gave a mutant considerably more potent in the peroxide shunt than the F87A mutant [118,119]. All thirteen methionine residues in the CYP102A1 heme domain were exchanged with norleucine to establish whether methionine oxidation could be the cause of P450 inactivation in peroxide-supported reactions [120], but the improvements were minor. Directed evolution has also been applied to the enhancement of CYP102A1's thermal stability [121] and its tolerance of the organic co-solvents often needed to improve substrate solubility in turnover experiments [122].

While the effects of the amino acid substitutions in the Schmid mutation are relatively easy to rationalize, the Arnold group's directed evolution variants are

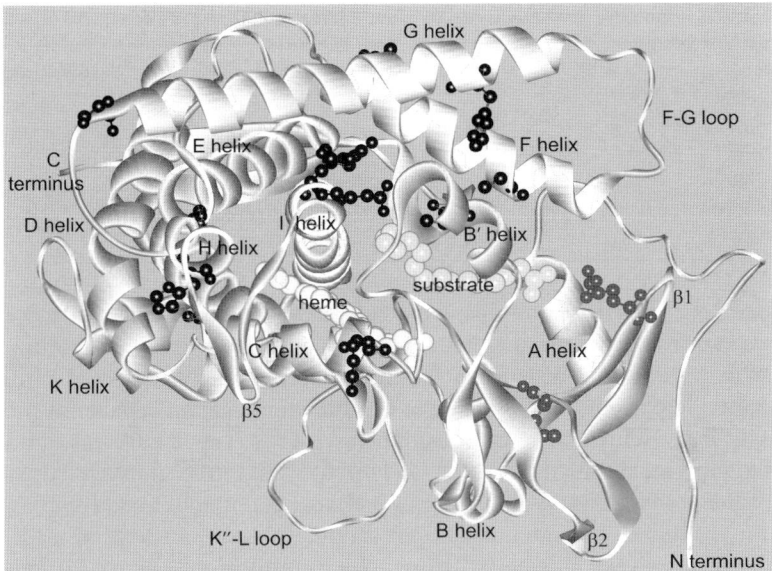

Figure 9. The substitution sites of the directed evolution mutant, 9-10A. The eleven substitutions in the α-domain are shown in black, the two in the β-domain (to either side of the A helix) in dark gray. Mutated residues occur at locations throughout the protein, with very little clustering. Few, if any, directly influence the active site, the heme, the substrate or the substrate access channel, meaning that secondary effects must be largely responsible for the dramatic changes in activity, specificity and other properties reported.

more difficult to understand as few if any of the substitutions directly affect the active site or substrate channel. The mutation sites of the CYP102A1 ethane hydroxylase, 9-10A, are depicted in Figure 9. With the exception of R47C (and possibly V78A), none of the mutations involved are at residues thought to have direct interactions with the substrate. Two occur on each of the F, G, and I helices, one on each of the B', C, E, H, and J helices and only two in areas of β strand – R47C and L353V. This is perhaps less surprising statistically than the fact that none are in randomly folded areas – though it is worth noting that three occur at structural transition points. Some amino acid changes are minor (Ala or Leu for Val in three cases), others dramatic (P142S, F205C and H236Q). Only three are on the proximal side of the heme, all in remote locations. The clustering of seven mutations in the region extending from the substrate terminus to the F/G helices may be significant.

Other members of the catalytically self-sufficient CYP102A1 family are also being engineered. The F88V/S189Q mutant of CYP102A3 from *B. subtilis* has been tested on a range of substrates [123]. Although only a double mutant, this variant is equivalent to the A74G/F87V/L188Q mutant of CYP102A1 since glycine occurs naturally at residue 75 in CYP102A3. Good activity towards

Table 3. Engineering the selectivity of octane hydroxylation in CYP102A3.

Mutant	F88V/S189Q	S189Q	F88V/S189Q/A330V	S189Q/A330V
Conversion rate (%)	49	43	7	5
% 1-octanol	0	0	11	48
% 2-octanol	0	13	42	36
% 3-octanol	68	41	40	16
% 4-octanol	32	46	7	0

substrates such as β-ionone and naphthalene was reported, and in particular towards capric acid and octane. Interestingly, a small quantity of ω-hydroxylated product was generated during the oxidation of capric acid. Directed evolution techniques were accordingly applied to engineering a mutant capable of producing significant yields of 1-octanol from octane [124]. The A330V mutation was found to promote terminal oxidation, and yields were further improved to almost 50% of the product mix by removing the original F87V mutation (Table 3). The similarity of this crucial alteration to the rational A328V substitution made during the engineering of mutant 9-10A to achieve terminal alkane oxidation using CYP102A1 is striking [114]. The P15S variant of *B. subtilis* CYP102A2 has also been engineered by error-prone PCR to improve activity towards substrates such as lauric acid and 1,4-naphthoquinone [125]. Although Pro15 is remotely located on the protein surface, it may have an influence over certain adjacent residues near the entrance to the access channel.

In summary, almost all the directed evolution work carried out to date in the bacterial field has been on CYP102 systems, largely because of their catalytic self-sufficiency. In principle it should be possible to augment the activity and selectivity enhancements that have been achieved by incorporating tried and tested rational mutations into the best variants, since there is typically little or no overlap between the residues involved. But in practice the effects are not always additive – perhaps not surprisingly, given that we cannot always explain at a structural level how mutations introduced by directed evolution alter reaction outcomes.

3. ENGINEERING MAMMALIAN CYTOCHROME P450 ENZYMES

This relatively new area has been reviewed [126–129]. Mammalian P450 enzymes are involved in the biosynthesis and degradation of endogenous compounds as well as xenobiotic metabolism [130–133]. The latter group of enzymes, by the nature of their function, is less substrate specific than most bacterial P450s which tend to carry out specific reactions. This broad substrate range enables a relatively small number of mammalian P450s with different, but overlapping, specificities to metabolize an extremely large range of molecules [1,2,127]. Among human P450 enzymes CYP1A1, CYP1A2, CYP2A6, CYP2B6, CYP2C8, CYP2C9, CYP2D6, CYP2E1, CYP3A4 and CYP3A5 are responsible for the oxidation of >90% of environmental toxicants, drugs and carcinogens (see also Chapter 16 of this volume). Mammalian P450s are also responsible for the biosynthesis of steroid hormones and bile salts, and the biosynthesis and degradation of vitamin D_3 (see also Chapter 12 of this volume) [130,134–138]. The potential for the application of mammalian P450s to bioconversion processes and preclinical drug development has been reviewed [128]. The broad specificity of degradative mammalian P450s makes them ideal starting points for designing chemical catalysts. However, mammalian P450 enzymes are more difficult to handle from a technical point of view as they are membrane-associated.

3.1. Rational Engineering

Rational re-engineering of mammalian P450 enzymes has been limited by a lack of structural information. Site-directed mutagenesis studies have relied largely on homology modeling based on the fact that the structurally characterized P450 enzymes have very similar tertiary structures. The known structures of bacterial P450 enzymes have been used to generate models of the mammalian enzymes [139]. Based on Gotoh's proposal of six substrate recognition sites, predicted substrate pocket amino acids could be targeted by site-directed mutagenesis [140]. The crystal structures of several mammalian P450 enzymes have recently been determined and no doubt more will become available in the near future [141–145]. The availability of these structures has led to an improved understanding of the functions of mammalian P450 enzymes and forms a basis for protein engineering [146,147]. It is also noteworthy that rational mutagenesis and deletions have played an important role in the heterologous production and crystallization of these enzymes [148,149].

The rational engineering of mammalian P450s to change the substrate specificity has tended to be limited to interchanging amino acid residues between closely related P450 family members. He, Szklarz, and Halpert have used site-directed mutagenesis at four key residues to interconvert the androstenedione hydroxylase activities of CYP2B4 and CYP2B5 [150]. The same group also engineered the regioselectivity for progesterone hydroxylation by CYP2B1 by replacing amino acid residues in this enzyme with the corresponding residues in CYP2C5. The I114A/F206V/F297G/V363L/I477F mutant was found to be the most efficient and also had the highest regioselectvity [146]. This work led to the generation of other CYP2B1 mutants which have increased activity for oxidizing the anticancer drugs cyclophosphamide (CPA) and ifosfamide (IFA) (Figure 10) [151]. The regio- and stereoselectivity of the mitochondrial steroid hormone hydroxylating P450 enzymes has been studied by swapping the residues

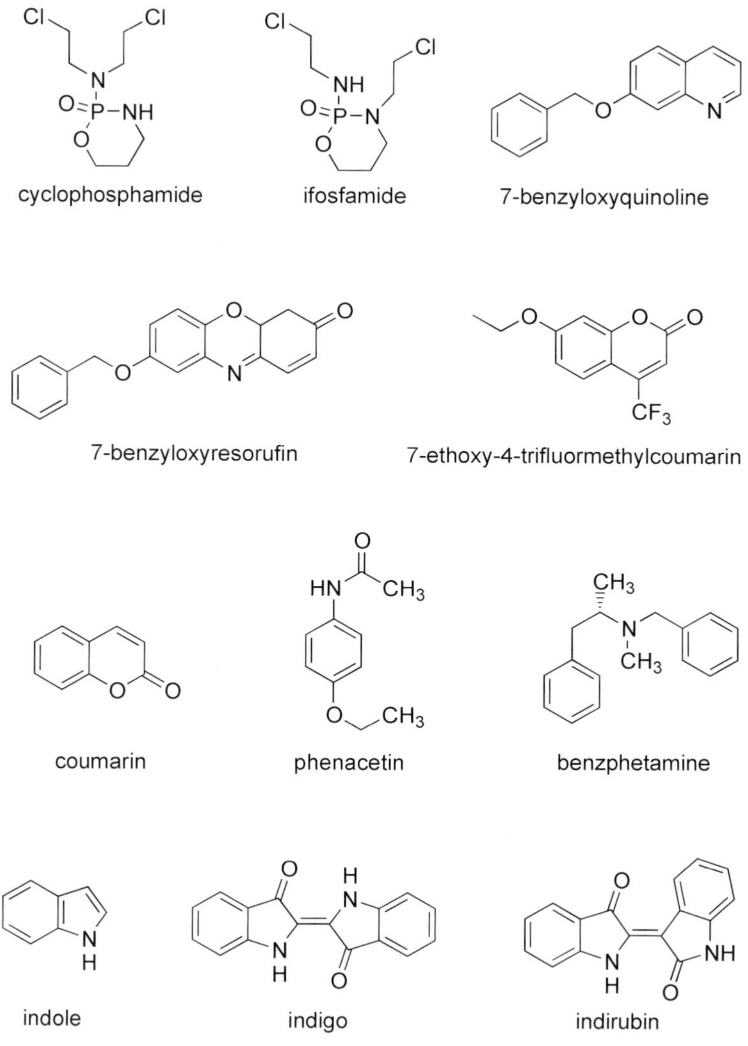

Figure 10. Substrates and probes of mammalian P450 substrates.

of human CYP11B1 and CYP11B2 enzymes [152,153]. Further residues involved in the stereoselectivity have also been found in the N-terminal region [154].

The two electrons from NADPH are transferred to mammalian cytochrome P450s via a membrane bound NADPH-reductase [155]. *In vivo* studies of mammalian P450 enzymes have been facilitated by the construction of fusions of the reductase protein and the P450 enzyme as well as bicistronic expression systems [156–159]. CYP2E1 has been engineered as a fusion protein to a bacterial reductase (from CYP102A1). Like CYP102A1 this protein is catalytically

DESIGN AND ENGINEERING OF CYTOCHROME P450 SYSTEMS

self-sufficient. The fusion protein maintains the activity of the wild-type CYP2E1 without the need for addition of lipids and detergents to stabilize the protein [160].

3.2. Random Mutagenesis

The development of random mutagenesis methods enables the generation of libraries containing large numbers of mutant enzymes. Some of these may have improved activity or other property of interest. This approach is particularly attractive for mammalian P450 enzymes due to the limited structural information available.

The development of high throughput screening/selection methods involving genotoxicity, colorimetric, fluorescent or luminescent based assays for mutant enzymes has led to rapid advances in random mutagenesis studies [161–163]. High-throughput luminescence-based assays for a wide range of human P450s have recently become commercially available.

3.2.1. CYP1A2

Guengerich and co-workers [164,165] have undertaken semi-random mutagenesis of the SRS of human CYP1A2. Using a gentotoxicity assay, single rounds of mutagenesis showed a 3- to 4-fold improvement in activity towards two test substrates, 7-ethoxyresorufin and phenacetin. Screening of larger random mutant libraries with mutations throughout the entire gene identified mutants with 10-fold enhanced activity for N-oxidation of heterocyclic amines [166]. A high throughput fluorescence assay based on 7-methoxyresorufin O-demethylation has been used to screen libraries of CYP1A2 mutants. A triple mutant with 5-fold enhancement of activity was identified using this method. A homology model of the CYP1A2 structure (based on the CYP2C5 structure) indicated that none of the mutations in the triple mutant were particularly close to the substrate binding site of the enzyme [167,168].

Abecassis, Pompon, and Truan [99] have used the CLERY family shuffling method to generate a library of mutants derived from CYP1A1 and CYP1A2. A colorimetric method based on the oxidation of naphthalene, a common substrate of both enzymes, was used to screen for functional progenies. Approximately 11% of the tested clones were active for naphthalene oxidation. The SHIPREC method was used to combine two unrelated P450 enzymes, CYP1A2 and the P450 domain of CYP102A1, to produce active chimeras. Screening CYP1A2 for 4-ethoxyresorufin deethylation activity resulted in the identification of two hybrids that were more soluble in the bacterial cytoplasm than wild-type CYP1A2. However, among correctly folded mutants there was a bias towards recombination at either terminus. The two mutants identified with improved solubility were comprised mostly of CYP1A2 with N-termini derived from CYP102 [100].

3.2.2. CYP2A6

Co-expression of P450 enzymes (including CYP2A family members) and cytochrome P450 reductase in bicistronic vector systems produced the blue pigment indigo from indole 3-oxidation [161]. Using semi-random mutagenesis techniques similar to those used with CYP1A2 Guenguerich and coworkers [169] produced variants of CYP2A6, including the mutant L240C/N297Q, that could oxidize indole more efficiently. Compared to the wild-type enzyme, these mutants also produced different colored products, including indirubin. Other mutants were found that oxidized indole less efficiently than the wild-type but could catalyze coumarin 7-hydroxylation up to 13 times faster [169]. These mutants were used to oxidize indole variants including 5-methoxy and 5-methylindole. Oxidation of 5-methoxyindole followed by coupling gives 5,5-dimethoxyindirubin which could inhibit human cyclin-dependent kinases (CDK-1 and -5) and glycogen synthase kinase-3 (GSK3) [170]. Further studies combining directed evolution using the entire open reading frame followed by site directed mutagenesis yielded the CYP2A6 mutant N297Q/I300V which could oxidize bulky 4- and 5-benzyloxyindoles to coloured products [171].

Random mutagenesis on the entire reading frame and screening with the coumarin 7-hydroxylation fluorescence assay resulted in mutations with altered activity. However, all the mutants selected for further study had lower catalytic efficiency than the wild-type enzyme [162].

3.2.3. CYP2B1

Halpert and co-workers [172,173] have used hydrogen peroxide (H_2O_2)-supported 7-ethoxy-4-trifluoromethylcoumarin (7-EFC) O-deethylation as a rapid screening procedure. An N-terminal truncated, C-terminal His-tagged form of the enzyme was used as a template for directed evolution as it was expressed more readily. The L209A mutant of this modified enzyme showed a five-fold higher activity for H_2O_2-driven turnover than the wild-type [172,173]. After two rounds of random mutagenesis a quadruple mutant V183L/F202L/L209A/S334P with a k_{cat} six-fold higher than the L209A mutant was isolated. This mutant also had improved hydrogen peroxide activity and higher stability to elevated temperatures and organic solvent concentration. Further rounds of random mutagenesis did not improve the activity. On the other hand removal of the V183L mutation increased the catalytic efficiency in the NADPH system. The resulting triple mutant had enhanced catalytic activity with several other substrates including

7-benzyloxyresorufin, benzphetamine and testosterone. Double mutants containing the individual mutations introduced by random mutagenesis with the L209A template mutation had increased activities with the anticancer drugs CPA and IFA [174].

3.2.4. CYP2C Family

Gillam has generated mutant libraries by restriction enzyme-mediated fragmentation and recombination of CYP2C9, CYP2C11, and CYP2C19 [126]. The recombinants were co-expressed with human cytochrome P450 reductase in *E. coli* and indole oxidation activity (to produce indigo) was used as a screening method. Results indicated no contamination of the library with parental forms. Sequencing of the mutants revealed an average of 6.9 crossovers and 2.6 spontaneous mutations per 1.5 kbp sequence. One mutant had a two-fold higher activity than the most active parent. Activities towards a second substrate, *p*-nitrophenol varied independently from indole oxidation across a range of mutants.

3.2.5. CYP3A4

A high-throughput H_2O_2-supported screening method for human CYP3A4 has been developed using whole cell suspension and 7-benzyloxyquinoline as a substrate [175]. Two mutants with more than twice the activity of the wild-type have been found but these did not show an increase in k_{cat} in the standard NADPH system.

Kumar and Halpert [129] have developed a high-throughput method for investigating CYP3A4 variants with altered cooperativity. NADPH-supported CYP3A4 activity for the debenzylation of 7-benzyloxyquinoline in the presence or absence of α-naphthoflavone was used to test for loss or gain in cooperativity. Initial results identified a mutant with decreased cooperativity but further analysis and screening is required [129].

4. ENGINEERING PLANT P450 ENZYMES

Cytochrome P450 enzymes are the largest family of plant proteins. They are involved in secondary metabolism and in the synthesis of a large number of complex natural products such as taxol, genistein, and artemisinin [176]. These products have important uses as flavors, aroma compounds, nutraceuticals, and pharmaceuticals. Plant P450s have also found uses in the production of unusual colors in transgenic plants [177,178]. As with mammalian P450s little structural information is available due to difficulties in handling the enzymes. Homology modeling coupled with mutagenesis has been used to study the structure/function of certain plant P450 enzymes [179]. As yet no extensive mutagenesis studies or engineering of plant P450s has been reported. The best examples of P450 engineering are for closely related enzymes. Two distinct cytochrome P450s that catalyze the regiospecific hydroxylation of (−)-4S-limonene exclusively at C3 or C6 are 70% identical at the amino acid level. A combination of domain swapping and reciprocal site-directed mutagenesis demonstrated that the exchange of a single residue (F363I) in the spearmint limonene-6-hydroxylase led to complete conversion to the regiospecificity and catalytic efficiency of the peppermint limonene-3-hydroxylase [180].

5. CONCLUSIONS AND OUTLOOK

Significant progress in re-engineering the substrate specificity and selectivity of P450 cytochromes has been reported, but we believe much more remains to be achieved. Research on human P450 cytochromes, which are ideal templates for exploring the synthesis and metabolism of pharmaceuticals and other molecules of potential clinical significance, is still in its early stages. Even less work has been carried out on the equivalent plant enzymes, whose exact role has yet to be unraveled, but their probable involvement in the biosynthesis of a number of secondary metabolites with medicinal properties makes the field a promising one.

Efforts to date have focused largely on bacterial enzymes. While these are better characterized, with many structures available, they are by no means fully understood. There have been numerous breakthroughs: CYP101A1 and the CYP102 family have been engineered to oxidize several classes of non-natural substrates; changes in regioselectivity such as shifting oxidation to terminal carbon positions have been accomplished; the stereoselectivity of the hydroxylation of linear and alicyclic compounds can be altered; aromatic compounds are readily oxidized to the phenols and catechols; ethane oxidation has been achieved. Yet the untapped opportunities remain considerable, particularly if effective new assay techniques can be developed. There is the potential to engineer variants with higher turnover activity towards both natural and non-natural substrates by accelerating the rates of electron transfer.

Directed evolution methods have explored only a fraction of the sequence space available. Investigations into the mechanisms by which mutations far removed from the active site exert their dramatic effects have barely begun, and the detailed kinetic consequences of mutations (whether introduced by design or by evolution) need to be explored.

Another attractive opportunity is chimera engineering – neither rational engineering nor random mutagenesis mimics nature's trick of inserting and/or deleting amino acids to transform the topology of an active site.

Finally, there remains considerable scope for harnessing new enzymes with properties that differ from those already known. By expanding the repertoire of

P450 systems under exploration, it should in time be possible to convert a far wider range of compounds to oxidation products with much greater ranges of specificity and selectivity.

ABBREVIATIONS

λ	reorganization energy
7-BQ	7-benzyloxyquinoline
CDK	cyclin-dependent kinases
CLERY	combinatorial libraries enhanced by recombination in yeast
CPA	cyclophosphamide
CYP	cytochrome P450
7-EFC	7-ethoxy-4-trifluoromethylcoumarin
FAD	flavin-adenine dinucleotide
FMN	flavin mononucleotide
GSK	glycogen synthase kinase
H_{AB}	electronic coupling
IFA	ifosfamide
ITCHY	incremental truncation for the creation of hybrid enzymes
kbp	kilobase pairs
k_{cat}	catalytic rate constant
k_{et}	rate of electron transfer
K_M	Michaelis constant
NAD(P)H	nicotinamide adenine dinucleotide (phosphate) hydride
PCR	polymerase chain reaction
PeCB	pentachlorobenzene
Por	Poryphyrin
SCRATCHY	combination of DNA shuffling and ITCHY
SHIPREC	sequence homology independent protein recombination
SRS	substrate recognition site
StEP	staggered extension process
TCB	trichlorobenzene

REFERENCES

1. F. P. Guengerich, *Curr. Drug Metab.*, 2, 93–115 (2001).
2. F. P. Guengerich, *Chem. Res. Toxicol.*, 14, 611–650 (2001).
3. A. Gutierrez, A. Grunau, M. Paine, A. W. Munro, C. R. Wolf, G. C. K. Roberts, and N. S. Scrutton, *Biochem. Soc. Trans.*, 31, 497–501 (2003).
4. R. E. Sharp and S. K. Chapman, *Biochim. Biophys. Acta, 1432*, 143–158 (1999).
5. I. Hanukoglu, *Adv. Mol. Cell Biol.*, 14, 29–56 (1996).
6. P. J. Loida and S. G. Sligar, *Biochemistry, 32*, 11530–11538 (1993).

7. P. J. Loida and S. G. Sligar, *Protein Eng.*, *6*, 207–212 (1993).
8. S. Kadkhodayan, E. D. Coulter, D. M. Maryniak, T. A. Bryson, and J. H. Dawson, *J. Biol. Chem.*, *270*, 28042–28048 (1995).
9. R. A. Marcus and N. Sutin, *Biochim. Biophys. Acta*, *811*, 265–322 (1985).
10. M. J. Honeychurch, H. A. O. Hill, and L. -L. Wong, *FEBS Lett.*, *451*, 351–353 (1999).
11. B. Zhao, F. P. Guengerich, M. Voehler, and M. R. Waterman, *J. Biol. Chem.*, *280*, 42188–42197 (2005).
12. B. Zhao, F. P. Guengerich, A. Bellamine, D. C. Lamb, M. Izumikawa, L. Lei, L. M. Podust, M. Sundaramoorthy, J. A. Kalaitzis, L. M. Reddy, S. L. Kelly, B. S. Moore, D. Stec, M. Voehler, J. R. Falck, T. Shimada, and M. R. Waterman, *J. Biol. Chem.*, *280*, 11599–11607 (2005).
13. R. Raag, S. A. Martinis, S. G. Sligar, and T. L. Poulos, *Biochemistry*, *30*, 11420–11429 (1991).
14. S. Nagano and T. L. Poulos, *J. Biol. Chem.*, *280*, 31659–31663 (2005).
15. T. Hishiki, H. Shimada, S. Nagano, T. Egawa, Y. Kanamori, R. Makino, S. Y. Park, S. Adachi, Y. Shiro, and Y. Ishimura, *J. Biochem. (Tokyo)*, *128*, 965–974 (2000).
16. S. G. Bell, X. Chen, F. Xu, Z. Rao, and L. -L. Wong, *Biochem. Soc. Trans.*, *31*, 558–562 (2003).
17. L.-L. Wong, A. C. G. Westlake, and D. P. Nickerson, *Structure and Bonding*, *88*, 175–207 (1997).
18. Z. Li and D. Chang, *Curr. Org. Chem.*, *8*, 1647–1658 (2004).
19. H. Li and T. L. Poulos, *Biochim. Biophys. Acta*, *1441*, 141–149 (1999).
20. C. S. Miles, T. W. B. Ost, M. A. Noble, A. W. Munro, and S. K. Chapman, *Biochim. Biophys. Acta*, *1543*, 383–407 (2000).
21. A. W. Munro, D. G. Leys, K. J. McLean, K. R. Marshall, T. W. Ost, S. Daff, C. S. Miles, S. K. Chapman, D. A. Lysek, C. C. Moser, C. C. Page, and P. L. Dutton, *Trends Biochem. Sci.*, *27*, 250–257 (2002).
22. V. B. Urlacher, S. Lutz-Wahl, and R. D. Schmid, *Appl. Microbiol. Biotechnol.*, *64*, 317–325 (2004).
23. A. J. Warman, O. Roitel, R. Neeli, H. M. Girvan, H. E. Seward, S. A. Murray, K. J. McLean, M. G. Joyce, H. Toogood, R. A. Holt, D. Leys, N. S. Scrutton, and A. W. Munro, *Biochem. Soc. Trans.*, *33*, 747–753 (2005).
24. I. C. Gunsalus and G. C. Wagner, *Methods Enzymol.*, *52*, 166–188 (1978).
25. T. L. Poulos, B. C. Finzel, I. C. Gunsalus, G. C. Wagner, and J. Kraut, *J. Biol. Chem.*, *260*, 16122–16130 (1985).
26. T. L. Poulos, B. C. Finzel, and A. J. Howard, *J. Mol. Biol.*, *195*, 687–700 (1987).
27. W. M. Atkins and S. G. Sligar, *J. Biol. Chem.*, *263*, 18842–18849 (1988).
28. D. Filipovic, M. D. Paulsen, P. J. Loida, S. G. Sligar, and R. L. Ornstein, *Biochem. Biophys. Res. Commun.*, *189*, 488–495 (1992).
29. J.-A. Stevenson, A. C. G. Westlake, C. Whittock, and L.-L. Wong, *J. Am. Chem. Soc.*, *118*, 12846–12847 (1996).
30. P. A. England, C. F. Harford-Cross, J.-A. Stevenson, D. A. Rouch, and L.-L. Wong, *FEBS Lett.*, *424*, 271–274 (1998).
31. A. Celik, R. E. Speight, and N. J. Turner, *Chem. Commun.*, 3652–3644 (2005).
32. J. J. de Voss, O. Sibbesen, Z. Zhang, and P. R. Ortiz de Montellano, *J. Am. Chem. Soc.*, *119*, 5489–5498 (1997).

33. Z. Zhang, O. Sibbesen, R. A. Johnson, and P. R. Ortiz de Montellano, *Bioorg. Med. Chem.*, 6, 1501–1518 (1998).
34. A. Verras, D. Kuntz Irwin, and R. Ortiz de Montellano Paul, *J. Med. Chem.*, 47, 3572–3579 (2004).
35. D. P. Nickerson and L.-L. Wong, *Protein Eng.*, 10, 1357–1361 (1997).
36. A. C. G. Westlake, C. F. Harford-Cross, J. Donovan, and L.-L. Wong, *Eur. J. Biochem.*, 265, 929–935 (1999).
37. K. J. French, M. D. Strickler, D. A. Rock, G. A. Bennett, J. L. Wahlstrom, B. M. Goldstein, and J. P. Jones, *Biochemistry*, 40, 9532–9538 (2001).
38. K. J. French, D. A. Rock, J. I. Manchester, B. M. Goldstein, and J. P. Jones, *Arch. Biochem. Biophys.*, 398, 188–197 (2002).
39. J.-A. Stevenson, J. K. Bearpark, and L.-L. Wong, *New J. Chem.*, 551–552 (1998).
40. S. G. Bell, J.-A. Stevenson, H. D. Boyd, S. Campbell, A. D. Riddle, E. L. Orton, and L.-L. Wong, *Chem. Commun.*, 490–491 (2002).
41. S. Yoshioka, S. Takahashi, K. Ishimori, and I. Morishima, *J. Inorg. Biochem.*, 81, 141–151 (2000).
42. S. Nagano, T. Tosha, K. Ishimori, I. Morishima, and T. L. Poulos, *J. Biol. Chem.*, 279, 42844–42849 (2004).
43. F. Xu, S. G. Bell, J. Lednik, A. Insley, Z. Rao, and L.-L. Wong, *Angew. Chem. Int. Ed. Engl.*, 44, 4029–4032 (2005).
44. D. P. Nickerson, C. F. Harford-Cross, S. R. Fulcher, and L.-L. Wong, *FEBS Lett.*, 405, 153–156 (1997).
45. S. G. Bell, C. F. Harford-Cross, and L.-L. Wong, *Protein Eng.*, 14, 797–802 (2001).
46. O. Sibbesen, Z. Zhang, and P. R. Ortiz de Montellano, *Arch. Biochem. Biophys.*, 353, 285–296 (1998).
47. C. F. Harford-Cross, A. B. Carmichael, F. K. Allan, P. A. England, D. A. Rouch, and L.-L. Wong, *Protein Eng.*, 13, 121–128 (2000).
48. J. P. Jones, E. J. O'Hare, and L.-L. Wong, *Chem. Commun.*, 247–248 (2000).
49. J. P. Jones, E. J. O'Hare, and L.-L. Wong, *Eur. J. Biochem.*, 268, 1460–1467 (2001).
50. X. Chen, A. Christopher, J. P. Jones, S. G. Bell, Q. Guo, F. Xu, Z. Rao, and L.-L. Wong, *J. Biol. Chem.*, 277, 37519–37526 (2002).
51. N. E. Jones, P. A. England, D. A. Rouch, and L.-L. Wong, *Chem. Commun.*, 2413–2414 (1996).
52. P. A. England, D. A. Rouch, A. C. G. Westlake, S. G. Bell, D. P. Nickerson, M. Webberley, S. L. Flitsch, and L.-L. Wong, *Chem. Commun.*, 357–358 (1996).
53. S. G. Bell, D. A. Rouch, and L.-L. Wong, *J. Mol. Catal. B: Enzymatic*, 3, 293–302 (1997).
54. S. G. Bell, R. J. Sowden, and L.-L. Wong, *Chem. Commun.*, 635–636 (2001).
55. S. G. Bell, X. Chen, R. J. Sowden, F. Xu, J. N. Williams, L.-L. Wong, and Z. Rao, *J. Am. Chem. Soc.*, 125, 705–714 (2003).
56. R. J. Sowden, S. Yasmin, N. H. Rees, S. G. Bell, and L.-L. Wong, *Org. Biomol. Chem.*, 3, 57–64 (2005).
57. K. G. Ravichandran, S. S. Boddupalli, C. A. Hasermann, J. A. Peterson, and J. Deisenhofer, *Science*, 261, 731–736 (1993).
58. H. Yeom and S. G. Sligar, *Arch. Biochem. Biophys.*, 337, 209–216 (1997).

59. H. Yeom, S. G. Sligar, H. Li, T. L. Poulos, and A. J. Fulco, *Biochemistry*, *34*, 14733–14740 (1995).
60. G. Truan and J. A. Peterson, *Arch. Biochem. Biophys.*, *349*, 53–64 (1998).
61. H. Li and T. L. Poulos, *Nat. Struct. Biol.*, *4*, 140–146 (1997).
62. S. Graham-Lorence, G. Truan, J. A. Peterson, J. R. Falck, S. Wei, C. Helvig, and J. H. Capdevila, *J. Biol. Chem.*, *272*, 1127–1135 (1997).
63. C. F. Oliver, S. Modi, M. J. Sutcliffe, W. U. Primrose, L. Y. Lian, and G. C. Roberts, *Biochemistry*, *36*, 1567–1572 (1997).
64. L. A. Cowart, J. R. Falck, and J. H. Capdevila, *Arch. Biochem. Biophys.*, *387*, 117–124 (2001).
65. M. A. Noble, C. S. Miles, S. K. Chapman, D. A. Lysek, A. C. MacKay, G. A. Reid, R. P. Hanzlik, and A. W. Munro, *Biochem. J.*, *339*, 371–379 (1999).
66. A. B. Carmichael and L.-L. Wong, *Eur. J. Biochem.*, *268*, 3117–3125 (2001).
67. Q. S. Li, J. Ogawa, R. D. Schmid, and S. Shimizu, *FEBS Lett.*, *508*, 249–252 (2001).
68. Q. S. Li, J. Ogawa, R. D. Schmid, and S. Shimizu, *Biosci. Biotechnol. Biochem.*, *69*, 293–300 (2005).
69. W. T. Sulistyaningdyah, J. Ogawa, Q.-S. Li, C. Maeda, Y. Yano, R. D. Schmid, and S. Shimizu, *Appl. Microbiol. Biotechnol.*, *67*, 556–562 (2005).
70. V. B. Urlacher, A. Makhsumkhanov, and R. D. Schmid, *Appl. Microbiol. Biotechnol.*, *70*, 53–59 (2006).
71. T. Kubo, M. W. Peters, P. Meinhold, and F. H. Arnold, *Chemistry*, *12* 1216–1220 (2006).
72. B. M. A. Lussenburg, L. C. Babel, N. P. E. Vermeulen, and J. N. M. Commandeur, *Anal. Biochem.*, *341*, 148–155 (2005).
73. C. F. Oliver, S. Modi, W. U. Primrose, L. Y. Lian, and G. C. Roberts, *Biochem. J.*, *327*, 537–544 (1997).
74. T. W. Ost, C. S. Miles, J. Murdoch, Y. Cheung, G. A. Reid, S. K. Chapman, and A. W. Munro, *FEBS Lett.*, *486*, 173–177 (2000).
75. A. B. Carmichael,*The Oxidation of Hydrocarbons by Variants of Cytochrome P450$_{BM3}$*, D.Phil. Thesis, University of Oxford, Oxford, 2002.
76. M. L. Klein and A. J. Fulco, *J. Biol. Chem.*, *268*, 7553–7561 (1993).
77. A. W. Munro, K. Malarkey, J. McKnight, A. J. Thomson, S. M. Kelly, N. C. Price, J. G. Lindsay, J. R. Coggins, and J. S. Miles, *Biochem. J.*, *303*, 423–428 (1994).
78. I. Sevrioukova and J. A. Peterson, *Biochimie*, *78*, 744–751 (1996).
79. T. W. Ost, C. S. Miles, A. W. Munro, J. Murdoch, G. A. Reid, and S. K. Chapman, *Biochemistry*, *40*, 13421–13429 (2001).
80. T. W. Ost, J. Clark, C. G. Mowat, C. S. Miles, M. D. Walkinshaw, G. A. Reid, S. K. Chapman, and S. Daff, *J. Am. Chem. Soc.*, *125*, 15010–14020 (2003).
81. H. M. Girvan, K. R. Marshall, R. J. Lawson, D. Leys, M. G. Joyce, J. Clarkson, W. E. Smith, M. R. Cheesman, and A. W. Munro, *J. Biol. Chem.*, *279*, 23274–23286 (2004).
82. M. G. Joyce, H. M. Girvan, A. W. Munro, and D. Leys, *J. Biol. Chem.*, *279*, 23287–23293 (2004).
83. R. Neeli, H. M. Girvan, A. Lawrence, M. J. Warren, D. Leys, N. S. Scrutton, and A. W. Munro, *FEBS Lett.*, *579*, 5582–5588 (2005).

84. H. Xiang, R. A. Tschirret-Guth, and P. R. Ortiz De Montellano, *J. Biol. Chem.*, 275, 35999–36006 (2000).
85. K. K. Khan, Y. A. He, Y. Q. He, and J. R. Halpert, *Chem. Res. Toxicol.*, 15, 843–853 (2002).
86. L. S. Koo, C. E. Immoos, M. S. Cohen, P. J. Farmer, and P. R. Ortiz de Montellano, *J. Am. Chem. Soc.*, 124, 5684–5691 (2002).
87. A. V. Puchkaev, L. S. Koo, and P. R. Ortiz de Montellano, *Arch. Biochem. Biophys.*, 409, 52–58 (2002).
88. S. A. Maves and S. G. Sligar, *Protein Sci.*, 10, 161–168 (2001).
89. J. D. Bloom, M. M. Meyer, P. Meinhold, C. R. Otey, D. MacMillan, and F. H. Arnold, *Curr. Opin. Struct. Biol.*, 15, 447–452 (2005).
90. E. T. Farinas, T. Bulter, and F. H. Arnold, *Curr. Opin. Biotechnol.*, 12, 545–551 (2001).
91. I. P. Petrounia and F. H. Arnold, *Curr. Opin. Biotechnol.*, 11, 325–330 (2000).
92. R. C. Cadwell and G. F. Joyce, *PCR Methods Appl.*, 2, 28–33 (1992).
93. C. Neylon, *Nucleic Acids Res.*, 32, 1448–1459 (2004).
94. Z. Shao, H. Zhao, L. Giver, and F. H. Arnold, *Nucleic Acids Res.*, 26, 681–683 (1998).
95. J. M. Joern, *Methods Mol. Biol.*, 231, 85–89 (2003).
96. O. Kagami, S.-H. Baik, and S. Harayama, *Methods Enzymol.*, 388, 11–21 (2004).
97. M. Ostermeier, *Trends Biotechnol.*, 21, 244–247 (2003).
98. H. Zhao, L. Giver, Z. Shao, J. A. Affholter, and F. H. Arnold, *Nat. Biotechnol.*, 16, 258–261 (1998).
99. V. Abecassis, D. Pompon, and G. Truan, *Nucleic Acids Res.*, 28, E88 (2000).
100. V. Sieber, C. A. Martinez, and F. H. Arnold, *Nat. Biotechnol.*, 19, 456–460 (2001).
101. K. E. Griswold, Y. Kawarasaki, N. Ghoneim, S. J. Benkovic, B. L. Iverson, and G. Georgiou, *Proc. Nat. Acad. Sci. USA*, 102, 10082–10087 (2005).
102. H. Joo, Z. Lin, and F. H. Arnold, *Nature*, 399, 670–673 (1999).
103. K. Matsuura, T. Tosha, S. Yoshioka, S. Takahashi, K. Ishimori, and I. Morishima, *Biochem. Biophys. Res. Commun.*, 323, 1209–1215 (2004).
104. S. A. Maves, H. Yeom, M. A. McLean, and S. G. Sligar, *FEBS Lett.*, 414, 213–218 (1997).
105. Q. S. Li, U. Schwaneberg, P. Fischer, and R. D. Schmid, *Chemistry*, 6, 1531–1536 (2000).
106. Q.-S. Li, J. Ogawa, R. D. Schmid, and S. Shimizu, *Appl. Environ. Microbiol.*, 67, 5735–5739 (2001).
107. D. Appel, S. Lutz-Wahl, P. Fischer, U. Schwaneberg, and R. D. Schmid, *J. Biotechnol.*, 88, 167–171 (2001).
108. W. T. Sulistyaningdyah, J. Ogawa, Q.-S. Li, R. Shinkyo, T. Sakaki, K. Inouye, R. D. Schmid, and S. Shimizu, *Biotechnol. Lett.*, 26, 1857–1860 (2004).
109. Q. S. Li, U. Schwaneberg, M. Fischer, J. Schmitt, J. Pleiss, S. Lutz-Wahl, and R. D. Schmid, *Biochim. Biophys. Acta*, 1545, 114–121 (2001).
110. U. Schwaneberg, C. Schmidt-Dannert, J. Schmitt, and R. D. Schmid, *Anal. Biochem.*, 269, 359–366 (1999).
111. O. Lentz, Q.-S. Li, U. Schwaneberg, S. Lutz-Wahl, P. Fischer, and R. D. Schmid, *J. Mol. Catal. B: Enzymatic*, 15, 123–133 (2001).

112. E. T. Farinas, U. Schwaneberg, A. Glieder, and F. H. Arnold, *Adv. Syn. Catal.*, *343*, 601–606 (2001).
113. A. Glieder, E. T. Farinas, and F. H. Arnold, *Nat. Biotechnol.*, *20*, 1135–1139 (2002).
114. M. W. Peters, P. Meinhold, A. Glieder, and F. H. Arnold, *J. Am. Chem. Soc.*, *125*, 13442–13450 (2003).
115. P. Meinhold, M. W. Peters, M. M. Y. Chen, K. Takahashi, and F. H. Arnold, *ChemBioChem*, *6*, 1765–1768 (2005).
116. D. F. Munzer, P. Meinhold, M. W. Peters, S. Feichtenhofer, H. Griengl, F. H. Arnold, A. Glieder, and A. de Raadt, *Chem. Commun.*, 2597–2599 (2005).
117. Q. S. Li, J. Ogawa, and S. Shimizu, *Biochem. Biophys. Res. Commun.*, *280*, 1258–1261 (2001).
118. P. C. Cirino and F. H. Arnold, *Adv. Syn. Catal.*, *344*, 932–937 (2002).
119. P. C. Cirino and F. H. Arnold, *Angew. Chem. Int. Ed. Engl.*, *42*, 3299–3301 (2003).
120. P. C. Cirino, Y. Tang, K. Takahashi, D. A. Tirrell, and F. H. Arnold, *Biotechnol. Bioeng.*, *83*, 729–734 (2003).
121. O. Salazar, P. C. Cirino, and F. H. Arnold, *ChemBioChem*, *4*, 891–893 (2003).
122. T. Seng Wong, F. H. Arnold, and U. Schwaneberg, *Biotechnol. Bioeng.*, *85*, 351–358 (2004).
123. O. Lentz, V. Urlacher, and R. D. Schmid, *J. Biotechnol.*, *108*, 41–49 (2004).
124. O. Lentz, A. Feenstra, T. Habicher, B. Hauer, R. D. Schmid, and V. B. Urlacher, *Chembiochem*, *7*, 345–350 (2005).
125. I. Axarli, A. Prigipaki, and N. E. Labrou, *Biomol. Eng.*, *22*, 81–88 (2005).
126. E. M. J. Gillam, *Clin. Exp. Pharmacol. Physiol.*, *32*, 147–152 (2005).
127. F. P. Guengerich, *Nat. Rev. Drug Discov.*, *1*, 359–366 (2002).
128. T. Sakaki and K. Inouye, *J. Biosci. Bioeng.*, *90*, 583–590 (2000).
129. S. Kumar and J. R. Halpert, *Biochem. Biophys. Res. Commun.*, *338*, 456–464 (2005).
130. F. P. Guengerich, *Molecular Interventions*, *3*, 194–204 (2003).
131. P. R. Ortiz de Montellano, in *Cytochrome P450: Structure, Mechanism, and Biochemistry* (P. R. Ortiz de Montellano, ed.), 2nd edn., Plenum, New York, 1995.
132. F. P. Guengerich, in *Human P450 Enzymes* (P. R. Ortiz de Montellano, ed.), New York, 1995.
133. F. P. Guengerich, *Drug Dev. Res.*, *49*, 4–16 (2000).
134. G. A. Kullak-Ublick and M.-B. Becker, *Drug Metab. Rev.*, *35*, 305–317 (2003).
135. R. W. Estabrook, *Biochem. Biophys. Res. Commun.*, *338*, 290–298 (2005).
136. M. Bureik, M. Lisurek, and R. Bernhardt, *Biol. Chem.*, *383*, 1537–1551 (2002).
137. D. E. Prosser and G. Jones, *Trends Biochem. Sci.*, *29*, 664–673 (2004).
138. F. P. Guengerich, Z.-L. Wu, and C. J. Bartleson, *Biochem. Biophys. Res. Commun.*, *338*, 465–469 (2005).
139. G. D. Szklarz and J. R. Halpert, *Life Sci.*, *61*, 2507–2520 (1997).
140. O. Gotoh, *J. Biol. Chem.*, *267*, 83–90 (1992).
141. M. R. Wester, E. F. Johnson, C. Marques-Soares, P. M. Dansette, D. Mansuy, and C. D. Stout, *Biochemistry*, *42*, 6370–6379 (2003).
142. P. A. Williams, J. Cosme, D. M. Vinkovic, A. Ward, H. C. Angove, P. J. Day, C. Vonrhein, I. J. Tickle, and H. Jhoti, *Science*, *305*, 683–686 (2004).
143. E. E. Scott, M. A. White, Y. A. He, E. F. Johnson, C. D. Stout, and J. R. Halpert, *J. Biol. Chem.*, *279*, 27294–27301 (2004).

144. G. A. Schoch, J. K. Yano, M. R. Wester, K. J. Griffin, C. D. Stout, and E. F. Johnson, *J. Biol. Chem., 279*, 9497–9503 (2004).
145. P. A. Williams, J. Cosme, A. Ward, H. C. Angove, D. M. Vinkovic, and H. Jhoti, *Nature, 424*, 464–468 (2003).
146. S. Kumar, E. E. Scott, H. Liu, and J. R. Halpert, *J. Biol. Chem., 278*, 17178–17184 (2003).
147. T. L. Domanski and J. R. Halpert, *Curr. Drug Metab., 2*, 117–137 (2001).
148. J. Cosme and E. F. Johnson, *J. Biol. Chem., 275*, 2545–2553 (2000).
149. C. W. Fisher, D. L. Caudle, C. Martin-Wixtrom, L. C. Quattrochi, R. H. Tukey, M. R. Waterman, and R. W. Estabrook, *FASEB J., 6*, 759–764 (1992).
150. Y. Q. He, G. D. Szklarz, and J. R. Halpert, *Arch. Biochem. Biophys., 335*, 152–160 (1996).
151. C.-S. Chen, J. T. Lin, K. A. Goss, Y.-a. He, J. R. Halpert, and D. J. Waxman, *Mol. Pharmacol., 65*, 1278–1285 (2004).
152. B. Bottner, H. Schrauber, and R. Bernhardt, *J. Biol. Chem., 271*, 8028–8033 (1996).
153. B. Bottner, K. Denner, and R. Bernhardt, *Eur. J. Biochem., 252*, 458–466 (1998).
154. S. Bechtel, N. Belkina, and R. Bernhardt, *Eur. J. Biochem., 269*, 1118–1127 (2002).
155. A. Gutierrez, O. Doehr, M. Paine, C. R. Wolf, N. S. Scrutton, and G. C. K. Roberts, *Biochemistry, 39*, 15990–15999 (2000).
156. G. R. Harlow and J. R. Halpert, *Arch. Biochem. Biophys., 326*, 85–92 (1996).
157. C. W. Fisher, M. S. Shet, D. L. Caudle, C. A. Martin-Wixtrom, and R. W. Estabrook, *Proc. Nat. Acad. Sci. USA, 89*, 10817–10821 (1992).
158. E. M. J. Gillam, R. M. Wunsch, Y.-F. Ueng, T. Shimada, P. E. B. Reilly, T. Kamataki, and F. P. Guengerich, *Arch. Biochem. Biophys., 346*, 81–90 (1997).
159. A. Parikh, E. M. Gillam, and F. P. Guengerich, *Nat. Biotechnol., 15*, 784–788 (1997).
160. M. Fairhead, S. Giannini, E. M. J. Gillam, and G. Gilardi, *J. Biol. Inorg. Chem., 10*, 842–853 (2005).
161. E. M. Gillam, A. A. Aguinaldo, L. M. Notley, D. Kim, R. G. Mundkowski, A. A. Volkov, F. H. Arnold, P. Soucek, J. J. DeVoss, and F. P. Guengerich, *Biochem. Biophys. Res. Commun., 265*, 469–472 (1999).
162. D. Kim, Z.-L. Wu, and F. P. Guengerich, *J. Biol. Chem., 280*, 40319–40327 (2005).
163. T. Shimada, E. M. J. Gillam, P. Sandhu, Z. Guo, R. H. Tukey, and F. P. Guengerich, *Carcinogenesis, 15*, 2523–2529 (1994).
164. A. Parikh, P. D. Josephy, and F. P. Guengerich, *Biochemistry, 38*, 5283–5289 (1999).
165. C.-H. Yun, G. P. Miller, and F. P. Guengerich, *Biochemistry, 39*, 11319–11329 (2000).
166. D. Kim and F. P. Guengerich, *Biochemistry, 43*, 981–988 (2004).
167. D. Kim and F. P. Guengerich, *Arch. Biochem. Biophys., 432*, 102–108 (2004).
168. D. Kim and F. P. Guengerich, *Annu. Rev. Pharmacol. Toxicol., 45*, 27–49 (2005).
169. K. Nakamura, M. V. Martin, and F. P. Guengerich, *Arch. Biochem. Biophys., 395*, 25–31 (2001).
170. F. P. Guengerich, J. L. Sorrells, S. Schmitt, J. A. Krauser, P. Aryal, and L. Meijer, *J. Med. Chem., 47*, 3236–3241 (2004).
171. Z.-L. Wu, L. M. Podust, and F. P. Guengerich, *J. Biol. Chem., 280*, 41090–41100 (2005).

172. E. E. Scott, Y. Q. He, and J. R. Halpert, *Chem. Res. Toxicol., 15*, 1407–1413 (2002).
173. E. E. Scott, M. Spatzenegger, and J. R. Halpert, *Arch. Biochem. Biophys., 395*, 57–68 (2001).
174. S. Kumar, S. Chen Chong, J. Waxman David, and R. Halpert James, *J. Biol. Chem., 280*, 19569–19575 (2005).
175. S. Kumar, D. R. Davydov, and J. R. Halpert, *Drug Metab. Dispos., 33*, 1131–1136 (2005).
176. M. Morant, S. Bak, B. L. Moller, and D. Werck-Reichhart, *Curr. Opin. Biotechnol., 14*, 151–162 (2003).
177. J. Ogata, Y. Kanno, Y. Itoh, H. Tsugawa, and M. Suzuki, *Nature, 435*, 757–758 (2005).
178. Y. Fukui, Y. Tanaka, T. Kusumi, T. Iwashita, and K. Nomoto, *Phytochemistry, 63*, 15–23 (2003).
179. G. A. Schoch, R. Attias, M. Le Ret, and D. Werck-Reichhart, *Eur. J. Biochem., 270*, 3684–3695 (2003).
180. M. Schalk and R. Croteau, *Proc. Nat. Acad. Sci. USA, 97*, 11948–11953 (2000).

15

Chemical Defense and Exploitation. Biotransformation of Xenobiotics by Cytochrome P450 Enzymes

Elizabeth M. J. Gillam and Dominic J. B. Hunter

School of Biomedical Sciences, The University of Queensland,
St. Lucia, Brisbane, Queensland, 4072, Australia
<e.gillam@uq.edu.au>
<d.hunter1@uq.edu.au>

1. INTRODUCTION. CHEMICAL DEFENSE	478
1.1. Evolution of P450-Mediated Chemical Defense	480
1.2. Scope of this Review	480
2. P450 SYSTEMS INVOLVED IN CHEMICAL DEFENSE	481
2.1. Mammals	481
2.1.1. Substrate Specificities of the Major Subfamilies	482
2.1.2. Regulation by Receptor-Mediated Systems and Other Mechanisms	486
2.2. Non-mammalian Vertebrates	487
2.2.1. Birds	488
2.2.2. Reptiles and Amphibians	490
2.2.3. Fish	493
2.3. Invertebrates	497
2.3.1. Non-vertebrate Chordates	498
2.3.2. Insects	498
2.3.3. Crustaceans	507
2.3.4. Echinoderms	508
2.3.5. *Caenorhabditis elegans* and Other Nematodes	509
2.3.6. Annelids	510

		2.3.7. Molluscs	511
		2.3.8. Cnidarians and Poriferans	512
	2.4.	Protista	513
	2.5.	Plants	513
		2.5.1. Metabolism of Herbicides and Other Xenobiotics	514
		2.5.2. Bioactivation of Xenobiotics by Plant P450s	517
		2.5.3. Induction of Xenobiotic Metabolizing P450s in Plants and Herbicide Safeners	519
		2.5.4. Implications of Immobility	520
	2.6.	Fungi	521
		2.6.1. Defense in Phytopathogenic Fungi	521
		2.6.2. Xenobiotic Mineralization	524
		2.6.3. Catabolism in Alkane-Assimilating Yeasts	524
	2.7.	Prokaryotes	525
		2.7.1. Catabolism of Xenobiotics for Carbon Substrate Exploitation: Alkane Oxidation and Other Biotransformations	525
3.	COMMON THEMES		529
4.	INDUSTRIAL APPLICATIONS OF P450 SYSTEMS FOR XENOBIOTIC DECOMPOSITION		530
	4.1. Pharmaceutical and Fine Chemical Synthesis		530
	4.2. Bacterial Systems for Bioremediation		532
	4.3. Phytoremediation and Herbicide Tolerance		533
	4.4. Biosensors		533
5.	CONCLUSIONS AND FUTURE PROSPECTS		534
	ACKNOWLEDGMENTS		535
	ABBREVIATIONS		535
	REFERENCES		537

1. INTRODUCTION. CHEMICAL DEFENSE

The task of reviewing the metabolism of xenobiotics by cytochrome P450 enzymes is a daunting one. P450s have been exploited throughout the biosphere, by almost all organisms so far investigated, because of their powerful ability to introduce an oxygen into an unactivated carbon center. It has been argued that the explosion in diversity seen in certain P450 families has been a direct response to the elaboration of environmental chemicals by other organisms, such as in animal-plant chemical warfare [1]. While P450s have also developed to fulfil specific physiological roles in most organisms, considerable genetic resources have been invested in P450s which serve primarily to enhance the clearance of xenobiotics with only limited roles in endobiotic clearance.

'Chemical defense', i.e., the response of an organism to foreign chemicals, can be viewed in evolutionary terms as a result of selection pressure imposed by the toxic effects of xenobiotics. Environmental chemicals pose a survival challenge for an organism, since they may exert adverse biological effects through interaction with receptors or interference with physiological processes. Alternatively such chemicals may react and form covalent adducts with endogenous molecules. Systems such as the mammalian kidney serve to non-specifically eliminate water-soluble chemicals. Chemicals that present the greatest problem are those that partition within lipid membranes, since these can accumulate in tissue and exert long-lasting effects.

For organisms to survive to reproduce in a chemically diverse environment, a system is needed to safely detoxify an almost limitless variety of chemicals, including those to which the species has not been previously exposed. The role of P450s in overcoming this problem is deceptively simple and elegant. Introduction of oxygen to a molecule serves to enhance water solubility, allowing chemicals to be eliminated via non-specific processes such as filtration in the kidney. Additionally monooxygenation serves to functionalize organic molecules for further metabolism (e.g., by conjugation). Since function is intimately linked with structure, biotransformation by P450s commonly alters the biological (i.e., receptor) or chemical reactivity of the molecule. While this will not result in detoxification in every case, enhanced water solubility generally facilitates clearance. Hence, a system for monooxygenation of diverse chemicals will, on the whole, serve to protect an organism from chemical challenge.

The remarkable chemistry effected by P450s is the subject of chapters elsewhere in this volume. The aspect critical to a discussion of xenobiotic metabolism is the *versatility* that the chemistry enacted at the P450 active site confers on these enzymes. By being able to introduce an oxygen into 'hard' nucleophiles, such as saturated carbon centers, but also heteroatoms and unsaturated carbons, P450s can biotransform an unequalled variety of organic molecules. This provides tremendous scope for specialization of these enzymes; modification of the protein binding site surrounding the heme prosthetic group allows adaptation to bind different types of xenobiotic substrates. However, fitness relies on the ability to survive challenge by numerous chemicals in the diet, air, water, and other environmental sources. Therefore, it is not in an organism's interest to overspecialize: effective chemical defense is a compromise between developing enzyme catalysts sufficiently efficient to detoxify chemicals before they can cause intolerable toxicity, while maintaining the versatility to respond to diverse chemical structures. Bearing in mind that the amount of genetic resources that can be invested in chemical defense is finite (i.e., it is not possible to have hundreds of relatively specialized genes to deal with the full variety of environmental toxins), it is not surprising that most organisms exploit a relatively small number of P450s for chemical defense, each with wide and overlapping substrate specificity but relatively poor catalytic rates on any one substrate. In these characteristics, P450s, along with other enzymes and transporters involved in xenobiotic clearance, contradict the classical view of enzymes as efficient and specialized biological catalysts.

1.1. Evolution of P450-Mediated Chemical Defense

From the earliest days of P450 investigation, it has been proposed that P450s functioned prior to the accumulation of oxygen in the atmosphere, possibly serving initially to scavenge oxygen that was toxic to early organisms and then later making use of the aerobic conditions for monooxygenations [2]. More recently, and in light of the accumulating mass of sequence information on P450s from diverse sources, P450s have been proposed to have arisen more than 3.5 billion years ago [3,4] from a putative common ancestor. As gene duplication and diversification progressed, certain ancestral P450 forms appear to have been recruited for chemical defense in each phylum resulting in lineage specific expansions of different P450 gene families and subfamilies. For example, CYP1-4 appear to have been exploited by vertebrates, with a very dramatic expansion in the CYP2 family, while insect P450s are derived from CYP4, CYP6, CYP9, and CYP12. Recent investigations in *Caenorhabditis elegans* suggest CYP13, CYP14, CYP33, CYP34, and CYP35 have diversified in nematodes (vide infra). The expansion in families such as CYP2 has been linked to the movement of animals onto land to exploit the new varieties of plants in the terrestrial environment as food [1]. Many of the families that have diversified in different groups of organisms cluster together phylogenetically, e.g., CYP6 and CYP28 in insects and CYP13 and CYP25 from nematodes cluster with CYP3 from vertebrates, while nematode CYP33, CYP34, CYP35, and CYP14 families are related to CYP2 in vertebrates [5].

As will be clear in subsequent sections, the last several years have seen an explosion of P450 sequence information for various organisms, especially as a result of the various genome projects. Unfortunately functional studies have lagged behind characterization of gene sequences and for no P450, not even the most heavily studied human forms, can we be entirely confident of defining a complete substrate range. The thermodynamic reality that some xenobiotics will be metabolized like endobiotics if they are structurally similar means that often it is difficult to determine whether a P450 has a principal or adventitious role in xenobiotic metabolism. For some systems (e.g., plants, vide infra) the separation between physiological roles and chemical defense is particularly unclear.

1.2. Scope of this Review

By far the most heavily studied P450s are those from mammals, and in particular those from humans. The study of mammalian xenobiotic-metabolizing

P450s has led to the discovery and characterization of many of the paradigms of xenobiotic metabolism throughout the biosphere. As such, mammalian xenobiotic-metabolizing P450s serve as comparators against which it is useful to evaluate what is known and understood about chemical defense in other types of organisms. Research on mammalian systems has also provided many of the essential tools for examination of other systems, such as simple assays for measuring activity towards xenobiotics (e.g., the alkoxyresorufin and alkoxycoumarin dealkylation assays), which are widely used for biomonitoring.

The purpose of this review is not to catalogue the full variety of xenobiotic clearance reactions in all organisms; the field is far too wide for such a Herculean effort. Rather, in the first part of the review, we will provide a general overview, compare and contrast different types of organisms, discuss some of the more informative and interesting examples and, in so doing, draw out general concepts that underpin this evolutionary response to chemical challenge. In particular, our principal focus will be on non-mammalian systems and we will intentionally avoid any significant discussion of a key group of xenobiotics, namely therapeutic drugs, since these have been given thorough and comprehensive coverage in Chapter 17 of this volume. However, in order to provide a framework within which to discuss other systems, a brief overview of mammalian xenobiotic metabolism will be provided. Likewise, while our focus will be on xenobiotic metabolism, we will make reference to certain endobiotic substrates where necessary. In the second part of the review, we will summarize how the properties of P450s have been exploited in various industrial applications.

2. P450 SYSTEMS INVOLVED IN CHEMICAL DEFENSE

2.1. Mammals

Xenobiotic metabolism by P450s in mammals, especially humans, has been the subject of intensive study for four decades or more, due to the importance of understanding the biotransformation of drugs and environmental chemicals for human health. More recently, particular groups such as marine mammals and marsupials have been studied for purposes of biomonitoring [6] and to explore evolutionary relationships in P450s or the response of animals to unusual dietary chemicals (e.g., folivorous marsupials and terpenes in eucalyptus leaves [7]).

Total levels of xenobiotic-metabolizing P450s are found at highest in the endoplasmic reticulum of the liver in mammals [8], where they metabolize chemicals delivered directly from the gut via the portal circulation and can facilitate excretion of products via the bile or release water-soluble metabolites into the bloodstream for elimination by the kidney. However, other organs directly exposed to the environment ('organs of entry and exit') also contain substantial P450 levels, and P450s have been demonstrated in almost all tissues examined [8–10].

Ding and Kaminsky have pointed out that P450s in the gastrointestinal tract serve as a primary barrier to the entry of chemicals by both detoxification and bioactivation of xenobiotics: those chemicals that are biotransformed to reactive products may covalently bind to nucleophilic components of intestinal cells and be sloughed off rather than taken into the body [9].

The functional importance of P450s has been underestimated in tissues such as brain and the vascular endothelium where expression is localized to discrete cell populations [8]. Rifkind has pointed out that the vascular endothelium, a major site of expression of forms such as P450 1A1, is actually the largest organ in the body [11]. Thus, the quantitative importance of extrahepatic forms may have been underestimated.

P450s and other xenobiotic-metabolizing enzymes have a particular role in the olfactory mucosa in addition to their normal, clearance roles in the liver and other tissues. P450s terminate olfactory stimuli by clearing chemicals from around olfactory neurons, thereby allowing the olfactory system to respond to new odorants [12]. Almost all mammalian subfamilies are represented in the olfactory epithelium [9] and some forms are relatively specific for this tissue, such as P450 2G1, rat P450 2A3, mouse P450 2a5, and human P450 2A13 (reviewed in [12]). Levels of P450 in the nasal mucosa have been reported to be higher than in any other extrahepatic tissue in most mammals with the exception of humans [9], perhaps due to the decreased importance of the sense of smell in our species. Interestingly, the conserved nasal specific form, P450 2G1, seems to be inactive in humans but functional in other species [9].

Families 1–4 catalyze almost all P450-mediated clearance of exogenous chemicals in mammals. At the time of writing (May 2006), subfamilies CYP2A, CYP2B, CYP2D, CYP2E, CYP2F, CYP2G, CYP2J, CYP2S, CYP2T, CYP2W, CYP2AB, CYP4A, CYP4B, CYP4X, and CYP4Z appear to be exclusively mammalian so far, while CYP1A, CYP1B, CYP2C, CYP2R, CYP2U, CYP2AC, CYP3A, CYP4F, and CYP4V are shared with other vertebrates (see http://drnelson.utmem.edu/P450.stats.all.2005.htm). Some families from which only mammalian forms are currently known may prove in time to contain fish, bird, reptile or amphibian sequences once more is discovered about the complement of P450s in these groups. Many of the newly characterized P450s in mammals appear to be expressed mostly outside the liver and their roles in xenobiotic metabolism remain to be established [13,14].

2.1.1. Substrate Specificities of the Major Subfamilies

The increasing availability of recombinant human P450s over the last fifteen years has reduced the interest in investigating P450s in other mammals. Thus, many informative reviews are available on the substrate specificities of human forms [15–17], but fewer are available for other animals [18–24]. The reader is referred especially to the database established by Rendic [25] for a comprehensive listing of human P450 substrates. In the following discussion, we will simply highlight some of the key aspects related to mammalian subfamilies and note characteristic substrate specificities including marker substrates that have been used extensively [19].

Guengerich has proposed a rough generalization of mammalian P450s into four groups based on the degree of extrapolation of substrate specificities that is possible among different species' P450 forms within a given subfamily [18]. For P450 2E, there is a high degree of conservation of substrate specificity between species. P450s 1A, 1B, and 4B show more differences necessitating some caution. P450 2D and P450 3A subfamily forms show more discrepancies between species but significant correlations are still obvious. Finally, for P450s 2A, 2B and 2C, extrapolations are very unreliable.

2.1.1.1. The P450 1 Family. P450s from the 1A and 1B subfamilies have well established roles in the metabolism of polycyclic aromatic hydrocarbons (PAH), aromatic (AAs) and heterocyclic amines (HAs) and diverse other chemicals including 17β-estradiol, many of which are bioactivated to proximate carcinogens or toxins [15,26,27]. Different but overlapping substrate ranges are seen for P450 1A1, 1A2, and 1B1. P450 1A1 has been more often associated with PAH metabolism and this has led to the proposal that the active site readily accommodates planar molecules [28].

Marker activities for P450 1A1 include benzo(a) pyrene (BaP) 3-hydroxylation (often abbreviated AHH for aryl hydrocarbon hydroxylase) and 7-ethoxyresorufin O-deethylation (EROD) [19] and these have served widely as measures of P450 1A induction in biomonitoring studies. P450 1A2 is responsible for most aromatic amine and heterocyclic amine metabolism in the liver. Prototypical P450 1A2 substrates are caffeine N-demethylation and phenacetin O-deethylation. In rats theophylline $N3$-demethylation and 7-methoxyresorufin O-demethylation (MROD) have been proposed as marker substrates whereas EROD is used for human P450 1A2 [19]. P450 1B1 marker reactions are 17β-estradiol 4-hydroxylation and EROD. Considerable overlap is seen between all three human enzymes in inhibitor sensitivity [29], and activities of each isoform are qualitatively comparable across species examined to date, although significant quantitative differences are evident [18]. All three forms appear to be induced via the aryl hydrocarbon receptor (AhR) by a variety of xenobiotic ligands such as the PAHs, 3-methylcholanthrene (3MC) and β-naphthoflavone (BNF) many of which are also substrates.

2.1.1.2. The P450 2 Family. The P450 2 family has diversified to the greatest extent in mammals of all P450 families. Forms from the 2A-2F subfamilies have clear roles in xenobiotic metabolism and other, as yet largely uncharacterized subfamilies may prove to be involved [14].

P450 2A6 from humans has recently been crystallized [30]. The active site appears small relative to other mammalian P450s for which structures are available, and this is in line with the substrate range characterized (including alkoxyethers and several nitrosamines; reviewed in [14]). Coumarin 7-hydroxylation is used as a marker activity for both P450 2A6 and 2A13 [19] but indole oxidation [31] is also useful. Testosterone 7α- and testosterone 15α- and 12α-hydroxylations are used as marker activities for rat P450s 2A1 and 2A2, respectively [19]. Both human P450 2A forms, plus rat P450 2A3 and mouse P450 2a5, are expressed in the respiratory tract and may play some role in olfaction (vide supra).

Importantly, P450s 2A6 and 2A13 metabolize nicotine and activate the tobacco procarcinogen, 4-(methylnitrosamino)-1-(3-pyridyl)-1-butanone (NNK) and may play key roles in cancer of the respiratory tract (reviewed in [9]). Interestingly, while most mammals characterized to date show two or three P450 2A forms, six hamster P450 2A enzymes are known (see http://drnelson.utmem.edu/ P450.stats.all.2005.htm). Another unusual feature of the hamster is that one gene CYP2A8 is induced by the classical AhR ligand, 3MC [32].

P450 2B forms in animals (e.g., rat P450 2B1 and 2B2) were first characterized as the major forms induced by phenobarbital (PB). Testosterone 16β-hydroxylation and 7-pentoxyresorufin O-dealkylation (PROD) are marker activities for rat P450 2B1 whereas (S)mephenytoin N-demethylation, 7-ethoxy-4-trifluoromethylcoumarin O-deethylation and buprorion hydroxylation are useful for the human form P450 2B6 [19]. P450 2B4 from rabbits has been the subject of extensive studies concerning the catalytic cycle of P450 and its interaction with redox partners. The recent crystallization of this form has provided evidence for the ability of P450s to adopt open and closed conformations, allowing access of large substrates to the heme center [33].

The P450 2C is the largest mammalian subfamily with at least 89 forms currently known. Forms in rats were originally isolated as male-specific or female-specific and pronounced sex differences in P450 2C expression exist in animals (but not apparently in humans). Marker reactions include [19]: progesterone 21-hydroxylation and S-warfarin 7-hydroxylation (rat P450 2C6); testosterone 16α-hydroxylation and progesterone 16α-hydroxylation (rat P450 2C7); 5α-androstane-3a,17β-diol, 3,17-disulfate 15β-hydroxylation and progesterone 2α-hydroxylation (rat P450 2C11); testosterone 6β-hydroxylation (rat P450 2C13); paclitaxel 6α-hydroxylation (human P450 2C8); diclofenac 4'-hydroxylation, tolbutamine methylhydroxylation, phenytoin 4-hydroxylation and (S)warfarin 7-hydroxylation (human P450 2C9); 3-[2,3-dichloro-4-(2-thenoyl)phenoxy]-propan-1-ol 5-hydroxylation and phenytoin 4-hydroxylation [34] (human P450 2C18); and (S)mephenytoin 4'-hydroxylation (human P450 2C19). Some clear differences in substrate specificity are evident, such as the ability of P450 2C8 amongst human forms, to accommodate larger substrates and the large proportion of negatively charged substrates metabolized by P450 2C9. Some of these differences have been rationalized by the available crystal structures [35–38].

P450 2D forms came to prominence due to the profound interindividual differences seen in drug metabolism due to the genetic polymorphism in P450 2D6. As a result, more is known about the substrate preferences of this form than any other mammalian P450, culminating the recent publication of the crystal structure [39]. P450 2D6 has a widely reported apparent preference for basic substrates, but a positive center is not an absolute requirement for substrate binding or metabolism, as evidenced by a variety of neutral substrates including steroids and the lack of pH effect on substrate binding [40]. Marker substrates include debrisoquine 4-hydroxylation, (+/−)bufuralol 1'-hydroxylation and dextromethorphan O-demethylation [19]. Many characteristic P450 2D6 activities are also shown by other P450 2D forms from rat, dog, etc. [19,23].

The substrate specificity of P450 2E1 forms is relatively well conserved across species in line with the single isoform designation. Originally characterized as the 'microsomal ethanol oxidizing system', P450 2E1 metabolizes low molecular weight organic compounds including solvents, nitrosamines, halogenated simple hydrocarbons, and lauric acid (by ω-1 hydroxylation) (reviewed by Guengerich [15]). Marker substrates include chlorzoxazone 6-hydroxylation, N-nitroso-N,N-dimethylamine N-demethylation and p-nitrophenol hydroxylation [19].

P450 2F forms are relatively poorly characterized to date, but the human form appears to play some role in the bioactivation of certain toxins such as 4-ipomeanol, 3-methylindole, styrene and naphthalene in the lung and also catalyzes ECOD, PROD and propoxycoumarin O-dealkylation (reviewed in [9]).

2.1.1.3. The P450 3 Family.

P450 3A forms appear to be both quantitatively and qualitatively the most important contributors to xenobiotic metabolism in mammals. Originally characterized as forms induced by glucocorticoids (e.g., dexamethasone (DEX)), antiglucocorticoids (e.g., pregnenolone-16α-carbonitrile), phenobarbital and some macrolide antibiotics (e.g., rifampicin, troleandomycin), they have subsequently been shown to metabolize a vast range of substrates of all sizes and natures, including macrolide drugs and fungal products (cyclosporin), alkylphenyl ether detergents (e.g., Triton N101), steroids (mostly by 6β-hydroxylation) (reviewed in [15]). P450 3A4 has also been associated with the activation of several procarcinogens including alkaloids, polyaromatic hydrocarbon diols, and fungal products such as aflatoxin B1 and G1 [27]. Marker substrates for rat and human enzymes include testosterone 6β-hydroxylation, erythromycin N-demethylation, nifedipine oxidation, midazolam 1'-hydroxylation [19] but significant differences in substrate specificity are seen across species [18].

P450 3A forms are the system in which cooperative phenomena and the associated 'atypical' kinetics have been best characterized, i.e., the ability of a substrate to alter its own metabolism (homotropic effects: positive cooperativity or substrate inhibition) or of one compound to affect the metabolism of a second compound (positive or negative heterotropic effects). In the case of human P450 3A4 at least, this has been proposed to result from the simultaneous binding of multiple molecules within the P450 active site (reviewed in [41]).

2.1.1.4. The P450 4 Family. P450 4 forms have been characterized in mammals as fatty acid hydroxylases, with predominant roles in the oxidation of such endobiotic substrates as lauric and arachidonic acid, and often producing important endogenous modulators such as the hydroxyeicosatrienoic acids [42]. Their involvement in xenobiotic metabolism in mammals appears limited to those substrates that resemble the endobiotic substrates, such as esters of fatty acids and lipophilic organic acids such as valproic acid. An exception appears to be P450 4B1, which in rabbits has been shown to metabolize 2-aminofluorene, 2-aminoanthracene, 4-ipomeanol and 3-methylindole amongst other substrates (reviewed in [15]). However, the human form has proven difficult to express, and its catalytic specificity is unclear.

2.1.2. Regulation by Receptor-Mediated Systems and Other Mechanisms

P450 forms from different subfamilies are upregulated in response to chemical exposure via nuclear receptor-mediated transcriptional changes. P450 1A and 1B forms are regulated by the aryl hydrocarbon receptor system [43], one of the PAS (Per, ARNT, Sim) family of receptors. Typical AhR ligands are the PAHs, 3MC, BNF and the halogenated aromatic hydrocarbon, 2,3,7,8-tetrachlorodibenzo-*p*-dioxin. The nuclear hormone receptors CAR (constitutive androstane or constitutively activated receptor) and PXR (pregnane X receptor otherwise known as the SXR, (steroid X receptor)) regulate the expression of many CYP2B, CYP2C and CYP3A genes in mammals. While the ligand specificities of PXR and CAR overlap to a large degree, CAR responds preferentially to PB and PXR responds to the typical CYP3A inducers: glucocorticoids (e.g., dexamethasone (DEX)), antiglucocorticoids (e.g., pregnenolone-16α-carbonitrile (PCN)) and rifampicin. Both PXR and CAR forms use the same heterodimerization partner, the retinoid X receptor (RXR).

The peroxisome proliferator activated receptor α (PPARα) controls the expression of many CYP4 genes in an analogous manner and responds to a variety of ligands including the classical CYP4 inducer, clofibrate (CF) [44]. In addition, the levels of several forms, principally P450 2E1, can be modulated by xenobiotic substrates by stabilization of the protein against degradation [45]. The formation of quasi-irreversible metabolite intermediate complexes from metabolism of methylenedioxy or amine-containing substrates, transformed ultimately into carbene and nitroso metabolites, respectively, is an extreme form of substrate stabilization but results in effective inactivation of the P450 [46].

2.2. Non-mammalian Vertebrates

The investigation of xenobiotic metabolism by P450s in non-mammalian vertebrate forms has been driven by several factors [47]:

(a) the need to assess the environmental impact of contamination by petrochemicals, pesticides, and other pollutants on growth, reproduction, behavior, and predation in wildlife. Induction of P450 activities, particularly those mediated by forms from the CYP1 and CYP2 families, is a sensitive way of quantifying the effects of exposure to certain chemicals and has been used in many biomonitoring studies in fish, birds, reptiles, and other wildlife;
(b) the development of avian or amphibian systems as test systems for toxicities relevant to humans, e.g., *in ovo* treatment of avian embryos for toxicity and genotoxicity testing, *Xenopus laevis* FETAX systems [48] and other test systems [47];
(c) the investigation of toxicities or chemical sensitivities particularly pronounced in certain animals, e.g., aflatoxicosis in poultry;
(d) for effective care of farmed animals (e.g., [49]); and
(e) to assess potential contamination of these animals as human foodstuffs (e.g., alligator and crocodile meat, frog legs, turtle eggs).

In the species of fish for which genomic data are available, and in the chicken (*Gallus gallus*) and *Xenopus laevis*, information on the functional capacities of individual forms has lagged behind the generation of sequence data, and transcriptome analysis may soon supercede measurement of activities as a means of assessing chemical exposure and impact. However, this will require more information to be obtained regarding the relationship of expression of 'sentinel' P450 forms to possible toxicity in monitored species. Other species, especially reptiles and amphibians, are largely unexplored with respect to both sequence and functional characterization of P450s.

All four P450 families thought to be responsible for xenobiotic metabolism in mammals are also represented in non-mammalian vertebrates and evidence has been obtained for a role for at least some members in the clearance of exogenous chemicals. However, different subfamilies are involved in more distantly related animals and orthologous relationships are generally not clear. Likewise, clear differences are seen in the nuclear receptor systems involved in the regulation of these P450s.

2.2.1. Birds

P450 levels are generally lower in birds than in most mammals examined [50–55] and appear to be inversely correlated with the degree of specialization of the diet [56]. This is consistent with the hypothesis that generalist feeders have a greater need to detoxify chemicals present in diverse prey than those with specialized diets. (e.g., [57].) Most studies on xenobiotic metabolism in birds have been done for the purpose of biomonitoring and have involved measurement of EROD or AHH activities in avian livers, induced in response to environmental contaminants such as polychlorinated biphenyls and pesticides.

2.2.1.1. Avian P450s Involved in Xenobiotic Metabolism.

At the time of writing (May 2006), almost all cloned avian P450s are from the domestic chicken (*Gallus gallus*). At least 41 forms have been described (see http://drnelson.utmem.edu/P450.stats.all.2005.htm). Amongst these are genes for chicken CYP1A4, CYP1A5, CYP1B1 and CYP1C like sequences, CYP2C45, at least five CYP2J genes, CYP2H1, CYP2H2, CYP2AC1, CYP2AC2, several other CYP2 related sequences, CYP3A37 and a second P450 3A form and multiple CYP4 genes. In addition, CYP1A5 has recently been cloned from turkey (*Meleagris gallopavo*) [58] and jungle crow [59]. P450 1, P450 2, and P450 3 forms have been shown to catalyze xenobiotic metabolism in chicken and certain other species.

2.2.1.1.1. P450 1 forms.

Two TCDD-induced CYP1As, CYP1A4 and CYP1A5, were cloned from chick embryo [60], with both showing more sequence similarity to mammalian CYP1A1 than CYP1A2. Recent sequence analysis suggests the CYP1A gene duplication occurred prior to the avian/mammalian divergence, and CYP1A4 and CYP1A5 appear to be orthologues of CYP1A1 and CYP1A2, respectively (D. R. Nelson, personal communication) [61]. This is consistent with functional studies, which suggest P450 1A5 is more P450 1A2-like and P450 1A4 is more P450 1A1-like [62]. P450 1A4 catalyzes EROD and AHH activities (P450 1A1 activities in rodents) as well as 3,3',4,4'-tetrachlorobiphenyl (TCB) metabolism in birds [63]. By contrast, P450 1A5 shows arachidonic acid epoxygenase activity (more pronounced in avian liver than mammalian) as well as some mammalian P450 1A2-like activities including the oxidation of uroporphyrinogen, an intermediate of heme biosynthesis, to uroporphyrin, a reaction (UROX) involved in the porphyria caused by halogenated aromatics.

Evidence exists for two P450 1A forms in at least four different orders of birds [64]. Marked differences are seen between species with respect to catalytic specificity and response to classical CYP1A inducers [65]. Turkey P450 1A5 was cloned and shown to metabolize aflatoxin B1 to both M1 and epoxide metabolites as well as catalyzing EROD and MROD, activities that could be inhibited by α-naphthoflavone [58].

2.2.1.1.2. P450 2 forms. The first avian P450 genomic clones [66] and cDNA and protein sequences [67] were obtained from PB-treated chickens. PB-induced P450 2H1 and P450 2H2 from chicken resemble P450 2B1 and P450 2B2 from rat and catalyze benzphetamine *N*-demethylation [68]. In addition, P450 2H1 mediates *p*-nitrophenol hydroxylation and acetaminophen oxidation [68]. Two more PB-inducible forms have been purified [68,69]. Chicken P450 2C45, the first avian P450 2C identified, metabolizes scoparone [70].

2.2.1.1.3. P450 3 forms. The observation of erythromycin *N*-demethylation, a typical CYP3A activity, in liver microsomes from PB-induced chickens led to the isolation of the first avian P450 3A form, P450 3A37. P450 3A37 catalyzed testosterone, progesterone, and androstenedione 6β-hydroxylation when expressed in COS cells [71].

Various other forms have been purified from avian liver but not fully characterized or sequenced, such as an acetone-inducible form functionally similar to rat P450 2E1 (reviewed in [56]).

2.2.1.2. Regulation of Avian P450s. Birds appear to respond to most of the same canonical inducers as mammals, such as TCDD, 3MC, PB, ethanol, acetone, DEX, and PCN. However, there are large interspecies differences in the relative response to inducers [54,65]. The patterns of TCDD induction of CYP1A4 and CYP1A5 differ to those seen with mammalian CYP1A induction [72]. Birds have a single form of AhR, like mammals, along with an AHRR [73]. Recently, the variation in responsiveness to TCDD has been attributed to two amino acid changes in the Ah receptor that affect the binding affinity for TCDD [74].

The chicken xenobiotic receptor (CXR) mediates induction of CYP2H1 in chickens by diverse chemicals including the prototypical inducer, PB [75]. CXR appears to be relatively promiscuous amongst the PXR and CAR-type receptors identified to date and to segregate in the same pharmacological group as mammalian PXR [76,77]. CYP2C45 can also be induced via CXR [70] and by murine PXR and CAR suggesting a degree of evolutionary conservation of regulation mechanisms. CYP3A37 was found to be inducible by CYP3A inducers, such as metyrapone, PB, DEX and PCN, but not rifampicin, in the same manner as CYP2H1.

In addition to the AhR and CXR systems, ergosterol biosynthesis-inhibiting fungicides such as prochloraz, propiconazole, and vinclozolin appear to induce forms from multiple P450 families (reviewed in [56]). In some cases fungicides have been shown to act via the CXR.

2.2.1.3. Bioactivation of Xenobiotics by Avian P450s. The bioactivation of xenobiotics by avian P450s has important implications for food production and management of animal health. The best-characterized example of veterinary importance is the sensitivity of poultry, and especially domesticated turkeys to the mycotoxin, aflatoxin B1 (AFB1). Upon exposure to even small quantities of AFB1 in contaminated feed, turkeys develop aflatoxicosis [78]. This profound sensitivity is due to two factors: the ability of turkey liver microsomal P450 1A5 to efficiently activate AFB1 coupled with a deficiency in the glutathione S-transferase activity that serves to protect other species [58,79]. Butylated hydroxytoluene appears to ameliorate AFB1 bioactivation by reducing P450-dependent metabolism [78].

Two adverse interactions between pesticides have been described in birds. The first involves the potentiation of organophosphate toxicity by ergosterol biosynthesis inhibiting fungicides. Prochloraz in particular appears to enhance the activation of organophosphates such as malathion by inducing the P450 responsible for generation of the active oxon species, and therefore leads to increased butyrylcholinesterase inhibition [80]. Evidence was obtained for increases in P450 1A, P450 2C, and P450 4A related forms in red-legged partridges (*Alectoris rufa*) as well as enhancement of total P450, EROD, and aldrin epoxidase activity [81]. In ketoconazole-treated bobwhite quail, malathion, parathion, and diazinon activation was increased [82].

In the second adverse interaction between pesticides, the organophosphate malathion acts as a mechanism based inactivator of a P450 when it undergoes P450-catalyzed oxidative desulfuration. The result is a decreased ability of affected birds (red legged partridges) to detoxify carbamate insecticides such as carbaryl, leading to enhanced cholinesterase inhibition [83].

Whilst the P450 1, P450 2 and P450 3 families appear to be principally responsible for xenobiotic metabolism in birds as in other vertebrates, an important exception is found in the bioactivation of the insecticide 1,1,1-trichloro-2,2-bis(4′-chlorophenyl)ethane (DDT) in chickens via metabolism of its primary metabolites 3-methylsulfonyl-DDE and o, p′-DDD by adrenal P450 11β-hydroxylase leading to covalent adduct formation and adrenocorticolytic effects [84,85].

2.2.2. Reptiles and Amphibians

Hepatic P450 levels in reptiles and amphibians are generally towards the lower end of the range seen in mammals [86,87]. In most cases, relative activities seem lower than in mammalian species such as rat with some exceptions, such as ECOD activities in frogs [88] and turtles [89]. The fact that reptiles and amphibians are poikilotherms may explain the typically lower temperature optima seen for reptilian and amphibian P450s [90–92], as well as some of the seasonal and life-stage variation in P450 activities and induction [93–95] (reviewed in [47]), factors that are crucial to understand in order to make sensible conclusions from biomonitoring studies [96].

P450-dependent monooxygenase activities demonstrated so far in amphibians (various species of frogs and the newt, *Pleurodeles waltyl*) are: MROD, EROD, PROD, BROD, MCOD, ECOD, ProCOD, BuCOD, AHH, aminopyrine N-demethylation, p-nitrophenetole O-deethylation, aldrin epoxidation, hydroxylation of the dieldrin analogue, HCE, chlordene epoxidation and hydroxylation, naphthalene hydroxylation, aniline hydroxylation, N-nitrosodimethylamine N-demethylation, laurate hydroxylation, p-nitroanisole O-demethylation, hydroxylation of polyunsaturated fatty acids, cypermethrin aryl- and C2-methyl hydroxylation, p-chloro-N-methylaniline N-demethylation, coumarin 7-hydroxylation and isoproturon N-demethylation, C-hydroxylation and formation of the olefin metabolite.

In reptiles the list of P450 mediated xenobiotic biotransformations includes: AHH, EROD, PROD, ECOD, benzphetamine N-demethylation, erythromycin N-demethylation, testosterone hydroxylation, aminopyrine N-demethylation, p-nitrophenetole O-deethylation and hydroxylation of polyunsaturated fatty acids in snakes; EROD, MROD, PROD, AHH, and oxidation of TCB in turtles; EROD, MROD, PROD, BROD, aminopyrine N-demethylation, ECOD, AHH, BaP bioactivation to mutagenic products in alligators; nifedipine oxidation and diazepam C3-hydroxylation in crocodiles; EROD, ECOD, AHH and ethylmorphine N-demethylation in caimans; p-chloro-N-methylaniline N-demethylation in lizards; and EROD, PROD and AHH in salamanders (reviewed in [47]).

While there is insufficient data to judge the relative scope of xenobiotic metabolism in these organisms, it has been speculated that amphibians with a largely terrestrial habitat have higher activities than those living predominantly in the water and that activities develop along with migration of certain amphibians to land upon maturation [97]. This idea is not new: prior to the discovery of P450s, Brodie et al. proposed that land animals may have developed more specialized mechanisms for xenobiotic metabolism whereas in aquatic amphibians and fish, clearance of foreign chemicals might be facilitated by diffusion through the gills and skin [98]. However, to date, insufficient information is available to make any generalizations regarding the scope of xenobiotic clearance in aquatic vs terrestrial organisms.

2.2.2.1. P450 Forms Involved in Xenobiotic Metabolism in Reptiles and Amphibians. As in the case of birds, most studies have been undertaken to assess the effects of environmental contamination. To date relatively few forms of reptile or amphibian P450 have been cloned and sequenced. However, at least 75 P450s are now known from *Xenopus laevis* and *X. tropicalis*. From *X. laevis*, 35 forms have been defined, including P450s: 1A6, 1A7 [99]; 2Q1 (a subfamily unique to *Xenopus* at present); 2AC1 (a subfamily shared with rat and chicken forms); 2R and at least ten other P450 2 forms; 4F42, 4T3, 4T4, 4T6, 4T7 (a subfamily also in fish); 4V4 (a subfamily with mammalian and fish forms) [100]; and at least one other CYP4 form (see http://drnelson.utmem.edu/P450.stats.all.2005.htm). While the *X. tropicalis* genome has not been analyzed fully, only a single CYP1A form is believed to exist suggesting the two forms in *X. laevis* arose via the genome duplication that occurred in this lineage [61]. Thus *X. laevis* may have considerably more P450 forms than other amphibians.

A P450, later assigned to the P450 2Q subfamily, was purified from *X. laevis* liver which reacted with anti-P450 2 form antibodies and catalyzed activities characteristic of mammalian P450 2E1 [101,102]. An alligator P450 2 protein has been *N*-terminally sequenced and P450 2 and P450 3-related immunoreactivity has been seen in alligator liver microsomes (reviewed in [47]). CYP3A42 has been identified in the African Ball or Royal python (*Python regius*) [103] and P450 3-related immunoreactivity and cortisol 6β-hydroxylation has been found in a *X. laevis* kidney cell line [104].

2.2.2.2. Regulation of P450s in Reptiles and Amphibians. In reptiles and amphibians induction of xenobiotic metabolism by classical inducers often reveals different patterns to those seen in mammals, e.g., 3MC and BNF, which both act through the Ah receptor in mammals, differentially induce multiple forms in reptiles and amphibians [47]. Part of the confusion here may be related to the different substrate and reaction specificity of reptilian and amphibian P450s compared to their mammalian counterparts. Typically, longer times are needed for induction and induced levels of activity are less than in mammals [105]. Induction in some species has been shown to depend upon ambient temperature (summarized in [47,106]). *X. laevis* has two AhRs (α and β) but these display much lower affinity for TCDD and lower activity in reporter gene assays compared to mammalian forms which explains the relative insensitivity of *X. laevis* to TCDD and the lower degree of induction of CYP1A6 and CYP1A7 [107]. This has important implications for the use of *X. laevis* as a model organism in the FETAX developmental toxicity assay, as teratogens acting through the AhR may not be effectively assessed [108].

PCBs, such as Aroclor 1254 and TCB, act as inducers in reptiles and amphibians [89,109–114]. The results of induction with PB are generally much less pronounced, if seen at all [115–117]. Glucocorticoids induced P450 3A-related immunoreactivity in a *Xenopus* cell line [104] and CYP2Q1 is DEX-inducible [102]. No PXR or CAR sequence is available to date from a reptile or amphibian. Intriguingly, however, novel nuclear receptors, BXRa and BXRb (benzoate X receptors) have been cloned from *X. laevis* that appear to be related to both mammalian PXR and CAR and avian CXR [76], but show considerably restricted ligand specificity compared to the mammalian receptors. No BXR homologue has been found in fish suggesting these originated in amphibians [77]. While these nuclear receptors (NRs) belong to the same NR1 family and likely share a common evolutionary origin, no data have been obtained to date to suggest BXRs have a role in regulation of xenobiotic-metabolizing enzymes but may rather serve an endogenous role [76].

2.2.2.3. Bioactivation of Xenobiotics by Reptilian and Amphibian P450s.

There is a general paucity of information regarding the role of P450s in detoxification versus toxification of xenobiotics in reptiles and amphibians. It will be advantageous to determine when detoxification became a driving factor in evolution of P450s [47]. Importantly, the jelly mass surrounding the eggs of amphibians appears not to exclude many xenobiotics [118] making it highly relevant to determine, for example, whether the decreasing populations of certain amphibians might be due to insecticide contamination of aquatic environments [119].

2.2.3. Fish

Fish from uncontaminated waters show hepatic P450 levels between $\sim$0.2–0.5 nmol/mg microsomal protein [120], i.e., towards the lower end of the range seen in most mammals studied. However, marked variation was seen in total P450 levels and certain activities in tropical fish and higher levels are seen in certain species feeding on Gorgonian corals [121–123]. P450 2B and P450 3A-related immunoreactivity was higher in species with higher overall P450 levels [121].

Enzyme expression and activities [124] and the effects of inducers have been shown to alter according to season [125,126]. Cyclical alterations have also been seen between activities in males and females that may be due to hormonal or other reproductive stage effects [125]. Temperature optima for fish P450s tend to vary markedly between fish from tropical and temperate climates. Fish from colder waters show optima typically lower than mammalian forms, at $\sim$18–25 °C rather than 37 °C (reviewed in [125,127–131]), and may be related to different organization of the membrane milieu [132]. Temperature optima for a single enzyme may differ between in recombinant expression systems due to alterations in membrane organization and lipid content [133]. However, enzyme concentrations also appear to alter with acclimatization temperature for at least one P450 in rainbow trout [134].

Metabolism occurs predominantly in the liver but extrahepatic tissues that are directly exposed to the external environment, such as the intestine and gill [136–139], plus certain other organs, such as the trunk kidney, can also metabolize xenobiotics [135–139]. The gill, like the respiratory epithelium in mammals is suited to a role in xenobiotic metabolism due to the extensive surface area it presents to both the environment and the systemic circulation [140]. Levels of total P450 are roughly an order of magnitude or more lower than in liver [139]. P450s are also expressed in the olfactory organ and likely play a role both in xenobiotic (odorant) clearance and maintenance of sensitivity in that tissue [141].

P450 1A levels or activity are used most widely as biomarkers of environmental contamination [127,142]; however recent evidence suggests induction of CYP2 and CYP3A related forms may also be observed in heavily polluted sites [143]. Biomonitoring of P450 3A forms has been proposed [144] based, in part, on their anticipated predominance in overall xenobiotic metabolism. Fish such as the estuarine killifish (*Fundulus heteroclitus*) are useful for examining the effects of environmental contamination in different sites since they present as discrete populations due to their limited home ranges [145]. Zebrafish are becoming increasingly important as models for developmental toxicity, due to the transparency of the embryo and the convenience of their size and growth requirements. Scup are useful since they appear to be both sensitive and responsive to environmental toxicants and toadfish are used for the opposite reason: they appear relatively insensitive to induction by environmental agents [146].

2.2.3.1. P450 Forms Involved in Xenobiotic Metabolism in Fish. The complete genomes are available for three fish. The P450 complement in the freshwater pufferfish (*Tetraodon nigroviridis*) has not been reported to date, but *Takifugu rubripes* (pufferfish) is known to have 54 P450s, and *Danio rerio* (zebrafish) at least 81 (see http://drnelson.utmem.edu/P450.stats.all.2005.htm). All mammalian P450 families have been found in *Fugu* now, and therefore all 18 appear to have predated the divergence of fish and tetrapods [147]. As in mammals, P450 families 1–4 are believed to be responsible for xenobiotic metabolism but important differences exist between mammals and fish in the substrate specificities of key P450 classes (reviewed in [148]). The CYP2 family is the most divergent between *Fugu* and humans with only 2R1 and 2U1 conserved [147]. CYP2D and CYP2J in mammals correspond to CYP2N, CYP2P, CYP2V, and CYP2Z in fish [147].

It is clear that the P450 1 family in fish is larger than in other organisms, with CYP1A1, CYP1A3, CYP1A9, CYP1B1, CYP1B2, CYP1C1, and CYP1C2 sequences currently known. The CYP2 family appears to have undergone a major expansion in fish, similar to that seen in mammals, with the following forms (including some pseudogenes) known to date: CYP2K1-22, CYP2M1, CYP2N1-13, CYP2P1-11, CYP2R1, CYP2U1, CYP2V1, CYP2X1-12, CYP2Y1-4, CYP2Z1-2, CYP2AA1-9, CYP2AD1-6, and CYP2AE1. The CYP3 family has three subfamilies in addition to the one found in mammals. CYP3A27, CYP3A30, CYP3A38, CYP3A40, CYP3A45, CYP3A48, CYP3A49, CYP3A50P, CYP3A56, CYP3A65, CYP3A68, CYP3A69, CYP3B1, CYP3B2, CYP3C1, and CYP3D1 are now known. Finally, CYP4F7, CYP4F28, CYP4F43, CYP4T1, CYP4T2, CYP4T5, CYP4T8, CYP4V1, and CYP4V5-7 have been identified to date.

2.2.3.1.1. *P450 1A forms.* Most fish are believed to have a single P450 1A form, probably orthologous to P450 1A1 in mammals, however salmonids and eels appear to have two [61]. P450 1A1 from rainbow trout catalyzed MROD, phenacetin *O*-deethylation, acetanilide *p*-hydroxylation, 2-methylnaphthalene hydroxylation, EROD, AHH, dimethylbenzanthracene (DMBA) metabolism to the 2- and 4- phenol metabolites, 7- and 12- methylhydroxylated derivatives and 8,9- and 5,6-diols and trace amounts of the 3,4-diol [149–152]. Rainbow trout P450 1A3 catalyzed EROD, MROD, AHH and phenacetin *O*-dealkylation as well as DMBA metabolism [149–152]. *Tilapia* P450 1A was shown to catalyze EROD, ECOD, and AHH [138]. P450 1A forms also detoxify AFB1 to AFM1 [153].

2.2.3.1.2. *P450 2 forms.* Rainbow trout P450 2K1 catalyzes 7-pentoxyresorufin *O*-dealkylation, testosterone 16B hydroxylation, E2 2-hydroxylation and progesterone-16α-hydroxylation, benzphetamine *N*-demethylation, AFB1 epoxidation and the ω-1 and ω-2 hydroxylation of laurate [127,154–157]. Some activity was also seen towards and 2-methylnaphthalene and testosterone 6β-, 7α-, 16α-, and 16β-hydroxylation [149,154]. P450 2K6 from zebrafish (*Danio rerio*) catalyzes AFB1 activation to the exo 8,9-epoxide [158].

P450 2M1 from rainbow trout catalyzes the ω-6 hydroxylation of laurate, an activity is not apparently present in mammalian liver microsomes [133]. P450 2M1 was shown to catalyze the metabolism of testosterone (6β-hydroxylation), E2 (2-hydroxylation) and progesterone (6β- and 16α-hydroxylation) at modest rates [133,154].

P450 2M and P450 2K forms have been proposed to catalyze the ω-, ω-1, and ω-2 hydroxylation of nonylphenol in Atlantic salmon (*Salmo salar*) due to the similarity of products formed to those from lauric acid hydroxylation [159]. Interestingly, just as certain mammalian P450s show positive heterotropic cooperativity, diethyldithiocarbamate was shown to enhance P450 2K1-mediated laurate (w-1)hydroxylation in trout liver microsomes [160].

P450 2P and P450 2N forms seem to be related to P450 2J from mammals [143] with overlap in activity and tissue specificity [161]. P450 2P3 from killifish (*Fundulus heroclitus*) catalyzed benzphetamine-*N*-demethylation and arachidonic acid metabolism to 14,15-, 11,12-, and 8,9-epoxyeicosatrienoic acids and 19-hydroxyeicosatetraenoic acid, with some activity towards EROD and MROD [143]. P450 2Ns cloned from *F. heteroclitus* were shown to metabolize arachidonic acid (to 8R,9S-epoxyeicosatrienoic acid 11R,12S-epoxyeicosatrienoic acid and hydroxyeicosatetraenoic acids), and benzphetamine (by *N*-demethylation) with poor, if any, activity towards MROD, EROD, PROD, and BROD [161].

2.2.3.1.3. *P450 3 forms.* P450 3A27 from rainbow trout was the first fish CYP3A cloned and catalyzes the 6β-hydroxylation of testosterone and progesterone and benzphetamine *N*-demethylation [154]. Limited activity was also seen towards testosterone 7α-, 16α-, and 16β-hydroxylation and 16β-estradiol 2-hydroxylation [154]. Medaka P450 3A38 and P450 3A40 catalyzes testosterone hydroxylation (6β-, 2β-, and 16β- for 3A38 with 6β- and 2β- only for 3A40) and BFC dealkylation [162]. In channel catfish, AHH and benzo(a)pyrene 7,8 diol monooxygenase activity (to triols and tetraols) was proposed to be mediated by P450 3A and was activated by aNF, but no effect of aNF was seen on testosterone 6β-hydroxylation [163]. P450 3C1 from zebrafish (*Danio rerio*) has been expressed as holoenzyme in yeast but not functionally characterized [164].

2.2.3.2. Regulation of Xenobiotic-Metabolizing P450s in Fish.

Constitutive P450 1A levels, as principally indicated by EROD or AHH activity are low or undetectable in the liver of most fish [125]. Many studies have shown AhR-coupled CYP1A induction in teleosts and elasmobranchs (cartilaginous fish); however, jawless fishes (lamprey, hagfish) appear to lack induction by typical PAH and HAH inducers of the AhR, while still expressing a P450 1A immunoreactive protein (reviewed in [165]). Bony fishes have at least two AhR genes, AhR2 being predominant in mediating the response to xenobiotics [73]. Some fish (e.g., zebrafish) lack TCDD binding and a functional transactivation domain in at least one form of AhR1 [73] (but zebrafish AhR1b retains responsiveness [166]). Cartilaginous fishes also have two AhRs and there may be more diversity still with an AhR3 reported in shark and dogfish [167]. Fish have also been shown to express AHRR which serves as a AhR-inducible negative regulator of AhR function by competing for binding to ARNT. Like mammalian AHRR, fish AHRR does not bind TCDD or BNF. Interestingly, polymorphisms of AhR1 have been found in Atlantic killifish populations associated with enhanced resistance to dioxin toxicity although a functional link between AhR1 and the mechanism of resistance is as yet unclear [168].

The classical induction of CYP2 forms by PB seen in mammals appears not to occur in fish [169,170]. Instead, PB appears to induce fish CYP1 via an AhR-mediated effect [171] rather than any direct effect on CYP2 forms [170,172,173]. The mechanism appears to differ from that of TCDD and may involve protein kinase-dependent signal transduction pathways [173].

To date no fish homologue of mammalian CAR has been characterized or shown to mediate P450 induction. The *Fugu* genome contains a nuclear receptor with significant homology to both mammalian PXR and CAR [77]. However, it has been defined as a PXR due to a characteristic insertion that expands the ligand binding site compared to the more restricted site in CAR [77].

PCN and DEX, typical CYP3A inducers in mammals acting through the PXR (or SXR) give contradictory effects in fish. No consistent increases are seen in any activity or total P450 levels. Rather decreases are seen in some studies (reviewed in [127]) whereas in others CYP1A induction by TCDD, TCB and BNF was observed to be increased by glucocorticoids but not other types of PXR ligands [174–176]. CYP3A65 [177] but not CYP3C1 [164] was induced by rifampicin and DEX. Intriguingly TCDD also induced CYP3A65 and an AHR2 knockdown abolished basal expression as well as blocking induction by TCDD [177], suggesting some crosstalk between these systems in fish. PXR sequences have been identified to date from zebrafish, *Fugu* and Atlantic salmon [76,77,178,179]. Zebrafish PXR has been shown to respond to nifedipine, clotrimazole, TCPOBOP and PB [76,178].

The peroxisome proliferators CF and ciprofibrate altered the regiospecificity of laurate hydroxylation in some fish but not others [180–182], consistent with the modest effect of peroxisome proliferators in trout [182,183]. CYP4T2 was induced in sea bass kidney by the peroxisome proliferators, CF, di(2-ethylhexyl) phthalate and 2,4-dichlorophenoxy acetic acid (2,4-D) [184].

Other inducers have received less study. Xenoestrogens have been shown to affect P450 levels (reviewed in [127]) and alkylphenols have been shown to have different effects in different systems [159,179,185,186]. In butterfly fish, one species feeding on terpene-rich Gorgonian corals showed markedly higher total P450 and P450 2- and P450 3-related immunoreactivity than congenic species that do not exploit this food source, suggesting induction of these forms [122,123]. P450 1 levels were relatively low [122].

2.2.3.3. Bioactivation of Xenobiotics in Fish. Many environmental contaminants are concentrated in fish by several orders of magnitude and fish appear to suffer hepatic cancers due to activation of PAHs [153]. AFB1 causes liver tumors in rainbow trout and other fish. Trout liver expresses little GST activity towards the AFB1 exo-8,9-epoxide. P450 2K1 activates AFB1 whereas P450 1A detoxifies to AFM1([153] and refs therein). P450 2K6 from zebrafish also catalyzed AFB1 activation to the exo 8,9-epoxide [158]. Zebrafish adults are relatively resistant but embryos and young animals are susceptible. P450 2K6 has lower activity relative to P450 2K1 from rainbow trout [157], which is in line with the resistance of zebrafish relative to rainbow trout[158].

Fish are also sensitive to the toxic effects of dioxins and PCBs (e.g., [187]). The toxicity of TCB may be in part due to the oxidative inactivation of P450 1As due to ROS generated in metabolism of TCB [188,189].

2.3. Invertebrates

Invertebrates comprise 95% of the animal kingdom [190] but, with the exception of certain insects, have been relatively poorly studied with respect to P450s and xenobiotic metabolism. Biomonitoring has been a driving force for much of the research on invertebrates, especially those in marine environments [191]. Rates of metabolism are generally lower in invertebrates than in fish [192]. Similarly, P450 levels tend to be lower, however this may be underestimated because of the difficulty of isolating specific organs in some cases.

2.3.1. Non-vertebrate Chordates

The ascidian *Ciona intestinalis* (sea squirt) genome is available and shows P450 sequences including some for P450 2-like forms but to date, however, no *Ciona* P450 has been functionally characterized.

2.3.2. Insects

Among eukaryotes, the greatest expansion of P450s has been in the insects. 43 families contain at least one insect form and almost 50 subfamilies are found in one family (CYP6) alone. To date the following P450 families (many of which contain multiple subfamilies) are found exclusively in class *Insecta*: CYP6, CYP9, CYP12, CYP15, CYP18, CYP28, CYP41, CYP48, CYP49, CYP301-319, CYP321, CYP324, CYP325, CYP329, CYP332-341. In CYP4, 27 subfamilies (CYP4D-E, CYP4G-H, 4J-N, 4P-Q, 4S, 4U, 4W, 4AA-AE, 4AJ, 4AQ, 4AU-AX) seem limited to insects. Notably, only one family, CYP4, and only one subfamily within that, CYP4C, is shared with other organisms, namely other invertebrates. This suggests that insect P450s have diverged markedly from other forms. However this inference may be biased by the greater degree of investigation of, and therefore information available for, insect P450s compared to those in other related invertebrate classes. What is clear is that within a species, forms generally show relatively low degrees of homology [193]. Genome projects have revealed 86 P450s in *Drosophila melanogaster*, and it has been estimated that each insect species will express ~100 forms [194]. While P450s catalyze many reactions important for normal physiology in insects, the major driving force for this explosion in P450 diversity is thought to have been plant versus animal chemical warfare over the last ~400 million years [1].

The elaboration of novel P450s is believed to have allowed insects to exploit plants as food sources, in the face of plants developing allelochemicals to repel or injure herbivores. P450 families found to be associated with xenobiotic metabolism in insects to date are: microsomal CYP6 and CYP28 (both related phylogenetically to CYP3 in vertebrates), CYP4 (containing a wider range of insect than vertebrate forms and therefore perhaps diversified for this role in insects) and mitochondrial CYP12. CYP4 has also been proposed to be involved in lipid metabolism and the mobilization of energy substrates [195] as in mammals. Evidence is accumulating for the involvement of newly-discovered families, e.g., CYP314, CYP321, and CYP325 in chemical defense.

Insect microsomal xenobiotic-metabolizing systems contain a typical NPR and cytochrome *b*5, and *b*5 appears to be essential in at least some P450-mediated xenobiotic biotransformations [196]. An notable difference to other animals is the clear importance of insect mitochondrial P450s, e.g., P450 12A1 of house fly [197], in xenobiotic metabolism [194]. Mitochondrial xenobiotic-metabolizing forms in insects use a redoxin and redoxin reductase as redox partners rather than the microsomal NADPH-cytochrome P450 reductase [197].

P450 expression varies according to life cycle, typically being substantial in larvae (correlating with feeding) and/or adult but often negligible in eggs and pupae. P450 activity in insect tissues tends to be greatest in the midgut but is also found in other tissues (fat body, Malpighian tubules (the excretory system), antennae, brain) [198,199]. P450s play roles in xenobiotic metabolism not only in the gut but also in odorant clearance in olfactory organs (e.g., P450 341A2 in the tarsal chemosensillae in *Papilio xuthus* [200], P450 4AW1 in the antennae of the Scarab beetle, *Phyllopertha diversa* [201] and P450s 9A13, 4G20, 4L4, and 4S4 in the sensilla trichodea of the antennae of *Mamestra brassicae* [202,203]). Based on the requirement for rapid inactivation of pheromones, the P450 4AW1 has been proposed to be located in the plasma membrane and oriented towards the extracellular environment [201].

Insects provide the best model system with which to examine both the evolution of P450s in response to natural selection pressures, i.e., allelochemicals produced by plants in animal-plant warfare, as well as the response to anthropogenic selection pressures, namely insecticides.

Indeed, evolution is proceeding apace in insects in response to xenobiotics from all sources, as evident from significant differences seen between insect and other animal P450s and the marginal degrees of sequence divergence seen between individual forms in subfamilies (suggesting recent duplication and which leads to difficulties in classification as allelic variants or different forms) [204]. In contrast to the situation with other invertebrates, a considerable amount of information is available on P450-dependent xenobiotic biotransformations in insects, and the reader is directed to the many excellent reviews written on the topic (e.g., [193,194,205–209]). Due to space limitations, this discussion will be limited to those examples where specific P450s have been identified and characterized.

Scott and colleagues [208] have pointed out that two criteria need to be fulfilled to positively assign a role for a P450 in the resistance to insecticides (and by extension, other xenobiotics): the P450 must be shown to metabolize the xenobiotic in question and the resistant strain should show higher activity of this P450 (due to any number of mechanisms). Table 1 lists P450s for which a role in xenobiotic metabolism is suspected. Both criteria have been met in only a very few cases (vide infra). Most work has been done on Lepidopteran and

Table 1. Insect P450s for which a role in xenobiotic metabolism is suspected.

P450 form	Insect	Substrates	Inducers	References
CYP4C27	Anopheles gambiae			[210]
CYP4D10	D. mettleri		SAG, PB	[211]
CYP4G8	Helicoverpa armigera			[212]
CYP4G18	Diabrotica virgifera virgifera			[213]
CYP4G19	Blattella germanica L.			[214]
CYP4H1, CYP4H15	A. gambiae			[210]
CYP4H15	A. gambiae			[210]
CYP4M1	M. sexta		CF, PB, UDO, NIC	[215]
CYP4M3	M. sexta		CF, TDO, UDO, NIC	[215]
CYP4AJ1, CYP4AK1	D. v. virgifera			[213]
CYP6A1	M. domestica	Aldrin, heptachlor, diazinon, JH (JH) I and III, JH analogues, methyl farnesoate, methyl farnesoate monoepoxide, methyl geranate, methyl nerate, (2E,6E)farnesal	ETH, PB, PBO	[194,216]–[218]
Cyp6a2	D. melanogaster	AFB1, 7,12-dimethylbenz[a]anthracene, and 3-amino-1-methyl-5H-pyrido[4,3-b]indole (Trp–P2), aldrin, diazinon, heptachlor, DDT	PB, BAR	[219]–[223]
CYP6A8	D. melanogaster	Lauric acid	BAR, PEP	[221,224]–[226]
CYP6A9	D. melanogaster		BAR	[221]

CYP6B1	*Papilio polyxenes*	Linear FCs (XAN, BER, IPN, psoralen, 4,5',8-trimethylpsoralen), angular FCs (ANG, SPH) to a lesser extent	XAN> BER>ANG >> FLA, coumarin, xanthonol, and psoralen	[227]–[231]
CYP6B2	*H. armigera*		PB, PM PIN, MEN, PBO	[205,232]
CYP6B3	*P. polyxenes*	XAN?	Linear and angular FCs (XAN, BER, ANG, SPH)	[233,234]
CYP6B4	*P. glaucus*	BER, IPN, IMP> XAN, psoralen> ANG, SPH, 4,5',8-trimethylpsoralen, ECOD	XAN, BER, ANG, IPN	[205,230,235,236]
CYP6B5	*P. glaucus*		XAN	[236]
CYP6B7	*H. armigera*		PIN, MEN, PM, CPM, fenvalerate, PB, PBO, BHA	[232,237]–[239]
CYP6B8	*Helicoverpa zea*	XAN, CGA, I3C, FLA, rutin, quercetin, CPM, aldrin, diazinon	XAN, CGA, I3C, PB, FLA, JAS, SAL	[240]–[243]
CYP6B9	*H. zea*		XAN, CGA, I3C, PB, FLA, JAS, SAL	[240,242]
CYP6B11	*P. canadensis*		XAN	[236]
CYP6B12	*P. glaucus*		XAN	[236]
CYP6B13	*P. canadensis*		XAN	[236]
CYP6B14–CYP6B16	*P. glaucus*		XAN	[236]
CYP6B17	*P. glaucus*	FCs, ECOD	XAN	[235,236,244]
CYP6B18, CYP6B19	*P. canadensis*		XAN	[236]

Table 1. (Continued)

P450 form	Insect	Substrates	Inducers	References
CYP6B20	P. glaucus		XAN	[236]
CYP6B21	P. glaucus	FCs, ECOD	XAN	[235,236,244]
CYP6B22	P. canadensis		XAN	[236]
CYP6B23	P. glaucus		XAN	[236]
CYP6B24	P. glaucus		XAN	[236]
CYP6B25	P. canadensis	FCs	XAN	[236,244]
CYP6B26	P. canadensis		XAN	[236]
CYP6B27	H. zea		XAN, CGA, I3C, PB, FLA, JAS, SAL	[240,242]
CYP6B28	H. zea		XAN, CGA, I3C, PB, FLA, JAS, SAL	[240,242]
CYP6D1	M. domestica	Phenoxybenzyl pyrethroids, e.g., CPM, deltamethrin, PM, organophosphates, chlorpyrifos, MROD, AHH, phenanthrene	PB, PBO, CF	[196,205,245]–[254]
CYP6D3	M. domestica		PB	[255]
CYP6E1	Culex quinquefasciatus			[256]
CYP6F1	C. p. quinquefasciatus, C. pipiens pallens			[257,258]
Cyp6g1	D. melanogaster	DDT		[226,259,260]
CYP6P7	A. minimus			[261]
CYP6X1	Lygus lineolaris			[262]
CYP6Z1	Anopheles gambiae	Pyrethroids?		[210,263]

CYP6Z2	A. gambiae		[210]
CYP6AA2	A. minimus		[264]
CYP6AA3	A. minimus		[261]
CYP6AB3	Depressaria pastinacella	IMP	[265,266]
CYP6AL1	Aedes aegypti	ECOD	[267]
CYP9A1	Heliothis virescens		[268]
CYP9A2	M. sexta		[269]
CYP9A4	M. sexta		[269]
CYP9A5	M. sexta		[269]
Cyp9b2	D. melanogaster		[224]
CYP12A1	M. domestica	Aldrin, heptachlor, diazinon, azinphosmethyl, amitraz, progesterone, testosterone, PCOD	[197]
Cyp12a4	D. melanogaster	Lufenuron	[270]
Cyp12d1	D. melanogaster		[224,226,271]
CYP12F1	A. gambiae		[210]
CYP28A1	D. mettleri	alkaloids	[272]
CYP28A2	D. mettleri		[272]
CYP28A3	D. nigrospiracula		[272]
CYP314A1	A. gambiae	XAN, ANG, CPM	[273]
CYP321A1	H. zea	XAN, ANG, CPM, ANF	[274]
CYP325A3	A. gambiae		[210]

CYP6Z2 (cont.)		FCs	
CYP6AA2		TLL	
CYP9A2		XAN, UDO, TDO, PB, I3C	
CYP9A4		XAN, CF, UDO	
CYP9A5		XAN, UDO	
Cyp9b2		PEP	
Cyp12a4		DDT, PEP	
CYP28A1		SEN, PB	
CYP28A2		SEN, SAG, PB	
CYP28A3		SAG, PB	
CYP314A1		XAN	
CYP321A1		XAN, CPM to a very small extent	

ANF, α-naphthoflavone; ANG, angelicin; BER, bergapten; BAR, barbital; BHT, butylated hydroxyanisole; CF, clofibrate; CGA, chlorogenic acid; CPM, α-cypermethrin; DDT: 1,1,1-trichloro-2,2,-bis(4′-chlorophenyl)ethane; DDA, dichlorodiphenyl acetic acid; DDD, dichlorodiphenyldichloroethane; ECOD, 7-ethoxycoumarin O-deethylase; ETH, ethanol; FCs, furanocoumarins; FLA, flavone; I3C, indole-3-carbinol; IMP, imperatorin; IPN, isopiminellin; JAS, jasmonate; JH, Juvenile hormone; LPR, Learn pyrethroid resistant; MEN, menthol; NIC, nicotine; PB, phenobarbital; PBO, piperonyl butoxide; PEP, piper negrum (black pepper) extract; PM, permethrin; PMB, pentamethyl benzene; PIN, a-pinene; Trp-P2, 3-amino-1-methyl-5H-pyrido[4,3-b]indole; SAG, Saguaro cactus isoquinoline alkaloids; SAL, salicylic acid; SEN Senita cactus isoquinoline alkaloids; SPH, sphondin; TDO, tridecanone; TLL, toxic leaf litter; UDO, 2-undecanone; XAN, xanthotoxin.

Dipteran species of economic importance as pests or disease vectors or on the insect version of the laboratory rat, *Drosophila*.

2.3.2.1. Detoxification of Plant Allelochemicals: Animal-Plant Warfare.
Plants produce allelochemicals as toxins to counter attack by herbivores such as insects. Herbivores may be oligophagous (feeding on a limited range of plants, e.g., typically ~three species) or polyphagous (feeding on many more) [243]. The advantage of polyphagy is that an insect can exploit a wider range of food sources. However, the cost of this strategy is that an insect must protect itself against a wider range of allelochemicals. By contrast, specialist feeders can develop sophisticated defenses against a limited number of allelochemicals, but will be more vulnerable should the supply of the plant to which they have specialized decline for any reason [243]. Plant allelochemicals have been shown to be substrates but also inducers of many insect P450s [215,229,234,241,275,276].

Furanocoumarins are photoactivated DNA crosslinking agents and can also damage other macromolecules and generate reactive oxygen species in the presence of O_2. These allelochemicals are found in at least 12 plant families but are most developed in *Rutaceae* and *Apiaceae (Umbelliferae)*, which are the principal host plant groups for *Papilio* species. Two groups exist: linear (e.g., xanthotoxin, bergapten, isopimpinellin, imperatorin) where the furan ring is attached to the 6,7-positions on the coumarin (benz-2-pyrone) nucleus, and angular (angelicin and sphondin) which have the furan ring attached to the 7,8-positions.

P450 6B1 has been shown to be constitutively expressed, and highly xanthotoxin-inducible in *P. polyxenes*, a species that feeds on xanthotoxin-containing plants [228,229]. By contrast P450 6B4 from *P. glaucus*, a species that encounters xanthotoxin less frequently, shows negligible constitutive expression but is still highly inducible [230,236]. Li et al. examined CYP6B sequences in different oligophagous (*P. polyxenes*) and polyphagous *Papilio* species (*P. glaucus* and *P. canadensis*), which encounter furanocoumarins at different frequencies in their host range. P450s in the polyphagous species metabolized a wider range of furanocoumarins at lower catalytic efficiency than the oligophagous ones, but worked on a wider range of furanocoumarins [235]. Within a limited comparison of a few P450s, the one functional P450 from *P. canadensis*, which only rarely encounters furanocoumarins, showed lower activities than those from *P. glaucus*, for which furanocoumarin exposure is more frequent. Li et al. proposed that the relaxation in selection pressure from furanocoumarins in *P. canadensis* allowed the accumulation of deleterious mutations in its P450s [235]. Despite rarely feeding on furanocoumarin-containing plants, xanthotoxin-mediated induction has been retained in *P. canadensis* [236].

Li et al. have also compared the substrate specificities of P450 6B1 from the specialist feeder *P. polyxenes* and P450 6B8 from the generalist *Helicoverpa zea* [243]. P450 6B1 showed a restricted substrate specificity compared to P450 6B8 on the substrates examined. Li et al. further proposed from modeling studies that the active site of P450 6B8 was more flexible and accommodating than P450 6B1 [243]. CYP6B8 was cloned from a xanthotoxin-exposed population resistant to cypermethrin, leading to the suggestion that the presence of broader specificity P450s in generalist feeders allowed rapid development of cross-resistance to insecticides [241].

An extra dimension to the animal-plant engagement was revealed by the demonstration that jasmonate and salicylate induce P450s in *H. zea* [242]. Jasmonate and salicylate act as chemical signals to plants to upregulate production of toxic allelochemicals such as furanocoumarins upon wounding (e.g., by insect attack). *H. zea* was shown to upregulate furanocoumarin detoxifying P450s in response to not only the furanocoumarin, but also to these prior signals. In such a way, the insect is prepared to defend itself against the allelochemicals as they are being generated [242].

An exquisite degree of specificity can be seen in the matching of different *Drosophila* species with host cacti. In the Saguaro cactus, *Carnegiae gigantea*, isoquinoline alkaloid phytotoxins, such as carnegine and gigantine can comprise up to ~1.5% of the plant dry weight with higher concentrations in the Senita cactus and in the soils surrounding the rotting cactus in which *D. mettleri* lives and breeds [272]. P450 4D10 appears to be the key form involved in the response of *D. mettleri* to the isoquinoline alkaloids produced by its principal host, the Saguaro cactus, based on the magnitude and specificity of its response [211]. In addition, several other P450s were detected classified on the basis of their differential responses to inducers [272]. P450 28A1 and P450 28A2 appear to confer resistance to alkaloids from cactus in *Drosophila mettleri* [272] while P450 28A3 appears to be the corresponding form in *D. nigrospiracula*.

Manduca sexta is able to feed on tobacco by virtue of induction of a P450 that metabolizes nicotine [277,278]. Nicotine-naïve larvae given a low dose of nicotine avoided feeding on tobacco plants for ~36 hours until the P450 could be induced [276]. Piperonyl butoxide (PBO) blocked the increase in consumption after 36 hours. CYP4M1 and CYP4M3 were found to be induced by nicotine, and these P450s have also been found to be upregulated by tridecanone and undecanone, allelochemicals produced by an alternative host plant, wild tomato [215].

2.3.2.2. Pesticide Resistance. Many synthetic organic pesticides are derived ultimately from plant-derived allelochemicals, e.g., the pyrethroids. Others share some structural (if not functional) similarity. Therefore overlap between those P450s responding to allelochemicals and insecticides is to be expected [209]. Upregulation of P450s responsible for the clearance of insecticides is commonly seen in resistant strains. In theory, resistance could also arise due to decreased P450-mediated activation of pro-insecticides, but in practice this is rare [207]. Resistance mediated by enhanced P450 activity is often demonstrated by the use of the synergist PBO which restores sensitivity to a greater or lesser extent by inhibiting the P450 involved [206]. Oakeshott et al. point out that the rapid transmission of resistance genes in a population following imposition of a strong selection pressure such as an insecticide may lead to replacement of a larger tract of a chromosome than simply the gene of interest, thereby reducing the overall genetic diversity of the population, at least temporarily [279].

The Learn pyrethroid-resistant (LPR) strain shows up to 5000-fold decreased sensitivity to cypermethrin compared to other flies. LPR flies show ~ 9 fold higher levels of mRNA and protein for P450 6D1. Resistance appears due to increased transcription of CYP6D1 in resistant flies. P450 6D1 mediates the 4'-hydroxylation of cypermethrin and other phenoxybenzyl pyrethroid insecticides. Cytochrome $b5$ is required. CYP6D1 also catalyzes the metabolism of several other xenobiotics (Table 1) [196,205,245–254].

Resistance mechanisms in *D. melanogaster* have been reviewed recently [206]. *D. melanogaster* Strain 91R was isolated in the field as a DDT-resistant strain and has higher levels of P450s 6a2, 6a8 and 6a9 than susceptible strains. While P450 6a2 levels correlated with resistance, other determinants were clearly involved [280]. Another DDT-resistant *D. melanogaster* strain, RDDT, also shows increased levels of P450 6a2 but in concert with enhanced P450 4e2 [281]. Regulation appears to be altered as evident from unresponsiveness to compounds that induced P450 in a sensitive strain [281]. Point mutations in cyp6a2 may also contribute to enhanced activity of the expressed protein (towards ECOD, benzylCOD and DDT detoxification by hydroxylation) [282].

Daborn et al. first used expression profiling to determine the P450 (P450 6g1) responsible for enhanced DDT resistance in *D. melanogaster* [260]. Brandt et al. extended the analysis of DDT resistance by crossing DDT resistant flies in the laboratory and showed that cyp12D1 was upregulated by DDT [271]. Cyp6a8 was identified by DDT exposure of flies in which cyp6g1 had been knocked out [226]. The authors propose that P450 6g1 confers resistance to a wider range of xenobiotics and is therefore more commonly selected in field exposure trials. While it has been proposed that a resistant allele may invoke a fitness cost in the absence of the cognate selection pressure, upregulation of furanocoumarin-detoxifying CYP6B forms proved not to compromise fitness in the absence of furanocoumarin selection in *H. zea* [242]. Indeed, McCart et al. have shown that the overexpression of cyp6g1 involved in DDT resistance may actually involve a fitness advantage when inherited through the maternal line, explaining the widespread retention of this gene despite discontinuation of DDT use for many years [283].

2.3.2.3. Regulation of Xenobiotic-Metabolizing P450s in Insects.

The canonical mammalian P450 inducers PB and clofibrate have been shown to induce several insect P450s (Table 1) but the mechanisms underlying these are unclear (reviewed in [207]). BaP induced EROD in adult *D. melanogaster* [281] and BaP induced its own metabolism in a *D. simulans* mutant [284,285]. While

AhR orthologues from insects and other invertebrates fail to bind typical mammalian ligands (reviewed in [73]), they still appear to play some role in mediating the response to xenobiotics [286].

The orthologues for the vertebrate AhR in insects appear to be spineless (Ss) and Sim. Tango (Tgo) is the ARNT orthologue. These transcription factors have clear roles in the development of antennae, tarsae, and the midline [287], and the Ss/Tgo heterodimer does not require ligand binding to activate transcription via XRE-AhR elements. However, XRE-AhR elements are also seen in some of the CYP6B genes where they may play a role in the upregulation in response to furanocoumarins. Petersen et al. [288] showed an XRE-Xan (xanthotoxin response element) was necessary but not sufficient for CYP6B1 expression. Both BaP and xanthotoxin had been shown to induce both the XRE-AhR and XRE-Xan *in vitro* in reporter gene assays with CYP6B1 and CYP6B4 promoters [286,289], but xanthotoxin and not BaP treatment induced xanthotoxin metabolism in midgut microsomes [286].

2.3.2.4. Bioactivation by Insect P450s. Most studies on insects have been biased towards the investigation of detoxifications, i.e., of plant allelochemicals or pesticides. However, at least one P450, P450 6a2 has been shown to activate several promutagens [219] and various toxicity and genotoxicity tests have been developed using *Drosophila* [290].The best example of bioactivation is that of the methylenedioxyphenyl compound, PBO, a widely used non-selective synergist that appears to pseudoirreversibly inhibit many P450s. It is converted into a reactive carbene that binds to the Fe^{2+} form of the heme iron. Sesamin and sesamolin from *Pyrethrum* species act in the same way [291], providing another dimension to animal-plant warfare. Another example is the bioactivation of organophosphate proinsecticides to their toxic forms by P450 6D1 amongst others [207,254].

2.3.3. Crustaceans

In crustaceans P450s are concentrated mainly in the endoplasmic reticulum of the hepatopancreas [190] and green gland [292](reviewed in [293]). In most species, levels of P450 are roughly an order of magnitude lower than for mammals such as rats, but some species show much higher levels than others [190]. P450 activities have been found to be better supported by organic hydroperoxides in some crustaceans [292]. Also, crustaceans had much lower ratios of reductase activity to P450 content than other marine invertebrates. Both of these differences may be due in part to differential lability of the reductase from crustaceans during preparation of microsomes [190].

Evidence has been found for the metabolism of BaP, PCBs, fenitrothion, heptachlor, tributyl tin, erythromycin, ethoxycoumarin, benzphetamine, aminopyrine, ethoxyresorufin, benzyloxyresorufin, pentoxyresorufin, ethylmorphine, bunitrol and imipramine in hepatopancreatic microsomes or purified P450s of various crustacean species [294] (reviewed in [293]). The endobiotics progesterone, testosterone and ecdysone were also metabolized [295]. James and Boyle concluded that typical mammalian P450 2 and P450 3 substrates were better metabolized by crustacean P450s than typical P450 1 substrates [293].

To date, P450s have been cloned from the lobsters: *Panulirus argus* (CYP2L1, CYP2L2 [296]), *Homarus americanus* (CYP4C18 [297] and CYP45 [298]); the crayfish, *Orconectes limosus* (CYPs 4C15 [299] and 4C24) and *Cherax quadricarinatus* (CYP4C29); the shrimp, *Penaeus setiferus* (CYP4C16 [297]); the crab *Carcinus maenas* (CYP330A1 and CYP4C39 [300]); and *Daphnia pulex* (CYP4C31-34, CYP4AN1-3 and CYP4AP1-2 [301]). However, functional studies have been limited to CYP2L1 which catalyzes testosterone and progesterone hydroxylation, but no xenobiotic substrates have been identified as yet [302].

No consistent data have been obtained suggesting induction of crustacean P450s by typical mammalian CYP1A inducers such as TCDD, BNF or 3MC. However, modest increases in total P450 and/or activity have been noted at lower but not higher concentrations (reviewed in [293]). Likewise lower but not higher doses of PB and BaP induced CYP330A1 in crabs [300]. Recently, oil effluent was shown to increase total P450 levels in shrimp [303]. Crabs, *Eriocheir japonicus*, from polluted environments showed higher total P450, ECOD and bunitrol hydroxylase activity than those from cleaner rivers and this was correlated with TEQs of PCBs and PCDDs [304,305]. CYP45 from lobster, a form that may be involved in regulation of molting hormone, was induced by PB and heptachlor [298]. Tributyl tin appears to induce one or more P450s but also acts as a mechanism-based inactivator in crabs [306,307].

2.3.4. Echinoderms

Levels of P450, reductase and $b5$ are roughly an order of magnitude lower in echinoderms such as sea stars, compared to rats but activities are not necessarily lower [190]. P450s are concentrated mainly in the endoplasmic reticulum of the pyloric caeca of sea stars (Asteroidea) [308]. Measurement of P450 spectra is complicated by the observation of an additional peak at 418 nm in the CO-Fe^{2+} versus Fe^{2+} difference spectra also found amongst crustaceans and molluscs. This has been proposed to be a protein breakdown product unrelated to P450 (or P420) (reviewed in [309]). The requirement for NADPH appears not to be as strict as in other systems. In fact, BaP metabolism in *A. rubens* was supported better by NADH and cumene hydroperoxide than NADPH, though a difference in product specificity accompanied this change in electron donor.

Activities proposed to be P450 mediated (reviewed in [309]) include *p*-nitroanisole demethylation in sea cucumber *Cucumaria miniata* and the sea urchin *Strongylocentrotus purpuratus*, 2,6-dimethyl-naphthalene hydroxylation by the sea urchin *S. droebachiensis*, AHH and biphenyl hydroxylation in sea stars including Asterias and *Ophiocomina nigra* and tributyltin metabolism by the sea star *Leptasterias polaris*. To date, only four echinoderm P450s have been cloned and sequenced: CYPs 4C19 and 4C20 from the white sea urchin (*Lytechinus anamesis*) [297], and CYPs 4C30 and 4FA1 from the Australian sea star (*Coscinasterias muricata*) (see http://drnelson.utmem.edu/P450.stats.all.2005.htm). To date, information about the substrates metabolized by these forms or their regulation is lacking.

Some evidence has been obtained for induction of AHH in echinoderms by PCBs and PAHs (reviewed in [309]).

2.3.5. *Caenorhabditis elegans and Other Nematodes*

Until relatively recently, there was little evidence for the existence or activity of cytochromes P450 in nematodes [310,311]. This was proposed to be due to the fact that the nematodes studied had largely been parasitic intestinal forms, which had either come to rely on host detoxification processes or were unable to utilize P450-dependent chemistry due to the oxygen-depleted environment of the intestine. An alternative explanation was that P450s represented a danger given the high degree of unsaturation in nematode lipid membranes [310]. However, the genome sequence of *C. elegans* revealed 84 P450 genes (76 intact, 8 pseudo genes; see http://drnelson.utmem.edu/P450.stats.all.2005.htm) that have been grouped in the CYP13, CYP14, CYP22, CYP23, CYP25, CYP29, and CYP31-CYP37, many of which contain several subfamilies and each of which has only been found in *C.elegans* to date.

The CYP13, 14, 33, 34, and 35 families in particular appear to have undergone a dramatic expansion (and CYP25 and CYP29 to a lesser degree), suggestive of a response to some environmental trigger, potentially chemical stress. Gotoh [5] has proposed that families 33, 34, 35, and 14 are related to CYP2 in vertebrates, while CYP13 and CYP25 are related to CYP3 in vertebrates and CYP6 in insects. CYP29, 31, 32, and 37 cluster with CYP4 but are sufficiently separate to be classified as different families. It is unclear which of the families unique so far to *C. elegans* will also be found in other nematodes, especially parasitic worms.

2.3.5.1. *Substrates and Activities.*

Monooxygenase activities have been described in few studies on parasitic nematodes (*Haemonchus contortus, Heligmosomoides polygyrus*, and *Trichostrongylus colubriformis*), specifically aldrin epoxidation, ECOD [312], aminopyrine dealkylation [314], and rotenone [315] and moxidectin metabolism [316]. P450-mediated xenobiotic metabolism has been suggested to be more important for larval, free-living stages of parasitic nematodes compared to adult forms living within the host [312,313].

Essentially nothing has been reported about the activities of *C. elegans* P450s [317]. Intriguingly, fluoranthene and PCB52 reproductive toxicity was ameliorated to some extent by RNAi knockdown of CYP35A and CYP35C forms suggesting these P450s may participate in the bioactivation or another aspect of the toxicity of these compounds [317]. RNAi knockdown experiments on other P450s showed phenotypic changes with CYP13A4, 13A8 and CYP44 (see http://drnelson.utmem.edu/C.elegans.RNAi.html). Work with *C. elegans* has led to the proposal that P450s and other xenobiotic-metabolizing enzymes are part of a longevity assurance system regulated by insulin/IGF-1 signalling pathways [318].

2.3.5.2. Regulation of Nematode P450s. In a large study with eighteen protypical (mammalian) inducers, CYP subfamilies 35A, 35C, 31A and 29A appeared to be the most subject to induction with the compounds chosen [319]. Three groups of CYP35 genes could be differentiated based on the degree and promiscuity of induction [319]. CYPs 35A4, 35A3, 35A2, and 35A1 were induced strongly by a limited range of compounds, CYPs 35C, 35C1, 35A5, 35B1, and 35B2 were induced by a broader range of compounds but to a lesser degree, while CYPs 35B3 and 35D1 appeared refractory to induction [319]. BNF, PCB52 and lansoprazole were the most effective inducers; however other compounds such as PB, ethanol, CF and atrazine also affected one or more forms [319].

In a more focused study, nine P450s were induced, as detected by gene array profiling, in response to either BNF (CYPs 35A1, 35A2, 35C1, 14A5), fluoranthene (CYP37B1), or atrazine (35B2, 35B1, 35A5, 22A) [320]. Interestingly, while CYP35 forms appear inducible by typical CYP1 family inducers, no sequence homology between these families is evident suggesting any orthologous relationship [317]. AhR and ARNT homologues have been detected in *C. elegans*, and XRE-like sequences are found in CYP35 genes, but the function of the AhR/ARNT pathway seems not to be associated with CYP35 induction [317]. Humic materials have also been shown to affect P450 expression in *C. elegans* [321]. Nothing is known about the mechanisms underlying regulation of these P450s but since the majority of NRs of *C. elegans* have no orthologues in vertebrates [322], it is unlikely that any direct parallels will be made between regulation in *C. elegans* and mammals.

Only limited studies have been done in other nematodes. PB was shown to induce aldrin epoxidation and ECOD [312] in *Haemonchus contortus* and a P450 difference spectrum could be measured in PB-treated larvae [312].

2.3.6. Annelids

P450s are concentrated mainly in endoplasmic reticulum of the intestine in annelids and levels are comparable to other invertebrates [190,323]. Measurement of P450 content in many annelids is complicated by the presence of the 'giant hemoglobin' [323,324].

AHH, EROD, BROD, ECOD, parathion desulfuration, aldrin epoxidation and carbofuran monooxygenation activities have been detected in the earthworm, *Lumbricus terrestris*, and polychaete *Nereis virens* ([323] and references therein). The principal substrates metabolized appear to be similar to those metabolized by P450 3 enzymes in mammals [325]. The PAH, fluoranthene was metabolized by *Capitella capitata* to 3- and 8-hydroxylated products [326]. EROD, PROD and BROD activities were catalyzed by microsomes from the worms, *Eisenia f. fetida* (tiger worm) and *Enchytraeus crypticus* (pot worm) [324]. To date only four annelid P450s have been sequenced. CYP4BB1 and CYP342A1 from *N. virens* [327] (shown to catalyze pyrene 1-hydroxylation) and CYP331A1 and CYP4AT1 from the aquatic polychaete, *Capitella capitata* [328].

The fact that many annelids ingest soil or sediment with food provides a clear means by which they may be exposed to environmental contaminants. The limited studies that have been done to date have provided some evidence for induction by PAHs and PCBs but only in certain species [329–331]. Benzanthracene, crude oil, and CF induced a *N. virens* P450 later identified as CYP342A1 [332]. CYP331A1 and CYP4AT1 from *Capitella capitata* were shown to be induced by BaP- and 3MC-containing sediments, respectively [328].

2.3.7. Molluscs

Mussels have been used extensively for biomonitoring, as a sedentary filter-feeding species (e.g., Mussel Watch [333]). P450s in molluscs are concentrated mainly in endoplasmic reticulum of the digestive gland [190]. Seasonal changes and lower temperature optima have been noted as for other poikilotherms [334–338].

EROD and ECOD activities have been detected in *Octopus pallidus* [338], and aminopyrine demethylase, PROD, AHH and biphenyl hydroxylation were observed in microsomes from the snail *Lymnaea* [339,340]. The common mussel shows EROD, ECOD, AHH, laurate hydroxylation and *N,N*-dimethylaniline *N*-demethylation amongst other activities [341–344]. In addition, NADPH-independent but P450-dependent activity towards AHH, ECOD, dimethylaniline demethylation, tributyltin dealkylation, testosterone hydroxylation and thiometon and disulfoton metabolism has been reported [190]. It has been speculated that molluscs use an endogenous peroxide or alternative source of reducing equivalents. Enhanced levels of NADPH-independent activity were seen in frozen and thawed versus freshly prepared microsomes from mussel, suggesting that lipid peroxides formed from membranes supported this activity [342]. The physiological relevance, if any, of this activity is unclear at present.

To date P450s have been cloned from: the pond snail, *Lymnaea stagnalis* (CYP10 [345]), the freshwater snail, *Biomphalaria glabrata* (CYP320A1) [346], the red abalone, *Haliotis rufescens* (CYP4C17 [297]), the blue mussel, *Mytilus galloprovincialis* (CYP4Y1 [297]), the common mussel, *Mytilus edulis planulatus* (CYP4AL1), the New Zealand green lipped mussel, *Perna canaliculus* (CYP322A1), the Pipi or Goolwa cockle (a clam) *Plebidonax deltoides* (CYP4AM1), the northern quahog (a clam), *Mercenaria mercenaria* (CYP30A1 [347]), and the Sydney rock oyster, *Saccostrea glomerata* (CYP323A1). CYP30A1 was found to be most closely related to mammalian CYP3 and the homologous insect CYP9 sequences [347].

Snyder has reviewed the induction of molluscan P450s [344]. Molluscan AhRs fail to bind typical mammalian AhR ligands but do bind to mammalian AHRE sequences [73]. Environmental contamination appears to increase levels of P450 1, P450 2 and P450 4-related immunoreactivity in *Mytilus* [348]. TCB, BaP and 3MC induced P450 and AHH in *M. galloprovincialis* [341,342] and AHH activity was increased in animals collected at polluted sites [342]. Divergent results have been obtained with BNF and various PCB mixtures in other studies [297,338,349]. PB induced total P450 and ECOD and laurate hydroxylase activities in *Mytilus* [342,343]. Atrazine induced AHH in the snail, *Lymnaea palustris* [340].

2.3.8. Cnidarians and Poriferans

Early studies failed to find evidence of P450 in Cnidaria (reviewed in [350]); however, recent studies have confirmed a role for P450s in xenobiotic metabolism. The first evidence of a P450 system in a poriferan was provided recently [190]. Measured P450 levels are within the range obtained for other invertebrates [190,350] but may underestimate the functional concentrations in Cnidarian tissues due to the inability to dissect specific organs [350]. P450s are expressed mainly in the endoplasmic reticulum. The '418' peak is also seen in cnidarians and has been proposed to represent another hemoprotein rather than P420 or another breakdown product of P450 [350].

EROD and BROD activities have been measured in *Hydra attenuata*, and BROD was induced by carbamazepine [351]. EROD has been measured in various species of sea anemone [350]. AHH has been detected in the coral, *Favia fragum* and was higher in animals collected close to shore suggesting some environmental chemical in runoff may have induced activity [352]. NADPH-independent AHH activity was detected for certain species [190] that occupy oxygen-poor environments.

To date, only one form has been cloned, namely CYP38A1, from the sponge, *Suberite domuncula* and from *Hydra*; however, other partial sequences from *Hydra* show similarity to known forms from other P450 families. CYP9- and CYP4-like sequences have been seen in the sponge, *Reniera*, genome.

2.4. Protista

Despite the increasing availability of P450 sequence data, very few functional studies have been undertaken on P450 forms from kingdom Protista. Evidence has been found for a role for P450s in xenobiotic metabolism in parasitic protozoa, and P450s have been implicated in drug resistance (reviewed in [311]). ECOD and EROD activities have been shown in Trypanosomes and PROD and ethanol oxidation in free-living *Euglena gracilis* [353,354]. In addition, the metabolism of herbicides and several prototypical P450 activities have been demonstrated in unicellular green algae [355–357]. PB and other compounds acted as inducers in *E. gracilis* and algae [355,358].

2.5. Plants

Plants are in a very different situation compared to most other organisms with respect to xenobiotics. The exposure of plants to xenobiotics is limited compared to heterotrophic animals: their 'ingestion' of toxins is only via water and air. They do not need to ingest substantial quantities of organic matter as food. In general, higher plants may be more involved in making toxins (allelochemicals for defense against herbivores or other plants) than in degrading them. Plants do not have mechanisms for excretion of large amounts of waste as do animals, consistent with their limited intake of chemicals and inability to move. Instead, 'removal' of chemicals is achieved principally by sequestration of toxic compounds away from sites of cellular metabolism (reviewed in [359]). Plants can also exclude certain chemicals efficiently, e.g., via epicuticular wax (increases of which may serve to confer resistance to e.g., herbicides like diclofop methyl [360]).

Sandermann coined the term the 'green liver' to describe the xenobiotic detoxification systems in plants [361] and its potential for use in environmental management (in the same manner as plants provide a 'green lung' for the biosphere). Parallels can be drawn with phase I, II and III processes in animals [361]. There are also three stages to the detoxification system in plants [362]: phase I is functionalization, e.g., by P450-mediated monooxygenation; phase II is conjugation with glucose, amino acids or glutathione; phase III in plants is a compartmentalization process, of soluble metabolites or conjugates in the vacuole and apoplasmic space, and of insoluble xenobiotic residues in cell structures, via incorporation into structural polymers, e.g., lignin in cell walls. As in other organisms, chemical efflux transporters, such as the ABC (ATP-binding cassette) transporters appear to play a major role in compartmentalization. In addition, evidence has been obtained that transgenic plants can exude xenobiotic metabolites from their roots [363].

While various genome sequencing projects are revealing the enormous numbers of P450s in plants relative to other organisms, only a small proportion of known P450s in plants have been ascribed functions to date (reviewed in [359]). The *Arabidopsis* functional genomics project website http://arabidopsis-P450.biotec.uiuc.edu/ lists the known functions of *A. thaliana* P450s, including several P450s of which the principal function is to degrade endobiotics, such as plant hormones or signalling molecules, and several of these show some activity towards xenobiotics [359]. To date, however, no plant P450s have been identified that show the exceptionally wide substrate specificity characteristic of the xenobiotic-metabolizing forms in mammals. Those P450s that serve to protect plants against xenobiotics such as herbicides appear to do so by catalysing similar activities that are physiologically relevant. It is unclear whether any plant P450s exist for which the substantial, evolutionarily important role is xenobiotic metabolism [364]. Nonetheless, chemical exposure is likely to exert the same selection pressures on plants, and this is probably more important than ever before in the evolutionary history of plants given the widespread use of herbicides since the last century. Interestingly, herbicide tolerance appears higher in monocots than dicots [365] and this may be related to higher GC content of monocot genomes which would potentiate recombination and gene duplication [364].

The study of xenobiotic metabolism in plants is still in its early stages. One hindrance in working with plants is the difficulty in measuring classical $Fe^{2+}\cdot CO$ versus Fe^{2+} P450 difference spectra due to interference from plant pigments. Most studies employ selective inhibitors and inducers to tease out differences in the P450 form specificity of different biotransformation reactions, approaches akin to early days of work with mammalian P450s [366]. Inhibitors commonly used are PBO, 1-aminobenzotriazole, malathion, palcobutrazole, tetcyclasis and SKF-525A, chemicals which also may be used as synergists to enhance the effect of herbicides by blocking their detoxification.

2.5.1. Metabolism of Herbicides and Other Xenobiotics

Studies of herbicide resistance provide the best information on xenobiotic metabolism in plants. Typically multiple P450s appear to play a role in the metabolism of a herbicide in any given plant and the K_m values for the forms induced may be higher than those found constitutively [364]. The first comprehensive evidence for P450-mediated xenobiotic metabolism in plants was presented for the *N*-demethylation of substituted 3-(phenyl)-1-methylureas in cotton [367,368]. Since then, a large number of compounds has been demonstrated to be substrates for plant P450s. Importantly, however, mediation by flavin-containing monooxygenases or peroxidases has not been excluded in many cases. Figure 1 illustrates the herbicides proposed to be metabolized by P450s in plants based on inhibitor studies or other indirect approaches [355,356,360,364,367–416]. Other xenobiotic substrates of plant P450s identified by the same methods are aminopyrine [377,417–421], benzo[a]pyrene [403,422], 7-methoxyresorufin [357,382],

BIOTRANSFORMATION OF XENOBIOTICS BY P450s

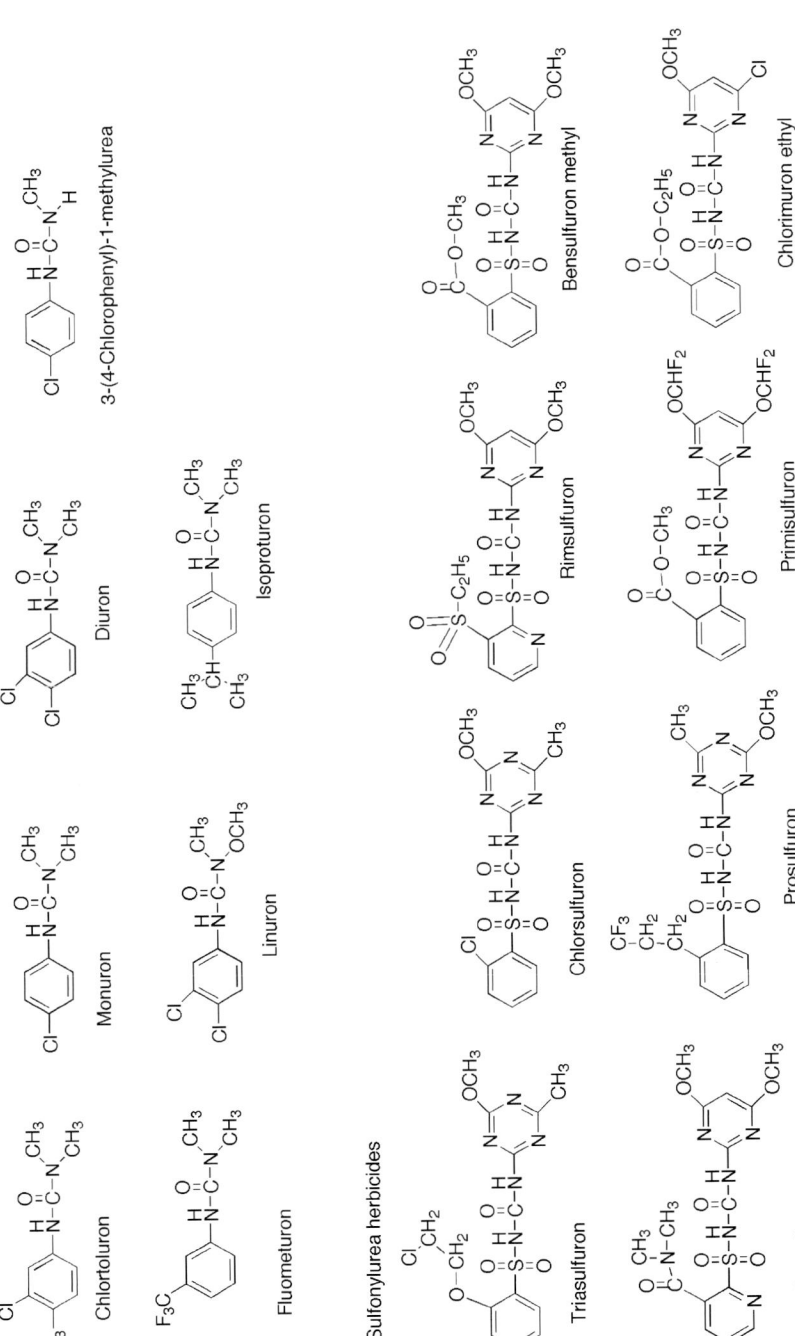

Figure 1. Structures of herbicides proposed to be substrates of plant P450s.

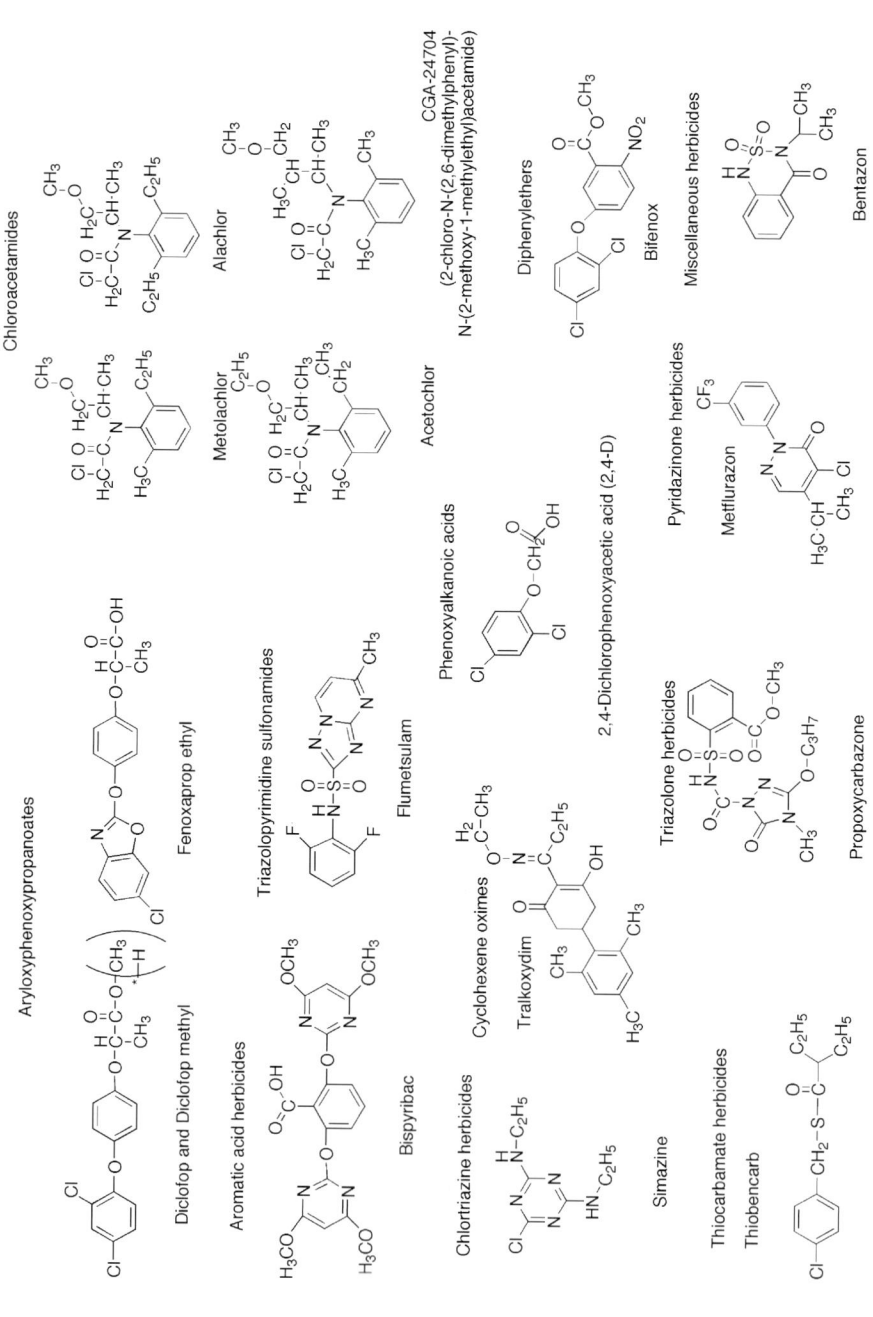

Figure 1. (Continued).

Table 2. Plant P450 forms identified in the metabolism of xenobiotics.

P450 form	Plant species	Xenobiotic substrate	Comments	Ref.
CYP71	Avocado (*Persea americana*)	*p*-chloro-*N*-methylaniline	A P450 involved in ripening of fruit	[425]
CYP71A10	Soybean (*Glycine max*)	Chlortoluron and other phenylurea herbicides linuron, diuron, and fluometuron	Rate ~10 fold lower than 76B1 commensurate with wider substrate specificity [364]	[379]
CYP71A11	Tobacco (*Nicotiana tabacum*)	Chlortoluron	Identified in cultured tobacco S401 cells treated with 2,4-D	[376]
CYP71B1	*Thlaspi arvense*			[431]
CYP71C6v1	Wheat (*Triticum aestivum*)	Chlorsulfuron and triasulfuron		[385]
CYP72	Maize (*Zea mays*)	Chlortoluron, bentazon, chlorimuron, malathion	Inhibited by glyphosate	[364]
CYP73	Jerusalem artichoke (*Helianthus tuberosus*)	Small planar aromatic molecules, e.g., 7-methoxycoumarin, 7-ethoxycoumarin, *p*-chloro-*N*-methylaniline plus chlorotoluron, aminopyrine, ethylmorphine erythromycin, 2-naphthoic acid	Enzyme efficiency and coupling was markedly better towards the endogenous substrate trans-cinnamate than any xenobiotic except 2-naphthoic acid	[377,430]
CYP76B1	Jerusalem artichoke (*Helianthus tuberosus*)	Phenylurea and other planar substrates, alkoxycoumarins and alkoxyresorufins	Inducible by chemical stress, i.e., Mn^{2+}, PB and aminopyrine	[382,426]
CYP81A	Maize (*Zea mays*)	Metolachlor and prosulfuron		[409]
CYP81B2	Tobacco (*Nicotiana tabacum*)	Chlortoluron	Identified in cultured tobacco S401 cells treated with 2,4-D	[376]

7-ethoxyresorufin [357,382,423], 7-pentoxyresorufin [357], 7-methoxycoumarin [357,382], 7-ethoxycoumarin [357,377,382,424–426], ethylmorphine [377], erythromycin [377], trichloroethylene [427], N-nitrosodimethylamine (NDMA) and N-nitrosomethylaniline [417–421], P-chloro-N-methylaniline [377,425,428], 3- and 4-chlorobiphenyl and 2,3-dichlorobiphenyl [355], benzoxazolin-2(3H)-one (BOA) [429], 1-phenylazo 2-hydroxynaphthalene (Sudan I) [417–421], 2-naphthoic acid [430], malathion (M. Barrett, unpublished cited in [364]), methidathion [410], diazinon [383,410] and isazofos [410]. Notably, this list includes at least one chemical, BOA [429], that is produced as an allelochemical by another plant. While indirect approaches have provided information on P450-dependent reactions, considerably less is known about particular forms catalyzing most of these biotransformations. Table 2 lists the specific P450 forms involved in xenobiotic metabolism that have been identified to date.

It is clear from this table and Figure 1 that, as a whole, plant P450s can deal with structurally and chemically divergent xenobiotic substrates. The fact that this appears somewhat less diverse than the substrate ranges for, e.g., mammalian hepatic P450s may be a function of the fact that plant P450s have received much less attention, and that e.g., herbicides are more limited in number and structural diversity compared to drugs. Given the small fraction of plant P450s that have been functionally characterized with respect to xenobiotics, and the clear variety in their physiological substrates and roles, it is not possible to predict at this stage whether plant xenobiotic-metabolizing P450s as a whole will show the same versatility as mammalian P450s. However, one can speculate that this is unlikely due to the more limited exposure of plants to xenobiotics.

2.5.2. *Bioactivation of Xenobiotics by Plant P450s*

While metabolism of herbicides typically leads to detoxification, bioactivation of xenobiotics has also been demonstrated in plants. Indeed this is exploited in the action of the pro-herbicide metflurazon, which is N-demethylated to the more phytotoxic metabolite, norflurazon [432]. P450s from unicellular green algae have been proposed to catalyze this bioactivation [356]; however the identity of metflurazon N-demethylases in plant species has not been established.

Evidence exists that the plants can bioactivate xenobiotics to mutagenic products (reviewed in [433]) and the products of metabolic activation may be mutagenic to plants [434]. In the best studied examples, *Persea americana* (avocado) S117 was shown to metabolize 2-aminofluorene [435] and *o*-, *m*-, and *p*-phenylenediamine [436] to mutagenic products in the Ames assay; P450s were proposed to dominate in the first activation and flavin-containing-monooxygenases in the second case [437]. Both P450s and peroxidases may also participate in the bioactivation of 2-aminofluorene by tobacco [433,438].

2.5.3. Induction of Xenobiotic Metabolizing P450s in Plants and Herbicide Safeners

Evidence for induction of xenobiotic metabolism was first reported by Reichart et al. [439,440], who showed that total P450 in Jerusalem artichoke (*Helianthus tuberosus*) slices could be weakly induced by classical mammalian P450 inducers such as PB (4 mM) and ethanol (300 mM) as well as two herbicides, monuron (3-(4-dichlorophenyl)-1,1-dimethylurea) and dichlobenil (2,6-dichlorobenzonitrile), with increases seen over changes due to wounding alone. *Trans*-cinnamic acid 4-hydroxylase activity, an activity now known to be important in the response to various stressors, was increased in parallel.

Since that time many inducers have been identified for the P450-mediated metabolism of herbicides, largely for exploitation in agriculture as herbicide safeners. Herbicide safeners are chemicals used to selectively protect crops from herbicides (reviewed by Davies and Caseley [441]) which exert their effects principally by upregulating enzymes and other processes for dealing with xenobiotics, e.g., P450s, GSTs, vacuolar transport, synthesis of glutathione, glutathione peroxidase and sulfate assimilation [364]. They are typically applied to seed prior to germination or mixed with the herbicide to be used in the field, depending on the degree of crop versus weed selectivity).

Many herbicides act as inducers of their own metabolism or that of other substrates, including 2,4-D [376], diclofop, chlortoluron, bentazon [397], bispyribac-sodium, fenoxaprop-ethyl, and thiobencarb [414], bensulfuron-methyl [389], triasulfuron [384,442], monuron and dichlobenil [439,440], and primisulfuron, prosulfuron, and bromoxynil octanoate [443].

Transcriptome (gene array) analysis has the potential to elucidate new pathways by which P450s and other genes may be induced by revealing distinct sets of genes transcriptionally upregulated in response to herbicides and other stressors [443]. CYP71B15, CYP71B19, CYP72A8, CYP73A5, CYP76C2, and CYP81D8, as well as a number of glycosyl transferases, glutathione transferases, and ABC transporters, were upregulated in response to one or more of the herbicides primisulfuron, prosulfuron, and bromoxynil octanoate. The functions of many of these P450s are unknown, while others appear to serve physiological roles in plants. It is unclear at this stage whether any are involved in xenobiotic metabolism. In another study, five P450s were shown to be induced by methanol treatment in *A. thaliana* [444]. Ekman et al. analyzed the response to 2,4,6-trinitrotoluene administration in *Arabidopsis* and found P450s featured prominently amongst the transcripts upregulated, including a putative CYP81D11, increased by > 14 fold as well as CYP71A12, a P450 706A2-like protein, a P450 73A5 type protein (a cinnamate hydroxylase), and a hypothetical CYP89A6 sequence [445]. Baerson et al. also found CYP81D11 featured among six P450s upregulated in reponse to the xenobiotic allelochemical BOA in *A. thaliana* [429] suggesting it may have a central role in the response to xenobiotics in *A. thaliana*.

While functional studies are lagging behind the identification of possible networks of genes turned up or turned down, microarray studies provide fertile areas for hypothesis generation and further studies to delineate the signal transduction pathways underpinning the response to inducers. To date, no xenobiotic-responsive receptor has been linked to upregulation of a P450 in plants that metabolizes foreign chemicals, in contrast again to the wealth of information available on mammalian nuclear receptors.

2.5.4. Implications of Immobility

Plants have developed sophisticated biochemistry to elaborate allelochemicals to defend themselves against herbivores and phytopathogenic fungi and bacteria. Wounding, chemical stress or microbial attack elicits responses including the upregulation of P450s and P450-dependent pathways. Moreover, allelopathy, a form of plant-plant chemical warfare, takes place between plants competing for the same light and other resources, in which toxins are released from one plant to injure another. However, such toxins may also be toxic to the plant of origin and therefore mechanisms must exist to prevent them disrupting endogenous physiological processes. In allelopathy, selection pressure may lead to the predominance of individuals in the population with the ability to detoxify the allelochemical.

FCs (psoralens) are one class of defense allelochemical which also act as mechanism-based inactivators of the P450 cinnamate-4-hydroxylase, a critical enzyme in the response to various stressors in plants. One mechanism for resistance to the toxic effects of these allelochemicals appears to be that the cinnamate-4-hydroxylases in plants producing psoralens are less prone to inactivation [446]. However, other studies suggest a more general response may be mounted to allelochemicals and potentially other xenobiotics.

Intriguingly, one study on tobacco has shown that a P450 is induced by the phytopathogenic bacterium *Pseudomonas solanacearum* that shows significant similarity with an avocado P450 implicated in metabolism of xenobiotic substrates [425,447]. It is unclear whether the elicitor in this case is a chemical or other type of stress imposed by the bacterial infection.

As noted above, the allelochemical BOA has been proposed to be metabolized by an as yet unidentified P450 in *A. thaliana*, a plant that does not synthesize it for defense [429]. Transcriptome studies showed six different P450s appeared to be upregulated by BOA exposure: CYP71B23, CYP72A8, CYP81D8, CYP81D11, CYP81F2 and CYP86A1. CYP81D11 and a battery of other genes were also markedly induced in response to several other xenobiotics of diverse structures, leading to the suggestion that plants could mount a coordinated response to diverse chemical insults similar to mammals [429]. This is the first real indication that plants may have a generalized response for chemical defense, rather than simply that xenobiotics are metabolized by enzymes with other physiological roles due to their structural similarity with endobiotic substrates. Future studies in this area, and particularly the functional characterization of P450s upregulated by multiple chemical stresses, should reveal the scope and versatility of P450-dependent xenobiotic metabolism in plants. Experiments can be performed conveniently with plant cell cultures in many cases, where the xenobiotic is supplied in, and the metabolite extracted from, the culture medium. Such studies have recently revealed wildtype tobacco cultures were capable of transforming loratidine and indole to their respective metabolites (H. Warzecha and M. Unger, personal communication).

2.6. Fungi

2.6.1. Defense in Phytopathogenic Fungi

The best-studied fungal phytopathogen is *Nectria haematococca (Fusarium solani)*, which metabolizes the phytoalexin pisatin (Table 3) via demethylation to produce 3,6α-dihydroxy-8,9-methylenedioxypterocarpan, which is far less toxic to most fungi tested [448]. Pisatin demethylase was identified as a P450, solubilized, reconstituted and partially purified [449–451] and six different loci were found coding for three pisatin demethylase phenotypes, distinguished by the amount of enzyme produced and the lag time for induction of the activity. Two genes, *PDA6-1* [452,453] and *PDAT9* [454], were identified and found to be associated with different levels of virulence [455]. Hybridisation of a fragment of the *PDAT9* gene to DNA fragments in *N. haematococca* isolates with known pisatin demethylating phenotypes identified other potential P450s, placed in the CYP57 family. The *PDAT9* gene was introduced into *N. haematococca* isolates lacking pisatin demethylase activity. Of 146 transformants, 52 showed an increase in virulence, though none became highly virulent. Three transformants were analyzed, differing in virulence, and showed approximately equal levels of pisatin tolerance and pisatin demethylase activity, suggesting that other genes are necessary for high virulence [456].

Pisatin demethylase expression is induced by pisatin and repressed by glucose and amino acids. A 40bp pisatin-responsive element was located 635bp upstream of the *PDA1* transcription start site and was shown to be required for pisatin regulation of *PDA1* expression. A binuclear zinc DNA-binding protein specific to this element was identified and suggested to be the transcription factor responsible for pisatin regulation of *PDA* expression [457].

P450-dependent pisatin demethylases are also found in *Ascochyta pisi, Mycosphaerella pinodes*, and *Phoma pinodella* [458]. Whilst the ability to demethylate pisatin is common amongst true fungi, virulence was linked to rapid pisatin demethylation. Only one fungal isolate showed hybridization to DNA specific for PDA genes, suggesting that the pisatin demethylase activity in the other isolates may be different to the *N. haematococca* enzyme [459].

Table 3. A selection of xenobiotics metabolized by fungi.

Organism	Compound metabolized	Refs
Nectria haematococca	Pisatin	[449–451]
Aspergillus niger	Benzoate	[464]
Aspergillus terreus	Pyrene, benzo[a]pyrene	[465]
Bjerkandera adusta DSM 3375	2,4,6-trinitrotoluene	[466]
Cladosporium sphaerospermum	Toluene	[467]
Coriolus versicolor	2,4,6-trichloro-4'-nitrophenyl ether, 2,4-dichloro-4'-nitrophenyl ether	[468, 469]
Cunninghamella bainieri	Benzo[a]pyrene	[470]
Cunninghamella elegans	Phenanthrene, Cortexolone	[471]
	Tributyltin	[472]
Exophiala jeanselmei	Styrene	[473]

Fusarium oxysporum	Fatty acids	[474–476]
Phanerochaete chrysosporium	Linear and substituted alkanes, PAHs, benzo[a]pyrene	[477–479]
	Biphenyl, biphenylene, diphenyl ether	[480]
	Benzoates, benzoic acid, camphor, cineole, cinnamic acid, *p*-coumaric acid, coumarin, cumene, 1,12-dodecanediol, 1-dodecanol, 4-ethoxybenzoic acid, 7-ethoxycoumarin	[481, 482]
	Nitrotoluenes, 4-nitrophenol	[483, 484]
	Lindane	[485]
	s-Triazine herbicides	[486]
	Endosulfan	[487]
Phlebia lindtneri	Chloronaphthalenes, dibenzo-*p*-dioxin, biphenyl, biphenylene, diphenyl ether, PAHs	[488, 489]
Pleurotus pulmonarius	Benzo[a]pyrene, phenanthrene, pyrene, anthracene, fluorene, dibenzothiophene	[490–492]

The cytochrome $P450_{11\alpha}$ of the filamentous fungus *Rhizopus nigricans* is involved in the transformation of steroidal compounds by 11α-hydroxylation. This activity is exploited commercially in corticosteroids production, but the physiological role of the P450 is uncertain. $P450_{11\alpha}$ alleviated the toxic effects of steroids on fungal growth; this effect was reversed when the P450 inhibitor metyrapone was present [460]. The P450 is induced by plant saponins and saponin aglycones [461] and has been purified [462]. Metabolism of the saponin α-tomatine by *Gibberella pulicaris* may also involve a P450 [463].

2.6.2. Xenobiotic Mineralization

Many fungi, such as the basidiomycetes, *Phanerochaete chrysosporium* (one of the white rot fungi) have an extraordinary ability to mineralize and detoxify xenobiotic pollutants and complex organic compounds such as the metabolites of lignin. The role of P450s in this metabolism has been inferred from the use of selective P450 inhibitors such as SKF-525-A, metyrapone, PBO and carbon monoxide. Some of these reactions are surveyed in Table 3.

2.6.3. Catabolism in Alkane-Assimilating Yeasts

P450alk, an alkane-inducible P450 capable of hydroxylating n-alkanes at the terminal position, was reported in *Candida tropicalis*, is distinct from the lanosterol demethylase (CYP51) [493,494]. Further work expanded the P450alk (CYP52) family [495–498] and showed that these genes were also present in other alkane-assimilating yeasts. For example, two genes (DH-ALK1 and DH-ALK2) have been isolated from the halotolerant yeast *Debaryomyces hansenii*, sequenced and shown to code for P450s (P450 52A12 and P450 52A13) [499].

Candida maltosa also possesses at least eight structurally related forms of CYP52 genes. The four major forms, P450S 52A3, A4, A5, and A9 were shown to have distinct substrate preferences with respect to chain length and substrate binding [500,501]. Comparison of P450cm2 and P450Alk3A, naturally occurring isoforms of P450 52A4 which differ in their fatty acid substrate preference, revealed that a single Val to Leu change was responsible, and that the fatty acid substrate preferences of these and other P450 52A enzymes could be altered by replacement of these, or equivalent residues [502]. Purified P450 52A3 was shown to be capable of catalysing a sequence of reactions to transform hexadecane to 1,16-hexadecanedioic acid [503], summarized in Figure 2.

Gene disruption of the major CYP52 genes in *C. maltosa* (ALK1, ALK 2, ALK 3, ALK 5) showed that a single isoform was sufficient to permit the yeast to grow on long-chain alkanes [504]. Though four other genes (ALK4, ALK6, ALK7, ALK8) are present, these are poorly induced by n-alkanes and were not capable of supporting growth on alkanes when the major genes were all disrupted.

A multigene family of alkane monooxygenases, YlALK1-YlALK8, was also found in the alkane assimilating yeast *Yarrowia lipolytica*, which has the advantage that it is haploid and has a sexual life cycle [505,506]. Gene disruption experiments suggested that YlALK1 functions to assimilate n-decane and longer alkanes, YlALK2 is involved in assimilation of n-dodecane and longer molecules

BIOTRANSFORMATION OF XENOBIOTICS BY P450s

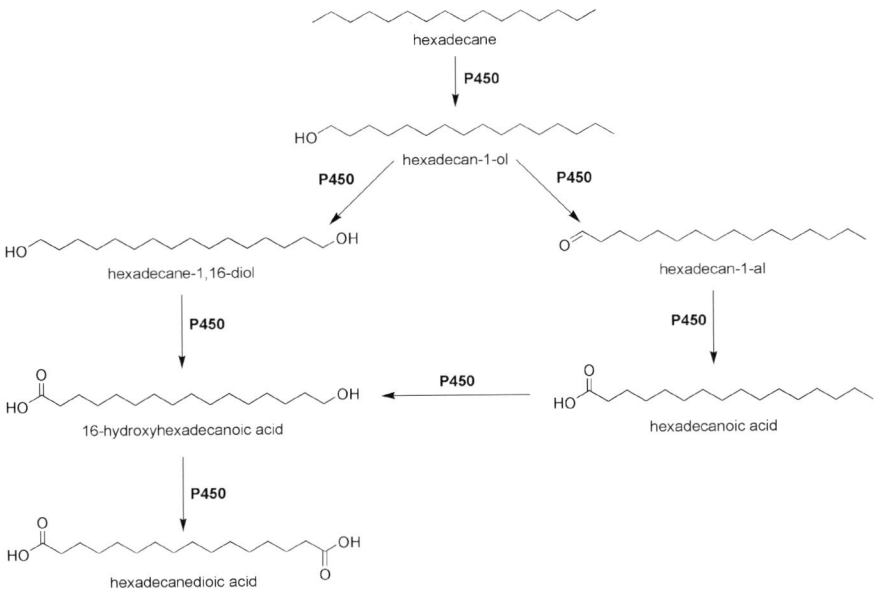

Figure 2. The biotransformation of hexadecane to 1,16 hexadecanedioic acid catalyzed by P450 52A3. Based on information provided in [503].

and that the remaining gene products are not significantly involved in *n*-alkane assimilation.

2.7. Prokaryotes

The number and type of prokaryotic P450 monooxygenases is very large, with over 150 families described. Many of these prokaryotic P450s are involved in the biosynthesis of complex molecules, and are outside the scope of this review. However, there are a large number of P450s that are involved in catabolic reactions, catalyzing reactions in the degradation pathways of xenobiotic molecules in order for their exploitation by the organism. Some of the compounds exploited in this way are illustrated in Figure 3.

2.7.1. Catabolism of Xenobiotics for Carbon Substrate Exploitation: Alkane Oxidation and Other Biotransformations

The best known and most intensively studied of the catabolic prokaryotic P450s is P450cam (CYP101), originally isolated from *Pseudomonas putida* [507,508]. P450cam catalyzes the stereospecific hydroxylation of (1R)-(+)-camphor to

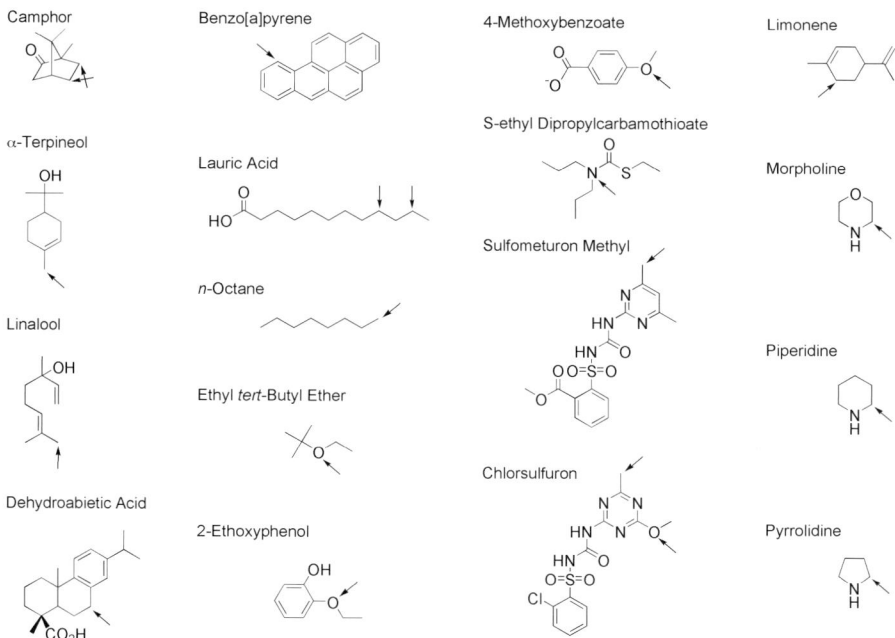

Figure 3. Xenobiotic substrates of bacterial P450s. Sites of metabolism are indicated by arrows.

5-*exo*-hydroxycamphor, allowing the bacterium to grow using camphor as a sole energy and carbon source. Detailed structural studies on this enzyme have revealed much regarding the binding of ligands at the active site of the P450, the determinants of its regio- and stereospecificity and the reaction pathway [509–513]. As discussed elsewhere in this volume, P450cam has been used widely as a model for P450s in general.

An interesting parallel to P450cam is a second camphor-hydroxylating P450, P450camr, isolated from *Rhodococcus* sp. NCIMB 9784 [514], which can use camphor as sole carbon source. In contrast to P450cam, which produces 5-*exo*-hydroxycamphor, P450camr hydroxylates (1R)-(+)-camphor at the 6-*endo* position, to produce 6-*endo*-hydroxycamphor.

P450terp (CYP108) was isolated from a *Pseudomonas* species capable of using a terpene natural product as its sole carbon and energy source [515]. P450terp makes up approximately 1% of the total protein in cell-free extracts, indicating its importance to the organism. It hydroxylates α-terpineol at the most sterically accessible, allylically activated position to form 7-hydroxyterpineol [516].

P450lin (P450 111) catalyzes the first committed step, the 8-methyl hydroxylation of linalool, in the use of this compound as a sole carbon source by *Pseudomonas incognita*. Despite the similarity of its function with P450terp, it only shares around 29% identity with that P450, and only 25% with P450cam (P450cam versus P450terp: 23% identity).

P450cin (P450 176A1) catalyzes the hydroxylation of cineole, a large component of the 'essential oil' of eucalypt trees, which is very similar structurally to camphor [517]. Despite the similarity of their substrates, P450cin shows notable differences to P450cam, using different redox partners (an FMN-containing protein versus a Fe_2S_2, ferredoxin-like protein), and with a substantially different active site, no B'-helix, no conserved active site threonine (T252 in P450cam) and other substantial changes [518].

More complex terpenes are also degraded by P450s. For example, P450TdtD is involved in the degradation of tricyclic diterpenes in *Pseudomonas diterpeniphila* [519]. P450terp is the closest homologue of this P450, but only shows 25.4% identity, so the relationship is not a close one.

Other aromatic compounds may be hydroxylated by bacterial P450s as part of biodegradative pathways. P450105D1 (P450 105D1) from *Streptomyces griseus* catalyzes the hydroxylation of xenobiotic aromatic substrates such as BaP, as well as other xenobiotics like testosterone, warfarin, and erythromycin [520].

P450 BM-3 (P450 102A1) from *Bacillus megaterium* requires no other proteins for activity, since the monooxygenase is fused to a diflavin reductase, and is thus self-sufficient. It catalyzes the β-hydroxylation of long-chain (c12 – c18) fatty acids; its native substrate is not known. Other fatty acid hydroxylating P450s have been characterized: these range from P450 BM-3 homologues in *B. subtilis* (P450 102A2 and P450 102A3) [521], to fatty acid α-hydroxylase from *Sphingomonas paucimobilis* (P450 152B1), which is involved in the catabolism of branched fatty acids [522].

Bacterial alkane hydroxylases include P450s, such as the *n*-octane hydroxylase of *R. rhodocrous* (f. *Corynebacterium* sp. Strain 7E1C) [523,524]. However, the best-characterized P450 family involved in alkane hydroxylation is P450 153, which appears to be common in bacteria lacking integral membrane alkane hydroxylases [525]. A P450 of this family was first characterized in *Acinetobacter* sp. EB104 [526], and was detected in cells growing on *n*-alkanes as the sole carbon source. In contrast to eukaryotic P450s which oxidize *n*-alkanes, it is soluble, though prone to aggregation, and is of an unusually high molecular weight. This is due to an N-terminal extension, containing a predicted amphipathic helix and a long F-G loop, closer in size to eukaryotic P450s than bacterial. Both these features may allow the apparently soluble P450 153 to associate with membranes and to bind its hydrophobic substrates. Homologues have been found in *Caulobacter crescentus* CB15 and *Bradyrhizobium japonicum* USDA 110 (P450 153A2 and P450 153A4) and in over twenty other alkane-degrading strains [525].

Sphingomonas sp. HXN200 possesses five different CYP153 genes (CYP153A7, 153A8, 153A11, 153D2, and 153D3) of which three have been confirmed to be functionally expressed [525]. In addition, *R. erythropolis* Strain

PR4, an alkane-degrading strain, has been shown to encode two possible homologues of CYP153 on a linear plasmid, with other alkane-degradation genes. Both homologues are closely related to the *Acinetobacter* P450 153A1, but lack the N-terminal extension [527]. Kubota et al. also reported the isolation of 16 novel CYP153 genes from various contaminated environments, including one from the *n*-alkane-degrading bacterium *Alcanivorax borkumensis* [528].

Two related pollutants, methyl *tert*-butyl ether (MTBE) and ethyl *tert*-butyl ether (ETBE), both used as gasoline additives, are degraded by P450s and used as sole carbon and energy sources for the bacteria expressing them. Two aerobic bacteria Strains E1 and E2, one tentatively identified as *Comamonas testosteroni*, were isolated and shown to degrade ETBE; the involvement of a P450 being inferred from the action of the specific P450 inhibitor metyrapone on ETBE degradation [529]. A gene cluster cloned from *R. ruber* IFP 2001, known to use ETBE, but not MTBE, as a sole carbon and energy source, encoded a P450 (P450 249A1), together with its ancillary redox proteins, presumed to be involved in the initial step of ETBE degradation [530]. This cluster also showed a remarkable similarity to the Thc system of *Rhodococcus* sp. Strain NI86/21 discussed below.

Two P450s ($P450_{RR1}$ and $P450_{RR2}$) have been isolated from a *R. rhodochrous* Strain (116) which can use 2-ethoxyphenol and 4-methoxybenzoate as sole carbon and energy sources, after enrichment for cultures able to degrade ethoxy-aromatics [531]. Similarly, a P450 was isolated from *R. erythropolis* K2-3 which catalyzes the ether bond cleavage of phenoxybutyrate herbicides [532], though in this case, the herbicide was not the sole carbon source [533]. Other herbicides have been investigated. In the case of *Rhodococcus* sp. Strain NI86/21, the degradation of thiocarbamate herbicides by *N*-dealkylation involves the Thc operon, ThcABCD. ThcB is known to be a P450 (P450 116) and is proposed to be responsible for the *N*-dealkylation reaction [534,535]. Sulfonylurea herbicides induce two P450s ($P450_{SU1}$ and $P450_{SU2}$, P450 105A1 and P450 105B1) in *Streptomyces griseolus* [536–538], which are involved in the metabolism of these herbicides. P450s of this class have been identified in other Streptomycetes as well as *Rhodococcus*, *Burkholderia*, and other bacteria.

Rhodococci are attractive targets, since they assimilate an unusually broad range of organic compounds. Further examples of catabolic P450s from Rhodococci include three CYP genes from *Rhodococcus* sp. Strain RHA1, which may be involved in limonene degradation [539] and the unusual P450 RhF (P450 116B2) which resembles P450 BM-3 in that it is a fused, self sufficient enzyme, but differs in that the reductase contains FMN and a 2Fe-2S cluster [540,541]. No function has yet been ascribed to the P450 RhF family.

Mycobacterium strains which use heterocyclic compounds such as morpholine, piperidine, and pyrrolidine as sole carbon, nitrogen and energy sources have been isolated and P450s implicated in the degradation of these compounds [542–544]. The P450 involved (P450 151) has been cloned from *M. smegmatis* mc^{2}155 [545] and isolated from *Mycobacterium* sp. Strain HE5 [546] and characterized, together with an associated ferredoxin.

3. COMMON THEMES

From the preceding discussion, several unifying themes emerge. The first is that P450s have been exploited to enable organisms to fill niches, whether this is in the case of bacteria living on the breakdown products of leaf litter, phytopathogenic fungi infecting pisatin-producing plants, Lepidopteran species consuming rutaceous plants, butterfly fish living on terpene-containing Gorgonian corals, or marsupials whose principal foodstuff is eucalyptus leaves. While in most examples, the role of the P450 is purely defensive, in bacteria and fungi the role of the P450 is to exploit the xenobiotic as a carbon source or assimilate it for some other purpose.

In diverse types of organism a correlation is evident between the degree of dietary specialization and the P450 complement. Polyphagous insects showed a wider range of P450 activities than related oligophagous species and amongst birds, generalist feeders showed higher P450 levels. A genetic investment in P450s appears important in the trade-off between specialization and versatility in feeding habits. While it is convenient to consider P450s in ecological terms as purely detoxifying enzymes, P450-mediated bioactivation is seen across all phyla depending on the xenobiotic under consideration.

Profound similarities are seen in regulatory mechanisms across organisms. In the case of the AhR, ligand binding and responsiveness appear to have been acquired later in evolution in vertebrates, in addition to or supplanting a primordial role in invertebrates [73]. While similarities are to some degree an inevitable result of evolution, e.g., it is not surprising that the AhRs in mammals, birds and fish respond to similar ligands, it is intriguing that chemicals identified as canonical mammalian inducers, e.g., PB and clofibrate, also act in plants and insects, organisms which appear to lack the cognate nuclear receptors. This raises the question of what is common about the regulatory systems in diverse organisms that enables certain chemicals to trigger inductive responses. Do the canonical inducers, for example, mimic naturally occurring xenobiotics, to which many types of organism have developed responses? Alternatively, is there, in most species, a sufficient degree of responsiveness to virtually any chemical, so that what we see is a reflection of this general capacity? A great number of chemicals are able to induce their own metabolism, albeit weakly, in a wide variety of organisms, which could be taken as an indication of some general capacity, mediated, for example, by stabilization of the P450 against degradation by the presence of a substrate.

Another common theme is the role P450s play in chemical senses. P450s that degrade chemicals in olfactory organs allow rapid recovery of this sense following stimulation. Examples exist currently in mammals and insects, and will most likely be found in other organisms if studied. One can speculate that the development and diversification of P450s facilitated the development of the chemical senses as well as, e.g., the elaboration of pheromone signalling.

Finally, cooperative effects on enzyme activity have been seen in humans and fish. In the case of several human P450s, cooperativity is proposed to result from multiple ligands binding within the P450 active site. This can be viewed as perhaps an inevitable consequence of a system for chemical defense that can deal with diverse chemicals. However, Atkins et al. have proposed that atypical kinetics carry a toxicological advantage [547]. P450s are involved in bioactivation as well as detoxification, and on any given chemical, either endpoint may result. Sigmoidal kinetics allow 'distributive catalysis' so that the potential for either the parent xenobiotic or its product exceeding a toxic threshold is minimized over a wide range of xenobiotic concentrations [547].

4. INDUSTRIAL APPLICATIONS OF P450 SYSTEMS FOR XENOBIOTIC DECOMPOSITION

4.1. Pharmaceutical and Fine Chemical Synthesis

The ability of P450s to introduce oxygen into non-activated carbon-hydrogen bonds in a regio- and stereoselective manner suggests roles in pharmaceutical and fine chemical synthesis which have been the subject of much investigation. Further, P450s comprise one of the largest superfamilies of enzymes and catalyze a diverse range of reactions in addition to simple hydroxylation, which makes them attractive targets as biocatalysts. However, significant problems exist: low activities and protein stability; the requirement for ancillary redox protein partners; the requirement for a costly cofactor, NAD(P)H, in cell-free processes; membrane association in many examples; which have thus far limited their industrial use. Nonetheless, the potential of P450s as efficient biological catalysts is under investigation by many groups.

The production of drug metabolites using P450s is one such area. Tens of milligrams of the metabolites 6β-hydroxytestosterone, 4'-hydroxydiclofenac, and acetaminophen have been purified from cultures of *E. coli* expressing human P450s [548] and similar metabolism has been demonstrated by Parikh et al. [549]. Otey et al. [550] used directed evolution to enhance the production of authentic human metabolites of the drug propranolol by bacterial cytochrome P450 BM-3.

Pravastatin, a potent, tissue-specific HMG-CoA reductase inhibitor, was originally isolated as a hydroxylated metabolite of mevastatin (compactin, ML-236B). Since the hydroxylation was difficult to perform chemically, microbial hydroxylation was adopted for the industrial production of pravastatin, using *Streptomyces carbophilus*, which expresses two P450s (P450sca1 and P450sca2) [551–555].

Terpene metabolism has been investigated with P450cam and other active site mutants with greatly enhanced activity for the oxidation of (+)-α-pinene and S-limonene have been produced [556]. The highly sought-after monooxygenated products verbenol and isopiperitenol were produced with high regio- and near total stereoselectivity. The biotransformation of (+)-valencene by P450 BM-3 and P450cam was investigated as a possible route to (+)-nootkatone, a fragrance molecule. Mutants of P450cam were obtained which gave (+)-nootkatone and (+)-*trans*-nootkatol as over 85% of the products [557]. A novel P450 from *Mycobacterium* sp. Strain HXN-1500 (P450 153A6) has been used to specifically hydroxylate L-limonene to form enantiomerically-pure (−)-perillyl alcohol, a proven anticancer agent [558], by expression in *P. putida*, a system which does not produce large amounts of difficult to separate side products.

P450s have also been investigated for alkane hydroxylations. P450cam has been engineered using a site-directed mutagenesis approach. Introduction of bulky amino acid side chains to reduce the active site volume was used to enhance the oxidation of hexane over 3-methylpentane [559] and to allow oxidation of butane and propane [560] and ethane [561], to their corresponding alcohols. Arnold et al. [562–565] have evolved P450 BM-3 to be regio- and stereoselective alkane hydroxylase, one variant of which was also an alkene epoxidase.

P450 BM-3 (and the newly discovered P450 RhF family) have the advantage of being soluble, self-sufficient enzymes [540,541,566], but they still require NADPH to function. Further evolution produced an enzyme capable of utilising peroxides as a source of oxygen, eliminating the need for NADPH, additional proteins, and eliminating rate-limiting electron transfer steps. Site-directed mutants of P450 BM-3 have allowed the natural substrate specificity to be altered towards shorter chain fatty acids [567].

An alternative approach to avoid the cost of supplying NAD(PH) is to use whole cell systems with their own intrinsic reduced nucleotide supply, at the expense of more difficult isolation and purification of products and with a number of limitations which pertain to whole cell systems, reviewed in [568].

A number of bioreactor systems have been described using recombinant cells from various sources, expressing a variety of P450s, for example, P450 1A1 in *Saccharomyces diastaticus* [569], P450s 3A4 and 2C9 in insect cell bioreactors [570] or immortalized human cells [571]. A bioreactor using liver microsomes has also been described [572]. However, here the microsomes were not immobilized but reacted with the substrate (ethoxyresorufin) in a temperature-controlled coil, the products being removed by an inline SPE column for analysis.

Electrocatalytic systems, using an antimony-doped tin oxide electrode [573], zinc dust [574] or a gold electrode [575], coupled to the P450 by the mediator cobalt sepulchrate, have been used to turn over P450, as well as coupling of rhodium-graphite electrode to P450 via a covalently bound riboflavin on the enzyme [576]. Light has also been used, by coupling light-driven NADPH synthesis in chloroplasts to a fused NADPH P450 reductase-P450 1A1 monooxygenase system immobilized together in a bioreactor [577,578]. A similar system coupled an $NADP^+$-dependent formate dehydrogenase from *Pseudomonas* sp. 101 to P450 BM-3 in a gel-sol matrix.

Many systems of immobilisation have been tried to stabilize the otherwise labile P450, such as DE52 (P450 105D1 [579]), reverse micelles or water/oil emulsions (P450cam [580,581]), polydimethylsiloxane on a microfluidic chip (PikC hydroxylase [582]), or thin-film silicate hydrogel using several P450s (P450 1A1, 3A4, 2C19, and 3A4/2C19) to create an array on a microassay plate for high-throughput assays [583].

4.2. Bacterial Systems for Bioremediation

The involvement of P450s in degradative pathways for environmental pollutants, toxins, and other xenobiotics points to potential roles in bioremediation. P450cam is a favored enzyme for these studies. In addition to binding its native substrate, camphor, a number of halogenated hydrocarbons, such as tri-, penta-, and hexachloroethane and 1,2-dichloropropane, are bound and reductively dehalogenated [584,585], but the products are inert to further reduction. Combination of P450cam with a toluene dioxygenase in a single recombinant *P. putida* strain permitted further metabolism to non-toxic products [586] and perhalogenated chlorofluorocarbons were degraded.

Wong et al. [587–590] have extensively investigated engineering of the active site of P450cam. One line of research has been directed towards enhancing the conversion of halogenated benzenes, which are relatively inert, to polychlorinated phenols, which are readily degraded. Mutations at F87, Y96, and L247 were used to increase active site hydrophobicity and force substrates to bind close to the heme to maintain a high degree of coupling, and were successful for 1,3,5-trichlorobenzene [587]. A more heavily halogenated compound, pentachlorobenzene, was less well handled. An additional mutation, F98W, increased coupling efficiencies for 1,3,5-trichlorobenzene [588]. The crystal structure of the F87W/Y96F/V247L mutant with 1,3,5-trichlorobenzene bound allowed a further mutant (L244A) to be designed, with more space for the additional chlorines of penta- and hexachlorobenzene. This improved their rates of oxidation markedly over the initial triple mutant [589]. Similar experiments with hexachloroethane were less successful, as a trade-off between tighter substrate binding and increased dehalogenation rate was noted [590].

Recombinant yeast expressing rat CYP1A subfamily P450s have been shown to degrade polychlorinated dioxins [591], suggesting a basis for a similar engineering scheme to that for P450cam above for these persistent environmental pollutants.

A similar approach has been taken to generate mutants to improve the oxidation of PAH by both P450cam [592] and P450 BM-3 [593]. In P450cam, the introduction of hydrophobic residues at Y96 improved the oxidation rates of naphthalene and pyrene by one to two orders of magnitude over the wild-type enzyme. Coupling was also improved. In P450 BM-3, substitution of hydrophobic for charged residues at the entrance to the substrate channel increased PAH oxidation up to 40-fold. These mutations were combined with active site mutations allowing more space for the PAH, leading to order of magnitude increases in activity, though the reactions were poorly coupled.

Contamination of munition sites by explosives, such as RDX (hexahydro-1,3,5-trinitro-1,3,5-triazine) has led to the isolation of a *Rhodococcus* strain (DN22) able to grow on RDX as its sole nitrogen source, using a plasmid-borne P450 [594], unusual in that it has a flavodoxin domain at the N-terminus [595]. Experiments to characterize this P450 used rabbit P450 2B4 to degrade RDX in a comparable manner [596], strengthening the evidence that RDX degradation *in vivo* proceeds via P450 and opening up the possibility of using recombinant mammalian enzymes in explosives bioremediation.

4.3. Phytoremediation and Herbicide Tolerance

Plants offer certain advantages for bioremediation (i.e., phytoremediation) of soil and water. In particular, plants do not excrete toxins; rather they conjugate and immobilise them in bound residues. This is useful in that residues can be removed by harvesting plant material [597].

A closely related application is the engineering of herbicide tolerance in plants. Bacterial (e.g., P450 105A1 from *Streptomyces griseolus* (see above) (reviewed in [364]), mammalian (e.g., human P450s 1A1, 2B6, 2C9, and 2C19 amongst others [598]) and other plant P450s (e.g., P450s 76B1 [599], and P450 71A10 [379]) have been expressed heterologously in plants. Werck-Reichhart et al. [364] point out that engineering of P450 expression to chloroplasts and specifically plastid DNA is desirable, so as to enhance coupling to photosynthesis and therefore the supply of NADPH, since the chloroplast is the major site of NADPH generation in plants in light. Localization in the plastid DNA would also serve to increase the number of copies per cell and avoid containment problems since plastid DNA would not be disseminated in pollen (e.g., to non-crop weeds).

A number of excellent recent reviews are available on the topic of engineering plants for enhanced xenobiotic metabolism [364,598,600] which will therefore not be covered in any further detail here.

4.4. Biosensors

Bistolas et al. [601] have surveyed biosensor types and P450 isoforms used therein. Attempts have been made to use ion-sensitive field effect transistor (ISFET) devices, often used as pH sensors. A fusion enzyme of rat P450 1A1 and yeast NADPH-P450 reductase was immobilized on the device in an agarose layer and the reaction was monitored by the local alkalinisation caused by proton consumption in the P450 turnover reaction [602].

A variety of methods for the immobilisation of P450s for direct reduction on electrode surfaces have been studied. Direct electron transfer was observed between P450cam in dimyristoyl-L-α-phosphatidylcholine (DMPC) or didodecyldimethylammonium bromide (DDAB) lipid films and a pyrolytic graphite electrode [603]. P450cam was also immobilized in a DDAB vesicular system and cross-linked on to a glassy carbon electrode using glutaraldehyde, in the presence of bovine serum albumin or methyltriethoxysilane to produce a functional biosensor [604]–[606].

Polycations have been used to immobilize P450s. P450 3A4 was used to construct a drug-screening biosensor by immobilization in a poly-(dimethyldiallylammonium chloride) multilayer on a gold electrode [607], which could detect P450 3A4 substrates in a concentration-dependent manner. Product analyzes confirmed that turnover of the substrates verapamil, midazolam, progesterone, and quinidine had taken place on the electrode. A polyaniline-doped glassy carbon electrode was used to immobilize P450 2D6 for the detection of the selective serotonin reuptake inhibitor drug fluorexidine, with potential applications in *in vivo* assessment of drug clearance [608].

Mixed colloidal clay/non-ionic detergent/protein films have been used to incorporate P450 2B4 into a biosensor [609]. The immobilized P450 2B4 turned over typical substrates such as aminopyrine and benzphetamine under aerobic conditions, producing bioelectrocatalytic reduction currents, which were blocked when metyrapone, a P450 inhibitor, was present. Shumyantseva et al. [610] have also used P450s 1A2, 2B4, and SCC conjugated with riboflavin, which acts as a covalently attached mediator of electron transfer, in biosensor fabrication. Gold nanoparticle films were also used to enhance activity of P450scc immobilized on rhodium-graphite electrodes [611].

One further application of P450s in biosensors is that of an ancillary protein. Phosphorothionate insecticides such as parathion show a considerably reduced inhibitory effect towards acetylcholinesterase (AChE), the usual enzyme used in organophosphorus pesticide detection, in their non-metabolized form. Chemical oxidation methods have disadvantages when used in food analysis, so a triple mutant of P450 BM-3 has been used to convert the pesticide residues to their oxo variants, which enabled detection of chlorpyrifos and parathion down to 20 μg/kg in infant food using a disposable AChE biosensor [612].

5. CONCLUSIONS AND FUTURE PROSPECTS

In this review we have attempted to provide an overview of the metabolism of xenobiotics across phyla and draw parallels between different organisms to identify interesting features of this versatile system for chemical defense. In the second part of the chapter, we have tried to give a brief insight into ways in which P450s have been exploited to date for anthropocentric ends.

It is clear that while mammalian, and especially human, P450s involved in xenobiotic metabolism are relatively well characterized, a great deal remains to be discovered about most other types of organism. While traditionally it has proven challenging to characterize such features as the substrate specificity of native P450 forms, especially from small animals such as insects, the ready availability of methods to express P450s as recombinants, coupled with the increased access to sequence information from many organisms, will facilitate further exploration of poorly characterized organisms.

Transcriptome studies using gene array technology have the potential to reveal further roles for P450s in the response to chemical and other stressors and therefore in xenobiotic metabolism (e.g., [613]). With the ready availability of sequence information for several plant species, it should be possible to determine genes for which the expression differs between strains that are sensitive or resistant to a particular toxic chemical (e.g., herbicide) or which are turned up or down in response to a chemical challenge, and which may therefore play a role in xenobiotic metabolism. Moreover, the development of microarray technology and transcriptome analysis shows great potential for advancing the understanding of regulatory and response systems involving P450s. Given that a significant proportion of what we know about the catalytic versatility and functional biology of P450s has come from human enzymes (as well as bacterial, fungal and plant forms), we stand to learn a great deal more about these exceptional enzymes by studying them in other organisms and ecological niches. This in turn will enable a better understanding of such aspects as interactions with accessory enzymes and control of the catalytic cycle, that will be of use in the engineering and exploitation of P450s for desired purposes.

ACKNOWLEDGMENTS

Grateful thanks are extended to Drs Wayne Johnston, Heribert Warzecha, Margaret James, Tara Sutherland, and Matthew Cheesman for constructive comments on drafts of this chapter. The amount of material published on this topic is enormous. We apologize to any colleagues whose work we have inadvertently omitted or failed to discuss due to space constraints.

ABBREVIATIONS

2,4-D	2,4-dichlorophenoxy acetic acid
3MC	3-methylcholanthrene
AA	aromatic amine

ABC	ATP-binding cassette
AChE	acetylcholinesterase
AFB1	aflatoxin B1
AHH	aryl hydrocarbon hydroxylase or benzo(*a*)pyrene 3-hydroxylase
AhR	aryl hydrocarbon receptor
AHRE	aryl hydrocarbon response element
AHRR	aryl hydrocarbon receptor repressor
aNF	α-naphthoflavone
ARNT	aryl hydrocarbon receptor nuclear translocator
BaP	benzo(*a*)pyrene
BFC	7-benzyloxy-4-trifluoromethylcoumarin
BNF	β-naphthoflavone
BOA	benzoxazolin-2(3H)-one
bp	base pair
BROD	7-benzyloxyresorufin *O*-debenzylation
BuCOD	7-butoxycoumarin 7-demethylation
BXR	benzoate X receptor
CAR	constitutive androstane or constitutively activated receptor
CF	clofibrate
CXR	chicken xenobiotic receptor
CYP	cytochrome P450 (gene, cDNA)
DDAB	didodecyldimethylammonium bromide
DDE	= 3-methylsulfonyl-DDE = 3-methylsulfonyl-2,2'-bis(4-chlorophenyl)-1,1'-dichloroethene)
DDT	1,1,1-trichloro-2,2-bis(4'-chlorophenyl)ethane
DEX	dexamethasone
DMBA	dimethylbenzanthracene
DMPC	dimyristoyl-L-α-phosphatidylcholine
E2	17β-estradiol
ECOD	7-ethoxycoumarin *O*-deethylase
EROD	7-ethoxyresorufin *O*-deethylase
ETBE	ethyl *tert*-butyl ether
FC	furanocoumarin
FETAX	frog embryo teratogenesis assay-xenopus
FMN	flavin mononucleotide
GST	glutathione S-transferase
HA	heterocyclic amine
HAH	halogenated aromatic hydrocarbon
HETEs	hydroxyeicosatrienoic acids
HMG-CoA	hydroxymethylglutarate-coenzyme A
IGF	insulin-like growth factor

ISFET	ion-sensitive field effect transistor
LPR	Learn pyrethroid-resistant
MCOD	7-methoxycoumarin 7-demethylation
MROD	7-methoxyresorufin O-demethylase
MTBE	methyl *tert*-butyl ether
NADH	nicotinamide adenine dinucleotide, reduced
NADPH	nicotinamide adenine dinucleotide phosphate, reduced
NDMA	N-nitrosodimethylamine
NNK	4-(methylnitrosamino)-1-(3-pyridyl)-1-butanone
NPR	NADPH-cytochrome P450 reductase
NR	nuclear receptor
o,p'-DDD	o,p'-dichlorodiphenyldichloroethane
P450	cytochrome P450
PAH	polycyclic aromatic hydrocarbon
PAS	Per-ARNT-Sim
PB	phenobarbital
PBO	piperonyl butoxide
PCB	polychlorinated biphenyl
PCN	pregnenolone-16α-carbonitrile
PPARα	peroxisome proliferator-activated receptorα
ProCOD	7-propoxycoumarin 7-demethylation
PROD	7-pentoxyresorufin O-dealkylase
PXR	pregnane X receptor (otherwise known as the SXR, steroid X receptor)
RDX	hexahydro-1,3,5-trinitro-1,3,5-triazine
ROS	reactive oxygen species
RXR	retinoid X receptor
SPE	Solid phase extraction
TCB	3,3',4,4'-tetrachlorobiphenyl
TCDD	2,3,7,8-tetrachlorodibenzo-p-dioxin
TEQ	toxic equivalent
UROX	uroporphyrinogen oxidation
XRE	xenobiotic response element
XRE-Xan	xanthotoxin response element

REFERENCES

1. F. J. Gonzalez and D. W. Nebert, *Trends Genet.*, 66, 164–168 (1990).
2. R. H. Wickramasinghe and C. A. Villee, *Nature*, 256, 509–511 (1975).
3. D. R. Nelson, T. Kamataki, D. J. Waxman, F. P. Guengerich, R. W. Estabrook, R. Feyereisen, F. J. Gonzalez, M. J. Coon, I. C. Gunsalus, O. Gotoh, K. Okuda, and D. W. Nebert, *DNA Cell Biol.*, 12, 1–51 (1993).

4. D. R. Nelson, *Arch. Biochem. Biophys.*, *369*, 1–10 (1999).
5. O. Gotoh, *Mol. Biol. Evol.*, *15*, 1447–1459 (1998).
6. I. Teramitsu, Y. Yamamoto, I. Chiba, H. Iwata, S. Tanabe, Y. Fujise, A. Kazusaka, F. Akahori, and S. Fujita, *Aquat. Toxicol.*, *51*, 145–153 (2000).
7. G. J. Pass, S. McLean, I. Stupans, and N. Davies, *Xenobiotica*, *31*, 205–221 (2001).
8. O. Pelkonen and H. Raunio, *Environ. Health Perspect.*, *105*, Suppl. 4, 767–774 (1997).
9. X. X. Ding and L. S. Kaminsky, *Annu. Rev. Pharmacol. Toxicol.*, *43*, 149–173 (2003).
10. L. Du, S. M. Hoffman, and D. S. Keeney, *Toxicol. Appl. Pharmacol.*, *195*, 278–287 (2004).
11. A. B. Rifkind, *Drug Metab. Rev.*, *38*, 291–331 (2006).
12. G. Y. Ling, J. Gu, M. B. Genter, X. L. Zhuo, and X. Ding, *Chem.-Biol. Interact.*, *147*, 247–258 (2004).
13. M. Karlgren, S. Miura, and M. Ingelman-Sundberg, *Toxicol. Appl. Pharmacol.*, *207*, 57–61 (2005).
14. F. P. Guengerich, Z. L. Wu, and C. J. Bartleson, *Biochem. Biophys. Res. Comm.*, *338*, 465–469 (2005).
15. F. P. Guengerich, in *Cytochrome P450* (P. R. Ortiz de Montellano, ed.), 3rd edn., Kluwer Academic/Plenum Publishing, New York, 2005, pp. 377–530.
16. F. P. Guengerich and T. Shimada, *Chem. Res. Toxicol.*, *4*, 391–407 (1991).
17. F. P. Guengerich, *Asia Pacific J. Pharmacol.*, *5*, 327–345 (1990).
18. F. P. Guengerich, *Chem.-Biol. Interact.*, *106*, 161–182 (1997).
19. M. A. Correia, in *Cytochrome P450* (P. R. Ortiz de Montellano, ed.), 3rd edn, Kluwer Academic/Plenum Publishers, New York, 2005, pp. 619–657.
20. V. Nedelcheva and I. Gut, *Xenobiotica*, *24*, 1151–1175 (1994).
21. F. P. Guengerich, in *Mammalian Cytochromes P-450* (F. P. Guengerich, ed.), Vol. 1, CRC Press, Boca Raton, Florida, 1987, pp. 1–54.
22. G. E. Schwab and E. F. Johnson, in *Mammalian Cytochromes P-450* (F. P. Guengerich, ed.), Vol. 1,CRC Press, Boca Raton, Florida, 1987, pp. 55–105.
23. M. Shou, R. Norcross, G. Sandig, P. Lu, Y. Li, Y. Lin, Q. Mei, A. D. Rodrigues, and T. H. Rushmore, *Drug Metab. Dispos.*, *31*, 1161–1169 (2003).
24. P. Anzenbacher, E. Anzerbacherova, R. Zuber, P. Soucek, and F. P. Guengerich, *Drug Metab. Dispos.*, *30*, 100–102 (2002).
25. S. Rendic, *Drug Metab. Rev.*, *34*, 83–448 (2002).
26. D. Kim and F. P. Guengerich, *Annu. Rev. Pharmacol. Toxicol.*, *45*, 27–49 (2005).
27. F. P. Guengerich and T. Shimada, *Mutation Res.*, *400*, 201–213 (1998).
28. S. K. Balani, H. J. C. Yeh, D. E. Ryan, P. E. Thomas, W. Levin, and D. M. Jerina, *Biochem. Biophys. Res. Commun.*, *130*, 610–616 (1985).
29. T. Shimada, H. Yamazaki, M. Foroozesh, N. E. Hopkins, W. L. Alworth, and F. P. Guengerich, *Chem. Res. Toxicol.*, *11*, 1048–1056 (1998).
30. J. K. Yano, M. H. Hsu, K. J. Griffin, C. D. Stout, and E. F. Johnson, *Nat. Struct. Mol. Biol.*, *12*, 822–823 (2005).
31. E. M. J. Gillam, L. M. Notley, H. L. Cai, J. J. De Voss, and F. P. Guengerich, *Biochemistry*, *39*, 13817–13824 (2000).
32. K. Kurose, M. Tohkin, and M. Fukuhara, *Mol. Pharmacol.*, *55*, 279–287 (1999).

33. E. E. Scott, Y. A. He, M. R. Wester, M. A. White, C. C. Chin, J. R. Halpert, E. F. Johnson, and C. D. Stout, *Proc. Natl. Acad. Sci. USA, 100,* 13121–13126 (2003).
34. R. T. Kinobe, O. T. Parkinson, D. J. Mitchell, and E. M. Gillam, *Chem. Res. Toxicol., 18,* 1868–1875 (2005).
35. M. R. Wester, E. F. Johnson, C. Marques-Soares, P. M. Dansette, D. Mansuy, and C. D. Stout, *Biochemistry, 42,* 6370–6379 (2003).
36. P. A. Williams, J. Cosme, A. Ward, H. C. Angove, D. M. Vinkovic, and H. Jhoti, *Nature, 424,* 464–468 (2003).
37. G. A. Schoch, J. K. Yano, M. R. Wester, K. J. Griffin, C. D. Stout, and E. F. Johnson, *J. Biol. Chem., 279,* 9497–9503 (2004).
38. M. R. Wester, J. K. Yano, G. A. Schoch, C. Yang, K. J. Griffin, C. D. Stout, and E. F. Johnson, *J. Biol. Chem., 279,* 35630–35637 (2004).
39. P. Rowland, F. E. Blaney, M. G. Smyth, J. J. Jones, V. R. Leydon, A. K. Oxbrow, C. J. Lewis, M. G. Tennant, S. Modi, D. S. Eggleston, R. J. Chenery, and A. M. Bridges, *J. Biol. Chem., 281,* 7614–7622 (2006).
40. G. P. Miller, I. H. Hanna, Y. Nishimura, and F. P. Guengerich, *Biochemistry, 40,* 14215–14223 (2001).
41. W. M. Atkins, *Annu. Rev. Pharmacol. Toxicol., 45,* 291–310 (2005).
42. R. T. Okita, and J. R. Okita, *Curr. Drug Metab., 2,* 265–281 (2001).
43. Q. Ma, *Curr. Drug Metab. 2,* 149–164 (2001).
44. O. Barbier, C. Fontaine, J. C. Fruchart, and B. Staels, *Trends Endocrinol. Metab., 15,* 324–330 (2004).
45. J. Y. Chien, K. E. Thummel, and J. T. Slattery, *Drug Metab. Dispos., 25,* 1165–1175 (1997).
46. P. F. Hollenberg, *Drug Metab. Rev., 34,* 17–35 (2002).
47. P. Ertl and G. W. Winston, *Comp. Biochem. Physiol. C, 121,* 85–105 (1998).
48. W. J. Davies and S. J. Freeman, *Methods Mol. Biol., 43,* 311–316 (1995).
49. M. H. Mayeaux and G. W. Winston, *J. Vet. Pharmacol. Therap., 21,* 274–281 (1998).
50. W. F. Khalil, T. Saitoh, M. Shimoda, and E. Kokue, *J. Vet. Pharmacol. Therap., 24,* 343–348 (2001).
51. J. T. Borlakoglu, J. P. G. Wilkins, M. P. Quick, C. H. Walker, and R. R. Dils, *Comp. Biochem. Physiol. C, 99,* 287–291 (1991).
52. R. P. Gupta and M. B. Abou-Donia, *Comp. Biochem. Physiol. C, 101,* 505–512 (1992).
53. M. B. Abou-Donia and A. A. Nomeir, *Fundam. Appl.Toxicol., 6,* 190–207 (1986).
54. R. P. Gupta, and M. B. Abou-Donia, *Comp. Biochem. Physiol. C, 121,* 73–83 (1998).
55. R. R. Dalvi, V. A. Nunn, and J. Juskevich, *Comp. Biochem. Physiol. C, 87,* 421–424 (1987).
56. C. H. Walker, *Comp. Biochem. Physiol. C, 121,* 65–72 (1998).
57. M. C. Fossi, A. Massi, L. Lari, L. Marsili, S. Focardi, C. Leonzio, and A. Renzoni, *Environ. Pollut., 90,* 15–24 (1995).
58. S. S. M. Yip, and R. A. Coulombe, *Chem. Res. Toxicol., 19,* 30–37 (2006).
59. M. X. Watanabe, H. Iwata, M. Okamoto, E. Y. Kim, K. Yoneda, T. Hashimoto, and S. Tanabe, *Toxicol. Sci., 88,* 384–399 (2005).

60. D. Gilday, M. Gannon, K. Yutzey, D. Bader, and A. B. Rifkind, *J. Biol. Chem.*, 271, 33054–33059 (1996).
61. H. M. H. Goldstone and J. J. Stegeman, *J. Mol. Evol.*, 62, 708–717 (2006).
62. P. R. Sinclair, N. Gorman, H. S. Walton, J. F. Sinclair, C. A. Lee, and A. B. Rifkind, *Drug Metab. Dispos.*, 25, 779–783 (1997).
63. A. Murk, D. Morse, J. Boon, and A. Brouwer, *Eur. J.Pharmacol.* 270, 253–261 (1994).
64. D. Gilday, G. D. Bellward, J. T. Sanderson, D. M. Janz, and A. B. Rifkind, *Toxicol. Appl. Pharmacol.*, 150, 106–116 (1998).
65. L. A. Verbrugge, J. P. Giesy, D. A. Verbrugge, B. R. Woodin, and J. J. Stegeman, *Comp. Biochem. Physiol. C*, 130, 67–83 (2001).
66. L. A. Mattschoss, A. A. Hobbs, A. W. Steggles, B. K. May, and W. H. Elliott, *J. Biol. Chem.*, 261, 9438–9443 (1986).
67. A. A. Hobbs, L. A. Mattschoss, B. K. May, K. E. Williams, and W. H. Elliott, *J. Biol. Chem.*, 261, 9444–9449 (1986).
68. J. F. Sinclair, S. Wood, L. Lambrecht, N. Gorman, L. Mende-Mueller, L. Smith, J. Hunt, and P. Sinclair, *Biochem. J.*, 269, 85–91 (1990).
69. R. P. Gupta, D. M. Lapadula, and M. B. Abou-Donia, *Comp. Biochem. Physiol.*, C 96, 163–176 (1990).
70. M. Baader, C. Gnerre, J. J. Stegeman, and U. A. Meyer, *J. Biol. Chem.*, 277, 15647–15653 (2002).
71. J. C. Ourlin, C. Handschin, M. Kaufmann, and U. A. Meyer, *Biochem. Biophys. Res. Comm.*, 291, 378–384 (2002).
72. S. S. Mahajan and A. B. Rifkind, *Toxicol. Appl. Pharmacol.*, 155, 96–106 (1999).
73. M. E. Hahn, *Chem.-Biol. Interact.*, 141, 131–160 (2002).
74. S. I. Karchner, D. G. Franks, S. W. Kennedy, and M. E. Hahn, *Proc. Natl. Acad. Sci. USA*, 103, 6252–6257 (2006).
75. C. Handschin, M. Podvinec, and U. A. Meyer, *Proc. Natl. Acad. Sci. USA*, 97, 10769–10774 (2000).
76. L. B. Moore, J. M. Maglich, D. D. McKee, B. Wisely, T. M. Willson, S. A. Kliewer, M. H. Lambert, and J. T. Moore, *Mol. Endocrinol.*, 16, 977–986 (2002).
77. J. M. Maglich, J. A. Caravella, M. H. Lambert, T. M. Willson, J. T. Moore, and L. Ramamurthy, *Nucl. Acids Res.*, 31, 4051–4058 (2003).
78. R. A. Coulombe, J. A. Guarisco, P. J. Klein, and J. O. Hall, *Animal Feed Sci. Technol.*, 121, 217–225 (2005).
79. P. J. Klein, R. Buckner, J. Kelly, and R. A. Coulombe, *Toxicol. Appl. Pharmacol.*, 165, 45–52 (2000).
80. G. Johnston, C. H. Walker, and A. Dawson, *Environ. Toxicol. Chem.*, 13, 615–620 (1994).
81. G. Johnston, A. Dawson, and C. H. Walker, *Environ. Pollution*, 91, 217–225 (1996).
82. M. J. J. Ronis, and T. M. Badger, *Toxicol. Appl. Pharmacol.*, 130, 221–228 (1995).
83. G. Johnston, C. H. Walker, and A. Dawson, *Pestic. Biochem. Physiol.* 49, 198–208 (1994).
84. I. Brandt, C. J. Jonsson, and B. O. Lund, *Ambio*, 21, 602–605 (1992).
85. C. J. Jonsson, B. O. Lund, B. Brunstrom, and I. Brandt, *Environ. Toxicol. Chem.*, 13, 1303–1310 (1994).

86. C. H. Walker, and M. J. Ronis, *Xenobiotica, 19*, 1111–1121 (1989).
87. R. J. Schwen and G. J. Mannering, *Comp. Biochem. Physiol. B, 71*, 431–436 (1982).
88. M. Noshiro and T. Omura, *Comp. Biochem. Physiol. B, 77*, 761–767 (1984).
89. A. Yawetz, B. R. Woodin, and J. J. Stegeman, *Biochim. Biophys. Acta, 1381*, 12–26 (1998).
90. M. J. Doherty and M. A. Khan, *Comp. Biochem. Physiol. C, 68C*, 221–228 (1981).
91. C. A. Herman and E. H. Oliw, *J. Exper. Zool., 280*, 1–7 (1998).
92. T. Andersson and E. Nilsson, *Comp. Biochem. Physiol. B, 94*, 99–105 (1989).
93. Y. Miura, *Comp. Biochem. Physiol. C, 82*, 63–67 (1985).
94. Y. Miura, H. Hisaki, and T. Nagai, *Comp. Biochem. Physiol. C, 85*, 353–355 (1986).
95. M. T. Rie, K. A. Lendas, B. R. Woodin, J. J. Stegeman, and I. P. Callard, *Biomarkers, 5*, 382–394 (2000).
96. R. P. Ertl and G. W. Winston, *Comp. Biochem. Physiol. C, 121*, 85–105 (1998).
97. M. A. Q. Khan, S. Y. Qadri, S. Tomar, D. Fish, L. Gururajan, and M. S. Poria, *Biochem. Biophys. Res. Comm., 244*, 737–744 (1998).
98. B. B. Brodie, J. R. Gillette, and B. N. La Du, *Annu. Rev. Biochem. 27*, 427–454 (1958).
99. Y. Fujita, H. Ohi, N. Murayama, K. Saguchi, and S. Higuchi, *Arch. Biochem. Biophys., 371*, 24–28 (1999).
100. Y. Fujita, H. Ohi, N. Murayama, K. Saguchi, and S. Higuchi, *Comp. Biochem. Physiol. B, 138*, 129–136 (2004).
101. H. Saito, H. Ohi, E. Sugata, N. Murayama, Y. Fujita, and S. Higuchi, *Arch. Biochem. Biophys., 345*, 56–64 (1997).
102. H. Ohi, E. Sugata, Y. Fujita, H. Saito, K. Saguchi, N. Murayama, and S. Higuchi, *Biochem. Mol. Biol. Int., 45*, 689–697 (1998).
103. A. G. McArthur, T. Hegelund, R. L. Cox, J. J. Stegeman, M. Liljenberg, U. Olsson, P. Sundberg, and M. C. Celander, *J. Mol. Evol., 57*, 200–211 (2003).
104. E. G. Schuetz, J. D. Schuetz, W. M. Grogan, A. Naray-Fejes-Toth, G. Fejes-Toth, J. Raucy, P. Guzelian, K. Gionela, and C. O. Watlington, *Arch. Biochem. Biophys., 294*, 206–214 (1992).
105. R. P. Ertl, J. J. Stegeman, and G. W. Winston, *Biochem. Pharmacol., 55*, 1513–1521 (1998).
106. D. Busbee, D. Colvin, I. Muijsson, F. L. Rose, and E. Cantrell, *Comp. Biochem. Physiol. C, 50*, 33–36 (1975).
107. J. A. Lavine, A. J. Rowatt, T. Klimova, A. J. Whitington, E. Dengler, C. Beck, and W. H. Powell, *Toxicol. Sci., 88*, 60–72 (2005).
108. A. A. Elskus, *Toxicol. Sci., 88*, 1–3 (2005).
109. A. M. Jelaso, C. DeLong, J. Means, and C. F. Ide, *Environ. Res., 98*, 64–72 (2005).
110. N. Gorman, H. S. Walton, J. F. Sinclair, and P. R. Sinclair, *Comp. Biochem. Physiol. C, 121*, 405–412 (1998).
111. Y. W. Huang, J. J. Stegeman, B. R. Woodin, and W. H. Karasov, *Environ. Toxicol. Chem., 20*, 191–197 (2001).
112. Y. W. Huang, M. J. Melancon, R. E. Jung, and W. H. Karasov, *Environ. Toxicol. Chem., 17*, 1564–1569 (1998).

113. A. Yawetz, M. BenedekSegal, and B. Woodin, *Environ. Toxicol. Chem.*, *16*, 1802–1806 (1997).
114. A. Yawetz, D. Goldman, J. J. Stegeman, and B. Woodin, *Water Sci. Technol.*, *27*, 457–464 (1993).
115. R. J. Schwen and G. J. Mannering, *Comp. Biochem. Physiol. B*, *71*, 445–453 (1982).
116. J. Marty, J. L. Riviere, M. J. Guinaudy, P. Kremers, and P. Lesca, *Ecotoxicol. Environ. Saf.*, *24*, 144–154 (1992).
117. Y. Gutman and M. Kidron, *Biochem. Pharmacol.*, *20*, 3547–3550 (1971).
118. K. Greulich, and S. Pflugmacher, *Arch. Environ. Contam. Toxicol.*, *47*, 489–495 (2004).
119. M. Berrill, S. Bertram, A. Wilson, S. Louis, D. Brigham, and C. Stromberg, *Environ. Toxicol. Chem.*, *12*, 525–539 (1993).
120. J. J. Stegeman, in *Polycyclic Hydrocarbons and Cancer* (H. V. Gelboin and P. O. P. Ts'o, eds)., Vol. 3, Academic Press, New York, 1981, pp. 1–60.
121. J. J. Stegeman, B. R. Woodin, H. Singh, M. F. Oleksiak, and M. Celander, *Comp. Biochem. Physiol. C*, *116*, 61–75 (1997).
122. N. H. Vrolijk, N. M. Targett, B. R. Woodin, and J. J. Stegeman, *Marine Biology*, *119*, 151–158 (1994).
123. N. H. Vrolijk, N. M. Targett, B. R. Woodin, and J. J. Stegeman, *Mar. Environ. Res.*, *39*, 11–14 (1995).
124. A. Arukwe and A. Goksoyr, *J. Exp. Zool.*, *277*, 313–325 (1997).
125. T. Andersson and L. Förlin, *Aquat. Toxicol.*, *24*, 1–19 (1992).
126. J. J. Schlezinger, C. Parker, D. C. Zeldin, and J. J. Stegeman, *Arch. Biochem. Biophys.*, *353*, 265–275 (1998).
127. D. R. Buhler and J. L. Wang-Buhler, *Comp. Biochem. Physiol. C*, *121*, 107–137 (1998).
128. K. L. Wall and J. Crivello, *Toxicol. Appl. Pharmacol.*, *151*, 98–104 (1998).
129. S. L. Levine, J. T. Oris, and T. E. Wissing, *Environ. Toxicol. Chem.*, *14*, 123–128 (1995).
130. M. Machala, K. Nezveda, M. Petrivalsky, A. B. Jarosova, V. Piacka, and Z. Svobodova, *Aquat. Toxicol.*, *37*, 113–123 (1997).
131. D. Ronisz, D. G. J. Larsson, and L. Forlin, *Comp. Biochem. Physiol. C*, *124*, 271–279 (1999).
132. D. E. Williams, R. R. Becker, D. W. Potter, F. P. Guengerich, and D. R. Buhler, *Arch. Biochem. Biophys.*, *225*, 55–65 (1983).
133. Y. H. Yang, J. L. Wang, C. L. Miranda, and D. R. Buhler, *Arch. Biochem. Biophys.*, *352*, 271–280 (1998).
134. H. M. Carpenter, L. S. Fredrickson, D. E. Williams, D. R. Buhler, and L. R. Curtis, *Comp. Biochem. Physiol. C*, *97*, 127–132 (1990).
135. D. E. Williams, B. S. S. Masters, J. J. Lech, and D. R. Buhler, *Biochem. Pharmacol.* *35*, 2017–2023 (1986).
136. P. Lindström-Seppä, U. Koivusaari, and O. Hänninen, *Comp. Biochem. Physiol. C*, *69*, 259–263. (1981).
137. M. R. Miller, D. E. Hinton, and J. J. Stegeman, *Aquat. Toxicol.*, *14*, 307–322 (1989).
138. Y. F. Ueng and T. H. Ueng, *Arch. Biochem. Biophys.*, *322*, 347–356 (1995).

139. I. Leguen, C. Carlsson, E. Perdu-Durand, P. Prunet, P. Part, and J. P. Cravedi, *Aquat. Toxicol.*, *48*, 165–176 (2000).
140. K. R. Olson, *Comp. Biochem. Physiol. A*, *119*, 55–65 (1998).
141. D. Saucier, A. K. Julliard, G. Monod, H. de Brechard, and L. Astic, *Fish Physiol. Biochem.*, *21*, 179–192 (1999).
142. A. Goksøyr and L. Förlin, *Aquat. Toxicol.*, *22*, 287–311 (1992).
143. M. F. Oleksiak, S. Wu, C. Parker, W. Qu, R. Cox, D. C. Zeldin, and J. J. Stegeman, *Arch. Biochem. Biophys.*, *411*, 223–234 (2003).
144. T. Hegelund and M. C. Celander, *Aquat. Toxicol.*, *64*, 277–291 (2003).
145. H. G. Morrison, E. J. Weil, S. I. Karchner, M. L. Sogin, and J. J. Stegeman, *Comp. Biochem. Physiol. C*, *121*, 231–240 (1998).
146. H. G. Morrison, M. F. Oleksiak, N. W. Cornell, M. L. Sogin, and J. J. Stegeman, *Biochem. J.*, *308*, 97–104 (1995).
147. D. R. Nelson, *Arch. Biochem. Biophys.*, *409*, 18–24 (2003).
148. A. Goksoyr, and A. M. Husoy, *Mar. Environ. Res.*, *34*, 147–150 (1992).
149. M. J. Melancon, D. E. Williams, D. R. Buhler, and J. J. Lech, *Drug Metab. Dispos.*, *13*, 542–547. (1985).
150. C. L. Miranda, M. C. Henderson, D. E. Williams, and D. R. Buhler, *Toxicol. Appl. Pharmacol.*, *142*, 123–132 (1997).
151. R. Gooneratne, C. L. Miranda, M. C. Henderson, and D. R. Buhler, *Xenobiotica*, *27*, 175–187 (1997).
152. D. E. Williams, and D. R. Buhler, *Biochem. Pharmacol.*, *33*, 3743–3753 (1984).
153. D. E. Williams, J. J. Lech, and D. R. Buhler, *Mutation Research-Fundamental and Molecular Mechanisms of Mutagenesis 399*, 179–192 (1998).
154. C. L. Miranda, J.-L. Wang, M. C. Henderson, and D. R. Buhler, *Arch. Biochem. Biophys.*, *268*, 227–238 (1989).
155. C. L. Miranda, Y.-H. Yang, M. C. Henderson, J.-L. Wang, and D. R. Buhler, *Federation Proc.*, *11*, A829 (1997).
156. D. E. Williams, and D. R. Buhler, *Cancer Res.*, *43*, 4752–4756 (1983).
157. Y. H. Yang, C. L. Miranda, M. N. Henderson, J. L. Wang-Buhler, and D. R. Buhler, *Drug Metab. Dispos.*, *28*, 1279–1283 (2000).
158. J. L. Wang-Buhler, S. J. Lee, W. G. Chung, J. F. Stevens, H. P. Tseng, T. H. Hseu, C. H. Hu, M. Westerfield, Y. H. Yang, C. L. Miranda, and D. R. Buhler, *Comp. Biochem. Physiol. C*, *140*, 207–219 (2005).
159. R. Thibaut, L. Debrauwer, E. Perdu, A. Goksoyr, J. P. Cravedi, and A. Arukwe, *Aquat. Toxicol.*, *56*, 177–190 (2002).
160. C. L. Miranda, M. C. Henderson, and D. R. Buhler, *Toxicol. Appl. Pharmacol.*, *148*, 237–244 (1998).
161. M. F. Oleksiak, S. Wu, C. Parker, S. I. Karchner, J. J. Stegeman, and D. C. Zeldin, *J. Biol. Chem.*, *275*, 2312–2321 (2000).
162. S. Kashiwada, D. E. Hinton, and S. W. Kullman, *Comp. Biochem. Physiol. C*, *141*, 338–348 (2005).
163. M. O. James, Z. Loua, L. Rowland-Faux, and M. C. Celander, *Aquat. Toxicol.*, *72*, 361–371 (2005).
164. G. E. Corley-Smith, H. T. Su, J. L. Wang-Buhler, H. P. Tseng, C. H. Hu, T. Hoang, W. G. Chung, and D. R. Buhler, *Biochem. Biophys. Res. Comm.*, *340*, 1039–1046 (2006).

165. M. E. Hahn, B. R. Woodin, J. J. Stegeman, and D. E. Tillitt, *Comp. Biochem. Physiol. C, 120*, 67–75 (1998).
166. S. I. Karchner, D. G. Franks, and M. E. Hahn, *Biochem. J., 392*, 153–161 (2005).
167. M. E. Hahn, S. I. Karchner, R. R. Merson, D. G. Franks, E. Montie, J. M. Lapseritis, E. W. Zabel, D. E. Tillitt, and M. Hannink, *Mar. Environ. Res., 58*, 132 (2004).
168. M. E. Hahn, S. I. Karchner, D. G. Franks, and R. R. Merson, *Pharmacogenetics, 14*, 131–143 (2004).
169. K. M. Kleinow, M. L. Haasch, D. E. Williams, and J. J. Lech, *Comp. Biochem. Physiol. C, 96*, 259–270 (1990).
170. A. A. Elskus, and J. J. Stegeman, *Comp. Biochem. Physiol. C, 92*, 223–230 (1989).
171. M. D. Sadar, R. Ash, J. Sundqvist, P. E. Olsson, and T. B. Andersson, *J. Biol. Chem., 271*, 17635–17643 (1996).
172. M. D. Sadar, R. Ash, and T. B. Andersson, *Biochem. Biophys. Res. Comm., 214*, 1060–1066 (1995).
173. M. D. Sadar, F. Blomstrand, and T. B. Andersson, *Biochem. Biophys. Res. Comm., 225*, 455–461 (1996).
174. M. Celander, M. E. Hahn, and J. J. Stegeman, *Arch. Biochem. Biophys., 329*, 113–122 (1996).
175. M. Celander, J. Bremer, M. E. Hahn, and J. J. Stegeman, *Environ. Toxicol. Chem., 16*, 900–907 (1997).
176. A. Devaux, M. Pesonen, G. Monod, and T. Andersson, *Biochem. Pharmacol., 43*, 898–901 (1992).
177. H. P. Tseng, T. H. Hseu, D. R. Buhler, W. D. Wang, and C. H. Hu, *Toxicol. Appl. Pharmacol., 205*, 247–258 (2005).
178. A. C. D. Bainy and J. J. Stegeman, *Mar. Environ. Res., 58*, 133–134. (2004).
179. V. Meucci and A. Arukwe, *Comp. Biochem. Physiol. C, 142*, 142–150 (2006).
180. M. L. Haasch, M. C. Henderson, and D. R. Buhler, *Comp. Biochem. Physiol. C, 121*, 297–303 (1998).
181. M. L. Haasch, M. C. Henderson, and D. R. Buhler, *Mar. Environ. Res., 46*, 37–40 (1998).
182. L. A. Baldwin, P. T. Kostecki, and E. J. Calabrese, *Ecotoxicol. Environ. Safety, 25*, 193–201 (1993).
183. L. Scarano, E. J. Calabrese, P. T. Kostecki, B. L. A., and D. A. Leonard, *Ecotoxicol. Environ. Safety, 29*, I3–19 (1994).
184. C. Sabourault, J.-B. Bergé, M. Lafaurie, J.-P. Girard, and M. Amichot, *Biochem. Biophys. Res. Comm., 251*, 213–219 (1998).
185. J. Sturve, L. Hasselberg, H. Fälth, M. Celander, and L. Förlin, *Aquat. Toxicol., 78, Suppl.* 1, S73–S78 (2006).
186. L. Hasselberg, S. Meier, A. Svardal, T. Hegelund, and M. C. Celander, *Aquat. Toxicol., 67*, 303–313 (2004).
187. T. R. Henry, J. M. Spitsbergen, M. W. Hornung, C. C. Abnet, and R. E. Peterson, *Toxicol. Appl. Pharmacol., 142*, 56–68 (1997).
188. J. J. Schlezinger, R. D. White, and J. J. Stegeman, *Mol. Pharmacol. 56*, 588–597 (1999).
189. J. J. Schlezinger, J. Keller, L. A. Verbrugge, and J. J. Stegeman, *Comp. Biochem. Physiol. C, 125*, 273–286 (2000).

190. M. Solé and D. R. Livingstone, *Comp. Biochem. Physiol. C, 141*, 20–31 (2005).
191. J. F. Narbonne, P. Mora, X. Michel, H. Budzinski, P. Garrigues, M. Lafaurie, J. P. Salaun, J. B. Berge, P. Den Besten, G. Pagano, C. Porte, D. Livingstone, P. D. Hansen, and A. Herbert, *J. Toxicol. Toxin Rev., 18*, 205–220 (1999).
192. D. R. Livingstone, *Comp. Biochem. Physiol. A, 120*, 43–49 (1998).
193. J. G. Scott and Z. M. Wen, *Pest Manag. Sci., 57*, 958–967 (2001).
194. R. Feyereisen, *Annu. Rev. Entomol., 44*, 507–533 (1999).
195. J. Y. Bradfield, Y. H. Lee, and L. L. Keeley, *Proc. Natl. Acad. Sci. USA, 88*, 4558–4562 (1991).
196. M. Zhang, and J. G. Scott, *Pestic. Biochem. Physiol., 55*, 150–156 (1996).
197. V. M. Guzov, G. C. Unnithan, A. A. Chernogolov, and R. Feyereisen, *Arch. Biochem. Biophys., 359*, 231–240 (1998).
198. E. Hodgson, *Insect Biochem., 13*, 237–246 (1983).
199. M. Maïbèche-Coisne, L. Monti-Dedieu, S. Aragon, and C. Dauphin-Villemant, *Biochem. Biophys. Res. Commun., 273*, 1132–1137 (2000).
200. H. Ono, K. Ozaki, and H. Yoshikawa, *Insect Biochem. Molec. Biol., 35*, 837–846 (2005).
201. M. Maïbèche-Coisne, A. A. Nikonov, Y. Ishida, E. Jacquin-Joly, and W. S. Leal, *Proc. Natl. Acad. Sci. USA, 101*, 11459–11464 (2004).
202. M. Maïbèche-Coisne, E. Jacquin-Joly, M. C. Francois, and P. Nagnan-Le Meillour, *Insect Molec. Biol., 11*, 273–281 (2002).
203. M. Maïbèche-Coisne, C. Merlin, M. C. Francois, P. Porcheron, and E. Jacquin-Joly, *Gene, 346*, 195–203 (2005).
204. H. Ranson, C. Claudianos, F. Ortelli, C. Abgrall, J. Hemingway, M. V. Sharakhova, M. F. Unger, F. H. Collins, and R. Feyereisen, *Science, 298*, 179–181 (2002).
205. M. R. Berenbaum, *J. Chem. Ecol., 28*, 873–896 (2002).
206. T. G. Wilson, *Annu. Rev. Entomol., 46*, 545–571 (2001).
207. J. G. Scott, *Insect Biochem. Molec. Biol., 29*, 757–777 (1999).
208. J. G. Scott, N. Liu, and Z. M. Wen, *Comp. Biochem. Physiol. C, 121*, 147–155 (1998).
209. M. A. Schuler, *Plant Physiol,. 112*, 1411–1419 (1996).
210. J. P. David, C. Strode, J. Vontas, D. Nikou, A. Vaughan, P. M. Pignatelli, C. Louis, J. Hemingway, and H. Ranson, *Proc. Natl. Acad. Sci. USA, 102*, 4080–4084 (2005).
211. P. B. Danielson, J. L. Foster, M. M. McMahill, M. K. Smith, and J. C. Fogleman, *Mol. Gen. Genet., 259*, 54–59 (1998).
212. B. Pittendrigh, K. Aronstein, E. Zinkovsky, O. Andreev, B. Campbell, J. Daly, S. Trowell, and R. H. ffrench-Constant, *Insect Biochem. Molec. Biol., 27*, 507–512 (1997).
213. M. E. Scharf, S. Parimi, L. J. Meinke, L. D. Chandler, and B. D. Siegfried, *Insect Molec. Biol., 10*, 139–146 (2001).
214. J. W. Pridgeon, L. Zhang, and N. Liu, *Gene, 314*, 157–163 (2003).
215. M. J. Snyder, J. L. Stevens, J. F. Andersen, and R. Feyereisen, *Arch. Biochem. Biophys., 321*, 13–20 (1995).
216. J. F. Andersen, J. G. Utermohlen, and R. Feyereisen, *Biochemistry, 33*, 2171–2177 (1994).
217. J. F. Andersen, J. K. Walding, P. H. Evans, W. S. Bowers, and R. Feyereisen, *Chem. Res. Toxicol., 10*, 156–164 (1997).

218. F. A. Carino, J. F. Koener, F. W. Plapp, Jr., and R. Feyereisen, *Insect Biochem. Molec. Biol.*, *24*, 411–418 (1994).
219. C. Saner, B. Weibel, F. E. Wurgler, and C. Sengstag, *Environ. Mol. Mutagen.*, *27*, 46–58 (1996).
220. B. C. Dunkov, V. M. Guzov, G. Mocelin, F. Shotkoski, A. Brun, M. Amichot, R. H. ffrench-Constant, and R. Feyereisen, *DNA Cell Biol.*, *16*, 1345–1356 (1997).
221. S. Maitra, S. M. Dombrowski, L. C. Waters, and R. Ganguly, *Gene*, *180*, 165–171 (1996).
222. A. Brun, A. Cuany, T. Le Mouel, J. Berge, and M. Amichot, *Insect Biochem. Molec. Biol.*, *26*, 697–703 (1996).
223. M. Amichot, S. Tares, A. Brun-Barale, L. Arthaud, J. M. Bride, and J. B. Berge, *Eur. J. Biochem.*, *271*, 1250–1257 (2004).
224. H. R. Jensen, I. M. Scott, S. Sims, V. L. Trudeau, and J. T. Arnason, *J. Agric. Food Chem.*, *54*, 1289–1295 (2006).
225. C. Helvig, N. Tijet, R. Feyereisen, F. A. Walker, and L. L. Restifo, *Biochem. Biophys. Res. Commun.*, *325*, 1495–1502 (2004).
226. G. Le Goff, S. Boundy, P. J. Daborn, J. L. Yen, L. Sofer, R. Lind, C. Sabourault, L. Madi-Ravazzi, and R. H. ffrench-Constant, *Insect Biochem. Molec. Biol.*, *33*, 701–708 (2003).
227. R. Ma, M. B. Cohen, M. R. Berenbaum, and M. A. Schuler, *Arch. Biochem. Biophys.*, *310*, 332–340 (1994).
228. H. Prapaipong, M. R. Berenbaum, and M. A. Schuler, *Nucl. Acids Res.*, *22*, 3210–3217 (1994).
229. C. F. Hung, H. Prapaipong, M. R. Berenbaum, and M. A. Schuler, *Insect Biochem. Molec. Biol.*, *25*, 89–99 (1995).
230. C. F. Hung, M. R. Berenbaum, and M. A. Schuler, *Insect Biochem. Molec. Biol.*, *27*, 377–385 (1997).
231. Z. M. Wen, L. P. Pan, M. R. Berenbaum, and M. A. Schuler, *Insect Biochem. Molec. Biol.*, *33*, 937–947 (2003).
232. X. P. Wang, and A. A. Hobbs, *Insect Biochem. Molec. Biol.*, *25*, 1001–1009 (1995).
233. R. A. Petersen, A. R. Zangerl, M. R. Berenbaum, and M. A. Schuler, *Insect Biochem. Molec. Biol.*, *31*, 679–690 (2001).
234. C. F. Hung, T. L. Harrison, M. R. Berenbaum, and M. A. Schuler, *Insect Molec. Biol.*, *4*, 149–160 (1995).
235. W. M. Li, M. A. Schuler, and M. R. Berenbaum, *Proc. Natl. Acad. Sci. USA*, *100*, 14593–14598 (2003).
236. W. Li, M. R. Berenbaum, and M. A. Schuler, *Insect Biochem. Molec. Biol.*, *31*, 999–1011 (2001).
237. C. Ranasinghe and A. A. Hobbs, *Insect Biochem. Molec. Biol.*, *28*, 571–580 (1998).
238. C. Ranasinghe, M. Headlam, and A. A. Hobbs, *Arch. Insect Biochem. Physiol.*, *34*, 99–109 (1997).
239. C. Ranasinghe and A. A. Hobbs, *Insect Molec. Biol.*, *8*, 443–447 (1999).
240. X. Li, M. R. Berenbaum and M. A. Schuler, *Insect Molec. Biol.*, *11*, 343–351 (2002).
241. X. Li, M. R. Berenbaum and M. A. Schuler, *Insect Biochem. Molec. Biol.*, *30*, 75–84 (2000).

242. X. C. Li, M. A. Schuler, and M. R. Berenbaum, *Nature*, *419*, 712–715 (2002).
243. X. Li, J. Baudry, M. R. Berenbaum, and M. A. Schuler, *Proc. Natl. Acad. Sci. USA*, *101*, 2939–2944 (2004).
244. W. Li, R. A. Petersen, M. A. Schuler, and M. R. Berenbaum, *Insect Molec. Biol.*, *11*, 543–551 (2002).
245. G. D. Wheelock and J. G. Scott, *J. Exp. Zool.*, *264*, 153–158 (1992).
246. G. D. Wheelock and J. G. Scott, *Pestic. Biochem. Physiol.*, *43*, 67–77 (1992).
247. M. Zhang and J. G. Scott, *Arch. Insect Biochem. Physiol.*, *27*, 205–216 (1994).
248. N. Liu and J. G. Scott, *Biochem. Genet.*, *34*, 133–148 (1996).
249. J. G. Scott, P. Sridhar, and N. Liu, *Arch. Insect Biochem. Physiol.*, *31*, 313–323 (1996).
250. P. J. Korytko and J. G. Scott, *Arch. Insect Biochem. Physiol.*, *37*, 57–63 (1998).
251. P. J. Korytko, R. J. MacLntyre, and J. G. Scott, *Insect Molec. Biol.*, *9*, 441–449 (2000).
252. P. J. Korytko, F. W. Quimby, and J. G. Scott, *J. Biochem. Mol. Toxicol.*, *14*, 20–25 (2000).
253. N. Liu and J. G. Scott, *Insect Biochem. Molec. Biol.*, *28*, 531–535 (1998).
254. R. Hatano and J. G. Scott, *Pestic. Biochem. Physiol.*, *45*, 228–233 (1993).
255. S. Kasai and J. G. Scott, *Insect Molec. Biol.*, *10*, 191–196 (2001).
256. S. Kasai, T. Shono, and M. Yamakawa, *Insect Molec. Biol.*, *7*, 185–190 (1998).
257. S. Kasai, I. S. Weerashinghe, T. Shono, and M. Yamakawa, *Insect Biochem. Molec. Biol.*, *30*, 163–171 (2000).
258. M. Q. Gong, Y. Gu, X. B. Hu, Y. Sun, L. Ma, X. L. Li, L. X. Sun, J. Sun, J. Qian, and C. L. Zhu, *Acta Biochim. Biophys. Sin. (Shanghai)*, *37*, 317–326 (2005).
259. P. Daborn, S. Boundy, J. Yen, B. Pittendrigh, and R. ffrench–Constant, *Mol. Genet. Genomics*, *266*, 556–563 (2001).
260. P. J. Daborn, J. L. Yen, M. R. Bogwitz, G. Le Goff, E. Feil, S. Jeffers, N. Tijet, T. Perry, D. Heckel, P. Batterham, R. Feyereisen, T. G. Wilson, and R. H. ffrench-Constant, *Science*, *297*, 2253–2256 (2002).
261. P. Rodpradit, S. Boonsuepsakul, T. Chareonviriyaphap, M. J. Bangs, and P. Rongnoparut, *J. Am. Mosq. Control Assoc.*, *21*, 71–79 (2005).
262. Y. C. Zhu and G. L. Snodgrass, *Insect Molec. Biol.*, *12*, 39–49 (2003).
263. D. Nikou, H. Ranson, and J. Hemingway, *Gene*, *318*, 91–102 (2003).
264. P. Rongnoparut, S. Boonsuepsakul, T. Chareonviriyaphap, and N. Thanomsing, *J. Vector Ecol.*, *28*, 150–158 (2003).
265. W. Li, A. R. Zangerl, M. A. Schuler, and M. R. Berenbaum, *Insect Molec. Biol.*, *13*, 603–613 (2004).
266. W. Mao, S. Rupasinghe, A. R. Zangerl, M. A. Schuler, and M. R. Berenbaum, *Insect Molec. Biol.*, *15*, 169–179 (2006).
267. J. P. David, S. Boyer, A. Mesneau, A. Ball, H. Ranson, and C. Dauphin-Villemant, *Insect Biochem. Molec. Biol.*, *36*, 410–420 (2006).
268. R. L. Rose, D. Goh, D. M. Thompson, K. D. Verma, D. G. Heckel, L. J. Gahan, R. M. Roe, and E. Hodgson, *Insect Biochem. Molec. Biol.*, *27*, 605–615 (1997).
269. J. L. Stevens, M. J. Snyder, J. F. Koener, and R. Feyereisen, *Insect Biochem. Molec. Biol.*, *30*, 559–568 (2000).
270. M. R. Bogwitz, H. Chung, L. Magoc, S. Rigby, W. Wong, M. O'Keefe, J. A. McKenzie, P. Batterham, and P. J. Daborn, *Proc. Natl. Acad. Sci. USA*, *102*, 12807–12812 (2005).

271. A. Brandt, M. Scharf, J. H. Pedra, G. Holmes, A. Dean, M. Kreitman, and B. R. Pittendrigh, *Insect Molec. Biol., 11*, 337–341 (2002).
272. P. B. Danielson, R. J. MacIntyre, and J. C. Fogleman, *Proc. Natl. Acad. Sci. USA, 94*, 10797–10802 (1997).
273. J. Vontas, C. Blass, A. C. Koutsos, J. P. David, F. C. Kafatos, C. Louis, J. Hemingway, G. K. Christophides, and H. Ranson, *Insect Molec. Biol., 14*, 509–521 (2005).
274. M. Sasabe, Z. Wen, M. R. Berenbaum, and M. A. Schuler, *Gene, 338*, 163–175 (2004).
275. M. B. Cohen, M. A. Schuler, and M. R. Berenbaum, *Proc. Natl. Acad. Sci. USA, 89*, 10920–10924 (1992).
276. M. J. Snyder and J. I. Glendinning, *J. Comp. Physiol., 179*, 255–261 (1996).
277. M. J. Snyder, E. L. Hsu, and R. Feyereisen, *J. Chem. Ecol., 19*, 2903–2916 (1993).
278. M. J. Snyder, J. K. Walding, and R. Feyereisen, *Insect Biochem. Molec. Biol., 24*, 837–846 (1994).
279. J. G. Oakeshott, I. Home, T. D. Sutherland, and R. J. Russell, *Genome Biol., 4*, 202 (2003).
280. J. M. Bride, A. Cuany, M. Amichot, A. Brun, M. Babault, T. L. Mouel, G. De Sousa, R. Rahmani, and J. B. Berge, *J. Econ. Entomol., 90*, 1514–1520 (1997).
281. M. Amichot, A. Brun, A. Cuany, G. De Souza, T. Le Mouel, J. M. Bride, M. Babault, J. P. Salaun, R. Rahmani, and J. B. Berge, *Comp. Biochem. Physiol. C, 121*, 311–319 (1998).
282. J. B. Berge, R. Feyereisen, and M. Amichot, *Philos. Trans. Roy. Soc. Lond. B Biol. Sci., 353*, 1701–1705 (1998).
283. C. McCart, A. Buckling, and R. H. ffrench-Constant, *Curr. Biol., 15*, R587–589 (2005).
284. S. Y. Fuchs, V. S. Spiegelman, R. I. Abdrjashitov, and G. A. Belitsky, *Exper. Oncol., 17*, 55–60 (1995).
285. S. Y. Fuchs, V. S. Spiegelman, and G. A. Belitsky, *Biochem. Pharmacol., 47*, 1867–1873 (1994).
286. R. P. Brown, C. M. McDonnell, M. R. Berenbaum, and M. A. Schuler, *Gene, 358*, 39–52 (2005).
287. R. B. Emmons, D. Duncan, P. A. Estes, P. Kiefel, J. T. Mosher, M. Sonnenfeld, M. P. Ward, I. Duncan, and S. T. Crews, *Development, 126*, 3937–3945. (1999).
288. R. A. Petersen, H. Niamsup, M. R. Berenbaum, and M. A. Schuler, *Biochim. Biophys. Acta, 1619*, 269–282 (2003).
289. C. M. McDonnell, R. P. Brown, M. R. Berenbaum, and M. A. Schuler, *Insect Biochem. Molec. Biol., 34*, 1129–1139 (2004).
290. U. Graf, F. E. Wurgler, A. J. Katz, H. Frei, H. Juon, C. B. Hall, and P. G. Kale, *Environ. Mutagen., 6*, 153–188 (1984).
291. C. B. Bernard, and B. J. Philogene, *J. Toxicol. Environ. Health, 38*, 199–223 (1993).
292. C. S. Jewell, M. H. Mayeaux, and G. W. Winston, *Comp. Biochem. Physiol. C, 118*, 369–374 (1997).
293. M. O. James and S. M. Boyle, *Comp. Biochem. Physiol. C, 121*, 157–172 (1998).
294. M. O. James, *Arch. Biochem. Biophys., 282*, 8–17 (1990).
295. M. O. James and K. T. Shiverick, *Arch. Biochem. Biophys., 233*, 1–9 (1984).

296. M. O. James, S. M. Boyle, H. G. TrapidoRosenthal, W. C. Smith, R. M. Greenberg, and K. T. Shiverick, *Arch. Biochem. Biophys.*, 329, 31–38 (1996).
297. M. J. Snyder, *Biochem. Biophys. Res. Comm.*, 249, 187–190 (1998).
298. M. J. Snyder, *Arch. Biochem. Biophys.*, 358, 271–276 (1998).
299. C. Dauphin-Villemant, D. Bocking, M. Tom, M. Maïbèche, and R. Lafont, *Biochem. Biophys. Res. Commun.*, 264, 413–418 (1999).
300. K. Rewitz, B. Styrishave, and O. Andersen, *Biochem. Biophys. Res. Comm.*, 310, 252–260 (2003).
301. P. David, C. Dauphin-Villemant, A. Mesneau, and J. C. Meyran, *Mol. Ecol.*, 12, 2473–2481 (2003).
302. S. M. Boyle, M. P. Popp, W. C. Smith, R. M. Greenberg, and M. O. James, *Mar. Environ. Res.*, 46, 25–28 (1998).
303. S. Arun, A. Rajendran, and P. Subramanian, *Ecotoxicology*, in press (2006).
304. M. Ishizuka, H. Hoshi, N. Minamoto, M. Masuda, A. Kazusaka, and S. Fujita, *Environ. Health Perspect.*, 104, 774–778 (1996).
305. S. Fujita, I. Chiba, M. Ishizuka, H. Hoshi, H. Iwata, A. Sakakibara, S. Tanabe, A. Kazusaka, M. Masuda, Y. Masuda, and H. Nakagawa, *Biomarkers*, 6, 19–25 (2001).
306. E. Oberdorster, D. Rittschof, and P. McClellan–Green, *Aquat. Toxicol.*, 41, 83–100 (1998).
307. E. Oberdorster, D. Rittschof, and G. A. LeBlanc, *Arch. Environ. Contam. Toxicol.*, 34, 21–25 (1998).
308. P. J. den Besten, H. J. Herwig, E. G. van Donselaar, and D. R. Livingstone, *Mar. Biol.*, 107, 171–177 (1990).
309. P. J. den Besten, *Comp. Biochem. Physiol. C*, 121, 139–146 (1998).
310. W. Y. Precious and J. Barrett, *Parasitol. Today*, 5, 156–160 (1989).
311. J. Barrett, *Comp. Biochem. Physiol. C*, 121, 181–183 (1998).
312. A. C. Kotze, *Int. J. Parasitol.*, 27, 33–40 (1997).
313. A. C. Kotze, *Int. J. Parasitol.*, 29, 389–396 (1999).
314. D. Kerboeuf, D. Soubieux, R. Guilluy, J. L. Brazier, and J. L. Riviere, *Parasitol. Res.*, 81, 302–304 (1995).
315. A. C. Kotze, R. J. Dobson, and D. Chandler, *Vet. Parasitol.*, 136, 275–282 (2006).
316. M. Alvinerie, J. Dupuy, C. Eeckhoutte, J. F. Sutra, and D. Kerboeuf, *Parasitol. Res.*, 87, 702–704 (2001).
317. R. Menzel, M. Rodel, J. Kulas, and C. E. W. Steinberg, *Arch. Biochem. Biophys.*, 438, 93–102 (2005).
318. D. Gems and J. J. McElwee, *Mech. Ageing Devel.*, 126, 381–387 (2005).
319. R. Menzel, T. Bogaert, and R. Achazi, *Arch. Biochem. Biophys.*, 395, 158–168 (2001).
320. K. Reichert and R. Menzel, *Chemosphere*, 61, 229–237 (2005).
321. R. Menzel, S. Sturzenbaum, A. Barenwaldt, J. Kulas, and C. E. W. Steinberg, *Environ. Sci. Technol.*, 39, 8324–8332 (2005).
322. E. Enmark, and J. Å. Gustaffson, *Trends Pharmacol. Sci.*, 21, 85–87 (2000).
323. R. F. Lee, *Comp. Biochem. Physiol. C*, 121, 173–179 (1998).
324. R. K. Achazi, C. Flenner, D. R. Livingstone, L. D. Peters, K. Schaub, and E. Scheiwe, *Comp. Biochem. Physiol. C*, 121, 339–350 (1998).

325. A. G. Berghout, E. Wenzel, J. Buld, and K. J. Netter, *Comp. Biochem. Physiol. C*, *100*, 389–396 (1991).
326. V. E. Forbes, M. S. Andreassen, and L. Christensen, *Environ. Toxicol. Chem.*, *20*, 1012–1021 (2001).
327. A. Jorgensen, L. J. Rasmussen, and O. Andersen, *Biochem. Biophys. Res. Comm.*, *336*, 890–897 (2005).
328. B. Li, H. C. Bisgaard and V. E. Forbes, *Biochem. Biophys. Res. Comm.*, *325*, 510–517 (2004).
329. R. F. Lee, S. C. Singer, and D. S. Page, *Aquat. Toxicol.*, *1*, 355–365 (1981).
330. C. R. Fries and R. F. Lee, *Mar. Biol.*, *79*, 187–193 (1984).
331. A. Jorgensen, A. M. B. Giessing, L. J. Rasmussen, and O. Andersen, *Environ. Toxicol. Chem.*, *24*, 2796–2805 (2005).
332. K. F. Rewitz, C. Kjellerup, A. Jorgensen, C. Petersen, and O. Andersen, *Comp. Biochem. Physiol. C*, *138*, 89–96 (2004).
333. E. D. Goldberg and K. K. Bertine, *Sci. Total Environ.*, *247*, 165–174 (2000).
334. M. A. Kirchin, A. Wiseman, and D. R. Livingstone, *Comp. Biochem. Physiol. C*, *101*, 81–91 (1992).
335. M. Solé, C. Porte, and J. Albaiges, *Environ. Toxicol. Chem.*, *14*, 157–164 (1995).
336. A. N. Wootton, P. S. Goldfarb, P. Lemaire, S. C. M. Ohara, and D. R. Livingstone, *Mar. Environ. Res.*, *42*, 297–301 (1996).
337. J. P. Shaw, A. T. Large, P. Donkin, S. V. Evans, F. J. Staff, D. R. Livingstone, J. K. Chipman, and L. D. Peters, *Aquat. Toxicol.*, *67*, 325–336 (2004).
338. D. M. Y. Cheah, P. F. A. Wright, D. A. Holdway, and J. T. Ahokas, *Aquat. Toxicol.*, *33*, 201–214 (1995).
339. M. Wilbrink, E. J. Groot, R. Jansen, Y. De Vries, and N. P. Vermeulen, *Xenobiotica*, *21*, 223–233 (1991).
340. W. Baturo, and L. Lagadic, *Environ. Toxicol. Chem.*, *15*, 771–781 (1996).
341. X. R. Michel, P. Suteau, L. W. Robertson, and J. F. Narbonne, *Aquat. Toxicol.*, *27*, 335–344 (1993).
342. X. Michel, J. P. Salaun, F. Galgani, and J. F. Narbonne, *Mar. Environ. Res.*, *38*, 257–273 (1994).
343. A. Galli, D. Del Chiaro, R. Nieri, and G. Bronzetti, *Mar. Biol.*, *100*, 69–73 (1988).
344. M. J. Snyder, *Aquat. Toxicol.*, *48*, 529–547 (2000).
345. Y. Teunissen, W. P. Geraerts, H. van Heerikhuizen, R. J. Planta, and J. Joosse, *J Biochem. (Tokyo)*, *112*, 249–252 (1992).
346. A. E. Lockyer, L. R. Noble, D. Rollinson, and C. S. Jones, *Mem. Inst. Oswaldo Cruz*, *100*, 259–262 (2005).
347. D. J. Brown, G. C. Clark, and R. J. Van Beneden, *Comp. Biochem. Physiol. C*, *121*, 351–360 (1998).
348. L. D. Peters, C. Nasci, and D. R. Livingstone, *Mar. Environ. Res.*, *46*, 295–299 (1998).
349. A. Binelli, F. Ricciardi, C. Riva, and A. Provini, *Chemosphere*, *62*, 510–519 (2006).
350. L. M. Heffernan, and G. W. Winston, *Comp. Biochem. Physiol. C: Pharmacol. Toxicol. Endocrinol.*, *121*, 371–383 (1998).
351. B. Quinn, F. Gagne, and C. Blaise, *Fresenius Environ. Bull.*, *13*, 783–788 (2004).
352. N. J. Gassman and C. J. Kennedy, *Bull. Marine Science*, *50*, 320–330 (1992).

353. J. Briand, H. Julistiono, P. Beaune, J. P. Flinois, I. de Waziers, and J. P. Leroux, *Biochim. Biophys. Acta, 1203*, 199–204 (1993).
354. H. Julistiono and J. Briand, *Comp. Biochem. Physiol. B, 102*, 747–755 (1992).
355. S. Pflugmacher and H. Sandermann, *Plant Physiol., 117*, 123–128 (1998).
356. F. Thies, T. Backhaus, B. Bossmann, and L. H. Grimme, *Plant Physiol., 112*, 361–370 (1996).
357. F. Thies and L. H. Grimme, *Arch. Microbiol., 164*, 203–211 (1995).
358. J. P. Barque, A. Abahamid, J. P. Flinois, P. Beaune, and J. Bonaly, *Biochem. Biophys. Res. Commun., 298*, 277–281 (2002).
359. M. A. Schuler and D. Werck-Reichhart, *Annu. Rev. Plant Biol., 54*, 629–667 (2003).
360. J. L. De Prado, M. D. Osuna, A. Heredia, and R. De Prado, *J. Agric. Food Chem., 53*, 2185–2191 (2005).
361. H. Sandermann, Jr., *Trends Biochem. Sci., 17*, 82–84 (1992).
362. D. Coupland, in *Herbicide Resistance in Weeds and Crops* (J. C. Caseley, G. W. Cussans, and R. K. Atkin, eds)., John Wiley & Sons, Inc., New York, 1991, pp. 263–278.
363. H. Kawahigashi, S. Hirose, H. Ohkawa, and Y. Ohkawa, *Plant Sci., 165*, 373–381 (2003).
364. D. Werck-Reichhart, A. Hehn, and L. Didierjean, *Trends Plant Sci., 5*, 116–123 (2000).
365. Y. Batard, A. Hehn, S. Nedelkina, M. Schalk, K. Pallett, H. Schaller, and D. Werck-Reichhart, *Arch. Biochem. Biophys., 379*, 161–169 (2000).
366. M. Barrett, *Drug Metabol. Drug Interact., 12*, 299–315 (1995).
367. D. S. Frear, *Science, 162*, 674–675 (1968).
368. D. S. Frear, H. R. Swanson, and F. S. Tanaka, *Phytochemistry, 8*, 2157–2169 (1969).
369. C. Mougin, N. Polge, R. Scalla, and F. Cabanne, *Pestic. Biochem. Physiol., 40*, 1–11 (1991).
370. J. Menendez, and R. De Prado, *Physiol. Plantarum, 99*, 97–104 (1997).
371. L. M. Hall, S. R. Moss, and S. B. Powles, *Pestic. Biochem. Physiol., 53*, 180–192 (1995).
372. C. Preston, F. J. Tardif, and S. B. Powles, *Molecular Genetics and Evolution of Pesticide Resistance, ACS Symposium Series 645*, 117–129 (1996).
373. J. T. Christopher, C. Preston, and S. B. Powles, *Pestic. Biochem. Physiol., 49*, 172–182 (1994).
374. R. DePrado, J. L. DePrado, and J. Menendez, *Pestic. Biochem. Physiol., 57*, 126–136 (1997).
375. Q. Yu, L. J. S. Friesen, X. Q. Zhang, and S. B. Powles, *Pestic. Biochem. Physiol., 78*, 21–30 (2004).
376. H. Ohkawa, N. Shiota, H. Imaishi, T. Yamada, H. Inui, and Y. Ohkawa, *Biotechnol. Biotech. Equip., 12*, 17–22 (1998).
377. M. A. Pierrel, Y. Batard, M. Kazmaier, C. Mignotte-Vieux, F. Durst, and D. Werck-Reichhart, *Eur. J. Biochem., 224*, 835–844 (1994).
378. R. Fonné-Pfister, J. Gaudin, K. Kreuz, K. Ramsteiner, and E. Ebert, *Pestic. Biochem. Physiol., 37*, 165–173 (1990).

379. B. Siminszky, F. T. Corbin, E. R. Ward, T. J. Fleischmann, and R. E. Dewey, *Proc. Natl. Acad. Sci. USA*, 96, 1750–1755 (1999).
380. D. S. Frear, H. R. Swanson, and F. W. Thalacker, *Pestic. Biochem. Physiol.*, 41, 274–287 (1991).
381. E. J. Belford, U. Dorfler, A. Stampfl, and P. Schroder, *Z. Naturforsch. C*, 59, 693–700 (2004).
382. T. Robineau, Y. Batard, S. Nedelkina, F. Cabello-Hurtado, M. LeRet, O. Sorokine, L. Didierjean, and D. Werck-Reichhart, *Plant Physiol.*, 118, 1049–1056 (1998).
383. D. E. Moreland, F. T. Corbin, and J. E. McFarland, *Pestic. Biochem. Physiol.*, 47, 206–214 (1993).
384. M. W. Persans and M. A. Schuler, *Plant Physiol.*, 109, 1483–1490 (1995).
385. W. S. Xiang, X. J. Wang, T. R. Ren, and X. L. Ju, *Pest Manag. Sci.*, 61, 402–406 (2005).
386. F. W. Thalacker, H. R. Swanson, and D. S. Frear, *Pestic. Biochem. Physiol.*, 49, 209–223 (1994).
387. M. K. Koeppe, C. M. Hirata, H. M. Brown, W. H. Kenyon, D. P. O'Keefe, S. C. Lau, W. T. Zimmerman, and J. M. Green, *Pestic. Biochem. Physiol.*, 66, 170–181 (2000).
388. R. S. Buker, B. Rathinasabapathi, G. MacDonald, and S. M. Olson, *Weed Sci.*, 52, 201–205 (2004).
389. F. Deng and K. K. Hatzios, *Pestic. Biochem. Physiol.*, 74, 102–115 (2002).
390. M. D. Osuna, F. Vidotto, A. J. Fischer, D. E. Bayer, R. De Prado, and A. Ferrero, *Pestic. Biochem. Physiol.*, 73, 9–17 (2002).
391. J. R. R. Hinz, M. D. K. Owen, and M. Barrett, *Weed Sci.*, 45, 474–480 (1997).
392. M. W. M. Burnet, B. R. Loveys, J. A. M. Holtum, and S. B. Powles, *Planta*, 190, 182–189 (1993).
393. D. S. Frear and H. R. Swanson, *J. Agric. Food Chem.*, 44, 3658–3664 (1996).
394. N. D. Polge and M. Barrett, *Pestic. Biochem. Physiol.*, 53, 193–204 (1995).
395. J. Menendez and R. De Prado, *Pestic. Biochem. Physiol.*, 56, 123–133 (1996).
396. C. Maneechote, C. Preston, and S. B. Powles, *Pestic. Sci.*, 49, 105–114 (1997).
397. N. Forthoffer, C. Helvig, N. Dillon, I. Benveniste, A. Zimmerlin, F. Tardif, and J. P. Salaun, *Eur. J. Drug. Metab. Pharmacokinet.*, 26, 9–16 (2001).
398. J. J. McFadden, D. S. Frear, and E. R. Mansager, *Pestic. Biochem. Physiol.*, 34, 92–100 (1989).
399. A. Zimmerlin and F. Durst, *Plant Physiol.*, 100, 874–881 (1992).
400. D. W. Bristol, A. M. Ghanuni, and A. E. Oleson, *J. Agric. Food Chem.*, 25, 1308–1314 (1977).
401. D. Scheel, and H. Sandermann, *Planta*, 152, 248–252 (1981).
402. A. Topal, N. Adams, W. C. Dauterman, E. Hodgson, and S. L. Kelly, *Pestic. Sci.*, 38, 9–15 (1993).
403. S. L. Kelly, A. Topal, I. Barnett, D. E. Kelly, and G. A. F. Hendry, *Pestic. Sci.*, 36, 27–30 (1992).
404. A. M. Makeev, A. Y. Makoveichuck, and D. I. Chkanikov, *Dokl. Akad. Nauk SSSR*, 233, 1222–1225 (1977).
405. A. J. Fischer, D. E. Bayer, M. D. Carriere, C. M. Ateh, and K. O. Yim, *Pestic. Biochem. Physiol.* 68, 156–165 (2000).
406. D. E. Moreland and F. T. Corbin, *Z. Naturforsch. C*, 46, 906–914 (1991).

407. D. E. Moreland, F. T. Corbin, W. P. Novitzky, C. E. Parker, and K. B. Tomer, *Z. Naturforsch. C*, 45, 558–564 (1990).
408. D. E. Moreland, F. T. Corbin, and J. E. McFarland, *Pestic. Biochem. Physiol.*, 45, 43–53 (1993).
409. S. Potter, D. E. Moreland, K. Kreuz, and E. Ward, *Drug Metabol. Drug Interact.*, 12, 317–327 (1995).
410. D. E. Moreland, F. T. Corbin, T. J. Fleischmann, and J. E. McFarland, *Pestic. Biochem. Physiol.*, 52, 98–108 (1995).
411. I. Jablonkai, and K. K. Hatzios, *Pestic. Biochem. Physiol.*, 48, 98–109 (1994).
412. D. S. Frear, *Drug Metabol. Drug Interact.*, 12, 329–357 (1995).
413. D. S. Frear, H. R. Swanson, and F. S. Tanaka, *Pestic. Biochem. Physiol.*, 45, 178–192 (1993).
414. M. S. Yun, Y. Yogo, R. Miura, Y. Yamasue, and A. J. Fischer, *Pestic. Biochem. Physiol.*, 83, 107–114 (2005).
415. K. W. Park, L. Fandrich, and C. A. Mallory-Smith, *Pestic. Biochem. Physiol.*, 79, 18–24 (2004).
416. J. J. McFadden, J. W. Gronwald, and C. V. Eberlein, *Biochem. Biophys. Res. Commun.*, 168, 206–213 (1990).
417. H. Hansikova, E. Frei, H. H. Schmeiser, P. Anzenbacher, and M. Stiborova, *Plant Sci.*, 110, 53–61 (1995).
418. H. Hansikova, E. Frei, P. Anzenbacher, and M. Stiborova, *Gen. Physio.l Biophys.*, 13, 149–169 (1994).
419. M. Stiborova and H. Hansikova, *Coll. Czech. Chem. Comm.*, 62, 1804–1814 (1997).
420. M. Stiborova, H. H. Schmeiser, and E. Frei, *Phytochemistry*, 54, 353–362 (2000).
421. M. Stiborova and H. Hansikova, *Coll. Czech. Chem. Comm.*, 61, 1689–1696 (1996).
422. C. A. L. Brady, R. A. Gill, and P. T. Lynch, *Environ. Geochem. Health*, 25, 131–137 (2003).
423. S. Pflugmacher, K. Geissler, and C. Steinberg, *Ecotoxicol. Environ. Saf.*, 42, 62–66 (1999).
424. M. Sugiura, T. Sakaki, Y. Yabusaki, and H. Ohkawa, *Biochim. Biophys. Acta*, 1308, 231–240 (1996).
425. D. L. Hallahan, J. H. A. Nugent, B. J. Hallahan, G. W. Dawson, D. W. Smiley, J. M. West, and R. M. Wallsgrove, *Plant Physiol.*, 98, 1290–1297 (1992).
426. Y. Batard, M. LeRet, M. Schalk, T. Robineau, F. Durst, and D. Werck-Reichhart, *Plant J.*, 14, 111–120 (1998).
427. T. Q. Shang, S. L. Doty, A. M. Wilson, W. N. Howald, and M. P. Gordon, *Phytochemistry*, 58, 1055–1065 (2001).
428. O. Young and H. Beevers, *Phytochemistry*, 15, 379–385 (1976).
429. S. R. Baerson, A. Sanchez-Moreiras, N. Pedrol-Bonjoch, M. Schulz, I. A. Kagan, A. K. Agarwal, M. J. Reigosa, and S. O. Duke, *J. Biol. Chem.*, 280, 21867–21881 (2005).
430. M. Schalk, M. A. Pierrel, A. Zimmerlin, Y. Batard, F. Durst, and D. Werck-Reichhart, *Environ. Sci. Pollution Res.*, 4, 229–234 (1997).
431. D. C. Lamb, D. E. Kelly, S. Z. Hanley, Z. Mehmood, and S. L. Kelly, *Biochem. Biophys. Res. Commun.*, 244, 110–114 (1998).

432. R. H. Strang and R. L. Rogers, *J. Agric. Food Chem.*, *22*, 1119–1125 (1974).
433. M. J. Plewa and E. D. Wagner, *Annu. Rev. Genet.*, *27*, 93–113 (1993).
434. J. Veleminsky and T. Gichner, *Mutation Res.*, *197*, 221–242 (1988).
435. C. Chiapella, P. Ysern, J. Riera, and M. Llagostera, *Mutation Res.*, *329*, 11–18 (1995).
436. C. Chiapella, J. A. Moreno, R. D. Radovan, N. Gaubert, and M. Llagostera, *Mutat. Res.*, *394*, 45–51 (1997).
437. C. Chiapella, R. D. Radovan, J. A. Moreno, L. Casares, J. Barbe, and M. Llagostera, *Mutat. Res.*, *470*, 155–160 (2000).
438. E. D. Wagner, J. M. Gentile, and M. J. Plewa, *Mutation Res.*, *216*, 163–178 (1989).
439. D. Reichhart, J. P. Salaun, I. Benveniste, and F. Durst, *Arch. Biochem. Biophys.*, *196*, 301–303 (1979).
440. D. Reichhart, J.-P. Salaün, I. Benveniste, and F. Durst, *Plant Physiol.*, *66*, 600–604 (1980).
441. J. Davies and J. C. Caseley, *Pestic. Sci.*, *55*, 1043–1058 (1999).
442. M. W. Persans, J. Wang, and M. A. Schuler, *Plant Physiol.*, *125*, 1126–1138 (2001).
443. S. Glombitza, P. H. Dubuis, O. Thulke, G. Welzl, L. Bovet, M. Gotz, M. Affenzeller, B. Geist, A. Hehn, C. Asnaghi, D. Ernst, H. K. Seidlitz, H. Gundlach, K. F. Mayer, E. Martinoia, D. Werck-Reichhart, F. Mauch, and A. R. Schaffner, *Plant Mol. Biol.*, *54*, 817–835 (2004).
444. A. Downie, S. Miyazaki, H. Bohnert, P. John, J. Coleman, M. Parry, and R. Haslam, *Phytochemistry*, *65*, 2305–2316 (2004).
445. D. R. Ekman, W. W. Lorenz, A. E. Przybyla, N. L. Wolfe, and J. F. Dean, *Plant Physiol.*, *133*, 1397–1406 (2003).
446. A. Gravot, R. Larbat, A. Hehn, K. Lievre, E. Gontier, J. L. Goergen, and F. Bourgaud, *Arch. Biochem. Biophys.*, *422*, 71–80 (2004).
447. P. Czernic, H. Chen Huang, and Y. Marco, *Plant Mol. Biol.*, *31*, 255–265 (1996).
448. H. D. Vanetten, S. G. Pueppke, and T. C. Kelsey, *Phytochemistry*, *14*, 1103–1105 (1975).
449. D. E. Matthews and H. D. VanEtten, *Arch. Biochem. Biophys.*, *224*, 494–505 (1983).
450. A. E. Desjardins, D. E. Matthews, and H. D. VanEtten, *Plant Physiol.*, *75*, 611–616 (1984).
451. A. E. Desjardins and H. D. VanEtten, *Arch. Microbiol.*, *144*, 84–90 (1986).
452. V. P. W. Miao, D. E. Matthews, and H. D. VanEtten, *Mol. Gen. Genet.*, *226*, 214–223 (1991).
453. V. P. Miao, S. F. Covert, and H. D. VanEtten, *Science*, *254*, 1773–1776 (1991).
454. A. P. Maloney and H. D. VanEtten, *Mol. Gen. Genet.*, *243*, 506–514 (1994).
455. C. Reimmann and H. D. VanEtten, *Gene*, *146*, 221–226 (1994).
456. L. M. Ciuffetti and H. D. VanEtten, *Mol. Plant Microbe Interact.*, *9*, 787–792 (1996).
457. R. Khan, R. Tan, A. G. Mariscal, and D. Straney, *Mol. Microbiol.*, *49*, 117–130 (2003).
458. H. L. George and H. D. VanEtten, *Fungal Genet. Biol.*, *33*, 37–48 (2001).
459. L. M. Delserone, K. McCluskey, D. E. Matthews, and H. D. VanEtten, *Physiol. Mol. Plant Pathol.*, *55*, 317–326 (1999).

460. K. Breskvar, Z. Ferencak, and T. Hudnikplevnik, *J. Steroid Biochem. Mol. Biol.*, 52, 271–275 (1995).
461. P. Znidarsic, M. Vitas, R. Komel, and A. Pavko, *Physiol. Mol. Plant Pathol.*, 55, 251–254 (1999).
462. T. Makovec and K. Breskvar, *Pflugers Archiv-Eur. J. Physiol.*, 439, R111–R112 (2000).
463. K. M. Weltring, J. Wessels, and G. F. Pauli, *Phytochemistry*, 48, 1321–1328 (1998).
464. B. W. Faber, R. F. M. van Gorcom, and J. A. Duine, *Arch. Biochem. Biophys.*, 394, 245–254 (2001).
465. G. Capotorti, P. Digianvincenzo, P. Cesti, A. Bernardi, and G. Guglielmetti, *Biodegradation*, 15, 79–85 (2004).
466. A. Eilers, E. Rungeling, U. M. Stundl, and G. Gottschalk, *Appl. Microbiol. Biotech.*, 53, 75–80 (1999).
467. D. Luykx, F. X. Prenafeta-Boldu, and J. A. M. de Bont, *Biochem. Biophys. Res. Comm.*, 312, 373–379 (2003).
468. N. Hiratsuka, H. Wariishi, and H. Tanaka, *Appl. Microbiol. Biotech.*, 57, 563–571 (2001).
469. H. Ichinose, H. Wariishi, and H. Tanaka, *Appl. Microbiol. Biotech.*, 58, 97–105 (2002).
470. J. P. Ferris, L. H. Macdonald, M. A. Patrie, and M. A. Martin, *Arch. Biochem. Biophys.*, 175, 443–452 (1976).
471. K. Lisowska and J. Dlugonski, *J. Steroid Biochem. Mol. Biol.*, 85, 63–69 (2003).
472. P. Bernat and J. Dlugonski, *Biotechnol. Lett.*, 24, 1971–1974 (2002).
473. H. H. J. Cox, B. W. Faber, W. N. M. VanHeiningen, H. Radhoe, H. J. Doddema, and W. Harder, *Appl. Environ. Microbiol.*, 62, 1471–1474 (1996).
474. N. Nakayama, A. Takemae, and H. Shoun, *J. Biochem.*, 119, 435–440 (1996).
475. T. Kitazume, N. Takaya, N. Nakayama, and H. Shoun, *J. Biol. Chem.*, 275, 39734–39740 (2000).
476. T. Kitazume, A. Tanaka, N. Takaya, A. Nakamura, S. Matsuyama, T. Suzuki, and H. Shoun, *Eur. J. Biochem.*, 269, 2075–2082 (2002).
477. H. Doddapaneni, R. Chakraborty, and J. S. Yadav, *BMC Genomics*, 6:92 (2005).
478. H. Doddapaneni, V. Subramanian, and J. Yadav, *Curr. Microbiol.*, 50, 292–298 (2005).
479. S. Masaphy, D. Levanon, Y. Henis, K. Venkateswarlu, and S. L. Kelly, *FEMS Microbiol. Lett.*, 135, 51–55 (1996).
480. N. Hiratsuka, M. Oyadomari, H. Shinohara, H. Tanaka, and H. Wariishi, *Biochem. Engin. J.*, 23, 241–246 (2005).
481. F. Matsuzaki and H. Wariishi, *Biochem. Biophys. Res. Comm.*, 334, 1184–1190 (2005).
482. F. Matsuzaki and H. Wariishi, *Biochem. Biophys. Res. Comm.*, 324, 387–393 (2004).
483. H. Teramoto, H. Tanaka, and H. Wariishi, *FEMS Microbiol. Lett.*, 234, 255–260 (2004).
484. H. Teramoto, H. Tanaka, and H. Wariishi, *Appl. Microbiol. Biotech.*, 66, 312–317 (2004).

485. C. Mougin, C. Pericaud, C. Malosse, C. Laugero, and M. Asther, *Pestic. Sci., 47,* 51–59 (1996).
486. C. Mougin, C. Laugero, M. Asther, and V. Chaplain, *Pestic. Sci., 49,* 169–177 (1997).
487. S. W. Kullman and F. Matsumura, *Appl. Environ. Microbiol., 62,* 593–600 (1996).
488. T. Mori and R. Kondo, *Appl. Microbiol. Biotech., 60,* 200–205 (2002).
489. T. Mori, S. Kitano, and R. Kondo, *Appl. Microbiol. Biotech., 61,* 380–383 (2003).
490. S. Masaphy, D. C. Lamb, and S. L. Kelly, *Biochem. Biophys. Res. Comm., 266,* 326–329 (1999).
491. L. Bezalel, Y. Hadar, P. P. Fu, J. P. Freeman, and C. E. Cerniglia, *Appl. Environ. Microbiol., 62,* 2547–2553 (1996).
492. L. Bezalel, Y. Hadar, P. P. Fu, J. P. Freeman, and C. E. Cerniglia, *Appl. Environ. Microbiol., 62,* 2554–2559 (1996).
493. D. Sanglard, O. Kappeli, and A. Fiechter, *Arch. Biochem. Biophys., 251,* 276–286 (1986).
494. D. Sanglard, C. Chen, and J. C. Loper, *Biochem. Biophys. Res. Comm., 144,* 251–257 (1987).
495. D. Sanglard and A. Fiechter, *FEBS Lett., 256,* 128–134 (1989).
496. D. Sanglard and J. C. Loper, *Gene, 76,* 121–136 (1989).
497. W. Seghezzi, D. Sanglard, and A. Fiechter, *Gene, 106,* 51–60 (1991).
498. D. L. Craft, K. M. Madduri, M. Eshoo, and C. R. Wilson, *Appl. Environ. Microbiol., 69,* 5983–5991 (2003).
499. J. S. Yadav and J. C. Loper, *Gene, 226,* 139–146 (1999).
500. T. Zimmer, M. Ohkuma, A. Ohta, M. Takagi, and W. H. Schunck, *Biochem. Biophys. Res. Comm., 224,* 784–789 (1996).
501. U. Scheller, T. Zimmer, E. Kargel, and W. H. Schunck, *Arch. Biochem. Biophys., 328,* 245–254 (1996).
502. T. Zimmer, U. Scheller, M. Takagi, and W. H. Schunck, *Eur. J. Biochem., 256,* 398–403 (1998).
503. U. Scheller, T. Zimmer, D. Becher, F. Schauer, and W. H. Schunck, *J. Biol. Chem., 273,* 32528–32534 (1998).
504. M. Ohkuma, T. Zimmer, T. Iida, W. H. Schunck, A. Ohta, and M. Takagi, *J. Biol. Chem., 273,* 3948–3953 (1998).
505. T. Iida, A. Ohta, and M. Takagi, *Yeast, 14,* 1387–1397 (1998).
506. T. Iida, T. Sumita, A. Ohta, and M. Takagi, *Yeast, 16,* 1077–1087 (2000).
507. J. Hedegaard and I. C. Gunsalus, *J. Biol. Chem., 240,* 4038–4043 (1965).
508. M. Katagiri, B. N. Ganguli, and I. C. Gunsalus, *J. Biol. Chem., 243,* 3543 (1968).
509. T. L. Poulos and R. Raag, *FASEB J., 6,* 674–679 (1992).
510. H. Y. Li, S. Narasimhulu, L. M. Havran, J. D. Winkler, and T. L. Poulos, *J. Am. Chem. Soc., 117,* 6297–6299 (1995).
511. T. L. Poulos, B. C. Finzel, and A. J. Howard, *J. Mol. Biol., 195,* 687–700 (1987).
512. I. Schlichting, J. Berendzen, K. Chu, A. M. Stock, S. A. Maves, D. E. Benson, B. M. Sweet, D. Ringe, G. A. Petsko, and S. G. Sligar, *Science, 287,* 1615–1622 (2000).
513. I. Schlichting, *FASEB J., 14,* A1307–A1307 (2000).
514. G. Grogan, G. A. Roberts, S. Parsons, N. J. Turner, and S. L. Flitsch, *Appl. Microbiol. Biotech., 59,* 449–454 (2002).

515. J. A. Peterson, J. Y. Lu, J. Geisselsoder, S. Grahamlorence, C. Carmona, F. Witney, and M. C. Lorence, *J. Biol. Chem.*, *267*, 14193–14203 (1992).
516. J. A. Fruetel, R. L. Mackman, J. A. Peterson, and P. R. O. Demontellano, *J. Biol. Chem.*, *269*, 28815–28821 (1994).
517. D. B. Hawkes, G. W. Adams, A. L. Burlingame, P. R. O. de Montellano, and J. J. De Voss, *J. Biol. Chem.*, *277*, 27725–27732 (2002).
518. Y. T. Meharenna, H. Y. Li, D. B. Hawkes, A. G. Pearson, J. De Voss, and T. L. Poulos, *Biochemistry*, *43*, 9487–9494 (2004).
519. C. A. Morgan and R. C. Wyndham, *Can. J. Microbiol.*, *48*, 49–59 (2002).
520. M. Taylor, D. C. Lamb, R. Cannell, M. Dawson, and S. L. Kelly, *Biochem. Biophys. Res. Comm.*, *263*, 838–842 (1999).
521. O. Lentz, V. Urlacher, and R. D. Schmid, *J. Biotechnol.*, *108*, 41–49 (2004).
522. I. Matsunaga, N. Yokotani, O. Gotoh, E. Kusunose, M. Yamada, and K. Ichihara, *J. Biol. Chem.*, *272*, 23592–23596 (1997).
523. G. Cardini and P. Jurtshuk, *J. Biol. Chem.*, *243*, 6070–6072 (1968).
524. G. Cardini and P. Jurtshuk, *J. Biol. Chem.*, *245*, 2789–2796 (1970).
525. J. B. van Beilen, E. G. Funhoff, A. van Loon, A. Just, L. Kaysser, M. Bouza, R. Holtackers, M. Rothlisberger, Z. Li, and B. Witholt, *Appl. Environ. Microbiol.*, *72*, 59–65 (2006).
526. T. Maier, H. H. Forster, O. Asperger, and U. Hahn, *Biochem. Biophys. Res. Comm.*, *286*, 652–658 (2001).
527. M. Sekine, S. Tanikawa, S. Omata, M. Saito, T. Fujisawa, N. Tsukatani, T. Tajima, T. Sekigawa, H. Kosugi, Y. Matsuo, R. Nishiko, K. Imamura, M. Ito, H. Narita, S. Tago, N. Fujita, and S. Harayama, *Environ. Microbiol.*, *8*, 334–346 (2006).
528. M. Kubota, M. Nodate, M. Yasumoto-Hirose, T. Uchiyama, O. Kagami, Y. Shizuri, and N. Misawa, *Biosci. Biotech. Biochem.*, *69*, 2421–2430 (2005).
529. M. Kharoune, L. Kharoune, J. M. Lebeault, and A. Pauss, *Appl. Microbiol. Biotech.*, *55*, 348–353 (2001).
530. S. Chauvaux, F. Chevalier, C. Le Dantec, F. Fayolle, I. Miras, F. Kunst, and P. Beguin, *J. Bacteriol.*, *183*, 6551–6557 (2001).
531. U. Karlson, D. F. Dwyer, S. W. Hooper, E. R. B. Moore, K. N. Timmis, and L. D. Eltis, *J. Bacteriol.*, *175*, 1467–1474 (1993).
532. H. Strauber, R. H. Muller, and W. Babel, *Biodegradation*, *14*, 41–50 (2003).
533. H. Mertingk, R. H. Muller, and W. Babel, *J. Basic Microbiol.*, *38*, 257–267 (1998).
534. I. Nagy, F. Compernolle, K. Ghys, J. Vanderleyden, and R. Demot, *Appl. Environ. Microbiol.*, *61*, 2056–2060 (1995).
535. I. Nagy, G. Schoofs, F. Compernolle, P. Proost, J. Vanderleyden, and R. Demot, *J. Bacteriol.*, *177*, 676–687 (1995).
536. J. A. Romesser and D. P. O'Keefe, *Biochem. Biophys. Res. Commun.*, *140*, 650–659 (1986).
537. D. P. O'Keefe, J. A. Romesser, and K. J. Leto, *Arch. Microbiol.*, *149*, 406–412 (1988).
538. C. A. Omer, R. Lenstra, P. J. Litle, C. Dean, J. M. Tepperman, K. J. Leto, J. A. Romesser, and D. P. O'Keefe, *J. Bacteriol.*, *172*, 3335–3345 (1990).
539. R. Warren, W. W. L. Hsiao, H. Kudo, M. Myhre, M. Dosanjh, A. Petrescu, H. Kobayashi, S. Shimizu, K. Miyauchi, E. Masai, G. Yang, J. M. Stott, J. E. Schein, H. Shin, J. Khattra, D. Smailus, Y. S. Butterfield, A. Siddiqui, R. Holt, M.

A. Marra, S. J. M. Jones, W. W. Mohn, F. S. L. Brinkman, M. Fukuda, J. Davies, and L. D. Eltis, *J. Bacteriol., 186*, 7783–7795 (2004).
540. G. A. Roberts, G. Grogan, A. Greter, S. L. Flitsch, and N. J. Turner, *J. Bacteriol., 184*, 3898–3908 (2002).
541. G. A. Roberts, A. Celik, D. J. B. Hunter, T. W. B. Ost, J. H. White, S. K. Chapman, N. J. Turner, and S. L. Flitsch, *J. Biol. Chem., 278*, 48914–48920 (2003).
542. P. Poupin, N. Truffaut, B. Combourieu, P. Besse, M. Sancelme, H. Veschambre, and A. M. Delort, *Appl. Environ. Microbiol., 64*, 159–165 (1998).
543. P. Poupin, J. J. Godon, E. Zumstein, and N. Truffaut, *Can. J. Microbiol., 45*, 209–216 (1999).
544. T. Schrader, G. Schuffenhauer, B. Sielaff, and J. R. Andreesen, *Microbiology-UK, 146*, 1091–1098 (2000).
545. P. Poupin, V. Ducrocq, S. Hallier–Soulier, and N. Truffaut, *J. Bacteriol., 181*, 3419–3426 (1999).
546. B. Sielaff, J. R. Andreesen, and T. Schrader, *Appl. Microbiol. Biotech., 56*, 458–464 (2001).
547. W. M. Atkins, W. D. Lu, and D. L. Cook, *J. Biol. Chem., 277*, 33258–33266 (2002).
548. R. B. Vail, M. J. Homann, I. Hanna, and A. Zaks, *J. Ind. Microbiol. Biotechnol., 32*, 67–74 (2005).
549. A. Parikh, E. M. J. Gillam, and F. P. Guengerich, *Nat. Biotech., 15*, 784–788 (1997).
550. C. R. Otey, G. Bandara, J. Lalonde, K. Takahashi, and F. H. Arnold, *Biotechnol. Bioeng., 93*, 494–499 (2006).
551. N. Serizawa and T. Matsuoka, *Biochim. Biophys. Acta, 1084*, 35–40 (1991).
552. Y. Kishida, A. Naito, S. Iwado, A. Terahara, and Y. Tsujita, *Yakugaku Zasshi–J. Pharm. Soc. Japan, 111*, 469–487 (1991).
553. I. Watanabe, F. Nara, and N. Serizawa, *Gene, 163*, 81–85 (1995).
554. N. Serizawa, *J. Synth. Org. Chem. Japan, 55*, 334–338 (1997).
555. M. Manzoni, and M. Rollini, *Appl. Microbiol. Biotech., 58*, 555–564 (2002).
556. S. G. Bell, R. J. Sowden, and L. L. Wong, *Chem. Comm.*, 635–636 (2001).
557. R. J. Sowden, S. Yasmin, N. H. Rees, S. G. Bell, and L. L. Wong, *Org. Biomol. Chem., 3*, 57–64 (2005).
558. J. B. van Beilen, R. Holtackers, D. Luscher, U. Bauer, B. Witholt, and W. A. Duetz, *Appl. Environ. Microbiol., 71*, 1737–1744 (2005).
559. J. A. Stevenson, J. K. Bearpark, and L. L. Wong, *New J. Chem., 22*, 551–552 (1998).
560. S. G. Bell, J. A. Stevenson, H. D. Boyd, S. Campbell, A. D. Riddle, E. L. Orton, and L. L. Wong, *Chem. Comm.*, 490–491 (2002).
561. F. Xu, S. G. Bell, J. Lednik, A. Insley, Z. H. Rao, and L. L. Wong, *Angew. Chemie-Int. Ed. Engl., 44*, 4029–4032 (2005).
562. A. Glieder, E. T. Farinas, and F. H. Arnold, *Nat. Biotech., 20*, 1135–1139 (2002).
563. M. W. Peters, P. Meinhold, A. Glieder, and F. H. Arnold, *J. Am. Chem. Soc. 125*, 13442–13450 (2003).
564. E. T. Farinas, M. Alcalde, and F. Arnold, *Tetrahedron, 60*, 525–528 (2004).
565. P. C. Cirino and F. H. Arnold, *Angew. Chem. Int. Ed. Engl., 42*, 3299–3301 (2003).
566. R. De Mot and A. H. A. Parret, *Trends Microbiol., 10*, 502–508 (2002).

567. D. Appel, S. Lutz-Wahl, P. Fischer, U. Schwaneberg, and R. D. Schmid, *J. Biotechnol.*, *88*, 167–171 (2001).
568. W. A. Duetz, J. B. van Beilen, and B. Witholt, *Curr. Opin. Biotech.*, *12*, 419–425 (2001).
569. Y. Liu, A. Kondo, H. Ohkawa, N. Shiota, and H. Fukuda, *Biochem. Engin. J.*, *2*, 229–235 (1998).
570. T. H. Rushmore, P. J. Reider, D. Slaughter, C. Assang, and M. Shou, *Metab. Engin.*, *2*, 115–125 (2000).
571. M. S. Zhu, F. Moulin, N. Aranibar, W. P. Zhao, B. Andrews, S. Callahan, J. Mitroka, and L. Klunk, *Drug Metab. Rev.*, *34*, 205–205 (2002).
572. S. M. van Liempd, J. Kool, J. Reinen, T. Schenk, J. H. N. Meerman, H. Irth, and N. P. E. Vermeulen, *J. Chromatography A*, *1075*, 205–212 (2005).
573. V. Reipa, M. P. Mayhew, and V. L. Vilker, *Proc. Natl. Acad. Sci. USA*, *94*, 13554–13558 (1997).
574. U. Schwaneberg, D. Appel, J. Schmitt, and R. D. Schmid, *J. Biotechnol.*, *84*, 249–257 (2000).
575. K. M. Faulkner, M. S. Shet, C. W. Fisher, and R. W. Estabrook, *Proc. Natl. Acad. Sci. USA*, *92*, 7705–7709 (1995).
576. V. V. Shumyantseva, T. V. Bulko, T. T. Bachmann, U. Bilitewski, R. D. Schmid, and A. I. Archakov, *Arch. Biochem. Biophys.*, *377*, 43–48 (2000).
577. M. Hara, H. Ohkawa, M. Narato, M. Shirai, Y. Asada, I. Karube, and J. Miyake, *J. Ferment. Bioeng.*, *84*, 324–329 (1997).
578. M. Hara, S. Iazvovskaia, H. Ohkawa, Y. Asada, and J. Miyake, *J. Biosci. Bioeng.*, *87*, 793–797 (1999).
579. M. Taylor, D. C. Lamb, R. J. P. Cannell, M. J. Dawson, and S. L. Kelly, *Biochem. Biophys. Res. Comm.*, *279*, 708–711 (2000).
580. H. Ichinose, J. Michizoe, T. Maruyama, N. Kamiya, and M. Goto, *Langmuir*, *20*, 5564–5568 (2004).
581. J. Michizoe, H. Ichinose, N. Kamiya, T. Maruyama, and M. Goto, *J. Biosci. Bioeng.*, *99*, 12–17 (2005).
582. A. Srinivasan, H. Bach, D. H. Sherman, and J. S. Dordick, *Biotechnol. Bioeng.*, *88*, 528–535 (2004).
583. K. Sakai-Kato, M. Kato, H. Homma, T. Toyo'oka, and N. Utsunomiya-Tate, *Anal. Chem.*, *77*, 7080–7083 (2005).
584. S. Y. Li and L. P. Wackett, *Biochemistry*, *32*, 9355–9361 (1993).
585. M. R. Lefever and L. P. Wackett, *Biochem. Biophys. Res. Commun.*, *201*, 373–378 (1994).
586. L. P. Wackett, M. J. Sadowsky, L. M. Newman, H. G. Hur, and S. Y. Li, *Nature*, *368*, 627–629 (1994).
587. J. P. Jones, E. J. O'Hare, and L. -L. Wong, *Chem. Comm.*, 247–248 (2000).
588. J. P. Jones, E. J. O'Hare, and L. -L. Wong, *Eur. J. Biochem.*, *268*, 1460–1467 (2001).
589. X. H. Chen, A. Christopher, J. P. Jones, S. G. Bell, Q. Guo, F. Xu, Z. H. Rao, and L. L. Wong, *J. Biol. Chem.*, *277*, 37519–37526 (2002).
590. M. E. Walsh, P. Kyritsis, N. A. J. Eady, H. A. O. Hill, and L. -L. Wong, *Eur. J. Biochem.* *267*, 5815–5820 (2000).

591. R. Shinkyo, M. Kamakura, S. Ikushiro, K. Inouye, and T. Sakaki, *Appl. Microbiol. Biotech.*, in press (2006).
592. P. A. England, C. F. Harford-Cross, J.-A. Stevenson, D. A. Rouch, and L.-L. Wong, *FEBS Lett.*, 424, 271–274 (1998).
593. A. B. Carmichael and L.-L. Wong, *Eur. J. Biochem.*, 268, 3117–3125 (2001).
594. N. V. Coleman, J. C. Spain, and T. Duxbury, *J. Appl. Microbiol.*, 93, 463–472 (2002).
595. H. M. B. Seth-Smith, S. J. Rosser, A. Basran, E. R. Travis, E. R. Dabbs, S. Nicklin, and N. C. Bruce, *Appl. Environ. Microbiol.*, 68, 4764–4771 (2002).
596. B. Bhushan, S. Trott, J. C. Spain, A. Halasz, L. Paquet, and M. Hawari, *Appl. Environ. Microbiol.*, 69, 1347–1351 (2003).
597. A. Schaffner, B. Messner, C. Langebartels, and H. Sandermann, *Acta Biotechnol.*, 22, 141–151 (2002).
598. H. Inui and H. Ohkawa, *Pest. Manag. Sci.*, 61, 286–291 (2005).
599. L. Didierjean, L. Gondet, R. Perkins, S. M. C. Lau, H. Schaller, D. P. O'Keefe, and D. Werck-Reichhart, *Plant Physiol.*, 130, 179–189 (2002).
600. M. Morant, S. Bak, B. L. Moller, and D. Werck-Reichhart, *Curr. Opin. Biotechnol.*, 14, 151–162 (2003).
601. N. Bistolas, U. Wollenberger, C. Jung, and F. W. Scheller, *Biosensors & Bioelectronics*, 20, 2408–2423 (2005).
602. M. Hara, Y. Yasuda, H. Toyotama, H. Ohkawa, T. Nozawa, and J. Miyake, *Biosensors & Bioelectronics*, 17, 173–179 (2002).
603. Z. Zhang, A. E. F. Nassar, Z. Q. Lu, J. B. Schenkman, and J. F. Rusling, *J. Chem. Soc., Faraday Trans.*, 93, 1769–1774 (1997).
604. E. I. Iwuoha, S. Joseph, Z. Zhang, M. R. Smyth, U. Fuhr, and P. R. Ortiz de Montellano, *J. Pharm. Biomed. Anal.*, 17, 1101–1110 (1998).
605. E. I. Iwuoha, S. Kane, C. O. Ania, M. R. Smyth, P. R. Ortiz de Montellano, and U. Fuhr, *Electroanalysi,s 12*, 980–986 (2000).
606. E. I. Iwuoha and M. R. Smyth, *Biosensors & Bioelectronics*, 18, 237–244 (2003).
607. S. Joseph, J. F. Rusling, Y. M. Lvov, T. Friedberg, and U. Fuhr, *Biochem. Pharmacol.*, 65, 1817–1826 (2003).
608. E. I. Iwuoha, A. Wilson, M. Howel, N. G. R. Mathebe, K. Montane-Jaime, D. Narinesingh, and A. Guiseppi-Elie, *Anal. Lett.*, 37, 929–941 (2004).
609. V. V. Shumyantseva, Y. D. Ivanov, N. Bistolas, F. W. Scheller, A. I. Archakov, and U. Wollenberger, *Anal. Chem.*, 76, 6046–6052 (2004).
610. V. V. Shumyantseva, T. V. Bulko, S. A. Usanov, R. D. Schmid, C. Nicolini, and A. I. Archakova, *J. Inorg. Biochem.*, 87, 185–190 (2001).
611. V. V. Shumyantseva, S. Carrara, V. Bavastrello, D. J. Riley, T. V. Bulko, K. G. Skryabin, A. I. Archakov, and C. Nicolini, *Biosensors & Bioelectronics*, 21, 217–222 (2005).
612. H. Schulze, R. D. Schmid, and T. T. Bachmann, *Anal. Chem.*, 76, 1720–1725 (2004).
613. Y. Narusaka, M. Narusaka, M. Seki, T. Umezawa, J. Ishida, M. Nakajima, A. Enju, and K. Shinozaki, *Plant Mol. Biol.*, 55, 327–342 (2004).

16

Drug Metabolism as Catalyzed by Human Cytochrome P450 Systems

F. Peter Guengerich

Department of Biochemistry and Center in Molecular Toxicology,
Vanderbilt University School of Medicine, Nashville, Tennessee 37232-0146, USA
<f.guengerich@vanderbilt.edu>

1. INTRODUCTION	562
2. IMPORTANCE OF P450 ENZYMES IN DRUG METABOLISM	563
2.1. Bioavailability	563
2.2. Toxicity	565
2.3. Drug-Drug Interactions	565
3. APPROACHES TO PREDICTING P450 ACTIVITY IN HUMANS	565
3.1. Reaction Phenotyping	565
3.2. High-throughput Screening, Including Genetic Polymorphism	567
3.3. Animal Models	568
3.4. Non-invasive Assays	570
4. P450s INVOLVED IN DRUG METABOLISM	570
4.1. General Considerations of Roles of P450s	570
4.2. Major P450s Involved in Drug Metabolism	572
4.2.1. P450 1A2	572
4.2.2. P450 2C9	573
4.2.3. P450 2C19	573
4.2.4. P450 2D6	574
4.2.5. P450 3A4	574
4.3. Other P450s with Contributions to Drug Metabolism	576

5. EXAMPLES OF MAJOR ISSUES INVOLVING DRUG
   METABOLISM BY P450   576
   5.1. Acetaminophen   576
   5.2. Terfenadine   578
   5.3. EE_2 579
   5.4. Tienilic Acid   580
   5.5. Perhexiline   581
6. SUMMARY   582
   ACKNOWLEDGMENTS   583
   ABBREVIATIONS   583
   REFERENCES   584

1. INTRODUCTION

The history of drug metabolism has its roots in work with xenobiotics in the 19th century [1]. The processes of oxidation and conjugation to form sulfates, mercapturic acids, hippuric acids, acetamides, glucuronides, and N-methyl compounds were all discovered, mainly through *in vivo* studies with simple chemicals. The pharmaceutical industry developed rapidly in the first half of the 20th century, and in 1947 Williams published the first edition of his book on 'detoxication' [2]. The next 20 years [3] involved a rapid development of *in vitro* approaches to these processes and the concepts of enzyme induction [4,5], activation to reactive products [6,7], and genetic polymorphism [8].

The modern biochemical phase of the work began, in one sense, with the separation and purification of the enzymes involved in drug metabolism, in the late 1960s. Important in this regard was the separation and reconstitution of a rabbit liver P450 system by Lu and Coon in 1968 [9]. The P450 purification studies progressed in the 1970s and were extended to human liver P450s in the 1980s [10,11]. The cloning and heterologous expression of human P450s occurred in the late 1980s and early 1990s [12], and the completion of the human genome at the turn of the century established the 57 P450s in the human genome.

Before 1980 the pharmaceutical industry emphasized *in vivo* animal studies in drug metabolism. As explained later in this chapter, there has been a major change in that *in vitro* metabolism studies with human systems are now done much earlier in the process, with the goal of selecting candidates that will not have bioavailability, metabolism, or toxicity problems in humans (Figure 1) [13].

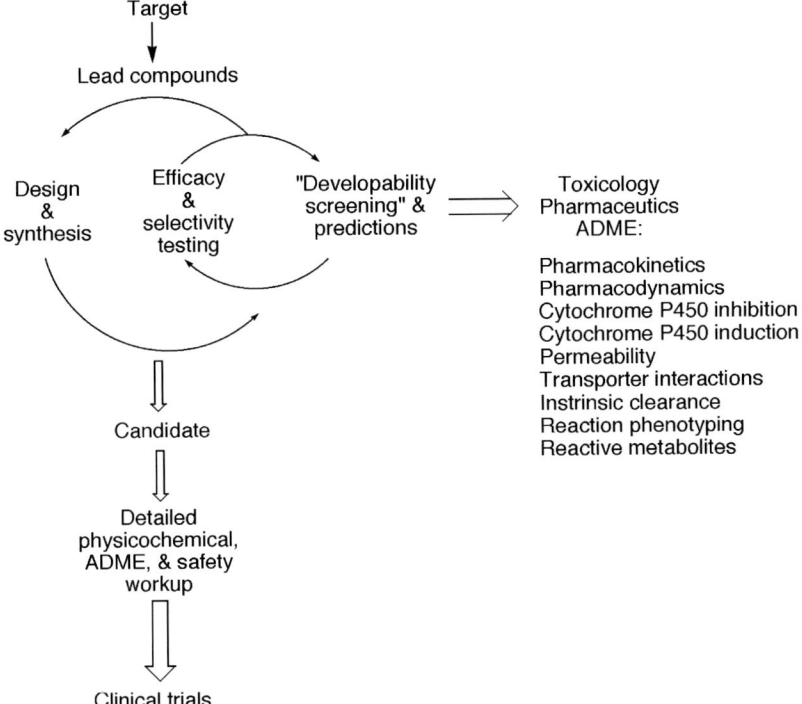

Figure 1. General strategy used in modern pre-clinical drug development. See text for discussion. ADME: absorption, distribution, metabolism, and excretion.

2. IMPORTANCE OF P450 ENZYMES IN DRUG METABOLISM

This chapter has been written with the perspective of introducing chemists to the biological and clinical significance of the P450s involved in drug oxidations.

Much of the understanding of the significance of P450s has come from studies with experimental animal models. With such systems, the importance of enzyme induction and genetic polymorphisms has been easy to demonstrate. Many early studies on genetic variability were done with congenenic animal lines [14] but can be done in more detail today with transgenic mice [15].

2.1. Bioavailability

Bioavailability is extremely important in the development of pharmaceuticals. Several factors are involved in bioavailability, including absorption, distribution, transport in and out of cells, and plasma protein binding. However, metabolism is often a controlling component.

One practical approach to analysis of bioavailability in humans involves the measurement of the concentration of the (parent) drug in the plasma, which is used as an indicator of the concentration in the target tissue, as a function of time (i.e., a pharmacokinetic analysis). After each dose of drug, the plasma level rises (to $C_{p,max}$, the maximum plasma concentration). The drug level decreases with time, and the integral is commonly termed the AUC.

The chart in the upper section of Figure 2 [16] shows a typical course expected for an 'extensive metabolizer' individual, who would probably be in the majority of the population. The lower panel shows what could happen in a 'poor metabolizer,' due to polymorphism or enzyme inhibition. The AUC is greatly increased in this case, and unexpected side effects might occur. However, if the AUC is decreased then adequate exposure to the drug might not occur.

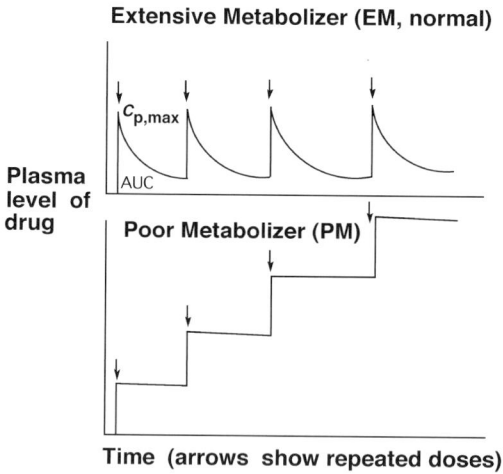

Figure 2. Significance of variations in P450 activity in drug clearance. The upper panel shows a typical pharmacokinetic course for a drug, with the plasma concentration of drug increasing after each dose to reach $C_{p,max}$. The 'AUC' is the integral of the curve(s) and represents the total exposure of the individual taking the drug. The lower panel shows the same doses of drug with an individual deficient in the P450 that catalyzes the oxidation of the drug. Much less oxidation occurs and after a few doses a much higher plasma level of the drug results, possibly leading to an exaggerated pharmacological response. Reprinted with permission of the American Society for Pharmacology and Experimental Therapeutics from [16].

2.2. Toxicity

The categories of drug toxicity have been reviewed elsewhere [17] and include (i) on-target toxicity, (ii) hypersensitivity and allergic reactions, (iii) off-target toxicity, (iv) reactive metabolites, and (v) idiosyncrasies. The molecular mechanisms for groups (iv) and (v) are largely unknown. With groups (i), (ii), and (iii), the effect of decreased metabolism (i.e., Figure 2) is to increase the amount of drug available to cause toxicity. Thus, a decreased P450 activity may contribute to the toxicity (e.g., see example of terfenadine in Section 5.2). In the case of bioactivation, the product of a P450 reaction can cause toxicity by covalently binding to macromolecules or perhaps by contributing to the load of oxidative damage (e.g., see example of acetaminophen in Section 5.1).

2.3. Drug-Drug Interactions

Many drug-drug interactions lead to adverse events and toxicity. Two major mechanisms involve enzyme induction and inhibition. Induction of a P450 by one drug might affect the ability of the P450 to oxidize another drug (Figure 2). For instance, the induction of P450s 2C9 and 3A4 by barbiturates can lead to enhanced clearance of another drug (e.g., see example of ethynylestradiol below). The inhibition of a P450 by a drug can lead to the accumulation of another drug (which is a substrate for that P450). Using Figure 2 as an example, the effect is to convert an extensive metabolizer to a poor metabolizer and increase the AUC for the drug substrate. An example is the accumulation of the drug terfenadine (see Section 5.2) that occurs when P450 3A4 is inhibited by compounds such as ketoconazole and erythromycin [18,19].

3. APPROACHES TO PREDICTING P450 ACTIVITY IN HUMANS

Since the early research on drug metabolism in laboratory animals, a major goal has been the application of the basic principals to humans and clinical medicine. Following the work with biochemistry and molecular biology, much progress has been made in this area. Following are brief descriptions of some of the approaches to assessing functions of individual 'drug-metabolizing' P450s in humans.

3.1. Reaction Phenotyping

This process is simply defining which P450 is the best catalyst of a reaction, using *in vitro* systems. Early research in this area utilized purified enzymes and antibodies in many cases [11,20]. Today the research is dominated by a few major approaches:

(i) The first is the direct use of recombinant P450 systems, the most popular of which are those derived from *Escherichia coli* and baculovirus systems. In many cases membranes containing the P450 and NADPH-P450 reductase are utilized instead of purified components. Today most studies require determination of the catalytic efficiency, i.e., k_{cat}/K_m, which is the most useful measure of comparison.
(ii) Use of human liver microsomes with diagnostic inhibitors. Some inhibitors are more specific than others. A very reasonable set for the major P450s (Figure 3) cited below follows. 1A2: α-naphthoflavone; 2C9: sulfaphenazole; 2D6: quinidine; 3A4: ketoconazole. A concentration of 1 μM is good for these inhibitors. A readily accessible inhibitor of P450 2C19 is not available.
(iii) Correlation analysis. A set of (10–12) human liver samples is analyzed (individually) for the catalytic activity of interest (i.e., new drug) and also a marker activity. For lists of useful reference activities of human P450s see [21].

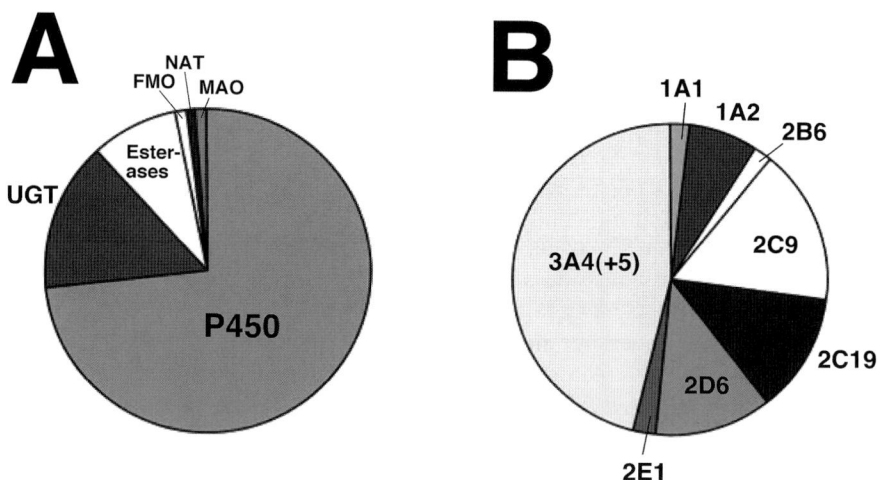

Figure 3. Contributions of enzymes to the metabolism of marketed drugs. The results are from a study of Pfizer drugs [44], and similar percentages have been reported by others [45]. A: Fraction of reactions on drugs catalyzed by various human enzymes. FMO: flavin-containing monoxygenase; NAT: *N*-acetyltransferase; MAO: monoamine oxidase; UGT: uridine dinucleotide phosphate glucuronosyl transferase. B: Fractions of P450 oxidations on drugs catalyzed by individual P450 enzymes. The segment labeled 3A4 (+3A5) is mainly due to P450 3A4, with some controversy about exactly how much is contributed by other subfamily 3A P450s. Reprinted with permission of the American Society for Pharmacology and Experimental Therapeutics from [44].

3.2. High-throughput Screening, Including Genetic Polymorphism

Many high-throughput methods are now used in drug metabolism in the pharmaceutical industry. These have different purposes.

One approach is to predict bioavailability. One of the most straightforward approaches is to incubate a low concentration of each of many drug candidates, individually, with human liver microsomes (or hepatocytes) for a short, fixed period of time (e.g., 15 min) and use rapid HPLC/mass spectrometry methods to determine how much of the parent molecule remains. At this stage there is no interest in defining and quantifying individual reaction products. The mass spectrometer provides a nearly universal detector for most drugs, with little if any optimization time required.

Another need is for screening of new drug candidates to determine if any will inhibit the metabolism of other drugs. The goal is trying to predict and avoid later problems with drug-drug interactions. Typically, either human liver microsomes or each of several major P450s is incubated with an individual drug candidate (at several concentrations) and the effect on a marked catalytic activity is measured. For instance, tolbutamide (methyl) hydroxylation or diclofenac 4-hydroxylation would be a marker for P450 2C9. These assays generally utilize high throughput HPLC or HPLC/mass spectrometry approaches. An alternative strategy is the use of fluorescence or luminescence assays [22]. These assays have some disadvantages in terms of the appropriate specificity for individual P450s but can be used in high-throughput assay formats [23].

P450 induction (Figure 4) is more difficult to adapt to high-throughput formats. However, drug candidates can be screened for affinity to some of the relevant receptors (e.g., AhR, PXR) using typical receptor technology. However, the point should be made that induction is a complex process and initial data on receptor binding is only part of the answer, so further tiers of assays are in order.

Genetic polymorphism has already been mentioned. The most common polymorphisms are SNPs, which can be screened rapidly. Thus, individuals participating in a clinical trial of a new drug can be genotyped rapidly. The cost per SNP measurement is reasonably low. The value is that knowledge of the gene status with a P450 involved in the metabolism of the drug may be coupled with information about drug efficacy and adverse events to better understand the issues. However, the approach is not simplistic. In large, late-stage clinical trials the number of individuals enrolled can reach 10,000–20,000 in some cases. In the case of P450 2D6, one of the most intensively studied polymorphisms in human P450s, 105 different allelic variants have been identified at the time of this writing (http://www.ki.se/CYPalleles/cyp2d6.htm). The effects of all of these variations (compared to the 'wild type' gene, *1A) have not been established, so using all of the information from patient genotyping is not a straightforward process. On the other hand, most of the aberrant P450 2D6 phenotypes can be

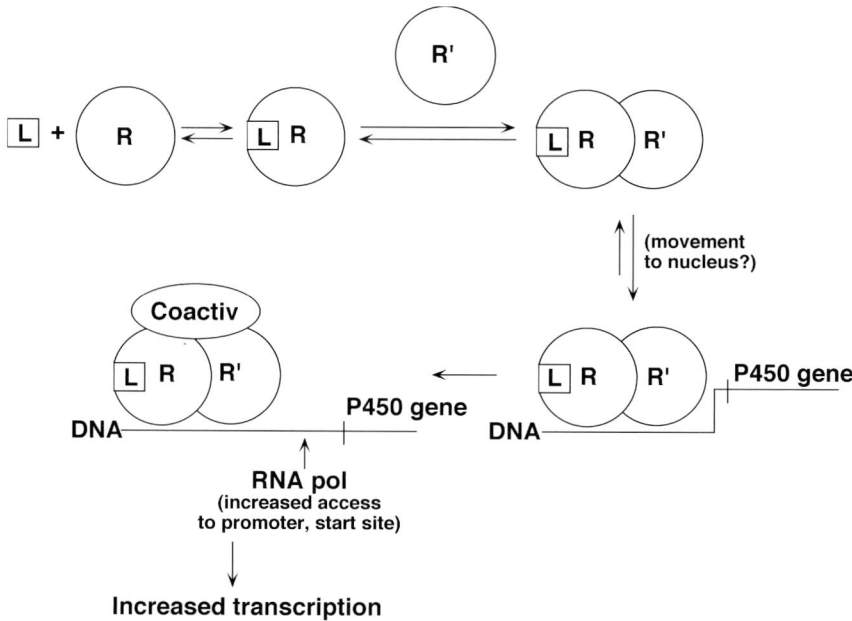

Figure 4. Generalized scheme for mechanism of induction of many P450s. A cytosolic receptor (R) binds a ligand (L), and this complex then forms a heterodimer, which moves to the nucleus to bind as an 'enhancer' site upstream of the P450 coding region, with the recognition due primarily to the nucleotide sequence. The complex may interact with a co-activator protein. The overall effect of the binding process is to change the gene/chromatin structure to make the promoter site more accessible to RNA polymerase and enhance transcription of the P450. Reprinted with permission of Springer-Verlag from [21].

accounted for with a relatively small number of the variant alleles [24,25]. The same statement applies for the P450 2C9 and 2C19 alleles [26–28]. Another problem in the application of the work is that the genetic polymorphisms identified to date explain very little of the phenotypic variability for P450s 1A2 and 3A4 [29,30].

3.3. Animal Models

Historically much of the drug metabolism research with pharmaceutical candidates involved pharmacokinetic work in animals. This is still the case, although *in vitro* assays with human systems now play an important role in making early predictions about the developability of drug candidates (Figure 1) [13]. *In vitro* systems have advantages in terms of speed, cost, and applicability. However, in *vivo* animal studies are still important for a number of reasons, primarily because these put any phenomena into a global and physiological context.

One reason for continuing the use of animal work is that efficacy and safety must be evaluated in some *in vivo* context before administration to humans. In some cases no animal models may be available for efficacy because of the nature of the disease. Some basic safety (toxicity) results must be obtained before administration to humans, for ethical reasons. Some problems are seen in animal models that are irrelevant to humans, and in certain cases decisions about biological significance can be made on the basis of past experience. Conversely, there is a small but finite risk that an adverse reaction will occur in humans that has not been seen in any of the animal models.

Another aspect of the use of animals involves transgenics. Four types of transgenic mice are currently available for use in drug metabolism research.

(i) The first is animals in which a regulatory gene has been deleted. For instance, $AhR^{-/-}$ mice do not show induction of P450 Family 1 enzymes, which may be of value in elucidating some mechanisms.

(ii) Systems are available in which the auxiliary enzyme NADPH-P450 reductase can be deleted, thus effectively eliminating all P450 function. If this is done in the germline, then the result is embryonic-lethal [31]. Conditional knock-outs have been made in which the reductase is ablated in a specific tissue, e.g., liver [32,33] or brain, heart, or lung (X. Ding, personal communication). Somewhat surprisingly, these animals are still viable and only show some problematic but not serious issues with fat metabolism. Another type of transgenic mouse is a hypomorphic animal, with the reductase down-regulated in many tissues [34,35]. It is of interest to note that strong down-regulation of many P450s in rats was achieved earlier with the chemical inhibitor 4-aminobenzotriazole without serious effects [36].

(iii) The third system involves the deletion of a particular P450 (structural gene). A caveat of this work is that defining exact orthologs is difficult with the metabolism of drugs. However, if a good match of function is possible, the approach is valuable in defining roles of individual P450s in physiological functions or toxicity. Mice (small letter in the nomenclature denotes a mouse gene/protein) have been developed devoid of P450s 1a1, 1a2, 1b1, 2a6, 2e1 [15], and others including 2g1, 2f2, 2s1, and the entire 2a/2b/2f/2g/2s gene cluster (X. Ding, personal communication). The animals have shown phenotypes that are generally normal.

(iv) The last type of mouse model is one that overexpresses a human P450 [37]. In a certain sense the results of studies done with such mice are more difficult to relate to other systems and to question about chemicals because of the potential supra-physiological effects. For instance, female mice expressing human P450 3A4 were found to have lactation problems due to the rapid metabolism of estradiol [38]. However, it is unclear as to whether this phenotype would be seen in an induced human.

A future goal is the removal of all of the apparently orthologous mouse genes, i.e., all members of a P450 subfamily, and replacement with the single human P450 from the same subfamily, and some systems have already been developed, e.g., P450 2E1 [39].

3.4. Non-invasive Assays

Non-invasive assays, for our purposes, are *in vivo* human studies in which the course of the metabolism of a drug may be monitored by using body fluids or breath, i.e. without surgery. Such systems have been developed for several human P450s and utilize highly specific substrates that can be ingested without risk to the individual. The following assay methods are generally considered to be well-validated measures of *in vivo* human P450 activity (Figure 3)—P450 1A2: caffeine N^3-demethylation, P450 2A6: coumarin 7-hydroxylation, P450 2C9: tolbutamide hydroxylation, P450 2C19: *S*-mephenytoin 4′-hydroxylation, P450 2D6: debrisoquine 4-hydroxylation, P450 2E1: chlorzoxane 6-hydroxylation, and P450 3A4: midazolam 1′-hydroxylation [21]. P450 3A4 is probably the most complex of these because of the inherent complexity of the enzyme itself (i.e., differential interactions with various substrates) and the problem of distinguishing between hepatic and intestinal metabolism. In most of the routine assays, extensive pharmacokinetic analysis is not done (i.e., many time points) but the assays rely on a single time point measurement.

The above assays can be used to address questions about induction and inhibition *in vivo*. For instance, increased clearance of caffeine following administration of a drug provides evidence that P450 1A2 has been induced at a particular drug dose. This information is more costly to obtain than *in vitro* work but can quickly put any *in vitro* results and questions of the relevance of particular inducer concentrations etc. into a more relevant biological scenario. In a similar way, the *in vivo* significance of P450 inhibition can be quickly established.

4. P450s INVOLVED IN DRUG METABOLISM

4.1. General Considerations of Roles of P450s

This chapter deals only with the human P450s except to mention some P450s in experimental animal models. The 57 human P450, or CYP, genes are listed in Table 1. The classification of the P450s is by function, as viewed by the author. Some variability exists in the system, e.g., P450s 1B1 and 27A1 could be in other categories [21]. Similar tables could be constructed for animal P450s, with considerable overlap. However, it should be pointed out that (i) mice have more P450s, 88, compared to humans, and (ii) some assignments of orthologues are

Table 1. Classification of human P450s based on major substrate class [21,40].

Sterols	Xenobiotics	Fatty acids	Eicosanoids	Vitamins	Unknown
1B1	1A1	2J2	4F2	2R1	2A7
7A1	1A2	4A11	4F3	24	2S1
7B1	2A6	4B1	4F8	26A1	2U1
8B1	2A13	4F12	5A1	26B1	2W1
11A1	2B6		8A1	26C1	3A43
11B1	2C8			27B1	4A22
11B2	2C9				4F11
17	2C18				4F22
19	2C19				4V2
21A2	2D6				4X1
27A1	2E1				4Z1
39	2F1				20A1
46	3A4				27C1
51	3A5				
	3A7				

not straightforward (e.g., humans have only one 4A subfamily gene but rodents have multiple ones).

About one-fourth of the human P450s are clearly involved in the metabolism of sterols. Most of these are essential and homozygous recessive individuals usually have serious problems [41]. Thus, levels of these enzymes usually do not vary much. In contrast, the P450s listed under the heading of xenobiotics vary considerably in the levels present in individuals, and effects are not seen unless an individual ingests inappropriate drugs. For example, 5–10% of many Caucasian populations are devoid of functional P450 2D6 but are not discernible except for problems with some medications [42]. Transgenic mice missing what appear to be the orthologues of these P450s are apparently very normal [15].

Many P450s can catalyze some oxidation of fatty acids but the four listed under this heading appear to do this as a true function [43]. The P450s involved in eicosanoid metabolism include the P450s involved in hydroxylation of the alkane chains and the unusual P450s that catalyze prostaglandin rearrangements (5A1 and 8A1, the thromboxane and prostacyclin synthases). The six P450s listed under Vitamins appear to be rather essential, in general, and selectively catalyze the oxidations of vitamins A and D [21].

Finally, about one-fourth of the human P450s have no clear substrates, either endogenous or exogenous, and are listed under the heading 'Unknown'. Little is currently known about these 'orphan' P450s, and some approaches to their study are presented elsewhere [40].

The roles of P450s in drug metabolism can be considered. The charts in Figure 3 [44] were published from a study by a major pharmaceutical company

(Pfizer) for company drugs [46]; others have published similar findings [45]. P450s are involved in the metabolism of ~ 75% of drugs, with the next categories of enzymes being the UGTs and esterases. Of the P450s, five account for ~ 95% of the metabolism: 1A2, 2C9, 2C19, 2D6, and 3A4 (with some possible contribution from 3A5 in the latter case, although the significance of the latter is debatable [47,48]).

Some of the 'minor' P450s do contribute to the metabolism of certain drugs and are important. However, the general point should be made that the situation with P450 is not so complicated as it might be. Of the 57 P450s, a proper understanding of a limited set of these enzymes provides considerable insight into the metabolism of most drugs.

4.2. Major P450s Involved in Drug Metabolism

The five P450s discussed here have been treated at length in a chapter that was written in Spring 2003; the reader is referred to this [21]. As one would imagine from the chart in Figure 3B and the number of drugs available, actual lists of substrates are quite extensive and will not be developed here; see [49].

4.2.1. P450 1A2

P450 1A2 was first characterized as the low K_m phenacetin O-deethylase [50]. This enzyme is expressed essentially only in liver, probably due to the involvement of factors such as HNF in its regulation [21,51]. The enzyme is highly inducible, by some drugs and also the polycyclic hydrocarbons and heterocyclic amines in cigarette smoke and charbroiled meat [52]. The induction process (Figure 4) involves the AhR receptor [53].

The amount of P450 1A2 in individuals varies 40-fold, as shown either with human liver microsomes (*in vitro*) [54] or with caffeine N^3-demethylation phenotyping assays (*in vivo*) (the latter reaction provides a reasonable approach for a non-invasive and safe assay [55]). The basis of this variability appears to be largely induction. SNPs in P450 1A2 are known (see http://www.imm.ki.se/CYPalleles/) but their relevance to function is still not established [29].

Some information is available about residues in the active site of P450 1A2 through site-directed and random mutagenesis studies [56–58]. The enzyme has not been crystallized, so all conclusions about the substrate binding site are based upon models, several of which have appeared [59,60].

Several inhibitors are known [21] but one of the most interesting of these is furafylline, a mechanism-based inactivator based on the purine pharmacophore. Furafylline was under development as a drug but the strong inhibition yielded a considerable contraindication with coffee and other caffeine-containing beverages, and the project was dropped [61].

4.2.2. P450 2C9

P450 2C9 is expressed primarily in liver and is one of the major P450 proteins present in human liver [21]. The protein appears to be absent in fetal liver [62].

The activity of P450 2C9 varies considerably among individuals because the enzyme is both inducible and because polymorphisms can have considerable effects on activity. The inducers are those in the barbiturate class, which acts through the CAR receptor in a manner which is not completely characterized [21,63]. The PXR receptor may also contribute to induction. Numerous SNPs are known (http://www.imm.ki.se/CYPalleles/) and the coding region SNPs appear to have different phenotypes depending upon the particular substrate [64]. The *2 and *3 variants (R144C and I359L) do impact the metabolism of (S)-warfarin and the balance between haemorrhaging (toxicity) and therapeutic effectiveness. This is an example in which the narrow therapeutic window of a drug renders the polymorphism and variability important.

Sulfaphenazole is one of the most widely used (in vitro) inhibitors of P450 2C9 [65]. Tienilic acid is a mechanism-based inhibitor (or at least a substrate that produces a reactive product that destroys the enzyme, if not necessarily exclusively) [66]. The modification of P450 2C9 by the tienilic acid product is involved in the production of autoantibodies to (unbound) P450 2C9, although the relationship of this phenomenon to the hepatotoxicity that develops is still unclear [67].

X-ray crystal structures of P450 2C9 have been published, with warfarin bound in one case [68,69]. However, the bound warfarin does not appear to be in a position amenable to hydroxylation, at least in the ferric enzyme [68].

4.2.3. P450 2C19

P450 2C19 is also primarily a liver enzyme. Historically the highly polymorphic 4'-hydroxylation of (S)-mephenytoin was recognized [70] and used as an assay to identify the enzyme, which was difficult because of its low abundance [71]. Although the protein is not abundant, it does contribute in the oxidations of a number of drug oxidations (Figure 3B).

Several SNPs contribute to the poor metabolizer phenotype, but the most common of these appears to be a variant with defective mRNA splicing [72]. In Western populations the incidence of the poor metabolizer phenotype is low (3–5%) but in Asian countries the incidence can be $\sim 20\%$ [26,73], giving rise to difficulties in development of drugs for different populations.

A crystal structure for P450 2C19 has not yet been obtained, although the success with other 2C subfamily enzymes suggests that this should be likely. Several residues important for substrate interaction have been defined through chimeric analysis and work with inhibitors [74].

4.2.4. P450 2D6

This enzyme was recognized early not as such, but as the polymorphic debrisoquine 4'-hydroxylase by R. L. Smith and his associates [75]. Other early *in vivo* work with polymorphic sparteine dehydrogenation [76] and metoprolol hydroxylation [77] also contributed to this field.

The protein was purified from human liver microsomes on the basis of its catalytic activity [50,78]; subsequent work led to the cloning of the cDNA and gene [79]. As in the case of P450 2C19, the major defective genotype involves aberrant mRNA splicing [79]. Other polymorphisms have other bases [80]. To date at least 105 allelic variants have been identified (http://www.imm.ki.se/CYPalleles/). Some of these yield altered activities. An interesting aspect is the existence of a genotype with gene duplication, yielding 13 copies of the 2D6 gene and thus 13 times more protein and catalytic activity (an 'ultra extensive metabolizer' phenotoype).

The list of substrates is considerable [49]. Most, but not all, have a basic nitrogen atom [81,82]. Although an X-ray crystal structure is not yet available, numerous homology models have been presented. Information about the active site is also available from site-directed mutagenesis studies. The anionic residues Asp301 and Glu216 have roles in binding basic substrates and Asp301 is a factor in protein stability [83]. Other residues with roles, probably hydrophobic stacking, include Phe120 and Phe483, which have been corroborated in a recent X-ray crystal structure [84].

Most of the drug substrates for P450 2D6 appear to have a wide margin for safety, or therapeutic index (i.e., the difference between the plasma concentration needed to produce beneficial *versus* adverse effects), or more problems would be encountered. In most pharmaceutical companies, the finding that a large fraction of the metabolism is due to a highly polymorphic enzyme (e.g., P450 2D6) is considered a serious liability and these drug candidates are generally not advanced, in the absence of a compelling reason. Pharmaceutical scientists would ideally like to have the metabolism of a drug shared by several enzymes so that a change in the level of one enzyme would not have major effects.

4.2.5. P450 3A4

P450 3A4 is the most abundant P450 in human liver and small intestine, accounting for a mean of $\sim 25\%$ of the total in the liver and being involved in the metabolism of about one-half of the drugs on the market (Figure 3B). Retrospective work indicates that P450s 3A4 and 2C9 were some of the first human P450s purified [10,85,86], but the definitive identification of P450 3A4 began with its characterization as nifedipine oxidase in purification studies [87]. Subsequent work quickly demonstrated the very broad substrate specificity of this catalyst [87–90]. An inherent problem in the reconstitution of this enzyme P450 has been the need for multiple protein and other components [91,92] and the sensitivity of the enzyme to residual non-ionic detergent (which is a substrate) [93].

P450 3A4 levels vary $\sim$ 20-fold among individuals [94,95]. Searches for functional SNPs in P450 3A4 have been negative [30]. The enzyme activity is largely a function of induction, which is mediated primarily through the PXR receptor [96], with some possible contribution from CAR and the glucocorticoid receptor [97,98]. An interesting but complex aspect of the regulation (and catalytic activity) of P450 3A4 is that the membrane transporter termed P-glycoprotein (*Mdr1*) uses many of the P450 substrates and inducers and can regulate the exposure to these in the cell [99,100]. Another point is that many drugs inhibit P450 3A4, and many of these compounds are mechanism-based inactivators [101]. One of the most interesting inhibitors of (intestinal) P450 3A4 is bergomattin (and related compounds), found in grapefruit juice [102–104].

X-ray crystal structures of P450 3A4 have been solved by Yano et al. [105] and Williams et al. [106]. In one of these [106] the substrate progesterone is present but is $\sim$ 17 Å away from the iron, too far for a facile chemical oxidation to occur. One possibility is that the enzyme undergoes a major movement upon reduction and O_2 addition. Another is that the substrate is bound adventitiously in this site. A third explanation is related to the rather unusual homotropic and heterotropic cooperativity (using terms from Kuby [107]), which is observed with some P450 3A4 ligands but not others. Several mechanisms have been proposed to account for this behavior in P450 3A4, but they can be generally categorized as (i) classic allosterism, (ii) multiple ligands (2 or 3) in the same (large) active sites, (iv) selection of pre-existing sub-population of P450 3A4 for ligand binding, and (iv) relatively slow movement of a ligand from the periphery of the protein to the heme [21,108].

Although high or low P450 3A4 activity is a problem in the metabolism of certain drugs, avoiding P450 3A4 action with all drug candidates is unrealistic. As the size of drug candidates becomes larger with the desire for more target selectivity, there is a tendency to 'steer' the drugs to the P450s with large active sites that can accommodate these, e.g., P450s 2C8 [109] and 3A4 [105,106].

One concept that is often expressed in the literature is that P450 3A4 is a large hole that substrates enter, followed by selection of atomic sites for oxidation based on the thermodynamic ease [110,111]. However, work in this laboratory has shown that testosterone, a substrate of medium size, shows very strong stereo- and regiospecificity in that only the β-hydrogen at C-6 is abstracted and that oxygen is transferred only back to the β-face [112]. Further, biomimetic metalloporphyrin models show very limited regioselectivity for oxidation, far from the extent that P450 3A4 does. Testosterone 6β-hydroxylation also shows a high intrinsic kinetic deuterium isotope effect, which is considerably attenuated in non-competitive intermolecular studies [112]. Exactly what steps are rate-limiting in various P450 3A4 reactions is not well understood at this point.

The above and other work indicate that P450 3A4 is a complex enzyme; however, the significance of this catalyst in drug metabolism argues that these biochemical complexities must be resolved. Currently all models for making predictions must be considered suspect, and even the choice of substrates to use in inhibition assays is problematic in that very variable results can be obtained [113].

4.3. Other P450s with Contributions to Drug Metabolism

More limited roles in drug metabolism are attributed to P450s 1A1, 2A6, 2B6, and 2E1 [21,49]. P450 1A1 has appreciable activity towards several drugs but the amount of this protein in liver and small intestine is very low. P450 2B6 is expressed mainly in liver but the concentration is ≤one-tenth that of P450 3A4. P450 2A6 has a modest level of expression in the liver but contributes to the metabolism of some drugs [114].

P450 2E1 and, to a more limited extent, P450 2A6 are involved in the metabolism of volatile anesthetics such as halothane, enflurane, etc. [115]. Both of these P450s have small active sites [116] and thus only small molecules tend to be good substrates. P450 2E1 shows a complex pattern of induction and inhibition with ethanol [117].

Some other human P450s have limited roles, e.g., P450 2C8 with paclitaxel [118], 3A5, and 3A7 with some of the P450 3A4 substrates [119]. The 'unknown' or 'orphan' P450s have not been examined but most are probably not major contributors to drug metabolism [40].

5. EXAMPLES OF MAJOR ISSUES INVOLVING DRUG METABOLISM BY P450

Five examples have been selected, in order to provide some insight into the variety of issues that occur in metabolism of drugs. Each of these sections is only a synopsis of the problem, and the reader is referred to the primary literature and critical reviews for each.

5.1. Acetaminophen

Acetaminophen (paracetamol, Tylenol®) is one of the most commonly used drugs in the USA and other Western countries. At low doses this is a very effective analgesic. However, high doses are toxic and nearly one-half the cases of acute liver failure can be attributed to this drug [120,121]. Thus, acetaminophen (also termed paracetamol in Europe and marketed as Tylenol® in the USA) is clearly an example of Paracelsus' axiom of 'the dose makes the poison' (and is the difference between a drug and a poison).

The pathways in Figure 5 [122,123] are involved in the metabolism of acetaminophen, which is known to be very relevant to toxicity. At low doses of acetaminophen the drug is conjugated to form the sulfate and glucuronide, and the reactive Michael acceptor (iminoquinone) that is formed reacts with GSH and is thus detoxicated. However, at higher acetaminophen doses the conjugation systems are depleted and more binding of the iminoquinone to proteins occurs. Depletion of the stores of the high-energy cofactors needed for conjugation (e.g., in malnutrition situations) also exacerbates the toxicity.

The relationship between bioactivation and covalent binding has been known since 1973 [124,125]. Exactly what proteins are critical targets for toxicity is still unclear, in that a non-toxic congener of acetaminophen (3-hydroxyacetanilide) gives a similar level of protein binding (but the set of modified proteins differs). An alternate hypothesis is that the removal of GSH removes a vital defense against oxidative stress and sensitizes cells. These hypotheses are both con-

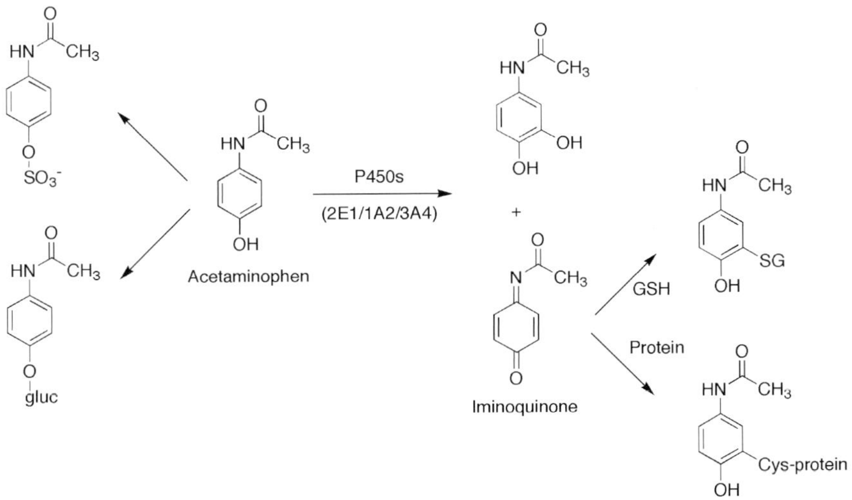

Figure 5. Metabolic pathways for acetaminophen. Acetaminophen can be conjugated to yield a sulfate or *O*-glucuronide. The parent drug is also oxidized (by several P450s, particularly 2E1 [122,123]) to an *ortho*-catechol and an iminoquinone. The iminoquinone reacts rapidly (and non-enzymatically) with GSH or proteins to yield aryl thioethers. The protein conjugates are generally considered to somehow initiate the process of toxicity.

sistent with the action of N-acetylcysteine as an antidote for acetaminophen poisoning.

Several P450s can activate acetaminophen [122]. P450 2E1 appears to be the most relevant. In a mouse model, elimination of P450 2E1 has a masked hepatoprotective effect against acetaminophen poisoning [123]. This result demonstrates not only the role of P450 2E1 but also the requirement for bioactivation as in the initiation of damage.

5.2. Terfenadine

Terfenadine (Seldane®) was the first commercial non-sedating antihistamine and was used widely throughout the world. However, some adverse effects were experienced, including approximately 20 deaths due to arrhythmias and abnormal QT intervals.

The relevant aspects of the metabolism are shown in Figure 6. Terfenadine is oxidized to two primary products, a deactivated cleavage product (of N-dealkylation) and the *tert*-butyl alcohol. The alcohol is further oxidized (by

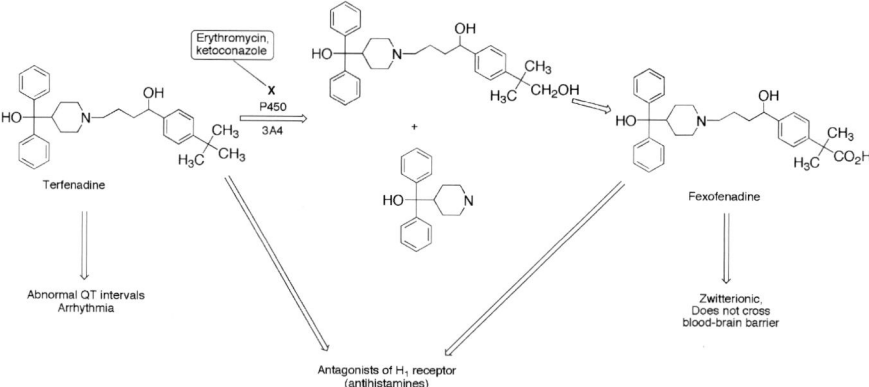

Figure 6. Metabolism of terfenadine and relevance to toxicity. Terfenadine is rapidly oxidized by P450 3A4 to produce roughly equal amounts of two products, an alcohol (*tert*-butyl group) and an N-dealkylation product [19]. The alcohol is rapidly oxidized (2 steps) to the carboxylic acid product, termed fexofenadine. In most individuals no terfenadine is found in the plasma, and fexofenadine is the dominant circulating form. Both terfenadine and fexofenadine have intrinsic antagonist activity with the H_1 receptor, which is involved in hay fever etc. Fexofenadine does not readily enter the central nervous system; thus terfenadine was the first non-sedating antihistamine marketed. Concurrent use of drugs that inhibit P450 3A4 (erthromycin, ketoconazole) leads to the accumulation of terfenadine. Terfenadine (but not fexofenadine) binds to hERG channels and can cause abnormal QT intervals and arrhythmias. In 1995 the FDA recalled terfenadine, and fexofenadine has replaced this drug.

P450 3A4 or dehydrogenases) to the carboxylic acid product, fexofenadine. In most individuals the process is rapid and the parent drug terfenadine is not seen in plasma. Both terfenadine and the carboxylic acid product have intrinsic activity as agonists of the H_1 receptor, which is involved in the histamine response. However, terfenadine (but not fexofenadine) also binds to hERG channels to produce the abnormal QT intervals and arrhythmia (this is an example of off-target drug toxicity mentioned earlier) [18,126].

When terfenadine was first marketed (1980s), no knowledge about the individual human P450s involved in its metabolism was available. Both oxidation steps were later shown to be catalyzed by P450 3A4, after the first reports of adverse incidents [19]. Many of the adverse reactions were traced to the concomitant use of P450 3A4 inhibitors, particularly the antibiotic erythromycin and the antifungal ketoconazole. For some time the FDA required labeling about the potential contraindications of using these inhibitory drugs (and other azoles) with terfenadine. Subsequently terfenadine was withdrawn from the market (1995) [127]. Fexofenadine is devoid of the cardiovascular problems of terfenadine and has been marketed as Allegra® since then.

5.3. EE_2

EE_2 (Figure 7) is the estrogenic component of most oral contraceptives. (Contraceptives also contain a progestin component, and this varies among the different preparations commercially available). The major oxidative route of metabolism of EE_2 is 2-hydroxylation, as in the case of the natural estrogens 17β-estradiol and estrone.

In the late 1960s several reports from Germany indicated that women who were using rifampicin or barbiturates experienced a loss of effectiveness of oral contraceptives and resumed bleeding and unexpected pregnancy [128–130]. The basis of the phenomenon was provided by Bolt and his associates, who demonstrated the enhanced clearance of EE_2 [131]. 2-Hydroxy EE_2 still has estrogenic activity but is more rapidly eliminated than EE_2.

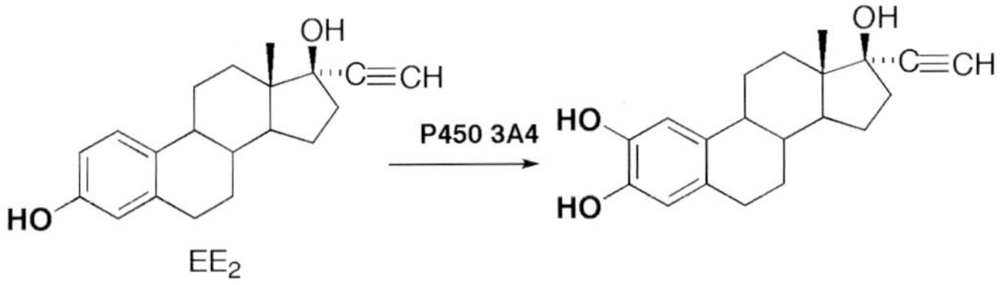

Figure 7. 2-Hydroxylation of EE_2 by P450 3A4 [95]. The *ortho*-catechol still binds the estrogen receptor and has contraceptive activity but is cleaved faster from the body. Thus, the effect of induction of P450 3A4 is to enhance this reaction and render oral contraceptives less effective.

Subsequent work demonstrated that P450 3A4 is the major catalyst of EE_2 2-hydroxylation [95] (EE_2 and several of the 17α-ethynyl-substituted progestins are also mechanism-based P450 inactivators, although this phenomenon is not relevant to the issue at hand here [95,132]). P450 3A4 is also inducible by barbiturates, rifampicin, and other compounds in hepatocyte cultures [133]. Indirect but convincing evidence of *in vivo* P450 3A4 inducibility by these compounds is also available [21]. Thus, the case of EE_2 provides a clear example of how P450 induction can decrease the bioavailability of a drug to make it ineffective. The phenomenon of loss of contraceptive effectiveness continues to be reported [134]. Another point to make in closing is the effectiveness of the herbal remedy St. John's Wort as a P450 3A4 inducer and produces the same effect [135].

5.4. Tienilic Acid

Tienilic acid (Figure 8) is another example that involves bioactivation but differs in several aspects from acetaminophen (Section 5.1). The drug had been used as a hypertensive agent but was withdrawn from use due to incidents of hepatitis.

A number of furans are activated by P450s [136], and today a furan ring, particularly an unsubstituted one, is considered suspect for problems in medicinal chemistry. In the case of tienilic acid, *S*-oxygenation is considered to be the initial activating event [137] (alternatively, epoxidation has also been proposed [138]). The product can react with thiols to yield covalent adducts. Tienilic acid is also a mechanism-based inactivator of P450 2C9, the enzyme that oxidizes it.

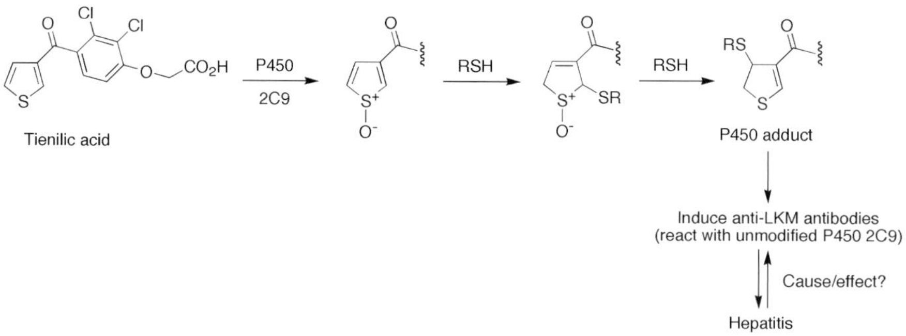

Figure 8. Metabolism of tienilic acid and proposed relevance to toxicity. The *S*-oxygenation is catalyzed by P450 2C9. A thiol reacts with the electrophilic *S*-oxide, and another attack of a thiol on the initial adduct can yield a more stable adduct. The thiol is thought to be a cysteine group of P450 2C9. The modified P450 2C9 is antigenic and induces the immune system to produce antibodies that recognize unmodified (and modified) P450 2C9. One hypothesis is that this autoimmune response is causal in the hepatitis that sometimes occurs with tienilic acid [67].

Radioactivity from tienilic acid is bound almost exclusively to P450 2C9, among the human liver microsomal proteins.

Some of the patients treated with tienilic acid also developed what are termed LKM (liver/kidney microsomal) antibodies in their sera [66]. These antibodies recognize P450 2C9, even the enzyme that has not been modified with tienilic acid [66], in human liver. Thus, we have a toxicity situation that incorporates aspects of bioactivation and hypersensitivity response. A similar phenomenon has been described for the drug dihydralazine, P450 1A2, other autoantibodies, and the associated hepatitis [139].

The chemical nature of the modification of P450 2C9 by the tienilic acid product has not been characterized in detail. However, the biological questions about this system are even more daunting. How does the immune system recognize the cellular protein as being foreign? Is the antibody causal for the hepatitis? Is the toxicity dose related? These idiosyncratic responses to drugs are very rare ($1/10^3$–$1/10^4$ patients) and the etiology has been difficult to understand. Further, animal models for these rare idiosyncratic problems are almost non-existent, and predicting which drugs may cause such problems and which patients might be at risk is one of the major issues in drug safety assessment today.

5.5. Perhexiline

Perhexiline (Figure 9) is an antianginal drug that works as a Ca^{2+} channel blocker, the use of which is complicated by instances of peripheral neuropathy and liver damage [141,142]. The metabolism of perhexiline appears relatively straightforward, although the point should be made that both the primary products, 'cis' and 'trans' hydroxyperhexiline, may exist as diastereomeric pairs [140]. The cis product is usually analyzed in clinical studies [140].

As pointed out earlier, work with P450 2D6 was based upon clinical work with debrisoquine 4-hydroxylation, which predated characterization of the enzyme [75]. In 1982, before the characterization of P450 2D6, Smith and his associates phenotyped 34 patients who had been treated with perhexiline [142]. Most of the 20 patients with neuropathy, but not the unaffected individuals, showed an impaired ability to oxidize debrisoquine. Further research by the group, working with Sherlock's group, showed a clear association between perhexiline liver injury and the deficiency of debrisoquine hydroxylation [143].

The role of P450 2D6 in perhexiline hydroxylation has been clearly demonstrated *in vivo* and *in vitro* [144]. Even before P450 2D6 was characterized or understood, phenotyping methods were used to help define subclinical neuropathy in a patient [145]. Genotyping can also be used to effectively predict poor metabolizer and intermediate phenotypes [146,147].

This is an example of a drug in clinical use where toxicity is clearly well associated with a genetic polymorphism in a P450 enzyme. This approach would be expected to be valuable in preventing side effects of drugs. However, the use

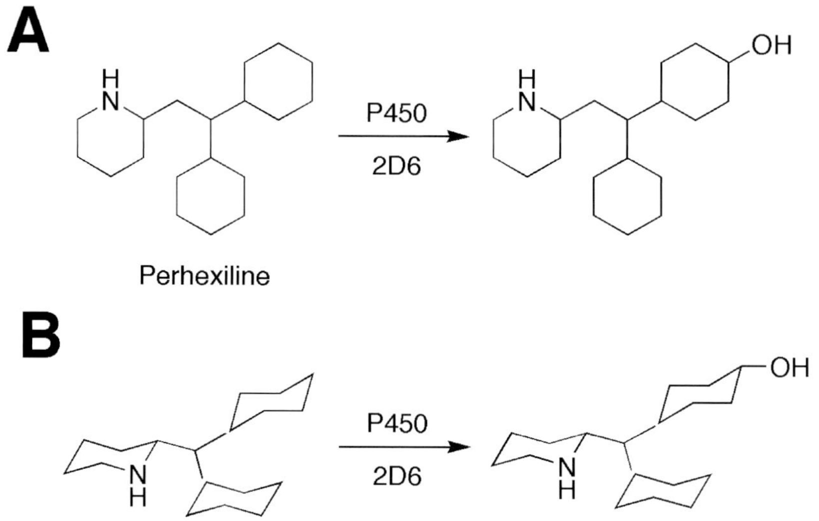

Figure 9. Hydroxylation of perhexiline by P450 2D6 [140]. A: Basic reaction. B: Reaction with stereochemical complexity shown. A main product, the '*cis*' alcohol, is shown. In perhexiline itself, the two cyclohexyl rings are prochiral. In the product there are three chiral centers.

of genotyping and phenotyping is still very limited. For instance, in the USA the FDA does not require genotyping for any drugs at this time. A recent survey of (639) Australian and New Zealand clinical laboratories (including hospitals and universities) indicated only one that was routinely phenotyping for perhexiline use and none were routinely genotyping. Some issues holding back the applications include a poor evidence base, perceived costs, unestablished clinical relevance, and a slow translation of basic research to clinical practice. Therefore, considerable opportunity still exists for use of P450 science in medical practice. The costs of simple assays may be easily recovered in decreased expenses of hospital stays.

6. SUMMARY

P450 enzymes are very important in the metabolism of drugs, a process that can markedly influence efficacy and toxicity. For these reasons, this research area is very relevant to individuals working in both the discovery and development phases of pharmaceutical research. A small subset of the human P450s (Figure 3) is responsible for most of the metabolism of drugs on the market, and this group of P450s has been extensively investigated. The information about these P450s (1A2, 2C9, 2C19, 2D6, 3A4) is considerable, and the drug development process and many of the regulatory issues are dealt with routinely. Progress has been made in the application of genomic analysis to issues, but this is still far from standard clinical practice. Recently the three-dimensional structures of several of the major human P450s involved in drug metabolism have been determined using X-ray crystallography. These structures are very useful in understanding the biochemistry of how these P450s function. The usefulness in predicting the course of reactions de novo will depend upon the flexibility of the enzymes and the extent to which they utilize induced fit with drugs and other non-physiological substrates that these enzymes deal with.

ACKNOWLEDGMENTS

The author is supported in part in this field of research by United States Public Health Service grants R37 CA090426 and P30 ES000267. Thanks are extended to K. Trisler for her assistance in preparation of the manuscript and to W. G. Humphreys (Bristol-Myers Squibb) for the concept of the paradigm in Figure 1.

ABBREVIATIONS

AhR	aryl hydrocarbon receptor
AUC	area under the curve
CAR	constitutive androgen receptor
$C_{p,max}$	maximum plasma concentration
EE_2	17α-ethynylestradiol
EM	extensive metabolizer
FDA	United States Food and Drug Administration
GSH	glutathione
hERG	human ether-a-go-go-related (gene and protein)
HNF	hepatocyte nuclear factor
HPLC	high performance liquid chromatrography
LKM	liver/kidney microsomal
NADPH	nicotinamide adenine dinucleotide phosphate, reduced
P450	cytochrome P450
PM	poor metabolizer
PXR	pregnane X receptor
QT	no definition (a descriptive cardiovascular term)
SNP	single nucleotide polymorphism
UGT	uridine dinucleotide phosphate glucuronosyl transferase

REFERENCES

1. F. P. Guengerich, in *Biotransformation, Vol. 3, Comprehensive Toxicology*, (F. P. Guengerich, ed.), Elsevier Science, Oxford, 1997, pp. 1–6.
2. R. T. Williams, *Detoxication Mechanisms*, John Wiley & Sons, Inc., New York, 1947.
3. R. T. Williams, *Detoxication Mechanisms*, John Wiley & Sons, Inc., New York, 1959.
4. H. Remmer, *Naunyn-Schmiedeberg's Arch. Exp. Pathol. Pharmakol., 237,* 296–307 (1957).
5. A. H. Conney, E. C. Miller, and J. A. Miller, *Cancer Res., 16,* 450–459 (1956).
6. G. C. Mueller and J. A. Miller, *J. Biol. Chem., 202,* 579–587 (1953).
7. B. B. Brodie, J. R. Gillette, and B. N. LaDu, *Annu. Rev. Biochem., 27,* 427–454 (1958).
8. W. Kalow, *Pharmacogenetics*, W. B. Saunders, Philadelphia, 1962.
9. A. Y. H. Lu and M. J. Coon, *J. Biol. Chem., 243,* 1331–1332 (1968).
10. P. Wang, P. S. Mason, and F. P. Guengerich, *Arch. Biochem. Biophys., 199,* 206–219 (1980).
11. L. M. Distlerath and F. P. Guengerich, in *Mammalian Cytochromes P-450, Vol. 1* (F. P. Guengerich, ed.), CRC Press, Boca Raton, Florida, 1987, pp. 133–198.
12. F. J. Gonzalez, *Pharmacol. Rev., 40,* 243–288 (1989).
13. F. P. Guengerich, *Int. J. Toxicol., 24,* 5–21 (2005).
14. D. W. Nebert, J. K. Heidema, H. W. Strobe, and M. J. Coon, *J. Biol. Chem., 248,* 7631–7636 (1973).
15. F. J. Gonzalez and S. Kimura, *Arch. Biochem. Biophys., 409,* 153–158 (2003).
16. F. P. Guengerich, *Mol. Interventions, 3,* 194–204 (2003).
17. D. C. Liebler and F. P. Guengerich, *Nature Rev. Drug Discov., 4,* 410–420 (2005).
18. P. K. Honig, D. C. Wortham, K. Zamani, D. P. Conner, J. C. Mullin, and L. R. Cantilena, *J. Am. Med. Assoc., 269,* 1513–1518 (1993).
19. C.-H. Yun, R. A. Okerholm, and F. P. Guengerich, *Drug Metab. Dispos., 21,* 403–409 (1993).
20. F. P. Guengerich and T. Shimada, *Chem. Res. Toxicol., 4,* 391–407 (1991).
21. F. P. Guengerich, in *Cytochrome P450: Structure, Mechanism, & Biochemistry* (P. R. Ortiz de Montellano, ed.), 3rd edn, Kluwer Academic/Plenum Press, New York, 2005, pp. 377–530.
22. M. D. Burke, S. Thompson, C. R. Elcombe, J. Halpert, T. Haaparanta, and R. T. Mayer, *Biochem. Pharmacol., 34,* 3337–3345 (1985).
23. C. L. Crespi, V. P. Miller, and B. W. Penman, *Anal. Biochem., 248,* 188–190 (1997).
24. A. K. Daly, S. Cholerton, W. Gregory, and J. R. Idle, *Pharmacol. Therapeut., 57,* 129–160 (1993).
25. E. U. Griese, U. M. Zanger, U. Brudermanns, A. Gaedigk, G. Mikus, K. Morike, T. Stuven, and M. Eichelbaum, *Pharmacogenetics, 8,* 15–26 (1998).
26. J. A. Goldstein, T. Ishizaki, K. Chiba, S. M. F. de Morais, D. Bell, P. M. Krahn, and D. A. P. Evans, *Pharmacogenetics, 7,* 59–64 (1997).
27. J. A. Goldstein and S. M. F. Demorais, *Pharmacogenetics, 4,* 285–299 (1994).
28. L. J. Dickmann, A. E. Rettie, M. B. Kneller, R. B. Kim, A. J. Wood, C. M. Stein, G. R. Wilkinson, and U. I. Schwarz, *Mol. Pharmacol., 60,* 382–387 (2001).

29. Z. Jiang, N. Dragin, L. F. Jorge-Nebert, M. V. Martin, F. P. Guengerich, E. Aklillu, M. Ingelman-Sundberg, G. J. Hammons, B. D. Lyn-Cook, F. F. Kadlubar, S. N. Saldana, M. Sorter, A. A. Vinks, N. Nassr, O.von Richter, L. Jin, and D. W. Nebert, *Pharmacogenetics Genomics*, *16*, 359–367 (2006).
30. J. K. Lamba, Y. S. Lin, E. G. Schuetz, and K. E. Thummel, *Adv. Drug Deliv. Rev.*, *54*, 1271–1294 (2002).
31. A. L. Shen, K. A. O'Leary, and C. B. Kasper, *J. Biol. Chem.*, *277*, 6536–6541 (2002).
32. C. J. Henderson, D. M. Otto, D. Carrie, M. A. Magnuson, A. W. McLaren, I. Rosewell, and C. R. Wolf, *J. Biol. Chem.*, *278*, 13480–13486 (2003).
33. J. Gu, Y. Weng, Q. Y. Zhang, H. Cui, M. Behr, L. Wu, W. Yang, L. Zhang, and X. Ding, *J. Biol. Chem.*, *278*, 25895–25901 (2003).
34. L. Wu, J. Gu, H. Cui, Q. Y. Zhang, M. Behr, C. Fang, Y. Weng, K. Kluetzman, P. J. Swiatek, W. Yang, L. Kaminsky, and X. Ding, *J. Pharmacol. Exp. Ther.*, *312*, 35–43 (2005).
35. Y. Weng, C. C. DiRusso, A. A. Reilly, P. N. Black, and X. Ding, *J. Biol. Chem.*, *280*, 31686–31698 (2005).
36. C. L. Meschter, B. A. Mico, M. Mortillo, D. Feldman, W. A. Garland, J. A. Riley, and L. S. Kaufman, *Fund. Appl. Toxicol.*, *22*, 369–381 (1994).
37. A. Galijatovic, D. Beaton, N. Nguyen, S. Chen, J. Bonzo, R. Johnson, S. Maeda, M. Karin, F. P. Guengerich, and R. H. Tukey, *J. Biol. Chem.*, *279*, 23969–23976 (2004).
38. A. M. Yu, K. Fukamachi, K. W. Krausz, C. Cheung, and F. J. Gonzalez, *Endocrinology*, *146*, 2911–2919 (2005).
39. C. Cheung, A. M. Yu, J. M. Ward, K. W. Krausz, T. E. Akiyama, L. Feigenbaum, and F. J. Gonzalez, *Drug Metab. Dispos.*, *33*, 449–457 (2005).
40. F. P. Guengerich, Z.-L. Wu, and C. J. Bartleson, *Biochem. Biophys. Res. Commun.*, *338*, 465–469 (2005).
41. D. W. Nebert and D. W. Russell, *Lancet*, *360*, 1155–1162 (2002).
42. J. R. Idle and R. L. Smith, *Drug Metab. Rev.*, *9*, 301–317 (1979).
43. J. H. Capdevila, V. R. Holla, and J. R. Falck, in *Cytochrome P450: Structure, Mechanism, and Biochemistry* (P. R. Ortiz de Montellano, ed.), 3rd edn., Kluwer Academic/Plenum Press, New York, 2005, pp. 531–551.
44. J. A. Williams, R. Hyland, B. C. Jones, D. A. Smith, S. Hurst, T. C. Goosen, V. Peterkin, J. R. Koup, and S. E. Ball, *Drug Metab. Dispos.*, *32*, 1201–1208 (2004).
45. L. C. Wienkers and T. G. Heath, *Nature Rev. Drug Discov.*, *4*, 825–833 (2005).
46. J. A. Williams, R. Hyland, B. C. Jones, D. A. Smith, S. Hurst, T. C. Goosen, V. Peterkin, J. R. Koup, and S. E. Ball, *Drug Metab. Dispos.*, *32*, 1201–1208 (2004).
47. I. Koch, R. Weil, R. Wolbold, J. Brockmoller, E. Hustert, O. Burk, A. Nuessler, P. Neuhaus, M. Eichelbaum, U. Zanger, and L. Wojnowski, *Drug Metab. Dispos.*, *30*, 1108–1114 (2002).
48. W. Huang, Y. S. Lin, D. J. McConn, 2nd, J. C. Calamia, R. A. Totah, N. Isoherranen, M. Glodowski, and K. E. Thummel, *Drug Metab. Dispos.*, *32*, 1434–1445 (2004).
49. S. Rendic, *Drug Metab. Rev.*, *34*, 83–448 (2002).
50. L. M. Distlerath, P. E. B. Reilly, M. V. Martin, G. G. Davis, G. R. Wilkinson, and F. P. Guengerich, *J. Biol. Chem.*, *260*, 9057–9067 (1985).

51. L. C. Quattrochi, T. Vu, and R. H. Tukey, *J. Biol. Chem.*, 269, 6949–6954 (1994).
52. E. J. Pantuck, K.-C. Hsiao, A. Maggio, K. Nakamura, R. Kuntzman, and A. H. Conney, *Clin. Pharmacol. Ther.*, 15, 9–17 (1974).
53. L. C. Quattrochi and R. H. Tukey, *Mol. Pharmacol.*, 36, 66–71 (1989).
54. M. A. Butler, M. Iwasaki, F. P. Guengerich, and F. F. Kadlubar, *Proc. Natl. Acad. Sci. USA*, 86, 7696–7700 (1989).
55. M. A. Butler, N. P. Lang, J. F. Young, N. E. Caporaso, P. Vineis, R. B. Hayes, C. H. Teitel, J. P. Massengill, M. F. Lawsen, and F. F. Kadlubar, *Pharmacogenetics*, 2, 116–127 (1992).
56. A. Parikh, P. D. Josephy, and F. P. Guengerich, *Biochemistry*, 38, 5283–5289 (1999).
57. H. Zhou, P. D. Josephy, D. Kim, and F. P. Guengerich, *Arch. Biochem. Biophys.*, 422, 23–30 (2004).
58. D. Kim and F. P. Guengerich, *Biochemistry*, 43, 981–988 (2004).
59. D. F. V. Lewis, B. G. Lake, S. G. George, M. Dickins, P. J. Eddershaw, M. H. Tarbit, A. P. Beresford, P. S. Goldfarb, and F. P. Guengerich, *Toxicology*, 139, 53–79 (1999).
60. D. Kim and F. P. Guengerich, *Arch. Biochem. Biophys.*, 432, 102–108 (2004).
61. D. Sesardic, A. Boobis, B. Murray, S. Murray, J. Segura, R. De La Torre, and D. Davies, *Brit. J. Clin. Pharmacol.*, 29, 651–663 (1990).
62. D. R. Umbenhauer, M. V. Martin, R. S. Lloyd, and F. P. Guengerich, *Biochemistry*, 26, 1094–1099 (1987).
63. Y. Yamamoto, T. Kawamoto, and M. Negishi, *Arch. Biochem. Biophys.*, 409, 207–211 (2003).
64. G. P. Aithal, C. P. Day, J. B. Leathart, and A. K. Daly, *Pharmacogenetics*, 10, 511–518 (2000).
65. M. E. Veronese, J. O. Miners, D. Randles, D. Gregov, and D. J. Birkett, *Clin. Pharmacol. Ther.*, 47, 403–411 (1990).
66. P. Beaune, P. M. Dansette, D. Mansuy, L. Kiffel, M. Finck, C. Amar, J. P. Leroux, and J. C. Homberg, *Proc. Natl. Acad. Sci. USA*, 84, 551–555 (1987).
67. P. Beaune, D. Pessayre, P. Dansette, D. Mansuy, and M. Manns, *Adv. Pharmacol.*, 30, 199–245 (1994).
68. P. A. Williams, J. Cosme, A. Ward, H. C. Angove, D. Matak Vinkovic, and H. Jhoti, *Nature*, 424, 464–468 (2003).
69. M. R. Wester, J. K. Yano, G. A. Schoch, C. Yang, K. J. Griffin, C. D. Stout, and E. F. Johnson, *J. Biol. Chem.*, 279, 35630–35637 (2004).
70. P. J. Wedlund, W. S. Aslanian, C. B. McAllister, G. R. Wilkinson, and R. A. Branch, *Clin. Pharmacol. Ther.*, 36, 773–780 (1984).
71. J. A. Goldstein, M. B. Faletto, M. Romkes-Sparks, T. Sullivan, S. Kitareewan, J. L. Raucy, J. M. Lasker, and B. I. Ghanayem, *Biochemistry*, 33, 1743–1752 (1994).
72. S. M. F. de Morais, G. R. Wilkinson, J. Blaisdell, K. Nakamura, U. A. Meyer, and J. A. Goldstein, *J. Biol. Chem.*, 269, 15419–15422 (1994).
73. H. H. Zhou, R. P. Koshakji, D. J. Silberstein, G. R. Wilkinson, and A. J. J. Wood, *New Engl. J. Med.*, 320, 565–570 (1989).
74. C. C. Tsao, M. R. Wester, B. Ghanayem, S. J. Coulter, B. Chanas, E. F. Johnson, and J. A. Goldstein, *Biochemistry*, 40, 1937–1944 (2001).

75. A. Mahgoub, J. R. Idle, L. G. Dring, R. Lancaster, and R. L. Smith, *Lancet, ii*, 584–586 (1977).
76. M. Eichelbaum, N. Spannbrucker, B. Steincke, and H. J. Dengler, *Eur. J. Clin. Pharmacol., 16*, 183–187 (1979).
77. M. S. Lennard, J. H. Silas, S. Freestone, G. T. Tucker, L. E. Ramsay, and H. F. Woods, *Brit. J. Pharmacol., 16*, 572P–573P (1982).
78. J. Gut, U. T. Meier, T. Catin, and U. A. Meyer, *Biochim. Biophys. Acta, 884*, 435–447 (1986).
79. F. J. Gonzalez, R. C. Skoda, S. Kimura, M. Umeno, U. M. Zanger, D. W. Nebert, H. V. Gelboin, J. P. Hardwick, and U. A. Meyer, *Nature, 331*, 442–446 (1988).
80. U. A. Meyer and U. M. Zanger, *Annu. Rev. Pharmacol. Toxicol., 37*, 269–296 (1997).
81. T. Wolff, L. M. Distlerath, M. T. Worthington, J. D. Groopman, G. J. Hammons, F. F. Kadlubar, R. A. Prough, M. V. Martin, and F. P. Guengerich, *Cancer Res., 45*, 2116–2122 (1985).
82. F. P. Guengerich, G. P. Miller, I. H. Hanna, M. V. Martin, S. Léger, C. Black, N. Chauret, J. M. Silva, L. Trimble, J. A. Jergey, and D. A. Nicoll-Griffith, *Biochemistry, 41*, 11025–11034 (2002).
83. I. H. Hanna, M.-S. Kim, and F. P. Guengerich, *Arch. Biochem. Biophys., 393*, 255–261 (2001).
84. P. Rowland, F. E. Blaney, M. G. Smyth, J. J. Jones, V. R. Leydon, A. K. Oxbrow, C. J. Lewis, M. G. Tennant, S. Modi, D. S. Eggleston, R. J. Chenery, and A. M. Bridges, *J. Biol. Chem., 281*, in press (2006).
85. P. P. Wang, P. Beaune, L. S. Kaminsky, G. A. Dannan, F. F. Kadlubar, D. Larrey, and F. P. Guengerich, *Biochemistry, 22*, 5375–5383 (1983).
86. P. B. Watkins, S. A. Wrighton, P. Maurel, E. G. Schuetz, G. Mendez-Picon, G. A. Parker, and P. S. Guzelian, *Proc. Natl. Acad. Sci. USA, 82*, 6310–6314 (1985).
87. F. P. Guengerich, M. V. Martin, P. H. Beaune, P. Kremers, T. Wolff, and D. J. Waxman, *J. Biol. Chem., 261*, 5051–5060 (1986).
88. D. J. Waxman, C. Attisano, F. P. Guengerich, and D. P. Lapenson, *Arch. Biochem. Biophys., 263*, 424–436 (1988).
89. R. H. Böcker and F. P. Guengerich, *J. Med. Chem., 29*, 1596–1603 (1986).
90. W. R. Brian, M.-A. Sari, M. Iwasaki, T. Shimada, L. S. Kaminsky, and F. P. Guengerich, *Biochemistry, 29*, 11280–11292 (1990).
91. S. Imaoka, Y. Imai, T. Shimada, and Y. Funae, *Biochemistry, 31*, 6063–6069 (1992).
92. E. M. J. Gillam, T. Baba, B.-R. Kim, S. Ohmori, and F. P. Guengerich, *Arch. Biochem. Biophys., 305*, 123–131 (1993).
93. N. A. Hosea and F. P. Guengerich, *Arch. Biochem. Biophys., 353*, 365–373 (1998).
94. R. W. Bork, T. Muto, P. H. Beaune, P. K. Srivastava, R. S. Lloyd, and F. P. Guengerich, *J. Biol. Chem., 264*, 910–919 (1989).
95. F. P. Guengerich, *Mol. Pharmacol., 33*, 500–508 (1988).
96. S. A. Kliewer, J. T. Moore, L. Wade, J. L. Staudinger, M. A. Watson, S. A. Jones, D. D. McKee, B. B. Oliver, T. M. Wilson, R. H. Zetterström, T. Perlmann, and J. M. Lehmann, *Cell, 92*, 73–82 (1998).
97. C. Calleja, J. M. Pascussi, J. C. Mani, P. Maurel, and M. J. Vilarem, *Nature Med., 4*, 92–96 (1998).

98. J. M. Pascussi, S. Gerbal-Chaloin, L. Drocourt, P. Maurel, and M. J. Vilarem, *Biochim. Biophys. Acta, 1619*, 243–253 (2003).
99. K. Yasuda, L. B. Lan, D. Sanglard, K. Furuya, J. D. Schuetz, and E. G. Schuetz, *J. Pharmacol. Exp. Ther., 303*, 323–332 (2002).
100. E. G. Schuetz, W. T. Beck, and J. D. Schuetz, *Mol. Pharmacol., 49*, 311–318 (1996).
101. F. P. Guengerich, in *Handbook of Drug Metabolism* (T. F. Woolf, ed.), Marcel Dekker, New York, 1999, pp. 203–227.
102. D. G. Bailey, J. D. Spence, C. Munoz, and J. M. O. Arnold, *Lancet, 337*, 268–269 (1991).
103. S. E. Rau, J. R. Bend, J. M. O. Arnold, L. T. Tran, J. D. Spence, and D. G. Bailey, *Clin. Pharmacol. Ther., 61*, 401–409 (1997).
104. T. C. Goosen, D. Cillie, D. G. Bailey, C. Yu, K. He, P. F. Hollenberg, P. M. Woster, L. Cohen, J. A. Williams, M. Rheeders, and H. P. Dijkstra, *Clin. Pharmacol. Ther., 76*, 607–617 (2004).
105. J. K. Yano, M. R. Wester, G. A. Schoch, K. J. Griffin, C. D. Stout, and E. F. Johnson, *J. Biol. Chem.*, 38091–38094 (2004).
106. P. A. Williams, J. Cosme, D. M. Vinkovic, A. Ward, H. C. Angove, P. J. Day, C. Vonrhein, I. J. Tickle, and H. Jhoti, *Science, 305*, 683–686 (2004).
107. S. A. Kuby, *A Study of Enzymes*, Vol. I, *Enzyme Catalysis, Kinetics, and Substrate Binding*. CRC Press, Boca Raton, FL, 1991.
108. E. M. Isin and F. P. Guengerich, *J. Biol. Chem., 281*, 9127–9136 (2006).
109. G. A. Schoch, J. K. Yano, M. R. Wester, K. J. Griffin, C. D. Stout, and E. F. Johnson, *J. Biol. Chem., 279*, 9497–9503 (2004).
110. D. A. Smith and B. C. Jones, *Biochem. Pharmacol., 44*, 2089–2098 (1992).
111. S. B. Singh, L. Q. Shen, M. J. Walker, and R. P. Sheridan, *J. Med. Chem., 46*, 1330–1336 (2003).
112. J. A. Krauser and F. P. Guengerich, *J. Biol. Chem., 280*, 19496–19506 (2005).
113. L. C. Wienkers, *J. Pharmacol. Toxicol. Methods, 45*, 79–84 (2001).
114. K. Nakamura, T. Yokoi, K. Inoue, N. Shimada, N. Ohashi, T. Kume, and T. Kamataki, *Pharmacogenetics, 6*, 449–457 (1996).
115. E. D. Kharasch, K. E. Thummel, D. Mautz, and S. Bosse, *Clin. Pharmacol. Ther., 55*, 434–440 (1994).
116. J. K. Yano, M. H. Hsu, K. J. Griffin, C. D. Stout, and E. F. Johnson, *Nat. Struct. Biol., 12*, 822–823 (2005).
117. J. Hakkola, Y. Hu, and M. Ingelman-Sundberg, *J. Pharmacol. Exp. Ther., 304*, 1048–1054 (2003).
118. A. Rahman, K. R. Korzekwa, J. Grogan, F. J. Gonzalez, and J. W. Harris, *Cancer Res., 54*, 5543–5546 (1994).
119. J. C. Gorski, S. D. Hall, D. R. Jones, M. VandenBranden, and S. A. Wrighton, *Biochem. Pharmacol., 47*, 1643–1653 (1994).
120. A. M. Larson, J. Polson, R. J. Fontana, T. J. Davern, E. Lalani, L. S. Hynan, J. S. Reisch, F. V. Schiodt, G. Ostapowicz, A. O. Shakil, and W. M. Lee, *Hepatology, 42*, 1364–1372 (2005).
121. J. G. O'Grady, *Hepatology, 42*, 1252–1254 (2005).
122. C. J. Patten, P. E. Thomas, R. L. Guy, M. Lee, F. J. Gonzalez, F. P. Guengerich, and C. S. Yang, *Chem. Res. Toxicol., 6*, 511–518 (1993).

123. S. S. T. Lee, J. T. M. Buters, T. Pineau, P. Fernandez-Salguero, and F. J. Gonzalez, *J. Biol. Chem., 271*, 12063–12067 (1996).
124. D. J. Jollow, J. R. Mitchell, W. Z. Potter, D. C. Davis, J. R. Gillette, and B. B. Brodie, *J. Pharmacol. Exp. Ther., 187*, 195–202 (1973).
125. J. R. Mitchell, D. J. Jollow, W. Z. Potter, J. R. Gillette, and B. B. Brodie, *J. Pharmacol. Exp. Ther., 187*, 211–217 (1973).
126. R. L. Woosley, Y. Chen, J. P. Freiman, and R. A. Gillis, *J. Am. Med. Assoc., 269*, 1532–1536 (1993).
127. S. C. Stinson, *Chem. Eng. News, 75*, 43–45 (1997).
128. D. Reimers and A. Jezek, *Prax. Pneumol., 25*, 255–262 (1971).
129. L. Nocke-Finck, H. Breuer, and D. Reimers, *Deutsch. Med. Wochenschr., 98*, 1521–1523 (1973).
130. D. Janz and D. Schmidt, *Lancet, i*, 1113 (1974).
131. H. M. Bolt, M. Bolt, and H. Kappus, *Acta Endocrinol., 85*, 189–197 (1977).
132. F. P. Guengerich, *Chem. Res. Toxicol., 3*, 363–371 (1990).
133. F. Morel, P. H. Beaune, D. Ratanasavanh, J.-P. Flinois, C.-S. Yang, F. P. Guengerich, and A. Guillouzo, *Eur. J. Biochem., 191*, 437–444 (1990).
134. K. Wilbur and M. H. Ensom, *Clin. Pharmacokinet., 38*, 355–365 (2000).
135. P. A. Murphy, S. E. Kern, F. Z. Stanczyk, and C. L. Westhoff, *Contraception, 71*, 402–408 (2005).
136. V. Ravindranath, L. T. Burka, and M. R. Boyd, *Science, 224*, 884–886 (1984).
137. P. M. Dansette, D. C. Thang, H. El Amri, and D. Mansuy, *Biochem. Biophys. Res. Commun., 186*, 1624–1630 (1992).
138. P. M. Dansette, G. Bertho, and D. Mansuy, *Biochem. Biophys. Res. Commun., 338*, 450–455 (2005).
139. M. Bourdi, D. Larrey, J. Nataf, J. Berunau, D. Pessayre, M. Iwasaki, F. P. Guengerich, and P. H. Beaune, *J. Clin. Invest., 85*, 1967–1973 (1990).
140. O. Beck, N. Stephanson, R. G. Morris, B. C. Sallustio, and P. Hjemdahl, *J. Chromatogr. B Analyt. Technol. Biomed. Life Sci., 805*, 87–91 (2004).
141. R. K. Roberts, D. Cohn, V. Petroff, and B. Seneviratne, *Med. J. Aust., 2*, 553–554 (1981).
142. R. R. Shah, N. S. Oates, J. R. Idle, R. L. Smith, and J. D. Lockhart, *Br. Med. J. (Clin. Res. Ed.), 284*, 295–299 (1982).
143. M. Y. Morgan, R. Reshef, R. R. Shah, N. S. Oates, R. L. Smith, and S. Sherlock, *Gut, 25*, 1057–1064 (1984).
144. L. B. Sorensen, R. N. Sorensen, J. O. Miners, A. A. Somogyi, N. Grgurinovich, and D. J. Birkett, *Br. J. Clin. Pharmacol., 55*, 635–638 (2003).
145. R. R. Shah, N. S. Oates, J. R. Idle, R. L. Smith, and J. D. Lockhart, *Am. Heart J., 105*, 159–161 (1983).
146. M. L. Barclay, S. M. Sawyers, E. J. Begg, M. Zhang, R. L. Roberts, M. A. Kennedy, and J. M. Elliott, *Pharmacogenetics, 13*, 627–632 (2003).
147. R. Hussein, B. G. Charles, R. G. Morris, and R. L. Rasiah, *Ther. Drug Monit., 23*, 636–643 (2001).

17

Cytochrome P450 Enzymes: Observations from the Clinic

Peggy L. Carver

University of Michigan College of Pharmacy, Clinical Sciences Department,
Ann Arbor, MI 48109-1065, USA
<peg@umich.edu>

1. INTRODUCTION	592
2. DRUG INTERACTIONS	593
3. DRUG METABOLISM BY CYTOCHROME P450 ENZYMES	594
3.1. Location of P450 Enzymes in Humans	594
3.2. The Role of the Gastrointestinal Tract in Drug Metabolism	594
3.3. Variability in the Activity of CYP3A4	595
4. ALTERATIONS IN P450 ENZYMES	596
4.1. Substrates, Inducers, and Inhibitors	596
4.2. Inhibitors of CYP Enzymes	598
4.2.1. Reversible Inhibition	598
4.2.2. Irreversible Inhibition	598
4.2.3. The 'Inhibition Index'	599
4.3. Inhibition of CYP and Adverse Effects of Drugs	599
4.4. *In Vitro-In Vivo* Scaling	600
4.5. Inducers of CYP Enzymes	600
4.6. Genetic Polymorphism	601
5. ACTIVE TRANSPORT OF DRUGS	603
6. ENZYME-TRANSPORTER COOPERATIVITY	603
7. USE OF PROBES TO QUANTITATE CYP ACTIVITY IN HUMANS	604

Metal Ions in Life Sciences, Volume 3   Edited by Astrid Sigel, Helmut Sigel and Roland K. O. Sigel
© 2007 John Wiley & Sons, Ltd

8.	PRODRUGS	604
9.	THE EFFECT OF INTRAVENOUS VERSUS ORAL ADMINISTRATION ON DRUG INTERACTIONS	605
10.	ADDITIONAL FACTORS AFFECTING DRUG INTERACTIONS	606
	10.1. Sex Differences	606
	10.2. Ethnicity Differences	606
	10.3. Diet	607
	10.4. Smoking	607
	10.5. Disease States	607
11.	HERBAL AND DIETARY EFFECTS ON CYP	608
	11.1. Flavonoids	608
	11.2. Foods	608
	11.2.1. Grapefruit Juice	608
	11.2.2. Garlic	609
	11.3. Herbal Interactions	609
	11.3.1. St. John's Wort	610
12.	INTERACTIONS WITH COMMONLY USED MEDICATIONS	610
	12.1. Terfenadine	610
	12.2. Drug Interactions with Macrolide Antibiotics	610
13.	BENEFICIAL EFFECTS OF DRUG INTERACTIONS	611
	13.1. Reducing the Cost of Expensive Medications	611
	13.1.1. Immunosuppressive Agents	611
	13.1.2. Antiretroviral Therapy	612
14.	FDA REGULATIONS REGARDING CYP450-MEDIATED DRUG INTERACTIONS	612
15.	CLINICAL SIGNIFICANCE OF DRUG INTERACTIONS	613
16.	SUMMARY	614
	ACKNOWLEDGMENT	615
	ABBREVIATIONS	615
	REFERENCES	615

1. INTRODUCTION

This chapter was written to offer chemists an overview of the clinical significance of cytochrome P450 (CYP) enzymes involved in drug metabolism, from the perspective of a clinical pharmacist practicing in the hospital and ambulatory setting. Clinicians involved in the day-to-day care of patients have benefited from the enormous growth of information in this area over the past two decades. An increasing number of drug interactions and undesirable effects that occur in the clinical use of drugs are being explained and avoided on the basis of knowledge about the specific CYP enzyme(s) involved in an agent's metabolism [1].

Most often, knowledge of the CYP enzyme(s) involved in the metabolism of drugs are utilized to predict and prevent drug interactions, minimize adverse effects, and to optimize the desired pharmacological response. Drug metabolism studies are increasingly used in the early stages of drug development for lead optimization and the selection of compounds for more extensive investigation, in the qualitative and quantitative prediction of drug biotransformation *in vivo*, and in the identification of likely determinants of metabolism following drug administration to humans, including possible drug interactions.

2. DRUG INTERACTIONS

A drug interaction occurs when the effectiveness or toxicity of a drug is altered by the administration of another drug or substance. The Institute of Medicine has identified medication-related errors as a leading cause of preventable death [2]. One area of particular concern has been adverse drug interactions. An adverse drug interaction occurs when two or more drugs are combined in a manner that alters either the effectiveness or the toxicity of one of the agents [3]. The incidence of adverse drug interactions has been estimated to be between 2.2% and 30% in hospitalized patients and between 9.2% and 70.3% in ambulatory patients [4,5]. Pharmacokinetic alterations are the most common and include changes in drug absorption (rate and/or extent), distribution (plasma protein binding displacement), metabolism and excretion (pulmonary, renal, and biliary) [6].

Drug interactions can be categorized as pharmacodynamic or pharmacokinetic. *Pharmacodynamic* interactions include additive, synergistic, or antagonistic interactions that can potentially affect efficacy or toxicity. *Pharmacokinetic* interactions result when one drug alters the absorption, distribution, metabolism, or excretion of another drug. The most common mechanisms underlying drug interactions are inhibition or induction of hepatic and extra-hepatic CYP enzymes and genetic polymorphism [5,7].

On a daily basis in the hospital or clinic, predicting and preventing cytochrome-mediated drug interactions is difficult as it is almost impossible to memorize all possible drug interactions. However, a thorough understanding of the mechanisms of drug interactions provides clinicians with the ability to predict their occurrence and devise strategies to minimize or avoid them. The challenge lies in recognizing among these drug interactions those of greatest clinical significance and in intervening to minimize or prevent them [7]. Fortunately, there are a number of online resources readily available to assist clinicians, including a very comprehensive list of cytochrome-mediated drug interactions maintained by the Division of Clinical Pharmacology at Indiana University [8].

To predict, evaluate, and manage drug interactions, clinicians must first identify the enzyme(s) responsible for an agent's metabolism, then consider the dosage and timing of administration of the drugs in question, the duration of therapy, baseline and steady state concentrations, the therapeutic index of each agent, and the potential for inter-individual variability in pharmacokinetic variables, including absorption, distribution, metabolism, and elimination [5].

3. DRUG METABOLISM BY CYTOCHROME P450 ENZYMES

The majority of oral medications are lipid soluble and nonpolar. Oral medications that are already hydrophilic escape this process of hepatic metabolism and are eliminated unchanged in the urine or feces [5,9]. The liver is the major site of drug metabolism, where two types of reactions occur. In phase I reactions, catalyzed by CYP enzymes, oxidation of a parent drug yields a more polar, hydrophilic moiety that may be pharmacologically active or inactive. CYP enzymes are heme proteins that catalyze phase I metabolism of many endogenous substrates, including steroids, fatty acids, prostaglandins, bile acids, and xenobiotics (including drugs, carcinogens, environmental pollutants, and many other synthetic chemicals). In phase II reactions, which are not mediated by CYP enzymes, conjugation of the parent drug or previously oxidized drug yields a more polar, hydrophilic moiety which is more readily excreted in the feces or urine [10].

3.1. Location of P450 Enzymes in Humans

In humans, CYP enzymes are found in the smooth endoplasmic reticulum of liver hepatocytes and in the villous columnar epithelium of the jejunum, lungs, kidney, and brain [11]. Although more than 40 different CYP enzymes have been identified in humans, only six (CYPs 3A4, 2D6, 1A2, 2C9/19, and 2E1) appear to be clinically significant in the metabolism of medications [11]. Enzymes belonging to the CYP3A subfamily are by far the most abundant of the human CYP isoforms. Since their substrate specificity is extremely broad, a variety of structurally diverse compounds are substrates of these enzymes. CYP3A enzymes are localized in organs of particular relevance to drug disposition, including the gastrointestinal tract, kidney, and liver; their catalytic activity is readily modulated by a variety of compounds [1].

3.2. The Role of the Gastrointestinal Tract in Drug Metabolism

Nearly 50% of all clinically used medications and endogenous steroids are metabolized by CYP3A4, which may explain why among the CYP 450 enzymes,

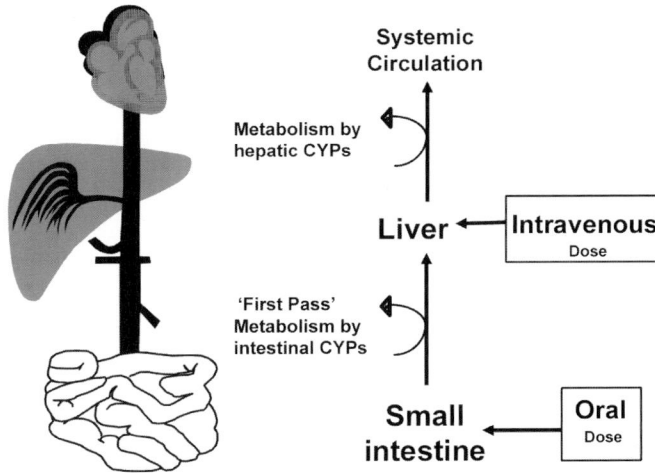

Figure 1. 'First pass' elimination refers to loss of drug as it passes through the gastrointestinal tract membrane and liver for the first time, during the absorption of an orally administered drug. Drugs administered intravenously, directly into the blood stream, go immediately to the liver for metabolism, avoiding metabolism by enzymes within the gastrointestinal tract membrane.

CYP3A4 is involved most often in drug metabolism and drug interactions [12]. CYP3A4 accounts for 29% and 70% of the total human hepatic and gastrointestinal tract CYP, respectively, which permits an effect on both pre-systemic and systemic drug distribution [10,13]. 'First pass' elimination refers to loss of drug as it passes through the gastrointestinal tract membrane and liver for the first time, during the absorption of an orally administered drug. First pass elimination results in variable (often very low) oral bioavailability of CYP3A4-metabolized drugs. Bioavailability refers to the percentage of an administered drug that is detected in the systemic circulation; by definition, an intravenously administered drug has a bioavailability of unity. Drugs administered intravenously, directly into the blood stream, go immediately to the liver for metabolism, avoiding metabolism within the gastrointestinal tract membrane (Figure 1).

3.3. Variability in the Activity of CYP3A4

An important characteristic of CYP3A is its large interindividual variability in activity, which reflects a genetic effect combined with modulation by environmental factors; this variability is often significantly increased by inhibition and induction of enzymes [1]. In humans, hepatic CYP3A4 varies at least 20-fold and enteric CYP3A4 by 10-fold among individuals; however, CYP3A4 content of each appear to be regulated independently. Thus, one patient might have relatively high CYP3A4 activity in the liver, and relatively low CYP3A4 activity in the intestine, and another patient the opposite quantities of enzyme, such that the liver could be the major site of drug metabolism in the first patient, versus the intestine in the second patient [14]. The relative contributions of hepatic and enteric metabolism can be estimated separately by determining differences in drug clearance following oral and intravenous administration of a CYP3A4-metabolized drug [13].

Studies using *in vitro* preparations are useful in identifying such potential interactions and possibly permitting extrapolation of *in vitro* findings to the likely *in vivo* situation. Even if accurate quantitative predictions cannot be made, several classes of drugs can be expected to result in drug interactions based on clinical experience. In many instances, the extent of such drug interactions is sufficiently pronounced to contraindicate the therapeutic use of the involved drugs [1].

4. ALTERATIONS IN P450 ENZYMES

4.1. Substrates, Inducers, and Inhibitors

Drugs can be substrates, inducers, and inhibitors of CYP enzymes. *Substrates* of CYP enzymes are moieties that undergo metabolism by one or more CYP enzymes. *Inducers* of CYP enzymes cause an increase in the amount or activity of a CYP enzyme. By contrast, drugs that are *inhibitors* of a CYP enzyme cause a decrease in the amount or activity of the enzyme. Enzyme inducers and inhibitors produce their effects in a dose- and concentration-dependent manner; higher doses of drugs produce greater induction or inhibition within a range of drug dosages or concentrations.

Some medications are a substrate for only one enzyme, while other drugs are metabolized by several CYP450 enzymes. Drugs may also act as inhibitors of enzymes responsible for the metabolism of other drugs [9]. For example, quinidine (used for the treatment of cardiac arrythmias) is a substrate for (and thus metabolized by) CYP2D6, but is an inhibitor of CYP3A4. CYP enzymes can be highly substrate specific and thus capable of metabolizing only a few substrates, or poorly specific, and capable of metabolizing a broad range of substrates. Enzymes capable of metabolizing a wide variety of substrates are more often involved in drug interactions than are those whose specificity limits their use to a few medications [7]. For example, CYP3A4 is poorly specific; it is responsible for metabolizing a large number of very commonly utilized medications, including many antibiotics, immunosuppressants, and lipid-lowering drugs. Table 1 lists common medications known to be substrates, inducers, and inhibitors of CYP enzymes.

Inducers and inhibitors differ in the onset and duration of their effects on CYP enzymes: while maximum enzyme induction is generally achieved over approximately two weeks, enzyme inhibition can generally be observed immediately after administration of the first dose of an inhibitor drug [10,11]. For

Table 1. Substrates, Inducers, and Inhibitors of Major Human Cytochromes.

Cytochrome	Substrates	Inducers	Inhibitors
CYP3A4	Alprazolam, astemizole, atorvastatin, carbamazepine, cisapride, cyclosporine, diltiazem, losartan, lovastatin, midazolam, nifedipine, simvastatin, terfenadine, verapamil	Carbamazepine, ethanol, phenobarbital, phenytoin, rifampin, St. John's Wort	Cimetidine, clarithromycin, cyclosporine, diltiazem, erythromycin, fluconazole, fluoxetine, fluvoxamine, grapefruit juice, itraconazole, ketoconazole, nifedipine, ritonavir, simvastatin, verapamil, voriconazole
CYP2D6	Codeine, dextromethorphan, flecainide, fluoxetine, haloperidol, metoprolol, paroxetine, perphenazine, propranolol, sertraline, timolol, tricyclic antidepressants, venlafaxine		Cimetidine, dextromethorphan, fluoxetine, haloperidol, paroxetine, sertraline
CYP2C9	Diclofenac, glipizide, ibuprofen, losartan, phenytoin, tolbutamide, torsemide, warfarin	Phenobarbital, rifampin	Amiodarone, cimetidine, fluconazole, fluoxetine, itraconazole, ketoconazole, metronidazole, ritonavir, voriconazole
CYP2C19	Diazepam, imipramine, nelfinavir, omeprazole pantoprazole, propranolol		Fluoxetine, fluvoxamine, omeprazole, ritonavir, pantoprazole, propranolol sertraline
CYP2E1	Acetaminophen, ethanol	Disulfiram	Ethanol, isoniazid
CYP1A2	Caffeine, theophylline, tricyclic antidepressants, warfarin	Charbroiled food, omeprazole, phenobarbital phenytoin, tobacco smoke	Fluvoxamine, ciprofloxacin, grapefruit juice

CYP3A4 substrates with *low* oral bioavailability (high presystemic elimination or first pass elimination), administration of a single dose of a CYP inhibitor will generally produce substantial increases in the substrate's area under the plasma-concentration time curve (AUC). By contrast, administration of a single dose of a CYP inhibitor concomitantly with a CYP substrate with *high* oral bioavailability does not produce a large effect; however, repeated administration of both may produce a cumulative increase in plasma substrate concentrations and a clinically significant interaction may be realized only during steady-state conditions [13]. Few data are available regarding the duration of enzyme inhibition or induction after discontinuation of inducers or inhibitors; however, they are affected by the half-lives and protein binding of inducers and inhibitors.

4.2. Inhibitors of CYP Enzymes

4.2.1. Reversible Inhibition

The most common type of inhibition is due to the reversible interaction of a drug or a stable metabolite with a CYP enzyme. This process is probably responsible for most pharmacokinetic drug-drug interactions *in vivo*. Essentially, reversible processes would occur when a drug or stable metabolite inhibits an early step in the CYP reaction cycle, usually substrate binding or oxygen coordination to the heme. Thus, substrate turnover is decreased [15]. The most common mechanism, reversible inhibition, occurs when a second substrate competes with a medication for the active site of the enzyme. Direct, rapidly reversible binding of an inhibitor or its metabolite to CYP results in either *competitive* or *noncompetitive* inhibition, the extent of which is determined by the relative binding constants of substrate and inhibitor for the enzyme and by the inhibitor's concentration. Clinically significant inhibitors include azole antifungals, erythromycin, quinidine, selective serotonin receptor inhibitors, and protease inhibitors. Diltiazem, a cardiac drug commonly used for the treatment of hypertension and angina, is a reversible inhibitor of CYP3A4, the enzyme responsible for the elimination of cyclosporine. Co-administration of diltiazem interferes with cyclosporine clearance, results in higher cyclosporine levels, and lengthens the cyclosporine dosing interval [16]. Reversible inhibition is transient, and the enzyme returns to normal activity once the inhibitor has been cleared.

4.2.2. Irreversible Inhibition

In irreversible inhibition, a substrate binds to a CYP enzyme, alters the enzyme structure, and produces permanent enzyme inactivation. Irreversible inhibition may occur either through a mechanism-based process or through the formation of a metabolite-intermediate complex. In mechanism-based inhibition, the substrate is metabolized by the enzyme and subsequently transforms and inactivates the enzyme. This suicide inhibition results in an enzyme that is irreversibly changed to a new compound and is essentially destroyed. In contrast, metabolite-intermediate complexation occurs when the metabolized substrate binds tightly to the enzyme to yield an inert enzyme. Although the enzyme is inactive, it is not completely destroyed [15].

Compared with reversible inhibition of CYP, mechanism-based inhibition of CYP more frequently causes pharmacokinetic-pharmacodynamic drug-drug interactions, as the inactivated CYP has to be replaced by newly synthesized protein. Clinically important mechanism-based CYP inhibitors include antibacterials (e.g., clarithromycin, erythromycin, and isoniazid), anticancer agents (e.g., tamoxifen and irinotecan), anti-HIV agents (e.g., ritonavir and delavirdine), and antihypertensives (e.g., hydralazine, verapamil, and diltiazem) [17].

4.2.3. The 'Inhibition Index'

The 'inhibition index', which consists of the ratio of the inhibited intrinsic clearance to that in the absence of inhibitor, is the inhibitor's concentration relative to its K_i value. The most potent reversible CYP3A inhibitors, which include azole antifungal agents and HIV protease inhibitors, have K_i values $< 1 \mu M$. There are a number of other commonly used drugs whose K_i values are in the low micromolar range which generally exhibit inhibitory effects on other CYP substrates. Clinically significant inhibition is uncommon for compounds with values $> 75-100 \mu M$ because sufficiently high concentrations are not clinically achieved [1,18]. By mimicking the substrate, a competitive inhibitor competes for binding to the active site of an enzyme. For a competitive inhibitor to fully inhibit a site, high concentrations of the inhibitor are needed. In fact, the higher the concentration, the larger the magnitude of the inhibition. The clinical significance of drug interactions resulting from competitive inhibition depends on the concentration of the inhibitor achieved at the site of inhibition, the relative dosages of the inhibitor and the substrate, the relative bioavailability, the relative affinity constants of inhibitor and substrate (i.e., how tightly they bind to the site), the inter-individual variability, and the therapeutic indices of the drugs [19]. A non-competitive inhibitor binds not to the active site, but to a different location on the enzyme, altering its conformation so that the active site is no longer fully functional. A non-competitive inhibitor needs only to be present in minimal amounts for inhibition to be effective.

4.3. Inhibition of CYP and Adverse Effects of Drugs

Inhibition of CYP450 is occasionally the intended pharmacological target of drugs. For example, the azole antifungal agents exert their antifungal effect by inhibiting fungal 14-α-demethylase, a CYP450 isoenzyme responsible for the conversion of lanosterol to ergosterol. Depletion of ergosterol lessens cell membrane integrity, and ultimately the viability of the pathogen. Unfortunately, azoles are poorly selective for human versus fungal CYP450. Currently available azoles are all inhibitors of CYP3A4, leading to adverse effects due to inhibition of CYP3A4-mediated metabolism of human steroid hormones. Inhibition of the conversion of lanosterol to ergosterol, and pregnenolone to a variety of adrenocorticoid and sex hormones, including cortisol, testosterone, and estrogen results in endocrine-related side effects such as the development of breasts in males, adrenocortical insufficiency, alopecia, and hypokalemia [5,20]. Newer azoles exhibit greater selectivity for fungal versus mammalian CYP3A4; consequently, they exhibit improved toxicity and drug interaction profiles.

4.4. In Vitro-In Vivo Scaling

The practice of using *in vitro-in vivo* scaling procedures to anticipate the effect of coadministration of CYP inhibitory agents on *in vivo* drug interactions has many limitations [21]. First, since *in vitro* studies are performed in human or animal preparations of liver microsomes, they do not account for the possible contributions of extrahepatic CYPs and transporter proteins [13]. Second, concentrations of the inhibitor at the enzyme active site *in vitro* may not mirror concentrations *in vivo* [22]. Third, *in vitro* experiments are usually conducted in media devoid of plasma proteins, which does not mimic the *in vivo* setting with respect to pH, ionic strength or protein binding, all of which can affect enzyme activity [23].

Finally, metabolites of the inhibitor may contribute to overall inhibitory effects on the enzyme. In general, the overall limitation is to make a given concentration of an inhibitor more potent *in vivo* than it is *in vitro* [22]. As such, human data should be obtained whenever possible [13]. However, *in vitro* drug interactions can be used as a screening tool to providing a rationale for further *in vivo* human investigations. In fact, case reports of interactions that occur in clinical practice often prompt more in-depth examination of drug interactions that can (and should) subsequently be confirmed in clinical trials [21].

4.5. Inducers of CYP Enzymes

Enzyme induction refers to an increase in enzyme activity, due to increased production of enzyme (through enhanced transcription and translation) or via a reduction in the natural rate of enzyme breakdown. Not all cytochrome enzymes are inducible; CYP1A2, CYP2C9, CYP2E1, and CYP3A4 are known to be inducible [15,24]. Unlike CYP inhibitors, which have an immediate effect, enzyme induction is a gradual process over time. Once an enzyme has been induced, removal of the inducing agent eventually will result in normalization of enzyme activity. Examples of common inducers include phenytoin, carbamazepine, rifampin, and ethanol. Predictably, these agents often results in increased metabolism of the other CYP substrates, which may result in therapeutic failures [25,26].

4.6. Genetic Polymorphism

A major cause of inter-individual variability of drug effects lies in the genetic polymorphism of CYP enzymes, in particular CYP2D6, CYP2C9, and CYP2C19, although genetic polymorphisms have been described for CYP1A2, CYP2D6, and CYP2E1 [9]. An individual's genotype is the combination of alleles on a chromosome that determines a specific trait or characteristic. A genetic polymorphism refers to a variation in the genotype in which a mutation in a gene results in an altered trait. A genetic polymorphism of a CYP gene results in a variation of enzyme activity. Polymorphism is typically a Mendelian trait with a population frequency of at least 1% (i.e., at least 1% of the population will demonstrate the varied form of the enzyme). The genetic polymorphism results in at least two distinct phenotypes, that is, at least two distinct biochemical characteristics of an enzyme [10].

Variant alleles, generally present in < 1% of individuals in a population, can result in enzyme activity that is increased, decreased, or absent; as compared to the rest of the population. Clinically, individuals with variant alleles are termed poor, intermediate, extensive or ultra rapid metabolizers. Individuals who are partially or completely deficient in an enzyme required for drug metabolism are termed 'poor metabolizers' (PMs); they have a decreased ability to metabolize the drug as compared to 'extensive metabolizers' (EMs), who possess a normal amount of enzyme. It is estimated that between 3–5% of Caucasians and African Americans and 15–20% of Asians are poor metabolizers of CYP2D6 and CYP2C19 (Figure 2). Individuals classified as PMs are homozygous for the autosomal recessive allele, while individuals classified as EMs are homozygous or heterozygous for the autosomal dominant allele. Ultrarapid metabolizers exhibit increased metabolic activity due to either duplicate or amplified alleles [5].

PMs generally have an increased response and a greater risk of toxicity after administration of small dosages of medications metabolized by the deficient enzyme [10]. If a drug is administered in its pharmacologically active form, and a CYP enzyme is required for its biotransformation, then a PM may experience more pronounced adverse effects and medication toxicity due to accumulation of the parent drug. For example, CYP2C9 is responsible for the metabolism of the anticoagulant S-warfarin, the active enantiomer of warfarin, to its inactive metabolite. The low but significant prevalence of PMs in the population contributes toward the variable anticoagulant response in patients taking warfarin:

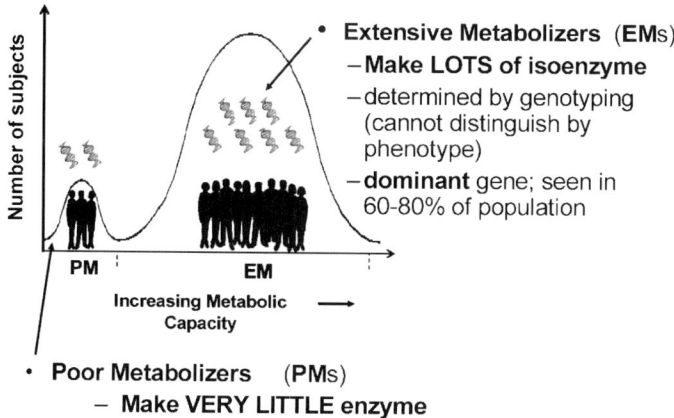

Figure 2. Genetic polymorphism of a CYP gene results in variations in the amount of enzyme activity in the population. 'Poor metabolizers' (PMs) have a decreased ability to metabolize a drug as compared to 'extensive metabolizers' (EMs), who possess a normal amount of enzyme.

PMs of CYP2C9 activity display an exaggerated response to normal doses of warfarin and are at increased risk of bleeding.

CYP2D6 is responsible for the metabolism of a wide variety of clinically important and widely utilized medications, including antidepressants, antipsychotics, and antihypertensives. Because routine screening for CYP2D6 phenotypic prevalence is not performed, PMs may be prescribed routine dosages of medications and risk the subsequent development of drug-related toxicity. Compared with EMs, PMs administered usual dosages of timolol ophthalmic solution experience higher plasma concentrations of medication and more pronounced cardiac effects [27]. Similar effects are of particular concern with usual dosages of neuroleptics (antipsychotic drugs, or tranquilizers, used to treat delusions and hallucinations) in PMs. PMs may be treated with inappropriately high doses, experience an increased risk of adverse effects, and comply less with the recommended course of therapy. However, in PMs, inhibition of CYP2D6 does not lead to significant effects because they lack an adequate amount of functioning enzyme. However, in EMs, inhibition of CYP2D6 can 'convert' an EM to a PM, in terms of their ability to metabolize drugs requiring CYP2D6. In the example described below (Section 6), inhibition of the conversion of codeine (an inactive drug administered orally to allow drug absorption into the bloodstream) to the active, pain-relieving drug morphine, is mediated rapidly via CYP2D6 in EMs. In PMs, or in EMs who have inhibition of CYP2D6, the conversion is blocked, and patients complain of a lack of pain relief despite 'normal' dosages of pain reliever. Although CYP3A4 drug metabolizing activity varies widely among individuals, significant polymorphisms have not yet been identified [13].

5. ACTIVE TRANSPORT OF DRUGS

In addition to CYP enzyme-mediated drug interactions, active transporters like P-glycoprotein (P-gp), the organic anion-transporting polypeptides (OATPs) such as the hepatic canalicular efflux transporter (MRP2; which transports drugs from hepatocytes into the bile ducts) play an important role in drug interactions. Active drug transporters play a key role by regulating access of drugs to the drug-metabolizing enzymes and controlling drug concentrations in the hepatocytes [16].

The most important of these proteins is P-gp, a plasma membrane-associated glycoprotein, which is a member of the ATP-binding cassette (ABC) transporter superfamily and a product of the multi-drug resistance 1 (MDR1) gene. Although variably expressed in the population, P-gp is present at high levels in the liver, kidney, pancreas, small intestine, colon, and adrenal gland as well as the capillary endothelium of the brain and testes. P-gp functions as an ATP-dependent efflux pump which excretes xenobiotics into bile, the gastrointestinal (GI) tract, and urine and prevents access to the central nervous system (CNS) by limiting transport across the blood-brain barrier. These actions result in lower plasma and cerebrospinal fluid (CSF) concentrations of xenobiotics, suggesting P-gp's role as a defense against xenobiotics [11]. In the gastrointestinal tract, P-gp is located within the brush border on the apical (luminal) surface of mature enterocytes; the colon has the highest P-gp expression, and the stomach and jejunum/ileum the lowest [28].

6. ENZYME-TRANSPORTER COOPERATIVITY

It has been observed that P-gp and CYP3A4 share some degree of substrate overlap and are colocalized in the small intestine, where they form a cooperative barrier that limits the oral bioavailability of xenobiotics and select drugs such as cyclosporine. Upon entering the enterocyte, a substrate of both P-gp and CYP3A4 may be absorbed directly into the systemic circulation, metabolized by CYP3A4 in the enterocyte, or secreted back into the intestinal lumen by P-gp. Drug pumped back into the lumen may be reabsorbed at a distal site and exposed again to any of the above three fates [14].

For drugs that are substrates of both CYP3A4 and P-gp, repeated pumping of substrate drugs out of the enterocytes by P-gp limits and regulates their access to metabolism by CYP enzymes, and prevents CYP3A4 from being overwhelmed by high drug concentrations in the enterocyte. The two proteins, therefore, function in concert to collectively reduce the intracellular concentration of xenobiotics [29,30].

P-gp can become saturated by high concentrations of P-gp substrate or inhibited by drugs; under these circumstances, passive diffusion becomes the rate-limiting step in drug absorption. Thus, increased doses of the P-gp substrate or inhibitor can lead to increases in the rate and extent of absorption of CYP substrate drugs. Drugs with high partition coefficients (highly lipophilic agents) can diffuse rapidly. Inhibition or saturation of P-gp has a greater effect on the oral bioavailability of more water soluble agents, as more lipophilic agents are able to rapidly diffuse across the enterocyte.

Coordinate up-regulation of CYP3A has been demonstrated in a cell culture model in response to some xenobiotics; however, other drugs have caused selective up-regulation of P-glycoprotein expression [31].

Despite this shared functionality, no correlation is found between intra-subject enterocyte P-gp levels and CYP3A4 in the small intestine and/or liver. Moreover, CYP3A4 and P-gp do not appear to be coordinately regulated [32].

7. USE OF PROBES TO QUANTITATE CYP ACTIVITY IN HUMANS

Given the wide interindividual variability in catalytic activity of CYP isoforms, a useful approach to characterizing such activity in an individual subject is phenotyping with an appropriate *in vivo* probe. This approach has been particularly valuable with CYP2C19 and CYP2D6, which exhibit genetic polymorphisms [5]. In principle, this approach should also be applicable to measuring CYP3A activity *in vivo* and alterations caused by drug interactions. In practice, however, there are several potentially complicating factors. A number of probes have been evaluated, and the FDA recently issued a 'white paper' outlining the preferred probes to use in the evaluation of new drugs. Probes capable of assessing the relative enzyme isoform activity in an individual could be utilized in the selection of optimal dosages of drugs, particularly in individuals who exhibit unusually high or low amounts of enzyme activity, due to genetic polymorphisms, induction or inhibition, or dietary or disease factors [21].

8. PRODRUGS

Although the process of drug metabolism normally causes medications to become less active and more readily removed from the body, metabolism occasionally results in a more active form of the drug. Common examples of this include the antihypertensive losartan and the analgesics codeine and acetaminophen.

Prodrugs are a form of pharmacotherapy in which the drug is administered as a pharmacologically inactive parent compound, which is then activated in the body, often catalyzed by CYP enzymes, to yield the active form responsible for the desired pharmacological effect. Prevention of this activation process can reduce or abolish the clinical effects of prodrugs. Coadministration of drugs that potentially result in inhibition of prodrug activation is a common problem in the clinical setting [12].

If an active metabolite is required for a clinical effect, then a PM may derive minimal benefit. For instance, codeine has relatively little pain-relieving ability on its own, but about 15% of absorbed codeine is changed through phase I metabolism via CYP2D6 into the pharmacologically active drug morphine. It is this transformation into morphine that accounts for most of the analgesic effects of codeine [4,6,33] Individuals with a poor analgesic response to codeine may be PMs with a low activity of CYP2D6. Approximately 5%–10% of white persons carry this CYP2D6 polymorphism and are classified as PMs.

9. THE EFFECT OF INTRAVENOUS VERSUS ORAL ADMINISTRATION ON DRUG INTERACTIONS

Because CYP3A4 is expressed both in the gastrointestinal tract wall and in the liver, inhibition of CYP3A4 can occur in both sites. Following oral drug administration, CYP3A in the enterocytes is exposed to all of the dose of a CYP inhibitor that is absorbed from the intestinal lumen, and intracellular drug concentrations are high. Subsequent drug transport from the gastrointestinal tract into the blood reduces this concentration, via metabolism in the enterocyte and dilution by distribution in the bloodstream. The extent of inhibition is dependent upon the route of administration (oral versus intravenous). Since inhibition of CYP3A4 within hepatocytes is probably less than that experienced within enterocytes, oral clearance (loss of drug due to the combined effects of hepatic and gastrointestinal metabolism) of a CYP substrate is likely to be affected to a greater extent than its systemic clearance (loss of the substrate solely from hepatic metabolism) [5,9].

For example, inhibition of CYP3A4-mediated metabolism of immunosuppressive agents (ISAs) is more pronounced when the ISA is administered orally rather than intravenously. Dual cooperativity of CYP enzymes and P-gp in the gastrointestinal tract contributes to the remarkable differences in effect on oral bioavailability of ISAs observed with administration of ketoconazole versus fluconazole: since ISAs are substrates of P-gp, and ketoconazole (but not fluconazole) is an inhibitor of P-gp, ketoconazole inhibition of P-gp-mediated ISA efflux results in greater increases in oral bioavailability of cyclosporine than fluconazole. For example, in five healthy subjects the clearance of oral or intravenous cyclosporine was decreased by 4.9- and 1.8-fold, with a 2.6-fold increase in overall oral bioavailability and 1.2-fold increase in hepatic bioavailability, before and after administration of five to nine days of oral ketoconazole 200 mg/day, respectively [34,35]. The use of fluconazole, a less potent inhibitor of CYP3A4, administered intravenously 400 mg/day, reduced the clearance of oral and intravenously administered cyclosporine by only 21 to 55% and 21%, respectively [31,36].

10. ADDITIONAL FACTORS AFFECTING DRUG INTERACTIONS

Both drug- and patient-related factors help determine susceptibility to potential drug interactions [32]. Factors such as genomic variability, gender, ethnicity, and the presence of disease states that alter plasma drug concentrations can affect the magnitude and clinical significance of drug interactions [13]. The effect of wide inter-patient variability in the absorption, distribution, metabolism, and clearance of drugs on the magnitude and clinical significance of drug interactions cannot be overemphasized [32].

10.1. Sex Differences

Sex difference can influence the magnitude of drug interactions between azoles and ISAs [32]. For example, tacrolimus (as immunosuppressant utilized by transplant patients to prevent organ rejection; it is a substrate of both CYP3A4 and P-gp) pharmacokinetics were compared with and without the concomitant administration of ketoconazole, a potent inhibitor of CYP3A4 and P-gp. Coadministration of oral ketoconazole and intravenous tacrolimus resulted in a significantly greater inhibition of tacrolimus clearance in the female than in the male subjects. By contrast, when both drugs were administered orally, a significantly greater increase in absolute bioavailability and higher blood levels were observed in the females [37]. Researchers theorize that these observed sex differences in drug levels could result from a lower metabolizing capacity of intestinal CYP3A4 microsomes in males than in premenopausal females or that higher P-gp activity in females results in more efficient CYP-mediated metabolism [32].

10.2. Ethnicity Differences

Ethnicity differences can affect drug pharmacokinetics due to the presence of increased CYP3A4 activity in the liver of certain ethnic groups [32]. For example, the oral bioavailability of tacrolimus in African-Americans is significantly lower than in non African-Americans, perhaps due to differences in intestinal P-gp and CYP3A4 metabolism rather than differences in hepatic metabolism [32,38]. These differences in oral bioavailability result in differences in dose-response: African-Americans require 37% higher daily dosages of tacrolimus to achieve plasma concentrations similar to those of Caucasian kidney transplant recipients [38].

10.3. Diet

Particular foods and overall nutrition play an important role in drug metabolism. Foods such as grapefruit juice, Sevilla (but not Valencia) oranges and watercress can inhibit certain metabolic enzymes, whereas other foods, such as the cruciferous vegetables (e.g., cauliflower, broccoli, etc.), can induce enzyme activity. Individuals with extremely poor nutritional status may lack the nutrients necessary for normal enzyme activity [39,40]. Further, diets excessively low or high in proteins or carbohydrates may alter the activity of specific enzymes [39–41].

10.4. Smoking

Aryl hydrocarbons in smoke are capable of inducing CYP1A2, which is thought to contribute to the activation of certain carcinogens, ultimately contributing to the formation of various cancers. Similar aryl hydrocarbons can also be found in charcoal-grilled foods, and other chemical toxins and pollutants in the environment can have similar effects, inducing and/or inhibiting particular enzymes [42,43].

10.5. Disease States

Finally, in addition to the above listed factors, the presence of inflammatory small bowel disease, cirrhosis, stress, infections, poor nutritional status and increased age are known to decrease the amount and activity of CYP3A4 present in tissues [5].

Liver disease, which can potentially reduce functional hepatic mass, can decrease drug metabolism. Patients with severe liver disease display decreased hepatic levels of total P450 and related enzyme activities, resulting in impairment of drug clearance and hepatic detoxification. P450s are clearly more susceptible to the effects of liver disease than are P450-reductase or phase II enzymes, which are preserved until the most advanced stages of liver failure [15]. In humans, significant impairment of P450 expression is found only in patients with severe liver disease, including hepatitis with liver failure and decompensated cirrhosis. Cirrhosis is characterized by densely connecting fibrosis and disruption of hepatic architecture by regeneration nodules. Changes in hepatic blood flow result from distortion of the hepatic microcirculation and eventually from portal hypertension, processes which lead to intrahepatic and extrahepatic 'shunting' of blood away from hepatocytes. This impairs the clearance of rapidly metabolized compounds by decreasing drug delivery to hepatocytes. Conversely, hepatic total P450 levels and related enzyme activities are within the broad normal range in uncomplicated cases of cholestasis, mild-moderate hepatitis, steatosis (fatty liver), and in cases of clinically compensated cirrhosis, and most cirrhotic patients remain responsive to the stimulation of hepatic drug metabolism by rifampicin and cigarette smoke [15].

11. HERBAL AND DIETARY EFFECTS ON CYP

11.1. Flavonoids

Due to an increased public interest in alternative medicine and disease prevention, the use of herbal preparations containing high doses of flavonoids for health maintenance has become very popular, raising the potential for interactions with conventional drug therapies. Flavonoids are present in fruits, vegetables and beverages derived from plants (tea, red wine), and in many dietary supplements or herbal remedies including Ginkgo biloba, soy isoflavones, and milk thistle. Flavonoids have been described as health-promoting, disease-preventing dietary supplements, and a high intake of flavonoids has been associated with a reduced risk of cancer, cardiovascular diseases, osteoporosis and other age-related degenerative diseases [44].

11.2. Foods

11.2.1. Grapefruit Juice

Grapefruit juice is an important dietary inhibitor of enteric (but not hepatic) CYP3A4. Grapefruit juice has been shown to contain nearly 200–850 mg/L of total flavonoids, among which naringin is the most abundant (145–638 mg/L). Acute or chronic consumption of as little as 4 oz of grapefruit juice in conjunction with orally administered substrates of CYP3A4 results in increased exposure to the CYP3A4 substrate [45] (Figure 3). Since the hepatic metabolism of an intravenously administered substrate is not affected, concentrations of the substrate are unaffected when the drug is administered with water or grapefruit juice. When the substrate is administered orally with water, plasma concentrations of the substrate are lower than observed with intravenous administration, due to first pass metabolism by gastrointestinal CYP3A4. However, oral administration of the substrate with grapefruit juice increases plasma concentrations of substrate due to inhibition of substrate metabolism by gastrointestinal tract membrane CYP3A4.

However, between patients, there is wide variability in the extent of the interaction with various CYP3A-metabolized drugs. For example, the bioavailability (i.e., the fraction of administered drug quantity which reaches the bloodstream) of cyclosporine may be unchanged, or as much as doubled following administration of a CYP3A4 inhibitor. Similarly, the bioavailability of felodopine (a drug used to treat high blood pressure) and terfenadine can increase insignificantly, or almost five-fold during concomitant administration with other CYP3A4-metabolized

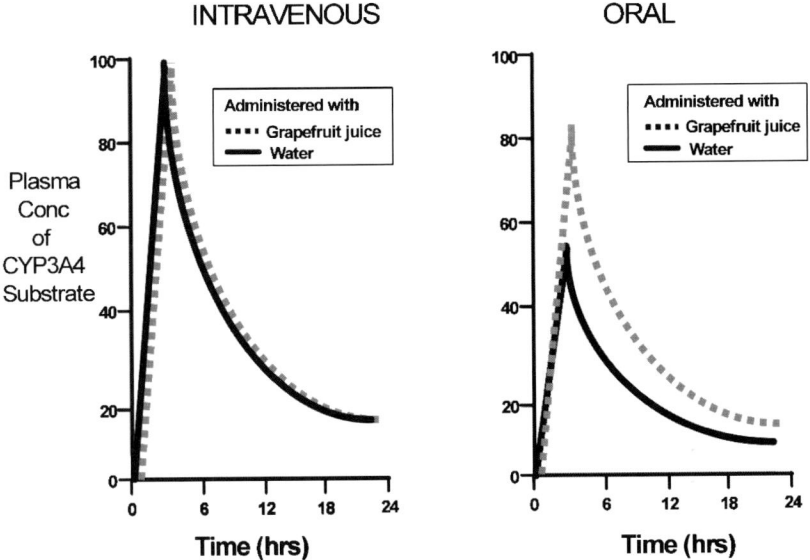

Figure 3. Example of the effect of grapefruit juice on the plasma concentrations of intravenously or orally administered CYP3A4 substrates. Grapefruit juice inhibits the metabolism of CYP3A4 in the gastrointestinal tract, but not the liver. Since the hepatic metabolism of an intravenously administered substrate is not affected, concentrations of the substrate are unaffected when the drug is administered with water or grapefruit juice. When the substrate is administered orally with water, plasma concentrations of the substrate are lower than observed with intravenous administration. However, oral administration of the substrate with grapefruit juice increases plasma concentrations of substrate due to inhibition of substrate metabolism as it crosses the gastrointestinal tract membrane.

drugs or inhibitors [1]. Since a predictable increase in drug exposure is not produced in each patient, they must be counseled to avoid consumption of any grapefruit juice while taking these medications.

11.2.2. Garlic

Garlic can induce CYP3A4 metabolism and/or P-gp. Administration of 2 garlic capsules twice daily can reduce the AUC of the protease inhibitor saquinavir by 51%; this could lead to resistance of HIV to saquinavir and loss of viral suppression [44].

11.3. Herbal Interactions

Many people mistakenly believe herbal medications to be free of adverse effects, including drug interactions. However, recent data suggests that many of these medications can cause clinically significant drug interactions, and it is important to remind patients, especially those taking CYP-metabolized drugs, of the potential for drug interactions with these 'benign' medications that are available over-the-counter, without a prescription [44].

11.3.1. St. John's Wort

St. John's Wort (*Hypericum perforatum*), a popular herbal medicine used for the treatment of depression, is a potent inducer of CYP3A4 following long-term (> 10 days) use. St. John's Wort has been implicated in decreased blood levels of cyclosporine (used to suppress organ rejection) and protease inhibitors (used for the treatment of HIV) resulting in several cases of organ transplant rejection and in HIV resistance [46].

12. INTERACTIONS WITH COMMONLY USED MEDICATIONS

12.1. Terfenadine

Terfenadine provides a classic example indicating the critical importance of drug metabolism and drug interactions in the development, regulation, and ultimate economic success of drugs. Terfenadine is normally metabolized by CYP3A4 to an active metabolite, fexofenadine. When this metabolism is inhibited, accumulation of terfenadine can result in 'torsades de pointe' a potentially fatal cardiac arrhythmia via blockade of cardiac potassium channels (Figure 4). Reaction with erythromycin was particularly common, as both medications tended to be co-prescribed for the treatment of pulmonary infections (erythromycin to treat the infection, and terfenadine as a nasal decongestant). As a mechanism-based (suicide) inhibitor of terfenadine, erythromycin resulted in serious adverse effects and even death in individuals, ultimately resulting in the withdrawal of terfenadine from world markets. However, an understanding of the metabolic mechanisms involved in these adverse effects led to the marketing of 'Allegra', an active metabolite of terfenadine which does not cause arrhythmia, as a safe and effective agent for the treatment of nasal congestion and allergic symptoms [47].

12.2. Drug Interactions with Macrolide Antibiotics

Macrolide-mediated CYP3A4 inhibition also has been implicated in adverse effects from benzodiazepines and 'statin' drugs used for the treatment of high cholesterol. CYP3A4 is responsible for the conversion of benzodiazepines, including midazolam and alprazolam to inactive metabolites. The inhibition of

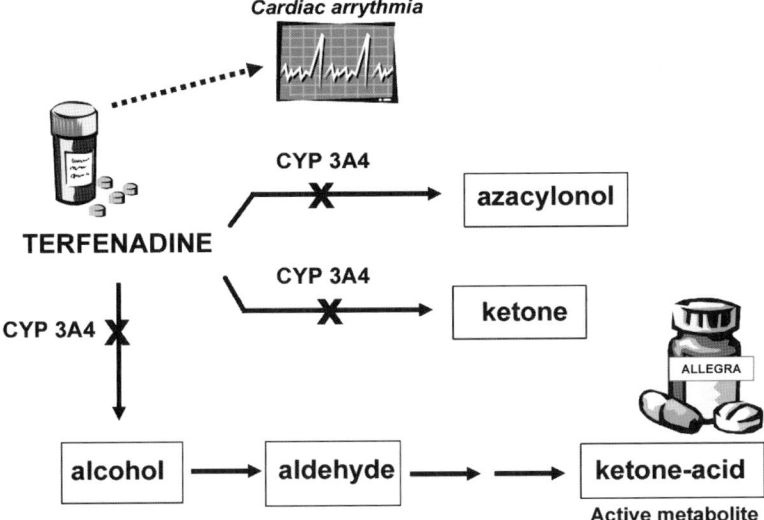

Figure 4. Interaction between CYP3A4 inhibitors and terfenadine. Terfenadine is normally metabolized by CYP3A4 to a number of inactive and active products, including the pharmacologically active ketone-acid metabolite, fexofenadine (Allegra®). When CYP3A4 metabolism is inhibited, accumulation of the parent drug, terfenadine can result in 'torsades de pointe' a potentially fatal cardiac arrhythmia via blockade of cardiac potassium channels.

CYP3A4 results in decreased clearance of the drug and increased sedation [48]. Similarly, inhibition of lovastatin metabolism produces dose-related toxic effects on skeletal muscle. In extreme cases, rhabdomyolysis (the breakdown of muscle fibers which can lead to severe pain and kidney damage) has been reported with the concomitant administration of lovastatin and macrolide antibiotics [49].

13. BENEFICIAL EFFECTS OF DRUG INTERACTIONS

13.1. Reducing the Cost of Expensive Medications

13.1.1. Immunosuppressive Agents

Although drug-drug interactions are usually considered undesirable events, they can also be beneficial. Therapy with immunosuppressive agents such as cyclosporine, tacrolimus, and sirolimus, which are used to prevent organ rejection following transplantation, can be cost-prohibitive. Reduction of daily dosages of ISAs by ketoconazole- or diltiazem-induced inhibition of their clearance can reduce therapeutic dosages of cyclosporine by 70–85%. This combination has proved to be well tolerated, effective and considered less costly than cyclosporine monotherapy [16,50].

A review of the clinical and economic saving potential of cyclosporine drug interactions concluded that ketoconazole appears to be the best candidate for reducing the financial pressures of chronic immunosuppressive therapy without sacrificing patients' well-being [50]. A five-year study of patients receiving a combination with ketoconazole and cyclosporine found no clinical difference in outcomes, including renal function, hepatic function, blood pressure, use of antihypertensive medications, or patient or graft survival. The authors concluded that the combined use of ketoconazole and cyclosporine could decrease the yearly per patient cost of cyclosporine by approximately $3,750, translating into a national annual savings of over $100 million in patients undergoing solid organ transplantation in the United States. Other reported advantages include a lower incidence of infections and acute rejection episodes in patients who received ketoconazole and cyclosporine versus placebo and cyclosporine [50].

13.1.2. Antiretroviral Therapy

Interactions between CYP enzymes have been widely exploited in the treatment of HIV infection. For example, the addition of low-dose ritonavir (a potent inhibitor of CYP3A4) significantly increases plasma concentrations of coadministered protease inhibitors (which are all CYP3A4 substrates), resulting in a decrease in the number of required pills and frequency of doses per day, fewer food and/or fluid restrictions, and a higher rate of virological suppression [51].

14. FDA REGULATIONS REGARDING CYP450-MEDIATED DRUG INTERACTIONS

Several commercially important drugs have been withdrawn from the US or worldwide market because of preventable adverse drug-drug interactions, including terfenadine, astemizole, mibefradil, and cisapride. Currently, a number of drugs have labeling based on the known interactions with other drugs metabolized by similar pathways. For example, many drugs have warning labels to the consumer regarding potential interactions with St. John's Wort or grapefruit juice, based on the metabolic and dispositional characteristics of the drugs being labeled despite a lack of actual *in vivo* studies characterizing the interaction [45,52].

As noted above, the FDA recently issued a 'white paper' outlining the preferred probes to use in the evaluation of new drugs. It is hoped that incorporation of these techniques early in the drug development process will assist in the identification of individuals at risk for altered efficacy or adverse drug effects due to CYP-mediated changes in drug metabolism resulting from drug interactions or genetic polymorphism [21].

15. CLINICAL SIGNIFICANCE OF DRUG INTERACTIONS

An important, but often overlooked, concept when discussing drug interactions is their clinical significance. Some drug interactions may be extremely serious, possibly resulting even in death, while other interactions are innocuous or even beneficial. Therefore, it is important to consider the significance of an interaction when assisting with a patient's care. Many things go into the determination of

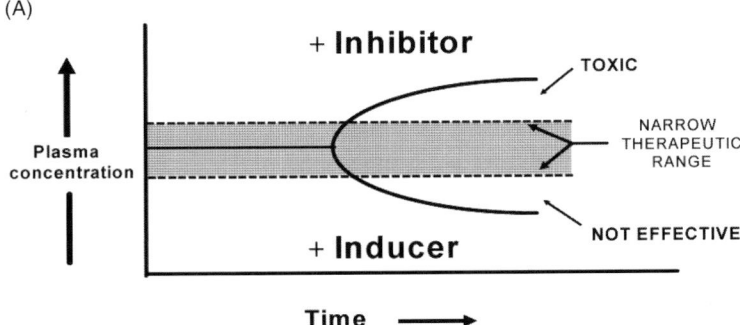

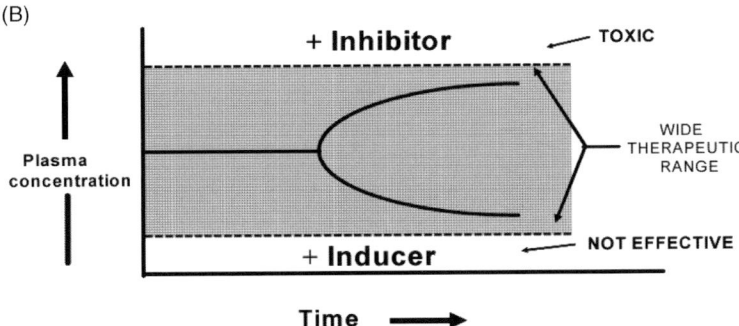

Figure 5. (A): The effect of an inhibitor or inducer of CYP enzymes on a drug with a narrow therapeutic range. Since the upper and lower limits of optimal plasma concentrations differ by a factor of only 2 or 3. Small increases or decreases in plasma concentrations result in decreased efficacy or increased toxicity. (B): The effect of an inhibitor or inducer of CYP enzymes on a drug with a wide therapeutic range. Larger variations in plasma concentrations usually occur before alterations in efficacy or toxicity are observed.

an interaction's clinical significance, and exactly how significant a particular interaction is may vary from one practitioner to another. Drugs for which CYP-mediated metabolism provides a major contribution (>30–50%) to the overall removal of drug from the body are more likely to be affected by changes in CYP metabolism than those for which CYP-mediated metabolism is limited [4,5].

One useful indicator of the significance of a change in drug metabolism which results in altered plasma concentrations is to evaluate the response in terms of the drug's therapeutic range. The therapeutic range refers to the range of plasma concentrations associated with effective pharmacologic effect, without undue toxicity. Some drugs (for example, digoxin, warfarin, cyclosporine) have a narrow therapeutic range; the upper and lower limits of optimal plasma concentrations differ by a factor of only 2 or 3. Small increases or decreases in plasma concentrations result in decreased efficacy or increased toxicity. For drugs with a wide therapeutic range, greater variation in plasma concentrations usually occur before alterations in efficacy or toxicity are noted (Figure 5). Although precise concentration limits are not usually definable, each drug has its own concentration-response range, although the range may differ with the specific disease being treated. Often, plasma concentration monitoring of drugs with narrow therapeutic ranges is necessary in order to maximize efficacy and minimize toxicity, particularly if potentially interacting drugs are introduced into a patient's drug regimen [4,5].

16. SUMMARY

As evidenced by both the large number of medications that are metabolized by and/or interact with metabolizing enzymes and the increasingly rapid rate at which new drugs are being approved, it is virtually impossible to predict and prevent all drug interactions. However, by understanding the mechanism by which most interactions occur and the large number of factors that can alter drug metabolism, many drug interactions can be minimized, prevented or detected early. Becoming familiar with commonly encountered medications that can alter enzyme activity allows the prediction and prevention of many interactions. Knowing which patients are at an increased risk for a drug interaction can also lead to closer monitoring as new drugs are added, allowing the prevention and/or early detection of drug interactions.

In the past, identification and evaluation of drug interactions was largely empirical, relying upon serendipitous clinical observations. However, recent knowledge about CYP-mediated metabolism provides opportunities for a more rational approach. The relative selectivity of currently available drugs to modulate CYP3A activity, based on clinical studies with model substrates, provides the basis for predicting their potential to interact with a new drug metabolized by the enzyme. Prediction is more challenging, however, when the goal is to identify and evaluate the potential *in vivo* inhibition/induction characteristics of a drug. Various *in vitro* approaches are being used, especially during drug development, to define inhibitor and induction potencies of a new chemical entity relative to other compounds. Minimally, such studies have the potential to identify interactions that are not likely to occur *in vivo* and, therefore, do not require clinical investigation. Such potency determinations should also allow an overall assessment of the probability of the new drug interacting with a CYP3A substrate. However, until that time, appropriately performed *in vivo* studies remain the ultimate means by which drug interactions and their clinical importance can be assessed.

ACKNOWLEDGMENT

Many thanks to Vincent L. Pecoraro for his thoughtful review and careful edits of this manuscript.

ABBREVIATIONS

ABC	ATP-binding cassette
AUC	area under the plasma-concentration time curve
CNS	central nervous system
CSF	cerebrospinal fluid
CYP	cytochrome P450
EM	extensive metabolizer
FDA	Food and Drug Administration
GI	gastrointestinal tract
ISA	immunosuppressive agents
MDR	multi-drug resistance
MRP2	hepatic canalicular efflux transporter
OATP	organic anion-transporting polypeptide
P-gp	P-glycoprotein
PM	poor metabolizer

REFERENCES

1. K. E. Thummel and G. R. Wilkinson, *Annu. Rev. Pharmacol. Toxicol.*, *38*, 389–430 (1998).
2. Committee on the Quality of Health Care in America. Institute of Medicine, in *To err is human: building a safer health system* (L. T. Kohn, J. M. Corrigan, and M. S. Donaldson, eds), National Academy Press, Washington, DC, 2000.
3. M. J. Ellenhorn and F. A. Sternad, *J. Am. Pharm. Assoc.*, *6*, 62–65 (1966).

4. M. Gosney and R. Tallis, *Lancet, 1*, 564–567 (1984).
5. A. Saad, D. D. DePestel, and P. L. Carver, *Pharmacotherapy*, (2006) in press.
6. J. Desmeules, M. P. Gascon, P. Dayer, and M. Magistris, *Eur. J. Clin. Pharmacol., 41*, 23–26 (1991).
7. J. G. Gillum, D. S. Israel, and R. E. Polk, *Clin. Pharmacokinet., 25*, 450–482 (1993).
8. Drug interactions: defining genetic influences on pharmacologic responses. Available at: http://medicine.iupui.edu/flockhart Accessed June 9, 2006.
9. R. Sikka, B. Magauran, A. Ulrich, and M. Shannon, *Acad. Emerg. Med., 12*, 1227–1235 (2005).
10. K. Venkatakrishnan, L. L. Von Moltke, and D. J. Greenblatt, *J. Clin. Pharmacol., 41*, 1149–1179 (2001).
11. Y. Zhang and L. Z. Benet, *Clin. Pharmacokinet., 40*, 159–168 (2001).
12. T. Tirkkonen and K. Laine, *Clin. Pharmacol. Ther., 76*, 639–647 (2004).
13. G. K. Dresser, J. D. Spence, and D. G. Bailey, *Clin. Pharmacokinet., 38*, 41–57 (2000).
14. S. D. Hall, K. E. Thummel, P. B. Watkins, K. S. Lown, L. Z. Benet, M. F. Paine, R. R. Mayo, D. K. Turgeon, D. G. Bailey, R. J. Fontana, and S. A Wrighton, *Drug Metab. Dispos., 27*, 161–166 (1999).
15. J. A. Hasler, R. Estabrook, M. Murray, I. Pikuleva, M. Waterman, J. Capdevila, V. Holla, C. Helvig, J. R. Falck, G. Farrell, L. S. Kaminsky, S. D. Spivack, E. Boitier, and P. Beaune, *Mol. Aspects Med., 20*, 1–137 (1999).
16. C. L. Smith, E. M. Hampton, and J. A. Pederson. *Pharmacotherapy, 14*, 471–481 (1994).
17. S. Zhou, S. Y. Chan, B. C. Goh, E. Chan, W. Duan, M. Huang, and H. L. McLeod, *Clin. Pharmacokinet., 44*, 279–304 (2005).
18. M. A. Gibbs, K. E. Thummel, D. D. Shen, and K. L. Kunze, *Drug Metab. Dispos., 27*, 180–187 (1999).
19. E. Albengres, H. Le Louet, and J. P. Tillement, *Drug Saf., 18*, 83–97 (1998).
20. C. A. Kauffman and P. L. Carver, *Adv. Pharmacol., 39*, 143–189 (1997).
21. K. A. Bachmann and J. D. Lewis, *Ann. Pharmacother., 39*, 1064–1072 (2005).
22. J. M. Neal, K. L. Kunze, R. H. Levy, R. A. O'Reilly, and W. F. Trager, *Drug Metab. Dispos., 31*, 1043–1108 (2003).
23. R. J. Bertz and G. R. Granneman, *Clin. Pharmacokinet., 32*, 210–258 (1997).
24. J. H. Lin and A. Y. H. Lu, *Clin. Pharmacokinet., 35*, 361–390 (1998).
25. G. Apseloff, D. M. Hilligoss, M. J. Gardner, E. B. Henry, P. B. Inskeep, N. Gerber, and J. D. Lazar, *Clin. Pharmacol., 31*, 358–361 (1991).
26. W. S. Tucker, B. B Snell, D. P. Island, and C. R. Gregg, *J. Am. Med. Assoc., 253*, 2413–2414 (1985).
27. T. I. Edeki, H. He, and A. J. Wood, *J. Am. Med. Assoc., 274*, 1611–1613 (1995).
28. A. Lo and G. J. Burckart, *J. Clin. Pharmacol., 39*, 995–1005 (1999).
29. U. Christians, T. Strom, Y. L. Zhang, W. Steudel, V. Schmitz, S. Trump, and M. Haschke, *Ther. Drug Monit., 28*, 39–44 (2006).
30. L. Z. Benet, C. L. Cummins, and C. Y. Wu, *Curr. Drug Metab., 4*, 393–398 (2003).
31. K. Venkatakrishnan, L. L. Von Moltke, and D. J. Greenblatt, *Clin. Pharmacokinet., 38*, 111–180 (2000).
32. U. Christians, W. Jacobsen, L. Z. Benet, and A. Lampen, *Clin. Pharmacokinet., 41*, 813–851 (2002).

33. L. Poulsen, K. Brosen, L. Arendt-Nielsen, L. F. Gram, K. Elbaek, and S. H. Sindrup, *Eur. J. Clin. Pharmacol., 51*, 289–295 (1996).
34. E. J. Wang, K. Lew, C. N. Casciano, R. P. Clement, and W. W. Johnson, *Antimicrob. Agents Chemother., 46*, 160–165 (2002).
35. D. Y. Gomez, V. J. Wacher, S. J. Tomlanovich, M. F. Hebert, and L. Z. Benet, *Clin. Pharmacol. Ther., 58*, 15–19 (1995).
36. C. L. Osowski, S. P. Dix, L. S. Lin, R. E. Mullins, R. B. Geller, and J. R. Wingard, *Transplantation 61*, 1268–1272 (1996).
37. S. Tuteja, R. R. Alloway, J. A. Johnson, and A. O. Gaber, *Transplantation, 71*, 1303–1307 (2001).
38. W. E. Fitzsimmons, I. Bekersky, D. Dressler, K. Raye, E. Hodosh, and Q. Mekki, *Transplant. Proc., 30*, 1359–1364 (1998).
39. I. Walter-Sack and U. Klotz, *Clin. Pharmacokinet., 31*, 47–64 (1996).
40. F. P. Guengerich, *Am. J. Clin. Nutr., 61(suppl)*, 651S–658S (1995).
41. J. Hong, J. Pan, F. J. Gonzalez, H. V. Gelboin, and C. S. Yang, *Biochem. Biophys. Res. Commun., 142*, 1077–1083 (1987).
42. D. L. Eaton, E. P. Gallagher, T. K. Bammler, and K. L. Kunze, *Pharmacogenetics, 5*, 259–274 (1995).
43. F. P. Guengerich and T. Shimada, *Mutat. Res., 400*, 201–213 (1998).
44. M. L. Chavez, M. A. Jordan, and P. I. Chavez, *Life Sci., 78*, 2146–2157 (2006).
45. A. Dahan and H. Altman, *Eur. J. Clin. Nutr., 58*, 1–9 (2004).
46. J. S. Markowitz, J. L. Donovan, C. L. DeVane, R. M. Taylor, Y. Ruan, J. S. Wang, and K. D. Chavin, *J. Am. Med. Assoc., 290*, 1500–1504 (2003).
47. R. L. Woosley, *Annu. Rev. Pharmacol. Toxicol., 36*, 233–252 (1996).
48. J. F. Westphal, *Br. J. Clin. Pharmacol., 50*, 285–295 (2000).
49. J. W. Grunden and K. A. Fisher, *Ann. Pharmacother., 31*, 859–863 (1997).
50. J. E. Martin, A. J. Daoud, T. J. Schroeder, and M. R. First, *Pharmacoeconomics, 15*, 317–337 (1999).
51. B. Motwani and W. Khayr, *Am. J. Ther., 13*, 57–63 (2006).
52. S. M. Huang, S. D. Hall, P. Watkins, L. A. Love, C. Serabjit-Singh, J. M. Betz, F. A. Hoffman, P. Honig, P. M. Coates, J. Bull, S. T. Chen, G. L. Kearns, M. D. Murray, and Center for Drug Evaluation and Research and Office of Regulatory Affairs, Food and Drug Administration, Rockville, MD, USA, *Clin. Pharmacol. Ther., 75*, 1–2 (2004).

Subject Index

A

Acenaphthene
  oxidation, 459
  structure, 454
Acetaminophen, 565, 580, 597, 604
  bioactivation, 577
  metabolism, *see* Metabolism
  oxidation, 489
  toxicity, 577
Acetone, 489
Acetylcholine esterase, 118, 119, 534
Acidity constants (of)
  coordinated water, 198
Acetinobacter sp., 527, 528
ACTH, *see* Adrenocorticotrophin
Activation (of)
  dioxygen, *see* Dioxygen
  enthalpy for carbon monoxide binding to P450, 210
  entropy for carbon monoxide binding to P450, 210
Active site of P450s, *see* P450s
Adamantane, 38, 44, 47, 174, 175, 204, 205, 209, 210
  acetamide, 178
  wire, 178
Adenosine monophosphate, *see* AMP
Adenosine 5'-triphosphate, *see* ATP
Adipose tissue
  steroid hormone biosynthesis, 370, 382
Adrenal cortex
  steroid hormone biosynthesis, 366–369, 372, 374–380
Adrenocorticosteroids, 600
Adrenocorticotrophin, 375–377, 380
Adrenodoxin, 8, 136, 364, 365, 367–369, 371, 372, 381, 382, 384
  phosphorylation, 372
  reductase, *see* Reductases
Aflatoxins, 418, 485, 487, 488, 490, 497
  biosynthesis, 417
  structure, 418
African Americans, 601, 606
Alcanivorax borkumensis, 528
Alcohols (*see also* individual names), 416–419
  keto, 403–408
  perillyl, 531
Aldehyde(s) (*see also* individual names), 408–415
  isocaproic, 398
  oxidation, 412
Aldosterone, 364, 367, 374, 384
  formation, 365–367, 375
  structure, 363
  synthase, *see* Synthases
Aldrin
  epoxidation, 509–511
Alectoris rufa, 490
Algae (*see also* individual names), 47
  blue green, 18

Algae (*see also* individual names) (Continued)
  green, 513
  toxins, 100, 119
Alkaloids, 417
  phytotoxins, 505
Alkanes (*see also* individual names), 446, 447
  hydroxylation, *see* Hydroxylation
  oxidation, 446, 459, 462, 525–529
Alkene
  epoxidation, 158, 343, 453
Allegra®, *see* Fexofenadine
Allelochemicals
  detoxification, 504, 505, 507
  plant, 504, 505, 507, 513, 520
Alopecia, 600
Amines (*see also* individual names)
  aromatic, 483
  heterocyclic, 483
Amino acid radicals, *see* Radicals
Aminopyrine, 514
AMP
  cyclic, *see* cAMP
Amphibians (*see also* individual names), 487
  chemical defense, 490, 491
  P450 regulation, 491–493
Anabaena variabilis, 18
Androstane diol, 44
Androstenedione, 323, 363, 369, 380, 381, 409–411
  hydroxylase, 463, 489
  structure, 410
Androsterone
  dehydroepi-, 363, 369, 380
  dehydroiso-, 403
  structure, 363
Angiotensin II, 383, 384
Animal(s) (*see also* individual names)
  models for drug metabolism, 586–570
  –plant chemical warfare, 478, 498, 499, 504, 505, 507
Annelids (*see also* individual names)
  chemical defense, 510, 511
  P450s, 101, 103
Anopheles gambiae
  genome, 4, 8
Anthracene
  9-methyl-, 454, 459
  dimethylbenz-, *see* Dimethylbenzanthracene
  oxidation, 459
  structure, 454
Anthropogenic chemicals, 103, 120
Antibacterials (*see also* individual names), 599
Antibiotic(s) (*see also* individual names), 579, 596
  antifungal treatment, 287
  macrolide, 485, 610–612
  resistance to azoles, 302
  synthesis, *see* Synthesis
Antibodies, 580, 581
Anticancer drugs (*see also* individual names), 467, 531, 599
  oxidation, 463
Anticoagulants (*see also* individual names), 601
Antidepressants, 602, 610
Antifungals (*see also* individual names), 300, 301, 303, 579
  azoles, 300–305, 598, 599
Antihistamine
  non-sedating, 578
Anti-HIV agents (*see also* individual names), 599
Antihypertensives (*see also* individual names), 599, 602, 604
Antimicrobial agents (*see also* individual names), 401
Antioxidants (*see also* individual names), 14
Antipsychotics, 602
Antisense oligonucleotides, 377
Antley-Bixler syndrome, 371
Apis mellifera
  genome, 4, 8
  P450, 4, 5
Apoptosis, 298, 369
Arabidopsis sp., 14
  P450 genes, 58
  thaliana, 4–6, 8, 9, 408, 418, 423, 514, 519, 520

SUBJECT INDEX

Arachidonic acid, 39, 247, 426, 486, 488
  metabolism, see Metabolism
  oxidation, 333
Archaea (see also individual names),
    86–91
  acidothermophilic, 87
  hyperthermophilic, 86
  P450s, 4, 86–90
Arenes
  hydroxylation, see Hydroxylation
Aromatase (see also CYP19), 104, 323,
    343, 381, 382, 409–412, 414, 429
  inhibition, 300, 301, 371, 374
  mechanism, 410, 411
Aromatic hydrocarbons, 14, 15, 99, 118,
    119, 453, 486
  halogenated, 496
  hydroxylation, see Hydroxylation
  oxidation, 447, 448, 459, 532
  polycyclic, 453, 483, 496, 497, 509,
    511, 532, 533
  structures, 448
Arrhythmias, 578, 596, 610
Artemisinin, 467
Arthrobacter globiformis, 171
Arthropods (see also individual names)
  P450s, 101, 105–108
Ascaris suum, 238
Ascidians (see also individual names), 498
Ascochyta pisi, 521
Ascorbate, 173, 352
  peroxidase, see Peroxidases
Asian population, 573, 601
Aspergillus parasiticus, 417
Association constants (see also Binding
    constants), 43, 47
Assays
  Comet, 118, 119
  fluorescence, see Fluorescence
  luminescence, see Luminescence
  micronucleus, 118, 119
Asterias rubens, 108, 508
Atomic force microscopy
  studies of P450s, 76
ATP, 475, 512, 607, 631, 637
  -binding cassette transporter, 603
  production, 320

Atrazine, 104, 510, 512
Autoxidation, 211
  rate constants, see Rate constants
Azoles
  as antifungal drugs, 300–305, 598, 599
  as inhibitors of P450s, 300–305, 382,
    606
  heme binding, 304
  structures, 301, 304

B

Bacillus
  megaterium, 7, 17, 87, 160–162, 291,
    306, 331, 332, 442, 453, 527
  subtilis, 7, 17, 291, 401, 461, 462, 527
  thuringiensis, 119
Bacteria(l) (see also individual names), 4,
    8, 59, 520
  P450 genes, 99
  P450s, see P450s
Balanus eburneus, 106
Baeyer-Villiger pathway, 404, 405, 408,
    409, 412, 413, 417, 418
Barbiturates, 309, 565, 573, 579, 580
Barium(II)
  crown ether complexes, 34
Barley, 418
Barnacles (see also individual names) 106
Benz[a]anthracene, 103, 511
Benzenes
  biodegradation, 447, 449
  ethyl-, 175, 176
  polychlorinated, 447–450
  trichloro-, see 1,3,5-Trichlorobenzene
Benzodiazepines, 610
Benzo[a]pyrenes, 14, 102–108, 447, 506,
    507, 511, 512, 514
  hydroxylation, see Hydroxylation
  metabolism, see Metabolism
Benzoic acid
  m-chloroperoxy-, see
    m-Chloroperoxybenzoic acid
Benzphetamine, 209, 331, 467
  demethylation, see Demethylation
  metabolism, 507
  structure, 330, 464

Bifonazole
  structure, 301, 304
Binding constants
  azoles/P450, 302
  heme/nitric oxide, 297
  imidazoles/P450, 300
  P450, 288
Bioavailability of
  cyclosporin, 608
  drugs, 563, 564, 567, 580, 595, 599, 604, 606, 608
  felodopine, 608
  imunosuppressive agents, 605
  terfenadine, 608
Biocatalysts, 530
Biodegradation of
  polychlorinated benzenes, 447, 449
  vitamin D_3, 463
Biomonitors or biomonitoring studies, 103, 106, 118–120, 481, 487, 488, 494, 497, 511
Biomphalaria glabrata, 105, 110, 511
Bioreaktor, 531, 532
Bioremediation, 165
  by bacterial P450s, 532
  of explosives, 533
Biosensors, 158, 165, 533, 534
Biosynthesis (of) (*see also* Synthesis)
  aflatoxins, 417
  aldosterone, 367, 372, 383, 385
  biotin, 401
  brassinolide, 15, 17, 408, 409
  brassinosteroids, 16, 17
  cAMP, 375
  cholesterol, 371, 400
  corticoid, 383
  cortisol, 308, 367, 372
  ecdysone, 16, 17
  erythromycin, 248
  estradiol, 16, 380
  estrogen, 11, 287, 300, 343, 370, 379, 381
  estrone, 323, 380
  gibberellins, 418, 419, 429
  hormones, 16, 17, 27, 158
  pregnonolone, 363, 365, 374–376, 379, 380, 398–400

  progesterone, 379–381, 403
  prostaglandins, 426–429
  psoralen, 407, 416, 417
  reaction types for steroid hormones, 365–371
  steroid hormones, 16, 17, 106. 107, 322, 361–386, 403, 409, 411, 463
  sterol, 4
  testosterone, 378–381
  thromboxane, 428, 429
  vitamin D, 384, 400, 463
Biotechnological application of P450s, 18, 286
Biotin, 401
Biphenyls
  polychlorinated, *see* PCBs
Birds
  chemical defense, 488–490
  P450s, 488–490
Blood pressure regulation, 384
Bombix mori, 8, 9
Bortezomib, 11
Bradyrhizobium japonicum, 527
Brain
  steroid hormone biosynthesis, 370, 382, 383
Brassinolide
  biosynthesis, 16, 17, 408, 409
Brassinosteroids
  biosynthesis, 16, 17
Breast
  cancer, 373, 374, 382
  cancer treatment, 300, 301, 371, 374, 382
Bunodosoma cavernata, 102
Buproprion
  hydroxylation, *see* Hydroxylation
Burkholderia, 528
Butler-Volmer theory, 130

C

Caenorhabditis elegans, 91, 480
  chemical defense, 509
  genome, 4, 8, 509

Caffeic acid, 14
Caffeine, 573, 597
 demethylation, see Demethylation
Calcitroic acid, see Vitamin D_3
Calcium(II)
 activation of fluxes, 290
 channel blocker (see also individual names), 581
Caldariomyces fumago, 81, 216, 307
Caldariomycin, 81
Callinectes sapidus, 106
Calmodulins, 290
Camalexin
 synthesis, 14
cAMP, 376, 378, 381, 382
 biosynthesis, see Biosynthesis
 -dependent phosphorylation, 372
 -dependent transcription of genes, 378
 response system, 377
Camphane, 199, 200, 205, 209, 210
Camphor, 29, 30, 145, 176, 197–199, 201, 204, 205, 212, 444
 1R-, 199, 201, 205, 209, 221, 323
 1S-, 199, 205
 5-hydroxy, 30, 238, 241
 bromo-, 205, 210
 hydroxylation, see Hydroxylation
 nor-, see Norcamphor
 P450 monooxygenase, see P450cam
 -quinone, 199
 structure, 73
Cancer
 breast, see Breast
 hormone-dependent, 371
 prostate, 373, 374
Candida
 maltosa, 18, 524
 tropicalis, 524
Capitella capitata, 103, 111, 511
Capric acid, 462
 oxidation, 459
 structure, 454
Carbamate(s)
 insecticides, 118, 490
 thio-, see Thiocarbamates
Carbon
 ^{13}C, 303
 radical, see Radicals
Carbon–carbon bond cleavage
 alpha to acids, 419–421
 alpha to alcohols, 416–419
 alpha to aldehydes, 408–415
 alpha to carbon bearing nitrogen, 422, 423
 alpha to ethers, 421, 422
 alpha to ketones, 408
 by P450s, 397–430
 diols, 398–403
 involving fatty acids, 423–425
 keto alcohols, 403–408
Carbon dioxide, 415, 420
Carbon monoxide, 12, 98, 192, 193, 203–211, 287, 289, 305–308, 424
 ^{13}CO, 32
 binding to Fe(II), 31, 38, 51, 82, 158, 306, 307, 322
 myoglobin complex, 203
 P450 complexes, see P450s
 Re complex, 175
 sensing transcriptional activators, 268
 sensor for the heme pocket, 203–211
 spectral properties of the P450 complexes, 203–208
Carcinogenesis, 99, 158, 182, 463
Carcinus
 aestuarii, 106
 maenas, 107, 110, 508
Cardiovascular diseases, 384
Carnegiae gigantea, 505
Carotenoids, 87
 epoxidation, see Epoxidation
Castasterone
 biosynthesis, 16
 oxidation, 408, 409
Catalases, 179, 320
Catfish, 496
Catharanthus roseus, 417
Caucasian population, 571, 601, 606
Caulobacter crescentus, 527
cGMP, 290, 299
Chemical defense (of), 477–535
 amphibians, 490, 491
 annelids, 510, 511

Chemical defense (of) (Continued)
  birds, 488–490
  Caenorhabditis elegans, 509
  cnidarians, 512
  crustaceans, 507, 508
  definition, 479, 480
  echinoderms, 508, 509
  evolution, 480
  fish, 493–497
  fungi, 521–525
  insects, 497–507
  invertebrates, 497–512
  mammals, 481–487
  molluscs, 511, 512
  nematodes, 509
  plants, 513–521
  poriferans, 512
  prokaryotes, 525–529
  protista, 513
  reptiles, 490, 491
Cherax quadricarinatus, 508
Chiral synthons, 286
Chlamydomonas reinhardtii, 4
Chloramphenicol, 309
m-Chloroperoxybenzoic acid, 39–42, 217, 218, 222, 325, 352
Chloroperoxidase, 12, 40, 50, 51, 66, 81, 169, 170, 179, 216–218, 252, 270, 273, 294, 307
  structure, 81, 268
Cholesterol, 209, 212, 295, 322, 610
  cleavage, 8, 365, 367
  conversion to pregnenolone, 363, 365, 374–376, 379, 380, 398–400
  ester hydrolase, 376
  structure, 362, 363
Chromatography
  high-performance liquid, *see* High-performance liquid chromatography
  ion exchange, 105
Chryptochiton stelleri, 105
Chymotrypsin, 70
Cigarette smoke (*see also* Tobacco), 572, 607
Cineole, 72, 239
  hydroxylation, *see* Hydroxylation
  structure, 73
Cinnamate hydroxylase, 519, 520
Ciona intestinalis, 498
Circular dichroism
  magnetic, *see* Magnetic circular dichroism
  P450s, 191, 192
Citrobacter braakii, 72
Citrulline, 82, 291
Clams (*see also* individual names), 512
  P450s, 111
Clarithromycin, 597, 599
Cleavage
  cholesterol, *see* Cholesterol
  diols, *see* Diols
  heterolytic, 279, 325, 333–335, 399, 439
  homolytic, 279, 325, 399, 404, 427
  O–O bond, 81, 213–215, 224, 251, 259, 273, 279, 325, 333–335, 337, 405, 414, 420, 427, 439, 446
  pregnenolone, 408
  progesterone, 408
  steroids, 408
Clofibrate, 103, 104, 486, 506, 510, 511
Clotrimazole, 497
  structure, 301
Clusters
  2Fe-2S (*see also* Ferredoxins), 7, 8, 74, 367, 368, 537, 538
  iron-sulfur, 8, 66
Cnidarians
  chemical defense, 512
  detoxification mechanism, 102, 103
  P450s, 101–103
Cobalt, 165, 167
  ^{60}Co, 349
  porphyrins, *see* Porphyrins
Cobalt(III)
  sepulchrate cage complex, 167
Cobaltocene, 167
Codeine, 602, 604, 605
Cofactors
  P450, *see* P450
Comamonas testosteroni, 528
Compound 0
  definition, 335

Compound I, 11, 29, 30, 32, 36, 39, 40, 52, 67, 68, 81, 82, 128, 164, 165, 171, 196, 217, 225, 320, 325, 326, 333, 334, 338, 341–343, 352, 352, 440
  analogues, 46
  characterization, 168
  definition, 169, 216, 218, 251, 345, 439
  $Fe^V=O$, 163, 346, 439
  formation, 439
  oxo iron(IV) porphyrin radical cation, 39–41, 159, 251, 345
  oxygen insertion, 37
  reactivity, 48
  structure, 345
  two states model, 347, 348
Compound II, 325
  definition, 169, 215
  $Fe^{IV}=O-H$, 169–171, 179, 180, 425
Congenital lipoid adrenal hyperplasia, 366, 367, 370, 376
Contraceptives (*see also* individual names)
  oral, 579
Copper(II), 46
Corals (*see also* individual names), 512
  Gorgonian, 493, 497
  P450s, 102, 103
Corticosteroids
  adreno-, 600
  commercial production, 523
Corticosterone, 369
  structure, 363
Cortisol, 364, 366, 369, 374, 600
  11-deoxy-, 365
  biosynthesis, *see* Biosynthesis
  hydroxylation, *see* Hydroxylation
Corynebacterium sp., 527
Coscinasterias muricata, 509
Cotton, 514
Coumarin(s), 72
  7-methoxy-, 518
  7-methoxy-4-trifluoromethyl-, 464, 466, 484, 496
  alkoxy-, 481
  demethylation, *see* Demethylation
  ethoxy-, 106, 507, 518
  furano-, 14, 17, 117, 416, 504, 505, 520

  hydroxylation, *see* Hydroxylation
  O-dealkylation, 485
  propoxy-, 485
  structure, 464
Crabs (*see also* individual names), 106, 110, 508
Crayfish (*see also* individual names)
  P450s, 102, 105–107, 508
Crocodile, 491
  meat, 487
Crown ethers
  barium(II) complex, 34
Crustaceans (*see also* individual names), 100, 118
  chemical defense, 507, 508
  P450s, 101, 102, 110, 111
Cryo crystallography, 191
Crystal structures of
  chloroperoxidase, 81
  CYP51, 69
  CYP119, 71, 72, 86–89, 304
  CYP121, 268, 269
  CYP158A2, 258, 442–444
  ferredoxin, 66
  P450 3A4, 78
  P450 BM3, 88
  P450 BSβ, 85
  P450 distal side, 275, 276
  P450cam, 88, 197
  P450nor, 83
  P450 models, 32, 36
  P450s, 6, 62–64, 269
  rubredoxin, 66
Cucumaria miniata, 508
Cushing's disease, 309
Cyanide
  as P450 inhibitor, 287–289, 308
Cyclic voltammetry (studies of) (*see also* Voltammetry), 140–144, 162
  P450 models, 32
β-Cyclodextrins
  Mn porphyrin, 44–46
  Ru porphyrin, 47
  substrate recognition, 43, 45, 46
Cyclohexane
  phenyl-, 448, 450
  structure, 448

Cyclohexene
  oxidation, 347
Cyclophosphamide, 463, 467
  structure, 464
Cyclosporin, 485, 597, 603, 605, 606,
  610–612, 614
  bioavailability, 608
  elimination, 598
Cylindrocarpon tonkinese, 296
CYP1 family, 105, 109–113, 463, 483,
  488, 492, 495, 508, 512
  A2, 243, 465, 466
CYP2 family, 105, 109, 110, 112–116,
  309, 463, 483–486, 488–490, 492,
  494, 495, 512
  inhibition, 302
CYP2A4 (= P450 15α), 72
CYP2A5 (= P450 COH), 72
CYP2A6, 466
CYP2B1, 466, 467
CYP2B4, 331, 333, 340, 344, 348, 410,
  413, 420, 463
  active site, 329
  dioxygen complex, 330
  mutants, 333, 347, 414
  redox potential, *see* Redox potentials
  structure, 329
CYP2B5, 442, 463
CYP2D6, 574, 582, 582, 596, 602
CYP3 family, 103–105, 109–111,
  113–116, 270, 485, 486, 489,
  494–497, 508
  mutation, 278
CYP3A4 (*see also* P450 3A4), 236, 278,
  279, 301, 302, 309, 422, 463, 467,
  574–576, 595–598, 600, 606,
  608–610
  activity, 595, 596
  inhibitors, 575, 606, 608, 610–612
CYP3A5, 463
CYP4 family, 4, 103, 105, 107, 108, 110,
  111, 114–116, 189, 270, 271, 486,
  498, 508, 512
CYP4A, 222, 453
CYP4B1, 268, 486
CYP5, 426–429
CYP8, 426–429

CYP11A1, *see* P450scc
CYP11B1 (= P450 11β), 364, 366, 367,
  372, 374, 375, 384, 385
  inhibition, 374
  mutation, 370
CYP11B2, *see* P450aldo
CYP17 (= P450c17), 363, 369, 370,
  373–375, 377–385, 413, 414
  inhibition, 374
CYP17A, 404, 405, 408, 410
CYP19 (*see also* Aromatase), 369–371,
  374, 380–383, 414
  inhibition, 374, 382
CYP21 (= P450c21), 363, 364, 369, 370,
  383
CYP24, 406, 408
CYP51 (= P450 14DM = lanosterol
  14α-demethylase), 4, 59, 60,
  304–306, 363, 409, 411, 412, 414,
  429
  crystal structure, *see* Crystal structures
  inhibition, 300, 302
CYP55A1, *see* P450nor
CYP74 family, 423–426
CYP85A family, 408
CYP88A, 418, 419
CYP93C, 421, 422
CYP101, *see* P450cam
CYP102, *see* P450 BM3
CYP107A1, *see* P450eryF
CYP107H1 (= P450 BioI), 400, 401
CYP108, *see* P450terp
CYP111 (= P450lin), 526
CYP119, 86, 238, 305, 456
  autoxidation rate constant, 212
  structure, 71, 72, 87, 88, 304
CYP121 (= Mtb P450), 268, 288, 302
  active site, 268
  crystal structure, 268, 269
  hydrogen bonding network, 276
CYP158A2, 253, 257, 258, 441, 456
  crystal structure, 258, 442–444
  hydrogen bonding network, 276–278,
    443
CYP165B1 (= OxyB), 239, 243
  structure, 241
CYP176A1, *see* P450cin

Cypermethrin, 506
Cysteine (and residues), 334
　dissociation, 308
　N-acetyl-, 578
　nitrosylation, 298
　oxidation to cysteic acid, 273
Cytochrome b_5, 8, 9, 102, 149, 189, 190, 291, 370, 373, 380, 499, 506, 508
　b_{562}, 275
　reductase, see Reductases
Cytochrome c
　oxidase, see Oxidases
Cytochrome c peroxidase, 273, 280, 335
　active site, 274
　hydrogen bonding network, 280
Cytochromes P450, see CYPs and P450s
Cytokines, 290

D

Danio rerio, 111–115, 494–496
Daphnia
　magna, 106
　pulex, 107, 508
DDT, 490
　detoxification, 506
　hydroxylation, see Hydroxylation
　resistance, 119, 506
Dealkylation (of) (see also Demethylation), 158
　7-pentoxyresorufin, 484, 485, 491, 513
　coumarins, 485, 496
　phenacetin, 495
　tributyltin, 511
Deamination
　oxidative, 99
Debaryomyces hansenii, 524
Debrisoquine, 581
Decarboxylation
　oxidative, 421
　P450-catalyzed, 419–421
O-Deethylases or deethylation
　ethoxycoumarin, 104, 105, 466, 485, 490, 491, 495, 506, 509–513
　ethoxyresorufin, 104, 106, 483, 488, 490, 491, 495, 496, 506, 511–513

　phenacetin, 483, 485
Deformylation, 344, 345
Dehalogenation, 99, 158
Dehydrogenases
　3β-hydroxysteroid, 376
　FAD, 7, 8
　formate, 166
　glyceraldehyde-3-phosphate, 298
Demethylation (of) (see also Dealkylation)
　androgen steroids, 343
　benzphetamine, 495
　caffeine, 483, 570, 572
　coumarin, 491
　erythromycin, 485, 489, 491
　lanosterol, 300
　methoxyresorufin, 488, 491, 495
　oxidative, 244
　theophylline, 483
Density functional theory calculations
　P450 models, 36, 273
　P450s, 67, 68, 198, 203, 213, 326, 327, 341, 342
6-Deoxyerythronolide, 248, 257
　hydroxylation, see Hydroxylation
Detoxification of
　DDT, 506
　genes, 120
　P450-based, 103
　pesticides, 507
　plant allelochemicals, 504, 505, 507
　xenobiotics, 9, 58, 102, 322
Dexamethasone, 485, 486, 489, 497
DFT calculations, see Density functional theory calculations
Diazepam, 597
　hydroxylation, see Hydroxylation
Diazinon, 490, 518
Dichlorodiphenyltrichloroethane, see DDT
Diclofenac, 597
　hydroxylation, see Hydroxylation
Dielectric
　constant, 135, 330
　environment, 137, 147–149
Digitalis, 362
Dihydralazine, 581

Diltiazem, 597–599, 611
Dimethylbenzanthracene
  metabolism, see Metabolism
Dinitrogen monoxide (nitrous oxide;
    N_2O), 7, 83, 249
  formation, 295–297
Dinitrogen trioxide
  formation, 298
Diols (see also individual names)
  androstane, 44
  cleavage, 29, 36, 398–403
Dioxins, 14, 15, 99, 106, 497
  dibenzo[p]-, 454, 459, 488, 489, 492, 496, 497, 508
  polychlorinated, 532
Dioxygen (see also Oxygen)
  activation, 68–70, 82, 243, 275, 341, 438, 439, 441, 443, 444
  atmospheric, 2, 7
  binding to P450s, see P450s
  cleavage, see Cleavage
  consumption, 216, 217, 241, 247
  heme-bound, 242, 328, 443
  iron-linked, 85
  labelling studies, 163–165, 168
  triplet state, 188
Dioxygenases, 2, 321
  definition, 320
Dipteran species, 504
Dolphins, 100
Drosophila
  genome, 4, 8, 13
  insecticide-resistant, 119
  melanogaster, 504–507
  mettleri, 505
  nigrospiracula, 505
  P450 genes, 58, 108, 110
  P450s, 9, 18, 60, 111, 118, 119, 498
  simulans, 506
Drug(s) (see also individual names), 98, 99, 158
  absorption, 593, 594
  administration, 605, 606
  azoles, see Azoles
  bioavailability, 563, 564, 567, 580, 595, 599, 604–606, 608
  candidates, 567

  clearance, 564
  distribution, 593, 594
  excretion, 593, 594
  high-throughput screening, 567, 568
  inhibition, 605, 606
  interactions, see Drug interactions
  metabolism, see Metabolism
  non-invasive assays, 570
  nonsteroidal anti-inflammatory, 7, 8
  oxidation, 463, 566
  P450 inducers, 600, 601
  preclinical development, 463, 563
  pro-, 604, 605
  properties in patients, 591–615
  targets, 373, 374
  therapeutic window, 573
  toxicity, 565
  transport, 603, 604
  triazole-based, 300
Drug interactions, 596
  adverse, 593, 599, 600
  beneficial, 611, 612
  clinical significance, 613, 614
  cytochrome-mediated, 593
  diet, 607
  disease state, 607
  –drug, 12, 565, 567, 598, 599, 610–612
  ethnicity differences, 606
  FDA regulations, 582, 612, 613
  herbs, 609, 610
  pharmacodynamic, 583
  pharmacokinetic, 583
  sex differences, 606
  smoking, 607
Dugong, 100

E

Earth
  life on, 85, 86
Earthworms (see also individual names), 511
Ecdysone, 106
  20-hydroxy-, 107
  biosynthesis, 16, 17

Echinoderms
  chemical defense, 508, 509
  P450s, 101, 108, 110, 508, 509
Econazole, 301
EE_2
  hydroxylation, see Hydroxylation
  metabolism, see Metabolism
  structure, 579
Eicosanoids, 98, 99, 571
  metabolism, see Metabolism
Eisenia fetida, 511
Electrochemistry of
  P450s, 127–151
  sensors, 171
Electrodes
  Ag/AgCl, 130, 162, 167, 171
  antimony-doped tin oxide, 168, 531
  basal-plane graphite, 147, 162, 166
  clay-modified, 147–149, 166
  edge-plane graphite, 145, 149, 150, 166
  glassy carbon, 145, 170, 531
  gold, 147, 166, 531, 534
  indium-tin oxide, 145
  list of modified electrodes, 148
  modifications, 144, 145, 147, 148
  montmorillonite-modified glassy carbon, 147
  normal hydrogen, 130, 158, 160, 171, 179
  polylysine-modified, 149
  pyrolytic graphite, 534
  rhodium graphite, 531, 534
  rotating disk, 165
  standard calomel, 130
  standard hydrogen, 129, 130
  surface-modified, 166
Electron density
  P450, 216
  P450 BM3, 79
  P450cam, 137, 207, 211, 335
Electron nuclear double resonance, see ENDOR
Electron paramagnetic resonance, see EPR
Electron potentials, see Redox potentials
Electron spin echo modulation, see ESEEM
Electron spin resonance, see EPR

Electron transfer (in)
  FAD, 290, 455
  FMN, 290, 455
  heme-dioxygen binding, 158
  mechanism, 331
  microsomal, 364
  mitochondrial, 364, 371–373
  P450s, 74, 75, 158–180, 245, 249, 338, 428, 440, 455, 456
  pathways, 455
  rate constant, 179
  regulation, 371–374
  second electron transfer, 245
Electron tunneling
  wires, see Wires
ELISA, 99, 118
Enchytraeus crypticus, 511
Endoplasmic reticulum
  steroid hormone biosynthesis, 369, 375, 379, 380
ENDOR spectroscopy (of)
  ^{1}H, 245
  P450 models, 33
  P450s, 191, 214, 245, 252, 349, 350
Enthalpy of activation, 190, 210
Entropy of activation, 210
Environment
  monitoring, 118, 119
  pollutants, 447, 463, 487, 494, 607
Enzyme-linked immunosorbent assay, see ELISA
Enzymes (see also Proteins and individual names)
  phenotyping, 565, 566
Eplerenone, 384
Epoxidation (of)
  aldrin, 509–511
  alkenes, 158, 343, 453
  by ferryl species, 169
  by manganese porphyrins, 42, 43
  carotenoids, 47
  double bonds, 246
  in P450s, 29, 36, 44, 286, 411
  mechanism, 342, 343
  olefins, 41, 48, 51, 324, 325, 342, 343, 439, 440
  P450-catalyzed, 319–353

Epoxidation (of) (Continued)
  propene, 36
  terpenes, 43
  α-pinene, 52
EPR studies (of)
  detection of amino acid radicals, 192
  P450 models, 31, 33, 38
  P450s, 191, 192, 197, 198, 214, 217–221, 302, 349, 350, 352
Ergosterol, 363
  formation, 300, 600
  structure, 362
Eriocheir japonicus, 508
Erythromycin, 212, 213, 248, 518, 527, 565, 578, 579, 597–599, 610
  -bound P450, *see* P450eryF
  demethylation, *see* Demethylation
  metabolism, *see* Metabolism
Escherichia coli (expression of)
  adrenodoxin, 368
  CYP11A1, 365
  CYP2C, 467
  CYP72A1, 417
  human P450s, 530, 566
  NADPH-cytochrome P450 reductase, 371
  P450 BioI, 401
ESEEM
  P450 models, 33
  P450s, 191, 192, 197
ESR, *see* EPR
Esterases (*see also* individual names), 376, 572
  acetylcholine, 118, 119, 534
Estradiol, 104
  17β-, 483, 579
  biosynthesis, *see* Biosynthesis
  ethynyl-, 565
  hydroxylation, *see* Hydroxylation
  metabolism, *see* Metabolism
  structure, 363, 410
Estrogens, 364, 370, 374, 409, 600
  biosynthesis, *see* Biosynthesis
  in breast cancer, 371
  structure, 410
Estrone, 409, 410, 579
  biosynthesis, *see* Biosynthesis

Ethanol (*see also* Alcohol), 489, 510, 601
Eucalyptus, 527
  leaves, 481
Euglena gracilis, 513
Eukaryotes (*see also* individual names), 59, 86, 98
Evolution of
  eukaryotes, 98
  life, 2
  P450s, 59, 60
EXAFS studies of
  P450s, 191, 192, 296
Extended absorption fine structure spectroscopy, *see* EXAFS
Extensive metabolizer, 564, 574, 601, 602

F

FAD (in), 76, 78, 90, 160, 161, 167, 190, 291, 323, 331, 368, 371
  dehydrogenase, *see* Dehydrogenases
  electron transfer, *see* Electron transfer
Fadrozole, 300
  structure, 301
Fatty acids, 98, 99, 175, 571
  hydroperoxy, 423–426
  hydroxylation, *see* Hydroxylation
  oxidation, 84, 453, 571
  oxygenated, *see* Eicosanoids
  transformation, 27
  unsaturated, 424, 425
Favia fragrum, 102, 512
FDA
  regulations for drug interactions, 582, 612, 613
Felodopine, 608
Fenchone, 199, 205, 210
Ferredoxins (*see also* individual names), 66, 86, 87, 149, 160, 291, 372
  Fe_2S_2, 7, 8, 74, 367, 368, 537, 538
  Fe_3S_4, 8
  reductase, *see* Reductases
Fexofenadine, 578, 579, 610, 611
Fish (*see also* individual names), 100, 487
  bioactivation of xenobiotics, 497

SUBJECT INDEX 631

Fish (Continued)
  chemical defense, 493–497
  Japanese puffer-, see Fugu and Takifugu rubripes
  killi-, see Killifish
  P450 families, 111–115
  zebra-, see Danio rerio
Flash photolysis of
  P450s, 191, 192, 208–210
Flavin adenine dinucleotide, see FAD
Flavin mononucleotide, see FMN
Flaviolin, 277–279
  binding, 257, 258, 442–444
Flavodoxins, 291
Flavonoids, 608
Fluconazole, 302, 605
  P450 binding, 304, 305
  structure, 301, 304
Fluoranthene, 103, 447, 511
  structure, 448
Fluorene oxidation, 459
Fluorescence
  assay, 165, 453, 567
  sensors, 171
FMN, 7, 74, 86, 90, 137, 167, 168, 190, 291, 323, 331, 371, 527
  electron transfer, see Electron transfer
  P450 BM3, 76, 79, 160, 161
  P450cin, 74, 76
  reductase, see Reductases
Food
  charbroiled, 572, 597, 607
  monitoring of toxicants, 119
  sea-, see Seafood
Food and Drug Administration of the United States, see FDA
Förster energy transfer quenching, 178
Fossils
  chemical, 86, 87
Fourier transform infrared spectroscopy (studies of)
  high pressure, 208
  P450s, 191, 192, 197, 201, 207, 209
Freeze-quench experiments, 221
Frogs, 487, 490–492
Fruit fly, see Drosophila melanogaster

FTIR, see Fourier transform infrared spectroscopy
Fugu (see also Takifugu rubripes), 497
  genome, 496
Fundulus heteroclitus, 494, 495
Fungi (see also individual names), 417, 418
  chemical defense, 521–525
  list of, 522, 523
  P450s, see P450s
  phytopathogenic, 520–524
  xenobiotic metabolism, 522
Fungicides (see also individual names), 489, 490
Furafylline, 572
Fusarium
  oxysporum, 7, 83, 295–297
  solani, 521

G

Gallus gallus, 487, 488
Garlic, 609
Gasoline additives, 528
Gastrointestinal tract
  drug metabolism, 594, 595
  P450s, 482
Genes
  P450s, see P450s
Genistein, 467
Genomes
  Anopheles gambiae, 4
  Apis mellifera, 4
  Caenorhabditis elegans, 4
  Chlamydomonas reinhardtii, 4
  Drosophila melanogaster, 4
  human, 4
  moss, 4
  Oryza sativa, 4
  yeast, 4
Genotoxicity of
  food toxicants, 119
  pesticides, 507
Gibberrella
  fujikuroi, 418, 419, 430
  pulicaris, 524

Gibberellins, 418, 419, 430
  biosynthesis, see Biosynthesis
Glucocorticoid remediable
  hyperaldosteronism, 367
Glucocorticoids, 364–366, 370, 374, 375,
  485, 486, 492, 497
Glutaraldehyde, 147
Glutathione, 577
Glycine
  N-palmitoyl-, 246, 248
Glycosides, 362
Glycosylation of
  P450s, 15
Gonads (see also Testis), 366, 369
  steroid hormone biosynthesis, 376,
  378–382
G-protein, 377
Grapefruit juice, 450, 575, 597, 607–609,
  612
GTP, 290
Guaiacols, 352
  peroxidation, 337
Guanosine monophosphate
  cyclic, see cGMP
Guanosine 5'-triphosphate, see GTP
Guanylate cyclase, 299
  nitric oxide binding, 290

H

Haemonchus contortus, 510
Haliotis rufescens, 512
Health of ecosystem, 118, 119
Heart
  disease, 374, 384
  failure, 384
  fibrosis, 373, 384
Helianthus tuberosus, 8, 9, 519
Helicoverpa zea, 504–506
Heme
  –azole complexes, 303–305
  degradation, 222
  dioxygen complex, see Dioxygen
  distal site, 12, 242–249
  electrochemical oxidation, 168–171
  electronic nature, 348

  pocket, 192, 195, 201–211, 221
  propionate interactions, 67, 68
  redox potentials, see Redox potentials
  solvent accessibility, 210, 221
  spin state, 192
  –thiolate systems, 12, 13, 28, 31, 51,
    66, 82, 158, 188, 216, 218, 268,
    273, 274, 306, 307
  water in pocket, 223
Heme oxygenase, 252, 277, 279
  homolytic dioxygen cleavage, 279
  hydrogen bonding network, 276, 277
Hemoglobin, 211, 320, 325, 337
Hepatotoxicity of
  tienilic acid, 573
Hepatitis, 580, 581, 607
Heptachlor, 107
Herbicides (see also individual names),
  513, 514, 528
  list of, 515, 516
  metabolism, see Metabolism
  resistance, 514
  safeners, 519, 520
  structures, 515, 516
Herbivores, 513, 520
Herbs (see also individual names)
  drug interactions, 609, 610
Hexadecane
  biotransformation, 523, 524
High-performance liquid
  chromatography, 567
HIV
  protease inhibitors, 599, 609, 610, 612
  treatment, 612
Homarus americanus, 107, 111, 508
Honeybee, see *Apis mellifera*
Hordeum vulgare, 418
Hormone(s) (see also individual names)
  biosynthesis, see Biosynthesis
  -dependent cancer, 371
  follicle stimulating, 381
  gonadal, see Gonads
  luteinizing, 379–381
  molting, 106, 107, 508
  plant, 418
  receptors, 486, 492, 493, 497, 575
  sex, 364, 369, 370, 379–381, 600

Hormone(s) (Continued)
    steroid, see Steroids and individual names
    treatment of disorders, 308
Horseradish peroxidase, 40, 169, 171, 252
Housefly (see also Musca domestica), 499
HPLC, see High-performance liquid chromatography
Human (see also Mammal)
    genome, 4
    immunodeficiency virus, see HIV
    in vivo drug studies, 570
    list of P450s, 571
    P450 genes, 58, 99
Hydra
    attenuata, 512
    magnipapillata, 110
Hydride transfer, 7, 84, 168, 296–298
3-Hydroacetanilide, see Acetaminophen
Hydrocarbons, 106
    aromatic, see Aromatic hydrocarbons
    formation in insects, 415
    halogenated, 532
    hydroxylation, see Hydroxylation
    metabolism, see Metabolism
Hydrogen bonding network in
    CYP121, see CYP121
    CYP158A2, see CYP158A2
    cytochrome c peroxidase, see Cytochrome c peroxidase
    heme oxygenase, see Heme oxygenase
    iron-sulfur clusters, 66, 67
    myoglobin, see Myoglobin
    nitric oxide synthase, see Nitric oxide synthase(s)
    P450cam, see P450cam
    P450eryF, see P450eryF
    P450nor, see P450nor
    P450s, see P450s
    proximal, 138
Hydrogen peroxide, 7, 49, 189, 211, 215, 217, 218, 222, 333, 344, 400, 414, 415, 466, 467
    formation, 164, 213, 214, 216, 222–224, 241, 242, 252, 286, 325, 335, 337, 440
    reduction, 321

scavenging, 179
Hydrolases (see also Proteases and individual names)
    androstenedione, 463
    cholesterol ester, 376
    peppermint limonene-3-, 468
    spearmint limonene-6-, 468
Hydroperoxide(s) (see also Peroxides), 101, 102, 106, 400
    ferric, see Iron(III)
Hydroxylases (see also individual names)
    11β-, 366
    androstenedione, 463
    aryl hydrocarbon, 488, 491, 495, 508, 509, 511, 512
    as drug targets, 374
    cinnamate, 519, 520
    laurate, 104, 243
    pregnenolone, 108
    steroid, 374–376
Hydroxylation (of), 30, 32, 67, 99, 286, 407, 409
    6-deoxyerythronolide, 248
    alkanes, 48, 338, 458, 460, 531
    androstenedione, 489
    arenes, 458
    aromatic, 29, 36
    aromatic hydrocarbons, 458
    aromatic hydrocarbons, 458
    benzo[a]pyrenes, 483
    benzylic, 29
    buproprion, 484
    camphor, 168, 214, 215, 217, 237, 239, 277, 323, 330, 340, 525, 526
    C–H bonds, 44, 49, 128
    cineole, 527
    cortisol, 492
    coumarin, 6, 466, 484, 491, 570
    DDT, 506
    debrisoquine, 581
    diazepam, 491
    diclofenac, 484, 567
    EE_2, 579, 580
    estradiol, 483, 496
    fatty acids, 7, 14, 17, 18, 86, 87, 246, 287, 292, 331, 401, 453, 458, 486, 491, 527

Hydroxylation (of) (Continued)
  hydrocarbons, 158
  isotope effect, 340
  L-arginine, 82, 291
  lauric acid, 167, 271, 485, 491, 495, 497, 511, 512
  limonene, 6, 468, 531
  linalool, 526
  mechanism, 337–342
  naphthalene, 491, 495, 508
  nitric oxide, 4
  octane, 462
  P450-catalyzed, 319–353
  paclitaxel, 484
  perhexiline, 581, 582
  phenytoin, 484
  p-nitrophenol, 489
  pregnenolone, 301, 369
  progesterone, 301, 369, 463, 464, 484, 489, 495, 508
  radical pathway, 339–341
  stereochemical scrambling, 340
  steroids, 45, 58, 365, 370–373, 485
  S-warfarin, 484
  testosterone, 6, 17, 484, 485, 489, 491, 495, 496, 508, 511, 576
  tolbutamine (methyl-), 484, 567, 570
  vitamin D_2, 408
Hydroxyl radical, *see* Radicals
Hydroxymethylglutarate, 530
Hyperaldosteronism
  type I, *see* Glucocorticoid remediable hyperaldosteronism
Hypericum perforatum, *see* St. John's wort
Hypertension, 373, 374
Hypokalemia, 600

I

Ifosfamide, 463, 467
  structure, 464
Imidazole (and moieties), 71, 72, 170
  as ligand, 335
  as P450 inhibitor, 287, 288, 300–305, 329, 333
  binding constants, 300
  phenyl-, 71, 72, 240, 300, 301, 304, 305, 330, 333
Immune function, 289
Immunosuppressive agents, 596, 605, 606, 611, 612
Imposex, 104
Indigo
  formation, 458, 466, 467
  structure, 464
Indirubin, 466
  structure, 464
Indole, 464
  3-methyl-, 485
  oxidation, 467, 484
Indomethacin
  structure, 421
Infrared (spectroscopy studies of)
  Fourier transform, *see* Fourier transform infrared spectroscopy
  P450s, 191, 192, 296
Inhibitors or inhibition (*see also* individual names)
  azoles, *see* Azoles
  cyanide, *see* Cyanide
  imidazole, *see* Imidazole
  index, 599
  of P450s, *see* P450s
  suicide, 287, 599, 610
Insects (*see also* individual names *and* species)
  chemical defense, 497–507
  hydrocarbon formation, 415
  list of P450s involved in xenobiotic metabolism, 500–503
  P450s, 8, 13, 15–18, 101, 110, 116, 117, 415, 480, 498
  steroid hormones, 16, 17, 111
  xenobiotic metabolism, 506, 507
Insecticides (*see also* individual names), 99, 119, 499, 505, 534
  carbamate, 118, 490
  organophosphate, 118
  pyrethroid, 505, 506
  resistance, 118–120
Insulin, 510
  -like growth factor, 510

Invertebrates (*see also* individual names)
  aquatic, 99–119
  chemical defense, 497–512
  P450 genes, 108–111
  physiology and ecology, 116, 117
Ion exchange chromatography, 105
β-Ionone, 459, 462
  structure, 454
IR, *see* Infrared
Iron
  ^{56}Fe, 303
  ^{57}Fe, 219
Iron(II), 50, 51, 166
  carbon monoxide binding, *see* Carbon monoxide
  redox potentials, *see* Redox potentials
  thiolate, 333
Iron(III), 29, 40, 41, 48–50, 169, 410, 411
  hydroperoxide, 164, 174, 341
  hydroperoxo intermediate, 333, 334, 342, 345, 349
  low-spin, 33
  peroxo species, 343–345, 349
  porphyrin, *see* Porphyrins
  redox potentials, *see* Redox potentials
  superoxide complex, 163, 164, 173
Iron(IV), 41, 216
  Fe^{IV}=O (oxoferryl species), 67, 68, 82, 159, 218, 219, 221, 244, 291, 399, 414–416, 421, 439
  Fe^{IV}=O–H, 169–171, 179, 180, 425
  Fe^{IV}–OH$_2$, 179
  Mössbauer parameters, 218
  -porphyrin π-cation radical, 216, 218, 251, 252, 325, 343, 352
Iron(V)
  Fe^{V}=O (compound I), 163, 169, 439
Iron-sulfur proteins (*see also* individual names), 67, 189
Isazofos, 518
Isoniazid, 599
Isotope effects, 245, 405, 409, 419
  kinetic, 338, 340, 345, 576
Itraconazole
  structure, 301

J

Jasmonic acid, 7, 8

K

Ketoconazole, 565, 566, 578, 579, 597, 605, 606, 611, 612
  P450 binding, 304, 305
  structure, 301, 304
Ketones, 408
Ketoprofen
  structure, 421
Killifish, 100, 494–496
Kinases
  cyclin-dependent, 466
  eIF2, 307, 308
  glycogen synthase, 466
  protein, *see* Protein kinase

L

Lactoperoxidase, 271, 294
Lactone
  formation, 429, 430
Lanosterol, 363, 411
  14α-demethylase, *see* CYP51
  conversion, 600
  metabolism, *see* Metabolism
Lauric acid, 167, 247, 462, 486
  hydroxylase, *see* Hydroxylases
  hydroxylation, *see* Hydroxylation
  oxidation, 333
Lepidopteran species, 499
Leptasterias polaris, 509
Letrozole, 300
  structure, 301
Leydig cells, 379–381
Ligase
  ubiquitin, 298
Lignin peroxidase, *see* Peroxidases
Limonene
  hydroxylation, *see* Hydroxylation
  oxidation, 450, 531
  peppermint, 468

Limonene (Continued)
  spearmint, 468
  structure, 448
Linoleic acid, 424
Linolenic acid, 424
Linum usitatissimum, 423
Lipid(s), 145, 146, 148
  metabolism, *see* Metabolism
Lipoxygenase, 171
Liver
  disease, 607
  P450s, 481, 607
Lobster (*see also* individual names)
  American, 107, 111
  P450s, 101, 107, 508
  spiny, 106, 110
Loganin
  structure, 417
Lufenuron, 111
Lumbricus terrestris, 511
Luminescence assay, 567
Lyase
  hydroperoxide, 7
Lycopersicon esculentum, 408, 423
Lymnaea, 511
  palustris, 104, 512
  stagnalis, 511
Lytechinus anamesus, 108, 509

M

Macroconstants, *see* Acidity constants *and* Binding constants
Macrolide antibiotics, *see* Antibiotic(s)
Magnetic circular dichroism studies of
  oxyferrous intermediate, 349
  P450s, 191, 192, 302, 303
Malathion, 106, 490, 518
Mamestra brassica, 499
Mammals (*see also* individual names *and* species)
  aquatic, 100, 115, 116
  chemical defense, 481–487
  marine, *see* individual names
  P450s, *see* P450s
Manduca sexta, 505

Manganese(II), 50
  propionate ligand, 67
Manganese(III), 40–45, 50
Manganese(IV), 41
Manganese(V), 41, 52
Marcus theory, 195
Marmesin, 416, 417
  conversion to psoralen, 416
  structure, 416
Marsupials, 481
Mass spectrometry, 567
MCD, *see* Magnetic circular dichroism
Meat
  charbroiled, 572, 597, 607
Meleagris gallopavo, 488
Mercenaria mercenaria, 111, 512
Metabolism
  acetaminophen, 576–578
  arachidonic acid, 495
  benzo[a]pyrenes, 102, 108, 507, 508
  benzphetamine, 507
  dimethylbenzanthracene, 495
  drugs, 27, 58, 69, 322, 365, 371, 485, 561–583, 593–596, 613, 614
  EE_2, 579, 580
  eicosanoids, 571
  erythromycin, 507
  estradiol, 104, 569
  fat, 569
  herbicides, 513–517
  hydrocarbons, 104
  lanosterol, 412, 413
  lipid, 498
  nicotine, 484, 505
  nitric oxide, 295–298
  olanexidine, 402, 403
  perhexiline, 581
  polychlorinated biphenyls, 507
  scoparone, 489
  steroids, 70, 106, 107, 111
  terfenadine, 578, 579
  terpene, 531
  tienilic acid, 580, 581
  testosterone, 104, 106
  tributyltin, 507, 509
  vitamin D_2, 408
  warfarin, 573

Metabolism (Continued)
  xenobiotics, 18, 102, 120, 365, 371, 413, 422, 462, 506–509, 514–517, 519, 533
Metabolizer
  extensive, 564, 574, 601, 602
  poor, 564, 573, 601, 602, 605
Methane oxidation, 447
3-Methylcholanthrene, 14, 15, 103, 104, 483, 484, 486, 489, 492, 508, 511, 512
Methylococcus capsulatus, 8
Methyl viologen, 173
Metyrapone, 308
Michaelis-Menten kinetics
  non-, 278
Miconazole, 302
  structure, 301
Microarray analysis, 13, 14
Microscopy
  atomic force, *see* Atomic force microscopy
Microsomes or microsomal
  electron transfer, *see* Electron transfer
  invertebrate, 101
  P450s, *see* P450s
Midazolam, 534
Mineralocorticoids, 364–366, 370, 374, 383, 385
Mitochondria(l)
  electron transfer, *see* Electron transfer
  P450s, *see* P450s
  steroid hormone synthesis, 379
Molecular dynamics simulation
  P450 3A4, 76, 77
  P450cam, 180, 202, 238, 259, 326
  random expulsion, 238
Molecular mechanics calculations
  P450eryF, 326
  P450s, 198, 215, 259
Molluscs (*see also* individual names)
  chemical defense, 511, 512
  gastropod, 104
  P450s, *see* P450s
Monooxygenases, 321
  cytochrome P450, *see* P450s
  definition, 320

steroid 11β-, 61
styrene, 168
Morphine, 602, 605
  ethyl-, 508, 518
Mosquitos (*see also* individual names)
  genome, 4, 8
  P450 genes, 108
  P450s, 5
Moss
  genome, 4
Mössbauer spectroscopy
  studies of P450s, 11, 191, 192, 211, 217–221, 352
Musca domestica
  P450s, 8, 118, 119, 415
Mussel (*see also* individual names), 119, 512
  as biomonitors, 511
  P450s, 105
Mutagenesis
  site-directed, *see* Site-directed mutagenesis
Mutation
  random, 457–459, 465–467, 572
  recombinatorial, *see* Recombinatorial mutation
  semi-random, 466
Mycobacterium, 531
  smegmatis, 528
  tuberculosis, 8, 90, 269, 288, 300–302, 304, 306
Mycosphaerella pinodes, 521
Mycotoxins (*see also* individual names), 417
Myeloperoxidase, 271
Myoglobin, 211, 320, 325, 337
  active site, 274
  carbon monoxide complex, 203
  dioxygen binding, 253
  docking sites for gases, 238
  hydrogen bonding network, 280
  mutation, 273
Myrtenol
  formation, 450
  structure, 448

Mytilus
  galloprovincialis, 104, 105, 512
  edulis, 104, 105, 512

N

NADH, 7, 84, 102, 106, 108, 128, 176, 295–298, 323, 365, 508
  consumption, 244
  dimer, 165, 166
  oxidation, 214, 215
  turnover rate, 447
NADP, 364
  oxidation, 216
NADPH, 7, 8, 30, 67, 82–84, 87, 106, 108, 128, 160, 162, 164, 165, 167, 169, 188, 190, 195, 202, 216, 217, 222, 249, 296, 322, 323, 325, 364, 365, 367, 368, 424, 439, 460, 464, 466, 467, 508, 511, 530, 531
  as electron donor, 291, 455
  consumption, 247, 398, 409, 415, 440, 446
  cytochrome P450 reductase, *see* Reductases
  dehydrogenation, 290
  dimer, 165, 166
  oxidase, *see* Oxidases
  oxidation, 216
Naphthalene, 462, 485
  oxidation, 459, 465, 533
Naphthoflavone
  α-, 106, 467, 488, 496, 497, 566
  β-, 14, 104, 106, 483, 486, 492, 496, 508, 510, 512
Naphthoic acid, 518
Nectria haematococca, 521, 522
Nematode(s), 108
  chemical defense, 509
  genome, 4, 8
  P450s, 480, 510
  parasitic, 509
Nereis virens, 103, 111, 511
Nernst
  equation, 132
  titration, 130, 133

Neuroleptics, 602
Neurotransmitters (*see also* individual names)
  nitric oxide, *see* Nitric oxide
Newt, 491
Nicotinamide adenine dinucleotide (reduced), *see* NADH
Nicotinamide adenine dinucleotide phosphate, *see* NADP
Nicotinamide adenine dinucleotide phosphate (reduced), *see* NADPH
Nicotine
  metabolism, *see* Metabolism
Nifedipine, 497
  oxidation, 485, 491, 575
Nitration
  P450s, 15
  tyrosine, 298, 299
Nitric oxide (NO), 4, 82, 84, 238, 307
  as neurotransmitter, 289
  as P450 inhibitor, 287, 288, 298, 299
  binding to guanylate cyclase, 290
  formation, 11, 82, 291, 293
  hydroxylation, *see* Hydroxylation
  in immune function, 289
  in vasodilation, 289
  metabolism, *see* Metabolism
  P450 interactions, 286–310
  reductase, *see* Reductases
  reduction, 236, 249, 295
  synthesis, *see* Synthesis
Nitric oxide synthase(s), 11, 12, 66, 77, 79, 82, 85, 225, 252, 270, 271, 279, 289, 298, 320
  active site structure, 268
  carbon monoxide binding, 306
  catalytic reactions, 291
  dioxygen complex, 82, 83
  endothelial, 83, 276, 277, 290
  heterolytic dioxygen cleavage, 279
  hydrogen bonding network, 276, 277
  inducible, 171, 189, 290, 294
  neuronal, 189, 290, 292, 294, 295, 303
  oxygenase domain, 189
Nitrite
  formation, 298
  peroxy-, 298, 299

SUBJECT INDEX

Nitrogen (*see also* Dinitrogen)
   ^{15}N, 303
   radical, *see* Radicals
Nitrogen monoxide, *see* Nitric oxide
Nitrogen dioxide, 298
p-Nitrophenol
   hydroxylation, *see* Hydroxylation
Nitrous oxide, *see* Dinitrogen monoxide
NMR
   studies of P450s, 191, 192
Nootkatol, 531
   oxidation, 451
   structure, 448
Nootkatone, 450, 451, 531
   structure, 448
Norbornane, 48, 198–200, 205, 209, 210, 221, 339, 340
Norcamphor, 198–200, 205, 209, 210, 221
Nuclear magnetic resonance, *see* NMR
Nucleotides (*see also* individual names)
   oligo, *see* Oligonucleotides
   pyridine, *see* Pyridine(s)

O

Obtusiferol, 363
Octane
   hydroxylation, *see* Hydroxylation
Octopus pallidus, 104, 105, 511
Olanexidine, 401
   metabolism, *see* Metabolism
   oxidation, 402, 403
   urinary metabolites, 402, 403
Olfactory or olfaction, 484
   neurons, 482
   organ, 493, 499
Oligonucleotides
   antisense, 377
Ophiocomina nigra, 509
Orconectes limosus, 107, 508
Oryza sativa
   genes, 99
   genome, 4, 8, 14
   P450, 9

Otter, 100
Ovary
   regulation of hormones, 381, 382
   steroid hormone biosynthesis, 370, 379–382
Oxidases (*see also* individual names)
   amine, 171
   cytochrome *c*, 299
   mixed-function, 2
   NADPH, 337
Oxidoreductase
   2-oxoacid-ferredoxin, 7, 87
Oximes, 423
Oxygen (*see also* Dioxygen)
   ^{17}O, 33
   ^{18}O, 38, 39, 48, 303, 410
   radical, *see* Radicals
   transport, 320
Oxygenases
   categories, 321
   di-, *see* Dioxygenases
   heme, *see* Heme oxygenases
   lip-, *see* Lipoxygenase
   mono-, *see* Monooxygenases
Oysters (*see also* individual names), 119, 512

P

P420
   carbon monoxide complex, 305, 306
   formation, 204, 294
   –P450 equilibrium, 294
   –P450 spectral transitions, 306
P450s (*see also* CYP *and* individual P450s)
   ^{57}Fe, 219
   access to the distal pocket, 237–239
   activation enthalpy, 190
   active site, 64, 70–74, 76, 80, 85, 164, 193, 214, 268, 269, 327–333
   active site analogues, 28, 32
   active site water, role of, 254–258, 279
   activity, 438, 439
   ancestral forms, 480

P450s (*see also* CYP and individual P450s) (*Continued*)
  animal, 4, 5
  antisera, 103
  aquatic, 97–121
  aquatic gene families, 108–116
  archaeon, 4, 86–90
  arthropodal, 101, 105–108
  axial cysteine, 303
  bacterial, 4, 6, 7, 17, 18, 60, 76, 160–162, 189, 190, 444–462
  binding constant, 288
  biocatalysis, 165–168
  biotechnological applications, 18, 286
  carbon–carbon bond cleavage, 397–430
  carbon monoxide binding, 192, 193, 203–211, 289, 305–308
  catalytic cycle, 28, 128, 129, 158–171, 175, 188, 189, 251, 287, 323–325
  -catalyzed epoxidations, *see* Epoxidation
  -catalyzed hydroxylations, *see* Hydroxylations
  -catalyzed substrate oxygenation, 163
  characteristic reactions, 3, 36
  clinical observations, 591–615
  cofactor, 28
  common themes, 529, 530
  competitive reactions, 222–225
  compound I, see Compound I
  compound II, see Compound II
  conformational changes, 70–74
  conservation of the fold, 62–64
  conserved acidic residues, 243
  conserved alcohol residues, 243, 259
  crustacean, *see* Crustaceans
  crystal structures, *see* Crystal structures
  cysteine ligand loop, 65–67
  cytosolic, 296
  designed protein models, 273–275, 280
  designing, 437–469
  dietary effects, 608–610
  dioxygen activation, *see* Dioxygen
  dioxygen binding, 10, 39, 68, 70, 190, 208, 211–215, 223, 224, 253–258, 286, 325–327, 349
  dioxygen reduction, 163–165
  directed evolution, 457–462, 466, 530
  distal acid–alcohol pair, 242–249
  distal pocket, 268, 275–279, 304, 331–333
  diversities, 1–18
  domain structure, 64, 65
  drug-metabolizing (*see also* Metabolism), 52, 58, 70, 71, 81, 329
  electrochemistry, 127–151, 167
  electron delivery, 188, 190, 202, 203
  electron shuttles, 167, 168
  electron transfer reactions, see Electron transfer
  energy landscape, 193
  engineering, 437–469
  eukaryotic, 4, 7, 294, 309, 332, 363
  evolution of common metabolic functions, 16–18
  evolution, *see* Evolution
  families and subfamilies, 60, 61
  fish, *see* Fish
  formation of protein radicals, 217–221, 224, 225
  fungal, 4, 6, 81, 83
  genes, 13, 58–60
  glycosylation, *see* Glycosylation
  heme–azole complexes, 303–305
  heme pocket, *see* Heme
  heme–propionate interactions, *see* Heme
  heme Soret band, 158
  heme–thiolate interactions, beyond, 267–280
  heme–thiolate systems, *see* Heme
  herbal interactions, 609
  high-spin state, 195–202
  high-valent metal-oxo intermediates, 11, 12
  high-valent species, 169
  Holy Grail, 165–168, 180
  human, 561–583
  human P450s, list, 571, 597
  human therapy, 287
  hydrogen bonding networks (*see also* individual P450s), 275–277, 348
  hydroperoxo complexes, 102, 213–215, 224, 243, 325

SUBJECT INDEX

P450s (*see also* CYP) (Continued)
  hydrophobic pocket, 348
  hydrophobic substrates, 239, 330
  hydrostatic pressure studies, 193–195
  inactivation, 221, 222
  inducers, 596, 597, 600, 601, 613
  induction mechanism, 568
  industrial applications, 530–534
  inhibitors (*see also* Inhibition), 288, 293–295, 297–310, 329, 333, 371, 374, 382, 415, 566, 573–600, 606, 608, 613
  insect, *see* Insect
  interaction with imidazoles, 300–305
  interactions with nitric oxide, 285–310
  invertebrate, 100–116
  iron-oxo intermediate, 215–221
  irreversible inhibitors, 309, 310
  kinetics of carbon monoxide binding, 208–211
  leakage in reactions, 12, 187–226
  location, 594
  low-spin state, 195–202
  major subfamilies, 482–486
  mammalian, 6, 13, 18, 59, 60, 100, 101, 160, 162, 323, 329, 403, 462–467
  mechanism, 6–13, 84, 85, 158–171, 258
  mechanism of catalysis, 337–342
  mechanism of epoxidation, 342, 343
  -mediated chemical defense, *see* Chemical defense
  membrane, associated, 58, 76, 77
  metal–diimine wires, 171–174
  microsomal, 73, 74, 136, 189, 294, 329, 332, 364, 365, 369–371, 373
  mimics, 27–52, 272, 273, 279, 280
  mitochondrial, 70, 111, 189, 296, 364–369, 371–373, 499
  molecular phylogeny, 57–91
  mollusc, 101, 104, 105, 110, 512
  monooxygenase catalytic mechanism, 9
  multimeric, 77–81
  multiple oxidants in catalysis, 343–347
  nematode, 480
  nitration, *see* Nitration
  nitric oxide binding, 293–295, 303
  nomenclature, 3–5, 60, 61, 99
  'orphan', 571, 576
  overview, 322–327
  oxidase reaction, 43, 216, 217
  oxidized species, 168–171
  oxy intermediates, 10, 11, 173
  -P420 equilibrium, 294
  -P420 spectral transitions, 306
  peppermint, 6
  peroxidase activity, 90
  peroxo complexes, 10, 11, 213–215
  pharmaceutical synthesis, 530–532
  phosphorylation, 15
  phylogenetic tree, 74, 98
  plant, 4–7, 13, 15, 18, 423, 467, 468, 513–520
  polymorphism, 567, 573, 601, 602, 604, 605, 613
  predicting drug activity in humans, 565–570
  prokaryotic, 70, 72, 74, 77, 87, 455, 525–529
  properties of human P450s, 597
  protein structural properties, 191–195, 222–225
  proximal cysteine, 290, 333–337
  proximal side, 268–275, 287, 304
  proximal surface, 86
  radical pathway, 339, 340
  rate constants for binding, *see* Rate constants
  reaction cycle, 9, 11, 188–190, 192, 195–222, 349
  reaction intermediates, 250–253
  reactivity landscape, 9–11, 193
  redox partners, 6–9
  redox potentials, list of, 134, 135, 146, 148
  redox titration, 131–139
  regulation by receptor-mediated systems, 486, 487
  regulation, 13–16
  resting state, 159
  reversible inhibitors, 308, 309
  secondary coordination sphere, roles of, 267–280
  second electron transfer, 245

Met. Ions Life Sci. **3**, 619–652 (2007)

P450s (*see also* CYP and individual P450s) (Continued)
  sequences, 3–5, 236
  side reactions, 189, 221, 222
  signature, 158
  similarities, 1–18
  site-directed mutagenesis, *see* Site-directed mutagenesis
  spectral properties of carbon monoxide complexes, 203–211
  spectroscopic techniques, list of, 191, 192
  steroid hormone biosynthesis, 361–386
  structural basis for binding and catalysis, 235–260
  structures (*see also* individual P450s), 5, 6, 57–91, 441–444
  substrate binding, 70–74, 199–211, 330, 331
  substrate entry, routes for, 202
  substrate hydroxylation, 225
  substrate-induced reactivity, 350, 351
  substrate interactions, 286–289
  substrate pocket, 449
  substrate recognition, 237–239
  substrate, role of, 439–441
  substrate specificity, 482–486
  superfamilies, 3, 4
  thermal-stable, 90, 91
  threonine, role of, 331–333
  transient oxygen intermediates, 351, 352
  turnover rates, 162
  ubiquitination, *see* Ubiquitination
  variant alleles, 567, 568, 574
  variation in function and fold, 81–86, 90
  vertebrate, 6, 8, 9, 15, 16, 480
  vesicle construct, 51
  voltammetry, *see* Voltammetry
  xenobiotic decomposition, 530–534
  yeast, *see* Yeast
P450 1A2, 160, 572, 573
  inhibition, 572
  mutagenesis studies, 572
  nitric oxide binding, 294
P450 15α (= CYP2A4), 72, 73

P450 2B4, 189, 198, 373
  autoxidation rate constants, 212
  azole binding, 304
  dioxygen binding, 209
  nitric oxide binding, 294
  spin states, 203
  structure, 80, 81
P450 2C5
  membrane bilayer, 77
  structure, 75
P450 2C9, 573, 581, 601, 602
P450 2C19, 573, 574, 601
P450 3A4
  active site pocket, 76–78
  structure, 76–78
P450 175A1, 87, 89
P450aldo (= CYP11B2), 364, 366, 367, 374, 375, 383–385
  inhibition, 374
P450arom, 149, 150, 364
P450 BM3 (CYP102), 7, 74–80, 84, 87, 143, 144, 147, 149, 160–162, 167, 174, 175, 189, 218–221, 237, 243, 246–248, 270, 271, 277, 291–293, 302, 303, 306, 331, 333, 464, 466, 527, 530–533
  ^{57}Fe, 219
  active site, 268, 452, 461
  autoxidation rate constants, 212
  carbon monoxide binding, 209, 306
  dioxygen binding, 452
  directed evolution, 458–462
  electron transfer, 160–162
  FMN domain complex, 78, 79
  mutation, 138, 214, 218–220, 247, 248, 452–456, 458–462
  nitric oxide binding, 293–295, 299
  redox potential, 224, 246, 247
  structure, 32, 70, 71, 74, 76, 78, 79, 88, 137, 140, 225, 246, 248, 268, 269, 292, 332
  structures of substrates, 454
P450 BSβ, 84, 86
  structure, 85, 269
P450cam (= CYP101), 68, 69, 70, 72–76, 87, 89, 162–164, 168, 170, 171, 175, 189, 190, 198, 218, 237–241, 243,

SUBJECT INDEX

P450cam (= CYP101) (Continued)
247–249, 259, 270, 279, 293, 296,
299, 302, 322, 323, 331, 340, 350,
372, 429, 444–452, 456, 525–527,
531, 532
- ^{57}Fe, 219
- active site, 201, 214, 239–241, 268, 274, 327–329, 445
- autoxidation rate constant, 212
- binding to wires, 176
- catalytic cycle, 28, 30, 324
- chain motions, 238
- conserved aspartic acid, 244–246
- conserved threonine, 242–244
- crystal structure of O_2 complex, 212
- dioxygen binding, 325–328, 335, 336
- directed evolution, 458
- Fe(II), 178, 179, 253, 326, 328, 336
- Fe(III), 178, 253, 326, 352
- ferrous dioxygen complex, 253–258
- hydrogen bonding network, 33, 204, 212–215, 223, 224, 239, 240, 242, 244, 245, 270, 275–278, 326, 328, 329, 334–336, 444, 445
- hydrogen bonding of thiolate, 33
- hydroperoxy intermediate, 68, 334
- hydrophobicity of active site, 138, 274
- modified electrodes, list of, 148
- mutants, 145, 180, 202, 214, 215, 219, 220, 223, 241, 244, 245, 253–256, 259, 274, 277, 278, 331, 333, 334, 336, 337, 343, 349, 444–452
- nitric oxide binding, 293–295, 303
- peroxo complex, 29, 30
- photoreduction, 178, 179
- rate constants for binding, 208–211
- rate of 5-OH-camphor formation, 241
- rate of dioxygen consumption, 241
- rate of hydrogen peroxide formation, 241
- redox potential, *see* Redox potentials
- redox states, 174
- resting state, 29, 34
- reversible potential, 132, 136, 138, 145–149

- structure, 32, 63, 64, 73, 74, 76, 88, 128, 197, 225, 240, 244, 257, 268, 269, 271, 278, 445, 449
- structures of substrates, 448
- structures of wire complexes, 178–180
- substrate complexes, 199, 200, 205, 208–211, 222, 449
- wire conjugates, list of, 177
- xenon binding, 238

P450cin (= CYP176A1), 68, 75, 76, 133, 139, 149, 239, 243, 527
- structure, 72–76, 241

P450 COH (CYP2A5), 72, 73

P450eryF (= CYP107A1), 68, 85, 189, 212, 213, 243, 248, 253, 276, 277, 456
- dioxygen binding, 256, 257, 336
- hydrogen bonding network, 326
- ketoconazole binding, 304, 305
- mutants, 256, 257, 278
- structure, 32, 257, 269, 304

P450nor (= CYP55A1), 7, 86, 236, 243, 249, 295–298
- hydrogen bonding network, 249, 296, 297
- mutants, 249
- nitric oxide binding, 294–298
- structure, 83, 84, 297

P450 RhF, 136

P450scc (= CYP11A1), 136, 145, 189, 363, 365–367, 372, 374–376, 379, 381–385, 398–401, 409, 413
- autoxidation rate constant, 212
- carbon monoxide binding, 209
- nitric oxide binding, 294

P450terp (= CYP108), 526, 527
- autoxidation rate constant, 212
- carbon monoxide binding, 209
- X-ray structure, 32, 269

Pacifastacus leniusculus, 106
Paclitaxel, 576
- hydroxylation, *see* Hydroxylation
Palmitate, 458
Palmitoleic acid, 331
Panulirus argus, 101, 106, 110, 508
Papilio sp., 504
- *canadensis*, 504
- *glaucus*, 504

Papilio sp. (Continued)
  polyxenes, 504
  xuthus, 499
Paracetamol, *see* Acetaminophen
Parathion, 490, 511, 534
Patinopectin yessoensis, 104
PCBs, 101, 104, 107, 118, 119, 488, 491, 497, 509–512
  metabolism, *see* Metabolism
Penaeus setiferus, 107, 508
Peppermint
  limonene-3-hydrolase, 468
  P450, 6
Perhexiline
  hydroxylation, *see* Hydroxylation
  metabolism, *see* Metabolism
  structure, 582
Perna canaliculus, 512
Peroxidases, 85, 215, 216, 320, 321
  ascorbate, 68
  chloro-, *see* Chloroperoxidase
  compound I, *see* Compound I
  compound II, *see* Compound II
  cytochrome *c*, 68
  heme, 65, 68
  horseradish, 40, 169, 171, 252
  lacto-, *see* Lactoperoxidase
  lignin, 49
  manganese(II), 68
  micro-, 171, 217
  myelo-, 171
Peroxide(s), 12, 85, 102
  formation, 82
  hydrogen, *see* Hydrogen peroxide
Peroxy acetic acid, 217–219, 352
Peroxy acids (*see also* individual names)
  m-chloroperoxybenzoic, 217, 218, 222, 325, 352
Persea americana, 518
Pesticides (*see also* individual names), 98, 101, 106, 111, 487, 488, 490
  detoxification, *see* Detoxification
  resistance, 505, 506
Phanerochaete chrysosporium, 524
Pharmaceutical industry, 562, 563, 567, 571

Pharmacodynamics or pharmacokinetics, 593, 598, 599
  analysis, 564, 568, 570
Pharmacological response, 564
Phenacetin, 465, 464
Phenobarbital, 14, 104, 106, 107, 329, 484–486, 489, 492, 496, 497, 506, 508, 510, 513
Phenytoin
  hydroxylation, *see* Hydroxylation
Pheromones, 120, 450, 499
Phoma pinodella, 521
Phosphate
  organic, 490, 507, 534
Phosphatidylcholine
  dimyristoyl-L-α-, 534
Phospholipid
  bilayer, 7
  membrane, 136
Phosphorus
  ^{32}P, 349
Phosphorylation of
  adrenodoxin, 372
  P450, 15
  serine, 15
  threonine, 15
Phyllopertha diversa, 499
Phylogenetic tree
  P450s, 74, 98
Physcomitrella patens
  genome, 4
Phytoalexins, 59
Phytotoxins
  alkaloid, 505
Pimelic acid, 401
Pinene
  binding modes, 451
  oxidation, 450, 531
  structure, 448
Piperidine, 423
  tetramethyl-, 422
Pisatin, 521
  structure, 522
Placenta, 365
  steroid hormone biosynthesis, 366, 370, 382, 383

Plant(s) (*see also* individual names *and* species)
  allelochemicals, 504, 505
  chemical defense, 513–521
  implications of immobility, 520, 521
  P450s, *see* P450s
  –plant chemical warfare, 520
  steroid hormone biosynthesis, 16
  versus animal chemical warfare, 498, 504, 505
Platelet aggregation, 426
Platyhelminths
  P450s, 101
Plebidonax deltoides, 512
Pleurodeles waltyl, 491
Pollutants, 102, 104, 447, 463, 487, 494, 607
Polychaete
  P450s, 101, 103, 111, 511
Polychlorinated biphenyls, *see* PCBs
Polycyclic aromatic hydrocarbons, *see* Aromatic hydrocarbons
Polymerase chain reaction
  error-prone, 457, 462
  real-time, 118, 120
Poor metabolizer, 564, 573, 601, 602, 605
Poriferans
  chemical defense, 512
  P450s, 101
Porphyria, 488
Porphyrins (*see also* Hemes *and* individual names)
  basket-handle, 28
  capped, 28
  cobalt, 46, 165
  electron deficient, 48–52
  epoxidation, *see* Epoxidation
  face-protected, 29
  Fe(II), 50, 51
  Fe(III), 29, 33, 40, 41, 48–50, 169, 410, 411
  Fe(IV), 41, 216
  Mn(III), 40–46
  Mn(IV), 41
  Mn(V), 41, 52
  Os, 46
  oxo Fe(IV), *see* Compound I
  picket-fence, 28, 37
  proto-, *see* Protoporphyrins
  radical, *see* Radicals
  Ru(II), 40, 46–48
  Ru(IV), 46
  Ru(VI), 40
  strapped, 28
  tetramesityl, 410, 411
  tetraphenyl, 40, 41
  thiolate-ligated, 273
  twin coronet, 279
  with a thiolate, 31–40
  Zn, 49, 50
Porpoise, 100
Potassium channel, 610, 612
Potatoes, 423
Potentials, *see* Redox potentials
Potentiometric titration of P450s, 130–139, 146, 147
Pravastatin, 530
Pregnonolone, 374–376, 404, 600
  16α-carbonitrile, 485, 486, 489
  biosynthesis, *see* Biosynthesis
  cleavage, 408
  hydroxylation, *see* Hydroxylation
  oxidation, 403, 405, 414
  structure, 363
  sulfate, 111
Pressure jump studies of P450s, 191, 192
Procambrus clarkii, 105, 106
Prochloraz, 489, 490
Progesterone, 278, 366, 367, 370, 534, 575
  cleavage, 408
  hydroxylation, *see* Hydroxylation
  oxidation, 404
  structure, 363
Prokaryotes (*see also* individual names), 59, 86
  chemical defense, 525–529
  P450s, *see* P450s
Propanolol, 530
Propiconazole, 489
Prostacyclin, 426–428
  synthase, *see* Synthases
Prostaglandin(s), 426–429, 571
  biosynthesis, *see* Biosynthesis

Prostaglandin(s) (Continued)
　endoperoxide synthase, see Synthases
　H synthase, see Synthases
　H_2, 7, 8
Prostate cancer, see Cancer
Proteases (see also individual names)
　HIV, 599, 609, 610, 612
　inhibitors, 309, 598, 599, 609, 610, 612
　serine, 70
Proteins (see also Enzymes and individual names)
　radical, see Radicals
　engineering, 463
　P-glyco-, 603–606, 609
　–protein interactions, 136, 371–373, 380
　sensor, 12
　SoxAX, 308
　steroidogenic acute regulatory, 376
Protein kinases
　A, 15, 376, 377
　C, 15
　cAMP-dependent, 15
Protista
　chemical defense, 513
Protonation constant, see Acidity constant
Proton transfer (in)
　cascade, 333
　P450s, 241, 243, 245, 252, 257–259, 279, 297, 326, 330, 331, 335, 428, 452
　relay system, 259, 298, 351
Protoprophyrin IX, 128, 320, 330
　Fe(III), 28, 160
　structure, 321
Pseudomonas sp., 532
　diterpeniphila, 527
　incognita, 526
　putida, 74, 162, 190, 237, 293, 322, 444, 525, 531
　solanacearum, 520
Psoralens, 520
　biosynthesis, see Biosynthesis
　structure, 416
Putidaredoxin, 29, 30, 74, 75, 137, 162, 168, 176, 192, 213, 270, 323, 351, 352, 372

reductase, see Reductases
Pyrene
　benzo[a]-, see Benzo[a]pyrene
　oxidation, 533
　structure, 448
Pyrethroids, 505, 506
Pyridine
　binding to P450s, 308
　N-oxides, 46
　nucleotide oxidation, 7, 12, 242, 244
Pyrogallol
　peroxidation, 337
Pyrrolidines, 422, 423
Pyruvate:ferredoxin oxidoreductase, 87
Python regius, 492

Q

Quantum mechanical calculations, 198, 215, 259
Quencher
　oxidative, 173–175, 179
　reductive, 173–175
Quinidine, 534, 566, 596, 598
Quinoline
　7-benzyloxy-, 456, 464, 467
Quinones (see also individual names)
　1,4-naphtho-, 454, 462
　tert-butylhydro-, 14
　imino-, 577

R

Radicals (see also individual names)
　alkoxy, 404, 408, 410
　allyl, 39, 425, 428
　amino acid, 192
　carbinyl, 428
　carbon, 339–341, 416, 420, 422, 428, 429
　carboxyl, 420
　clock, 341, 348
　porphyrin π-cation, see Iron(IV)
　hydroxyl, 325, 341
　indolyl, 224

SUBJECT INDEX

Radicals (Continued)
  intermediates, 36, 37, 339
  nitrogen, 423
  olefin cation, 343
  oxygen, 428, 429
  pairs, 224
  pathway in P450s, 339, 340
  porphyrin, 67, 68, 159, 179, 216–218
  protein, 188, 189, 217–221
  tryptophan, 189, 192, 221, 224, 225
  tyrosine, 39, 180, 189, 192, 219–221, 224, 225, 352
Radiolysis
  cryo-, 252
  for redox reactions, 250–253
Rainbow trout, 493, 495, 497
Raloxifene, 309
Raman spectroscopy (studies of)
  nitric oxide synthase, 295
  P450 models, 279
  P450s, 191, 192, 211, 246, 275, 278, 296, 299, 302, 303, 326, 348
Random expulsion molecular dynamics, 238
Rate constants for
  autoxidation, 212
  carbon monoxide binding, 208–210
  dioxygen binding, 208–210, 212, 325
  hydroperoxy complexes, 217
  P450s, 199, 208–212, 217, 346
Ravuconazole, 301, 302
Reaction rates, see Rate constants
Reaction volume, 199
Reactive oxygen species, 10, 12, 369, 497, 504
Recombination mutation methods
  combinatorial libraries enhanced by recombination in yeast (CLERY), 457, 465
  DNA family shuffling, 457, 458
  DNA shuffling and ITCHY (SCRATCHY), 458
  incremental truncation for the creation of hybrid enzymes (ITCHY), 458
  sequence homology independent protein recombination (SHIPREC), 458, 465

  staggered extension process (StEP), 457
Redox potentials, 440
  Co(III)/Co(II), 167
  compound I/compound II, 224
  CYP2B4, 331
  Fe(II)/Fe(I), 166
  Fe(III)/Fe(II), 29, 32, 33, 35, 38, 48, 50, 67, 158, 160, 162, 166, 170, 173
  Fe(IV)/Fe(III), 170, 171
  heme, 161, 270, 286, 324
  indolyl radical pair, 224
  list of, 134, 173
  Mn(III)/Mn(II), 50
  P450 BM3, 246, 247, 455
  P450cam, 202, 273, 331, 334
  pH dependence, 139
  potentiometrically determined, 130–139, 146, 147
  practical considerations, 143, 144
  Re(I)/Re(0), 173
  Re(II)/Re(I), 173
  reversible, 128–151
  Ru(III)/Ru(II), 173, 179
  theoretical considerations, 140–143, 195
  tyrosine phenoxy radical pair, 224
  voltammetrically determined, 130, 131, 139–150
Reductases
  adrenodoxin, 364, 365, 368, 369, 371, 372
  artificial, 162
  cytochrome b_5, 8, 9
  ferredoxin, 160
  flavin containing, 160
  FMN, 137
  FMN/FAD, 77, 90
  methionine synthase, 290
  NADPH, 464
  NADPH-cytochrome P450, 101, 102, 104, 371, 373, 414, 499, 569
  NADPH-dependent ferredoxin, 8, 9
  nitric oxide, 249
  P450, 364
  putidaredoxin, 176, 323

Reduction potentials, *see* Redox potentials
Reptiles, 487
  chemical defense, 490, 491
  P450 regulation, 491–493
Resonance Raman spectroscopy, *see* Raman spectroscopy
Resorufin
  7-benzyloxy-, 464, 467, 508
  alkoxy-, 481
  demethylation, *see* Demethylation
  ethoxy-, 465, 508, 518
  methoxy-, 514
  O-dealkylation, *see* Dealkylation
  O-deethylation, *see* *O*-Deethylation
  pentoxy-, 484, 508, 518
  structure, 464
Respiratory chain
  inhibition, 299
Respiratory tract
  cancer, 484
Rhenium (0), 173
Rhenium(I)
  wires, 173, 175–177
  redox potentials, *see* Redox potentials
Rhizopus nigricans, 523
Rhodium(I)
  diimine complex, 171
Rhodococcus sp., 8, 526, 528, 533
  erythropolis, 527, 528
  rhodochrous, 527, 528
Rhodospirillum rubrum, 307
Rhodotorula minuta, 419
Rhodovulum sulfidophilum, 308
Rice, *see Oryza sativa*
Rifampicin, 485, 486, 497, 579, 580, 608
Ritonavir, 309, 599, 612
Rubredoxin, 149
  crystal structure, 66
Ruthenium(I)
  2,2'-bipyridyl complex, 178
Ruthenium(II), 40, 46–48, 172, 173
  2,2'-bipyridyl complex, 175, 178, 179
  diimine, 162, 171, 174
  porphyrins, *see* Porphyrins
  redox potential, *see* Redox potentials
  wires, 176, 177

Ruthenium(III), 172, 173
  2,2'-bipyridyl complex, 171, 173, 178
  redox potential, *see* Redox potentials
Ruthenium(IV), 46
Ruthenium(VI), 40

S

Saccharomyces
  cerevisiae, *see* Yeast
  diastaticus, 531
  genome, 4
Saccharopolyspora erythraea, 304
Saccostrea glomerata, 512
Salmo salmar, 495, 497
Salt bridge in P450s, 89, 239, 246
Saponines, 362
Saxitoxin
  red-tide, 119
Scallop (*see also* individual names), 104, 119
Schistosoma mansoni, 105
Schistosomiasis, 105
Schizosaccharomyces pombe, 8
Scoparone
  metabolism, *see* Metabolism
Sea anemone (*see also* individual names), 512
  P450s, 102, 103
Seal, 100
Sea lion, 100
Sea squirts (*see also* Ascidians), 498
  P450 genes, 108
Sea stars (*see also* individual names), 108, 508, 509
  P450s, 102
Sea urchins (*see also* individual names), 108, 508, 509
Secologanin
  structure, 417
Seldane®, *see* Terfenadine
Serine
  phosphorylation, 15
Serotonin
  receptor inhibitor, 598
Shark, 100

Shrimps (see also individual names), 107, 508
  grass, 106
Shunt pathway, 32, 37, 40, 196, 217, 224, 280, 325
Site-directed mutagenesis of P450s, 72, 137, 138, 218–221, 242, 270, 271, 370, 444–452, 459, 463–465, 468, 531, 572, 574
Skin
  expression of P450s, 384
  fungal infections, 301
Snails (see also individual names), 105, 110, 511
  atrazine-exposed, 104
Snake (see also individual names), 491, 492
Solanum tuberosum, 423
Soret band, 179
  compound I, 217
  metal-diimine wires, 175
  P420, 306
  P450-carbon monoxide complex, 203–208, 275, 305–307, 325
  P450-dioxygen complex, 211, 213
  P450-nitric oxide complex, 293–295, 308
  P450s, 158, 197, 223, 287–300, 302, 349, 352
Spearmint
  limonene-6-hydrolase, 468
  P450, 6
Spectroscopy (of)
  azole drug binding to P450s, 302, 303
  list of techniques, 191, 192
Sphingomonas paucimobilis, 527
Sphingosine, 378
Spironolactone, 384
Sponges (see also individual names), 110, 111, 512
Stability constants, see Binding constants
Steroids (see also individual names), 98, 99, 287, 600
  acute regulation of biosynthesis, 375, 376
  androgens, 300, 343, 370, 374, 375, 381, 409
  biosynthesis, see Biosynthesis
  chlorinated, 309
  chronic regulation of biosynthesis, 376–379
  ecdy-, 107, 108, 111
  hydroxylation, see Hydroxylation
  neuro-, 383
  overproduction, 373
  plant, see individual names
  receptors, 492
  structures, 362
Steroidogenic factor 1, 376–379, 382, 384, 385
Sterols, 571
  biosynthesis, see Biosynthesis
  demethylase activity, 302
  phyto-, 362, 363
  structures, 362
Stigmasterol
  structure, 362
St. John's wort, 612
  as P450 3A4 inducer, 580, 597, 610
Stopped flow studies of P450s, 192, 208–210, 213, 217–221
Streptomyces
  carbophilus, 530
  coelicolor, 275, 441, 456
  griseus, 527, 528, 533
  P450s, 4, 257
Strongylocentrotus
  droebachiensis, 509
  purpuratus, 110, 508
Structure–function relationships of P450s, 370
Styrene, 485
  3-chloro-, 453, 454
  structure, 454, 522
Suberite domuncula, 512
Sulfaphenazole, 566
  as inhibitor, 573
Sulfide oxidation, 322
Sulfite oxidation, 308
Sulfolobus solfataricus, 7, 87, 304, 456
Sulfones, 322
Sulfoxides, 322

Sulfur
  ^{34}S, 303
Superoxide, 38, 189, 211, 298
  ferric complex, see Iron(III)
  formation, 12, 216, 286, 325
Superoxide dismutase, 298
Surfactants, 145–148
Synthases
  aldosterone, 61
  allene oxide, 7, 8
  cystathionine β-, 12, 307
  divinyl ether, 7
  isoflavone, 421
  prostaglandin H, 8, 39, 40
  prostacyclin, 299, 426, 571
  thymidylate, 87
  thromboxane, 7, 426, 571
Synthesis of
  antibiotics, 286
  bio-, see Biosynthesis
  camalexin, 14
  nitric oxide, 290–293

T

Tacrolimus, 606, 611
Takifugu rubripes (*see also Fugu*), 111–115, 494
Tamoxifen, 599
Tanabe-Sugano diagram, 198
Taxol, 467
Temperature jump studies of P450s, 191, 192, 199
Teratogens, 492
Terfenadine, 565, 597, 610–612
  bioavailability, 608
  metabolism, see Metabolism
  toxicity, see Toxicity
Terpenes, 481, 526, 527
  metabolism, see Metabolism
  oxidation, 450
  sesqui-, 450
Testis (*see also* Gonads), 369, 379
  hormone biosynthesis, 370, 387–381
  regulation of hormones, 379, 380
Testosterone, 72, 248, 278, 364, 409, 456, 467, 527, 575, 600

17-*O*-acetyl-, 404
6β-, 278
biosynthesis, see Biosynthesis
hydroxylation, see Hydroxylation
isotope labelling studies, 278
metabolism, see Metabolism
structure, 363, 410
3,3',4,4'-Tetrachlorobiphenyl, 488, 492, 497, 512
  oxidation, 491
Tetradecanoic acid, 401
Tetrahydrobiopterin, 82, 291
3,3,5,5-Tetramethylcyclohexanone, 199, 205, 210
Tetraodon nigroviridis, 494
Theophylline, 483
Therapeutic index, 574
Thermus thermophilus, 87
Thioanisole, 171
Thiocarbamates
  di-, see Dithiocarbamates
  diethyldi-, 495
Thiols (and thiolate groups) (*see also* individual names), 333–337
  as porphyrin ligands, 31–40, 128
  heme–, see Heme
  proximal, 136
Thiosulfate
  oxidation, 308
Threonine
  hydroxyl group, 331–333
  phosphorylation, 15
  role in P450 catalysis, 331–333
Thromboxanes, 426–429
  A$_2$, 7
  biosynthesis, see Biosynthesis
  synthase, see Synthases
Ticlopidine, 309
Tienilic acid, 309
  as inhibitor, 573
  bioactivation, 580
  hepatotoxicity, 573, 580
  metabolism, see Metabolism
  structure, 580
Tin
  tributyl, 104, 106, 111, 508, 511, 522

SUBJECT INDEX

Tobacco, 505, 518, 520, 607
  procarcinogen, 484
  smoke, 597
Tolbutamide
  hydroxylation, see Hydroxylation
Tomato, 408, 423
Toxicity (of)
  acetaminophen, 577
  geno-, see Genotoxicity
  hepato-, see Hepatotoxicity
  terfenadine, 578
  test systems, 487
Toxins (see also individual names)
  afla-, see Aflatoxins
  algal, 100, 119
  myco-, 417
  phyto-, 505
  saxi-, 119
  xantho-, see Xanthotoxin
Transferases
  glutathione-S-, 14, 118–120, 519
  glycosyl, 519
  uridine dinucleotide phosphate glucuronosyl, 571
Tributyltin, 104, 106, 111, 508
  metabolism, see Metabolism
  structure, 522
1,3,5-Trichlorobenzene, 532
  oxidation, 449
  structure, 448
Trypsin, 70
  chymo-, see Chymotrypsin
Triazoles
  as P450 inhibitors, 288
1,1,1-Trichloro-2,2-bis(4-chlorophenyl)ethane, see DDT
Trichosporon cutaneum, 296
Troleandomycin, 485
Trout
  rainbow, 493, 495, 497
Trypanosomes, 513
Tryptophan
  radical, see Radicals
Tulipa gesneriana, 423
Tuna, 100
Turkey, 488, 490

Turtles, 487, 490, 491
Tylenol®, see Acetaminophen
Tyrosine
  nitration, 298, 299

U

Ubiquitination of
  P450, 15
Unio tumidus, 105
UV-Vis spectrophotometry of
  oxyferrous adducts, 349, 350
  P450s, 191, 192, 197, 245, 252

V

Valencene, 450, 451, 453, 531
  structure, 448
Vancomycin, 236
Vasoconstriction, 426
Vasodilation, 426
  nitric oxide, see Nitric oxide
Verapamil, 534, 597, 599
Vertebrates (see also individual names)
  non-mammalian, 487–498
  P450s, see P450s
Vinclozolin, 489
Vitamin A, 571
Vitamin C, see Ascorbate
Vitamin D, 384
  oxidation, 571
Vitamin D_2
  hydroxylation, see Hydroxylation
  metabolism, see Metabolism
Vitamin D_3, 400, 406
  oxidation, 417
  reactions, 406, 407
Voltammetry (studies of)
  comparison of data, 150, 151
  cyclic, see Cyclic voltammetry
  P450s, 130, 131, 138–151, 166, 170
  surface-confined, 142–144, 146, 151
Voriconazole, 302
  structure, 301

W

Warfarin, 527, 597, 601, 602, 614
  hydroxylation, *see* Hydroxylation
  metabolism, *see* Metabolism
Western population, 573, 576
Whale, 100, 101
  P450 genes, 115, 116
Wires
  biphenyl-tethered, 179
  electrun-tunneling, 171–180
  -mediated P450 oxidation, 179, 180
  -mediated P450 reduction, 178, 179
  metal-diimine, 171–174
  P450-wire conjugates, 174–178

X

XAS, *see* X-ray absorption spectroscopy
Xanthotoxin, 504, 507
Xenobiotics, 14, 98, 99, 101, 116, 158, 321, 344, 571
  bioactivation, 518
  biotransformation, 477–537
  catabolism, 17
  chemical defense against, 477–535
  decomposition, 530–534
  detoxification, *see* Detoxification
  effects of P450s, 102–108, 117
  excretion, 603
  hydrophobic, 330
  list of, 522, 523, 526
  metabolism, *see* Metabolism
  mineralization, 524
  receptor, 489
  resistance, 111, 118, 119
  structures, 522, 523, 526
Xenopus
  laevis, 487, 491, 492
  tropicalis, 491, 492
X-ray absorption spectroscopy of P450 models, 273

Y

Yarrowia lipolytica, 524
Yeast (*see also* individual names), 419, 532
  alkane assimilation, 524, 525
  genome, 4
  P450, 4, 18, 373

Z

Zebrafish (*see also Danio rerio*), 494, 497
Zinc(II), 49, 50